The Geology of Somalia: a Selected Bibliography of Somalian Geology, Geography and Earth Science.

AD

By R. Lee Hadden

Topographic Engineering Center
US Army Corps of Engineers
7701 Telegraph Road
Alexandria, Virginia 22315

February 2007

Introduction

This bibliography on the geology, geography and earth sciences of Somalia was gathered from a variety of different abstracting, bibliographical and cartographical resources. They include citations from agriculture, botany, engineering, geology, geography, medical, military science, soils, transportation and other subject resources. These citation resources are provided by a number of scientific societies, such as the American Geographical Society; from government resources, such as the Defense Technical Information Center; non-governmental organizations such as the United Nation's Food and Agricultural Organization (FAO); and from commercial databanks such as GeoRef, WorldCat and GeoBase. Many unique citations were collected from the catalogs and resources of major research libraries, such as the Library of Congress and the US Geological Survey Library.

Within this bibliography, the article retrieval information is given as much as possible. These include specific ISSN, ISBN, OCLC and Library of Congress numbers that allow the electronic borrowing or copying of these items through library networks. Alternately, the citations also include information on acquiring these items through document delivery companies and commercial services. Very often, scientific publications in less developed countries are not published in large numbers, and it is very difficult to retrieve reports or maps more than even a few years old. This bibliography is intended to be a resource for those scientific citations on Somalia that can still be retrieved.

Within these citations are many variations in spelling and place names. Many scientific and cartographic investigations were done in Cushitic, European and Semitic languages, such as Arabic, English, French, Italian and Somali. Thus, the same name may be spelled differently according to the language(s) used. Variations on single and doubled consonants (geminated consonants) and single and doubled vowels (diphthongs) are common. Mogadisco, Mogadiscio, Mogadishu, Mogadischu and Muqdisho are all variations on the name of the capital city. The river Juba can be spelled Jubba, Juba, Giuba, etc., according to the language used, as can the port city of Chisimaio, Chisimayo, Kismayo, Kisimayo, Kisymao, etc. So any search for authors, place names and locations in this bibliography should take into account spelling variations.

Abbreviations and links to resources used:

(All links and URLs in this bibliography are current as of January 2007)

AGI: American Geological Institute, Alexandria, VA. See: www.agiweb.org

AGS: American Geographical Society Library, University of Wisconsin, Milwaukee Campus. See: http://www.amergeog.org

AS&T: Applied Science & Technology from H.W. Wilson is a bibliographic database that indexes articles of at least one column in length. English-language periodicals published in the United States and elsewhere are covered; non-English language articles are included if English abstracts are provided. Periodical coverage includes trade and industrial publications, journals issued by professional and technical societies, and specialized subject periodicals, as well as special issues such as buyers' guides, directories, and conference proceedings. See: http://www.hwwilson.com/Databases/applieds.htm#Abstracts

ASFA: Aquatic Sciences and Fisheries Abstracts Input to ASFA is provided by a growing international network of information centers monitoring more than 5,000 serial publications, books, reports, conference proceedings, translations, and limited distribution literature. ASFA is a component of the Aquatic Sciences and Fisheries Information System (ASFIS), formed by four United Nations agency sponsors of ASFA and a network of international and national partners. Aquatic Sciences and Fisheries Abstracts are produced by CSA under contract to FAO. See: http://www.csa.com/

British Library: The British Library Document Supply Service can supply many of the article citations and reports given in this bibliography, especially those maps and other materials owned by the British Library. See: http://www.bl.uk/services/document/dsc.html

CISTI: Canada Institute of Scientific and Technical Information. This is a Canadian document supply service for scientific and technical literature. "Through Global Service, CISTI can obtain any document for you, from anywhere in the world. Most documents are supplied within four weeks. You can specify the level of service you prefer at the time of ordering by choosing the appropriate line from the drop down menu on any of the CISTI order forms." See: http://cisti-icist.nrc-cnrc.gc.ca/

CSA Technology Research Database: This comprehensive database provides a single mega-file of all the unique records available through its 3 components: the CSA Materials Research Database with METADEX, CSA High Technology Research Database with Aerospace, and the CSA Engineering Research Database. The database content represents the most comprehensive and current coverage of the relevant serial and non-serial literature available. Sources covered include over 4,000 periodicals, conference proceedings, technical reports, trade journal/newsletter items, patents, books, and press releases. See: http://www.csa.com/

DTIC: Defense Technical Information Center, Alexandria, VA. See: www.dtic.mil

ESPM: The *CSA Environmental Sciences and Pollution Management* database offers access to the international literature in the environmental sciences. Abstracts and citations are drawn from over 6000 serials including scientific journals, conference proceedings, reports, monographs, books and government publications. See: http://www.csa.com/

FAO: Food and Agriculture Organization Library, United Nations, Rome, Italy. See: www.fao.org

GeoBase: GEOBASE is a unique multidisciplinary database supplying bibliographic information and abstracts for development studies, the Earth sciences, ecology, geomechanics, human geography, and oceanography. The database provides current coverage of almost 2,000 international journals, including both peer-reviewed titles and trade publications, and provides archival coverage of several thousand additional journal titles and books. GEOBASE is unequalled in its coverage of international literature of the core scientific and technical periodicals. Papers are selected, read, and classified using a unique classification scheme that is versatile and updated annually to adapt coverage to current research trends. The material covered includes refereed scientific papers; trade journal and magazine articles, product reviews, directories and any other relevant material. GEOBASE has a unique coverage of non-English language and less readily available publications including books, conference proceedings and reports, making this the best resource available for multidisciplinary searches of international literature. The content crosses over subject, language, and cultural boundaries, providing a unique research tool to users. All material in *GEOBASE* is also available as print in the following Elsevier/Geo Abstracts journals: *Geographical Abstracts, Physical Geography, Human Geography, Geological Abstracts, Ecological Abstracts, International Development Abstracts* and *Oceanographic Literature Review, Geomechanics Abstracts*. See: www.elsevier.com

GeoRef: American Geological Institute, Alexandria, VA. The American Geological Institute not only identified materials for the abstracting database, GeoRef, but also locates and supplies materials as a document delivery service. See: www.agiweb.org

ISBN: International Standard Book Number. This unique number can be used to identify and locate library holdings of a particular book or report title. See: http://www.isbn.org/standards/home/index.asp

ISSN: International Standard Serial Number. This unique number can be used to locate libraries which have subscriptions to this journal, magazine or serial. See: http://www.issn.org/

Library of Congress Control Number – LCCN: This is a unique number applied by the Library of Congress to identify individual publications. This number can be used to identify copies of this item in libraries held in the US and abroad. See: http://www.loc.gov/marc/lccn_structure.html

LC or LOC: Library of Congress, Geography and Map Division, Washington, DC. The Geography and Map Division has the largest collection in the world, with 5.4 million maps, 75,000 atlases, 500 globes, 3,000 three-dimensional objects and thousands of digital files. Recently, the Library of Congress has digitally scanned and mounted its 10,000[th] map online. See: http://www.loc.gov/rr/geogmap/

Linda Hall Library: "Our Document Delivery Services Department allows students, researchers, and businesses to request copies of journal articles, conference proceedings, historical documents, or many other documents housed at the Linda Hall Library. We are committed to filling every in-scope, properly cited request within 24-48 hours. Requests are processed during the local working hours of 8 am – 5 pm, U. S. Central Time, Monday through Friday. Our fee is a cost recovery fee intended to support a strong collection and dedicated services." See: http://www.lhl.lib.mo.us/services/document_delivery/index.shtml

Northwestern University Transportation Library: The Transportation Library was founded in 1958 to support the curricula and research programs of the Transportation Center and the Center for Public Safety of Northwestern University, including the School of Police Staff and Command. Containing over 400,000 items, the Transportation Library of Northwestern University is one of the largest transportation information centers in the world, encompassing information on all transportation modalities, including: air, rail, highway, pipeline, water, urban transport and logistics. Its collection of environmental impact statements is one of the most complete in the world. See: http://www.library.northwestern.edu/transportation/

NTIS: National Technical Information Service, Alexandria, VA. See: www.ntis.gov

OA: Oceanic Abstracts. For over 32 years, Oceanic Abstracts from Cambridge Scientific Abstracts has been focused exclusively on worldwide technical literature pertaining to the marine and brackish-water environment. The journal has long been recognized as a leading source of information on topics relating to oceans. It focuses on and is totally comprehensive in its coverage of marine biology and physical oceanography, fisheries, aquaculture, non-living resources, meteorology and geology, plus environmental, technological, and legislative topics. See: http://www.csa.com/factsheets/oceanic-set-c.php

OCLC: Founded in 1967, OCLC Online Computer Library Center is a nonprofit, membership, computer library service and research organization dedicated to the public purposes of furthering access to the world's information and reducing information costs. More than 41,555 libraries in 112 countries and territories around the world use OCLC services to locate, acquire, catalog, lend and preserve library materials. Researchers, students, faculty, scholars, professional librarians and other information seekers use OCLC services to obtain bibliographic, abstract and full-text information when and where they need it. See: http://www.oclc.org/

SWRA: Selected Water Resources Abstracts (1967-94). *SWRA* provides more than 271,138 abstracts compiled by the Water Resources Scientific Information Center (WRSIC) of the USGS. *SWRA* provides thorough coverage of worldwide technical literature across the life, physical, and social-science aspects of water resources as well as U.S. Government documents produced by the USGS's many research facilities. Records are drawn from journals, monographs, conference proceedings, reports, court cases, and other federal and state publications. *SWRA*, and now *Water Resources Abstracts*, are your best sources for issues pertaining to groundwater, water quality, water planning, and water law and rights.

TRIS: TRIS is a bibliographic database funded by sponsors of the Transportation Research Board (TRB), primarily the state departments of transportation and selected federal transportation agencies. TRIS Online is hosted by the National Transportation Library under a cooperative agreement between the Bureau of Transportation Statistics and TRB. See: http://ntlsearch.bts.gov/tris/index.do

UN: United Nations Library, New York, NY. See: www.un.org

USGS: US Geological Survey Library, Reston, VA. See: www.usgs.gov/library

University of Texas at Austin. Perry-Castañeda Library Map Collection: "Many of these maps have been scanned and are available for downloading and other uses." See: http://www.lib.utexas.edu/maps/

WorldCat: Among other things, this a free database from OCLC showing local library holdings of desired publications. See: http://www.worldcat.org/

Acknowledgements

The author also wants to acknowledge the help of Mr. Abdirahman S. Ali, a native of Somalia, for his help in resolving some of the multi-lingual questions in this bibliography and for his help in interpreting the cultural geography of Somalia.

Mr. Matt Marks and Mr. Willie Sayyad of Refworks.com has kindly provided the bibliographer with a temporary subscription to their reference management software, and this bibliography would have been much more difficult to produce without this valuable resource. Refworks.com allowed the bibliographer to place the paper and electronic citations from very different resources and different libraries into one large file, and to allow for the automatic formatting and preparation of the bibliography. "RefWorks supports hundreds of online databases and output styles covering a broad range of subject areas. RefWorks collaborates with some of the world's most prestigious online information service providers, including CSA Illumina, BioOne, EBSCOhost, Elsevier, H.W. Wilson, ISI Web of Science, OCLC, Ovid and ProQuest, to name a few." Their support for this project is gratefully acknowledged. See: http://www.refworks.com/

Somalia[1]

Somalia ("Soomaaliya" in the Somali language[2]; "As-Suumaal" in Arabic), formerly known as the Somali Democratic Republic, is a coastal nation at the "Horn of Africa" in East Africa. Somalia is shaped at an angle somewhat like the number "7", and this country is also called "The Horn of Africa" due to its shape somewhat like the horn of a rhinoceros.

Somalia has about 637,540-637,657 square kilometers (estimates vary), of which only 10,320 km² is covered with water, and the country is slightly smaller in land area than the state of Texas.

Inland, Somalia has borders 58 kilometers long with the country of Djibouti to the north; Ethiopia shares a border 1,626 km long with Somalia in the north and west, and in the south and west, Kenya shares a border with Somalia of some 682 km.

Seaward, the country lies on the coast of the Gulf of Aden and the Indian Ocean, and has a coastline of some 3,025 km. Somalia occupies a strategic location on Horn of Africa along southern approaches to Bab el Mandeb and sea route through the Red Sea and the Suez Canal. Somalia's long coastline has been of strategic and tactical importance chiefly in permitting trade with the Middle East and the rest of East Africa. The exploitation of the shore and the continental shelf for fishing and other purposes had barely begun by the early 1990s. Sovereignty was claimed over territorial waters up to 200 nautical miles.

A generation ago, Somalia had an estimated population of around 8,591,000 people. However, population estimates are very difficult to verify because of the political situation and the mostly nomadic nature of the Somalis. The last accurate census was taken almost a generation ago in 1975. Currently, Somalia is believed to be one of the fastest growing countries in Africa and the world. Recent population estimates from 2005 range greatly between 15 and 25 million inhabitants. About 60% of all Somalis are nomadic or semi-nomadic pastoralists who raise cattle, camels, sheep, and goats. About 25% of the populations are settled farmers who live mainly in the fertile agricultural region between the Juba and Shebelle rivers in southern Somalia. The remainder of the population (15%-20%) is urban.

Because of the wars and conflict, Somalia has a large dispersed community, one of the largest of the whole continent. There are over a million Somalis living outside of Africa. This "Diaspora" adds stress to the society, since the Somali clan structure is extremely important to their identity and way of life. Breaking the clan cohesion by distance or displaced refugees induces stress on both the refugees and the family members left behind. The US Department of State says on their website: "The status of expatriate Somalis has been an important foreign and domestic issue. The Somali-populated region of the Horn of Africa stretches from the Gulf of Tadjoura in modern-day Djibouti through Dire Dawa, Ethiopia, and down to the coastal regions of southern Kenya. Unlike many countries in Africa, the Somali nation extends beyond its national

---

[1] This review of Somalia has been taken from a wide number of open source information, including the US Department of State; the CIA Factbook; the American University's Foreign Area Study, "Somalia, A Country Study"; the US Geological Survey's commodities reports, geological maps and other data; and data acquired from various general reference sources such as soil maps, regional encyclopedias, geographical gazetters and geological dictionaries.

[2] "Soomaal" in Somali means "to milk," as in "to milk a goat or to milk a camel."

borders. Since gaining independence in 1960, the goal of Somali nationalism, also known as Pan-Somalism, has been the unification of all Somali populations, forming a Greater Somalia. This issue has been a major cause of past crises between Somalia and its neighbors--Ethiopia, Kenya, and Djibouti."[3]

Ethnic groups in the country include the Bajuni people who live in the coastal settlements and islands south of Kismayo and are of East African Swahili origin. The Bajuni people in Somalia speak Kibajuni and the Bravanese people speak Chimwiini. Both of these languages are dialects of Swahili.

The Bravanese live in Brawa and are believed to be of mixed Arab, Portuguese and other descent.

The Benadiri (Reer Hamar), are an urban people of East African Swahili origin who live in cities in the "gobolka Benadir" coastal region.

The Bantu are agricultural workers who live along the Juba River.

Several thousand Arabs also live in Somalia, along with some hundreds of Indians and Pakistanis.

A small population of people of Italian descent, which dated back to Somalia's colonial era, began to emigrate following independence, and by the outbreak of war most Italian Somalis had left the country.

Nearly all the inhabitants speak Somali, the official language of the country. The Somali language is based on various dialects from the Mudug province. The Somali language was not used extensively for writing until 1973, when a standard orthography using the Latin alphabet was decreed by the Supreme Revolutionary Council (SRC). Somali is now the language of instruction in schools, although there are few schools operating today. Arabic, English, and Italian languages are also used extensively. Maay, a language closely related to Somali, is spoken by the Rahanweyn.

Politically, Somalia is divided into 18 regions (singular "gobolka"; plural is "gobollada"), and subdivided into districts. The regions are: Awdal; Bakool (Bakol); Banaadir (Benadir); Bari; Bay; Galguduud (Galgudud); Gedo; Hiiraan (Hiran); Jubbada Dhexe (Middle Jubba); Jubbada Hoose (Lower Jubba); Mudug; Nugaal (Nugal); Sanaag; Shabeellaha Dhexe (Middle Shabelle); Shabeellaha Hoose (Lower Shabelle); Sool; Togdheer (Togder) and Woqooyi Galbeed.

The northwestern area has declared its independence as the "Republic of Somaliland"; the northeastern region of Puntland is a semi-autonomous state; and the remaining southern portion is riddled with the struggles of rival factions.

In May of 1991, the major northern Somaliland clan, the "Isaqs,"declared an independent Republic of Somaliland (Somali: *Soomaaliland*), that now includes the administrative regions of Awdal, Woqooyi Galbeed, Togdheer, Sanaag, and Sool. Although not recognized by any other government, this entity has maintained a stable existence, aided by the overwhelming dominance of a ruling clan and economic infrastructure left behind by British, Russian, and American military assistance programs. According to the US Department of State, "The self-declared "Republic of Somaliland"

---

[3] See: http://www.state.gov/r/pa/ei/bgn/2863.htm, "US Department of State, Background Note: Somalia," accessed January 10, 2007.

consists of a regional authority based in the city of Hargeisa, including a President, Vice President, parliament, and cabinet officials."[4]

Somaliland now covers roughly the region of the former British Somaliland protectorate, an area of about 137,600 square kilometers (53,128 sq mi), which was also briefly an independent country for about five days in 1960. Somaliland is bordered by Ethiopia on the south, Djibouti on the west, the Gulf of Aden on the north, and the autonomous region Puntland in Somalia on the east. The capital of Somaliland is Hargeisa. The majority of the people are Sunni Moslems, and Somali and Arabic are the languages used mostly, with English taught in the schools.

Yet another secession from Somalia took place in the northeastern region. The regions of Bari, Nugaal, and northern Mudug now comprise a neighboring self-declared autonomous state of Puntland. Puntland is centered on the capital, Garowe, in the Nugaal Gobolka. Puntland has also made strides toward reconstructing a legitimate, representative government, but has suffered some civil strife. Puntland disputes its border with Somaliland as it also claims portions of eastern Sool and Sanaag.

In 1998 Puntland seceded and declared itself to be an autonomous state. This semi-secession is apparently an attempt to reconstitute Somalia as a federation of semi-autonomous states. Unlike neighboring Somaliland, Puntland does not seek outright independence from Somalia. The self-proclaimed state took the name "Puntland" after declaring "temporary" independence in 1998, with the intention that it would participate in any Somali reconciliation to form a new central government. The name is derived from the "Land of Punt" mentioned by ancient Egyptian sources, which is believed by some to have existed in what is now Somalia.

A third secession occurred in 1998 with the declaration of the state of Jubaland. The territory of Jubaland is now encompassed by the state of Southwestern Somalia and its status is unclear. Jubaland (Somali: *Jubbaland*) or Juba Valley (Somali: *Dooxada Jubba*), formerly Trans-Juba (Italian: *Oltre Giuba*), is the southwestern most part of Somalia, on the far side of the Juba River (thus "Trans"-Juba), bordering on Kenya.

It has a total area of 87,000 km² (33,000 sq mi), and the main city is Kismayo, on the coast near the mouth of the Juba. The area was mainly inhabited by the Ogaden nomads sub clan with other Dir and Hawiye clans

Jubaland is divided among the administrative regions of Gedo, Middle Juba, and Lower Juba and is technically part of the Transitional Government of Somalia. Despite this, until recently the actual administration of Jubaland was carried out by the Juba Valley Alliance led by Barre Adan Shire Hiiraale, also known as "Barre Hiiraale".

A fourth self-proclaimed entity led by the Rahanweyn Resistance Army (RRA) was set up in 1999. This "temporary" secession was reasserted in 2002, leading to de facto autonomy of Southwestern Somalia. The RRA had originally set up an autonomous administration over the Bay and Bakool regions of south and central Somalia in 1999.

The central and southern areas of Somalia are flat, with an average altitude of less than 180 metres (600 ft). South and west of the central area, extending to the Shabeelle River, lays a plateau whose maximum elevation is 685 meters (2,250 feet). The southwestern part of Somalia is largely savanna, or grassland. Only about 13 percent of

[4] See also: http://www.state.gov/r/pa/ei/bgn/2863.htm The self-declared "Republic of Somaliland" consists of a regional authority based in the city of Hargeisa, including a President, Vice President, parliament, and cabinet officials.

the country is suitable for farming. The region between the Juba and Shabeelle Rivers is low agricultural land, and the area that extends southwest of the Jubba River to Kenya is low pastureland.

The Jubba and Shabeelle Rivers originate in Ethiopia and flow toward the Indian Ocean. They provide water for irrigation but the rivers are not navigable by commercial vessels. The Shabeelle, with a total length of 2,011 kilometers (1,250 miles), dries up before reaching the ocean, but it is still the longest river in the country. Most of Somalia's food crops are grown in the fertile region in the south between Somalia's two permanent rivers, the Juba and the Webi Shabeelle. The amount of water in both rivers is highly variable, depending on rainfall. In periods of heavy rainfall there are also a number of seasonal streams, none of which can be used for irrigation or as dependable sources of drinking water.

Despite its lengthy shoreline, Somalia has only one natural harbor, Berbera. There is also a limited area around Baidoa that receives an average of 500–600 mm of rainfall each year.

At present there are still many species of wildlife found throughout the country, especially in the far southern part of the nation: small antelopes, a large variety of birds, foxes, hyenas, leopards, lions, ostriches and warthogs. Unfortunately, many of the large game mammals have been nearly wiped out: the giraffes, the hippopotami, the oryx, the rhinoceri, zebras, and, above all, the elephants. The elephants are slaughtered chiefly by ivory poachers. Some measures to protect the endangered animals have recently been taken by creating a national wildlife reserve in the lower Shabeelle swamp region.

Soils

The types of soil vary according to climate and parent rock. The arid regions of northeastern Somalia have mainly thin and infertile desert soils. The limestone plateaus of the interfluvial area have fertile, dark gray to brown, calcareous residual soils that provide good conditions for rain-fed agriculture. The most fertile soils in Somalia are found on the alluvial plains of the Jubba and Shabeelle rivers. These deep vertisols (black cotton soils) have a high water-retention capacity and are mainly used for irrigation agriculture. Irrigated land in Somalia is between 1,800 km² and 2,000 sq km², according to 1993 estimates.

There are also large areas of dark cracking clays (vertisols) in the southern part of Somalia that appear to have a higher water-holding capacity than the generally sandy soils found elsewhere. According to both the FAO and USDA soil taxonomy, a vertisol is a soil in which there is a high content of expansive clay known as montmorillonite. This soil forms deep cracks in drier seasons or years. Alternate shrinking and swelling of the soil causes self-mulching, where the soil material consistently mixes itself, causing vertisols to have an extremely deep A horizon and with no B horizon. (A soil with no B horizon is called an A/C soil). This heaving of the underlying material to the surface often creates microrelief known as "gilgai".

Vertisols typically form from highly basic rocks such as basalt in climates that are seasonally humid or subject to erratic droughts and floods. Depending on the parent material and the climate, they can range from grey or red to the more familiar deep black. Vertisols are found between 50° N and 45° S of the equator. In their natural state, vertisols are covered with grassland or grassy woodland. The heavy texture and unstable behavior prevents forest from growing.

The shrinking and swelling of vertisols can damage buildings and roads, leading to extensive subsidence. Vertisols are generally used for grazing of cattle or sheep. It is not unknown for livestock to be injured through falling into cracks in dry periods. However, the shrink-swell activity allows rapid recovery from compaction.

When irrigation is available, crops such as cotton, wheat, sorghum and rice can be grown. Vertisols are especially suitable for rice because they are almost impermeable when saturated. Rainfed farming is very difficult because vertisols can be worked only under a very narrow range of moisture conditions: they are very hard when dry and very sticky when wet.

The Somali farmers have developed their own system of making a checker-board of small ridges to retain the limited rainfall. Sorghum is grown in the basins that are formed, and in a year of average or above-average rainfall a reasonable crop can be harvested. Where they are well maintained, these basins probably also help to prevent soil erosion, but the crop-failure rate in seasons of below average rainfall can be high. One way in which the cultivators have attempted to respond to this crop uncertainty is by ratooning[5] their first "gu" rainy season sorghum crop in the hope that the re-growth may produce some grain in the second "dayr" rains. Yields obtained by this method appear to be generally low, and it may also contribute to the carry over of moth stemborers such as Busseola fusca[6], Chilo partellus[7] and Sesamia calamistis[8], that often cause severe damage to the crop.

The cultivators usually try to plant a second crop during the second rainy period, the "dayr", but the yields are usually low and the crop may fail as often as once in every 3 years. In 1986, for example, much of this area did not receive any rain at all in the second rainy season, so that all crops failed and severe hardship was caused to many people in early 1987.

Population pressure is increasing on the better soil regions, and deficiencies of phosphate are widespread. Some of these soils have a high phosphate-fixation capacity, so that heavy and uneconomic broadcast applications were sometimes found to be necessary to increase yields. Recently, it has been found that good responses can be obtained from small phosphate applications placed near the seed at the time of planting.

Few other crops except sorghum are grown, although some cowpeas may be planted in pure stands or intercropped, and a little maize, some groundnuts, watermelon, sesame, and tomato are sometimes grown in low-lying areas. Attempts have been made to introduce cotton with some success in limited areas.

---

[5] Ratoon: A shoot sprouting from a plant base, as in the banana, pineapple, or sugar cane.
[6] *Busseola fusca* is indigenous to Africa and is present in high and mid altitude areas (3500 ft and above sea level*)*. Adult moths lay eggs on maize plants, larvae emerge and after feeding on leaves for two to three days enter inside the maize stems. *Busseola fusca* lays its eggs between the stem and the leafsheaths
[7] *Chilo partellus* accidentally came to Africa from Asia in the 1930s. *Chilo partellus* is present on low and mid-altitude areas (zero to 4000 ft above sea level). *Chilo partells* lays its eggs on plant surface in form of egg batches.
[8] The Sesamia calamistis, aka "Pink Stalk Borer" or "Pink Maize Borer", is indigenous to Africa, occurring in savanna areas with a distinct dry season. Host plants include a wide variety of native grass species but also include cultivated grasses such as sugarcane, wheat and maize.

As elsewhere in the arid and subarid zones, the most widespread agricultural activity in Somalia is livestock keeping. Many of these stocks belong to nomadic herdsmen, who often move to and from the Ogaden area of Ethiopia or other areas with their herds, depending on the availability of grazing and water supplies. However, although some Somalis have cultivated sorghum for many years, many relied mainly on their livestock for their food supplies. In recent years, in suitable growing areas, an increasing number have taken to both growing some sorghum while also maintaining their livestock, which are taken to the traditional seasonal grazing areas by some of the young men of the family. Animal traction is used by some farmers, but this is not widespread. Manure is not generally used to fertilize the soil.

The two important rivers in the south of the country, the Shebelle and the Juba, are being increasingly used for irrigation, particularly of maize and sorghum. There are also limited areas of plantation crops, principally bananas for export, sugar cane, and some citrus and other crops.

Terrain, Vegetation, and Drainage

Somalia's terrain consists mainly of plateaus, plains, and highlands. In the far north, however, the rugged east-west ranges of the Karkaar Mountains lie at varying distances from the Gulf of Aden coast.

"Somalia consists of the eastern fringe of the Ethiopian plateau, consisting of a series of tablelands sloping to the southeast, the natural continuation of the great highlands that rise to the east of the Gall-Danakil tectonic trench. The mountains along the Coast of Aden are an extension of the raised rim forming the interface between the highlands and the Danakil depression, and contain the country's highest elevations…
…which drop down to the sea in steep cliffs all the way to rocky Ras Asir. From the mountainous rim, the highlands slope away toward the southeast, merging into the plateaus of Ogaden in Ethiopia and Mijertins regions, which in turn descend to the vast peneplain of Mudug and the coastal plain of Benadir, surrounded by long chains of sand dunes that block the passage of the Webi Shabelle, Somalia's largest river, as it flows to the ocean. Geologically, these lowlands constitute an ancient substructure of crystalline rocks that, as they sank, were invaded by the sea and blanketed with sediments of Cenozoic and Recent age, which were then overlain by alluvial deposits caused by surface erosion. Somalia can therefore be divided into two major natural regions: the north, characterized by highly incised highlands sloping toward the ocean; and the south, where the low tablelands fall away to the wide coastal plains."[9]

The far northern part of Somalia consists of hills and low mountains, which reach to about 8,000 feet (2,440 meters) in height. The northern region is somewhat mountainous, with plateaus reaching between 900 and 2,100 meters (3,000 and 7,000 feet). The highest point in the country, Mount Shimbiris, is at an elevation of 2,450 meters (8,000 feet). To the northeast there is an extremely dry, dissected plateau that reaches a maximum elevation of nearly 2,450 meters (8,000 feet).

The weather is hot throughout the year, except at the higher elevations in the north. Rainfall is sparse, and most of Somalia has a semiarid-to- arid environment suitable only for the nomadic pastoralism practiced by well over half the population. Only in limited areas of moderate rainfall in the northwest, and particularly in the

---

[9] World Geographical Encyclopedia. 1994. Volume I "Africa". Milan. Page 114.

southwest, where the country's two perennial rivers are found, is agriculture practiced to any great extent.

Somalia has problems with recurring droughts; frequent dust storms over the eastern plains in summer; floods during rainy season. It also has famine; the use of contaminated water contributes to human health problems; deforestation; overgrazing; soil erosion and desertification. Owing to inappropriate land use, the original vegetation cover, especially in northern Somalia, has been heavily degraded and in various places has been entirely destroyed. This progressive destruction of plant life also has impaired animal habitats and reduced forage, affecting not only Somalia's greatest resource, its livestock (chiefly goats, sheep, camels, and cattle), but also the country's wildlife.

Most of Somalia consists of dry savanna, suitable only for limited pasturage and occasional cultivation. A good deal of Somalia is arid and able to sustain only limited numbers of people and animals. The long coastline has a number of ports and some fishing villages but is for the most part uninhabitable. A coastal plain on the Gulf of Aden is backed by the mountains of the Karkaar Range. The tallest peak is Shimber Beris (Surad Ad; 2,408 m/7,900 ft). Below the mountains is a plateau region averaging 915 m (3,000 ft) in elevation, the western part of which stretches into the Ogaden of Ethiopia, which is considered the Somali "heartland."

Physiographically, Somalia is a land of limited contrast. In the north, a maritime plain parallels the Gulf of Aden coast, varying in width from roughly twelve kilometers in the west to as little as two kilometers in the east. Scrub-covered, semiarid, and generally drab, this plain, known as the "guban" or "scrub land", is crossed by broad, shallow watercourses that are beds of dry sand except in the rainy seasons. When the rains arrive, the vegetation, which is a combination of low bushes and grass clumps, is quickly renewed, and for a time the guban provides some grazing for nomad livestock.

Inland from the gulf coast, the plain rises to the precipitous northward-facing cliffs of the dissected highlands. These form the rugged Karkaar mountain ranges that extend from the northwestern border with Ethiopia eastward to the tip of the Horn of Africa, where they end in sheer cliffs at Caseyr. The general elevation along the crest of these mountains averages about 1,800 meters above sea level south of the port town of Berbera, and eastward from that area it continues at 1,800 to 2,100 meters almost to Caseyr. The country's highest point, Shimber Berris, which rises to 2,407 meters, is located near the town of Erigavo.

Southward the mountains descend, often in scarped ledges, to an elevated plateau devoid of perennial rivers. This region of broken mountain terrain, shallow plateau valleys, and usually dry watercourses is known to the Somalis as the "Ogo."

In the Ogo's especially arid eastern part, the plateau is broken by several isolated mountain ranges, and gradually slopes toward the Indian Ocean and in central Somalia constitutes the Mudug Plain. A major feature of this eastern section is the long and broad Nugaal Valley, with its extensive network of intermittent seasonal watercourses. The eastern area's population consists mainly of pastoral nomads. In a zone of low and erratic rainfall, this region was a major disaster area during the great drought of 1974 and early 1975.

The western part of the Ogo plateau region is crossed by numerous shallow valleys and dry watercourses. Annual rainfall is greater than in the east, and there are flat areas of arable land that provide a home for dry land cultivators. Most important, the western area has permanent wells to which the predominantly nomadic population returns during the dry seasons. The western plateau slopes gently southward and merges into an area known as the Haud, a broad, undulating terrain that constitutes some of the best grazing lands for Somali nomads, despite the lack of appreciable rainfall more than half the year. Enhancing the value of the Haud are the natural depressions that during periods of rain become temporary lakes and ponds.

The Haud[10] zone continues for more than sixty kilometers into Ethiopia, and the vast Somali Plateau, which lies between the northern Somali mountains and the highlands of southeast Ethiopia, extends south and eastward through Ethiopia into central and southwest Somalia. The portion of the Haud lying within Ethiopia was the subject of an agreement made during the colonial era permitting nomads from British Somaliland to pasture their herds there. After Somali independence in 1960, it became the subject of Somali claims and a source of considerable regional strife.

Southwestern Somalia is dominated by the country's only two permanent rivers, the Jubba and the Shabeelle (Shabele). With their sources in the Ethiopian highlands, these rivers flow in a generally southerly direction, cutting wide valleys in the Somali Plateau as it descends toward the sea; the plateau's elevation falls off rapidly in this area. The adjacent coastal zone, which includes the lower reaches of the rivers and extends from the Mudug Plain to the Kenyan border, averages 180 meters above sea level.

The Jubba River enters the Indian Ocean at Chisimayu (Kismaayo). Although the Shabeelle River at one time apparently also reached the sea near Merca, its course is thought to have changed in prehistoric times. The Shabeelle now turns southwestward near Balcad (about thirty kilometers north of Mogadishu) and parallels the coast for more than eighty-five kilometers. The river is perennial only to a point southwest of Mogadishu; thereafter it consists of swampy areas and dry reaches and is finally lost in the sand east of Jilib, not far from the Jubba River. During the flood seasons, the Shabeelle River may fill its bed to a point near Jilib and occasionally may even break through to the Jubba River farther south. Favorable rainfall and soil conditions make the entire riverine region a fertile agricultural area and the center of the country's largest sedentary population.

In most of northern, northeastern, and north-central Somalia, where rainfall is low, the vegetation consists of scattered low trees, including various acacias, and widely scattered patches of grass. This vegetation gives way to a combination of low bushes and grass clumps in the highly arid areas of the northeast and along the Gulf of Aden.

A broad plateau encompassing the northern city of Hargeysa, which receives comparatively heavy rainfall, is covered naturally by woodland (much of which has been degraded by overgrazing) and in places by extensive grasslands. Parts of this area have been under cultivation since the 1930s, producing sorghum and corn; in the 1990s it constituted the only significant region of sedentary cultivation outside southwestern Somalia.

The Haud south of Hargeysa is covered mostly by a semi-arid woodland of scattered trees, mainly acacias, underlain by grasses that include species especially favored by livestock as forage. As the Haud merges into the Mudug Plain in central Somalia, the aridity increases and the vegetation takes on a sub-desert character. Farther

---

[10] Merriam-Webster's Geographical Dictionary: Haud or Hawd, a region in SE Ethiopia and SW Somalia; a semi-desert plateau that also has grassy plains.

southward the terrain gradually changes to semiarid woodlands and grasslands as the annual precipitation increases.

Along the Indian Ocean from Mereeg, about 150 kilometers northeast of Mogadishu, southwestward to near Chisimayu (Kismaayo) lies a stretch of coastal sand dunes. This area is covered with scattered scrub and grassy clumps. Overgrazing, particularly in the area between Mogadishu and Chisimayu, has resulted in the destruction of the protective vegetation cover and the gradual movement of the once-stationary dunes inland. Beginning in the early 1970s, efforts were made to stabilize these dunes by replanting.

Land

Estimates vary, but from about 46 to 56 percent of Somalia's land area can be considered permanent pasture. About 14 percent of the land area is classified as forest. Approximately 13 percent of the land is suitable for cultivation, but most of that area would require additional investments in wells and transportation for it to be usable. The remaining land is not economically feasible for exploitation. In the highlands around Hargeysa, relatively high rainfall has raised the organic content in the sandy calcareous soils characteristic of the northern plains, and this soil has supported some dry farming. South of Hargeysa begins the Haud, whose red calcareous soils continue into the Ethiopian Ogaden. This soil supports vegetation ideal for camel grazing. To the east of the Haud is the Mudug Plain, leading to the Indian Ocean coast; this region, too, supports a pastoral economy. The area between the Jubba and Shabeelle rivers has soils varying from reddish to dark clays, with some alluvial deposits and fine black soil. This is the area of plantation agriculture and subsistence agro-pastoralism.

Practices concerning land rights varied from rural to urban areas. In pre-colonial times, traditional claims and inter-clan bargaining were used to establish land rights. A small market for land, especially in the plantation areas of the south, developed in the colonial period and into the first decade of Somalia's independence. The socialist regime sought to block land sales and tried to lease all privately owned land to cooperatives as concessions. Despite the government's efforts, a de facto land market developed in urban areas; in the bush, the traditional rights of clans were maintained.

The Siad Barre regime also took action regarding the water system. In northern Somalia from 1988 to 1991, the government destroyed almost all pumping systems in municipal areas controlled by the Somali National Movement, or failing that, stole the equipment. In rural areas, the government poisoned the wells by either inserting animal carcasses or engine blocks that leaked battery acid. As a result, northern Somalis had to rely on older gravity water systems, use poor quality water, or buy expensive water. Following the declaration of the independent Republic of Somaliland in the north in May 1991, the government of the republic began ongoing efforts to reconstruct the water system.

In the south, in the late 1980s onward, as a result of war damage and anarchy, the water situation in the towns tended to resemble that in the north. Few pumping systems were operational in early 1992. Conditions in rural areas varied. Many villages had at least one borehole from which poor quality water could be obtained in buckets; pumps generally were nonfunctioning. Somalis who lived near the Jubba or Shabeelle rivers could obtain their water directly from the river.

Climate

Due to Somalia's location on the equator, there is relatively little seasonal variation in climate. The weather is hot throughout the year, with mean maximum temperatures of 30–40° C (86–104° F) except at higher elevations and along the Indian Ocean coast. Mean daily minimums usually vary from about 15°C to 30°C (60°F-85°F). Somalia has some of the highest mean annual temperatures in the world. At Berbera on the northern coast the afternoon high averages more than 100° F (38° C) from June through September. Temperature maxima are even higher inland, but along the coast of the Indian Ocean temperatures are considerably lower because of a cold offshore current. The average afternoon high at Mogadishu, for example, ranges from 83° F (28° C) in July to 90° F (32° C) in April. Northern Somalia experiences the greatest temperature extremes, with readings ranging from below freezing in the highlands in December to more than 45° C in July in the coastal plain skirting the Gulf of Aden. The north's relative humidity ranges from about 40 percent in mid-afternoon to 85 percent at night, varying somewhat with the season.

Along the coast and in lowland areas it is very humid. During the colder months, December to February, visibility at higher elevations is often restricted by fog. In summation, major climatic factors are a year-round hot climate, seasonal monsoon winds, and irregular rainfall with recurring droughts.

Climate is the primary factor in much of Somali life. For the large nomadic population, the timing and amount of rainfall are crucial determinants of the adequacy of grazing and the prospects of relative prosperity. During droughts such as occurred during 1974-75 and 1984-85, widespread starvation can occur. There are some indications that the climate has become drier in the last century and that the increase in the number of people and animals has put a growing burden on water and vegetation.

Unlike typical climates at this latitude, conditions in Somalia range from arid in the northeastern and central regions to semiarid in the northwest and south. Long-term mean annual rainfall is less than 4 inches (100 millimeters) in the northeast and approximately 8 to 12 inches in the central plateaus. The southwest and northwest receive an average of 20 to 24 inches a year. While the coastal areas experience hot, humid, and unpleasant weather year-round, the interior is most often dry and hot. The problem in the dry land areas is that neither rainy season is sufficiently reliable in quantity or distribution to produce a crop regularly, so that production is uncertain at the best of times.

The climatic year in Somalia comprises of four seasons. The "gu," or main rainy season, lasts from April to June. The second rainy season, the "day" or "dayr", lasts from October to December. Each rainy season is followed by a dry season: the main one (jilaal) lasts from December to March and the second dry season, the "hagaa" or "xagaa", lasts from June to September.

The gu rains, the Southwest monsoons, begin in April and last until June, producing a fresh supply of pasture and for a brief period turning the desert into a flowering garden. Lush vegetation covers most of the land, especially the central grazing plateau where grass grows tall. Milk and meat abound, water is plentiful, and animals do not require much care. The clans, reprieved from four months' drought, assemble to engage alternately in banter and poetic exchange or in a new cycle of hereditary feuds. They also offer sacrifices to Allah and to the founding clan ancestors, whose blessings they seek. Numerous social functions occur: marriages are contracted, outstanding disputes are settled or exacerbated, and a person's age is calculated in terms of the number of gus he or she has lived. The monsoon season lasts from May until October, and is torrid in the north and hot in the south, with irregular rainfall.

The gu season is followed by the "hagaa" or "xagaa" drought (July-September) and the hagaa by the day rains (October-November). The southwest monsoon, a sea breeze, makes the period from about May to October the mildest season at Mogadishu. The southwest monsoons blow parallel to the coast and brings in maritime air masses, but quickly loses its moisture and brings little rain. During the second dry season, showers fall in the coastal zone.

The periods that intervene between the two monsoons (October through November and March through May) are hot and humid, and are called "tangambili" by the Somalians.

The second rainy season, called the "dayr," extends from October to December.

Next is the second dry season, the jiilaal, which lasts from December through March. This is the harshest season for the pastoralists and their herds. From December until February is also the northeast monsoon season, with moderate temperatures in north and very hot temperatures in the south. Temperatures in the south are less extreme, ranging from about 20° C to 40° C. The December-February period of the northeast monsoon is also relatively mild, although prevailing climatic conditions in Mogadishu are rarely pleasant. The dry but cool winter monsoon blows from the Asian continent, gets weaker as it moves south, and often is imperceptible in Mogadishu.

The hottest months are February through April. Coastal readings are usually five to ten degrees cooler than those inland. The coastal zone's relative humidity usually remains about 70 percent even during the dry seasons.

Most of the country receives less than 500 millimeters of rain annually, and a large area encompassing the northeast and much of northern Somalia receives as little as 50 to 150 millimeters. Certain higher areas in the north, however, record more than 500 millimeters a year, as do some coastal sites. The southwest receives 330 to 500 millimeters. Generally, rainfall takes the form of showers or localized torrential rains and is extremely variable.

Forestry

There is little forestland, much of that diminishing. Nearly 14 percent of Somalia's land area was covered by forest in 1991. Frankincense and myrrh, both forest products, generated some foreign exchange. In 1988, myrrh exports alone were valued at almost 253 million shillings. A government parastatal (a semi-autonomous, quasi-governmental, state-owned enterprise) in 1991 no longer had monopoly rights on the sale of frankincense and myrrh, but data on sales since privatization were not available.

In accordance with rainfall distribution, southern and northwestern Somalia has a relatively dense thornbush savanna, with various succulents and species of acacia. By contrast, the high plateaus of northern Somalia have wide, grassy plains, with mainly low formations of thorny shrubs and scattered grass tussocks in the remainder of the region. Northeastern Somalia and large parts of the northern coastal plain, on the other hand, are almost devoid of vegetation. Exceptions to this are the wadi areas and the moist zones of the northern coastal mountains, where the frankincense tree (Boswellia) grows. The myrrh tree (Commiphora) thrives in the border areas of southern and central Somalia. More efficient and careful handling of Boswellia, Commiphora, and other resin-exuding trees could increase yields of aromatic gums.

Savanna trees had been Somalia's principal source of fuel, but desertification had rapidly eroded this fuel source, especially because refugees from the Ogaden War had foraged the bush in the vicinity of refugee camps for fuel. The acacia species of the thorny savanna in southern Somalia supply good timber and are the major source of charcoal, but charcoal production has long exceeded ecologically acceptable limits.

The government's 1988 development report stated that its sand dune stabilization project on the southern coast remained active: 265 hectares of a planned 336 hectares had been treated. Furthermore, thirty-nine range reserve sites and thirty-six forestry plantation sites had been established. Forestry amounted to about 6 percent of the Gross Domestic Product.

As elevations and rainfall increase in the maritime ranges of the north, the vegetation becomes denser. Aloes are common, and on the higher plateau areas of the Ogo are woodlands. At a few places above 1,500 meters, the remnants of juniper forests (protected by the state) and areas of candelabra euphorbia (a chandelier-type cactus) occur. In the more arid highlands of the northeast, boswellia and commiphora trees are sources, respectively, of the frankincense and myrrh for which Somalia has been known since ancient times.

The region encompassing the Shabeelle and Jubba rivers is relatively well watered and constitutes the country's most arable zone. The lowland between the rivers supports rich pasturage. It features arid to sub-arid savanna, open woodland, and thickets that include frequently abundant underlying grasses. There are areas of grassland, and in the far southwest, near the Kenyan border, some dry evergreen forests are found.

Other vegetation includes plants and grasses found in the swamps into which the Shabeelle River empties most of the year and in other large swamps in the course of the lower Jubba River. Mangrove forests are found at points along the coast, particularly from Chisimayu to near the Kenyan border. Uncontrolled exploitation appears to have caused some damage to forests in that area. Other mangrove forests are located near Mogadishu and at a number of places along the northeastern and northern coasts.

Transportation

Inadequate transport facilities are a considerable impediment to Somalia's economic development. In 1988 the total expenditure for transportation and communications was $57.8 million. Nearly 55 percent of this amount was for new infrastructure; 28 percent was for rehabilitation and maintenance of existing infrastructure. This activity must be understood in the context of the ongoing civil war in Somalia; much of the infrastructure particularly bridges in the north, either had deteriorated or been destroyed, as a result of the fighting. As of early 1992, no systematic study existed of the infrastructural costs of the ongoing civil war.

At independence, Somalia inherited a poorly developed transportation system consisting of a few paved roads in the more populated areas in the south and northwest, four undeveloped ports equipped only with lighterage facilities, and a handful of usable airstrips. During the next three decades, some improvement was made with the help of substantial foreign aid. By 1990 all-weather roads connected most of the important towns and linked the northern and southern parts of the country. Three ports had been substantially improved, eight airports had paved runways, and regular domestic air service also was available. But in early 1992, the country still lacked the necessary highway infrastructure to open up undeveloped areas or to link isolated regions, and shipping had come to a virtual halt because of the security situation.

Roads and Railroads

There are no railways in Somalia; internal transportation is by truck and bus. The national road system nominally comprises 22,100 kilometers (13,702 mi.) of roads that include about 2,600 kilometers (1,612 mi.) of all-weather roads, although most roads have received little maintenance for years and have seriously deteriorated. In 1990 Somalia had more than 21,000 kilometers of roads, of which about 2,600 kilometers were paved, 2,900 kilometers were gravel, and the remainder was improved and unimproved dirt roads. The country's principal highway was a 1,200-kilometer two-lane paved road that ran from Chisimayu in the south through Mogadishu to Hargeysa in the north. North of Mogadishu, this route ran inland, roughly paralleling the border with Ethiopia; a 100-kilometer spur ran to the Gulf of Aden at Berbera. By early 1992 much of this road, especially the northern part between Hargeysa and Berbera, was relatively unsafe because of land mines. Somalia's 1988 plan provided for another connection from this main route to Boosaaso on the Gulf of Aden. Somalia had only one paved road that extended from north of Mogadishu to Ethopia; all other links to neighboring countries were dirt trails that are impassable in rainy weather.

Only about 1,800 miles of the paved roads are passable year-round, and in the rainy seasons most rural settlements are not accessible by motor vehicle. Buses, trucks, and minibuses are the main means of transport for the population. In rural areas camels, cattle, and donkeys are still used for personal transportation and as pack animals. Motor vehicles registered in 1995 numbered only 24,000.

Currently, there are no operating railways in Somalia; the railroad between Jawhar and Mogadishu has been closed since 1988.

Ports and Airports

Four ports handled almost all of Somalia's foreign trade. Berbera, Mogadishu, and Chisimayu were deepwater ports protected by breakwaters. Merca, just south of Mogadishu, was a lighterage port that required ships to anchor offshore in open roadsteads while loading and unloading. Mogadishu was the principal port of entry for most general cargo. Berbera received general cargo for the northern part of the country and handled much of the nation's livestock exports. United States aid enabled the doubling of the berths at the port of Barkera and the deepening of the harbor, completed in 1985. Maydh, northwest of Erigaro, was the only other and much smaller northern port. Chisimayu's main function was the export of bananas and meat; the meat was processed and packed at the port. The United States also financed the development of Chisimayu port in the latter half of the 1980s. Merca was an export point for bananas. In 1986 the Somali Ports Authority launched a modernization project for all ports, with concentration on Mogadishu.

Ports and terminals are located in Boosaaso, Berbera, Chisimayu (Kismaayo), Merca and Mogadishu. The three larger cities, Mogadishu, Berbera, and Kismaayo, also have deep-water harbors, but dangerous coral reefs keep coastal traffic to a minimum.

There are no commercial water transport facilities. The ports of Mogadishu, Chisimayu, and Berbera are served by vessels from many parts of the world.

The major airfields are in Mogadishu and Berbera. Mogadishu International Airport was the nation's principal airfield. In the 1980s, the runway was extended to 4,500 meters, and is one of Africa's longest runways. The runway was built with the United States' financial aid. The airport was further expanded in 1989 by Italy's contribution from its emergency aid fund for Africa. Only Mogadishu's airport offered international flights.

Somali Airlines, the nation's flag carrier, was partially owned by Alitalia, the Italian national airline. Somali Airlines in 1989 replaced its fleet of five aging 707 airplanes with one Airbus 310, making it a one-plane international airline. In 1990 domestic service linked Mogadishu with Berbera and six other Somali cities; flights were scheduled at least once a week. As of April 1992, Somali Airlines had no scheduled flights, domestic or international, and no other regular flights existed.

According to statistics from 2005, there are 64 airports in Somalia. Of these, 6 are paved, and 58 are not. Of the paved airports, only 4 have runways in excess of 3,000 meters; of the unpaved airports, only 4 are over 3,000 meters, and 6 more are less than 914 meters. During peacetime, the state-owned Somali Airlines operated national routes as well as on international routes to Kenya, Arabia, and Europe. Mogadishu, Berbera, and Kismaayo; all have airports with runways long enough for major aircraft takeoff and landing.

In 1989, before the collapse of the government, the national airline had only one airplane. According to a World Bank report, the "private airline business in Somalia is now thriving with more than five carriers and price wars between the companies." The owner of Daallo Airlines says, "Sometimes it's difficult without a government and sometimes it's a plus, but corruption is not a problem, because there is no government."[11]

Economy

Somalia's economic fortunes are driven by its deep political divisions. Economic life continues, in part because much activity is local and relatively easily protected. Agriculture is the most important sector, with livestock normally accounting for about 40% of Gross Domestic Product (GDP) and about 65% of export earnings, but Saudi Arabia's ban on Somali livestock, due to Rift Valley Fever[12] concerns, has severely hampered the sector. Nomads and semi-nomads, who are dependent upon livestock for their livelihood, make up a large portion of the population.

Somalia's agricultural products include bananas, sorghum, corn, coconuts, rice, sugarcane, mangoes, sesame seeds, beans; cattle, sheep, goats and fish. Livestock, hides, fish, charcoal, and bananas are Somalia's principal exports, while sugar, sorghum, corn,

---

[11] *Africa Open for Business*, World Bank, March 18, 2005.

[12] According to the World Health Organization, "The virus, which causes Rift Valley Fever, is a member of the *Phlebovirus* genus, one of the five genera in the family *Bunyaviridae*. Since 1930, when the virus was first isolated during an investigation into an epidemic amongst sheep on a farm in the Rift Valley of Kenya, there have been outbreaks in sub-Saharan and North Africa. In 1997-98, there was a major outbreak in Kenya and Somalia. In September 2000, RVF was for the first time reported outside of the African Continent. Cases were confirmed in Saudi Arabia and Yemen. This virgin-soil epidemic in the Arabian Peninsula raises the threat of expansion into other parts of Asia and Europe." See: http://www.who.int/mediacentre/factsheets/fs207/en/

qat, and machined goods are the principal imports. Somalia's industries include a few light industries, including sugar refining, textiles and wireless communication. However, the small industrial sector, based mostly on the processing of agricultural products, has largely been looted and sold as scrap metal.

Despite the seeming anarchy, Somalia's service sector has managed to survive and grow. The ongoing civil disturbances and clan rivalries, however, have interfered with any broad-based economic development and international aid arrangements. Somalia's arrears to the IMF continued to grow in 2005. Statistics on Somalia's GDP, growth, per capita income, and inflation should be viewed skeptically. In late December 2004, a major tsunami caused an estimated 150 deaths and resulted in destruction of property in coastal areas.

Before the start of civil war in the early 1990s, the manufacturing sector was beginning to develop. However, all industries suffered major losses during the civil war. In 2000 industry and manufacturing accounted for only 10% of GDP. Industries mainly serve the domestic market and, to a lesser extent, provide some of the needs of Somalia's agricultural exports, such as the manufacture of crates for packing bananas.

The most important industries were petroleum refining (as of 2000 shut down), the state-owned sugar plants at Jowhar and Gelib, an oilseed-crushing mill, and a soap factory. Other industries manufactured corrugated iron, paint, cigarettes and matches, aluminum utensils, cardboard boxes and polyethylene bags, and textiles. A cement plant at Berbera was completed in 1985.

The fish- and meat-canning export industries operate below capacity. Textiles are produced at the SOMALTEX plant, which supplies virtually the entire domestic market. Most major enterprises were government-owned, but private plants produce food, beverages, chemicals, clothing, and footwear. There are also plants for milk processing, vegetable and fruit canning, and wheat flour and pasta manufacturing, as well as several grain mills. The country's first pharmaceuticals factory, near Mogadishu, opened in 1986. Local craft industries produce sandals and other leather products, cotton cloth, pottery, baskets, and clay or meerschaum vessels.

Since the collapse of the state, Somalia has transformed from what Mohamed Siad Barre referred to as "scientific socialism" to a free market economy. It has long been one of the world's poorest and least developed countries and has relatively few natural resources. Somalia's poverty was even further aggravated by the hostilities of the civil war started in 1991. Agriculture is the most important sector, with livestock accounting for about 40% of GDP and about 65% of export earnings. Somalia continues to have one of the highest child mortality rates in the world, with 10% of children dying at birth and 25% of those surviving birth dying before age five. The international aid group, Medecins Sans Frontieres (Doctors Without Borders) has further stated that the level of daily violence due to the lack of government is "catastrophic."

There are some signs of growth in Somalia. The CIA Factbook says: "Despite the seeming anarchy, Somalia's service sector has managed to survive and grow. Mogadishu's main market offers a variety of goods from food to the newest electronic gadgets. Hotels continue to operate, and militias provide security." Infrastructure, such as roads are as numerous as those in neighboring countries but of much lower quality. A World Bank report states that the private sector has found it too hard to build roads due to high transaction costs and the fact that those who pay road fees are not the only ones using the road, presenting a problem with recuperation of investment.

The thriving telecommunications industry is private, offering wireless service and internet cafés. Competing phone companies have agreed on interconnection standards, which were brokered by the United Nations funded Somali Telecom Association. Electricity is furnished by entrepreneurs, who have purchased generators and divided cities into manageable sectors.

The private sector also supplies clean water. However, a statistic from 2000 indicated that only 21% of the population had access to safe drinking water at that time.

The main problem affecting economic growth is the lack of stability, or the perception of it. For businesses to operate, it is necessary to provide some level of security and internationally recognized governments are widely perceived as being more reliable in this than the traditional tribal leadership that currently holds sway in Somalia. However, investors are feeling more comfortable lately; for example, a Coca-Cola bottling plant opened in Mogadishu in 2004. Remittance services has become a large industry in Somalia. Successful people from the world-wide diaspora who fled because of the war contribute to the economy around $2 billion annually. In the absence of a formal banking sector, money exchange services have sprouted throughout the country, handling between $500 million and $1 billion in remittances annually. Wireless communications has also become a giant economic force in Somalia. Because of the war, nobody really knows the size of the economy or how much it is growing.

Somalia's economy is based on agriculture; however, the main economic activity is not crop farming but livestock raising. Between 1969 and the early 1980s, the military government imposed a system of "Scientific Socialism," which featured the nationalization of banks, insurance firms, oil companies, and all large industrial firms. This "Scientific Socailism" also included the setting up of state-owned enterprises, farms, and trading companies, and the organizing of state-controlled cooperatives. In the end, this experiment weakened the Somalian economy considerably, and since the collapse of the military regime the economy has suffered even more as a result of civil war. Generally speaking, at the present, the Somalian economy cannot survive without foreign aid.

Somalia's most valuable resources are the natural pastures that cover most of the country. Another resource that has scarcely been exploited is the abundant fish life in the coastal waters, still unpolluted by industrial waste. A potential source of hydroelectricity is the Jubba River.

By far the most important sector of the economy is agriculture, with livestock raising surpassing crop growing fourfold in value and earning about 90 percent of Somalia's foreign exchange. Agriculture in Somalia can be divided into three subsectors. The first is nomadic pastoralism, which is practiced outside the cultivation areas. This sector, which raises goats, sheep, camels, and cattle, has become increasingly market-oriented. The second sector is the traditional, chiefly subsistence, agriculture practiced by small farmers. This traditional sector takes two forms: rain-fed farming in the south and northwest, which raises sorghum, often with considerable livestock; and small irrigated farms along the rivers, which produce corn (maize), sesame, cowpeas, and, near towns, vegetables and fruits. The third sector consists of market-oriented farming on medium-

and large-scale irrigated plantations along the lower Jubba and Shabeelle rivers. Here the major crops are bananas, sugarcane, rice, cotton, vegetables, grapefruit, mangoes, papayas, and other fruits.

In the small fishing sector, tunny and mackerel are caught and canned in the north, sharks are often caught and sold dried by artisanal inshore fishers, and, in southern Somalia, choice fish and shellfish are processed for export. A small fishing industry exists in the north where tuna, shark, and other warm-water fish are caught, although fishing production is seriously affected by poaching.

In the late 1980s, industry was responsible for just under 10 percent of Somalia's gross domestic product. Mogadishu was the chief industrial center, with bottling plants, factories producing spaghetti, cigarettes, matches, and fishing boats, a petroleum refinery, a small tractor assembly workshop, and small enterprises producing construction materials. In Kismaayo there were a meat-tinning factory, a tannery, and a modern fish factory. There were two sugar refineries, one near Jilib on the lower reach of the Jubba and one at Jawhar (Giohar) on the middle reach of the Shabeelle.

Even before the destruction caused by Somalia's civil wars of the 1980s and '90s, the productivity of Somalian factories was very low. Often entire works did not operate at full capacity or produced nothing at all over long periods. The few existing power stations, located at Mogadishu, Hargeysa (Hargeisa), and Kismaayo, were often out of order, resulting in frequent power cuts with adverse effects on factory production. (Rural areas have no power plants at all.) A significant portion of commodities necessary for daily life is produced by small workshops in the informal sector.

Somalia has a large trade deficit. Its chief export commodities are livestock (to Arab countries, mainly Saudi Arabia) and bananas (to Italy and Arab countries). Other, much less important exports are hides and skins, fish, and frankincense and myrrh. Almost everything is imported, even food for an urban population no longer accustomed to the traditional diet.

Besides the official market, there is also a flourishing black market, by means of which tens of thousands of Somali workers in Arab countries provide commodities missing on the Somali market while avoiding the duties levied on imports. Since wages in Somalia are very low, almost every family is directly or indirectly involved in informal trading.

Energy

Somalia relied principally on domestic wood and charcoal and on imported petroleum to meet its energy needs. Attempts to harness the power of the Jubba River at the proposed Baardheere Dam had not come to fruition as of early 1992. Electrical utilities had been state owned since 1970, when foreign-owned enterprises were nationalized. Throughout the country, about eighty different oil-fired thermal and diesel power plants relied on imported petroleum. With aid from Finland, new plants were constructed in the Chisimayu and Baidoa areas in the mid-1980s.

Somalia relied on foreign donors, first the Soviet Union and then Saudi Arabia, to meet its petroleum needs. In the late 1970s, Iraq helped Somalia build a refinery at Jasiira, northeast of Baraawe, that had a capacity of 10,000 barrels a day. But when the Iran-Iraq War broke out in 1980, deliveries were suspended, and Somalia again required refined oil imports. As of mid-1989, Somalia's domestic requirements were again being met by this refinery, but deliveries of Iraqi crude oil were erratic. In May 1989, Somalia signed an agreement with the Industrial Export, Import, and Foreign Trade Company of Romania by which the company was to construct an oil refinery on the outskirts of Mogadishu. The project was to cost US$500 million and result in a refining capacity of 200,000 barrels per day. Because of subsequent events in Romania and Somalia, the refinery project had not materialized as of early 1992.

Throughout the 1980s, various international oil companies explored for oil and natural gas deposits in Somalia. In October 1991, the World Bank and the UN Development Programme announced the results of its hydrocarbon study in the countries bordering the Red Sea and the Gulf of Aden. The study indicated the potential for oil and gas in northern Somalia was good. However, exploitable oil and natural gas have not yet been found. In view of the civil war in Somalia following the fall of Siad Barre, however, various foreign oil exploration plans were canceled.

There is some evidence that both oil and natural gas are present, and the now-defunct transitional government installed in 2000 and two of the breakaway republics signed exploration agreements with foreign oil companies. But exploration and extraction are likely to be difficult and expensive for the foreseeable future.

The oil refinery at Mogadishu, with a production capacity of 10,000 barrels per day, has been out of operation since 1991. There is one natural gas field, but exploration and exploitation of oil and natural gas has been suspended since political conflict began.

A successful innovation was the completion of a wind energy utilization project. Four wind turbines, each rated at 50 kilowatts, were embedded in the Mogadishu electrical grid. In 1988 these turbines produced 699,420 kilowatt hours of energy. Total electric energy produced in 1988, the latest year for which figures were available in early 1992, was 257 million kilowatt hours. Five self-contained wind energy conversion systems in rural centers also were planned, but as of May 1992 there was no information that these had been built.

Mining

The local geology suggests the presence of valuable mineral deposits, but both quantity and quality are too low for mining to be commercially worthwhile. No commercially exploitable quantities of minerals have been found in Somalia, with the possible exception of uranium (for which at this time there is no market). However, as of 1992, only a few significant mining sites had been located, and mineral extraction played a very minor role in the economy. Natural resources in Somalia include uranium and largely unexploited reserves of bauxite, coal, copper, guano, gypsum, iron ore, phosphate, salt, tin, uranium, natural gas and possibly some oil reserves.

Tin was commercially mined by the British in Somalia before World War II. Sea salt is collected at several sites on the coast. The deposits of the clay mineral sepiolite, or meerschaum, in south-central Somalia are among the largest known reserves in the world.

The Somali minerals sector, which was not a significant economic force before the 1991 overthrow of the government, failed to expand in the ensuing years of political and economic instability. In 2003, small quantities of gypsum, marine salt, and sepiolite (meerschaum) were exploited, and the country also presumably produced clays, sand and gravel, crushed and dimension stone, and limestone (for lime manufacture and agriculture). Officially reported mineral and trade data have been unavailable owing to lack of a central government from 1991 to 2000, and because of the secession of Somaliland and Puntland.

In 2003, Somalia's output of gypsum, marine salt and sepiolite were estimated at 1,500 metric tons, 1,000 metric tons and 6 metric tons, respectively. The civil war forced the closure of Somalia's cement plant and oil refinery (a leading industry), and halted exploration for natural gas and other resources. There were unexploited deposits of anhydrite, bauxite, columbite, feldspar, natural gas, iron ore, kaolin, quartz, silica sand, tantalum, thorium, tin, and uranium. The mining outlook shows little change for the short run.

Recent discoveries of amethyst, aquamarine, emerald, garnet, opal, ruby, and sapphires indicate a possible source of gemstones. But current mining of the gemstones in Somaliland has been limited by a lack of modern equipment, civil strife, and damage to the infrastructure. An EU-funded non-governmental organization was working with Somaliland's government to exploit these gemstone resources.

History

From their connection with the Ethiopian hinterland, their proximity to Arabia, and their export of precious gums, ostrich feathers, ghee (clarified butter), and other animal products as well as slaves from farther inland, the northern and eastern Somali coasts have for centuries been open to the outside world. This area probably formed part of Punt, "the land of aromatics and incense," mentioned in ancient Egyptian writings. Between the 7th and 10th centuries, immigrant Muslim Arabs and Persians developed a series of trading posts along the Gulf of Aden and Indian Ocean coasts. Many of the early Arab geographers mentioned these trading posts and the sultanates that grew out of them, but they rarely described the interior of the country in detail.

Intensive European exploration really began only after the occupation of Aden by the British in 1839 and the ensuing scramble for Somali possessions by Britain, France, and Italy. In 1854, while Richard Burton was exploring the country to the northwest in the course of his famous journey from Berbera to Harer, his colleague, John Hanning Speke, was making his way along the Makhir coast in the northeast.

This region had previously been visited by Charles Guillain, captain of the brig Ducouedid, between 1846 and 1848. Guillain also sailed down the Indian Ocean coast and went ashore at Mogadishu, Marka, and Baraawe, penetrating some distance inland and collecting valuable geographic and ethnographic information.

In 1865 the German explorer Karl Klaus von der Decken sailed up the Jubba River as far as Baardheere in the small steamship Welf, which foundered in rapids above the town. Decken was killed by a Somali, but much valuable information collected by his expedition survived.

In 1883 a party of Englishmen, F.L. and W.D. James, G.P.V. Aylmer, and E. Lort-Phillips, penetrated from Berbera as far as the Shabeelle, and between 1886 and 1892 H.G.C. and E.J.E. Swayne surveyed the country between the coast and the Shabeelle and also reached farther east toward the Nugaal valley.

During 1894–95, A. Donaldson-Smith explored the headwaters of the Shabeelle in Ethiopia, reached Lake Rudolf, and eventually descended the Tana River to the Kenyan coast.

In 1891 the Italian Luigi Robecchi-Bricchetti trekked from Mogadishu to Hobyo and then crossed the Ogaden to Berbera. About the same time, further explorations were made by another Italian, Captain Vittorio Bottego.

In the 20th century several extensive surveys were made, especially in the British protectorate, by J.A. Hunt between 1944 and 1950, and much of the country was mapped by aerial survey.

During World War II the British protectorate evacuated their European citizens in 1940. But the British Protectorate was recaptured along with Italian Somalia in 1941, at the same time when Ethiopia also was liberated. Just before and during World War II, a number of maps were made by the British, French, US and Italian forces. Many of these maps and geological surveys, now de-classified, are now available for public use from the British Library and the Library of Congress.

With the exception of French Somaliland, all the Somali territories were united under one British military administration. In 1948 the protectorate reverted to the Colonial Office; the Ogaden and the Hawd regions were gradually surrendered to Ethiopia; and by 1950 the Italians had returned to southern Somalia with a mandate of 10 years to prepare the country for independence under a United Nations trusteeship.

Taking advantage of the modest progress that the British military administration had effected, the Italians rapidly pursued social and political advancement, although commercial development and infrastructure improvement was much more difficult. The independence of the British Somaliland Protectorate from the United Kingdom was proclaimed on June 26, 1960. A few days later, on July 1, 1960, Italian Somalia followed suit, and the two territories were unified as the Somali Republic. The government was formed by Abdullahi Issa. Aden Abdullah Osman Daar was appointed as President and Abdirashid Ali Shermarke as Prime Minister. Later, in 1967, Mohammed Ibrahim Egal became Prime Minister in the government appointed by Abdirishid Ali shermarke. Egal was later chosen as President of the self-declared independent Somaliland. He died in a hospital in Pretoria on May 3, 2002.

The politics of the new republic were conditioned by clan allegiances, but the first major problems arose from the last-minute marriage between the former Italian trust territory and the former British protectorate. Urgent improvements in communication between the two areas were necessary, as were readjustments in their legal and judicial systems. The first independent government was formed by a coalition of the southern-based Somali Youth League (SYL) and the northern-based Somali National League (SNL).

While modest developments were pursued internally with the help of mainly Western aid, foreign policy was dominated by the Somali unification issue and by the campaign for self-determination of adjoining Somali communities in the Ogaden, French Somaliland, and northern Kenya. The Somalian government strongly supported the Kenyan Somali community's aim of self-determination (and union with Somalia); when this failed in the spring of 1963, after a commission of inquiry endorsed Somali aspirations, Somalia broke off diplomatic relations with Britain, and a Somali guerrilla war broke out in northern Kenya, paralyzing the region until 1967. By the end of 1963 a Somali uprising in the Ogaden led to a brief confrontation between Ethiopian and Somalian forces. Since the United States and the West provided military support to Ethiopia and Kenya, Somalia turned to the Soviet Union for military aid. Nevertheless, the republic maintained a generally neutral but pro-Western stance, and, indeed, a new government formed in June 1967 under the premiership of Maxamed Xaaji Ibrahiim Cigaal (Muhammad Haji Ibrahim Egal) embarked on a policy of détente with Kenya and Ethiopia, muting the Pan-Somali campaign.

In March 1969 more than 1,000 candidates representing 64 parties (mostly clan-based) contested the 123 seats in the National Assembly. After these chaotic elections, all the deputies (with one exception) joined the SYL, which became increasingly authoritarian. The assassination of President Cabdirashiid Cali Sherma'arke (Abdirashid Ali Shermarke) on October 15, 1969, provoked a government crisis, of which the military took advantage to stage a coup on October 21.

The overthrow of Cigaal brought to power as head of state and president of a new Supreme Revolutionary Council the commander of the army, Major General Maxamed Siyaad Barre (Muhammed Siad Barre). At first the new regime concentrated on consolidating its power internally. Siyaad quickly adopted "Scientific Socialism," which, he claimed, was fully compatible with his countrymen's traditional devotion to Islam. Leading a predominantly military administration, Siyaad declared a campaign to liberate the country from poverty, disease, and ignorance. The president was soon hailed as the "Father" of the people (their "Mother" was the "Revolution," as the coup was titled). Relations with socialist countries (especially the Soviet Union and China) were so greatly strengthened at the expense of Western connections that, at the height of Soviet influence, slogans proclaiming a trinity of "Comrade Marx, Comrade Lenin, and Comrade Siyaad" decorated official Orientation Centres throughout the land. Siyaad's authoritarian rule was reinforced by a national network of vigilantes called Victory Pioneers, by a National Security Service headed by his son-in-law, and by National Security Courts notorious for ruthless sentencing. Rural society was integrated into this totalitarian structure through regional committees on which clan elders (now renamed "peace-seekers") were placed under the authority of a chairman, who was invariably an official of the state apparatus. Clan loyalties were officially outlawed, and clan-inspired behavior became a criminal offense. Of the government's many crash programs designed to transform society, the most successful were mass literacy campaigns in 1973 and 1974, which made Somali a written language (in Latin characters) for the first time.

After 1974, Siyaad Barre turned his attention to external affairs. Somalia joined the Arab League, gaining much-needed petrodollar aid and access to political support from those Persian Gulf states to which Somali labor and livestock were exported at a growing rate. Following Haile Selassie's overthrow in September 1974, Ethiopia began to fall apart, and guerrilla fighters of the Western Somali Liberation Front (WSLF) in the Ogaden pressed Siyaad (whose mother was an Ogaadeen) for support. When in June 1977 France granted independence to Djibouti (under a Somali president), the WSLF, backed by Somalia, immediately launched a series of fierce attacks on Ethiopian garrisons. By September 1977 Somalia had largely conquered the Ogaden region and the war was at the gates of Harer. Then the Soviet Union turned to fill the superpower vacuum left in Ethiopia by the gradual withdrawal of the United States. In the spring of 1978, with the support of Soviet equipment and Cuban soldiers, Ethiopia re-conquered the Ogaden, and hundreds of thousands of Somali refugees poured into Somalia.

In late 1969, a military government assumed power following the assassination of Shermarke, who had been chosen, and served as, President from 1967–1969. Mohamed

Siad Barre, a General in the armed forces became the President in 1969 following a coup d'état. The revolutionary army leaders, headed by Major General Mohamed Siyad Barre, who came from the majority tribe in Somalia, established large-scale public works programs. They also successfully implemented an urban and rural literacy campaign, in which they helped to dramatically increase the literacy rate from a mere 5% to 55% by the mid-1980s.

In the meantime, Barre assassinated a major figure in his cabinet, Major General Gabiere, and two other officials. Intermittent civil war has been a fact of life in Somalia since 1977. In 1991, first insurgent forces led by Abdullahi Yusuf Ahmed, leader of the (SSDF), and President Ali Mahdi Mohamed officially unrecognised, ousted Siad Barre's government. Turmoil, factional fighting, and anarchy have followed in the years since.

The same year, the northern portion of the country declared its independence as Somaliland; although de facto independent and relatively stable compared to the tumultuous south, it has not been recognized by any foreign government.

Many years of conflict, severe drought, and famine have left Somalia in a continuing state of crisis. Hundreds of thousands of Somali have been displaced by warfare and civil unrest. Chronic food shortages have led to high rates of malnutrition in many parts of the country. Much of Somalia is without adequate water supplies or sanitation. Cholera, measles, tuberculosis, and malaria are widespread. The absence of health or welfare infrastructure in the country, which are largely destroyed after years of conflict, has left international relief organizations struggling to provide essential services normally offered by the government. However, their efforts are hindered by continuing violence, and most Somali have little or no access to health care.

Conditions in the Republic of Somaliland and Puntland are somewhat better than in the rest of the country but still fall short of ideal. Because of the overall level of stability enjoyed by the two self-governing regions, they have been able to rebuild much of their health care infrastructure.

Somalia was also one of the many countries affected by the tsunami which struck the Indian Ocean coast following the 2004 Indian Ocean earthquake, destroying entire villages and killing an estimated 300 people.

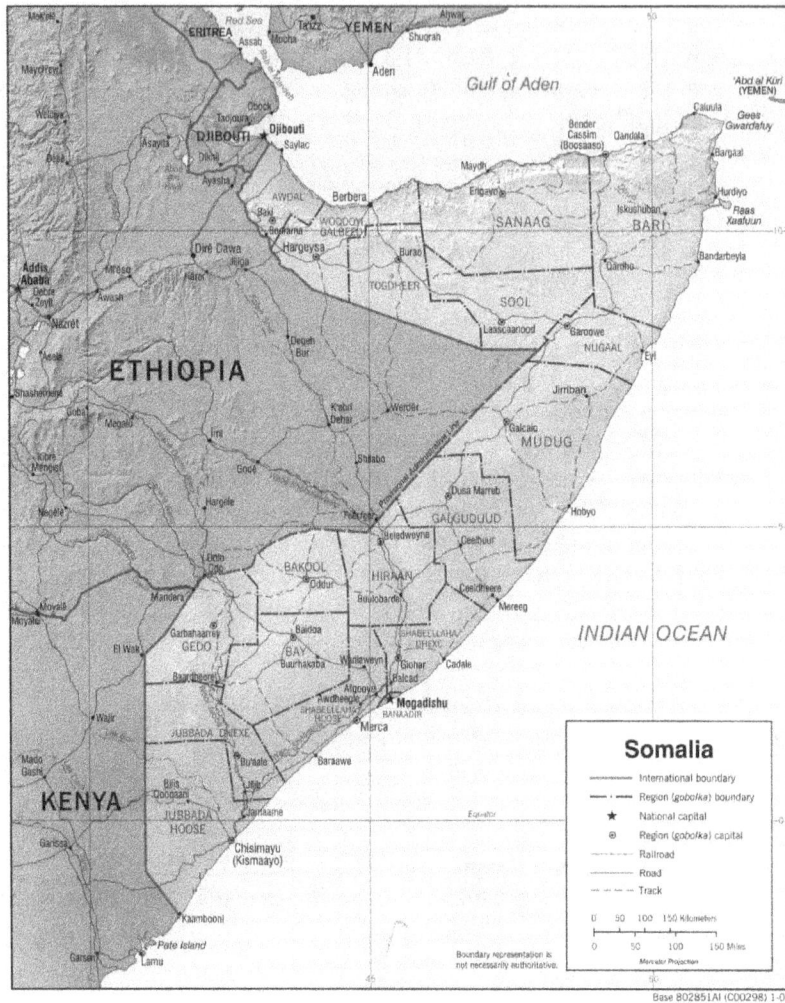

A

Abbate, E. 1985. "Costituzione Geologica Del Bacino dell'Uebi Scebeli in Relazione Al Contenuto Salino Delle Acque Superficali e Sotterranee." Translated title: "Geological Constitution of Scebeli River Basin in Connection to the Saline Contents of the Surface Water and Ground Water." Somalia: Universita Nazionale Somala, Facolt Agraria, Mogadishu, Somalia. Descriptors: Africa; agriculture; East Africa; ground water; saline composition; Scebeli River basin; Somali Republic; surface water; water resources. Database: GeoRef. Accession Number: 1996-040304.

Abbate, E., Abdi Salah, H., Bruni, P. and Sagri, M. 1993. Pre- and Synrift Sedimentation in a Tertiary Basin along the Gulf of Aden (Daban Basin, Northern Somalia). Firenze; Istituto Agronomico L'Oltremar; 1993. Relazioni E Monografie Agrarie Subtropicali E Tropicali Nuova Serie113 A, pages: 211-240. International Meeting; 1st- 1987 Nov: Mogadishu. Notes: Geology of Somalia and surrounding regions: Geology and mineral resources of Somalia and surrounding regions. OCLC Accession Number: 34175029.

Abbate, E. Abdi, Saleh H. Bruni, P. and Sagrim, M. 1987. Geology of the Daban Basin; Excursion B, Guidebook. Somalia: Universita Nazionale Somala, Mogadishu, Somalia. Descriptors: Africa; areal geology; Daban Basin; East Africa; guidebook; northern Somalia; Somali Republic. Notes: Guidebook for GeoSom 87, Nov. 23-30, 1987. Database: GeoRef. Accession Number: 1996-040305.

Abbate, E. Bigazzi, G. Norelli, P. and Quercioli, C. 1993. Fission-Track Ages and the Uplift of the Northern Somali Plateau. Firenze; Istituto Agronomico L'Oltremar; 1993. Relazioni E Monografie Agrarie Subtropicali E Tropicali Nuova Serie113 A, pages: 369-378. International Meeting; 1st- 1987 Nov: Mogadishu. Notes: Geology of Somalia and surrounding regions: Geology and mineral resources of Somalia and surrounding regions. OCLC Accession Number: 34175029.

Abbate, E. Bruni, P. Fazzuoli, M. and Sagri, M. 1987. The Continental Margin of Northern Somalia and the Evolution of the Gulf of Aden; GeoSom 87; International Meeting; Geology of Somalia and Surrounding Regions; Abstracts. Somalia: Somali Natl. Univ., Mogadishu, Somalia. GeoSom 87; Geology of Somalia and Surrounding Regions, Mogadishu. Somalia Conference: Nov. 23-30, 1987. Descriptors: Africa; Arabian Plate; Arabian Sea; Cenozoic; continental margin; East Africa; en echelon faults; evolution; faults; Gulf of Aden; Indian Ocean; intrusions; magnetic anomalies; Miocene; Neogene; Nubian Plate; Oligocene; Paleogene; plate tectonics; reactivation; rifting; sea-floor spreading; Somali Republic; structural geology; tectonics; tectonophysics; Tertiary; upper Miocene. OCLC Accession Number: 34175029.

Abbate, E. Dal Piaz, G. V. and Del Moro, A., et al. 1987. Review of Late Proterozoic-Cambrian Evolution of Northeastern Somali Continental Crust; GeoSom 87; International Meeting; Geology of Somalia and Surrounding Regions; Abstracts. Somalia: Somali Natl. Univ., Mogadishu, Somalia. GeoSom 87; Geology of Somalia and Surrounding Regions, Mogadishu. Somalia Conference: Nov. 23-30, 1987. Descriptors: Africa; age; areal geology; Basaso; Cambrian; clastic rocks; correlation; Crinozoa; East Africa; Echinodermata; facies; greenschist facies; Heis; Inda Ad Complex; intrusions; Invertebrata; Mait; Mait Complex; metamorphic rocks; metasedimentary rocks; molasse; Neoproterozoic; northeastern Somali Republic; Paleozoic; Pan-African Orogeny; passive margins; plate tectonics; Porifera; Precambrian; Proterozoic; review; sedimentary rocks;

sedimentary structures; shallow-water environment; Somali Republic; upper Precambrian. OCLC Accession Number: 34175029.

Abbate, E. Hussein, Abdi Salah; Bruni, P. Fazzuoli, M. and Sagri, M. 1987. Pre- and Syn-Rift Deposits in a Tertiary Basin Along the Gulf of Aden (Daban Basin, Northern Somalia); GeoSom 87; International Meeting; Geology of Somalia and Surrounding Regions; Abstracts. Somalia: Somali Natl. Univ., Mogadishu, Somalia. GeoSom 87; Geology of Somalia and Surrounding Regions, Mogadishu. Somalia Conference: Nov. 23-30, 1987. Descriptors: Africa; basins; Baulder Beds; carbonate sediments; Cenozoic; Chordata; clastic sediments; coastal environment; cyclic processes; Daban Basin; deltaic environment; depositional environment; East Africa; Eocene; fluvial environment; folds; Foraminifera; fresh-water environment; grain size; intertidal environment; Invertebrata; lacustrine environment; lagoonal environment; microfossils; middle Eocene; Mollusca; northern Somali Republic; Oligocene; Paleogene; Paleosols; Pisces; Protista; rifting; sebkha environment; sedimentary basins; sediments; Somali Republic; stratigraphy; subtidal environment; synclines; Tertiary; upper Oligocene; Vertebrata. OCLC Accession Number: 34175029.

Abbate, E. Sagri, M. Sassi, F. P. and Istituto agronomico per l'oltremare. 1993. Geology and Mineral Resources of Somalia and Surrounding Regions: With a Geological Map of Somalia 1:1,500,000. Firenze Italy: Istituto agronomico per l'oltremare. Descriptors: 2 v. Geology- Somalia; Geology- Somalia- Maps; Mines and mineral resources- Somalia; Hydrogeology- Somalia; Groundwater- Somalia; Water resources development- Somalia. Abstract: A. Regional geology- B. Mineral and water resources. Notes: " ... 43 papers presented at the First International Meeting on the Geology of Somalia and Surrounding Regions (GEOSOM '87), held in Mogadishu, November 23-30, 1987" Two maps on 2 folded leaves in pocket in volume 2. Includes bibliographical references. Funding: Supported by Istituto agronomico per l'oltremare. Other Titles: Geological map of Somalia 1:1,500,000; Papers presented at the First International Meeting on the Geology of Somalia and Surrounding Regions, 1987; GEOSOM '87. First International Meeting on the Geology of Somalia and Surrounding Regions, 1987. OCLC Accession Number: 34175029.

Abdalla, J. A. Cortiana, G. Frizzo, P. and Jobstraibizer, P. 1987. Black Heavy- Mineral Beach Sands from Batalale (Berbera, Northern Somalia); GeoSom 87; International Meeting; Geology of Somalia and Surrounding Regions; Abstracts. Somalia: Somali Natl. Univ., Mogadishu, Somalia. GeoSom 87; Geology of Somalia and Surrounding Regions, Mogadishu. Somalia Conference: Nov. 23-30, 1987. Descriptors: Africa; amphibole group; Batalale; Berbera; chain silicates; coastal environment; composition; East Africa; economic geology; epidote; epidote group; garnet group; genesis; granulometry; heavy mineral deposits; ilmenite; magnetite; monazite; nesosilicates; northern Somali Republic; orthosilicates; oxides; phosphates; placers; pyroxene group; rutile; silicates; Somali Republic; sorosilicates; titanite. OCLC Accession Number: 34175029. See also: Abdalla, J. A. Cortiana, G. Frizzo, P. and Jobstraibizer, P. G. 1993. Black Heavy-Mineral Beach Sands from Batalaleh (Berbera, N. Somalia). Firenze; Istituto Agronomico L'Oltremar; 1993. Relazioni E Monografie Agrarie Subtropicali E Tropicali Nuova Serie113 B, pages: 541-550. International Meeting; 1st- 1987 Nov: Mogadishu. Notes: Geology of Somalia and surrounding regions: Geology and mineral resources of Somalia and surrounding regions.

Abdalla, J. A., Said, A. A. and Visona, D. 1996/10. "New Geochemical and Petrographic Data on the Gabbro-Syenite Suite between Hargeysa and Berbera-Shiikh (Northern Somalia)." Journal of African Earth Sciences. Volume 23, Issue 3, Pages 363-373. Notes: ID: 2327. Abstract: The Somali crystalline basement in northern Somalia contains a Neoproterozoic igneous suite essentially composed of gabbros with minor syenitic and granitic bodies. This Gabbro-Syenite Suite (GSS) was emplaced at a relatively high crustal level and is related to Late Precambrian crustal extension. The basic plutons are composed mostly of gabbro with petrographic and geochemical characteristics ranging from N-type to T- and P-type MORB. This compositional variability points to heterogeneous sub-continental mantle sources. The plutons do not show significant in situ differentiation, suggesting that the cumulitic rocks may be related to cooling processes developed in deeper magma chambers and/or during magma ascent. ISSN: 1464-343X.

Abdel-Gawad, M. 1970. "Interpretation of Satellite Photographs of the Red Sea and Gulf of Aden; A Discussion on the Structure and Evolution of the Red Sea and the Nature of the Red Sea, Gulf of Aden and Ethiopia Rift Junction." Philosophical Transactions of the Royal Society of London, Series A: Mathematical and Physical Sciences. Volume 267, Issue 1181, Pages 23-40. Descriptors: Apollo 7; Arabian Sea; displacement; faults; folds; Gemini XI; Gulf of Aden; Indian Ocean; interpretation; photo geology; Red Sea; regional; rift zones; satellite photographs; Somali plateau; structural geology; systems; tectonics. Notes: With discussion; illustrations (incl. sketch maps); Abstract: Large-scale translational movements, displacement structures, left-lateral displacement of Arabia some 40 km to the north-northeast since Plio-Pleistocene. ISSN: 0080-4614.

Abdi Mohamed M. 1989. "Croyances Populaires Et Religions Classiques En Somalie." Mondes En Developpement. Volume 17, Issue 66, Pages 49-75. Descriptors: International Development Abstracts- 74; religion; Islam; divination. Notes: Somalia. Abstract: Demonstrates that the Somali people today still use divination and worship some of their ancient gods, despite the spread of Islam. Christianity and Judaism have also had an influence, and together they constitute a complex mystic amalgam. -English summary. ISSN: 0302-3052.

Abdi Salah, H. 1993. Paleocene to Eocene Planktonic Foraminiferal Biozones in Coriolei 1 Well (Mogadishu Coastal Basin, Somalia). Firenze; Istituto Agronomico L'Oltremar; 1993. Relazioni E Monografie Agrarie Subtropicali E Tropicali Nuova Serie113 A, pages: 181-196. International Meeting; 1st- 1987 Nov: Mogadishu. Notes: Geology of Somalia and surrounding regions: Geology and mineral resources of Somalia and surrounding regions. OCLC Accession Number: 34175029.

Abdi, Abdullahi Ahmed. 1987. Preliminary Report on the Investigation of the El Bur Sepiolite Project; GeoSom 87; International Meeting; Geology of Somalia and Surrounding Regions; Abstracts. Somalia: Somali Natl. Univ., Mogadishu, Somalia. GeoSom 87; Geology of Somalia and Surrounding Regions, Mogadishu. Somalia Conference: Nov. 23-30, 1987. Descriptors: Africa; Bur; Cenozoic; clay minerals; clays; East Africa; economic geology; Eocene; Indho Qabyo; lagoonal environment; Miocene; Neogene; Paleogene; possibilities; reserves; sepiolite; sheet silicates; silicates; Somali Republic; Tertiary. OCLC Accession Number: 34175029.

Abdillahi, O.H. 1991. "Somaliassa Osa-Aikatiesto" Translated title: "Somalia's Part-Time Roads." Tie Ja Liikenne. Vol. 51 No. 4. 32 pages. Finnish. English Summary: Comments of Somali Highway Engineer Osman Hagi Abdillahi. Available from Northwestern University Transportation Library through interlibrary loan or document delivery.

Abdirahim, M. M. Abdirahman, H. Ali Kassim, M. Carmignani, L. and Fantozzi, P. 1987. Geological Mapping of Juba Valley; GeoSom 87; International Meeting; Geology of Somalia and Surrounding Regions; Abstracts. Somalia: Somali Natl. Univ., Mogadishu, Somalia. GeoSom 87; Geology of Somalia and Surrounding Regions, Mogadishu. Somalia Conference: Nov. 23-30, 1987. Descriptors: Africa; areal geology; Caanoole Formation; Cambar Formation; Cretaceous; East Africa; Fanweyn Formation; folds; Garbahaarey Formation; Gedo; Giuba Valley; Jubba Valley; Jurassic; Kuredka Formation; Luuq-Mandera Basin; Mesozoic; reactivation; regional; Somali Republic; synclinoria; Tomalo Syncline; transpression; Waajid Formation. OCLC Accession Number: 34175029.

Abdirahim, M. M. Ali Kassim, M. Carmignani, L. and Coltorti, M. 1993. The Geomorphological Evolution of the Upper Jubba Valley in Southern Somalia. Firenze; Istituto Agronomico L'Oltremar; 1993. Relazioni E Monografie Agrarie Subtropicali E Tropicali Nuova Serie113 A, pages: 241-250. International Meeting; 1st- 1987 Nov: Mogadishu. Notes: Geology of Somalia and surrounding regions: Geology and mineral resources of Somalia and surrounding regions. OCLC Accession Number: 34175029.

Abdulla, Ali D. 1996. "Somalia's Reconstruction: An Opportunity to Create a Responsive Information Infrastructure." The International Information & Library Review. Volume 28, Issue 1, Pages 39-57. Abstract: This article envisages the development of an effective information infrastructure in future Somalia that would contribute to the economic and social renewal of that country. The article highlights the role of information in Somalia's reconstruction. It then reviews the state of the information infrastructure in Somalia before the collapse of the government in 1990, examining both existing and proposed components of the infrastructure and identifying associated problems. In its final section, the article delineates an "appropriate" information infrastructure which would be consistent with the conditions and needs of information users in that environment. This contains an outline of the attributes of such an infrastructure and offers specific recommendations to future Somali development planners and international donors supporting information development activities.

Abdulla, Omer Haji Dualeh. 1980. Characterization and Pedogenic Classification of the Soils of the Central Agricultural Research Station, Afgoi, Somali Democratic Republic. Descriptors: 3, iii, 72 i.e. 74 leaves; Soils- Somalia; Soils- Classification. Notes: Includes bibliographical references (leaves 66-67). Microfilm copy of typescript. Ann Arbor: University Microfilms, 1980.- 1 reel; 35 mm. Dissertation: Thesis (M.S.)-- Michigan State University. Dept. of Crop and Soil Sciences, 1979. Responsibility: by Omer Haji Dualeh Abdulla. OCLC Accession Number: 6564078.

Abdulla, Omer Haji Dualeh. 1979. Characterization and Pedogenic Classification of the Soils of the Central Agricultural Research Station, Afgoi, Somali Democratic Republic. Descriptors: 3, iii, 72 i.e. 74 leaves; Soils- Somalia; Soils- Classification. Notes: Includes bibliographical references (leaves 66-67). Dissertation: Thesis (M.S.)--

Michigan State University. Dept. of Crop and Soil Sciences, 1979. Responsibility: by Omer Haji Dualeh Abdulla. OCLC Accession Number: 6564026.

Abdullahi A.M and Jahnke H.E. 1990. "Some Aspects of Pastoral Supply Behaviour and Rural Livestock Marketing in Africa: The Case of Central Somalia." Q. J. Int. Agric. Volume 29, Issue 4, Pages 341-359. Descriptors: Social, economic and political aspects of rural change; rural area; developing country; pastoralism; nomads; supply behavior; livestock marketing; market mechanism; pastoral supply behavior Species Term: Somalia; Animalia. Notes: Africa Somalia. Abstract: The paper shows the main relationships between demand and supply and livestock prices, between prices and season and between demand and supply and season. These individual segments are subsequently summarized to characterize the underlying economic principle determining quantities and prices of marketed livestock. The paper illustrates the existence of a market equilibrium at three levels, in each case prices and quantities of livestock are determined by their supply and demand. While seasonal flexibilities in the supply of animals from herds to markets can be observed, depending on the condition of stock and a herding household's particular needs and aspirations, the general market mechanism shows the same prevalent interdependencies between demand and supply as in any other free market system. Social behavior reinforces this development rather than being an obstacle as can be seen during the religious feasts where the supply response of the nomads is to sell more livestock.

Abdurahman O.S and Bornstein S. 1991. "Diseases of Camels (Camelus Dromedarius) in Somalia and Prospects for Better Health." Nomadic Peoples. Volume 29, Pages 104-112. Descriptors: Crop and livestock production- technical aspects- 74.1.3; camel disease research; veterinary treatment. Notes: Somalia. Abstract: Diseases diagnosed in camels in Somalia are described, together with an outline of traditional knowledge of such diseases and healing practices. The difficulties of interpreting the significance of sero-epidemiological studies of camels using methods originally designed e.g. for cattle are discussed. Although camel disease research is still in its infancy, there is ample scope to improve the health and production of the animals by introducing extension services including disease prophylaxis, treatment, and better management. - Authors. ISSN/ISBN: 0822-7942.

Abid, S. M. and Jibril, S. A. 1986. Assistance to the Rehabilitation and Effective Utilization of Rural Water Reservoir in Somalia; Implementation of the Project. Somalia: FAO, Mogadishu, Somalia. Descriptors: Africa; East Africa; reservoirs; rural environment; Somali Republic; surface water; water resources. Database: GeoRef. Accession Number: 1996-067440.

Aboukhaled, A. and Saouma, E. (director). 1973. "Progress Report on the Regional Applied Research Programme on Land and Water use in the Near East; Calcareous Soils; Report of the FAO/UNDP Regional Seminar on Reclamation and Management of Calcareous Soils." Soils Bulletin. Issue 21, Pages 133-151. Descriptors: Africa; Asia; characterization; Cyprus; East Africa; Egypt; genesis; Iran; Iraq; Jordan; leaching; Lebanon; Middle East; morphology; Near East; North Africa; profiles; soil management; soil surveys; soils; Somali Republic; Sudan; surveys; Syria; utilization; water regimes. Notes: References: 16; tables. ISSN/ISBN: 0532-0437.

Abucar, Moallim Hassan; Corazza, M. Corsini, F. Malesani, P. G. Sagri, M. and Tanelli, G. 1987. Carnotite and Sepiolite Mineralizations in the Galgudug Region

(Central Somalia); GeoSom 87; Geology of Somalia and Surrounding Regions; Late Abstracts. Somalia: Somali Natl. Univ., Dep. Geol., Mogadishu, Somalia. GeoSom 87; Geology of Somalia and Surrounding Regions, Mogadishu. Somalia Conference: Nov. 24-Dec. 1, 1987. Descriptors: Africa; basalts; breccia; calcite; carbonate rocks; carbonates; carnotite; Cenozoic; chlorides; clastic rocks; clay minerals; coal; East Africa; economic geology; El-Bur; electron microscopy data; framework silicates; Galgudug; goethite; gypsum; halides; halite; igneous rocks; iron oxides; lacustrine environment; lignite; limestone; manganese oxides; metal ores; mineral deposits, genesis; Miocene; Mudug Series; Neogene; Oligocene; organic compounds; organic materials; organic residues; oxides; Paleogene; palygorskite; quartz; sedimentary rocks; SEM data; sepiolite; sheet silicates; silica minerals; silicates; Somali Republic; spectra; spinel; sulfates; Tertiary; upper Miocene; upper Oligocene; uranium ores; vanadates; vanadium ores; volcanic rocks; Waabo; X-ray diffraction data; X-ray fluorescence spectra. OCLC Accession Number: 34175029.

Acocella, V., Korme, T., Salvini, F. and Funiciello, R. 2002. "Elliptic Calderas in the Ethiopian Rift: Control of Pre-Existing Structures." Journal of Volcanology and Geothermal Research. Volume 119, Issue 1-4, Pages 189-203. Descriptors: Geothermal prospecting; Tectonics; Fracture; Remote sensing. Abstract: The Ethiopian Rift is characterized by several Quaternary calderas. Remote sensing and field analyses were used to investigate the regional structural control on three calderas (Fantale, Gariboldi, Gedemsa) in the axial part of the rift. These calderas are located along the Wonji Fault Belt (WFB), a zone of Quaternary NNE-SSW normal faults and extensional fractures. The three calderas show E-W elongation and major E-W vent alignments, oblique with regard to the mean NW-SE extension direction. No significant evidence of E-W tectonic structures has been found near the calderas, the only relevant systems being those of the WFB. Conversely, left-lateral E-W-trending faults are present at the rift borders and on the Nubia and Somalia plateaus, implying a predominant pre-rift activity. The E-W fractures were partly reactivated during rifting, possibly controlling the development of the magma chambers. Thus, the E-W elongation of the calderas would be the surface expression of such a control, rather than the result of regional extension. An evolutionary model on the role of different structures on magmatism at different crustal levels within the rift is proposed. ISSN: 0377-0273.

Adam H.M. 1994. "Formation and Recognition of New States: Somaliland in Contrast to Eritrea." Rev. Afr. Polit. Econ. Volume 59, Pages 21-38. Descriptors: Political; democratic form; developing country; state formation; political comparison; comparative studies. Notes: Somalia- Somaliland. Abstract: "Somaliland' has reasserted the separate existence it had as the colony of British Somaliland before independence and union with the former Italian Somalia in 1960. It has avoided the devastation of warlordism that has afflicted the rest of Somalia through compromise politics between clan elders. However, its de facto statehood since 1991 has not received the international recognition accorded Eritrea in 1993. The experiences of Somaliland and Eritrea in the circumstances of their post-colonial union with other entities, in their liberation movements and in their current politics are contrasted. It is suggested that there can be mutual learning from Somaliland's consociational, ethnic democracy and Eritrea's radical social democracy of an eventual, orchestrated multipartism that eschews ethnic and religious divides. ISSN: 0305-6244.

Aden A.J and Frizzo P. 1996. "Geochemistry and Origin of Low and High TiO2 Mafic Rocks in the Barkasan Complex: A Comparison with Common Neoproterozoic Gabbros of Northern Somali Crystalline Basement." Journal of African Earth Sciences. Volume 22, Issue 1, Pages 43-54. Descriptors: Igneous geochemistry- 72.2.1; magma composition; igneous geochemistry; Proterozoic; gabbro; Barkasan Complex. Notes: Somalia. Abstract: The Barkasan complex comprises numerous metagabbroic bodies intruded along a Neoproterozoic shear zone in the crystalline basement of northern Somalia. The constituting rocks can be subdivided into a low TiO2 (0.41-1.39 wt.%) group, and a high TiO2 (2.13-6.75 wt.%) group, typical of the upper portion of the Ago Marodi intrusion. The compositional and mineralogical changes result from fractionation and differentiation in a cooling tholeiitic magma. By comparison, the Sheikh gabbro (one of the principal Neoproterozoic gabbroic intrusions in northern Somalia), despite its common contents in phases of the first stage of fractional crystallization, has lower nickel, chromium and Mg. Significant differences also exist in terms of Zr/Y, Zr/Rb and Zr/Nb ratios, reflecting a different parental magma composition. ISSN: 0899-5362.

Aden, A. I. 1985. "Somali Democratic Republic: Solar Desalination Experience." Non-Conventional Water Resources use in Developing Countries.Proceedings of the Interregional Seminar. Volume Netherlands Antilles April 22-28, Pages 2 fig. Descriptors: Desalination; Somali Democratic Republic; Solar distillation; Water supply development; Distillation; Solar energy; Developing countries; Corrosion; Maintenance; Pilot plants; Biomass; Saline water intrusion. Abstract: High evaporation rates in the Somali Democratic Republic are responsible for saline water conditions in many groundwater supplies. To alleviate the problems associated with high costs of imported oil, the government of Somali embarked on research into low-cost renewable energies such as solar, wind and biomass. The United Nations Childrens Fund and the United Nations Industrial Development Organization helped design and build the Kudha solar desalination plant. The pilot plant consists of 15 blocks of 9 solar still units each. Three storage tanks store raw seawater, rain water and distilled water, and the water supply for the village, respectively. Deficiencies associated with plant operation include poor reliablity, an inadequate labor force, and corrosion and maintenance problems. The 3,600 sq m plant serves a village population of about 2,000 people. (See also W88-07850) (Geiger-PTT). Database: Environmental Sciences and Pollution Mgmt.

Aerospace Center (U.S.). 1995. "Tactical Pilotage Chart, TPC. K-5C, Djibouti, Ethiopia, Somalia." St. Louis, Mo; Riverdale, MD: The Center; NOAA Distribution Branch (N/CG33), National Ocean Service distributor. Descriptors: Aeronautical charts- Djibouti; Aeronautical charts- Ethiopia; Aeronautical charts- Somalia; Government publication; National government publication. Notes: Description: 1 map; color; on sheet 106 x 146 cm. folded to 27 x 37 cm. Scale 1:500,000; Lambert conformal conic proj., standard parallels 9020' and 14040'; (E 370--E 430/N 120--N 80). Cartgrph Code: Category of scale: a Constant ratio linear horizontal scale: 500000 Coordinates-- westernmost longitude: E0370000 Coordinates--easternmost longitude: E0430000; Notes: Relief shown by contours, shading, tints, and spot heights. "Air information current through 24 February 1995." "Compiled September 1982; revised February 1995." "Printed by DMA 10-95." "Copyright 1995 by the United States Government. No copyright claimed under Title 17 U.S.C." Includes elevation tints and interchart relationship diagrams. Other Titles: Djibouti, Ethiopia, Somalia; Tactical pilotage chart; TPC. OCLC Accession Number: 33387842.

Aerospace Center (U.S.). 1993. "Tactical Pilotage Chart, TPC. L-6A, Ethiopia, Somalia." St. Louis, Mo; Riverdale, MD: The Center; NOAA Distribution Branch (N/CG33), National Ocean Service distributor. Descriptors: Aeronautical charts-Ethiopia; Aeronautical charts- Somalia; Government publication; National government publication. Notes: Description: 1 map; color; on sheet 106 x 146 cm. folded to 27 x 37 cm. Scale 1:500,000; Lambert conformal conic proj., standard parallels 1020' and 6040'; (E 420--E 480/N 80--N 40). Cartgrph Code: Category of scale: a Constant ratio linear horizontal scale: 500000 Coordinates--westernmost longitude: E0420000 Coordinates--easternmost longitude: E0480000; Notes: Relief shown by contours, shading, tints, and spot heights. "Air information current through 3 January 1985." Shipping list no. 99-2018-S. "Compiled February 1985." "Lithographed by DMAAC 3-93." Includes elevation tints and interchart relationship diagrams. Other Titles: TPC, Ethiopia, Somalia; Ethiopia, Somalia. OCLC Accession Number: 42386026.

Aerospace Center (U.S.). 1985. "Tactical Pilotage Chart, TPC. L-5B, Ethiopia, Kenya, Somalia." St. Louis, Mo; Riverdale, MD: The Center; NOAA Distribution Branch (N/CG33), National Ocean Service distributor. Descriptors: Aeronautical charts-Ethiopia; Aeronautical charts- Kenya; Aeronautical charts- Somalia; Government publication; National government publication. Notes: Description: 1 map; color; on sheet 106 x 146 cm., folded to 27 x 37 cm. Scale 1:500,000; Lambert conformal conic proj., standard parallels 1020' and 6040', convergence factor .06979; (E 360--E 420/N 80--N 40). Cartgrph Code: Category of scale: a Constant ratio linear horizontal scale: 500000 Coordinates--westernmost longitude: E0360000 Coordinates--easternmost longitude: E0420000; Notes: Relief shown by contours, shading, tints, and spot heights. "Air information current through 22 January 1985." Shipping list no. 99-2024-S. "Compiled June 1984." "Lithographed by DMAAC 3-85." (Printing distributed by GPO lacks this note.). Includes elevation tints and interchart relationship diagrams. Other Titles: TPC, Ethiopia, Kenya, Somalia; Ethiopia, Kenya, Somalia. OCLC Accession Number: 30798125.

Aerospace Center (U.S.). 1985. "Tactical Pilotage Chart, TPC. L-5C, Ethiopia, Kenya, Somalia." St. Louis, Mo; Riverdale, MD: The Center; NOAA Distribution Branch (N/CG33), National Ocean Service, distributor. Descriptors: Aeronautical charts-Ethiopia; Aeronautical charts- Kenya; Aeronautical charts- Somalia; Government publication; National government publication. Notes: Description: 1 map; color; on sheet 106 x 146 cm., folded to 27 x 37 cm. Scale 1:500,000; Lambert conformal conic proj., standard parallels 1020' and 6040', convergence factor .06979; (E 360--E 420/N 40--N 00). Cartgrph Code: Category of scale: a Constant ratio linear horizontal scale: 500000 Coordinates--westernmost longitude: E0360000 Coordinates--easternmost longitude: E0420000; Notes: Relief shown by contours, shading, tints, and spot heights. "Air information current through 15 February 1985." DMAAC 3-93 lithograph distributed to depository libraries on shipping list no. 99-2017-S. "Compiled December 1984." "Lithographed by DMAAC 7-85." Includes elevation tints and interchart relationship diagrams. Other Titles: TPC, Ethiopia, Kenya, Somalia; Ethiopia, Kenya, Somalia. OCLC Accession Number: 30798169.

Aerospace Center (U.S.). 1984. "Operational Navigation Chart, ONC. K-5, Djibouti, Ethiopia, Somalia, Sudan, Yemen (San`a), Yemen (Aden)." St. Louis, Mo; Riverdale, MD: The Center; Distribution Division (C-44), National Ocean Service. Descriptors: Aeronautical charts- Djibouti; Aeronautical charts- Ethiopia; Aeronautical charts- Somalia; Aeronautical charts- Sudan; Aeronautical charts- Yemen; Aeronautical charts- Yemen (People's Democratic Republic); Government publication; National government publication. Notes: Description: 1 map; color; on sheet 105 x 146 cm. Scale 1:1,000,000; Lambert conformal conic proj., standard parallels 9020' and 14040'; (E 310--E 430/N 160--N 80). Cartgrph Code: Category of scale: a Constant ratio linear horizontal scale: 1000000 Coordinates--westernmost longitude: E0310000 Coordinates--easternmost longitude: E0430000; Notes: Relief shown by shading, tints, and spot heights. "Elevations in feet." "Air information current through 1 February 1984." "Compiled April 1977. Revised February 1984." "Lithographed by DMAAC 8-84." Includes terrain characteristic tints and interchart relationship diagrams. Other Titles: Djibouti, Ethiopia, Somalia, Sudan, Yemen (San`a), Yemen (Aden); Operational navigation chart. ONC, K-5. Responsibility: prepared and published by the Defense Mapping Agency Aerospace Center. OCLC Accession Number: 30152224.

Aerospace Center (U.S.). 1981. "Operational Navigation Chart, ONC. M-6, Seychelles, Somalia." St. Louis, Mo; Riverdale, MD: The Center; Distribution Division (C-44), National Ocean Service. Descriptors: Aeronautical charts- Seychelles; Aeronautical charts- Somalia; Government publication; National government publication. Notes: Description: 1 map; color; on sheet 105 x 146 cm. Scale 1:1,000,000; Lambert conformal conic proj., standard parallels 1020' and 6040'; (E 430--E 550/S 00--S 80). Cartgrph Code: Category of scale: a Constant ratio linear horizontal scale: 1000000 Coordinates--westernmost longitude: E0430000 Coordinates--easternmost longitude: E0550000; Notes: Relief shown by shading, tints, and spot heights. "Elevations in feet." "Air information current through 11 June 1981." "Compiled January 1970. Revised June 1981." "Lithographed by DMAAC 8-81." Other Titles: Seychelles, Somalia. Operational navigation chart. ONC, M-6. Responsibility: prepared and published by the Defense Mapping Agency Aerospace Center. OCLC Accession Number: 30304512.

Aerospace Center (U.S.) and Defense Mapping Agency Aerospace Center (U.S.). 1982. "Tactical Pilotage Chart, TPC. M-5B, Kenya, Somalia, Tanzania." St. Louis, MO: DMAAC. Descriptors: Aeronautical charts- Tanzania; Aeronautical charts- Kenya; Aeronautical charts- Somalia; Government publication; National government publication. Notes: Description: 1 map; color; on sheet 105 x 146 cm., folded to 27 x 37 cm. Scale 1:500,000; Lambert conformal conic proj., standard parallels 1020 and 6040, convergence factor .06979; (E 370--E 430/S 00--S 40). Cartgrph Code: Category of scale: a Constant ratio linear horizontal scale: 500000 Coordinates--westernmost longitude: E0370000 Coordinates--easternmost longitude: E0430000; Notes: Scale 1:500,000. Relief shown by shading, contours, and spot heights. "Air information current through 16 July 1982." "Lithographed by DMAAC 3-83." Other Titles: Kenya, Somalia, Tanzania. TPC M-5B. Responsibility: produced and published by the Defense Mapping Agency Aerospace Center. OCLC Accession Number: 30798330.

AFIS, Mogadishu, Somalia. 1957. Situazione Dei Pozzi Trivellati Ed a Gola Aperta Alla Data Del 20-27 Luglio 1957; Carta Geografica. Situation of the Drill Wells at Open Gorges on July 20-27, 1957; Geographic Map. Somalia: AFIS, Mogadishu,

Somalia. Descriptors: Africa; drilling; East Africa; Somali Republic; water resources; water wells; wells. Database: GeoRef. Accession Number: 1996-067503.

AFIS, Mogadishu, Somalia. 1956. Bonifica Di Bulo Mererta, Basso Uebi Scebeli; Informazioni e Dati Tecnici (Planimetrie, Relazioni e Foto). Reclamation of Bulo Mererta, Lower Scebele River; Information and Technical Data, Planimetry, Relationships and Photos. Somalia: AFIS, Mogadishu, Somalia. Descriptors: Africa; Bulo Mererta; East Africa; fluvial features; reclamation; rivers; Scebele River; Somali Republic; southern Somali Republic; water resources. Database: GeoRef. Accession Number: 1996-067502.

AFIS, Mogadishu, Somalia. 1954. Programmi Di Collaborazione Tecnica Con La FAO Per Lo Sviluppo Economico Della Somalia, Accordo Di Progetto n.2-Risanamento e Valorizzazione Dei Descek. Programs of Technical Collaboration with FAO for Economic Development of Somalia, Agreement of Project N.2-Reformation and Evaluation. Somalia: AFIS, Mogadishu, Somalia. Descriptors: Africa; East Africa; Somali Republic; water resources. Database: GeoRef. Accession Number: 1996-067498.

AFIS, Mogadishu, Somalia. 1954. Studio Per La Valorizzazione Delle Risorse Idriche Del Medio e Basso Scebeli. Study about the Exploitation of Water Resources of the Middle and Lower Scebele River. Somalia: AFIS, Mogadishu, Somalia. Descriptors: Africa; East Africa; exploitation; Scebele River; Somali Republic; southern Somali Republic; water resources. Database: GeoRef. Accession Number: 1996-067499.

AGIP MINER, Mogadishu, Somalia. 1977. Reports Diversi Dei Periodi 1957-1963 e 1976-1977, Numerati Da 1 a 56, Con Diverse Carte Geol.. several Reports of the Periods 1957-1963 and 1976-1977, Numbers from 1 to 56, with several Geologic Maps. Somalia: MRMI, Mogadishu, Somalia. Descriptors: Africa; East Africa; geologic maps; maps; petroleum; petroleum exploration; Somali Republic. Database: GeoRef. Accession Number: 1996-067505.

Agrar Hydrotech, Essen, Federal Republic of Germany. 1985. Hydrology and Water Management; Juba Valley; Progress Report; Annexes 1-5. Federal Republic of Germany: AHT/GTZ, Essen, Federal Republic of Germany. Descriptors: Africa; East Africa; hydrology; Juba River; report; Somali Republic; southern Somali Republic; water management; water resources. Database: GeoRef. Accession Number: 1996-067508.

Agrar Hydrotech, Essen, Federal Republic of Germany. 1985. Hydrology of the Juba River; Main Report; Annex 1-4. Federal Republic of Germany: AHT/GTZ, Essen, Federal Republic of Germany. Descriptors: Africa; East Africa; hydrology; Juba River; report; rivers and streams; Somali Republic; southern Somali Republic; surface water. Database: GeoRef. Accession Number: 1996-067509.

Agrar Hydrotech, Essen, Federal Republic of Germany. 1972. Study of Water Resources and Planning of Water Supply. Federal Republic of Germany: Agrar Hydrotech, Essen, Federal Republic of Germany. Descriptors: Africa; East Africa; Somali Republic; water management; water resources; water supply. Database: GeoRef. Accession Number: 1996-067507.

Agrisystems (EA) Ltd; IUCN East Africa Regional Office; Somali Natural Resources Management Programme and IUCN Eastern Africa Programme. 1997. Environmental Impact Assessment Manual and Guidelines for the Somali Water Sector. Nairobi, Kenya: The Office. Descriptors: 39; Water-supply engineering- Environmental aspects- Somalia; Water resources development- Environmental aspects- Somalia; Water-supply, Rural- Somalia- Management; Environmental impact analysis- Somalia-Handbooks, manuals, etc. Notes: At head of title: Somali Natural Resources Management Programme. "Project no. 6/SO-82/95 + 6/SO-83/04." "IUCN Eastern Africa Programme" taken from the cover. Responsibility: by Agrisystems (EA) Ltd for IUCN--the World Conservation Union, Eastern Africa Regional Office. LCCN: 00-284252. OCLC Accession Number: 45248559.

Ahmad, M. M. Coughlin, Terry L. and Hampton, D. L., et al. 1982. Somali Democratic Republic, East Africa: Coriole Basin, Interpretation of GECO 1980 Non-Exclusive Seismic Survey. Bartlesville: Phillips Petroleum Co., Frontier Exploration Section: SEAGAP. Descriptors: 1 case; Petroleum- Geology- Somalia; Geology-Somalia. Notes: COMPANY CONFIDENTIAL. Case includes 2 spiral binders and assorted maps. Other Titles: Coriole Basin, interpretation of GECO 1980 non-exclusive seismic survey. Interpretation of GECO 1980 non-exclusive seismic survey. Responsibility: by M.M. Ahmad, T.L. Coughlin, D.L. Hampton, W.R.E. Kelly. OCLC Accession Number: 14223068.

Ahmed, Ali H. 1986. Pozzo Sperimentale Di Ceel Baqal; Raccolta Ed Interpretazione Dei Dati. Experimental Wells of Ceel Baqal; Collection and Interpretation of the Data. Somalia: Universita Nazionale Somalia, Mogadishu, Somalia. Descriptors: Africa; Benadir; Ceel Baqal; East Africa; experimental studies; ground water; observation wells; Somali Republic; water wells; wells. Database: GeoRef. Accession Number: 1996-069170.

Ahmed, M. Abukar; Worsame, Ali; Sheikh, Abdel Ghaffar A. A. Sheikh-Ali, Bashir M. and Jobin, William. 1990. Water Quality of Jubba River in Somalia. Durango, CO, USA: Publ by ASCE, New York, NY, USA. pages: 559-566. Irrigation and Drainage- Proceedings of the 1990 National Conference, Jul 11-13 1990. Descriptors: Rivers- Somalia; Water Pollution- Water Quality; Environmental Testing; Irrigation. Abstract: Monitoring of water quality in the Jubba River in Somalia of East Africa, led to indications of a potential impact of toxic trace elements, salinity and biocides in restricting agricultural development. A multipurpose dam near the Ethiopian border and irrigation systems in the alluvial plain near the Indian Ocean were the main components in the proposed development. ISSN/ISBN: 0-87262-769-1.

Ahrens, Thomas P. [Ahrenz, T. P. and Azzaroli, A.] 1951. "A Reconnaissance Ground-Water Survey of Somalia, East Africa." Descriptors: ground water; Italian Somaliland; water supply; water, ground and surface. Notes: illustrations (incl. maps), Italy, Com. Interministeriale Ricostruzione, Rome; Abstract: The ground-water resources of Italian Somaliland are limited in ocurrence, and the waters are frequently mineralized. Study of the aquifers and the hydrology of the various provinces leads to the conclusion that additional development is possible, however. Database: GeoRef. Accession Number: 1996-069217.

Aimed, O. S.; Yusuf, O. S.; Hassan, A. D. 1991. "Standard Penetration Tests of Eolian Sands of Somalia." In: Proceedings of the 6th International Congress of the International Association of Engineering Geology, Amsterdam, 6-10 August 1990. Volume 1, Pages 323-327. Publ Rotterdam: A A Balkema, 1990. Also: 1991. International Journal of Rock Mechanics and Mining Science & Geomechanics Abstracts. Volume 28, Issue 4, Page A232.

Air Survey & Development GmbH; Clyde Surveys Limited; United States and Agency for International Development. 1985. "Mogadishu Town Plan." Mogadishu, Somalia: USAID. Descriptors: Government publication; National government publication. Notes: Description: 1 map; color; 57 x 74 cm., folded to 30 x 13 cm. Geographic: Mogadishu (Somalia)- Maps. Scale 1:18,000. Cartgrph Code: Category of scale: a Constant ratio linear horizontal scale: 18000; Notes: Panel title. Includes indexes, inset of "Town center," and location map. Responsibility: concept and design, Air Survey & Development GmbH; cartography of town plan, Clyde Surveys Limited. LCCN: 90-685829. OCLC Accession Number: 23047185.

Alaily F. 1995. "Soil Properties of Quaternary Deposits in the Arid Part of Northeastern Somalia." Arid Soil Research & Rehabilitation. Volume 9, Issue 1, Pages 39-50. Descriptors: SOILS- 71.5; mineralogy; carbonate; gypsum; Vertisol; saline soil. Notes: Somalia- Darror Valley. Abstract: The investigated soils are found in the Darror Valley in NE Somalia. Investigations indicate that all soils developed from Quaternary sediments are rich in carbonates and gypsum, are often saline, and often have a color of 7.5 YR, which becomes more red in the subsoil. Compared with Vertisols of more humid climatic conditions, the investigated Vertisols have cracks that penetrate the soil to a greater depth. They have a moderate content of clay and an unusual mineralogical composition. The clay fraction consists mostly of palygorskite, kaolinite, and chlorite and is poor in smectite. The palygorskite as well as most of the soluble salts, gypsum, and carbonates are lithogenous. According to the framework for land evaluation (Food and Agriculture Organization, 1976), the major part of the valley is classified as N1, N2, and NR. ISSN: 0890-3069.

Alaily F. 1993. "Soil Formation on Limestones in the Arid Region of Northeast-Somalia." Catena. Volume 20, Issue 3, Pages 227-246. Descriptors: Soils; alteration intensity; landscape history; carbonate weathering; arid soil; limestone Species Term: Somalia. Notes: Somalia- Darror Valley. Abstract: In order to clarify the history of landscape in the arid part of northern Somalia, the genesis of soils in the Darror Valley and its surroundings has been investigated. Four of the sampled soils are developed from limestone and are situated on levelled plateaus whereas the fifth soil is developed from alluvial material in a shallow depression. The alteration intensity of these soils are estimated by means of the Feo content and the Sil:All-ratio. By balancing the carbonate amount in the soils of the elevated area, the primary height of their parent rocks was calculated. The age of the soils was also calculated through determination of the amount of carbonate lost by weathering. The results of these calculations indicate that the oldest soil in the landscape is relatively young, and that most soils developed since the late Pleistocene, as a result of a continuous soil erosion since Tertiary. ISSN: 0341-8162.

Alaily F. 1990. "Soils of the Darror Valley (Northeast Somalia)." Catena. Volume 17, Issue 1, Pages 13-24. Descriptors: soil; arid region; evaporite; gypsum; saline; Tertiary evaporite formation Species Term: Somalia. Notes: Somalia- Darror Valley. Abstract: The Darror Valley is located in the arid part of NE Somalia. Its soils are highly influenced by their parent material, which consists mainly of evaporites, limestone, conglomerate and alluvial deposits. The most common soils in this area are calcic Solonchaks, gypsic Solonchaks, gypsic Calcisols, haplic Calcisols, eutric Leptosols and calcic Vertisols. All soils have a neutral to weakly alkaline pH-value and are very rich in carbonates, often rich in gypsum, often saline, and their moisture content at the end of the dry season is much lower than the moisture content of the permanent wilting point. The soil salinization is mainly due to Tertiary evaporite formations.ISSN: 0341-8162.

Alaily F, Lassonczyk B, Huth A and Gensior A. 1990. "Genesis of Soils in the Arid Part of Northeast Somalia." Berliner Geowissenschaftliche Abhandlungen, Reihe A. Volume 120, Issue 2, Pages 695-704. Descriptors: fluvial terrace; pedogenesis; parent material; arid soil; basaltic plateau; alluvial deposit. Notes: Somalia- Darror Valley Somalia- Tog-Damal-Cad Valley. Abstract: The genesis of soils of landscapes representing 3 geomorphological units has been investigated. These consist of 2 plateaus (near Elayo and Candala), and the Darror and Tog-Damal-Cad Valley. The parent material of the soils in the investigated areas consists of: basalt and alluvial sediments on two basalt plateaus; 5 fluvial terraces bearing coarse gravel in Tog-Damal-Cad Valley; and evaporites, limestones, conglomerates and alluvial deposits in the Darror Valley. According to field and laboratory investigations the Solonchaks and the saline soils in the area of the Gulf of Aden and in the southern part of the Darror Valley are developed by translocation of lithogenic and/or atmospheric salt. Skeletic soils from fluvial terraces, conglomerates and basalts as well as the upper part of most soils are poor in salt and have a weak alkaline pH that might be due to residual enrichment of sodium. Also caliche seems to develop only in skeletic soils, such as soils from fluvial gravel or conglomerates. In addition all soils are influenced by alteration processes and only the oldest ones are also influenced by clay migration. The oldest landscape surfaces are found under a dense stony cover blackened with desert varnish. The intensity of desert varnish increases with the surface age as well as with the Si:Al ratio. ISSN: 0172-8784.

Alaily, F. 1993. "Soil Formation on Limestones in the Arid Region of Northeast Somalia." Catena (Giessen). Volume 20, Issue 3, Pages 227-246. Descriptors: Africa; aluminum; atmospheric precipitation; calcareous composition; carbonate rocks; Cenozoic; chemical composition; chemically precipitated rocks; climate; color; Cretaceous; density; duricrust; East Africa; electrical conductivity; evaporites; gypsum; Halomorphic soils; horizons; iron; limestone; Mesozoic; metals; parent materials; pedogenesis; Pleistocene; Quaternary; relief; sedimentary rocks; silicon; size distribution; soil profiles; soils; Solonchak soils; Somali Republic; sulfates; Tertiary; water regimes. Notes: 28 anals., 7 tables, sketch map; Reference includes data from Geoline, Bundesanstalt fur Geowissenschaften und Rohstoffe, Hanover, Germany. ISSN/ISBN: 0341-8162.

Alaily, Fayez. 1996. "Palygorskitreicher Vertisol Im Ariden Bereich NO-Somalias." Translated title: "Palygorskite Rich Vertisol in the Arid Part of NE-Somalia." Z. Pflanzenernaehr. Bodenkd. Volume 159, Issue 2, Pages 163-167. Descriptors: Africa; arid environment; atmospheric precipitation; bedding plane irregularities; calcareous composition; chlorite; chlorite group; clay mineralogy; clay minerals; climate; color; East Africa; electrical conductivity; evapotranspiration; fines; horizons; hydrology; illite; kaolinite; major elements; moisture; mudcracks; palygorskite; pedogenesis; pH; relief; sedimentary structures; sheet silicates; silicates; smectite; soil profiles; soils; Somali Republic; temperature; terrestrial environment; tropical environment; Vertisols; water content; X-ray analysis. Notes: References: 24; illustrations incl. 16 anals., sect., 3 tables, sketch map; CY: GeoRef in Process, Copyright 2006, American Geological Institute. After editing and indexing, this record will be added to Georef. Reference includes data from Geoline, Bundesanstalt fur Geowissenschaften und Rohstoffe, Hanover, Germany. Database: GeoRef In Process. ISSN: 0044-3263.

Aldhous, Peter. 1995. "Doomed Somalis Saved with Special Diet." New Scientist. August 5. Volume 147, Pages 5. Descriptors: Famines; Public health- Somalia; Malnutrition; General Science; Applied Science & Technology. Notes: Illustration. Abstract: Aid agencies should make greater efforts to save the adult victims of famine, according to the author of a study of starving people in Somalia. Steve Collins of the Irish relief agency Concern Worldwide discovered that severely emaciated adults who have previously been written off as certain to die can make a full recovery if they receive the right treatment. ISSN/ISBN: 0262-4079.

Alemannin, Mazzocchi. 1925. "Il Problema Del Giuba." Translated title: "Juba River Problem." La Rassegna Italiana Del Mediterraneo. Volume 1925, Descriptors: Africa; East Africa; hydrology; Juba River; Somali Republic; surface water; water resources. Database: GeoRef.

Alexander, Julius. 1992. "Creating a Country's Infrastructure: Relief Organizations Operating in Somalia Have Built Distribution and Communication Networks in the Absence of a National Government." Air Cargo World. Vol. 81, No. 11 (Nov. 1992). Air transportation; Freight traffic. ISSN: 0745-5100.

Alexander, Julius. 1992. "The Logistics of Hope: Airlifting Food and Relief Supplies into Civil War-Torn Somalia is a Herculean Logistics Feat: Here's How Inbound Managers Cope With Crude Landing Conditions, No Security, and Other Obstacles." Inbound Logistics. Volume 12, No. 11 (Nov. 1992), Pages 24-27. Descriptors: Agricultural products; Air cargo; Disaster relief. Geographical Terms: Somalia. ISSN: 0888-8493.

Alexander, LM. 1998. Somali Current Large Marine Ecosystem and Related Issues. Malden, MA: Blackwell Science. Large Marine Ecosystems of the Indian Ocean: Assessment, Sustainability, and Management. Pppages: 327-331. Descriptors: Water currents; Fishery resources; Continental shelves; Resource development; Aquatic environment; Living resources; Man-induced effects; Resource conservation; Ecosystems; Somali Current; Marine. Abstract: The Somali Current LME measures approximately 700,000 km super(2). It extends for approximately 800 km southwest to northeast along the coast, from the vicinity of Dar es Salaam in the south to just north of the island of Socotra. The LME is fairy narrow here, because both the continental shelf and the bulk of the fisheries resources, excluding tuna, are close to the coast. In the north, the LME performs and abrupt turn to the west, as does the coastline itself, and occupies the southern part of the Gulf of Aden, westward to the vicinity of Bab-el-Mandeb. If some future management program is anticipated for the Somali Current LME, it should be approached cautiously, because of the unfortunate situation in Somali and because more general data will probably be required before the scientific needs of the ecosystem are fully understood. However, tentative plans could be formulated, including identification of the parameters of the ecosystem's health and preferred species in the interests of sustained development. Notes: TR: KE0000159. ISSN/ISBN: 0632043180. Database: ASFA: Aquatic Sciences and Fisheries Abstracts. Accession Number: 4674890.

Ali Kassim, M., Fantozzi, P.L, Carmignani, L. and Conti, P. 2002. "Geology of the Mesozoic-Tertiary Sedimentary Basins in Southwestern Somalia." Journal of African

Earth Sciences. Volume 34, Issue 1-2, Pages 3-20. Descriptors: Tertiary; Mesozoic; basin evolution; geological mapping; sedimentary basin. Notes: Additional Info: United Kingdom; References: Number: 89; Geographic: Somalia.

Abstract: Two main sedimentary basins can be recognized in southern Somalia, the NE-SW trending Mesozoic-Tertiary Somali coastal basin, and the NNE-SSW Mesozoic Luuq-Mandera basin. The two basins are separated by the Bur region where the Proterozoic-Early Paleozoic Metamorphic basement of southern Somalia outcrops. The investigated area covers part of the Metamorphic basement of southern Somalia and of the Luuq-Mandera basin, although this basement is not described in details in this paper. In the Bur region the basement outcrops discontinuously near inselbergs and monadnocks, which stand out of a blanket of recent sediments. Because of this patchy distribution and the limited areal extent of the outcrops, the structure of the metamorphic basement is difficult to reconstruct. A NW-SE trend of structures prevails and two metamorphic complexes (the Olontole and Diinsor complexes) can be recognized. The Luuq-Mandera basin is a wide NNE-SSW synclinorium, delimited to the SE by the basement high of the Bur region, and to the west by the crystalline basement high of NE Kenya (Northern Frontier district). The extreme thickness of Triassic sediments in the axial part of the basin, and the thinner and younger succession on both sides of the basin suggest that the Luuq-Mandera basin was a subsiding elongated area that was invaded by the sea in the early Mesozoic, during the dismembering of Gondwana. The Jurassic-Cretaceous succession that followed comprises two main cycles of transgression and regression: the carbonate sediments that lie at the bottom pass up section into shales, evaporites and sandstone deposits. Since late Cretaceous, continental contition prevaled, with a long phase of peneplanation, and then a general uplift, which brought about the creation of lake depressions and the capture of the Dawa river, with formation of the present Jubba valley. The main tectonic events in the study area, and throughout SW Somalia, are represented by strike-slip movements along vertical faults in the Sengif and Garbahaarrey belt. Deformation is localized within a narrow belt that extends for more than a 100 km in a NE-SW direction. The near parallelism between the fold axes and the regional orientation of faults indicates a right-lateral movements along faults. The structure of the Garbahaarrey belt consists of an anastomosing fault system that delimits elongated folded blocks, arranged in anticline-syncline structures, with subvertical axial surfaces and fold axes parallel to the main wrench faults. The orientation of folds and the typical "positive flower structur&equot; profile of the anticlines indicate that shortening was perpendicular to the strike of the wrench, i.e. in a SE-NW direction. In the Garbahaarrey belt, strike-slip and shortening, therefore, occurred contemporaneously and led to a relative transpression between the NW and SE blocks. The observed parallelism between fold and fault orientation cannot be explained with a simple rotation of pre-existing fold axes during transpression, but can be regarded as an example of folding and strike-slip movements that occurred simultaneously but independently along frictionless faults. The faults delimiting the anticlines accommodated the strike-slip component of transpression only, whereas the compressive component led to the generation of fold axes parallel to the wrench zone. Results of the field work are summarized in two geological maps of the Gedo, Bakool, and Bay regions (1:250,000) which accompany this report (maps are attached with this issue). ISSN: 0899-5362.

Ali Kassim, M. Carmignani, L. and Fantozzi, P. 1993. Tectonic Transpression in the Gedo Region. Firenze; Istituto Agronomico L'Oltremar; 1993. Relazioni E Monografie Agrarie Subtropicali E Tropicali Nuova Serie113 A, pages: 379-388. International Meeting; 1st- 1987 Nov: Mogadishu. Notes: Geology of Somalia and surrounding regions: Geology and mineral resources of Somalia and surrounding regions. OCLC Accession Number: 34175029.

Ali Kassim, M. Carmignani, L. and Fantozzi, P. 1987. Tectonic Transpression in the Gedo Region; GeoSom 87; International Meeting; Geology of Somalia and Surrounding Regions; Abstracts. Somalia: Somali Natl. Univ., Mogadishu, Somalia. GeoSom 87; Geology of Somalia and Surrounding Regions, Mogadishu. Somalia Conference: Nov. 23-30, 1987. Descriptors: Africa; East Africa; en echelon folds; faults; fold belts; folds; Gedo; grabens; Luuq-Mandera Basin; normal faults; reactivation; shear zones; Somali Republic; structural geology; synthetic faults; tectonics; transcurrent faults; transpression; wrench faults. OCLC Accession Number: 34175029.

Ali Kassim, M. Carmignani, L. Fantozzi, P. and Ferrara, G. 1993. Fissural Basalts of the Luuq Area (Central-Southern Somalia): Geology, Petrology and Isotopic Geochemistry. Firenze; Istituto Agronomico L'Oltremar; 1993. Relazioni E Monografie Agrarie Subtropicali E Tropicali Nuova Serie113 A, pages: 251-258. International Meeting; 1st- 1987 Nov: Mogadishu. Notes: Geology of Somalia and surrounding regions: Geology and mineral resources of Somalia and surrounding regions. OCLC Accession Number: 34175029.

Ali Kassim, M. Schneider, M. and Farah, Ibrahim Mohamed. 1987. Lineament Map of Northeastern Somalia; GeoSom 87; International Meeting; Geology of Somalia and Surrounding Regions; Abstracts. Somalia: Somali Natl. Univ., Mogadishu, Somalia. GeoSom 87; Geology of Somalia and Surrounding Regions, Mogadishu. Somalia Conference: Nov. 23-30, 1987. Descriptors: Africa; Bari; Cenozoic; East Africa; faults; Jurassic; lineaments; Mesozoic; Middle Jurassic; Miocene; Neogene; northeastern Somali Republic; Pliocene; reactivation; Somali Republic; structural geology; tectonics; Tertiary. OCLC Accession Number: 34175029.

Ali, Kassim M. Carmignani, L. and Fazzuoli, M. 1987. Geology of the Luuq-Mandera Basin. Somalia: Somali National University, Mogadishu, Somalia. Descriptors: Africa; areal geology; East Africa; expeditions; Gedo Somali Republic; guidebook; Luuq Somali Republic; Luuq-Mandera Basin; Somali Republic. Notes: Guidebook for GEOSOM '87; Excursion A. Database: GeoRef. Accession Number: 1996-070024.

Ali, M. Ashraf. 1984. A Compendium of Land Resource Studies in Somalia: Preparatory Assistance in Land use Planning and Management, Somalia. Mogadishu: Ministry of Agriculture: Food and Agriculture Organisation of the United Nations. Descriptors: 61 leaves; Land use- Research- Somalia; Soils- Research- Somalia; Hydrology- Research- Somalia. Notes: "May 1984." "TCP/SOM/2309(T)." Includes bibliographical references (leaves 56-60). LCCN: 90-980171; OCLC Accession Number: 29390284.

Ali, M. Y. 2005. Petroleum Geology and Hydrocarbon Potential of Somaliland. Feria de Madrid, Spain: Society of Petroleum Engineers, Richardson, TX 75083-3836, United States. pages: 3087-3090. 67th European Association of Geoscientists and Engineers, EAGE Conference and Exhibition, Incorporating SPE EUROPEC 2005-Extended Abstracts, Jun 13-16 2005. Descriptors: Crude petroleum; Geology;

Hydrocarbons; Geographical regions; Wells; Tectonics; Sediments; Deposition; Seismology. Abstract: Fewer than 10 exploration wells have been drilled in Somaliland (Northern Somalia), mostly in the northern onshore and offshore areas, making it one of the least explored countries in the world. Although tectonic developments have controlled the accommodation space available for sediment deposition since Early Jurassic, the wells drilled on both onshore and offshore have encountered source, reservoir and seal rocks. In this study seismic, well and outcrop data have been used to determine the petroleum system of the country. The results show that Upper Jurassic and Cretaceous units, and possibly Oligocene-Miocene units in the offshore area, show potential for hydrocarbon generation. Traps are provided by rollover anticlines associated with listric growth faults and rotated basement fault and horst blocks. These are controlled by Upper Jurassic to Lower Cretaceous tentional stresses. Basins which have potential for hydrocarbons include Berbera (Bihendual) basin (conjugate side of the Balhaf basin in Yemen), Las Durch, Raguda, and Al Mado basin between Eragavo and Puntland (conjugate side of the Masilah basin in Yemen). ISBN: 9073781981.

Allen, R. B., Abouzakhm, A. G. and Sikander, A. H. 1991. "Petroleum Geology of the Gulf of Aden; AAPG International Conference; Abstracts." AAPG Bull. Aug. Volume 75, Issue 8, Pages 1402. Descriptors: Africa; anhydrite; Arabian Peninsula; Arabian Sea; Asia; carbonate rocks; Cenozoic; clastic rocks; color alteration index; cuttings; East Africa; economic geology; Eocene; extension faults; faults; geothermal gradient; Gulf of Aden; Habshiya Limestone; Indian Ocean; Jurassic; limestone; macerals; marl; maturity; Mesozoic; natural gas; offshore; onshore; Paleogene; petroleum; reflectance; reservoir rocks; sedimentary rocks; shale; Somali Republic; source rocks; strike-slip faults; structural geology; subsidence; sulfates; Tertiary; transform faults; vitrinite; Wanderer Limestone; wells; Yemen. ISSN/ISBN: 0149-1423.

Al-Najim M and Briggs J. 1992. "Livestock Development in Somalia- a Critical Review." GeoJournal. Volume 26, Issue 3, Pages 357-362. Descriptors: Agriculture; Crop and livestock production- general; agricultural development; livestock production; developing country; range productivity; overgrazing; livestock development. Notes: Somalia. Abstract: Livestock development in Somalia has mostly focused on the provision of improved veterinary services, water supplies and marketing facilities, whilst range productivity, expressed in terms of the availability of range forage and improved methods and techniques of animal raising, has largely been mismanaged or ignored. This partial development by the government has had some positive consequences. However, there have also been negative consequences, in particular increased animal pressures on the range resources, with the result that Somali rangelands have been increasingly exposed to overgrazing. ISSN: 0343-2521.

Al-Najim M and Briggs J. 1990. "Overstocking in Somalia: A Study of the Carrying Capacity of Bay Region's Rangelands." Occasional Papers- University of Glasgow, Department of Geography. Volume 28, Pages 23p. Descriptors: Information systems, climatic and soil conditions; overstocking; overgrazing; communal range management; livestock; carrying capacity; pastoralism. Notes: Somalia- Bay Region. Abstract: The paper investigates the carrying capacity of Somali rangelands, and aims to identify the consequences of increased stocking rates of livestock on the availability of palatable plants. This is achieved by examining the relationship between the total annual amounts of consumable dry matter available under different levels of annual rainfall and the existing livestock units. The area selected for study is located in the centre of the southern part of Somalia, between the Juba and Shebelle rivers, and covers about 40 000 square kilometres. The results suggest that overstocking has induced overgrazing in Somalia in general, and in Bay Region in particular. Each pastoral household tends to keep as many animals as possible in order to maximise its gain from the available communal range resources. In addition, Somali rangelands have been subjected to only partial economic development by the Somali government: increases in watering points and improvement in veterinary services and livestock marketing facilities have been made, while the availability of range forage has not been sufficiently increased nor have pastoral practices been properly controlled or managed. British Library: GU-DG-OPS--28 6224.905.

Al-Thour, Khalid A. 1997. "Facies Sequences of the Middle-Upper Jurassic Carbonate Platform (Amran Group) in the Sana'a Region, Republic of Yemen; Special Issue on Mesozoic Rift Basins of Yemen." Mar. Pet. Geol. Sep. Volume 14, Issue 6, Pages 643-660. Descriptors: Africa; Al-Khothally Formation; Amran Group; Anthozoa; Arabian Peninsula; Asia; biozones; carbonate platforms; carbonate rocks; clastic rocks; Coelenterata; cycles; depositional environment; East Africa; Ethiopia; Invertebrata; Jurassic; lithofacies; marine environment; marl; Mesozoic; Middle Jurassic; paleogeography; Raydah Formation; regression; Sanaa Yemen; sea-level changes; sedimentary rocks; sedimentation; shallow-water environment; Somali Republic; Stromatoporoidea; structural controls; Tawilah Group; transgression; unconformities; Upper Jurassic; Wadi Al-Ahjur Formation; Yemen. Notes: References: 36; illustrations incl. block diags., sects., strat. cols., sketch maps; Latitude: N151000, N161000 Longitude: E0442000, E0432000. ISSN/ISBN: 0264-8172.

Amadei, R. 1960. "Notizie Sull Programma Di Ricerche Idriche Per La Pastorizia Somala. Information on the Water Research Program of Pastoral Somalia." Riv. Agric. Subtrop. Trop. Volume 54, no.4-9, Descriptors: Africa; East Africa; exploration; programs; Somali Republic; water resources. ISSN: 0035-6026.

"Amoco Units Sign Exploration Pacts in Somalia, Off Ghana." 1987. Oil & Gas Journal. May 11. Volume 85, Pages 20. Descriptors: Petroleum prospecting; Applied Science & Technology. Map. ISSN: 0030-1388.

Amore, C. and Xasan, Axmed Cali. 1982. "La Diga Di Baardheere Sul Fiume Jubba (Somalia); Indagini Geologico-Tecniche." Translated title: "The Baardheere Dam on the Jubba River, Somalia; Geological-Technical Study." Memorie Di Scienze Geologiche. Volume 35, Pages 1-20. Descriptors: Africa; atmospheric precipitation; dams; East Africa; engineering geology; geophysical methods; geophysical surveys; hydrogeology; hydrology; Jubba River; mechanical properties; permeability; properties; seismic methods; site exploration; Somali Republic; surveys; water balance. Notes: References: 1 p. illustrations incl. portrs. Reference includes data from PASCAL, Institute de l'Information Scientifique et Technique, Vandoeuvre-les-Nancy, France. ISSN: 0391-8602.

"Ancoraggio di Mogadiscio." 1941? EAF 257. First ed. Scale 1:7 500. Nairobi: [Survey Directorate, East Africa Command], 1941?. Maps. Note: Copy of Italian hydro chart published in 1934. Subject: Somalia- Maps and charts 1941. Added name: Great Britain. War Office. General Staff. Geographical Section. 1 sheet. British Library.

Anderson, D. M. W. and Weiping, Wang. 1990. "Acacia Gum Exudates from Somalia and Tanzania: The Acacia Senegal Complex." Biochemical Systematics and Ecology. October 1990, Volume 18, Issue 6, Pages 413-418. Descriptors: Acacia leucospira; A. cheilanthifolia; A. senegal; Acacia sp. nov.; gum exudates; amino acids; sugar composition; 13C NMR spectroscopy; chemotaxonomy. Abstract: Analytical data are presented for the polysaccharide and proteinaceous components of the gum exudates from three Somalian Acacia spp. viz. Acacia sp.nov. (Fagg & Styles 85) and A. cheilanthifolia Chiov. (members of the A. senegal complex) and A. leucospira Brenan. In addition, data are given for a Tanzanian specimen of gum arabic and, for comparative purposes, for a reference specimen of Sudanese gum arabic from A. senegal (L.) Willd. The gum from A, leucospira is dextrorotatory and has a low rhamnose, high uronic acid and high methoxyl content; it therefore has affinities to the sources of commercial gum talha (A. seyal Del. and relatives). In contrast, the gums from A. cheilanthifolia and Acacia sp. nov. are laevorotatory and of special interest as they provide the first opportunity to study positively identified members of the A. senegal complex backed by properly documented herbarium voucher specimens. Although undoubtedly closely related species, their chemical and spectroscopic data indicate that their gum exudates, and that from the Tanzanian sample, are analytically and structurally distinct from that of A. senegal (L.) Willd.

Angelis D'Ossat, G. and Millosevich, F. 1900. "Seconda Spedizione Bottego; Studio Geologico Del Materiale Raccolto Da M. Sacchi." Translated title: "The Second Bottego Expedition; Geologic Study of the Materials Collected by M. Sacchi." R.Soc.Geogr.Ital. Volume 1900, Descriptors: Africa; areal geology; Bottego Program; East Africa; expeditions; Somali Republic. Database: GeoRef.

Angellucci A, De'gennaro M, De Magistris M.A and Di Girolamo P. 1994. "Economic Aspects of Red Sands from the Southern Coast of Somalia." Int. Geol. Rev. Volume 36, Issue 9, Pages 884-889. Descriptors: Non-metals; glass sand; mineral resource; mineralogy; Quaternary; sand. Notes: Somalia. Abstract: Analysis of Quaternary sands from coastal Somalia indicates the potential for economically useful glass-sand components and heavy minerals. The sands occur in four distinct stratigraphic levels and form subparallel ridges. Four components are present in all sand specimens: calcium carbonate, quartz, feldspars, and green hornblende; mineral compositions vary along a seaward-landward progression. The red sands are characterized by silica dissolution-precipitation features and alteration to laterite products. ISSN: 0020-6814.

Angelucci A, De Gennaro M, De Magistris M.A and Di Girolamo P. 1995. "Mineralogical, Geochemical and Sedimentological Analysis on Recent and Quaternary Sands of the Littoral Region between Mogadishu and Merka (Southern Somalia) and their Economic Implication." Geologica Romana. Volume 31, Pages 249-263. Descriptors: Non-metals- 72.15.2; mineralogy; geochemistry; sedimentology; littoral deposit; mineral resources; sand. Notes: Somalia. Abstract: Mineralogical, geochemical and electron microscopy-sedimentology investigations were performed on samples from sandy complexes (siliceous 'red dune' and carbonaceous 'white dune') along an approximately 80 km stretch of the Indian Ocean coastline between Mogadishu and Merka (southern Somalia). After an introduction on the geological characteristics of the Merka 'red dune' complex, we will discuss the following points: 1) Mineralogical composition of the dune and gradual variations of the silicate (quartz feldspar and accessory minerals) and carbonate fractions with respect to the distance from the present-day coastline to the white and red dunes. 2) The strong processes of dissolution and precipitation of silica, and subordinately, other mineral salts in the 'red dune'. 3) Laterization within the 'red dune' and the selective enrichment of trace elements of 'transition period 4'. 4) Quartz grain micromorphology and sedimentology characteristics. 5) Origin of the sands from the metamorphic PreCambrian complex in the Bur Region. 6) Discussion on the use of the name 'dune' for this complex, a term which is not appropriate for a large part of the stratigraphic sequence of the 'red dune' complex. Furthermore, some aspects of economic interest are discussed, including: the use of sand for glassworks; the concentration process of useful 'transition' elements in laterites (Co, Ni, Sc, Fe, V, Cr, Ti); and the possible concentration of heavy mineral phases like zircon, kyanite, ilmenite, and rutile etc. Finally, the potential is discussed for the use of Rare Earth Elements to indicate the protolith of generated sands and laterites. ISSN/ISBN: 0435-3927.

Angelucci, A., Arush, Mahamud A., Cabdulqaadir, M. M., Boccaletti, M., Piccoli, G. and Robba, E. 1986. "Geological History of Central and Southern Somalia since the Triassic." Memorie Della Societa Geologica Italiana. Volume 31, Descriptors: Africa; central Somali Republic; East Africa; historical geology; Mesozoic; Somali Republic; southern Somali Republic; stratigraphy; Triassic. ISSN/ISBN: 0375-9857.

Angelucci, A., Boccaletti, M., Cabdulqaadir, M. M., Arush, Mahamud A., Piccoli, G. and Robba, E. 1981. "Evoluzione Geologica Della Somalia Dal Mesozoico; Nota Preliminare. Geologic Evolution of Somalia from the Mesozoic; Preliminary Note." Quaderni Di Geologia Della Somalia (Mogadiscio). Volume 5, Descriptors: Africa; areal geology; East Africa; evolution; Mesozoic; Somali Republic. LCCN: 93093377.

Antonides, L. E. Dolley, T. P. Newman, H. R. Panulas, J. G. and Rabchevsky, G. A. 1988. "Minerals Yearbook: The Mineral Industries of East Africa and the Indian Ocean. 1988 International Review." Bureau of Mines, Washington, DC. pages: 22. Descriptors: Production; Imports; Exports; Tables Data; Mineral economics; Mineral industry; Comoros; Djibouti; Ethiopia; Kenya; Madagascar; Mauritius; Reunion; Seychelles; Somalia; Sudan; Tanzania; Uganda; Foreign technology; Natural resources and earth sciences Mineral industries; Business and economics Foreign industry development and economics. Abstract: The report discusses the mineral industries of the following: Comoros, Djibouti, Ethiopia, Kenya, Madagascar, Mauritius, Reunion, Seychelles, Somalia, Sudan, Tanzania, and Uganda. Notes: Also available from Supt. of Docs. NTIS Accession Number: PB91142620XSP.

Arcangelo, H. C. 1844. A Sketch of the River Juba or Gochob or Gowin from a Trip up the Stream in 1844. International (III): United Service Journal, International (III). Descriptors: Africa; areal geology; East Africa; Gochob; Gowin; Juba River; Somali Republic; southern Somalia. Database: GeoRef. Accession Number: 1996-069234.

Archer, L. S. 1955. Report on the Hargeisa Water Supply. Somalia: Govern. Somaliland Protectorate, Hargeisa, Somalia. Descriptors: Africa; East Africa; Hargeisa Somali Republic; northern Somali Republic; Somali Republic; Waqooyi Galbeed Somali Republic; water resources; water supply. Database: GeoRef. Accession Number: 1996-069196.

Arfaa, F. 1975. "Studies on Schistosomiasis in Somalia." American Journal of Tropical Medicine and Hygiene. March 1975. Volume 24, Issue 2, Pages 280-283.

Descriptors: Animals; Bulinus/parasitology; Cricetinae; Disease Reservoirs; Female; Geography; Gerbillinae; Humans; Male; Schistosoma haematobium/pathogenicity; Schistosomiasis/epidemiology/transmission; Somalia; Urinary Tract Infections; epidemiology; parasitology. Abstract: A survey was made on the prevalence and intensity of schistosomiasis in 14 localities in 4 areas in Somalia has revealed the presence of urinary bilharziasis among the inhabitants of all these localities, this being much higher in the two areas where water development has been accomplished. The mean prevalence of infection was found to be 27.2% and 58.1% in the two areas where water development has only been planned, while it was 58.7% and 75.6% in the two areas where the extension of irrigation was achieved years ago. Snails were collected from the habitats visited and these were identified. The role of Bulinus abyssinicus in the transmission of infection was proven. ISSN: 0002-9637.

Arre, N. How Safe are Somali Water Sources? Lawrenceville, NJ: The Red Sea Press. 2004. pages: 453-460. 8th- 2001 Jul: Hargeisa, Somalia. Notes: Congress of the Somali studies international association; war destroys: Peace nurtures: Somali reconciliation and development; sponsor: Somali studies international association. ISSN/ISBN: 1569021864.

Arecchi, Alberto. 1984/2. "Mogadishu." Cities. Volume 1, Issue 3, Pages 221-228. Abstract: Muqdisho (Somali spelling), the capital of Somalia, now has 1 million inhabitants, about 20% of the population of the entire country. In the last ten years in particular its growth has been chaotic and uncontrolled, with the annual rate reaching 10% or more, caused largely by the immigration of nomadic families hit by famines and wars. This profile examines the problems caused by Mugadishu's rapid growth and offers some suggestions for controlling the town's development.

Arrou-Vignod, Jean-François; United Nations Development Programme and International Committee of the Red Cross. 1990s. Somalia, Biyo, that's Life Zimbabwe, Life Lessons; Djibouti, PK12, a Herculean Task. Geneve: Azimuths-UNDP. Volume: 1 videocassette (ca. 30 min.), Descriptors: Water-supply- Somalia; Rural unemployment- Zimbabwe; Rural development- Zimbabwe; Human settlements- Djibouti; Infrastructure (Economics)- Djibouti; LC: H85. Abstract: [Pt. 1]. Crippled by war and successive droughts, Somalia is one of the poorest countries in the world. Water supply facilities have been destroyed in the wars between the clans that have divided up the country and access to drinking water has become a major issue for all. In this struggle for survival, a trade that was almost forgotten until now has returned: water carriers. In the Somalian language, the word for water is iyo like life. (ca. 10 min.). [Pt. 2]. Rural unemployment is a major problem in Zimbabwe. David Jura, the principal of the primary school in Chenaka in eastern Zimbabwe, sees a clear link between rural poverty and environmental degradation. Through activities such as a highly productive nursery and an eco-tourism project he is setting up, David is restoring hope and a life not only to his pupils, but to the community as a whole. (ca. 10 min.). [Pt.3]. Hasna, a director at Djibouti Radio and Television, has chosen to produce a story on Point Kilometre 12, i.e. a place known only by its distance from Djibouti. This lunar desert is home to several thousand people, who have been "dehoused", as the local expression goes. In the harsh environment imposed on them, young and old have taken their destiny into their own hands by building the infrastructure they need by themselves. (ca. 10 min.). Notes: System Info: VHS. PAL. Notes: Video of three different documentaries originally broadcast in the television programme Azimuths. On cassette label: Azimuths Productions. Series statement from container. Participants: Narration: Julian Finn. Responsibility: made by the TV Unit of the United Nations Development Programme, with the aid of ICRC, the International Committee of the Red Cross ... [et al.]; conceived by Jean-François Arrou-Vignod. Editor, Jean-François Arrou-Vignod. OCLC Accession Number: 44699917.

Arush, M. A. and Basu, A. 1993. Tectonic Significance of Quartz Types in Yesomma Sandstone, Somalia. Firenze; Istituto Agronomico L'Oltremar; 1993. Relazioni E Monografie Agrarie Subtropicali E Tropicali Nuova Serie113 A, pages: 169-180. International Meeting; 1st- 1987 Nov: Mogadishu. Notes: Geology of Somalia and surrounding regions: Geology and mineral resources of Somalia and surrounding regions. OCLC Accession Number: 34175029.

Arush, Mohamud A. 1993. Sedimentological and Geochemical Studies of the Tisje-Yesomma Successions (Cretaceous/Tertiary of Somalia) and their Source Rocks. Berlin: Selbstverlag Fachbereich Geowissenschaften, FU Berlin. Descriptors: 116; Sediments (Geology)- Somalia; Geochemistry- Somalia; Geology, Stratigraphic-Cretaceous; Geology, Stratigraphic- Tertiary. Notes: Summary also in German. Includes bibliographical references (p. 90-104). Responsibility: Mohamud A. Arush. ISSN/ISBN: 3927541575.

Asfaw, Laike M. 1992. "Constraining the African Pole of Rotation." Tectonophysics. Volume 209, Issue 1-4, Pages 55-63. Abstract: In the absence of well defined transform faults in the East African rift system for constraining the plate kinematic reconstruction, the pole of relative motion for the African (Nubian) and Somalian plates has been determined from residual motion.If Africa and Somalia are to continue to drift apart along the East African rift system (which would then evolve into a series of ridges offset by transform faults) then incipient transform faults that may reflect the direction of relative motion should already be in place along the East African rift system. The incipient transforms along the East African rift system are characterized by shear zones, such as the Zambezi shear zone in the south and the Aswa and Hamer shear zones in the north. Some of these shear zones have been associated with recent strike-slip faulting in the NW-SE direction during periods of earthquakes. Provided that these, consistently NW-SE oriented, strike-slip movements in the shear zones give the direction of relative motion of the adjacent plates, then they can be used to constrain the position of the Africa-Somalia Euler pole. Due to the fact that identifying transform faults in the East African rift system is difficult and because the genesis of transform faults characterizing a plate boundary at an inception stage is not well known, the discussion here is limited to the northern segment of the East African rift system where shear zones are better characterized by the existing geophysical data. The characterizing features vary with latitude, indicating the complexity of the problem of the genesis of transform faults. I believe, however, that the relatively well defined intra-continental transform fault in the northern East African rift system, which is characterized by strike-slip faulting and earthquakes, constrains the pole of relative motion for the African and Somalian plates to a position near 1.5° S and 29.0° E.

Ashkenazi, I. and Shemer, J. 2005. "Tsunami--the Death Waves." Harefuah. Mar. Volume 144, Issue 3, Pages 154-9, 232. Descriptors: English Abstract; Homeless Persons-statistics & numerical data; Humans; Indian Ocean; Indonesia; Mortality; Natural Disasters/statistics & numerical data. Abstract: On December 26, 2004, the fourth strongest earthquake over the past century struck in the Indian Ocean off the western coast of northern Sumatra, Indonesia. Measuring 9.0 in magnitude, the earthquake triggered massive tsunamis that struck the Indian Ocean countries and Somalia, and killed more than tens of thousands and destroyed entire villages, leaving over a million homeless. Tsunamis are water waves that are caused by sudden vertical movement of a large area of the sea floor during an undersea earthquake. Tsunami speed can exceed 800 kilometers per hour, and as it reaches shallow water the height of the wave drastically increases. There are two natural warning signs of a possible tsunami: the earthquake itself and later, in the minutes preceding a tsunami strike, the sea often recedes temporarily from the coast. Despite these warning signs and despite a lag of up to several hours between the earthquake and the impact of the tsunamis, nearly all of the victims were taken completely by surprise. One of the most common myths associated with natural disasters is that dead bodies are responsible for the spread of epidemics. This article discusses the myths that often lead authorities and others to take inappropriate action, and presents valuable lessons to be learned from this catastrophic disaster. ISSN: 0017-7768.

Atmaoui, N. and Hollnack, D. 2003. "Neotectonics and Extension Direction of the Southern Kenya Rift, Lake Magadi Area." Tectonophysics. March 27. Volume 364, Issue 1-2, Pages 71-83. Abstract: The Magadi area, located in the southern part of the Kenya Rift, is a seismically active region where rifting is still in progress. The recent tectonic activity has been investigated through a seismological survey and the study of neotectonic joints found in Lake Magadi sediments, which were deposited some 5000 years ago. The structural analysis of these open fractures was combined with a quantitative analysis of the orientation and size characteristics of imagery faults. The gathered data demonstrate (1) that the majority of the systematic joints have straight and parallel trajectories with a common en echelon mode of propagation displayed through a rich variety of patterns, and (2) that there is a self-similarity in fault and joint principal directions recognised at the different telescopic scales. SPOT image (1: 125,000), aerial photos (1:76,000), and outcrop fieldwork reveal two important structural orientations which are N015°E and N015°W The N015°E regional direction is consistent with the orientation of the southern segment of the Kenya Rift. Structural analysis is supported by results of a joint microseismic investigation in the Lake Magach area. Obtained focal mechanism solutions indicate an E-W to ESE-WNW normal faulting extension direction.

Aubry, A. 1886. "Observations Géologiques sûr les Pays Danakils, Somalis, le Royaume Du Choa et les Pays Gallas. Geologic Observations on the Countries of Danakil, Somalia and the Kingdom of Choa and Gallas." Bull. Soc. Geol. Fr. Volume 14, Pages 201-222. Descriptors: Afar; Africa; areal geology; Choa; East Africa; Galla; Somali Republic. ISSN: 0037-9409.

Audin, L., Manighetti, I., Tapponnier, P., Metivier, F., Jacques, E. and Huchon, P. 2001. "Fault Propagation and Climatic Control of Sedimentation on the Ghoubbet Rift Floor- Insights from the Tadjouraden Cruise in the Western Gulf of Aden." Geophysical Journal International. Feb. Volume 144, Issue 2, Pages 391-413. Descriptors: Geological Faults; Climate; Sediments; Sea Floor Spreading; Volcanoes; Mid-Ocean Ridges; Somalia; Earth Mantle; Plates (Tectonics). Abstract: A detailed geophysical survey of the Ghoubbet AI Kharab (Djibouti) clarifies the small-scale morphology of the last submerged rift segment of the propagating Aden ridge before it enters the Afar depression. The bathymetry reveals a system of antithetic normal faults striking N130 deg E, roughly aligned with those active along the Asal rift. The 3.5 kHz sub-bottom profiler shows how the faults cut distinct layers within the recent, up to 60 m thick, sediment cover on the floor of the basin. A large volcanic structure, in the center of the basin, the 'Ghoubbet' volcano, separates two sedimentary flats. The organization of volcanism and the planform of faulting, with en echelon subrifts along the entire Asal-Ghoubbet rift, appear to confirm the westward propagation of this segment of the plate boundary. Faults throughout the rift have been active continuously for the last 8400 yr, but certain sediment layers show different offsets. The varying offsets of these layers, dated from cores previously retrieved in the southern basin, imply Holocene vertical slip rates of 0.3-1.4 mm/yr and indicate a major decrease in sedimentation rate after about 6000 yr BP. The two central sub-basins E and W of the volcano have distinct depositional histories, and the Ghoubbet may have been isolated from the sea between 10 and 72 kyr BP. Database: CSA Technology Research Database. ISSN/ISBN: 0956-540X.

Augusti, G. Blasi, C. Ceccotti, A. Fontana, A. and Sacchi, G. 1983. Tests and Technical Specifications for Concretes in Somalia. Nairobi, Kenya: E. &F. N. Spon, London, Engl. pages: 426-436. Appropriate Building Materials for Low Cost Housing: African Region, Proceedings of a Symposium. Descriptors: Concrete Testing- Somalia; Cement- Testing; Concrete Aggregates- Somalia; Sand and Gravel- Somalia; Concrete Construction- Standards; Mineral Industry and Resources- Somalia. Abstract: After a brief description of the materials commonly used in the Mogadishu area (Benadir), a standard mortar is proposed to allow strength determination of the cement. Then, results of destructive and non-destructive tests of concrete, made with the materials available in Somalia, are reported; finally, a cost-strength comparison is made between commonly used concretes, obtained from coral aggregates, and concretes made with siliceous materials, which can only be found outside the Mogadishu area and so are seldom applied in practice. The present work is intended as a first step towards the establishment of national Somali standards for reinforced concrete constructions. ISSN/ISBN: 0-419-13280-5.

Autori, Vari. 1962. "Carte Geologiche Della Somalia Settentrionale a Scale Diverse." Translated title: "Geologic Maps of various Scales of Northern Somalia." Somalia: Somali Republic Geological Survey, Hargeisa, Somalia. Descriptors: Africa; East Africa; geologic maps; maps; northern Somali Republic; Somali Republic. Database: GeoRef. Accession Number: 1996-069202.

Awal, A. and Egal, A. H. 1985. Assisted Rural Water Supply Programme in the N-W Region, Situation in the District of Gebiley, Hargeisa. International (III): UNICEF, International (III). Descriptors: Africa; East Africa; Gebiley Somali Republic; Hargeisa Somali Republic; Somali Republic; water resources; water supply. Database: GeoRef. Accession Number: 1996-069204.

Ayele, A. 2000. "Normal Left-Oblique Fault Mechanisms as an Indication of Sinistral Deformation between the Nubia and Somalia Plates in the Main Ethiopian Rift." Journal of African Earth Sciences. Volume 31, Issue 2, Pages 359-367. Abstract: Four regional earthquakes were studied for source parameters and focal mechanisms in the Main Ethiopian Rift. The results show normal faulting with significant left-lateral strike-slip components for all of the studied earthquakes. The horizontal component of the T-axis shows northwest-southeast/east-west orientation. From the mode of deformation of all the earthquakes studied, it is inferred that the Somalia Plate is experiencing sinistral oblique deformation with respect to the Nubia Plate in the Main Ethiopian Rift. One of the studied events was followed by two aftershocks with similar rupture processes as the main shock, deduced from waveform comparison at station AAE. ISSN: 1464-343X.

Aylmer, G. P. V. 1898. "Two Recent Journeys in Northern Somaliland; II." Geogr. J. Volume 11, Pages 34-48, 113. Descriptors: Africa; areal geology; East Africa; field trips; Somali Republic. Notes: sketch map. ISSN: 0016-7398.

Aylmer, L. 1911. "The Country between the Jubba River and Lake Rudolf." Geogr. J. Volume 38, Pages 289-296. Descriptors: Africa; areal geology; East Africa; East African Lakes; Ethiopia; Juba River; Lake Turkana; Somali Republic. ISSN: 0016-7398.

Aylmer, L. 1908. "Captain Aylmer's Journey in the Country South of the Tana River, East Africa." Geogr. J. Volume 32, Pages 55-59, 120. Descriptors: Africa; areal geology; East Africa; field trips; Somali Republic; Tana River. ISSN: 0016-7398.

Azzali, S. and Menenti, M. 1997. "Fourier Analysis of Temporal NDVI in the Southern African and American Continents." Soil and Water Research, Wageningen (Netherlands). Agricultural Research Dept; Netherlands: Winand Staring Centre for Integrated Land. Volume: PB97161889, pages: 153. Descriptors: Fourier analysis; Vegetative index; Zambia; Somalia; Argentina; Satellite imagery; Image analysis; Image classification; Vegetation growth; Environmental monitoring; Ground cover; Agroclimatology; Ecosystems; Quantitative analysis; Time series analysis; Harmonic analysis; Mapping; Remote sensing; Advanced very high resolution radiometer; Fuzzy sets; Fast fourier transforms; Normalizing (Statistics); Foreign technology; NDVI (Normalized Difference Vegetation Index); Normalized Difference Vegetation Index. Abstract: This report summarizes the results of two years investigation on the application of the Fourier analysis of temporal NDVI in Southern Africa and Southern America. The project provided the first detailed and quantitative climatology of the response of vegetation cover at the continental scale of Africa over a period of 10 years. The objectives of this project are: Extend the application of Fourier analysis of time series of NDVI, after the first very promising study of Zambia and Somalia, to longer time series and other climates and ecosystems; and Improve methods applied to map iso-growth zones in that first study. Notes: Color illustrations reproduced in black and white. Also pub. as Winand Staring Centre for Integrated Land, Soil and Water Research, Wageningen (Netherlands). Agricultural Research Dept. rept. no. REPT-108. Database: NTIS. Accession Number: PB97161889.

Azzaroli, A. and Fois, V. 1970. "Geological Outlines of the Northern End of the Horn of Africa; Rock Deformation and Tectonics." Report of the ...Session- International Geological Congress. Volume Part 4, Pages 293-314. Descriptors: Africa; areal geology; Cenozoic; East Africa; lithofacies; Mesozoic; north; paleogeography; regional; sedimentary rocks; Somali Republic; tectonics. Notes: SP: IGC, International Geological Congress; With discussion. Part 4; illustrations (incl. sketch maps); Abstract: General treatment of the stratigraphy, physiography, tectonics and paleogeography of the area. ISSN/ISBN: 1023-3210.

Azzaroli, A. and Merla, G. 1960. "Carta Geologica Della Penisola Somala, Scala 1:4,000,000." Translated title: "Geologic Map of the Somali Peninsula, Scale 1:4,000,000." Italy: AGIP Mineraria- C.N.R., Litogr. Artist. Cartogr., Florence, Italy.

Descriptors: Africa; East Africa; geologic maps; maps; Somali Peninsula; Somali Republic. Database: GeoRef. Accession Number: 1996-071061.

Azzaroli, A. and Merla, G. 1959. Carta Geologica Della Somalia e Dell'Ogaden, Scala 1:500,000. Translated title: Geologic Map of Somalia and of Ogaden, Scale 1:500,000. Italy: AGIP Mineraria- C.N.R., Litogr. Artist. Cartogr., Florence, Italy. Descriptors: Africa; East Africa; Ethiopia; geologic maps; maps; Ogaden; Somali Republic. Database: GeoRef. Accession Number: 1996-071060.

Azzaroli, Augusto. 1958. "L'Oligocene e Il Miocene Della Somalia; Stratigrafia, Tettonica, Pateontologia (Macroforaminiferi. Coralli, Molluschi)." Translated title: Oligocene and Miocene of Somalia; Stratigraphy, Tectonics, Paleontology (Macroforaminiferi, Corals, Mollusks)." Palaeontographia Italica. Volume 52, Pages 143. Descriptors: Africa; Anthozoa; Cenozoic; Coelenterata; East Africa; Foraminifera; geologic history; Invertebrata; microfossils; Miocene; Mollusca; Neogene; Oligocene; Paleogene; paleontology; Protista; Somali Republic; tectonics; Tertiary. Notes: illustrations (incl. geol. sketch maps). Abstract: The Oligocene and Miocene are represented in Somalia by neritic marine and inland lagoonal sediments, capped by coarse conglomerate composed of material derived by erosion of fault scarps. Details of the stratigraphy and structure are reported, and included foraminiferal, coral, and molluscan faunas are systematically described, including one new genus (Somalidacna) and many new species. ISSN: 0373-0972.

Azzaroli, Augusto. 1958. "Carta altimetrica della pensisola Somala" Translated title: "Altimetric map of the Somalan Peninsula." Compilata da A. Azzaroli e G. Merla. Scale: 1:4 000 000. [S.l.]: AGIP Mineraria: Consiglio Nazionale delle Ricerche. 1 map: color Subjects: Somalia. Physical geography- maps and charts; Ethiopia. Physical geography- maps and charts. Added Name: Merla, G. British Library: Maps X.5444.

Azzaroli, Augusto. 1957. "Missione Geologica in Migiurtinia." Ricerca Scientifica (1948). Volume 2, Pages 301-346. Descriptors: Cenozoic; faults; geologic mission; Italian Somaliland; Migiurtinia; report; Tertiary. Notes: illustrations (incl. sk. maps). Abstract: A report, mainly geologic, of an expedition to Migiurtinia, northern Italian Somaliland, in 1956. The area is underlain by a thick series of Mesozoic and Tertiary sedimentary formations, predominantly limestones, resting on a basement of sedimentary and crystalline rocks of uncertain age, but undoubtedly much older than the covering deposits. Over most of the region, the sedimentary sequence ends with the Eocene, but along the coast of the Gulf of Aden and the Indian Ocean, discontinuous strips of Oligocene and Miocene deposits occur, which permit dating of the successive movements responsible for the present position of regional structural elements. ISSN: 0392-7482.

Azzaroli, Augusto. 1952. "I Macroforaminiferi Della Serie Del Carcar (Eocene Medio e Superiore) in Somalia e La Loro Distribuzione Stratigrafica." Palaeontographia Italica. Volume 47, Issue 17), Pages 99-131. Descriptors: Africa; Carcar series; Cenozoic; East Africa; Foraminifera; historical geology; Invertebrata; Italian Somaliland; microfossils; paleontology; Protista; Somali Republic; Tertiary. Notes: illustrations (incl. sketch map). Abstract: On the basis of the foraminiferal fauna, defines five subdivisions of the Carcar beds in Italian and British Somaliland and correlates the series with the middle- upper Eocene of Europe and India. Descriptions of new species and varieties are included. ISSN: 0373-0972.

Azzaroli, Augusto; Canavari, Mario and Stefanini, Giuseppe. 1958. L'Oligocene e Il Miocene Della Somalia: Stratigrafia, Tetonica, Paleontologia (Macroforaminiferi, Coralli, Molluschi). Pisa: Tipografia Moderna. pages: 143. Descriptors: Paleontology-Somalia; Paleontology- Oligocene; Paleontology- Miocene; Mollusks, Fossil. Notes: At head of title: Raccolta di monografie paleontologiche, fondata da Mario Canavari, continuata e accresciuta da Giuseppe Stefanini. Includes bibliographical references (p. 135-139). Responsibility: A. Azzaroli. Summary in English. Database: Worldcat. OCLC Accession Number: 4286801.

Azzaroli, Augusto and Fois, V. 1964. "Geological Outlines of the Northern End of the Horn of Africa (Somali Region)." Report of the ...Session- International Geological Congress. Pages 36-37. Descriptors: Africa; geology; Horn of Africa; Italian Somaliland; north. Notes: IGC, International Geological Congress; IGCGAO; 22d, India, Rept., Vol. Abs. ISSN: 1023-3210.

B

Bab-Ttat, M. A. Windley, B. F. Al-Mishwit, A. T. and Almond, D. C. 1993. Geology of the Lowder-Mudiah Area, Yemen. Firenze; Istituto Agronomico L'Oltremar; 1993. Relazioni E Monografie Agrarie Subtropicali E Tropicali Nuova Serie113 A, pages: 143-152. International Meeting; 1st- 1987 Nov: Mogadishu. Notes: Geology of Somalia and surrounding regions: Geology and mineral resources of Somalia and surrounding regions. OCLC Accession Number: 34175029.

Babadzhanov, P. B., Rubtsov, L. N. and Solovey, B. G. 1976. "Drift and Anisotropy of Small-Scale Inhomogeneities in the Lower Ionosphere in the Equatorial Region." Vsesoyuznoye Soveshchaniye Po Issledovaniyu Dinamicheskikh Protsessov v Verkhney Atmosfere, Obninsk, Nov. 1973, Trudy. Edited by L. A. Kataseva and by I. A. Lysenko. Moscow, 1976. Pages 96-99. Descriptors: Ionospheric inhomogeneities; Ionospheric drift measurements; Mogadishu, Somalia. Refs. Russian summary. Abstract: Parameters of anisotropy of small-scale inhomogeneities in the lower equatorial region, and summarized data of investigations of the drifts of nonhomogeneities in this region, are presented on the basis of results of the Soviet equatorial meteor expedition in Somalia (Mogadishu, magnetic lat. Phi similar to 10°S). The measurement procedure is described with a graph showing the distribution of useful information and interfering factors in the course of a day, and a diagram of the orientation of the measurement triangle of the receiving antenna field. The drift records were analyzed by means of similitude and correlation analysis. During the entire measurement cycle of ionospheric drifts of Mogadishu, the prevailing movement was westward. The range of most probable apparent velocities of drift was 40-80 m/sec. The spatial characteristics of the small-scale inhomogeneities were determined by the method of total correlation. The large and small axes of the anisotropy ellipses were, respectively, 391 plus or minus 151 and 163 plus or minus 55 m, with an axial ratio of 2.35 plus or minus 0.09. The ellipsoidal elongation of the inhomogeneities of electron density were orientated predominantly at an angle of similar to plus or minus 20 degrees from the direction of the magnetic meridian. The ratio of actual and apparent drift was equal to 1.35 plus or minus 0.20. The variability of the drift picture in the course of regular drift was estimated by the parameter of random variability that was equal to 34 plus or minus 7 m/sec in Mogadishu. The ratio of this parameter to actual drift velocity is equal to 0.68 plus or minus 0.10. The life of small-scale nonuniformities is equal to 3.48 plus or minus 0.63 sec. The dynamic and spatial characteristics of the ionospheric nonhomogeneities in Mogadishu are compared with those of near equatorial regions. Database: Meteorological & Geoastrophysical Abstracts.

Baker, B. H. 1969. "Structural Evolution of the Rift Zones of the Middle East; a Comment." Nature (London). Volume 224, Issue 5217, Pages 359-360. Descriptors: Africa; Arabian Peninsula; Asia; East Africa; evolution; Middle East; Nubia; rift zones; sea-floor spreading; Somali Republic; structural geology; tectonics. Abstract: Rotation pole location-number, Red Sea-Aden gulf spreading rate, role of shearing along rift zones reaffirmed (discussion of paper by D. G. Roberts; for reference, see this Bibliography Vol. 34, No. 2, 21 U70-06442). ISSN: 0028-0836.

Baldacci, R. 1920. Cenni Sulle Condizione Geologico-Minerarie Delle Colonie Italiane. Translated title: Geology and Mining in the Italian Colony. Italy: Atti Conv. Naz. per il dopo-guerra delle Colonie, Rome, Italy. Descriptors: Africa; East Africa; mineral resources; mining; mining geology; Somali Republic. Database: GeoRef. Accession Number: 1998-015981.

Ballu, V., Diament, M., Briole, P. and Ruegg, J. C. 2003. "1985-1999 Gravity Field Variations Across the Asal Rift: Insights on Vertical Movements and Mass Transfer." Earth & Planetary Science Letters. MAR 15. Volume 208, Issue 1-2, Pages 41-49. Abstract: Repeated gravity measurements across the Asal Rift zone reveal that the central part of the northern rift might be subject to mass input in addition to vertical movements. Precise relative gravity measurements were made in 1985, 1988 and 1999. Temporal variations in the gravity field over the 11- and 14-yr periods reveal a probable relative gravity increase on the order of 30 muGal (300 nm/s(2)) in the central zone of the rift (active zone since the Ardoukoba rifting episode) with respect to both sides of the rift. The gravity data are compared with vertical deformation data collected by the, French National Geographical Institute (IGN) in 1984 and 2000 along 50 km of the across-axis leveling profile. The comparison shows that vertical movements alone cannot explain the observed gravity variations and that, unlike in many other volcano studies, our gravity and elevation changes cannot be modeled by a single process such as subsidence, inflation or localized mass injection. ISSN: 0012-821X.

Banakar, V. K., Nair, R. R., Tarkian, M. and Haake, B. 1993. "Neogene Oceanographic Variations Recorded in Manganese Nodules from the Somali Basin." Mar. Geol. Mar. Volume 110, Issue 3-4, Pages 393-402. Descriptors: absolute age; Antarctic bottom water; Antarctic Ocean; carbonate compensation depth; Cenozoic; Indian Ocean; lower Pliocene; manganese composition; Miocene; Neogene; nodules; paleo-oceanography; paleocirculation; Pliocene; Somali Basin; stratigraphic boundary; Tertiary; Th/Th; upper Miocene; X-ray diffraction data. Notes: SP: DSDP, Deep Sea Drilling Project; References: 38; illustrations incl. 1 table, sketch map. Abstract: A detailed study of a nodule from the Somali Basin dated by (super 230) Th (sub excess) was correlated with the paleoceanographic events recorded in Site 236 (Leg 24) Deep Sea Drilling Project (DSDP) cores. Tentative indications are that the phase of nodule accretion starting with the development of pillar structure at a depth of 20mm in the nodule around 13 Ma coincides with increased Antarctic Bottom Water (AABW) flow and an elevated calcium carbonate compensation depth (CCD). The Late Miocene lowering of the CCD is represented by the mottled zones between 8 and 18mm in the nodule is characterized by an abundant silicate component (>20%) of Aeolian origin. The

Miocene/Pliocene boundary (5 Ma) occurs at a depth of about 8mm and is represented by the development of pillar structure and a minimum of Aeolian dust (10.3%). The increased biological productivity of the Somali surface water since the Middle Miocene is demonstrated by the increasing C (sub org) content of the nodule (from 0.11 to 0.19%) towards its surface. ISSN: 0025-3227.

Bankwitz, P. and Bankwitz, E. 1988. "Photo-tectonic Interpretation of Selected Areas on the Western Margin of the Indian Ocean." Zeitschrift Fuer Geologische Wissenschaften. Volume 16, Issue 7, Pages 635-646. Descriptors: aerial photography; Africa; Arabian Peninsula; Asia; East Africa; geophysical surveys; image analysis; Indian Ocean; Indian Ocean Islands; Indian Peninsula; Madagascar; Oman; Pakistan; remote sensing; Somali Republic; structural geology; Sultan Massif; superposed tectonics; surveys; tectonic units; tectonics. Notes: References: 16; illustrations incl. 8 plates, geol. sketch maps; Reference includes data from Geoline, Bundesanstalt fur Geowissenschaften und Rohstoffe, Hanover, Germany. ISSN: 0303-4534.

Barberi, F., Bonatti, E., Marinelli, G. and Varet, J. 1974. "Transverse Tectonics during the Split of a Continent: Data from the Afar Rift." Tectonophysics. Volume 23, Issue 1-2, Pages 17-29. Descriptors: Afar Depression; Africa; concepts; crust; Djibouti; east; East Africa; emplacement; Ethiopia; fractures; genesis; magmas; models; oceanic; rift zones; rifting; Somali Republic; structural geology; tectonics; tectonophysics; transverse; upwelling. Notes: Univ. Miami, Rosenstiel School, Contrib. illustrations (incl. sketch map); Abstract: Magmatism, splitting of continent, creation of new oceanic crust; East Africa. ISSN: 0040-1951.

"Bardera." EAF Bardara 500k. 1941? Scale: 1:500 000. [Nairobi]: Map production section, [19--] Maps. Note(s):These maps do not have EAF numbers. Originally classified Secret. Subjects: Kenya- Maps and charts; Somalia- Maps and charts. Added Name: Great Britain. War Office. General Staff. Geographical Section. 1 sheet. British Library: Maps MOD EAF Bardara 500k.

"Bardera." 1941. Great Britain. Army. East Africa Command. East African Survey Group. EAF 228. First ed. Scale: 1:5 000. [Nairobi]: [Survey Directorate, East Africa Command], 1941. Maps. Subjects: Somalia- Maps and charts 1941. Added Name: Great Britain. War Office. General Staff. Geographical Section. 1 sheet. British Library: Maps MOD EAF 228.

Barbieri C, Vianini V and Fea M. 1985. "Use of Meteosat Data to Assess the Suitability of Sites for Astronomical Observations." ESA Bulletin. Volume 43, Pages 32-36. Descriptors: Remote Sensing, GIS and Mapping; sites for astronomical observations; Canaries; Spain; Sardinia; Sicily; Somalia; Italy. Abstract: Three years' worth of Meteosat images have been analyzed to investigate the suitability of six sites for astronomical observations. The six sites are: the Canaries (with particular attention to La Palma), the Sierras near Almeria (Spain), the Gennargentu (Sardinia), the northern coastal region of Sicily, the northern coastal region of Somalia, and finally that of Asiago Observatory (Italy) to provide a comparison with ground data. The analysis conducted has highlighted both the great value of this novel application of Meteosat data and also the limitations. ISSN: 0376-4265.

Barbour, E. K., Nabbut, N. H., Al-Nakhli, H. M. and Haile, H. T. 1986. "O-Serotypes of Pseudomonas Aeruginosa from Animal and Inanimate Sources in Saudi Arabia." Zentralbl. Bakteriol. Mikrobiol. Hyg. [A]. Jul. Volume 261, Issue 4, Pages 400-

406. Descriptors: Agglutination Tests; Animal Feed; Animals; Chickens/microbiology; Feces/microbiology; Food Microbiology; Nasal Cavity/microbiology; Poultry; Poultry Diseases/microbiology; Pseudomonas Infections/microbiology/veterinary; Pseudomonas aeruginosa/classification/isolation & purification; Saudi Arabia; Serotyping; Sewage; Sheep/microbiology; Sheep Diseases/microbiology; Water Microbiology. Abstract: A total of 1012 samples were examined for Pseudomonas aeruginosa and 257 (25.4%) were positive. The incidence of Ps. aeruginosa in samples collected from animal sources (N = 730) was significantly higher (28.7%) than that in 282 samples of inanimate sources (16.7%). The percentage of samples infected with these organisms was lowest in poultry feed (2.8%) and highest in sewage effluent (57.1%). Nine serotypes were defined from all sources. P5 was the common predominant individual O type in infected chicken navels and in the nasal cavities of Najdi sheep (a Saudi Arabian sheep breed), while P3 and P6 were predominate in the nasal cavities of Somali sheep (a breed imported from Somalia). No Ps. aeruginosa serotype was predominant in sheep faeces. In inanimate sources, P4 was predominant in water and sewage effluent. The isolate from the animal feed was untypeable. In using the slide agglutination technique for serotyping, most of the unusual agglutination reaction types of Ps. aeruginosa (70%) were of strains isolated from Somali sheep. ISSN: 0176-6724.

Barker, J.R., Thurow, T.L. and Herlocker, D.J. 1990. "Vegetation of Pastoralist Campsites within the Coastal Grassland of Central Somalia." Afr. J. Ecol. Volume 28, Issue 4, Pages 291-297. Descriptors: General Microbial Ecology; camp site; livestock; soil fertility; grassland; grazing Species Term: Somalia; Cynodon dactylon; Cenchrus ciliaris; Cleome; Indigofera; Gisekia; Cynodon; Cenchrus; Cleome tenella; Gisekia pharnaceoides; Indigofera intricata. Notes: Somalia. Abstract: Seasonal occupancy campsites where livestock are bedded down at night has created islands of soil fertility within the coastal grassland of central Somalia. The fast-growing, grazing-tolerant, stoloniferous grass, Cynodon dactylon, and the non-palatable, annual forbs, Cleome tenella and Gisekia pharnaceoides, occupied the centre of the campsites. The campsite perimeters were dominated by the slow-growing, grazing-sensitive, perennial plants Cenchrus ciliaris and Indigofera intricata. ISSN: 0141-6707.

Barker, J. R., Herlocker, D. J. and Young, S. A. 1989. "Vegetal Dynamics in Response to Sand Dune Encroachment within the Coastal Grasslands of Central Somalia." Afr. J. Ecol. Volume 27, Issue 4, Pages 277-282. Descriptors: grasslands; coastal environments; sand; soil characteristics; species diversity; Somalia; vegetation patterns; salt. Abstract: Sand dune encroachment within the coastal grasslands of central Somalia influenced vegetal dynamics as a result of sand saltation and deposition. This impact was greatest near the dune bodies and decreased with distance. The soil was an Ustic Torripsamment gypsic, isohyperthermic that differed with respect to surface sand deposition. Vegetal cover and species diversity were low near the dunes and increased with distance in front of them. Cyperus chordorrihzus Chiov. and Cenchrus ciliaris L., both members of the original plant community, were the last species to succumb to the encroaching dunes. ISSN: 0141-6707.

Barkhadle, A. M. I., Ongaro, L. and Pignatti, S. 1994. "Pastoralism and Plant Cover in the Lower Shabelle Region, Southern Somalia." Landscape Ecology. Volume 9, Issue 2, Pages 79-88. Descriptors: Evolution And Palaeoecology; rangeland; remote sensing; Somalia; vegetation. Notes: Geographic: Somalia- Shabelle. Abstract: Four vegetation types were recognized and mapped using remote sensing techniques; general vegetation characteristics, mainly floristic and physiognomic aspects, are described. Dynamic relationships between these vegetation types are outlined. ISSN: 0921-2973.

Barnard, P. C., Thompson, S., Bastow, M. A., Ducreux, C. and Mathurin, G. 1992. "Thermal Maturity Development and Source-Rock Occurrence in the Red Sea and Gulf of Aden." J. Pet. Geol. Volume 15, Issue 2, Pages 173-186. Descriptors: Petroleum geology; Sedimentology; Geochemistry; Geothermal energy; Tectonics; Petroleum prospecting. Abstract: Detailed geochemical and geothermal studies have been carried out in the Red Sea and Gulf of Aden in order to understand the regional distribution of source rocks through time, and the effects of changing heat flow through time. These studies have shown the presence of generally thin developments of fair to very good quality source rocks. Most of the identified source rocks occur in syn-rift sediments throughout the Red Sea area, but pre-rift source-rock occurrences have been identified in Yemen (Gulf of Aden), Somalia, and to a lesser extent in Egypt. Isolated occurrences of post-rift source rocks have also been identified. The data can be interpreted in the regional context of the sedimentary facies of the region in order to predict possible geographic distribution of source rocks. Maturity gradients determined using vitrinite reflectivity and spore coloration range from low (often in post-rift sections) to high (often in syn- and pre-rift sections). The maturity gradients in many of the sections analyzed show intersecting bilinear trends, suggesting very high palaeogeothermal gradients in sediments close to the rifting centers. In such areas, the oil and gas 'windows' are relatively shallow and thin. In areas where crustal thickening has reduced heat flux, hydrocarbon generation may have occurred in the past, but has since ceased. The models derived during this project for source-rock distribution and heat-flow variations are consistent with the tectonic evolution of the basin, and show that there is good potential in parts of the study area for oil generation, accumulation and preservation in pre-rift, syn-rift as well as in post-rift sediments. ISSN: 0141-6421.

Barnes, Sydney U. 1976. "Geology and Oil Prospects of Somalia, East Africa." AAPG Bulletin (American Association of Petroleum Geologists). Volume 60, Issue 3, Pages 389-413. Descriptors: Petroleum Geology; Petroleum Prospecting- Somalia. Abstract: Regional setting, stratigraphy and six major structural features of Somalia are described. The most promising region for oil and gas prospecting in Somalia is believed to be the Mesozoic shelf and reef area around the Somali embayment and around the Nogal uplift. Lithofacies-isopach maps are of much assistance in determining areas of limestone buildup for subsequent geophysical surveys. Secondary prospective oil and gas regions in Somalia are the coastal and offshore marine Tertiary sedimentary rocks which have had gas shows. Stratigraphic traps in clastic sedimentary rocks caused by facies changes or overlaps against the uplift regions of Somalia, and monoclinal porosity pinchouts in carbonate sedimentary rocks are other possibilities. It is the opinion of the writer that commercial oil and gas deposits are present in Somalia. Refs. ISSN: 0149-1423.

Barnes-Dean, Virginia Lee; Dean, Andrew G. and International Rescue Committee. 1981. The Value of the Cross-Culturally Expert Team in Disaster Situations: A Needs Assessment Survey in Suriye Refugee Camp, Lower Gedo, Somalia, June 1981. Descriptors: Refugees- Somalia- Lower Gedo; Disaster relief- Somalia; Water-supply-Somalia; Emergency relief programmes; Health services; Research; Surveys. Notes:

Suriye (Somalia: Refugee camp); Notes: Includes bibliographical references. Responsibility: Virginia Lee Barnes-Dean, Andrew G. Dean. Worldcat Accession Number: 68764299.

Barrington, Brown C. 1931. "The Geology of North-Eastern British Somaliland." Quarterly Journal of the Geological Society of London. Volume 87, Descriptors: Africa; areal geology; Somali Republic; East Africa; northern Somali Republic. ISSN: 0370-291X.

Bartolet, Jeffrey. 1992. "The Road to Hell." Newsweek. 9/21/92, Vol. 120 Issue 12, p52, 3p, 3c. Subject Terms: Somalia- Social conditions. Abstract: Examines the destructive situation in Somalia. Starvation and the 100,000 lives claimed already; The need for international intervention; Bleak future; Breakdown of the government into clan-based fiefdoms; Freelance gangs, warlords and gunmen hired by local merchants; The hundreds of tons of food looted by these groups; Fears that rainfalls will bring major cholera and typhoid epidemics; Current international action; Lesson learned by the international community. ISSN: 0028-9604.

Bashir, A. A. and Abdirisaq, B. S. 1988. Stima Di Fabbisogni Idrici Del Mais, Sesamo e Cotone in Somalia. Translated title: Evaluation of Water Demand of Maize, Sesame and Cotton in Somalia. Somalia: Universita Nazionale Somala, Mogadishu, Somalia. Descriptors: Africa; agriculture; East Africa; evaluation; hydrology; irrigation; Somali Republic; water resources; water supply; water use. Database: GeoRef. Accession Number: 1998-015566.

Basile, C. 1935. Uebi Scebeli Nella Spedizione Di S.A.R. Luigi Di Savoia (Diaro Di Tenda e Di Cammino). Translated title: Expedition to the Shebelle River of S.A.R. Luigi Di Savoia, Diary of the Trip. Italy: Cappeli Editore, Bologna, Italy. Descriptors: Africa; areal geology; East Africa; expeditions; Luigi di Savoia; Shebelle River; Somali Republic; southern Somali Republic. Database: GeoRef. Accession Number: 1998-011093.

Bassias, Yannis, Clement, Pierre and Giannesini, Pierre-Jean. 1993. "Les Rapports Des Campagnes a La Mer MD 64 SOMIRMAS (Somalie, Amirantes, Mascareignes) a Bord Du "Marion Dufresne" 5 Juillet- 14 Aout 1990. Report of the MD 64/SOMIRMAS Cruise (Somalia, Amirante Islands, Mascarene Islands) Aboard the "Marion Dufresne", July 5-August 14, 1990." Publications De l'Institut Francais Pour La Recherche Et La Technologie Polaires. Volume 94-1, Pages 58. Descriptors: cores; Indian Ocean; marine geology; marine sediments; Mascarene Basin; ocean basins; sediments; Somali Basin; stratigraphy. Notes: illustrations incl. 1 table. ISSN: 1246-7375.

Bastow I.D, Stuart G.W, Kendall J.-M and Ebinger C.J. 2005. "Upper-Mantle Seismic Structure in a Region of Incipient Continental Breakup: Northern Ethiopian Rift." Geophysical Journal International. Volume 162, Issue 2, Pages 479-493. Descriptors: Applied seismology; continental breakup; seismic velocity; upper mantle; mantle structure Species Term: Somalia; Nubian. Notes: Additional Info: United Kingdom; References: Number: 68; Geographic: Ethiopia East Africa Sub-Saharan Africa Africa Eastern Hemisphere World. Abstract: The northern Ethiopian rift forms the third arm of the Red Sea, Gulf of Aden triple junction, and marks the transition from continental rifting in the East African rift to incipient oceanic spreading in Afar. We determine the P- and S-wave velocity structure beneath the northern Ethiopian rift using independent tomographic inversion of P- and S-wave relative arrival-time residuals from teleseismic earthquakes recorded by the Ethiopia Afar Geoscientific Lithospheric Experiment (EAGLE) passive experiment using the regularised non-linear least-squares inversion method of VanDecar. Our 79 broad-band instruments covered an area 250 × 350 km centred on the Boset magmatic segment ∼70 km SE of Addis Ababa in the centre of the northern Ethiopian rift. The study area encompasses several rift segments showing increasing degrees of extension and magmatic intrusion moving from south to north into the Afar depression. Analysis of relative arrival-time residuals shows that the rift flanks are asymmetric with arrivals associated with the southeastern Somalian Plate faster than the northwestern Nubian Plate. Our tomographic inversions image a 75 km wide tabular low-velocity zone beneath the less-evolved southern part of the rift in the uppermost 200-250 km of the mantle. At depths of >100km, north of 8.5°N, this low-velocity anomaly broadens laterally and appears to be connected to deeper low-velocity structures under the Afar depression. An off-rift low-velocity structure extending perpendicular to the rift axis correlates with the eastern limit of the E-W trending reactivated Precambrian Ambo-Guder fault zone that is delineated by Quaternary eruptive centres. Along axis, the low-velocity upwelling beneath the rift is segmented, with low-velocity material in the uppermost 100km often offset to the side of the rift with the highest rift flank topography. Our observations from this magmatic rift zone, which is transitional between continental and oceanic rifting, do not support detachment fault models of lithospheric extension but instead point to strain accommodation via magma assisted rifting. ISSN: 0956-540X.

Bauchot, M. L. and Blache, J. 1979. "(Presence of Ariosoma Balearicum (De La Roche, 1809) in Red Sea (Pisces, Teleostei, Congridae))." Bull. Mus. Natl. Hist. Nat. (France) (4e Ser.) (A.Zool.). Volume 1, Issue 4, Pages 1131-1137. Descriptors: synonymy; new records; geographical distribution; Ariosoma balearicum; Red Sea; Somali Dem. Rep. morphology (organisms); Ariosoma somaliense; Ariosoma scheeli; Marine. Abstract: Ariosoma balearicum (de la Roche, 1809), until now known in Western and Eastern Atlantic Ocean and in Mediterranean Sea, is recorded from Red Sea and North West Indian Ocean. A. somaliense Kotthaus, 1968, described from coast of Somalian is, pro parte, a junior Synonym of A. balearicum and, pro parte, a junior synonym of A. scheeli (Stroemann, 1896). Database: ASFA: Aquatic Sciences and Fisheries Abstracts.

BBC Somali Service; Indiana University, Bloomington and Archives of Traditional Music. 1967-1981. [Germany; England, London; Wales, Cardiff, Somalis, 1967-1981]. Volume: 1 sound cassette (52 min.), Descriptors: Formations (Geology)-Somalia; International law- Somalia; Interviews- Somalia; Medicine- Somalia; Science and state- Somalia; BBC Somali Speech Archive; Non-musical recording; Cassette recording. Abstract: Somalis in UK, interview Canab Trabi, interviewee; Cabdullaahi Xaaji, interviewer (6:38)- International law and development commented by Yuusuf Cali Haaruun (14:48)- Exploration [of] the solar system Osman Sugulle and Saciid Cabdulqaadir, commentators (11:00)- Science policy, traditional medicine, geographic formation commented by specialists Yuusuf Cumar Cali, Dr. Xassan Ismaaciil, Maxamuud; presented by Zainab M. Jamac, Moxamuud Cabdi Cali Dauleh (18:53). Notes: Outer space- Exploration- Somalia. Somalia. Notes: Somali general. Deposited by the BBC at the Archives of Traditional Music in 1994, under option 3. Recorded in

Cardiff, Wales, 1970 (item 1); at the BBC Bush House in London, England, 1981 (item 2); 1967 (item 3); and in Germany, no date provided (item 4). Responsibility: collected by British Broadcasting Corporation, Somali Service. Worldcat Accession Number: 33498631.

BBC Somali Service; Indiana University, Bloomington and Archives of Traditional Music. 1958-1981. [Germany; England, London, Somalis, 1958-1981]. Volume: 1 sound cassette (48 min.), Descriptors: Commerce; Formations (Geology)-Somalia; Interviews- Somalia; Water- Djibouti; BBC Somali Speech Archive; Non-musical recording; Cassette recording. Abstract: Djibouti water, African film symposium Omar A. Axmed, Abdul-Nejid Hussein, interviewees; Cabdullaahi Xaaji, interviewer (14:42)- First anniversary of the BBC Somali Service commented by Xaaji Cabdi Dauleh and Xaaji Osman (13:53)- Science policy, trade, geological formation, Somalia commented by Yuusuf Cumar Cali, Xassan Ismaaciil, Moxamad Cabdi Kaarshe (18:57). Notes: Somalia. Notes: Somali general. Deposited by the BBC at the Archives of Traditional Music in 1994, under option 3. Recorded at the BBC Bush House in London, England, 1981 (item 1); 1958 (item 2); and in Germany, 1978 (item 3). Responsibility: collected by British Broadcasting Corporation, Somali Service. Worldcat Accession Number: 33487473.

BCI Geonetics, New Hampshire, United States. 1985. Ground Water Exploration in Northwest Somalia; Regional Groundwater Potential and Identification of Potential Test Drilling Sites in Fractured Bedrock Aquifer. United States: BCI Geonetics Int., New Hampshire, United States. Descriptors: Africa; aquifers; drilling; East Africa; exploration; fractured materials; ground water; northwestern Somali Republic; Somali Republic; water resources. Notes: Report 1 (complete report). Database: GeoRef. Accession Number: 1998-011095.

Beatty, W. H. Bruce, J. G. and Guthrie, R. C. 1981. "Circulation and Oceanographic Properties in the Somali Basin as Observed during the 1979 Southwest Monsoon. Technical Rept." Woods Hole Oceanographic Institution, MA. Jun. pages: 87. Descriptors: Somalia; Eddies Fluid mechanics; Hydrographic surveying; Water analysis; Ocean currents; Monsoons; Oceanographic data; Ocean basins; XBT Expendable Bathythermograph; STD Salinity Temperature Depth; AGS 33 Vessel; Somali current; Ocean circulation; Ocean technology and engineering Physical and chemical oceanography; Ocean technology and engineering Dynamic oceanography. Abstract: Shipboard expendable bathythermograph (XBT), salinity-temperature-depth (STD), and sea surface temperature and salinity observations were taken from USNS WILKES (T-AGS 33) from 16 August until 5 September 1979 in the area of the Somali Current off the coast of Northeast Africa. Analysis of the data indicated the presence of two large anti-cyclonic gyres. The larger of the two gyres, called the Great Whirl or Prime Eddy, was centered at about 7° N and 55° E with a diameter of approximately 350 nautical miles. The smaller gyre, known as the Socotra Eddy, was centered at approximately 12° N and 57° E with an approximate diameter of 200 nautical miles. The two eddies were separated by a trough of cold water advected from the region of upwelling off the Somali coast between 9 N and 11 N. Studies of TIROS-N satellite infrared photographs and XBT cross-sections taken from tankers transiting the area during July and early August 1979 indicated the presence of a southern eddy separated from the Great Whirl by a trough of cold, upwelled water between 3° N and 5° N. During the early part of the WILKES survey, the southern eddy and the Great Whirl coalesced. DTIC Accession Number: ADA2762383XSP.

Beauchamp, J., Omer, M. K. and Perriaux, J. 1990. "Provence and Dispersal of Cretaceous Elastics in Northeastern Africa: Climatic and Structural Setting." Journal of African Earth Sciences. Volume 10, Issue 1-2, Pages 243-251. Abstract: During Cretaceous times, NE Africa was covered by clastic sediments. These sandy deposits correspond to the so-called "Nubian Sandstones" of Sudan, and the equivalent series of Egypt, Ethiopia and Somalia. In Central Sudan, the sandstone are alluvial, deposited from braided rivers under a dry tropical climate. They grade into alluvial plain and beach deposits in northern and eastern Sudan, Egypt, Ethiopia and Somalia. The source province was a north-south basement high, the Butana Massif, which extended northward into Egypt and eastward into Ethiopia and Somalia (Ethiopia-Sudan High, Harar-Nogal Swell). Nubian Sandstones were deposited in extensional tectonic framework. Old lineaments were rejuvenated as normal and strike-slip faults. Several geodynamics event could have interfered: northward drift of the African craton and downwarping of its northern margin as evidenced by Tethys transgressions, Indian Ocean opening and progressive uplift of the eastern margin of Africa, eastern updoming preceding the Red Sea and Aden Gulf opening. ISSN: 1464-343X.

Beaumont T.E. Kennie T.J.M and Matthews M.C. 1985. "Interpretation of Landsat Satellite Imagery for Regional Planning Studies (North Africa)." Surrey University Press, Blackie Group; US distributors Halsted Press. Descriptors: Remote Sensing, Gis and Mapping; Benin; water resources; Somalia; agricultural; Sudan; highway; Niger. Abstract: Three major themes are detailed: 1) applications in planning and regional engineering: highway transportation, urban and development planning, hazard studies and geographical databases; 2) the interpretation of engineering data from Landsat imagery, 3) examples of applications in feeder road planning in Benin, water resources in Somalia, agricultural development in Sudan, and highway maintenance planning in Niger.-after Author. In: Remote sensing in civil engineering (1985) p. 162-203; Notes: Special Features: 6 figs, 23 photos, 65 refs. OCLC Accession Number: 0550160.

Beaumont, T. E. 1982. "Land Capability Studies from Landsat Satellite Data for Rural Road Planning in North-East Somalia; Terrain Evaluation and Remote Sensing for Highway Engineering in Developing Countries; Proceedings of a Symposium at OECD." Supplementary Report- Transport and Road Research Laboratory. Volume 690, Pages 86-95. Descriptors: Africa; developing countries; East Africa; engineering geology; geophysical surveys; highways; imagery; Landsat; northeastern Somali Republic; planning; remote sensing; satellite methods; Somali Republic; surveys; terranes. Notes: References: 8; sketch maps. ISSN: 0305-1315.

Beazeley, G. A. and Somaliland Field Force. 1904. "Compass Road Traverse. Galkayu to Bohotele." Somaliland Field Force; Africa; Eastern. Descriptors: Manuscript (mss). Notes: Description: 1 map. Geographic: Gaalkacyo (Somalia)- Maps. Bohatele (Somalia)- Maps. Notes: Manuscript. Relief shown as form lines. Shows: roads with distances, tracks. "Copied by S.S. Bhamba Ram." Scale 1:253,440. Cartgrph Code: Category of scale: a Constant ratio linear horizontal scale: 253440; General Info: Coverage: Gaalkacyo to Bohotele, passing through Ethiopia. Other Titles: Galkayu to Bohotele. Responsibility: Capt. G.A. Beazeley. Worldcat Accession Number: 48597974.

Beazeley, G. A. and Somaliland Field Force. 1903. "Compass Road Traverse. Galkayu to Bohodle." Somaliland Field Force; Africa; Eastern. Descriptors: Manuscript (mss). Notes: Description: 1 map. Geographic: Gaalkacyo (Somalia)- Maps. Bohodle (Somalia)- Maps. Notes: Manuscript. Relief shown as form lines. Shows: roads with distances, tracks. With pin pricks following the route. Scale 1:253,440. Cartgrph Code: Category of scale: a Constant ratio linear horizontal scale: 253440; General Info: Coverage: Gaalkacyo to Bohodle, passing through Ethiopia. Other Titles: Galkayu to Bohodle. Responsibility: Capt. G.A. Beazeley. OCLC Accession Number: 48597955.

Behi, M., Mohamed, S. and Abukar, A. K. 1991. "Somalian Gulf of Aden Petroleum Geology and Regional Geophysical Evaluation; AAPG International Conference; Abstracts." AAPG Bull. Aug. Volume 75, Issue 8, Pages 1405. Descriptors: Africa; Arabian Sea; biostratigraphy; Cenozoic; cores; Cretaceous; cuttings; East Africa; economic geology; geophysical methods; geophysical surveys; Gulf of Aden; Indian Ocean; Jurassic; lithostratigraphy; Mesozoic; Miocene; Neogene; Oligocene; paleoenvironment; Paleogene; petroleum; petroleum exploration; seismic methods; Somali Republic; structural traps; surveys; Tertiary; traps; well-logging; wells. ISSN: 0149-1423.

Bellahsen, N., Faccenna, C., Funiciello, F., Daniel, J. M. and Jolivet, L. 2003. "Why did Arabia Separate from Africa? Insights from 3-D Laboratory Experiments." Earth and Planetary Science Letters. November 30. Volume 216, Issue 3, Pages 365-381. Abstract: We have performed 3-D scaled lithospheric experiments to investigate the role of the gravitational force exerted by a subducting slab on the deformation of the subducting plate itself. Experiments have been constructed using a dense silicone putty plate, to simulate a thin viscous lithosphere, floating in the middle of a large box filled with glucose syrup, simulating the upper mantle. We examine three different plate configurations: (i) subduction of a uniform oceanic plate, (ii) subduction of an oceanic-continental plate system and, (iii) subduction of a more complex oceanic-continental system simulating the asymmetric Africa-Eurasia system. Each model has been performed with and without the presence of a circular weak zone inside the subducting plate to test the near-surface weakening effect of a plume activity. Our results show that a subducting plate can deform in its interior only if the force distribution varies laterally along the subduction zone, i.e. by the asymmetrical entrance of continental material along the trench. In particular, extensional deformation of the plate occurs when a portion of the subduction zone is locked by the collisional process. The results of this study can be used to analyze the formation of the Arabian plate. We found that intraplate stresses, similar to those that generated the Africa-Arabia break-up, can be related to the Neogene evolution of the northern convergent margin of the African plate, where a lateral change from collision (Mediterranean and Bitlis) to active subduction (Makran) has been described. Second, intraplate stress and strain localization are favored by the presence of a weakness zone, such as the one generated by the Afar plume, producing a pattern of extensional deformation belts resembling the Red Sea-Gulf of Aden rift system.

Bellahsen, N., Fournier, M., d'Acremont, E., Leroy, S. and Daniel, J. M. 2006. "Fault Reactivation and Rift Localization: Northeastern Gulf of Aden Margin." Tectonics. FEB 17. Volume 25, Issue 1, Pages TC1007. Abstract: [1] The Gulf of Aden and the Sheba spreading ridge (Gulf of Aden) forms the southern boundary of the Arabian Plate. Its orientation (075°E) and its kinematics (about 030°E divergence) are interpreted as the result of an oblique rifting. In this contribution, a field study in the northeastern Gulf of Aden allows us to confirm the Oligo-Miocene synrift directions of extension and to precise the normal fault network geometry. The synrift extensions are 020° E and 160°E (possibly in this chronological order); the normal faults strike 070°E, 090°E, and 110°E. The results show that some characteristics are consistent with oblique rifting analogue models, while some others are not. Especially, fault reactivation of Mesozoic structures is shown to have occurred significantly at the beginning and during rifting. These data are therefore compared to analogue models of oblique reactivation, and this comparison demonstrates that fault reactivation played a key role during the early stage of the Gulf of Aden rifting. Finally, scenarios of the lithospheric evolution during the eastern Gulf of Aden opening (preexisting weaknesses in the lithosphere or not) are discussed to better constrain the deformation history of the northern margin. Especially, we show that rift localization processes may imply stress rotations through time.

Bellieni, G., Brotzu, P., Morbidelli, L., Nicoletti, M., Piccirillo, E. M. and Traversa, G. 1983. "Pre-Rift and Rift Stage Volcanism in East Africa (SE Ethiopia-Somalia and East Kenya); Tezisy, 27-y Mezhdunarodnyy Geologicheskiy Kongress--Abstracts, 27th International Geological Congress." Report of the ...Session- International Geological Congress. Volume 27, Issue 3, Pages 125-126. Descriptors: Africa; East Africa; eastern Kenya; Ethiopia; Kenya; rift zones; Somali Republic; southeastern Ethiopia; structural geology; tectonics; volcanoes. Notes: IGC, International Geological Congress. ISSN: 1023-3210.

Beltrandi, M. and Pyre, A. 1973. "Geologic Evolution of South-West Somalia." France: Assoc. Serv. Geol. Afric., Paris, France. Descriptors: Africa; East Africa; Somali Republic; southwestern Somali Republic; structural geology. Database: GeoRef. Accession Number: 1998-011099;1998-011100.

Beltrando G and Camberlin P. 1993. "Interannual Variability of Rainfall in the Eastern Horn of Africa and Indicators of Atmospheric Circulation." Int. J. Climatol. Volume 13, Issue 5, Pages 533-546. Descriptors: Geographical Abstracts: Physical Geography; Clouds, fog, precipitation; rainfall; Southern Oscillation; interannual variability; rainy season; teleconnections; atmospheric circulation; ENSO. Notes: Ethiopia Indian Ocean Somalia Africa- (East). Abstract: The results indicate significant negative correlations between northern autumn rains in Somalia and the Southern Oscillation during the same season. These rains are also negatively correlated with pressure in the Western Indian Ocean, and positively correlated in the Eastern Indian Ocean. The reverse pattern is shown with sea-surface temperature. Central Ethiopia summer rains, in contrast, indicate significant positive correlations with the Southern Oscillation, at zero and 3 months lag. El Nino years often correspond to drought years in this region. Evidence of an inverse relationship between the amount of rainfall in Ethiopia during summer (especially during September) and both the pressure and sea temperature over the Indian Ocean is also given. Correlations with the Arabian Sea are particularly strong. These results suggest that, although summer rains over Ethiopia are said to be related to the monsoon air flow from the Congo Basin and the Atlantic Ocean, there also exist quite strong connections with the surface conditions prevailing in the Indian and Pacific Oceans. ISSN: 0899-8418.

Belyavskiy, G. A. 1977. Fiziko-Mekhanicheskiye Svoystva Donnykh Osadkov Severnoy Chasti Indiyskogo Okeana. Translated title: Physical and Mechanical Properties of Bottom Sediments from the Northern Part of the Indian Ocean. USSR: Nauk. Dumka, Kiev, USSR. Descriptors: active specific surface; Arabian Sea; Bay of Bengal; chemical composition; data; density; depth; engineering geology; grain size; Gulf of Aden; Indian Ocean; materials; Mid-Indian Ridge; moisture; north; ocean floors; oceanography; petrology; physical properties; plasticity; pore water; porosity; properties; sediments; Somali Basin. Notes: References: 67; illustrations incl. tables, strat. cols., sketch maps; Latitude: N220000, E1100000 Longitude. Database: GeoRef. Accession Number: 1977-047784.

Bendavid-Val A. 1989. "Rural-Urban Linkages: Farming and Farm Households in Regional and Town Economies." Review of Urban & Regional Development Studies. Volume 1, Issue 2, Pages 89-97. Descriptors: Regional And Spatial Development And Planning; regional development; farm household income; nonfarm economy; rural urban exchange. Notes: Kenya Somalia. Abstract: While it is known that agricultural expansion is critical to growth in nonfarm sectors of rural regions, the extent and mechanisms of economic interdependency between agriculture and other sectors remains an inadequately understood aspect of the rural-urban economic growth dynamic. Least well understood are the linkages between household farm income and development of town economies. This paper presents findings from rural-urban exchange research in rural regions of Kenya and Somalia that provide insights into relationships between agriculture and the nonfarm economy, and offers conclusions for rural regional development strategies. ISSN: 0917-0553.

Bendick R, Klemperer S, McClusky S, Bilham R and Asfaw L. 2006. "Distributed Nubia-Somalia Relative Motion and Dike Intrusion in the Main Ethiopian Rift." Geophysical Journal International. Volume 165, Issue 1, Pages 303-310. Descriptors: Regional structure and tectonics; viscoelasticity; rifting; tectonic evolution Species Term: Somalia. Notes: Additional Info: United Kingdom; References: Number: 50; Geographic: Ethiopian Rift East African Rift Sub-Saharan Africa Africa. Abstract: The Main Ethiopian Rift (MER) in central Ethiopia extended in the rift-normal direction at a mean rate of 4.0 ± 0.9 mm yr-1(1σ) during the period 1992-2003, nearly a factor of two slower than the opening rate estimated from global plate motion inversions. Rift opening near a geodetic array during this period was accommodated by a single dike injection event in 1993, spatially coincident with active magmatic segments, probably triggered by observed seismicity. Following dike injection, the crust in the rift relaxed as a layered medium, with a ∼15-km-thick elastic lid over a viscous half space of 1018 Pa s. Diking, rather than normal faulting on rift-bounding faults, appears to be the predominant mechanism of extension in the MER, explaining the very low regional rates of moment release. The length scale and temporal behaviour of surface displacements require viscoelastic rheology in the rift. ISSN: 0956-540X; 1365-246X.

Benedini, M. 1983. E' Possible Costruire War Di Grandi Dimensioni? Translated title: Is Construction Possible on Large Scales? Somalia: Somali National University, Mogadishu, Somalia. Descriptors: Africa; catchment hydrodynamics; construction; dimensions; East Africa; hydrodynamics; hydrology; Somali Republic; water resources. Database: GeoRef. Accession Number: 1998-015572.

Benini, G. Baraldi, G. Falciai, M. and Farah, A. A. 1985. Irrigazione e Drenaggio Nei Terreni Salsi. Translated title: Irrigation and Drainage in Saline Soils. Somalia: Universita Nazionale Somala, Mogadishu, Somalia. Descriptors: Africa; agriculture; drainage basins; East Africa; hydrology; irrigation; salinity; soils; Solonchak soils; Somali Republic; water. Database: GeoRef. Accession Number: 1998-015573.

Benny, P. N. 2002. "Variability of Western Indian Ocean Currents." West.Indian Ocean J.Mar.Sci. Volume 1, Issue 1, Pages 81-90. Descriptors: Water circulation; Water currents; Dynamic topography; Variability; Somali Current; Indian Ocean; Marine. TR: KE0300038. Abstract: In the study reported, an attempt was made to understand the intra-annual variability of the western Indian Ocean circulation by estimating the monthly dynamic topography with respect to 400db. The major currents in the western Indian Ocean are clearly depicted in the topography. Among the currents, the Somali Current exhibits strong annual variability. Eddy circulation is prominent in the northern part of the Somali Current during the southwest monsoon period. Seasonal variability is also noticed in the North Equatorial Current. Slight spatial and temporal changes are noticed in the South Equatorial Current and Equatorial Counter Current. The Equatorial Jet flow occurs in the monsoon transition periods of May and November between the equator and 3 alpha South. ISSN: 0856-860X.

Benvenuti, G., Vallario, A., Salad, M. H. and Shire, Y. O. 1987. "Problematiche Idrogeologiche e Stima Delle Risorse Idriche Della Formazione Di Baidoa Nella Regione Di Bay (Somalia Sud-Occidentale)." Geologia Applicata e Idrogeologia. Volume 22, Pages 83-103. Descriptors: Geographical Abstracts: Physical Geography; water resource estimation; permeability; hydrologic balance; rainfall; evapotranspiration. Notes: Somalia- Bay Region- Baidoa Formation. Abstract: This research considered the hydrogeologic conditions of the various lithotypes outcropping over a very large area roughly including Baidoa, Oddur and Dinsor. Data collected at the Baidoa meteorological station, roughly representative of the climatic conditions of the studied area gave average annual rainfall as 581.5 mm; average annual temperature as 26.3°C, and annual evapotranspiration as 582 mm. Water falling on the limestones was calculated to be about 9900; 106 m3/year. A conservative estimate of evapotranspiration was considered to be 99% of the annual rainfall. The average water available in the carbonatic structure thus turns out to be approximately 98; 106 m3/year or 5900 m3/year per square kilometer of outcropping limestone. The average outflow values (wells 3 l/s; springs 0.6 l/s) are about 8.5- 9; 106 m3/year, or about 10% of the average annual infiltration. ISSN: 0435-3870.

Benvenuti, G. Hussein Salad, M. Omar Shire Yusuf and Vallario, A. 1993. Preliminary Hydrogeologic Balance of the Baidoa Formation (Bay Region, South West Somalia). Firenze; Istituto Agronomico L'Oltremar; 1993. Relazioni E Monografie Agrarie Subtropicali E Tropicali Nuova Serie113 B, pages: 665-670. International Meeting; 1st- 1987 Nov: Mogadishu. Notes: Geology of Somalia and surrounding regions: Geology and mineral resources of Somalia and surrounding regions. OCLC Accession Number: 34175029.

Benvenuti, G. Hussein Salad, M. Omar, Shire Y. and Vallario, A. 1987. Preliminary Hydrogeological Balance of the Baidoa Formation in the Bay Region, Southwest Somalia; GeoSom 87; International Meeting; Geology of Somalia and Surrounding Regions; Abstracts. Somalia: Somali Natl. Univ., Mogadishu, Somalia. GeoSom 87; Geology of Somalia and Surrounding Regions, Mogadishu. Somalia

Conference: Nov. 23-30, 1987. Descriptors: Africa; Baidoa Formation; Bay Somali Repblic; East Africa; ground water; Somali Republic; surface water; water balance; water resources. OCLC Accession Number: 34175029.

Benvenuti, G., Mantovani, F., Mase, G., Pozzi, R. and Maxamed, Huseen Salaad. 1988. "Hydrogeological Observations of the Tug Wajale Area (Hargeisa, Northwest Somalia)." Memorie Di Scienze Geologiche. Volume 40, Pages 173-183. Descriptors: Africa; aquifers; clastic rocks; Cretaceous; dams; East Africa; geomorphologic maps; granites; ground water; ground-water dams; Hargeisa Somali Republic; igneous rocks; international cooperation; maps; Mesozoic; metamorphic rocks; morphostructures; northwestern Somali Republic; plutonic rocks; remote sensing; rockfill dams; sandstone; sedimentary rocks; Somali Republic; Tug Wajale Somali Republic; water resources; water supply; water wells; wells. Notes: References: 16; illustrations incl. sect. ISSN: 0391-8602.

Benvenuti, G., Salad, M. Hussein, Shire, V. Omar and Vallario, A. 1987. "Problematiche Idrogeologiche e Stima Delle Risorse Idriche Della Formazione Di Baidoa Nella Regione Di Bay (Somalia Sud-Occidentale). Hydrogeologic Problems and Estimating Water Resources of the Baidoa Formation in the Bay Region, Southwestern Somalia." Geologia Applicata e Idrogeologia. Volume 22, Pages 83-103. Descriptors: Africa; aquifers; atmospheric precipitation; Baidoa Formation; Bay Somali Republic; carbonate rocks; East Africa; ground water; hydrogeologic maps; infiltration; international cooperation; Jurassic; limestone; maps; Mesozoic; oolitic limestone; sedimentary rocks; Somali Republic; springs; water resources; water wells; wells. Notes: References: 15; illustrations incl. 4 tables. ISSN: 0435-3870.

Berg, Hans Dietmar. 1977. "Grundwasserneubildung Durch Fluss-Infiltration In Somalia/Ostafrika." Translated title: "Left Bracket Ground Water Replenishment by River Infiltration in Somalia, East Africa Right Bracket." GWF, Das Gas- Und Wasserfach: Wasser/Abwasser. Volume 118, Issue 9, Pages 411-414. Descriptors: Water Resources; Hydrology. Abstract: Within the framework of a development and project undertaken by the German Society for Technical Cooperation, a search was made for new water supplies for a number of cities in Somalia. Three of the sites investigated lie in the vicinity of the Scebelli River, the second largest river in the country, which, however, carries water for only two-thirds of the year. In this connection an attempt was made to study the infiltration from the river and to determine quantitatively the volume of water infiltrating into the ground water, in order to be able to establish the amounts required for replenishment with the aid of a long-term ground water balance.

Berger, Carol. 1985. "Aid Wrangle Brings Cholera to Somalia." New Scientist. May 23. Volume 106, Pages 9. Descriptors: Cholera; Refugees; Public health/Somalia; General Science. Notes: Illustration; Map. ISSN: 0262-4079.

Bernardi, A. 1973. "Ricerca Di Acqua Nelle Formazioni Cristalline e Metaforiche Dello Zoccolo Africano; Possibilita' e Limiti Dei Metodi Sismico a Rifrazione, Elettrico e Magnetico. Translated title: The Water from the Alteration Zone of the Metamorphic and Crystalline Formation from the African Basement; Possibilities and Limits of the Seismic Refraction, Electrical and Magnetic Methods." Conv. Int. Acque Sotterranee, Atti. Issue 2, Pages 386-392. Descriptors: Africa; basement; Benin; Burkina Faso; East Africa; electrical methods; geophysical methods; geophysical surveys; ground; ground water; hydrogeology; igneous rocks; Kenya; magnetic methods; metamorphic rocks; movement;

Precambrian; refraction; seismic methods; Somali Republic; surveys; Togo; West Africa. Notes: With discussion; illustrations. Database: GeoRef.

Bertani, G. V. 1961. Data on New Mogadishu Pumping Station. Somalia: Publisher unknown, Mogadishu, Somalia. Descriptors: Africa; East Africa; ground water; Mogadishu Somali Republic; pumping; Somali Republic; water resources; water wells; wells. Database: GeoRef. Accession Number: 1998-015575.

Bertonelli, P. 1937. Cenni Monografici Sul Paese Del Gherire. Translated title: Monographic Outlines on Gherire City. Somalia: Publisher unknown, Mogadishu, Somalia. Descriptors: Africa; areal geology; East Africa; Gherire Somali Republic; Somali Republic. Notes: illustrations incl. sketch map. Database: GeoRef. Accession Number: 1998-015576.

Besairie, Henri. 1949. "La Côte Française des Somalias." Pages 116. Descriptors: Physical geography- Somaliland, French; Somaliland, French- Description and Travel. Notes: At Head of Title: Haut Commissariat de Madagascar et Dependences. Database: USGS Library.

Besairie, Henri. 1946. Notice Explicative de la Carte Géologique de la Côte Des Somalis Au 1:50 000. Translated title: Explanatory Note of the Geologic Map of the Somali Coast, 1:50 000. France: Imprimerie Nationale, Paris, France. Descriptors: Africa; East Africa; geologic maps; maps; Somali Republic. Database: GeoRef. Accession Number: 1996-040100.

Besairie, Henri. 1941. "Notes Hydrologiques et Pedologiques sûr la Somalie Française." Bull. Soc. Hist. Nat. Toulouse. Volume 76, Pages 305-338. Descriptors: Africa; Djibouti; East Africa; French Somaliland, hydrogeology and soils; soils; water supply. Notes: sk. Maps. Abstract: Notes on the regional ground-water aquifers and soils of French Somaliland, in relation to the basalt terrain and overlying lacustrine, alluvial, and coastal deposits. Chemical analyses of waters and soils are included. ISSN: 0366-3477.

Besteman, C. and Mabry, J.B. 1996. "Dhasheeg Agriculture in the Jubba Valley, Somalia." University of Arizona Press. Descriptors: 53-68; Irrigated agriculture- 74.1.8; flood recession agriculture; state intervention; resource management; social organisation. Abstract: Drawing on data collected during twelve months of fieldwork in a village in the mid-valley, discusses the importance of dhasheeg agriculture to subsistence, analyzing risk minimization strategies used by local farmers. The point is to demonstrate how flexibility is maintained and stratification stifled in the social organization of flood recession agriculture. Concludes with some remarks on the nature of the conflict between local management of resources and state intervention in local resource management (in the form of new tenure laws, titling programs, development projects), and the place of irrigable land in contemporary war-torn Somalia. Notes: In: Canals and communities: small-scale irrigation systems (1996) p. 53-68; References: 38; Geographic: Somalia-Jubba Valley; Notes: Special Features: 1 map, 38 references, 1 table; Update: A 05 JAN 1998. ISBN: 0816515921; OCLC Accession Number: 2063389.

Beydoun, Z. R. 1970. "Southern Arabia and Northern Somalia; Comparative Geology; A Discussion on the Structure and Evolution of the Red Sea and the Nature of the Red Sea, Gulf of Aden and Ethiopia Rift Junction." Philosophical Transactions of the Royal Society of London, Series A: Mathematical and Physical Sciences. Volume 267, Issue 1181, Pages 267-292. Descriptors: Africa; Arabian Peninsula; areal geology; Asia;

East Africa; explanatory text; geologic; maps; north; regional; Somali Republic; South; Yemen. Notes: illustrations (incl. colored geol. maps under separate cover). ISSN: 0080-4614.

Beydoun, Z. R. 1969. "Southern Arabia and Somalia, Comparative Geology; the Structure and Evolution of the Red Sea and the Nature of the Red Sea, Gulf of Aden and the Ethiopian Rift Junction." Descriptors: Africa; Arabian Peninsula; Asia; continental drift; East Africa; regional; Saudi Arabia; Saudi Arabia-Somalia relations; Saudi Arabia-Somalia structural relations; Somali Republic; South; structural geology; tectonics. Notes: The Royal Society, London; Abstract: Reconstruction of pre-drift configuration. Database: GeoRef.

Beydoun, Z. R. and Royal Society (Great Britain). 1970. "Geological Map, Northern Somalia and Socotra, Gulf of Aden." Philosophical Transactions of the Royal Society of London. Volume 267. London: Royal Society. Descriptors: Geology- Aden, Gulf of- Maps; Geology- Somalia- Maps; Geology- Yemen (Republic)- Socotra- Maps; Coasts- Somalia- Maps. Notes: Description: 1 map; color; 26 x 123 cm. Scale [ca. 1:1,200,000]; Lambert conical orthomorphic proj. (E 430--E 550/N 130--N 100). Cartgrph Code: Category of scale: a Constant ratio linear horizontal scale: 1200000 Coordinates--westernmost longitude: E0430000 Coordinates--easternmost longitude: E0550000; Notes: Issued with text: A discussion on the structure and evolution of the Red Sea and the nature of the Red Sea, Gulf of Aden and Ethiopia Rift Junction organized by H.L. Falcon, Red Sea Discussion Meeting, 1969. (Philosophical transactions of the Royal Society of London, A.). In upper right margin: Beydoun (Red Sea Discussion). Phil. trans. Roy. Soc. A. "Figure 3." Responsibility: compiled by Z.R. Beydoun. ISSN: 0261-0523. Library of Congress; LCCN: 89-691382.

Beydoun, Z. R. and Sikander, A. H. 1992. "Red Sea- Gulf of Aden: Re-Assessment of Hydrocarbon Potential." J. Pet. Geol. Volume 15, Issue 2, Pages 245-246. Descriptors: Petroleum geology; Tectonics; Stratigraphy; Sedimentology; Geochemistry; Petroleum prospecting. Abstract: The results and conclusions of the UNDP/World Bank Red Sea- Gulf of Aden Regional Hydrocarbons Study Project (June, 1989- January, 1992) are summarised. These results were derived from detailed examination of pooled public-sector data provided to the project by the participating bordering states in the region; these included samples and logs from ca. 56 wells drilled in the two basins (out of a total of 71 wells), and some 30,000 line-kms of reflection seismic (out of 110,000 km available). Well logs were digitised and correlated, and basin-wide biostratigraphic, lithostratigraphic and paleoenvironmental studies, source-rock organic richness, quality and maturity studies and sedimentological analyses were performed by contracted specialist staff in the UK and France, and integrated with basin-wide seismic structural and isopach mapping for selected regional reflectors/intervals by national task forces in Cairo under project supervision; academically derived crustal seismic and other geophysical data, and industrially derived velocity and other borehole data were also integrated for the Red Sea. Modelling, utilising recently acquired geophysical data, indicates crustal asymmetry for the Red Sea rift, with oceanic crust of pull-apart basin type flooring some portions of offshore Egypt and Sudan on the west, and stretched continental crust underlying the eastern Arabian flank. These results suggest that sinistral strike-slip processes controlled the initial break-up of the Arabian Plate, shaping the African flank as a sharp plate boundary. Similar processes appear to have initially shaped the Somalia flank of the Gulf of Aden rift. Only later did true sea-floor spreading propagate: 10-12 million years ago in the eastern Gulf of Aden; and < ca. 5 million years ago in the western Gulf of Aden, Afar and the southern/central Red Sea. Biostratigraphic work shows deep subsidence and ocean-water inundation commenced in the late middle Oligocene in the Gulf of Aden, in the late Oligocene in the southern Red Sea, and in the early Miocene in the Central Red Sea/Gulf of Suez. ISSN: 0141-6421.

Beydoun, Ziad R. 1982. "The Gulf of Aden and Northwest Arabian Sea." United States: Plenum Press, New York, United States. Descriptors: Afar; Africa; Arabian Peninsula; Arabian Sea; Asia; Dhofar; Djibouti; East Africa; Ethiopia; geologic maps; Gulf of Aden; Indian Ocean; maps; northwestern Arabian Sea; Oman; Phanerozoic; plate tectonics; Precambrian; Socotra; Somali Republic; stratigraphy; tectonophysics; Yemen. Notes: References: 174; illustrations incl. 2 tables, sects., geol. sketch maps. ISBN: 0306377764.

Beyene A and Abdelsalam M.G. 2005. "Tectonics of the Afar Depression: A Review and Synthesis." Journal of African Earth Sciences. Volume 41, Issue 1-2, Pages 41-59. Descriptors: Plate tectonics; rifting; tectonics Species Term: Nubian; Somalia. Notes: Additional Info: United Kingdom; References: Number: 101; Geographic: Afar East Africa Sub-Saharan Africa Africa Eastern Hemisphere World. Abstract: This article outlines geomorphological and tectonic elements of the Afar Depression, and discusses its evolution. A combination of far-field stress, due to the convergence of the Eurasian and Arabian plates along the Zagros Orogenic Front, and uplift of the Afar Dome due to a rising mantle plume reinforced each other to break the lithosphere of the Arabian-Nubian Shield. Thermal anomalies beneath the Arabian-Nubian Shield in the range of 150°C-200°C, induced by a rising plume that mechanically and thermally eroded the base of the mantle lithosphere and generated pulses of prodigious flood basalt since ∼30 Ma. Subsequent to the stretching and thinning the Afar Dome subsided to form the Afar Depression. The fragmentation of the Arabian-Nubian Shield led to the separation of the Nubian, Arabian and Somalian Plates along the Gulf of Aden, the Red Sea and the Main Ethiopian Rift. The rotation of the intervening Danakil, East-Central, and Ali-Sabieh Blocks defined major structural trends in the Afar Depression. The Danakil Block severed from the Nubian plate at 20 Ma, rotated anti-clockwise, translated from lower latitude and successively moved north, left-laterally with respect to Nubia. The westward propagating Gulf of Aden rift breached the Danakil Block from the Ali-Sabieh Block at ∼2 Ma and proceeded along the Gulf of Tajura into the Afar Depression. The propagation and overlap of the Red Sea and the Gulf of Aden along the Manda Hararo-Gobaad and Asal-Manda Inakir rifts caused clockwise rotation of the East-Central Block. Faulting and rifting in the southern Red Sea, western Gulf of Aden and northern Main Ethiopian Rift superimposed on Afar. The Afar Depression initiated as diffused extension due to far-field stress and area increase over a dome elevated by a rising plume. With time, the lithospheric extension intensified, nucleated in weak zones, and developed into incipient spreading centers. ISSN: 0899-5362.

Bibliography on Desert Soils of Egypt, Eritrea, Somaliland, Aden, (1964-1931). 1964. Harpenden: Descriptors: 15 pages. ISBN: 000294476. OCLC Accession Number: 67333319.

Bibolini, A. 1922. "Contributions a l'Etude De La Geologie De l'Afrique Orientale Italienne. » Translated title : « Contributions to the Study of the Geology of the

Italian East Africa." Report of the ...Session- International Geological Congress. Volume 13, Pages 798-817. Descriptors: Africa; areal geology; East Africa; Eritrea; Ethiopia; Somali Republic. Notes: SP: IGC, International Geological Congress; Volume 2. ISSN: 1023-3210.

Bigazzi, G. Bonadonna, F. P. Di Paola, G. M. and Giuliani, A. 1993. K-Ar and Fission Track Ages of the Last Volcano-Tectonic Phase in the Ethiopian Rift Valley, (Tullu Moye Area). Firenze; Istituto Agronomico L'Oltremar; 1993. Relazioni E Monografie Agrarie Subtropicali E Tropicali Nuova Serie113 A, pages: 311-322. International Meeting; 1st- 1987 Nov: Mogadishu. Notes: Geology of Somalia and surrounding regions: Geology and mineral resources of Somalia and surrounding regions. OCLC Accession Number: 34175029.

Bigi, F. Funaioli, U. and Gatti, V. 1970. L'Opera Della SAIS in Somalia; Significato e Valore Delle Realizzazioni, Delle Esperienze e Degli Studi Compiuti Della SAIS Nei Suoi 44 Anni Di Vita. Work of SAIS in Somalia; Significance and Value of the Realization, of the Experience and of the Complete Studies of SAIS in their 44 Years of Existence. Italy: SAIS, Milan, Italy. Descriptors: Africa; agriculture; East Africa; Giuba River; SAIS; Shebelle River; Somali Republic; water resources. Database: GeoRef. Accession Number: 1998-015578.

Bigioggero, B. Dal Piaz, G. V. Del Moro, A. and Ibrahim, H. A. 1993. Pan-African Foliated Granites and Post-Tectonic Granitoids from North-Eastern Somalia. Firenze; Istituto Agronomico L'Oltremar; 1993. Relazioni E Monografie Agrarie Subtropicali E Tropicali Nuova Serie113 A, pages: 91-118. International Meeting; 1st- 1987 Nov: Mogadishu. Notes: Geology of Somalia and surrounding regions: Geology and mineral resources of Somalia and surrounding regions. OCLC Accession Number: 34175029.

Bile, K., Isse, A. and Mohamud, O., et al. 1994. "Contrasting Roles of Rivers and Wells as Sources of Drinking Water on Attack and Fatality Rates in a Hepatitis E Epidemic in Somalia." American Journal of Tropical Medicine and Hygiene. Oct. Volume 51, Issue 4, Pages 466-474. Descriptors: Acute Disease; Adolescent; Adult; Age Factors; Child; Child, Preschool; Disease Outbreaks; Female; Fresh Water; Hepatitis Antibodies/blood; Hepatitis E/epidemiology/mortality; Hepatitis E virus/immunology; Humans; Incidence; Infant; Male; Middle Aged; Pregnancy; Pregnancy Complications, Infectious; epidemiology; mortality; Rain; Research Support, Non-U.S. Gov't; Seasons; Sex Factors; Somalia-epidemiology; Water Supply. Notes: LR: 20041117; (Hepatitis Antibodies). Abstract: In early 1988, an increased incidence of acute hepatitis was observed in villages along the Shebeli River in the Lower Shebeli region of Somalia. This was followed by a large epidemic that lasted until late 1989. In a survey of 142 villages with a population of 245,312 individuals, 11,413 icteric cases were recorded, of which 346 died, corresponding to an attack rate and a case fatality rate of 4.6% and 3.0%, respectively. The etiologic role of hepatitis E virus (HEV) in this epidemic was proven by demonstrating anti-HEV in 128 of 145 sampled cases as a sign of recent infection with HEV. In three villages, where a special study protocol was implemented, the attack rate was found to increase significantly with age from 5% in the group 1-4 years of age to 13% in the group 5-15 years of age and to 20% for persons older than 15 years of age. Among cases 20-39 years of age, the female-to-male ratio was 1.5:1, which was a significant predominance of females. As in other hepatitis E outbreaks, there was a high fatality rate in pregnant females, estimated to be 13.8%. The epidemic peaked with the rise in the level of the river during rainfall, suggesting that the disease was waterborne. The attack rate was higher (6.0%) in villages supplied with river water, while fewer cases were recorded in those relying on wells or ponds for their water supply, 1.7% and 1.2%, respectively. ISSN: 0002-9637.

Binda, P. L. Omenetto, P. and Warden, A. J. 1993. Mineral Deposits and Occurrences in the Precambrian of Northeast Africa and Arabia: A Review. Firenze; Istituto Agronomico L'Oltremar; 1993. Relazioni E Monografie Agrarie Subtropicali E Tropicali Nuova Serie113 B, pages: 429-516. International Meeting; 1st- 1987 Nov: Mogadishu. Notes: Geology of Somalia and surrounding regions: Geology and mineral resources of Somalia and surrounding regions. OCLC Accession Number: 34175029.

Bird, P. 2003. "An Updated Digital Model of Plate Boundaries." Geochemistry Geophysics Geosystems. MAR 14. Volume 4, Pages 1027. Abstract: [1] A global set of present plate boundaries on the Earth is presented in digital form. Most come from sources in the literature. A few boundaries are newly interpreted from topography, volcanism, and/or seismicity, taking into account relative plate velocities from magnetic anomalies, moment tensor solutions, and/or geodesy. In addition to the 14 large plates whose motion was described by the NUVEL-1A poles (Africa, Antarctica, Arabia, Australia, Caribbean, Cocos, Eurasia, India, Juan de Fuca, Nazca, North America, Pacific, Philippine Sea, South America), model PB2002 includes 38 small plates (Okhotsk, Amur, Yangtze, Okinawa, Sunda, Burma, Molucca Sea, Banda Sea, Timor, Birds Head, Maoke, Caroline, Mariana, North Bismarck, Manus, South Bismarck, Solomon Sea, Woodlark, New Hebrides, Conway Reef, Balmoral Reef, Futuna, Niuafo'ou, Tonga, Kermadec, Rivera, Galapagos, Easter, Juan Fernandez, Panama, North Andes, Altiplano, Shetland, Scotia, Sandwich, Aegean Sea, Anatolia, Somalia), for a total of 52 plates. No attempt is made to divide the Alps-Persia-Tibet mountain belt, the Philippine Islands, the Peruvian Andes, the Sierras Pampeanas, or the California-Nevada zone of dextral transtension into plates; instead, they are designated as "orogens" in which this plate model is not expected to be accurate. The cumulative-number/area distribution for this model follows a power law for plates with areas between 0.002 and 1 steradian. Departure from this scaling at the small-plate end suggests that future work is very likely to define more very small plates within the orogens. The model is presented in four digital files: a set of plate boundary segments; a set of plate outlines; a set of outlines of the orogens; and a table of characteristics of each digitization step along plate boundaries, including estimated relative velocity vector and classification into one of 7 types (continental convergence zone, continental transform fault, continental rift, oceanic spreading ridge, oceanic transform fault, oceanic convergent boundary, subduction zone). Total length, mean velocity, and total rate of area production/destruction are computed for each class; the global rate of area production and destruction is 0.108 m(2)/s, which is higher than in previous models because of the incorporation of back-arc spreading.

Bisson, Robert A. and Lehr, Jay H. 2004. Modern Groundwater Exploration; Discovering New Water Resources in Consolidated Rocks using Innovative Hydrogeologic Concepts, Exploration, Drilling, Aquifer Testing, and Management Methods. United States: John Wiley & Sons, Hoboken, NJ, United States. Descriptors: Africa; Antilles; aquifers; Caribbean region; case studies; East Africa; exploration; future; global; ground water; history; hydrology; Lesser Antilles; regional; remote sensing; reservoir rocks; Somali Republic; Sudan; techniques; technology; Tobago; Trinidad; Trinidad and Tobago; water quality; water resources; water supply; watersheds; West Indies. Notes: References: 27; illustrations. ISBN: 0471064602.

Biswas, Asit K., Masakhalia, Y. F. O., Odero-Ogwel, L. A. and Pallangyo, E. P. 1987. "Land use and Farming Systems in the Horn of Africa." Land use Policy. October 1987, Volume 4, Issue 4, Pages 419-443. Abstract: Efficient land use and farming systems are essential if agricultural production is to be increased substantially in Africa on a sustainable basis. Land use and farming systems in the Horn of Africa- Sudan, Somalia, Ethiopia and Djibouti- are analysed in this article. Policy options and recommendations to improve the situation in the four countries are outlined. ISSN: 0264-8377.

Blenkey, Nick. 2005. "Dealing with the Pirates." Mar. Log. Volume 110, Issue 12, Pages 4. Descriptors: Economic and social effects; Coastal zones; Ships; Military operations; Naval warfare. Abstract: The problem of piracy of Somalia coast is discussed. The foiled attack on the Seabourn ship includes the hijacking of two ships carrying world food program aid. The attackers opened fire on ships and the vessels slowed down or stopped close to the Somali coast. Vessels protected themselves by staying out of the region. ISSN: 0897-0491.

Blok, H. P. 1959. "Il Somalo Della Somalia. Grammatica e Testi." Translated title: "Somali of Somalia: Grammar and Use." Ed. Istituto Poligrafico Dello Stato (Roma, 1955). A Cura Dell' Amministrazione Fiduciaria Italiana Della Somalia: Pp. VIII + 404. Lingua. Volume 8, Pages 111-111.

Blondel, F. 1935. "La Geologie Et Les Ressources Minerales De l'Ethiopie, De La Somalie Et De l'Eritree." Chronique Des Mines Coloniales. Oct. Volume 43, Pages 306-317. Descriptors: Africa; bibliography; East Africa; economic geology; Eritrea; Ethiopia; geologic maps; historical geology; maps; mineral resources; Somali Republic. Notes: (Bur. d'Études Geol. et Min. Coloniales), an. 4; 5 figs. (incl. geol. map); Abstract: Stratigraphy and mineral resources of Ethiopia, Somaliland, and Eritrea; bibliography. ISSN: 0366-7324.

Blondel, F. b. 1894. La Géologie Et Les Ressources Mínérales De l'Ethiopie, De La Somalie Et De l'Eritrée. Translated title: The Geology and mineral resources of Ethiopia, Somalia and Eritrea. Paris: Bureau d'études géologiques et miniéres coloniales. Descriptors: p. 306-317. maps. 27 cm. Geology- Ethiopia; Geology- Somalia; Mines and mineral resources- Ethiopia; Mines and mineral resources- Somalia. Notes: Caption title. Extracted from la Chronique des mines coloniales, 4me année, no. 43, 1er oct. 1935. "Cette étude n'est aucune façon originale: elle dérive tout particuliérement des deux ouvrages principaux de E. Krenkel (1926) et de G. Stefanini (1933) cités dans la bibliographie." "Bibliographie": p. 314-317. OCLC Accession Number: 20939775.

Boccaletti, M., Dainelli, P. and Angelucci, A., et al. 1988. "Folding of the Mesozoic Cover in Sw Somalia: A Compressional Episode Related to the Early Stages of Indian Ocean Evolution." J. Pet. Geol. Volume 11, Issue 2, Pages 157-168. Descriptors: Petroleum Geology- Somalia; Geology- Tectonics. Abstract: The tectonic-sedimentary evolution of SW and central Somalia is characterized by two main depositional cycles. The first cycle (Triassic to Early Cretaceous) is characterized by subsiding basins related to a process of crustal thinning, and is associated with the separation of Madagascar from Africa, between 165 and 121 million years ago. The second cycle, starting in the Late

Cretaceous with a regional unconformity, is related to the separation and northeastward drift of India, at approximately 80 million years ago. The folds that affect the Jurassic and Early Cretaceous formations of the Lugh-Mandera basin in SW Somalia are the result of a compressive phase connected with dextral transcurrent movements along NE-SW trending fracture zones. These zones are developed parallel to the oceanic system of transform faults, in connection with the change in the stress regime intervening at the shift of the direction of spreading between the first and the second stages of evolution of the Indian Ocean. ISSN: 0141-6421.

Boeckelmann, K., Schreiber, W and Schroeder, J.H. 1990. "Sedimentology and Stratigraphy of Taleh and Karkar Formations (Eocene) in Northern Somalia." Berliner Geowissenschaftliche Abhandlungen, Reihe A. Volume 120, Issue 2, Pages 595-620. Descriptors: Tertiary; sedimentation; transgression/regression; Tertiary; Palaeocene; Eocene; sea level variation. Notes: Somalia. Abstract: Detailed measuring of sections and preliminary microfacies analysis combined with biostratigraphical investigations of the Eocene sediments of N Somalia facilitate the recognition of transgressive and regressive successions. The Auradu Fm. (Maastrichtian/Lower Eocene) is mainly composed of uniform, massive platform carbonates. In the upper part of this formation cyclic shelf carbonates have been deposited; they grade upward into a lagoonal transitional facies. The Taleh Fm. (Lower/Middle Eocene) is composed of evaporites and carbonates of sub- to supratidal zones. Following the arid, regressive or offlap model of a vast tidal flat area with interfingering sabkha and lagoonal environments, vertical and lateral facies variations can be observed. The generally open marine sediments of Karkar Fm. (Middle/Upper Eocene) start with a relatively rapid transgression. The distinct regression in the upper Karkar Fm. confirms a retreat of the sea towards the east in the Upper Eocene. ISSN: 0172-8784.

Bohannon, R. G. 1989. "Style of Extensional Tectonism during Rifting, Red Sea and Gulf of Aden." Journal of African Earth Sciences. Volume 8, Issue 2-4, Pages 589-602. Abstract: Models describing the development of the Red Sea and the Gulf of Aden, prior to the present periods of sea-floor spreading, include those that use block faulting on steep normal faults, uniform diffuse shear in continental crust, simple shear on large detachment faults that cut the entire lithosphere, combinations involving detachment faults/ductile deformation/plutonic inflation, and ones that minimize the role of mechanical extension in favor of an earlier stage of sea-floor spreading. Geologic and geophysical studies from the Arabian continental margin in the southern Red Sea and LANDSAT analysis of the northern Somalia margin in the Gulf of Aden suggest that the early continental rifts were long narrow features that formed by extension on closely spaced normal faults above moderate- to shallow-dipping detachments with break-away zones defining one rift flank and root zones under the opposing rift flank. The rift flanks presently form the opposing continental margins across each ocean basin. The detachment on the Arabian margin dips gently to the west, with a breakaway zone now eroded above the deeply dissected terrain of the Arabian escarpment. The Arabian detachment projects westward to middle crustal levels beneath the sediment of the southern Red Sea coastal plain. Strata in the upper plate dip as steeply as 60° to the west, and the beds are repeated by numerous planar and listric normal faults that dip to the east. Most of the faults truncate downward at the detachment. Thus, the upper plate is highly extended and the rocks in its eastern part have been translated about 20 km westward and

21/2- to 5-km downward relative to the rest of Arabia. A prominent detachment surface, with a north dip, is evident in northernmost Somalia where it breaks away north of the Somalian escarpment in an otherwise undeformed section of cratonic strata of Jurassic to Eocene age. The upper plate of the Somalian detachment consists of a highly faulted collage of the cratonic strata. This fault projects to middle crustal levels in the opposing Arabian margin to the northeast.A model is proposed in which upper crustal breakup occurs on large detachment faults that have a distinct polarity. Most of the intense faulting is confined to the upper plates of the detachments, which root in the middle crust. Deeper extension is accommodated chiefly by ductile shear that results in vertical uplift, thinning, and lateral expansion of the rocks of the lower crust and the lithospheric mantle. Thus, lithospheric thinning, the thermal anomaly associated with it, and nearly all isostatic effects of rifting are confined to the area of the rift, its flanks, or directly beneath it. Tholeiitic magmatism began, on the Arabian margin, once the crust had thinned to 20-25 km. It is proposed that most of the continental divergence was accommodated by disorganized plutonic expansion after that, rather than by mechanical extension, until the present system of organized sea-floor spreading was established. ISSN: 1464-343X.

Bohra, D. M. 1979. "Patterns of Urban Settlements In Somalia." GeoJournal. Volume 3, Issue 1, Pages 63-68. Descriptors: Urban Planning; Regional Planning- Land Use. Abstract: The paper analyzes the distribution of 68 urban settlements in 14 regions of the country. The value of near-neighbor statistic (R) shows nearly random pattern of urban settlements in some regions and approachingly uniform distribution of urban settlements in other regions. Regional distribution of urban settlements and their growth have been explained in context of region's geographical, economic and industrial frameworks. ISSN: 0343-2521.

Bonavia, Franco F., Chorowicz, Jean and Collet, Bernard. 1995. "Have Wet and Dry Precambrian Crust Largely Governed Cenozoic Intraplate Magmatism from Arabia to East Africa?" Geophysical Research Letters. 01 Sep. Volume 22, Issue 17, Pages 2337-2340. Descriptors: Afar; Africa; African Plate; Arabian Peninsula; Arabian Plate; Arabian Shield; Asia; basalts; Burundi; Cenozoic; Central Africa; Congo Democratic Republic; continental crust; crust; East Africa; East African Rift; Ethiopia; flood basalts; igneous activity; igneous rocks; intraplate processes; Kenya; lithosphere; mantle plumes; models; moisture; Nubia; Nubian Shield; plate boundaries; plate tectonics; Precambrian; Somali Republic; Tanzania; underplating; uplifts; volcanic rocks; water; Yemen; Zaire. Notes: References: 32; illustrations incl. geol. sketch maps. ISSN: 0094-8276.

Bonini M, Mazzarini F and Abebe T, et al. 2005. "Evolution of the Main Ethiopian Rift in the Frame of Afar and Kenya Rifts Propagation." Tectonics. Volume 24, Issue 1, Pages 1-21. Descriptors: Regional structure and tectonics; plate tectonics; faulting; tectonic evolution; rift zone Species Term: Somalia. Notes: Additional Info: United States; References: Number: 92; Geographic: Ethiopian Rift East African Rift Sub-Saharan Africa Africa Eastern Hemisphere World. Abstract: The Main Ethiopian Rift (MER) has a complex structural pattern composed of southern, central, and northern segments. Ages of onset of faulting and volcanism apparently indicate a heterogeneous time-space evolution of the segments, generally referred to as a northward progression of the rifting process. New structural, petrological, and geochronological data have been used to attempt reconciling the evolution of the distinct MER segments into a volcano-tectonic scenario accounting for the propagation of the Afar and the Kenya Rifts. In this evolutionary model, extension affected the Southern MER in the early Miocene (20-21 Ma) due to the northward propagation of the Kenya Rift-related deformation. This event lasted until 11 Ma, then deformation decreased radically and was resumed in Quaternary times. In the late Miocene (11 Ma), deformation focused in the Northern MER forming a proto-rift that we consider as the southernmost propagation of Afar. No major extensional deformation affected the Central MER in this period, as testified by the emplacement at 12-8 Ma of extensive plateau basalts currently outcropping on both rift margins. Significant rift opening occurred in the Central MER during the Pliocene with the eruption of voluminous ignimbritic covers (Nazret sequence) exposed both on the rift shoulders and on the rift floor. The apparent discrepancy between the heterogeneous propagation of the three MER segments could be reconciled by considering the opening of Central MER and the later reactivation of the Southern MER as due to a southward propagation of rifting triggered by counterclockwise rotation of the Somalian plate starting around 10 Ma. ISSN: 0278-7407.

Bosellini A, Russo A and Schroeder R. 1999. "Stratigraphic Evidence for an Early Aptian Sea-Level Fluctuation: The Graua Limestone of South-Eastern Ethiopia." Cretaceous Res. Volume 20, Issue 6, Pages 783-791. Descriptors: Mesozoic; Aptian; Cretaceous; eustacy; sea level change Species Term: Somalia; Palorbitolina lenticularis; Socotra; Orbitolinidae; Foraminifera. Notes: Additional Info: United Kingdom; References: 53; Geographic: Ethiopia. Abstract: The occurrence of a thin band of marine limestone (Graua Limestone) within a thick succession of fluviatile sandstones in south-eastern Ethiopia is direct evidence of flooding of part of the East African craton (Horn of Africa). According to the presence of abundant orbitolinid foraminifers (Palorbitolina lenticularis Blumenbach and Praeorbitolina cormyi Schroeder), the age of the Graua Limestone can be referred to the Early Aptian. Stratigraphy and palaeogeographic reconstructions for the Early Cretaceous in the surrounding regions (Kenya, Somalia, Yemen, Socotra, Oman, Syria, Lebanon, Israel, Egypt and Libya) show that the Early Aptian transgression was of regional extent. Our data seem to confirm that this transgression was of relatively short duration. This pulse cannot be related to tectono-eustatic mechanisms, which are too slow. A short-lived event should be invoked: either a regional tectonic pulse or the desiccation of the proto-South Atlantic. ISSN: 0195-6671.

Bosellini, Alfonso. 1987. The Continental Margins of Somalia; their Structural Evolution and Sequence Stratigraphy; GeoSom 87; International Meeting; Geology of Somalia and Surrounding Regions; Abstracts. Somalia: Somali Natl. Univ., Mogadishu, Somalia. GeoSom 87; Geology of Somalia and Surrounding Regions, Mogadishu. Somalia Conference: Nov. 23-30, 1987. Descriptors: Afar; Africa; Arabian Sea; carbonate rocks; Cenozoic; clastic rocks; compression tectonics; continental margin; Cretaceous; deltaic environment; East Africa; evolution; Gulf of Aden; Indian Ocean; Jurassic; Karroo Supergroup; Kenya; magnetic anomalies; Mesozoic; Miocene; Neogene; paleogeography; plate tectonics; rifting; sandstone; sea-level changes; sedimentary rocks; sedimentation; Somali Republic; stratigraphy; structural geology; tectonics; Tertiary; transgression; uplifts; Upper Cretaceous. OCLC Accession Number: 34175029.

Bosellini, A., Russo, A., Arush, M. A. and Cabdulqadir, M. M. 1987. "The Oligo-Miocene of Eil (NE Somalia): A Prograding Coral-Lepidocyclina System." Journal of African Earth Sciences. Volume 6, Issue 4, Pages 583-593. Abstract: The Oligo-Miocene succession of Eil is the product of a depositional regression and constitutes a 120-150 m thick depositional sequence that prograded seaward for at least 20-25 km. Its time-transgressive stratigraphy is documented physically by well exposed tangential clinoforms (previously considered as evidence of a tectonic coastal flexure) and biostratigraphically by the occurrence of calcareous nannoplankton, planktonic and benthonic foraminifera, and a rich coral fauna. The upper boundary of the sequence is indicated by a reefal toplap, which constitutes the flat surface of the Nogal Plateau. Age (Chattian to Burdigalian) and toplap relationships of the sequence indicate clearly that progradation took place after the Late Oligocene flooding which followed the strong fall of sea-level during the Chattian.Because of the horizontal geometry of the entire sedimentary system, it has been possible to make a clear environmental reconstruction and a facies model with original water depths. A worldwide Tertiary facies--the Lepidocyclina beds-- was confined to the front of the reef, at depths ranging from 35-40 to 120-130 m. ISSN: 1464-343X.

Bosellini, A., Russo, A. and Assefa, G. 2001. "The Mesozoic Succession of Dire Dawa, Harar Province, Ethiopia." Journal of African Earth Sciences. Volume 32, Issue 3, Pages 403-417. Abstract: The Mesozoic succession of Dire Dawa, Harar Province, Ethiopia, consists of a lower fluviatile sandstone (Adigrat Sandstone); an intermediate carbonate-marly unit, formerly called Antalo Limestone; and an upper fluviatile sandstone (Amba Aradam Formation). This study has shown that the intermediate unit consists of four different formations grouped into two depositional sequences. These sequences and their boundaries, Middle-Late Jurassic in age, are well correlated with sequences recognised throughout East Africa and a large part of Yemen. The base of the lower sequence (Antalo Supersequence) is time-transgressive (Pliensbachian to Oxfordian) and is the result of the first flooding of this sector of the Gondwana continent during the Mesozoic. The second major sequence boundary is also time-transgressive and corresponds to an abrupt deepening of East Africa and southern Arabia shallow water ramps and carbonate platforms, a collapse most probably related to the separation of Madagascar from Africa. A major tectonic event occurred in Early Cretaceous from northern Ethiopia to Yemen, and southern Ethiopia and Somalia. This vast uplift, testified by faults and angular uncomformities, was followed by deposition of fluviatile sediments over the entire region. ISSN: 1464-343X.

Bosellini, Alfonso, Russo, Antonio and Schroeder, Rolf. 1999. "Stratigraphic Evidence for an Early Aptian Sea-Level Fluctuation: The Graua Limestone of South-Eastern Ethiopia." Cretaceous Research. Volume 20, Issue 6, Pages 783-791. Descriptors: sea-level fluctuations; eustacy; Aptian; Cretaceous; East Africa. Abstract: The occurrence of a thin band of marine limestone (Graua Limestone) within a thick succession of fluviatile sandstones in south-eastern Ethiopia is direct evidence of flooding of part of the East African craton (Horn of Africa). According to the presence of abundant orbitolinid foraminifers (Palorbitolina lenticularis Blumenbach andPraeorbitolina cormyi Schroeder), the age of the Graua Limestone can be referred to the Early Aptian. Stratigraphy and palaeogeographic reconstructions for the Early Cretaceous in the surrounding regions (Kenya, Somalia, Yemen, Socotra, Oman, Syria, Lebanon, Israel, Egypt and Libya) show that the Early Aptian transgression was of regional extent. Our data seem to confirm that this transgression was of relatively short duration. This pulse cannot be related to tectono-eustatic mechanisms, which are too slow. A short-lived event should be invoked: either a regional tectonic pulse or the desiccation of the proto-South Atlantic. ISSN: 0195-6671.

Bossolasco, M. 1935. Resume Des Travaux Executes a La Station Geographique Temporaire De Mogadiscio. Summary of the Studies done by the Temporary Geographic Station in Mogadishu. France: Publisher unknown, Paris, France. Descriptors: Africa; areal geology; Benadir; East Africa; geodesy; geography; geophysical surveys; Mogadishu Somali Republic; Somali Republic; surveys. Database: GeoRef. Accession Number: 1998-015582.

Bosworth, W., Huchon, P. and McClay, K. 2005. "The Red Sea and Gulf of Aden Basins." Journal of African Earth Sciences. OCT. Volume 43, Issue 1-3, Pages 334-378. Abstract: We here summarize the evolution of the greater Red Sea-Gulf of Aden rift system, which includes the Gulfs of Suez and Aqaba, the Red Sea and Gulf of Aden marine basins and their continental margins, and the Afar region. Plume related basaltic trap volcanism began in Ethiopia, NE Sudan (Derudeb), and SW Yemen at similar to 31 Ma, followed by rhyolitic volcanism at similar to 30 Ma. Volcanism thereafter spread northward to Harrats Sirat, Hadan, Ishara-Khirsat, and Ar Rahat in western Saudi Arabia. This early magmatism occurred without significant extension, and continued to similar to 25 Ma. Much of the Red Sea and Gulf of Aden region was at or near sea level at this time. Starting between similar to 29.9 and 28.7 Ma, marine syn-tectonic sediments were deposited on continental crust in the central Gulf of Aden. At the same time the Horn of Africa became emergent. By similar to 27.5-23.8 Ma a small rift basin was forming in the Eritrean Red Sea. At approximately the same time (similar to 25 Ma), extension and rifting commenced within Afar itself. At similar to 24 Ma, a new phase of volcanism, principally basaltic dikes but also layered gabbro and granophyre bodies, appeared nearly synchronously throughout the entire Red Sea, from Afar and Yemen to northern Egypt. This second phase of magmatism was accompanied in the Red Sea by strong rift-normal extension and deposition of syn-tectonic sediments, mostly of marine and marginal marine affinity. Sedimentary facies were laterally heterogeneous, being comprised of inter-fingering siliciclastics, evaporite, and carbonate. Throughout the Red Sea, the principal phase of rift shoulder uplift and rapid syn-rift subsidence followed shortly thereafter at similar to 20 Ma. Water depths increased dramatically and sedimentation changed to predominantly Globigerina-rich mart and deepwater limestone. Within a few million years of its initiation in the mid-Oligocene the Gulf of Aden continental rift linked the Owen fracture zone (oceanic crust) with the Afar plume. The principal driving force for extension was slab-pull beneath the Urumieh-Doktar arc on the north side of the narrowing Neotethys. Drag of Arabia by the northward-moving Indian plate across the partially locked northern Owen fracture zone and the position of the Carlsberg oceanic ridge probably also influenced the geometry of the Aden rift. The trigger for the onset of rifting, though, was the impingement of the Afar plume at similar to 31 Ma. The Red Sea propagated away from the plume head, perpendicular to the extensional stresses then operating in Arabia, and arrived at the bend in the African-Levant margin, which itself may have been a stress concentration ripe for rifting. The local geometry of the early Red Sea rift was strongly influenced by pre-existing basement structures, and as a consequence followed a complex path from Afar to Suez. Each segment of the rift was initially an asymmetric half graben, with well-defined accommodation zones between sub-basins. In the Gulf of Aden, the positions of accommodation zones were strongly influenced by older Mesozoic rift basins. Early rift structures can be restored to their original contiguous geometries along both the Red Sea and Gulf of Aden conjugate margins. In both basins, present-day shorelines restore to a separation of 40-60 km along most of their lengths. The initial rift basins were 60-80 km in width. Oceanic spreading initiated on the Sheba Ridge east of the Alula-Fartaq fracture zone at similar to 19-18 Ma. After stalling at this fracture zone, the ridge probably propagated west into the central Gulf of Aden by similar to 16 Ma. This matches the observed termination of syn-tectonic deposition along the onshore Aden margins at approximately the same time. At similar to 14 Ma, a transform boundary cut through Sinai and the Levant continental margin, linking the northern Red Sea with the Bitlis-Zagros convergence zone. This corresponded with collision of Arabia and Eurasia, which resulted in a new plate geometry with different boundary forces. Red Sea extension changed from rift normal (N60°E) to highly oblique and parallel to the Aqaba-Levant transform (N15° E). North of Suez in Egypt the rift system became emergent, perhaps due to minor compression of the Sinai sub-plate, and the marine connection to the Mediterranean Sea became restricted but not terminated. Red Sea sedimentation changed from predominantly open marine to evaporitic, although deep water persisted in many regions. A third phase of magmatism commenced, locally in Ethiopia but predominantly in western Saudi Arabia and extending north to Harrat Ash Shama and Jebel Drusc in Jordan, Lebanon, and Syria. At similar to 10 Ma, the Sheba Ridge rapidly propagated west over 400 kin from the central Gulf of Aden to the Shukra al Sheik discontinuity. Oceanic spreading followed in the south-central Red Sea at similar to 5 Ma. This corresponded in time to an important unconformity throughout the Red Sea basin and along the margins of the Gulf of Aden, coeval with the Messinian unconformity of the Mediterranean basin. A major phase of pull-apart basin development also occurred along the Aqaba-Levant transform. In the early Pliocene the influx of marine waters through Bab al Mandeb increased and Red Sea sedimentation thereafter returned to predominantly open marine conditions. By similar to 3-2 Ma, oceanic spreading moved west of the Shukra al Sheik discontinuity, and the entire Gulf of Aden was an oceanic rift. During the last similar to 1 My, the southern Red Sea plate boundary linked to the Aden spreading center through the Gulf of Zula, Danakil Depression, and Gulf of Tadjoura. Presently, the Red Sea spreading center appears to be propagating toward the northern Red Sea to link with the Aqaba-Levant transform. Alkali basaltic volcanism continues within the Younger Harrats of western Saudi Arabia and Yemen and offshore southern Red Sea islands. Most of the Arabian plate is now experiencing N-S upper crustal compression, whereas the maximum horizontal stress is oriented E-W in NE Africa. Arabia and Africa, now on separate plates, are therefore completely decoupled in terms of regional, far-field stresses. ISSN: 1464-343X.

Boudra, William F. 1993. "Engineers Restore Hope." Military Engineer. Volume 85, Issue 557, Pages 4-9. Descriptors: Military engineering; Military operations; Contracts; Planning; Engineers. Abstract: Operation Restore Hope in Somalia was supported by an engineering contract. During the planning phase, specified contingency missions are examined, requirements are identified, and a sourcing study is conducted to determine how assets will mobilize. The execution phase consists of mobilizing assets accomplishing specified tasks when a contingency arises. ISSN: 0026-3982.

Bowden, Jared Heath. 2004. Recent and Projected Climate Variability during the Seasonal Rains of the Greater Horn of Africa. Raleigh, NC: North Carolina State

University. 2004-08-06. Abstract: Bowden, Jared Heath. Recent and Projected Climate Variability during the Seasonal Rains of the Greater Horn of Africa. (Under the direction of Dr. Frederick H.M. Semazzi). Abstract: The objective of this study is to investigate the recent climate variability on intra-seasonal, interannual, and decadal time scales for the Greater Horn of Africa for the seasons of October, November, December (OND) and March, April, May (MAM). We use Empirical Orthogonal Functions (EOFs) to separate the variability. The observed climate variability is used to characterize the historical Parallel Climate Model variability. We then investigate the projected climate variability from the Parallel Climate Model business-as-usual run. Specifically, we demonstrate the observed rainfall-circulation relationships for ENSO during the OND and MAM seasons in relation to the rainfall over GHA using EOFs and the weighted wind composites. We find that during the recent climate of the OND season ENSO and the Indian Ocean Zonal Mode (IOZM) occur simultaneously with strong anticyclonic flow near Sumatra. As for the MAM season, there is no sign of the IOZM. During the OND season, the historical EOFs separate ENSO and the IOZM, suggesting the two can behave independently. Overall, the GHA region experiences positive anomalous rainfall during the OND season with the positive phase of the IOZM. The MAM season has no IOZM and demonstrates a complicated spatial temporal pattern because ENSO is in a transition phase during this season. We also demonstrate the significance of a trend mode for both OND and MAM seasons. The trend of the OND season is highly correlated to the tropical South Atlantic Index. The trend is seen in all months but strongest during October. The MAM characteristically favors the tropical South Atlantic Index, but the correlations are much lower. Overall, the combined affect of the MAM and OND season would generate an increasingly wetter northwest GHA and drier southern GHA. Somalia is the only region that compensates the increasing trend through an opposite loading relationship between the OND and MAM seasons. Notes: Department of Marine, Earth and Atmospheric Sciences. URL: http://www.lib.ncsu.edu/theses/available/etd-08152004-100319/unrestricted/etd.pdf

Bowen, M. R. 1988. "Survey of Tree Planting in Somalia 1925-1985. Occasional Paper." Oxford Forestry Inst. (England). Performer: National Range Agency, Mogadishu (Somalia). Volume: OFIOP36, pages: 49. Descriptors: Research projects; History; Developing countries; Sites; Area; Surveys; Somalia; Trees Plants; Afforestation; Reforestation; Species diversity; Biomass plantations; Foreign technology; Nurseries; Silviculture; Natural resources and earth sciences Forestry. Abstract: The names and locations of past and present organizations involved in tree planting in Somalia are listed. in total, about 30 projects of both government and non-governmental agencies have planted perhaps ten million trees and tested some 215 species during the past 60 years. Many of these plantings have taken place as shelterbelts or windbreaks (equivalent to about 1,600 ha) or for amenity purposes. Block plantations have been established to stabilize both coastal and inland sand dunes on about 1,500 ha (a lower estimate than the previously published figures), while about 1,800 ha have been established as community wood lots. Fruit trees form a sizeable proportion of amenity plantings. There has generally been excessive optimism as to what species, especially exotics, will grow successfully. It is premature to make firm species recommendations for the different Somali ecological zones, most of which are arid or semi-arid and frequently have saline soils. Notes: Prepared in cooperation with National Range Agency, Mogadishu

(Somalia). NTIS Accession Number: PB90229725XSP. Library of Congress; LCCN: 88-980818. , ISBN: 0850741068.

Bower, A. S., Fratantoni, D. M., Johns, W. E. and Peters, H. 2002. "Gulf of Aden Eddies and their Impact on Red Sea Water." Geophysical Research Letters. Nov. Volume 29, Issue 21, Descriptors: Mesoscale oceanic eddies; Oceanic eddies-oceanic circulation relationships; Water mass movements; Anticyclone oceanic eddies; Oceanographic surveys; Hydrographic surveys; Mesoscale features; Water masses; Current observations; Oceanic eddies; Current rings; Transport processes; Ocean circulation; Indian Ocean, Aden Gulf; Indian Ocean, Somali Current; Red Sea; Marine. Notes: 2025; TR: CS0404051. Abstract: New oceanographic observations in the Gulf of Aden in the northwestern Indian Ocean have revealed large, energetic, deep-reaching mesoscale eddies that fundamentally influence the spreading rates and pathways of intermediate-depth Red Sea Water (RSW). Three eddies were sampled in February 2001, two cyclonic and one anticyclonic, with diameters 150-250 km. Both cyclones had surface-intensified velocity structure with maxima similar to 0.5 m s super(-1), while the equally-energetic anticyclone appeared to be decoupled from the surface circulation. All three eddies reached nearly to the 1000-2000 m deep sea floor, with speeds as high as 0.2-0.3 m s super(-1) extending through the depth range of RSW. Comparison of salinity and direct velocity measurements indicates that the eddies advect and stir RSW through the Gulf of Aden. Anomalous water properties in the center of the anticyclonic eddy point to a possible formation site in the Somali Current System. Database: Oceanic Abstracts. ISSN: 0094-8276.

Bowman, Martin. 1994. "Mission to Mogue." Air Progress. February. Volume 56, Pages 42-45. Descriptors: Somalia/History/1991- (Civil War)/Relief work; Applied Science & Technology. Notes: Illustration. Abstract: An account of the writer's experiences in Mogadishu during Operation Provide Relief, established as an immediate short-term response to sickness and starvation in Somalia. Civil war broke out in Somalia in 1990, with an estimated 30,000 people being killed in the first 4 months of fighting. Ten times that figure died in the resulting famine. ISSN: 0002-2500.

Bowman, Martin. 1993. "German Air Force 467 to Mandera." Air Progress. July. Volume 55, Pages 68-70. Descriptors: Somalia/History/1991- (Civil War)/Relief work; Military transport airplanes; Applied Science & Technology. Notes: Illustration. Abstract: The writer describes his trip on German Air Force flight 467 to Mandera, Somalia. The flight service is part of Operation Provide Relief that began in late August 1992 to supply food to Somalia. The German government has made airlifts for the World Food Program a priority for its Air Force. In Somalia, food aid is delivered using 3 C-160D Transalls. ISSN: 0002-2500.

Box, T., Hadis, W. 1971. "Nomadism and Land Use in Somalia." Economic Development and Cultural Change, Volume 19, no 2, Pages 222-228. January 1971. 13 ref. Descriptors: grazing; arid lands; livestock; land management; water holes; geographical regions; social aspects; economic impact; rainfall; human population; foreign trade; ecosystems; land tenure; nomadism; somalia; camels. Abstract: the climate of Somalia is dry tropical. Annual rainfall is extremely variable, ranging from1-24 inches per year, divided between spring and autumn rains. Only about 12% of the land is cutivatable, and livestock production, involving camels, sheep, goats and cattle, supports about 73% of the population. The Somali herdsmen have evolved a system of nomadic rotation of rangeland that results in a balance between range vegetation and animal needs. In the wet season, herds are scattered onto the ranges receiving rainfall. During the dry season clans are concentrated near their home wells over which they have primary rights. Scattering again when rains reappear relieves the grazing pressures around the waterholes. Recent studies indicate that numbers of herd animals are increasing and that a large percentage of livestock production is being sold outside the country. There are various estimates indicating that up to 90% of the hard currency coming into the country may come through sales of nomadic livestock. Recently a number of foreign studies have suggested the elimination of nomadic pastoralists and their resettlement as sedentary farmers. The author contends that under the ecologic, economic and cultural conditions now existing, nomadic land use is the best utilization of the existing resource. A system of land tenure protecting and encouraging nomadism should be initiated in conjunction with improved nomadic range management programs. ISSN: 0013-0079.

Bradbury, M. 1994. "The Case of the Yellow Settee: Experiences of Doing Development in Post-War Somaliland." Community Dev. J. Volume 29, Issue 2, Pages 113-122. Descriptors: Water; war; aid project. Notes: Somalia- Somaliland. Abstract: In order to illustrate the complexities of doing development work in a situation of armed conflict, describes the background to and aftermath of an incident when armed youths attacked and looted an ActionAid office in Somaliland. ISSN: 0010-3802.

Bradbury, M., Abokor, A.Y. and Yusuf, H.A. 2003. "Somaliland: Choosing Politics Over Violence." Rev. Afr. Polit. Econ. Volume 30, Issue 97, Pages 455-478. Descriptors: Political; Political systems and political change; political development; democratization; violence; territorial disintegration. Notes: Additional Info: United Kingdom; References: 31; Geographic: Somalia East Africa Sub-Saharan Africa Africa. Abstract: Since breaking away from Somalia in 1991, the people of Somalliand have charted a different path from Somalia away from violent conflict towards constitutional politics. Unrecognised by the International community, political reconstruction in Somaliland has largely been an internal affair. While lack of formal recognition has had its costs, it has also has given Somallianders the opportunity to craft a system of government rooted in their local culture and values that is appropriate to their needs. For the past decade this has comprised a system of government that uses traditional forms of social and political organisation with Western-style Institutions of government. In December 2002 Somalliand took the first step towards changing this system by holding multi-party elections for district councils. These were followed in April 2003 by presidential elections. This paper describes the process of political transition in Somalliand and the first democratic elections in this region for 33 years. ISSN: 0305-6244.

Brady, M. J., Harms, J. C., Lowell, J. D. and MacKenzie, D. B. 1994. "Somalia; a Future Exploration Objective; AAPG Annual Convention." Annual Meeting Abstracts-American Association of Petroleum Geologists and Society of Economic Paleontologists and Mineralogists. Volume 1994, Pages 109. Descriptors: Africa; Arabian Sea; arches; basins; carbonate platforms; carbonate rocks; Cenozoic; chemically precipitated rocks; clastic rocks; deep-water environment; deformation; deltaic environment; East Africa; evaporites; evolution; faults; Gondwana; Gulf of Aden; Indian Ocean; limestone; marine environment; Mesozoic; Miocene; Neogene; petroleum exploration; plate tectonics;

rifting; sedimentary rocks; shale; shelf environment; Somali Republic; source rocks; Tertiary; thickness; unconformities. ISSN: 0094-0038.

Branca, Francesco, Abdulle, Ali and Persson, L. A., et al. 1993/6/5. "Famine in Somalia." The Lancet. Volume 341, Issue 8858, Pages 1478-1479.

Brandt, S.A. 1988. "Early Holocene Mortuary Practices and Hunter-Gatherer Adaptations in Southern Somalia." World Archaeology. Volume 20, Issue 1, Pages 40-56. Descriptors: Historical Geography; holocene; mortuary practices; hunter-gatherer; adaptations; archaeological excavations; rockshelter. Notes: Somalia- Buur Heybe. Abstract: Recent archaeological excavations of a large rockshelter at Buur Heybe, southern Somalia, resulted in the discovery of fourteen human burials of early Holocene age. The mortuary dates are examined in the light of an ecological model of hunter/gatherer socio/territorial organization which predicts that when critical human resources are spatio/temporally unpredictable and scarce, hunter/gatherers are unlikely to bury their dead in formal burial areas or build grave monuments. Conversely, when resources are abundant and predictable across time and space, conditions will arise that favour the construction of grave monuments and/or formal burial areas, possibly as a means of ritualizing corporate lineal descent. ISSN: 0043-8243.

Brandt, S. A. 1983. "Early Stone Age Occupation of the Horn of Africa: The Evidence from Somalia." Journal of Human Evolution. Volume 12, Issue 8, Pages 693-693. ISSN: 0047-2484.

Brassier, M. and Geleta, S. 1993. "A Planktonic Marker and Callovian-Oxfordian Fragmentation of Gondwana: Data from Ogaden Basin, Ethiopia." Palaeogeogr. , Palaeoclimatol. , Palaeoecol. Volume 104, Issue 1-4, Pages 177-184. Descriptors: Mesozoic; foraminifera; Jurassic; marker event; Callovian/Oxfordian boundary; Gondwana; continental breakup Species Term: Foraminifera; Dinophyceae; Somalia; Bacteria (microorganisms). Notes: Ethiopia- Ogaden Basin. Abstract: This paper outlines the Triassic to Jurassic history of the eastern Ogaden Basin as obtained from micropalaeontological studies. Continental sediments bearing Upper Triassic spores and pollen are overlain by Tethyan shallow water limestones yielding foraminifera and dinoflagellates of Callovian aspect. Flooding of the carbonate platform close to the Callovian Oxfordian boundary brought about replacement of larger foraminiferid assemblages. A brief influx of early planktonic foraminifera provides evidence for a connection with oceanic waters of the western Tethys. These changes are attributed to a series of major transgressions which, in turn, coincided with the initiation of sea floor spreading during the opening of the Somalia Basin of the western Indian Ocean. ISSN: 0031-0182.

Bricchetti-Robecchi, Luigi (Robecchi-Bricchetti, Luigi). 1899. Somalia e Benadir; Viaggio Di Esplorazione Dell'Africa Orientale. Translated title: Somalia and Benadir; Exploration Journey in East Africa. Italy: Publisher unknown, Milan, Italy. Descriptors: Africa; areal geology; Benadir; East Africa; field trips; Somali Republic. Notes: With analysis of soil from Somaliland by E. Zenoni, and notes and catalogue of rocks and minerals collected by T. Taramelli, p. 488-492. Database: GeoRef. Accession Number: 1998-011357.

Brima, E. I., Haris, P. I., Jenkins, R. O., Polya, D. A., Gault, A. G. and Harrington, C. F. 2006. "Understanding Arsenic Metabolism through a Comparative Study of Arsenic Levels in the Urine, Hair and Fingernails of Healthy Volunteers from

Three Unexposed Ethnic Groups in the United Kingdom." Toxicol. Appl. Pharmacol. Jun 5. Abstract: Very little is known about arsenic (As) metabolism in healthy populations that are not exposed to high concentrations of As in their food or water. Here we present a study with healthy volunteers from three different ethnic groups, residing in Leicester, UK, which reveals statistically significant differences in the levels of total As in urine and fingernail samples. Urine (n = 63), hair (n = 36) and fingernail (n = 36) samples from Asians, Somali Black-Africans and Whites were analysed using inductively coupled plasma mass spectrometry (ICP-MS) and graphite furnace atomic absorption spectroscopy (GF-AAS). The results clearly show that the total concentrations of As in urine and fingernail samples of a Somali Black-African population (urine 7.2 mug/g creatinine; fingernails 723.1 mug/kg) are significantly (P 0.05) in the level of As in the hair samples from these three groups; Somali Black-Africans (116.0 mug/kg), Asians (117.4 mug/kg) and Whites (141.2 mug/kg). Significantly different levels of total As in fingernail and urine and a higher percentage of urinary DMA in the Somali Black-Africans are suggestive of a different pattern of As metabolism in this ethnic group. ISSN: 0041-008X.

Briot, P. and Toens, P. D. 1984. "Surficial Uranium Deposits in Somalia; Surficial Uranium Deposits; Report of the Working Group on Uranium Geology, Organized by the International Atomic Energy Agency." Austria: Volume: IAEA-TECDOC-322, Descriptors: Africa; calcrete; carbonate rocks; carnotite; Cenozoic; Dusa Mareb; East Africa; economic geology; El Bur; Mercia Series; metal ores; mineral deposits, genesis; Miocene; Mudugh; Neogene; possibilities; resources; sedimentary processes; sedimentary rocks; Somali Republic; source rocks; surficial deposits; Tertiary; uranium ores; vanadates. Notes: References: 6; illustrations incl. geol. sketch map; Latitude: N043000, N063000 Longitude: E0464000, E0461000. Database: GeoRef. Accession Number: 1985-052799.

Briot, Philippe. 1982. "Données Isotopiques Sûr Les Carbonates Et Les Sulfates Des Calcretes De Somalie, Mauritanie, Namibie Et Australie Occidentale." Translated title: "Isotopic Data on the Carbonates and Sulfates of Calcretes from Somalia, Mauritania, Namibia and Western Australia." Bulletin Du Bureau De Recherches Géologiques Et Minieres, Deuxieme Serie, Section 2: Geologie Des. Issue 1, Pages 57-65. Descriptors: Mineral Exploration. Abstract: This isotopic study of uraniferous or non-uraniferous calcretes was performed for Somalian, Mauritanian, Namibian and Western Australian samples. The isotopic compositions of the samples are relatively similar and characterize all these carbonate-rich formations. The data are characteristic of continental origin and are close to those of mildly evaporated meteoritic waters. ISSN: 0153-6540.

Brisou, J., Courtois, D. and Denis, F. 1974. "Microbiological Study of a Hypersaline Lake in French Somaliland." Appl. Microbiol. May. Volume 27, Issue 5, Pages 819-822. Descriptors: Aerobiosis; Alcaligenes- isolation & purification; Bacillus-isolation & purification; Bacteria/isolation & purification; Flavobacterium-isolation & purification; Halobacterium- isolation & purification; Micrococcus- isolation & purification; Minerals; Pseudomonas/isolation & purification; Sarcina- isolation & purification; Sodium Chloride; Soil Microbiology; Somalia; Staphylococcus- isolation & purification; Vibrio- isolation & purification; Water Microbiology; Xanthomonas-isolation & purification. ISSN: 0003-6919.

"British Somaliland." 1929. S.l: s.n.; Somalia. Notes: Description: 7 maps. Geographic: Somalia- Maps. Scale 1:250,000. Cartgrph Code: Category of scale: a Constant ratio linear horizontal scale: 250000; Notes: Bromide prints of original survey sheets. Title supplied by cataloguer. Relief shown as form lines. Shows: international boundaries, tracks. General Info: Coverage: Somalia. OCLC Accession Number: 48599586.

British Somaliland. Geological Survey. 1958. "Somaliland Protectorate: Geological Survey." Geological Survey; Somalia. Descriptors: Geology- Somalia- Maps. Notes: Description: maps. Geographic: Somalia- Maps. Scale 1:125,000. Cartgrph Code: Category of scale: a Constant ratio linear horizontal scale: 125000; Notes: Accompanied by reports and cross section. Shows roads and geological information. Based on D.C.S. map. General Info: Coverage: Somalia. OCLC Accession Number: 48598980.

British Somaliland. Geological Survey. 1957. "Somaliland Protectorate. Topographical Sketch Map." Geological Survey; Somalila. Notes: Description: maps. Geographic: Somalia- Maps, Topographic. Scale 1:125,000. Cartgrph Code: Category of scale: a Constant ratio linear horizontal scale: 125000; Notes: Relief shown as contours (interval 500 ft) and heights. Shows: roads, tracks. Based on D.C.S. map. General Info: Coverage: Somalia. Other Titles: Topographical sketch map. OCLC Accession Number: 48598935.

British Somaliland. Geological Survey. 1952-1960. "Annual Report of the Geological Survey." Descriptors: Geology- Somalia- Periodicals; LC: QE339.S6. Notes: Frequency: Annual; 1952-1960. v.; 22-33 cm; Notes: Title from cover. At head of title: Somaliland Protectorate. Library of Congress; LCCN: 59-42758.

Brook, G.A., Burney, D.A. and Cowart, J.B. 1990. "Desert Paleoenvironmental Data from Cave Speleothems with Examples from the Chihuahuan, Somali-Chalbi, and Kalahari Deserts." Palaeogeogr. , Palaeoclimatol. , Palaeoecol. Volume 76, Issue 3-4, Pages 311-329. Descriptors: Geographical Abstracts: Physical Geography; Oceans; General Theory And Methods; palaeoenvironmental data; cave; speleothem; pollen spectra; dating; desert palaeoenvironment; palaeoenvironmental; pollen spectrum; desert; age determination; pollen; savannah Species Term: Somalia. Notes: Mexico- Chihuahuan Desert Africa- Kalahari Desert Africa- Somali- Chalbi Desert Botswana- Kalahari Desert Somalia- Somali-Chalbi Desert USA- New Mexico- Chihuahuan Desert Zaire- Matupi Cave. Abstract: Pollen-bearing lake, bog, and spring sediments are relatively scarce in many arid and semiarid regions of the world, and few are dateable beyond the 14C range. We have obtained pollen spectra from speleothems collected from caves in the Somali-Chalbi and Kalahari deserts suggesting that these deposits may be an important future source of desert paleovegetation data. As cave speleothems can be dated by the 230Th/234U method to c.350 000 yr BP, and by the TL and ESR methods potentially to 1 Myr BP, and can sometimes give paleotemperature and paleohydrologic data, they could provide a first glimpse of desert paleoenvironments during isotope stages 4-9. Pollen from three northern Somalia speleothems indicate more mesic conditions in the Horn of Africa at 10 000, 11 800, and 176 500 yr BP, while speleothem pollen spectra from Matupi Cave in northeastern Zaire, presently surrounded by tropical rainforest, suggest a savanna grassland at this cave c.14 000 yr BP. ISSN: 0031-0182.

Brook, G.A., Cowart, J.B., Brandt, S.A. and Scott, L. 1997. "Quaternary Climatic Change in Southern and Eastern Africa during the Last 300 Ka: The Evidence from

Caves in Somalia and the Transvaal Region of South Africa." Zeitschrift Fur Geomorphologie, Supplementband. Volume 108, Pages 15-48. Descriptors: the quaternary; climate change. Notes: South Africa Somalia. Abstract: Evidence from Sudwala, Echo, and Sterkfontein Cave in the Transvaal region of South Africa suggests that speleothems were deposited at times of lower temperature when precipitation was 140170% of present values. Clastic sediments were washed into the caves at times of climatic/environmental transition when surface soils, accumulated under wetter conditions, became unstable. Speleothem, tufa and sand dune age data for Namibia, Botswana and the Transvaal and northern Cape, South Africa, have provided a detailed record of wet and dry periods in the southern African summer rainfall zone to 300 ka B.P. Three sites record moister conditions at 202-186 ka B.P. The period 50-14 ka B.P. was characterized by five wet periods at 50-43, 38-35, 31-29, 26-21, and 19-14 ka B.P. There is also strong evidence for increased moisture during the mid and late Holocene from 6.9-2.6 and 1.8-0.5 ka B.P. Ages for submerged speleothems in Namibia and for dune activity at Etosha Pan and in the Kalahari Desert indicate significant dry periods at 35-31, 29-28, and 11-8 ka B.P. Speleothem, tufa and rock shelter sediment data for Somalia show wet conditions at 260-250, 176-160, 116-113, 87-75, 13, 10, 7.5, 5, and 1.5 ka B.P. The most obvious differences in the last 35 ka, between the two records, are that southern Africa experienced wet conditions in late glacial times, when eastern Africa was dry, and eastern Africa experienced wet conditions in the early Holocene, when southern Africa was dry. ISSN: 0044-2798.

Brook, G.A., Cowart, J.B. and Ford, D.C. 1996. "Raised Marine Terraces Along the Gulf of Aden Coast of Somalia." Physical Geography. Volume 17, Issue 4, Pages 297-312. Descriptors: Sea level; Sea-level and river terraces; marine terrace; sea level change; isotopic analysis; coral. Notes: Somalia- Gulf of Aden. Abstract: Mapping in the Galweda-Elayu area of northern Somalia has revealed depostional and erosional marine terraces at elevations of approximately 16m, 8m, and 2m. These terraces vary from 0-2300m, 200-2200m, and 0-800m in width, respectively. Sediments exposed in stream-valley walls demonstrate that the two higher terraces were formed by marine transgressions followed later by regressions to below present sea level. Beach ridges on the terraced alluvial fan at the mouth of togga Galweda imply that sea level and/or land elevation varied by at least 6m during the formation of the 16m terrce and by at least 3m during the formation of the 8m terrace. 230Th/234U ages of corals suggest that the 8m terrace was formed during deep-sea isotope substage 5c (105 kyr BP) and the 2m terrace during substage 5a (80 kyr BP). A 7-kyr-old coral from above the present storm beach on the outer flanks of the 2m terrace suggests that sea level in the Gulf of Aden was close to its present level by the middle Holocene. No material suitable for dating was recovered from the 16-m terrace, but on morphological grounds and based on marine-terrace elevations elsewhere in the Red Sea-Gulf of Aden rift zone, we believe that the 16m terrace, but on morphological grounds and based on marine-terrace was formed during isotope substage 5e (132-120 kyr BP) when global sea level was about 6m above present. ISSN: 0272-3646.

Brook, George A. Alsharhan, A. S. and Glennie, K. W. 1995. Quaternary Environments in the Somali-Chalbi and Kalahari Deserts of Africa; Abstracts of the International Conference on Quaternary Deserts and Climatic Change. United Arab Emirates: United Arab Emirates University, Desert and Marine Environment Center, Al

Ain, United Arab Emirates. International Conference on Quaternary Deserts and Climatic Change, Al Ain. United Arab Emirates Conference: Dec. 9-11, 1995. Descriptors: Africa; algae; archaeology; artifacts; biogeography; biostratigraphy; Botswana; C-14; carbon; Cenozoic; Central Africa; Chordata; clastic rocks; climate change; Congo Democratic Republic; cores; diatoms; Drotsky's Cave; East Africa; East African Lakes; Etosha Pans; fresh-water environment; geochronology; Gewihaba Valley; Holocene; humid environment; IGCP; interglacial environment; Invertebrata; isotopes; Ituri Rainforest; Kalahari Desert; lacustrine environment; Lake Albert; lower Holocene; Matupi Cave; middle Holocene; Mollusca; Namibia; Ngamiland; northeastern Congo Democratic Republic; northern Somali Republic; Otavi Mountains; paleohydrology; palynomorphs; Pisces; Plantae; Pleistocene; Quaternary; radioactive isotopes; red beds; relative age; savannas; sedimentary rocks; sediments; shallow-water environment; shells; solution features; Somali Republic; Somali-Chalbi Desert; Southern Africa; speleothems; stalagmites; terrestrial environment; thermoluminescence; Tsodilo Hills; upper Pleistocene; upper Quaternary; Vertebrata; White Paintings rock shelter; Xeidum Valley. Notes: IGCP, international geological correlation programme; IGCP project no. 349. Database: GeoRef. Accession Number: 2000-008975.

Brown, Barnum and Currie, Ethel Dobbie. 1943. Palaeontology of Harrar Province, Ethiopia. Part 1, the Dudley Expedition. New York: American Museum of Natural History. Descriptors: 29 pages; Paleontology- Ethiopia- Harerge Kifle Hager; Paleontology- Somalia; Geology- Ethiopia- Harerge Kifle Hager; Geology- Somalia; Sea urchins, Fossil- Ethiopia- Harerge Kifle Hager; Sea urchins, Fossil- Somalia. Abstract: pt. 1. The Dudley Expedition / by Barnum Brown. The Harrar Plateau. Crystalline basement. The Adigrat sandstone. Upper Jurassic limestones. Cretaceous deposits. Nubian sandstone(?). Tertiary and Quaternary deposits. The Dirre Daua region. The Hargeisa, British Somaliland, region- pt. 2. Echinoidea by Ethel D. Currie. Jurassic Echinoidea from Ethiopia. Cretaceous Echinoidea from Ethiopia. Eocene Echinoidea from British Somaliland. Notes: Named Conf: Dudley Expedition of the Anglo-American Oil Company (1920-1921); Note(s): "Issued December 30, 1943"--T.p. verso. Includes bibliographical references (pt. 1, p. 13; pt. 2, p. 28-29).; Other Titles: Dudley Expedition; Echinoidea; Palaeontology of Harrar Province, Ethiopia; Part 2; Echinoidea; Responsibility: Barnum Brown. Part 2, Echinoidea by Ethel D. Currie. OCLC Accession Number: 31841007. URL: http://digitallibrary.amnh.org/dspace/handle/2246/1758

Brown, G. A. 1962. "Groundwater Geology in the Vicinity of Mogadishu." Somalia: Descriptors: Africa; Benadir; East Africa; ground water; Mogadishu Somali Republic; Somali Republic; water resources. Database: GeoRef. Accession Number: 1998-015583.

Brown, Nick. 2005. "Somalia Calls for Piracy Help." Jane's Defence Weekly. Issue CT, Pages 2. Descriptors: Military operations; Security systems; Coastal engineering. Abstract: Senior officials from the fledging Somali Transitional Federal Government (TFG) have asked for international assistance to combat piracy. The Somalian Prime Minister Ali Mohamed Gedi called for help to combat the armed raiders operating in the country's waters. The country has had no effective central government and is bereft of any traditional security apparatus and with absolutely no coastal patrol capability. The respect for territorial integrity of Somalia's boundary claims has complicated international efforts to counter the pirates. ISSN: 0265-3818.

Brown, Nick. 2004. "France Assumes Command of CTF 150." Jane's Navy International. Issue JUE, Pages 1. Descriptors: Law enforcement; Shipyards; Security systems; Drug products; Societies and institutions. Abstract: Rear Adm Tony Rix, UK Royal Navy (RN), has handed over control of the multinational Combined Task Force (CTF) 150- one of the key forces undertaking the global 'war on terror'- to the French Navy's Rear Adm Jean-Pierre Teule. CTF 150 has responsibility for ensuring maritime security in support of the US-led war on terror, and for generally policing around two million square miles of sea around the Horn of Africa, the Gulf of Aden, the Somalia Basin and extending up into the Red Sea and east through the Gulf of Oman round to the Straits of Hormuz. The task force regularly carries out maritime interdiction operations and, as well as interdicting arms shipments bound for embargoed countries, has seized a number of drug runners and delivered them to local authorities. A wide range of countries have made rotations through the CTF 150 force structure, but it is currently made up of forces mainly from Australia, Canada, France, Germany, Italy, New Zealand, Pakistan, Spain, the UK and the USA. Rear Adm Rix handed over command to Rear Adm Teule on 1 June 2004, aboard HMS Cumberland, one of the RN's Type 22 Batch 3 frigates, which arrived in the Red Sea to join CTF 150 in early May. ISSN: 1358-3719.

Brown, O. B. Schott, F. and Miami Univ., FL. 1981. MONEX Oceanographic Observations Along the East African Coast (Global Atmospheric Research Program). WMO Intern. Conf. on Early Results of FGGE and Large-Scale Aspects of its Monsoon Expt. 16 p (SEE N82-23851 14-47); International Organization; 1981; WMO Intern. Conf. on Early Results of FGGE and Large-Scale Aspects of its Monsoon Expt. 16 p (SEE N82-23851 14-47); International Organization. Descriptors: Coastal Currents; Monsoons; Oceanographic Parameters; Somalia; Air Water Interactions; Density (Mass/Volume); Mesoscale Phenomena; Ocean Data Acquisitions Systems; Salinity; Surface Temperature. Abstract: Ship, buoy, and satellite current, temperature and air pressure data were used in order to study the Somali current. Wind field development, near surface circulation and upwelling were noted. Temperature, salinity, and potential density were recorded. The salinity profiles show several layers which deviate by more than 0.1 ppt from the mean. The density profile is smooth. So-called anomalous flow and expected flow often occur in the same year. Results suggest a mesoscale ocean-atmosphere interaction driven by the large horizontal difference in ocean surface temperature along the East African coast. Notes: Available from MF E06; HC WMO. Database: CSA Technology Research Database. Accession Number: N82-23908 (AH).

Brown, Ronald J. 1998. "Operation Eastern Exit, Somalia Evacuation January 1991" (174K). From U.S. Marines in the Persian Gulf, 1990-1991: With Marine Forces Afloat in Desert Shield and Desert Storm by Ronald J. Brown. U.S. Marine Corps, History and Museums Division, Washington, D.C. 1998. See: http://www.lib.utexas.edu/maps/historical/somalia_evac_1991.jpg

Bruce, J. B. Wright, M. M. Hollister, D. T. and Birch, L. E. 1981. "Construction Training Manual for the Pilot Housing Project, Kurtunwaare, Somalia." Florida Agricultural and Mechanical Univ., Tallahassee. Experimental Low-Cost Construction Unit. Performer: Berger (Louis), Inc., East Orange, NJ. Jul. Volume: PUB4, AIDPNAAM637, pages: 129. Descriptors: Construction; Houses; Manuals; Project management; Methodology; Planning; Floors; Supports; Walls; Tools; Roofs; Doors; Windows; Somalia; Training; Foundation engineering; Mechanical industrial civil and marine engineering Structural engineering; Behavioral and social sciences Personnel selection training and evaluation; Building industry technology Construction management and techniques. Abstract: A training manual in basic construction methods for use by settlers in the Kurtunwaare, Somalia housing project is herein presented. The manual begins with an overview of project administration, including training objectives, the organization of construction teams, training facilities, the training program and methods employed, records and files maintenance, and construction plans. The succeeding chapter focuses on foundation building and describes foundation layout, floor construction, and the use of piles, perimeter beams, partition grade beams, and subfloor fill. Next, masonry operations are explained, with reference to building layout and the construction of corners, exterior/interior walls, doors and windows, wall finishes, and roof anchor blocks. Lastly, carpentry operations are described, with an emphasis on roof design, truss fabrication and erection, roofing, and construction of temporary door/window frames and wooden doors and windows. Appropriate tools and their use are described in each section. One hundred forty-eight illustrations are included. Notes: Funder: Agency for International Development, Washington, DC. NT: Prepared in cooperation with Berger (Louis), Inc., East Orange, NJ. NTIS Accession Number: PB85150167XSP.

Bruce, J. G. 1987. "Seasonal Heat Content Variations in the Northwestern Indian Ocean." NSTL Station, MS; United States: Naval Oceanographic Office. Volume: ADA1964238, pages: 7. Descriptors: Heat transfer; Indian Ocean; Monsoons; Thermoclines; Advection; Autumn; Basins(Geographic); Dynamic response; Heat; Heat loss; Horizontal orientation; North(Direction); Ocean surface; Reprints; Somalia; Spring season; Temperature; Vertical orientation; West(Direction); Wind; Bathythermograph data; Seasonal variations; Sea water; Wind stress; Ocean basins. Abstract: The changes of heat storage of the Somali Basin in the western Indian Ocean, as shown from 55 detailed temperature sections along the western sea lane (tanker XBT program 1975-1979), are only partly accounted for by the net heat gain at the sea surface. Vertical movement of the thermocline and horizontal advection appear to be the cause of major changes both between and during the monsoons. During the southwest monsoon the large amount of heat stored in the 0 to 400-m layer during northern spring (up to 31 x 10 to the 8th power J/sq m) is redistributed with a heat loss of approximately 13 x 10 to the 8th power J/sq m in the upper 0 to 100-m layer and the gain of a similar amount in the 100 to 400-m layer, caused to a large extent by the dynamic response of these layers to the monsoon wind stress. In late autumn this heat is then advected from the region. Reprints. (EDC). Notes: Pub. in Oceanol. Acta, p77-83 1987. Original contains color plates: All DTIC/NTIS reproductions will be in black and white. Abstract in English and French. Proceedings Int. Symp. on Equatorial Vertical Motion, Paris, 6-10 May 85. NU: Contract N00014-76-C-0071. Database: NTIS. Accession Number: ADA1964238.

Bruce, J. G. 1977. "Somali Current: Recent Measurements during the Southwest Monsoon." Science (Wash.). Volume 197, Issue 4298, Pages 51-53. Descriptors: Water temperature; Monsoons; Somali Current; Marine. Notes: Records keyed from 1977 ASFA printed journals. TR: 1977. Abstract: The Somali Current was measured at approx maximum strength for the year during the southwest monsoon on 11 and 12 Aug 1975 (~1.5 days) at stations spaced 20-35 km apart. These measurements permit a nearly synoptic detailed mapping of the temp structure of a major western boundary current. Database: ASFA: Aquatic Sciences and Fisheries Abstracts.

Bruce, J. G. 1974. "Some Details of Upwelling Off the Somali and Arabian Coasts." J. Mar. Res. Volume 32, Issue 3, Pages 419-423. Descriptors: Physical oceanography; Upwelling; Water temperature; Salinity; Geographical distribution; Arabian Sea; Somali Dem. Rep. Notes: Records keyed from 1975 ASFA printed journals. TR: 1975. Abstract: Surface temp and salinity maps from measurements during the period of maximum coastal upwelling in the Arabian Sea are given. The region of coldest, temp <14°C, and freshest, salinity <35.5 ppt, surface water off Ras Mabber (9°N Somali coast) shifted northeastward during a 10-day period. Off the Arabian coast the upwelled water was not as cold (minimum, 18°C) or fresh (minimum, 35.7ppt) and was found between 16°N and 20°N. Database: ASFA: Aquatic Sciences and Fisheries Abstracts. ISSN: 0022-2402.

Bruce, J. G. and Beatty, W. H. 1985. "Some Observations of the Coalescing of Somali Eddies and a Description of the Socotra Eddy. Technical Rept." Woods Hole Oceanographic Institution, MA. Volume: WHOICONTRIB4960, pages: 15. Descriptors: Eddies Fluid mechanics; Circulation; Indian ocean; Upwelling; Monsoons; Seasonal variations; Tropical regions; Offshore; Somalia; Coastal regions; Water flow; Sea water; Ocean surface; Surface temperature; Expendable; Reprints; Ocean currents; Somali Basin; Socotra eddy circulation; Boundary currents; Ocean circulation; Earth sciences and oceanography Dynamic oceanography; Ocean technology and engineering Dynamic oceanography. Abstract: The eddy circulation pattern in the Somali Basin was observed to shift during the latter part of the 1979 southwest monsoon. Measurements were obtained from research ships and from tankers taking repeated XBT sections along the standard offshore sea lane. Initially the current turned strongly offshore between 4 deg N-6 deg N forming a loop or eddy to the south of the northern eddy which turned offshore at about 10 deg N. This pattern resulted at the surface in two wedges of relatively cold and fresh upwelled water along the coast similar to the type of circulation observed during a 1970 survey. However by late August the location of the southern turn off point had shifted northward and the coastal boundary current then merged and flowed into the northern eddy. This flow pattern produced a relatively fresh and cold cell in the near surface water on the western portion of the northern eddy. By October the circulation appeared to have reverted back to a pattern somewhat similar to that observed during the earlier (June, July) part of the monsoon. Observations of the Socotra eddy circulation indicate that it develops each southwest monsoon approximately between 10 deg N-14 deg N and during some years its horizontal and vertical scale approaches that of the northern Somali eddy. Notes: Pub. in Oceanologica ACTA, v8 n2 p207-219 1985. DTIC Accession Number: ADA1635770XSP.

Bruchon, R. 1972. Comparaison Des Dispositions Géologiques De Part Et d'Autre Du Golfe d'Aden. Translated title: Comparison of Geological Aspects of the Shores of the Gulf of Aden. France: Universite de Picardie, Amiens, France. Descriptors: Africa; Arabian Peninsula; Arabian Sea; areal geology; Asia; correlation; Djibouti; East Africa; Gulf of Aden; Indian Ocean; lithologic controls; paleogeography; petroleum; petroleum exploration; plate tectonics; rifting; Somali Republic; stratigraphy; structural controls; tectonics; Yemen. Database: GeoRef. Accession Number: 1998-010517.

Brummer, G. J. A. Kloosterhuis, H. T. and Helder, W. 2002. Monsoon-Driven Export Fluxes and Early Diagenesis of Particulate Nitrogen and its Delta Across the Somalia Margin. Geological Society; 2002. Special Publication- Geological Society of London 195, pages: 353-370. Meeting- 2001 Apr: London. Notes: The tectonic and climatic evolution of the arabian sea region; sponsor: Geological society. geological society. geological society. ISBN: 1862391114.

Bruni, P. Abbate, E. Abdi, Salah H. Fazzuoli, M. and Sagri, M. 1987. Geological Map of the Daban Basin, Northern Somalia. Italy: S.E.L.C.A., Florence, Italy. Descriptors: Africa; cartography; Daban Basin; East Africa; geologic maps; maps; northern Somali Republic; Somali Republic. Database: GeoRef. Accession Number: 1998-015584.

Bruni, P. and Fazzuoli, M. 1976. "Sedimentological Observation on Jurassic and Cretaceous Sequences from Northern Somalia; Preliminary Report." Bollettino Della Societa Geologica Italiana. Volume 95, Issue 6, Pages 1571-1587. Descriptors: Africa; Arabian Sea; carbonate rocks; clastic rocks; Cretaceous; East Africa; environment; Gulf of Aden; Indian Ocean; Jurassic; lagoons; limestone; lithofacies; lithostratigraphy; Mesozoic; sandstone; sedimentary petrology; sedimentary rocks; sedimentation; shale; shore features; Somali Republic; stratigraphy; structural analysis; structural geology; tectonics. Notes: References: 19; sects., sketch map. ISSN: 0037-8763.

Bruno, V. Shaie, M. Balley, O. A. and Ismail, A. 1985. Salinit Dei Suoli in Somalia. Saline Soils in Somalia. Translated title: Somalia: Universita Nazionale Somala, Mogadishu, Somalia. Descriptors: Africa; East Africa; geochemistry; Kismaio Somali Republic; saline composition; salinity; soils; Solonchak soils; Somali Republic; water resources. Database: GeoRef. Accession Number: 1998-015585.

Bryzgolov, L. 1967. Results of Hydrogeological Work Carried Out in 1966 on the Modu-Mode Apatite Occurrence. Somalia: United Nations Development Programme, Mogadishu, Somalia. Descriptors: Africa; Bay; East Africa; hydrology; Modu-Mode; occurrence; phosphate deposits; phosphorus; Somali Republic; surface water. Database: GeoRef. Accession Number: 1998-015586.

Buettner, W. G. 1966. "Tribal and Resident Boundary Map." Mogadiscio, Somali Republic. Descriptors: Tribes- Somalia- Maps. Notes: Description: color map; 87 x 71 cm. Scale 1:2,000,000. Notes: Reproduction: Photocopy. Hand colored. Responsibility: Drafted by W.G. Buettner, August 1966. Library of Congress; LCCN: gm 69-1382.

Burans J.P and Et Al. 1990. "HIV Infection Surveillance in Mogadishu, Somalia." East Afr. Med. J. Volume 67, Issue 7, Pages 466-472. Descriptors: International Development Abstracts- volume 74; sexually transmitted disease; HIV infection Species Term: Somalia; Human immunodeficiency virus. Notes: Somalia- Mogadishu. Abstract: A group of 89 prostitutes and 45 patients attending sexually transmitted disease clinics in Mogadishu were examined for evidence of HIV infection. There was a significant amount of sexually transmitted diseases (STDs) in these two groups, with 11.2% and 6.7% respectively being culture positive for N. gonorrhoea. All study participants were negative for antibodies to HIV, suggesting an extremely low prevalence of HIV in high risk behaviour groups in the capital city of Somalia. ISSN: 0012-835X.

Burbridge, Peter R. 1987. Coastal And Marine Resource Management Prospects in Somalia. Seattle, WA, USA: ASCE, New York, NY, USA. pages: 4556-4570. Coastal Zone '87, Proceedings of the Fifth Symposium on Coastal and Ocean Management.

Descriptors: Coastal Zones- Management; Environmental Protection- Somalia; Water Pollution- Marine Pollution. Abstract: The main coastal resource management issues in Somalia are centered in the terrestrial component of the coastal zone. There are no significant problems of marine pollution or the overexploitation of marine living resources. The development problems of primary significance are: natural hazards associated with floods and destabilization of sand dunes, land based industrial and municipal pollution and low standards of resource management and planning. ISBN: 0-87262-602-4.

Burgess, N. D., Clarke, G. P. And Rodgers, W. A. 1998/7. "Coastal Forests of Eastern Africa: Status, Endemism Patterns and their Potential Causes." Biological Journal of the Linnean Society. Volume 64, Issue 3, Pages 337-367. Descriptors: endemic species; evolution; centres of endemism; conservation. Abstract: Eastern African coastal forests are located within the Swahili regional centre of endemism and Swahili-Maputaland regional transition zone in eastern Africa, between 1° North and 25° South, and 34-41° East. Approximately 3167 km² coastal forest remains: 2 km² in Somalia, 660 km² in Kenya, 697 km² in Tanzania, 16 km² in Malawi, 3 km² in Zimbabwe and perhaps 1790 km² in Mozambique. Most forests are small (2), and all but 19 are under 30 km² area. Over 80% of coastal forest is located on government land, principally Forest Reserves; only 8.3 km² found in National Parks (6.2 km² in Kenya (Arabuko-Sokoke), 2 km² in Tanzania (Mafia Island) and tiny patches in Zimbabwe). Coastal forests are an important and highly threatened centre of endemism for plants (c550 endemic species), mammals (6 species), birds (9 species), reptiles (26 species), frogs (2 species), butterflies (79 species), snails (>86 species) and millipedes (>>20 species). Endemic species are concentrated in the forests of the Tana River, between Malindi in Kenya to Tanga in northern Tanzania, and in southern Tanzania. Forests with highest numbers of endemics are: lower Tana River, Arabuko-Sokoke, Shimba Hills (Kenya); lowland East Usambara, Pugu Hills, Matumbi Hills, Rondo and Litipo and other plateaux near Lindi (Tanzania); the Tanzanian offshore island of Pemba; Bazaruto archipelago (Mozambique), and tiny forest remnants of southern Malawi, eastern Zimbabwe and Mozambique. Most coastal forest endemics have a narrow distributional range, often exhibiting single-site endemism or with scattered or disjunct distributional patterns. They are best interpreted as relicts and not the result of recent evolution. Relictualization probably started with the separation of the ancient Pan African rainforest into two parts during the Miocene. The coastal forests are interpreted as a > with the endemic species gradually becoming more and more relict (and presumably extinct) due historically to climatic desiccation and more recently to human destruction. ISSN: 0024-4066.

Burgett, Charles L., Hoover, Constance S., Rousseau, Jacqueline J., United States Board on Geographic Names, United States and Defense Mapping Agency. 1987. "Gazetteer of Somalia: Names Approved by the United States Board on Geographic Names." Pages 519. Descriptors: GovDoc; Geographic: Somalia- Gazetteers. Note(s): "June 1987." Includes bibliographical references (p. ix). Prepared by Charles L. Burgett, Constance S. Hoover, Jacqueline J. Rousseau. LCCN: 88-600915.

Burnham, L. 1986. "Population of Angola by Major Subdivisions. Final Rept." Library of Congress, Washington, DC. Federal Research Div. 30 Apr. pages: 5. Descriptors: Demography; Regions; Somalia; Census; Population; Angola; Behavior and society Education law and humanities. Abstract: Provides the population of Angola.

Entries are listed alphabetically by administrative districts according to the current subdivisions. DTIC Accession Number: ADA3034162XSP.

Burroughs, R. H., I.I.I. 1974. The Structural and Sedimentological Evolution of the Somali Basin; Paleooceanographic Interpretations. United States: Woods Hole Oceanographic Institution, Woods Hole, MA, United States. Descriptors: basins; bottom features; carbonate compensation depth; Cenozoic; changes of level; evolution; fracture zones; Indian Ocean; marine geology; northwest; ocean circulation; ocean floors; oceanography; paleo-oceanography; plate tectonics; productivity; rates; ridges; sedimentation; Somali Basin; thermal upwelling. Database: GeoRef. Accession Number: 1977-028828.

Buscaglione, L. Fazzuoli, M. Chiocchini, M. and Pavia, G. 1993. Contributions to the Stratigraphy of the Early to Middle Jurassic Formations of the Eastern Side of the Luuq-Mandera Basin, Bay and Gedo Region, Southwestern Somalia. Firenze; Istituto Agronomico L'Oltremar; 1993. Relazioni E Monografie Agrarie Subtropicali E Tropicali Nuova Serie113 A, pages: 153-167. International Meeting; 1st- 1987 Nov: Mogadishu. Notes: Geology of Somalia and surrounding regions: Geology and mineral resources of Somalia and surrounding regions. OCLC Accession Number: 34175029.

C

C.S. Hammond & Company. 1936. "Map of Ethiopia and Adjoining Territories." New York: C.S. Hammond & Co. Inc. Notes: Description: 1 map; color; 47 x 40 cm., on sheet 63 x 43 cm., folded to 15 x 24 cm. Geographic: Ethiopia- Maps. Somalia- Maps. Djibouti- Maps. Scale [ca. 1:6,336,000]; Cartgrph Code: Category of scale: a Constant ratio linear horizontal scale: 6336000; Notes: Relief shown by shading and spot heights. Insets: Eritrea- Chart of distances in statute miles- Africa- Comparative populations and areas. Includes text and advertisement on panel for Northwest Motors. Other Titles: Panel title; Map of Ethiopia; OCLC Accession Number: 35032276.

Cabdulqaadir, M. M. 1981. "Geologia Della Zona Di Buqda Caqable Nella Regione Di Hiran." Translated title: "Geology of the Buqda Zone in the Hiran Region." Quaderni Di Geologia Della Somalia (Mogadiscio). Volume 5, Descriptors: Africa; areal geology; Buqda zone; East Africa; Hiran Somali Republic; Somali Republic. LCCN: 93093377.

Cabdulqaadir, M. M. and Carush, Maxamud C. 1982. "The Indo-Mediterranean Characters of the Somali Shallow Marine Benthic Faunas from Jurassic Up to Oligocene." Bollettino Della Societa Paleontologica Italiana. Volume 21, Issue 2-3, Pages 243-253. Descriptors: Africa; Bivalvia; Brachiopoda; Cenozoic; continental drift; East Africa; Echinodermata; Foraminifera; Gastropoda; Invertebrata; Mediterranean region; Mesozoic; microfossils; Mollusca; paleoenvironment; paleogeography; paleontology; Protista; shallow-water environment; Somali Republic; stratigraphy; Tethys. Notes: References: 50; sketch maps. ISSN: 0375-7633.

Cabdulqaadir, Maxamuud Maxamed. 1987. Some Thoughts on the Tethyan Characteristics of the Somali Shallow Benthic Faunas; Shallow Tethys 2. Netherlands: A. A. Balkema, Rotterdam, Netherlands (NLD). International Symposium on Shallow Tethys 2, Wagga Wagga, N.S.W. Australia Conference: Sept. 15-17, 1986. Descriptors: Africa; benthonic taxa; Bivalvia; Brachiopoda; Cenozoic; correlation; East Africa; Echinodermata; Echinoidea; Echinozoa; Foraminifera; Gastropoda; Invertebrata;

Mesozoic; microfossils; Mollusca; numerical analysis; Protista; shallow-water environment; Somali Republic; stratigraphy; Tertiary; Tethys. ISBN: 906191647X.

Cabdulwaaxid, S. M. 1981. Rilevamento Geologico Della Zona Di Dinsoor. Translated title: Geological Surveying of Dinsor. Somalia: Universita Nazionale Somala, Mogadishu, Somalia. Descriptors: Africa; areal geology; Bay Somali Repblic; Dinsor Somali Republic; East Africa; mapping; Somali Republic. Database: GeoRef. Accession Number: 1998-015594.

Cadotsch, A. F. 1986. "Does Selective Intervention Affect Other Areas of Primary Health Care? Evaluation of a Vaccination Program in Somali Refugee Camps." Soz. Praventivmed. Volume 31, Issue 6, Pages 302-307. Descriptors: Child, Preschool; Developing Countries; English Abstract; Humans; Nutritional Status; Primary Health Care/organization & administration; Refugees; Sanitation; Somalia; Vaccination, water. Abstract: The discussion, whether poor countries of the third world should work towards comprehensive or selective primary health care has been based on political and economical arguments. The present study examines how far a selective strategy with one target problem influences other areas of primary health care. The study is based on the experiences with a vaccination program in the Somalia refugee camps. The program produced for the first time a reliable census of the refugee population which produced much smaller figures than had been reported previously. Data on the nutritional status of all children under 5 was collected at the vaccination places. These data provided a new basis for the general food distribution. At the same time the data served to evaluate the supplementary feeding program for vulnerable groups. The study also showed that general food distribution was more important for the reduction of malnutrition than other interventions, such as curative health care, community water supplies, sanitation, supplementary feeding program and vaccination program. Furthermore it illustrates the conditions in which sectorial actions can become a useful asset to community oriented health programmes. ISSN: 0303-8408.

Calderoni, G. Masi, U. and Petrone, V. 1993. Chemical Features of Springwaters from the East African Rift. A Reconnaissance Study. Firenze; Istituto Agronomico L'Oltremar; 1993. Relazioni E Monografie Agrarie Subtropicali E Tropicali Nuova Serie113 B, pages: 699-710. International Meeting; 1st- 1987 Nov: Mogadishu. Notes: Geology of Somalia and surrounding regions: Geology and mineral resources of Somalia and surrounding regions. OCLC Accession Number: 34175029.

Cameron, J. 1970. "The Alio Ghelle Radioactive Mineral Occurrence in the Bur Region of the Republic of Somalia; a Brief Summary of the Principal Features; Uranium Exploration Geology." Panel Proceedings Series- International Atomic Energy Agency. Pages 169-175. Descriptors: actinides; Africa; Alio Ghelle; Bur region; East Africa; economic geology; metals; Somali Republic; uranium. Notes: illustrations (incl. sketch maps). ISSN: 0074-1876.

Cancelliere, G. 1987. Belet Weyne Gypsum Stone Quarry; GeoSom 87; Geology of Somalia and Surrounding Regions; Late Abstracts. Somalia: Somali Natl. Univ., Dep. Geol., Mogadishu, Somalia. GeoSom 87; Geology of Somalia and Surrounding Regions, Mogadishu. Somalia Conference: Nov. 24-Dec. 1, 1987. Descriptors: Africa; alabaster; Belet Weyne; breccia; clastic rocks; Cretaceous; East Africa; economic geology; FerFer Formation; gypsum deposits; Hiiraan; Mesozoic; production; reserves; sedimentary rocks; Somali Republic; sulfates. OCLC Accession Number: 34175029.

Cannon, R. T., Simiyu Siambi, W. M. N. and Karanja, F. M. 1981. "The Proto-Indian Ocean and a Probable paleozoic/mesozoic Triradial Rift System in East Africa." Earth and Planetary Science Letters. Volume 52, Issue 2, Pages 419-426. Abstract: A revised Paleozoic-Mesozoic stratigraphy of coastal Kenya (including, in particular, the Karroo) based on current geological mapping near Mombasa is briefly described. This stratigraphy provides the geological framework for proposals concerning the Proto-Indian Ocean and the tectonic setting of the Karroo depositional basins. Recent geophysical evidence suggests that, within Gondwanaland, Madagascar was situated off East Africa near Kenya/Tanzania. The southern limits of the marine Lower Jurassic and southern limits of the marine Middle and Upper Jurassic are in similar positions in mainland Africa and Madagascar using the latter reconstruction. These paleogeographic limits also define the position, during the Jurassic, of an embayment from an ocean to the north. Regional geological similarities also support this reconstruction and are reinforced by paleocurrent data from the Karroo of Kenya indicating drainage north-northeast during the Permian and Triassic and possibly the Lower Jurassic. Marine connections during Karroo times appear to be of different ages in Kenya, Tanzania, Somalia, and Madagascar, probably reflecting physical limitations to marine access in fault-separated basins. The above embayment encroached across the Karroo depositional basins from northeast Kenya to southern Tanzania during the Lower and Middle Jurassic, i.e. from the direction towards which the Karroo drainage had been previously directed. Marine conditions remain to the present day so this embayment can be considered the Proto-Indian Ocean for East Africa. The marine incursion took place before the breakup of Gondwanaland suggesting that during the Jurassic the Proto-Indian Ocean in East Africa was an epicontinental sea and not a true ocean (i.e. floored by simatic crust). The epicontinental nature of this sea is confirmed by the lithologies of the associated sediments. Paleontological data indicate that this sea was an arm of Tethys. True oceanic conditions could not have been established until the displacement of Madagascar away from Africa, probably in the Cretaceous. Accepting the above northern position of Madagascar, the writers also postulate that in East Africa the fault-bounded Karroo depositional basins (troughs) were located within a major triradial rift system extending from Lake Malawi at least as far as eastern Kenya (some 1600 km). This rift system, if valid, was established within Gondwanaland over a period ~100 m.y. in the Paleozoic/Mesozoic (pre-breakup) in marked contrast to the East African Rift System (classical rift valleys) which is mainly a Cainozoic phenomenon (post-breakup). It is, therefore, considered that there is a fundamental difference in origin between the two rift systems.

Cantone, G. 1976. "Research on the Littoral Zone of Somalia. Annelida, Polychaeta from Bender Mtoni and Sar Uanle." Monit.Zool.Ital., Suppl. Volume 8, Issue 9, Pages 223-254. Descriptors: New records; New species; Geographical distribution; Annelida; Polychaeta; Scyphoproctus somalus; Somali Dem.Rep; Marine. Notes: Records keyed from 1977 ASFA printed journals. TR: 1977. Abstract: The author describes 77 spp of benthonic polychaetes found in southern Somalia, south of the equator, in the mangrove swamps of Bender Mtoni (approximately 1km north of Sar Uanle) and along the beach of Sar Uanle; 8 of them: Phyllodoce capensis Day, 1960, Branchiosyllis uncinigera (Hartmann-Schroder, 1960), Glycera tesselata Grube var. minor La Greca, 1946, Rhynchospio glutaea (Ehlers, 1897), Paraspio mecznikowianus

(Claparede, 1869), Cirriformia capensis (Schmarda, 1861), Pherusa laevis (Stimpson, 1856), Fabricia filamentosa Day. 1963 had never been found in the Indian Ocean. A description of a new sp of Capitellidae,, Scyphoproctus somalus sp.nov.is also given. Database: ASFA: Aquatic Sciences and Fisheries Abstracts.

Canuti P, Fazzuoli M, Ficcarelli G and Venturi F. 1983. "Occurrence of Liassic Faunas at Waaney (Uanei), Province of Bay, South-Western Somalia." Rivista Italiana Di Paleontologia e Stratigrafia. Volume 89, Issue 1, Pages 31-46. Descriptors: Sedimentology; Species Term: Somalia. Notes: Special Feature: 2 figs, 40 refs. Abstract: The sequence exposed at Waaney consists of seven horizons. Sedimentological features give evidence that the environment evolved from a restricted lagoon, with fine terrigenous supply and probably evaporitic episodes, to a shallow sea with varying energy conditions and eventually to a deeper but partially restricted sea with oligotypic fauna. The palaeontological and sedimentological data agree with previous findings of Liassic faunas in Arabia, in N Somalia and in NE Kenya. ISSN: 0035-6883.

Canuti, P., Fazzuoli, M. and Tacconi, P. 1980. "A Geological Map of the Lugh-Bardera-Baidoa-Ted Area (S.W. Somalia)." Atti Dei Convegni Lincei, Accademia Nazionale Dei Lincei. Volume 1980, Descriptors: Africa; Baidoa Somali Republic; Bardera Somali Republic; East Africa; geologic maps; Lugh Somali Republic; maps; Somali Republic; southwestern Somali Republic; Ted Somali Republic. ISSN: 0391-805X.

Canuti, P., Fazzuoli, M., Tacconi, P. and Carrelli, Antonio (president). 1980. "Photogeological Map of the Region between Juba and Webi Shebeli Rivers (Southwest Somalia); Geodynamic Evolution of the Afro-Arabian Rift System." Atti Dei Convegni Lincei, Accademia Nazionale Dei Lincei. Issue 47, Pages 209-217. Descriptors: Africa; areal geology; Bur Uplift; Cretaceous; East Africa; East African Rift; erosion features; explanatory text; fault zones; geologic maps; geophysical surveys; Juba River; Jurassic; maps; Mesozoic; Oddur Arch; peneplains; remote sensing; Somali Republic; southwestern Somali Republic; surveys; Webi Shebeli Homocline; Webi Shebeli River. Notes: References: 24. ISSN: 0391-805X.

Caponera, D. A. 1964. "Report to the Government of Somalia on Water Legislation and Administration with Draft Water Code." Italy: Descriptors: Africa; drawdown; East Africa; legislation; policy; Somali Republic; water management; water resources; water rights; water use. Database: GeoRef. Accession Number: 1998-015598.

Caponera, D. A. 1973. "Water Laws in Moslem Countries." Food and Agriculture Organization of the United Nations. Pages 118. Refs. Descriptors: Water Resources; Water Law; Legislation; Water Rights; Reviews; Legal Aspects; Water Resources Development; Arid Lands; Asia; Africa; Irrigation; Moslem Water Law; Afghanistan; Bahrain; Brunei; Iran; Jordan; Kuwait; Morocco; Qatar; Somalia; Tunisia; The Peoples' Democratic Republic Of Yemen; Yemen Arab Republic. Abstract: this is a revision of FAO agricultural development paper no. 43, 'Water Laws in Moslem Countries' published in 1954. The earlier paper has been outdated by changes in water law and by the emergence of many new nations in the Moslem world. In this, volume one of a two volume update, a general introduction is given and twelve nations (Afghanistan, Bahrain, Brunei, Iran, Jordan, Kuwait, Morocco, Qatar, Somalia, Tunisia, the Peoples' Democratic Republic of Yemen, and the Yemen Arab Republic) are described. Since Moslem law is based upon revelations transmitted from Allah to man through the agency of the prophet

Mohammed, its religious overtones are unmistakable and western planners must expect plans to be suspect which ignore or contradict this divine law. Moslem law has faced the practical problems of dividing a limited resource among the populace; however, it should be remembered that the scarcer the water is, the more complicated and elaborate the regulations are. Modern tendencies in Moslem water law make water a national resource under central control and ownership. This is compatible with traditional Islamic thinking which has always viewed the concept of community interest as the basis of moslem water law. Database: Environmental Sciences and Pollution Mgmt.

Caponera, D. A. and Grigg, N. S. 1982. "Assistance in Water Legislation; Mission Report, 7-10 June, 1982." Italy: Volume: TCP/SOM/0105, Descriptors: Africa; East Africa; legislation; policy; Somali Republic; water management; water resources. Database: GeoRef. Accession Number: 1998-015600.

Cappellini, V., Faenza, V. and Fondelli, M. 1982. "Il Telerilevamento e l'Informatica Per Lo Sviluppo Del'Agricoltura Nei Paesiterzo Mondo." Riv. Agric. Subtrop. Trop. Volume 76, Issue 1-2, Pages 5-17. Descriptors: Remote Sensing, Gis And Mapping; International Development Abstracts- volume 74; techniques of remote sensing; vegetation cover; hydrological conditions; Amazonia; Somalia; Sudan; Tanzania; Kenya. Notes: Special Feature: 2 tables, 3 refs. Abstract: Provides an introduction to the techniques of remote sensing, and shows how systematic information can be used as a data bank for subsequent analysis not only for agricultural purposes, though emphasis is placed on the value of Landsat in defining vegetation cover in relation to hydrological conditions. Examples are noted from Amazonia, Somalia, Sudan, Tanzania and Kenya. ISSN: 0035-6026.

Cappola, V. and Manning, R. B. 1994. "Research on the Coast of Somalia. Crustacea Stomatopoda." Trop. Zool. Volume 7, Issue 2, Pages 271-291. Descriptors: new records; distribution records; geographical distribution; new genera; check lists; Stomatopoda; Pseudosquilla; Somali Dem. Rep. Mesacturoides brevisquamatus; Marine. Notes: TR: CS9516082. Abstract: Seventeen species of gonodactyloid stomatopods representing nine genera and four families are recorded from shore habitats in Somalia. All but one are new records for Somalia. Mesacturoides brevisquamatus (Paulson 1875) is recorded from below Djibouti for the first time. Two new genera are erected for species previously assigned to the genus Pseudosquilla Dana 1852, Pseudosquillana for P. megalophthalma Bigelow 1893, Pseudosquillisma for P. oculata (Brulle 1837) and two other species. A new host, Mesacturoides fimbriatus (Lenz 1905), is reported for the parasitic gastropod, Caledoniella montrouzieri Souverbie 1869. Database: ASFA: Aquatic Sciences and Fisheries Abstracts. ISSN: 1121-919X.

Caraci, G. 1927. "La Migiurtinia Ed Il Territorio Del Nogal Secondo Recenti Studi. » Translated title : « Recent Studies of Migiurtania and Nugal Territories." Rivista Geografica Italiana. Volume 34, Pages 117-123. Descriptors: Africa; areal geology; East Africa; Migiurtinia; Nugal; Somali Republic. ISSN: 0035-6697.

Carbone F and Accordi G. 2000. "The Indian Ocean Coast of Somalia." Mar. Pollut. Bull. Volume 41, Issue 1-6, Pages 141-159. Descriptors: Coastal and Marine Management; Regional Geology; Marine ecology: general studies; Marine and coastal environment; anthropogenic effect; coastal zone; continental shelf; coral reef; environmental degradation; geological structure; mangrove; marine ecosystem; seagrass meadow Species Term: Somalia; Anthozoa; Halophila; Rhizophoraceae; Socotra;

Thalassodendron; Cheloniidae; Testudines; Aves; Neotonchus. Notes: Additional Info: United Kingdom; References: Number: 83; Geographic: Indian Ocean- Somali Shelf Somalia. Abstract: Somalia has the longest national coastline (3025 km) in Africa with an estimated shelf area (depth 0-200 m) of 32500 km2. The country is divided into the northern coastal plain of Guban, which has a semi-arid terrain; the northern highlands with rugged mountain ranges containing the country's highest peak (2407 m); and the Ogaden region which descends to the south from the highlands and which consists of shallow plateau valleys, wadis and broken mountains. The latter region continues to the Mudug plain in central Somalia. From Ras Caseyr to the Kenya border, the coast runs north-east to south-west, coinciding with the displacement caused by the Mesozoic marginal subsidence. This general structure is complicated by sedimentary troughs crossing the Horn of Africa, and by large sedimentary basins, cutting the coastline and extending inland into Southern Somalia and Northern Kenya (Juba-Lamu embayment, Mogadishu basin). Offshore, the western Somali Basin extends from Socotra to the Comores. The open shelf environments developed along the Somali coast are a consequence of an extensive marine transgression, connected to coastal subsidence or inland uplift. The rocks along the southern coastal belt are Pliocene-Pleistocene, and are characterized by a sequence of both marine and continental deposits of skeletal sands, coral build-ups, eolian sands and paleosols. As well as eolian and biogenic sedimentary processes, sea-level fluctuations, Holocene climatic changes and neotectonic movements have combined to produce the modern coastline. A notable feature is an ancient dune ridge complex, known as the Merka red dune, which rims the coast extending beyond the Kenyan border and which separates the narrow coastal belt from the Uebi Shebeli alluvial plain. Two features of note are the Bajuni Archipelago, which consists of islands, islets and skerries, forming a barrier island separated from the coast by a narrow marine sound, and a braided, channelized coastal area, which originated from the drowning of a paleofluvial net. The southern Somali coast, with that of Kenya and Tanzania, forms part of the Somali Current Large Marine Ecosystem, encompassing 700000 km2, and extending 800 km between Dar es Salaam and Ras Hafun. Abundant biomass develops here due to upwelling. The shelf area has a wide variety of coral reefs, mangroves, seagrass meadows, beaches and estuaries. In shallow water areas the abraded flats are colonized by scattered coral communities with variable cover. A true fringing reef is achieved in places only in the Bajuni archipelago. All along the southern Somali coastal shelf there are spreading meadows of Thalassodendron seagrass, and benthic communities typical of mobile sandy substrates are limited to beach ridges and shoals developed along the coastline. Around the Bajuni barrier island and the channelized area there is more diversity. Mangroves grow on the tidal belts of the channels, and there are expanses of salt flats. Large-scale alteration produced by man on the Somali coast is relatively recent, but has accelerated in the last few decades, especially around major cities. This alteration affects especially backshore areas where the Pleistocene coral reefs are quarried. At present, the continental shelf is not adequately monitored or protected, so coastal habitats are being degraded, living marine resources are overexploited, and pollution levels are increasing, all of which affect natural resources and biodiversity. Somalia is one of the world's poorest and least developed countries, with few resources and devastated by civil war, but since 1993 it has been part of the Common Market for Eastern and Southern Africa (COMESA). This will affect fisheries and aquaculture in terms of the investment, production, trade and fish consumption of the member states. There are currently no marine protected areas and no legislation concerning their establishment and management, although the World Conservation Monitoring Centre (WCMC) Protected Areas Database lists Busc Busc Game Reserve as an MPA. In 1992, The WCMC also listed the following coastal sites as proposed protected areas: Zeila (important sea bird colonies on offshore islets), Jowhar-Warshek, Awdhegle-Gandershe. The area from Kisimayo to Ras Chiambone is probably of highest priority, as it is important for coral reefs, marine turtles, and mangrove resources, although it is still poorly known. ISSN: 0025-326X.

Carbone, F. and Accordi, G. 2000. "The Indian Ocean Coast of Somalia." United Kingdom. Descriptors: 63-82; Regional and general; biomass; coastal morphology; environmental assessment; industrial development; sedimentary structure. Abstract: Somalia has the longest national coastline (3025 km) in Africa with an estimated shelf area (depth 0-200 m) of 32,500 km2. The country is divided into the northern coastal plain of Guban, which has a semi-arid terrain, the northern highlands with rugged mountain ranges containing the country's highest peak (2407 m), and the Ogaden region whichdescends to the south from the highlands and which consists of shallow plateau valleys, wadis and broken mountains; this region continues to the Mudug plain in central Somalia. From Ras Caseyr to the Kenya border, the coast runs northeast to southwest, coinciding with the displacement caused by the Mesozoic marginal subsidence. This general structure is complicated by sedimentary troughs crossing the Horn of Africa, and by large sedimentary basins, cutting the coastline and extending inland into southern Somalia and northern Kenya (Juba-Lamu embayment, Mogadishu basin). Offshore, the western Somali Basin extends from Socotra to the Comores. The open shelf environments developed along the Somali coast are a consequence of an extensive marine transgression, connected to coastal subsidence or inland uplift. The rocks along the southern coastal belt are Pliocene-Pleistocene, and are characterized by a sequence of both marine and continental deposits of skeletal sands, coral build-ups, eolian sands and paleosols. As well as eolian and biogenic sedimentary processes, sea-level fluctuations, Holocene climatic changes and neotectonic movements have combined to produce the modern coastline. A notable feature is an ancient dune ridge complex, known as the Merka red dune, which rims the coast extending beyond the Kenyan border and which separates the narrow coastal belt from the Uebi Shebeli alluvial plain. Two features of note are the Bajuni Archipelago which consists of islands, islets and skerries, forming a barrier island separated from the coast by a narrow marine sound, and a braided, channelized coastal area which originated from the drowning of a paleofluvial net. The southern Somali coast, with that of Kenya and Tanzania, forms part of the Somali Current Large Marine Ecosystem, encompassing 700,000 km2, and extending 800 km between Dar es Salaam and Ras Hafun. Abundant biomass develops here due to upwelling. The shelf area has a wide variety of coral reefs, mangroves, seagrass meadows, beaches and estuaries. In shallow water areas the abraded flats are colonized by scattered coral communities, with variable cover. A true fringing reef is achieved in places only in the Bajuni archipelago. All along the southern Somali coastal shelf there are spreading meadows of Thalassodendron seagrass, and benthic communities typical of mobile sandy substrates are limited to beach ridges and shoals developed along the coastline. Around the Bajuni barrier island and the channelized area there is more diversity. Mangroves grow on the tidal belts of the channels, and there are expanses of salt flats. Large-scale alteration produced by man on the Somali coast is relatively recent, but has accelerated in the last few decades, especially around major cities. This alteration affects especially backshore areas where the Pleistocene coral reefs are quarried. At present, the continental shelf is not adequately monitored or protected, so coastal habitats are being degraded, living marine resources are overexploited, and pollution levels are increasing, all of which impact on natural resources and biodiversity. Somalia is one of the world's poorest and least developed countries, with few resources and devastated by civil war, but since 1993 it has been part of the Common Market for Eastern and Southern Africa (COMESA). This will affect fisheries and aquaculture in terms of the investment, production, trade and fish consumption of the member states. There are currently no Marine Protected Areas and no legislation concerning their establishment and management, although the WCMC (World Conservation Monitoring Centre) Protected Areas Database lists Busc Busc Game Reserve as an MPA. In 1992, The WCMC also listed the following coastal sites as proposed protected areas: Zeila (important sea bird colonies on offshore islets), Jowhar-Warshek, Awdhegle-Gandershe. The area from Kisimayo to Ras Chiambone is probably of highest priority, as it is important for coral reefs, marine turtles, and mangrove resources, although it is still poorly known. Notes: In: Seas at the millennium- an environmental evaluation- Volume 2 (2000) p. 63-82; References: 83; Geographic: Indian Ocean Somalia. ISBN: 0080432077.

Carbone, F., Matteucci, R., Pignatti, J.S. and Russo, A. 1993. "Facies Analysis and Biostratigraphy of the Auradu Limestone Formation in the Berbera-Sheikh Area, Northwestern Somalia." Geologica Romana. Volume 29, Pages 213-235. Descriptors: Tertiary; Palaeogene; foraminifera; facies analysis; diagenesis; Auradu Limestone Formation; biostratigraphy. Notes: Somalia- Berbera-Sheikh. Abstract: The main stratigraphic and sedimentological features of the Auradu Limestone Formation in NW Somalia, close to its type area, are outlined. The present study of the depositional and diagenetic characteristics, associated with the investigation of the fossil assemblages and their biostratigraphy and paleoecology, focuses on some significant features useful for a better definition of the paleoenvironmental setting and the stratigraphic evolution of the formation. Biostratigraphic evidence gathered from the larger foraminiferal assemblages allows a correlation of the lower Paleogene faunal associations of Somalia with the existing peri-Mediterranean biozonations, thus extending their validity to the Horn of Africa, with implications for the dating of neighbouring formations of the Arabic peninsula. ISSN: 0435-3927.

Carbone, F. and Accordi, G. 2000. "The Indian Ocean Coast of Somalia." Marine Pollution Bulletin. Jan-Jun. Volume 41, Issue 1-6, Pages 141-159. Descriptors: Coral reefs; Sea grass; Mangrove swamps; Marine parks; Nature conservation; Man-induced effects; Ecosystem management; Geology; Coastal zone management; Marine pollution; Overexploitation; Coastal morphology; Living resources; Environmental protection; Environment management; Coastal zone; Seagrasses; Resources Management; Coastal Waters; Corals; Sea Grasses; Estuaries; Marine Environment; Environmental Quality; Degradation; Resources; Swamps; Coelenterates (Corals); Sea-grass; Marine environment (see also Sea water); Environmental quality standards; Decomposition; Thalassodendron; Somalia; Somalia; Thalassodendrom. Notes: Special Issue: Seas at the Millennium: An Environmental Evaluation. TR: CS0102716. Abstract: Somalia has the longest national coastline (3025 km) in Africa with an estimated shelf area (depth 0-200 m) of 32 500 km super(2). The country is divided into the northern coastal plain of Guban, which has a semi-arid terrain; the northern highlands with rugged mountain ranges containing the country's highest peak (2407 m); and the Ogaden region which descends to the south from the highlands and which consists of shallow plateau valleys, wadis and broken mountains. Offshore, the western Somali Basin extends from Socotra to the Comores. The open shelf environments developed along the Somali coast are a consequence of an extensive marine transgression, connected to coastal subsidence or inland uplift. The rocks along the southern coastal belt are Pliocene-Pleistocene, and are characterized by a sequence of both marine and continental deposits of skeletal sands, coral build-ups, eolian sands and paleosols. A notable feature is an ancient dune ridge complex, known as the Merka red dune, which rims the coast extending beyond the Kenyan border and which separates the narrow coastal belt from the Uebi Shebeli alluvial plain. The southern Somali coast, with that of Kenya and Tanzania, forms part of the Somali Current Large Marine Ecosystem, encompassing 700 000 km super(2), and extending 800 km between Dar es Salaam and Ras Hafun. Abundant biomass develops here due to upwelling. The shelf area has a wide variety of coral reefs, mangroves, seagrass meadows, beaches and estuaries. In shallow water areas the abraded flats are colonized by scattered coral communities with variable cover. A true fringing reef is achieved in places only in the Bajuni archipelago. All along the southern Somali coastal shelf there are spreading meadows of Thalassodendron seagrass, and benthic communities typical of mobile sandy substrates are limited to beach ridges and shoals developed along the coastline. Around the Bajuni barrier island and the channelized area there is more diversity. Mangroves grow on the tidal belts of the channels, and there are expanses of salt flats. Large-scale alteration produced by man on the Somali coast is relatively recent, but has accelerated in the last few decades, especially around major cities. This alteration affects especially backshore areas where the Pleistocene coral reefs are quarried. At present, the continental shelf is not adequately monitored or protected, so coastal habitats are being degraded, living marine resources are overexploited, and pollution levels are increasing, all of which affect natural resources and biodiversity. There are currently no marine protected areas and no legislation concerning their establishment and management, although the World Conservation Monitoring Centre (WCMC) Protected Areas Database lists Busc Busc Game Reserve as an MPA. Database: Environmental Sciences and Pollution Mgmt. ISSN: 0025-326X.

Carmignani, Luigi; Università di Siena and Consiglio nazionale delle ricerche (Italy). 1994. "Carta Geologica Della Somalia Nord-Orientale = Geological Map of Northeastern Somalia." Firenze: Litografia Artistica Cartografica. Descriptors: Geology-Somalia- Maps; Mapa geológico- Somália. Notes: Description: 1 map; color; 90 x 137 cm. Scale 1:200,000. Cartgrph Code: Category of scale: a Constant ratio linear horizontal scale: 200000; Notes: Relief shown by contours and spot heights. Includes location map, tectonic sketch map of Gulf of Aden, 2 color cross-sections, stratigraphic chart, and index to mapping. Other Titles: Geological map of northeastern Somalia; Responsibility: Authors, Pier Lorenzo Fantozzi, Abdirahman Hilowle Mohamed, Ali Kassim Mohamed; Cartographic Supervisor, Luigi Carmignani; Dipartimento di Scienze della Terra Università di Siena; Consiglio Nazionale delle Ricerche. OCLC Accession Number: 58676288.

Carrara, E., Rapolla, A., De Florentis, N. and Dorre, Abdulkadir S. 1985. "Indagini Geofisiche Applicate Al Rinvenimento Di Falde Acquifere Ed Allo Sbarramento Di Torrenti Nella Regione Di Hiran (Somalia)." Translated title: "Geophysical Investigation in the Discovery of the Aquifers and in the Obstruction of the Flood in Hiran Region, Somalia." Quaderni Di Geologia Della Somalia (Mogadiscio). Volume 8, Descriptors: Africa; aquifers; dams; East Africa; floods; geologic hazards; geophysical surveys; ground water; Hiran Somali Republic; Somali Republic; surveys. LCCN: 93093377.

Carter, S. and McCormack, D.P. 2006. "Film, Geopolitics and the Affective Logics of Intervention." Political Geography. Volume 25, Issue 2, Pages 228-245. Descriptors: Political; geography, geopolitics; war; media role; culture Species Term: Buteogallus; Somalia. Notes: Additional Info: United Kingdom; References: Number: 81; Geographic: Somalia East Africa Sub-Saharan Africa Africa. Abstract: This paper explores the way in which questions of affect are implicated in the relation between film and popular articulations of geopolitics. Recent work in political and cultural geography has foregrounded the role of affect in the performative enactment of space and spacing. Drawing upon such work, in this paper we explore the particular role of film as an affective assemblage through which geopolitical sensibilities emerge and are amplified. More specifically, we argue that the relation between cinema and enactments of geopolitical intervention must be understood not only in terms of the way one reproduces or subverts the discursively framed codes and scripts of the other but also in terms of the amplification and anchoring of particular affects through specific tactics and techniques. We illustrate this through a brief discussion of how the relations between the affective and geopolitical logics of intervention are implicated in U.S. involvement in Somalia in 1993 and its depiction in the 2002 film Black Hawk Down. In moving towards a conclusion, we draw upon this engagement with film in order to point to the possibilities for a more expansive engagement with the role played by the logics of affect in contemporary geopolitical cultures. ISSN: 0962-6298.

Carter, S. 2004. "Two New Species of Euphorbia Subsp Euphorbia (Euphorbiaceae) from East and Northeast Somalia." Nord. J. Bot. Volume 23, Issue 3, Pages 295-297. Abstract: Two small species of succulent, spiny Euphorbia are described as new, both with limited distributions, E. ammophila from the coastal plain of eastern Somalia, and E. densispina from the mountain ranges of the northeast.

Castagno, A. A. 1964. Controversies in the Horn of Africa. Descriptors: 50 leaves. Notes: Geographic: Somalia- Boundaries- Kenya. Kenya- Boundaries- Somalia. Somalia- Boundaries- Ethiopia. Ethiopia- Boundaries- Somalia. Note(s): Caption title.; Responsibility: A.A. Castagno, Jr. OCLC Accession Number: 38056865.

Castagno, A. A. United States and Agency for International Development. 1969. The Somali National Police Force and A.I.D. Assistance. Descriptors: 33 leaves; Police- Somalia; Police training- Somalia. Notes: Cover title. Bibliography: leaves 31-33. Responsibility: A.A. Castagno. OCLC Accession Number: 11228752.

Castagno, Margaret. 1999. Historical Dictionary of Somalia. Bosano: Hagi. Descriptors: 216. Notes: Met lit. opg. Original: Oorspr. uitg. Metuchen: Scarecrow, 1975. Responsibility: Margaret Castagno. OCLC Accession Number: 66048266.

Castagno, Margaret. 1975. Historical Dictionary of Somalia. Metuchen, N.J: Scarecrow Press. Descriptors: 213. Notes: Somalia- History- Dictionaries. Notes:

Includes bibliographical references (p. 165-213). Reproduction: Microfiche. Ann Arbor, Mich.: Filmed by University Microfilms International for Indiana University, Bloomington, 1995. 3 microfiches. ISBN: 0810808307.

Castellani E, Gullino G and Mohamed M.I. 1983. "Macchie Fogliari Del Conocarpus Lancifolius." Riv. Agric. Subtrop. Trop. Volume 77, Issue 4, Pages 509-514. Descriptors: Combretaceae; Somalia; Pseudocercospora afgoiensis. Notes: Special Feature: 2 figs, 5 refs. Abstract: Describes a leaf spot disease of Conocarpus lancifolius (Combretaceae), a timber and shade tree appreciated in Somalia for its drought resistance. Technical description and diagnosis of its causal agent, Pseudocercospora afgoiensis sp. n. are also given. English summary. ISSN: 0035-6026.

Castellani, E. and Mohamed, M. I. 1981. "(Downy Mildew of Maize in Somalia.)." Riv. Agric. Subtrop. Trop. Volume 75, Issue 2-3, Pages 223-230. Descriptors: downy mildew; Zea mays; Somalia; infection; Sclerospora sorghi. Abstract: Downy mildew of maize caused by Sclerospora sorghi was observed for the first time in Somalia in the summer of 1980, near Afgoy (Low Scebeli). The attacks were enhanced by exceptionally high rainfall. A description of the disease is given. Seed treatments with Metalaxyl are suggested for the disease control. ISSN: 0035-6026.

Castellano, G. 1931. Analisi Di Alcuni Campioni Di Acque Prelevati Dai Pozzi Del Comprensorio Della SAIS. Translated title : Analysis of Water Samples Drawn from Wells of the SAIS District. Italy: SAIS, Milan, Italy. Descriptors: Afgoi Somali Republic; Africa; chemical composition; East Africa; Genale Somali Republic; geochemistry; Giohar Somali Republic; ground water; hydrochemistry; samples; Shabeellaha Dhexe Somali Republic; Somali Republic; water wells; wells. Database: GeoRef. Accession Number: 1998-015611.

Castellano, G. 1931. Variazione Della Composizione Dell'Acqua Dell'Uebi Scebeli Dal 10.3.1930 Al 25.4.1931. Translated title: Variation of the Water Composition of the Shabelle River from March 10, 1930 to April 25, 1931. Italy: SAIS, Milan, Italy. Descriptors: Africa; chemical composition; East Africa; geochemistry; hydrochemistry; hydrology; Shebelle River; Somali Republic; southern Somali Republic; surface water. Database: GeoRef. Accession Number: 1998-015612.

Castelli, R. J. and Secker, N. 1986. Port of Kismayo (Somalia) Rehabilitation. Oakland, CA, USA: ASCE, New York, NY, USA. pages: 954-965. Ports '86, Proceedings of a Specialty Conference on Innovations in Port Engineering and Development in the 1990's. Descriptors: Ports And Harbors- Kismayo, Somalia; Port Structures- Repair; Docks- Design; Piles- Steel; Foundations. Abstract: Extensive deterioration of the 2070 ft long four-berth, marginal wharf at the Port of Kismayo necessitated a major rehabilitation to maintain operations. The selected scheme was a steel sheet pile bulkhead, reinforced with steel cover plates and anchored by a continuous sheet pile deadman. Variable subsurface conditions require the removal and replacement of soft soils along a portion of the bulkhead, while in another area the sheet piles will penetrate a shallow zone of coral and dense gravel. Deep compaction of underwater fill will be performed to reduce surface settlement and potential damage to concrete pavement. A primary criteria for the rehabilitation was to minimize future maintenance requirements. ISBN: 0-87262-538-9.

Cattellani, Giorgio. 1897. "l'Avvenire coloniale d'Italia nel Benadir, Somalia. Manuale ... corredato di carte geographiche, etc." 183 pages. Napoli, 1897. Octovo. British Library: 10097.a.46.

Cattin, R., King, G. and Vigny, C, et al. 2005. "Numerical Modelling of Quaternary Deformation and Post-Rifting Displacement in the Asal-Ghoubbet Rift (Djibouti, Africa)." Earth & Planetary Science Letters. 15 NOV. Volume 239, Issue 3-4, Pages 352-367. Descriptors: Regional structure and tectonics; fault displacement; Quaternary; igneous intrusion; rifting; crustal evolution; numerical model Species Term: Somalia. Notes: Additional Info: Netherlands; References: Number: 47; Geographic: Djibouti [East Africa] East Africa Sub-Saharan Africa Africa Eastern Hemisphere World. Abstract: Over the last three decades a host of information on rifting process relating to the geological and thermal structure, long-time scale deformation (Quaternary and Holocene) and rifting cycle displacement across the Asal-Ghoubbet rift has been made available. These data are interpreted with a two-dimensional thermo-mechanical model that incorporates rheological layering of the lithosphere, dyke inflation and faulting. Active fault locations and geometry are mainly controlled by both thermal structure and magma intrusion into the crust. The distributed slip throughout the inner rift is related to the closeness of magma chamber, leading to additional stress into the upper thinned crust. Assuming a constant Arabia-Somalia motion of 11 mm/year, the variation of subsidence rate between the last 100 and 9 ka is associated with a decrease of the average injection rate from 10 to 5 mm/year. These values, about equal to the regional opening rate, suggest that both volcanism and tectonic play an equivalent role in the rifting process. Our modelled sequence of events gives one possible explanation for both vertical and horizontal displacements observed since the 1978 seismovolcanic crisis. Although part of the post-rifting deformation could be due to viscous relaxation, the high opening rate in the first years after the event and the abrupt velocity change in 1984-1986 argue for a large dyke inflation of 12 cm/year ending in 1985. The asymmetric and constant pattern of the GPS velocity since 1991 suggests that present post-rifting deformation is mainly controlled by fault creep and regional stretching. This study demonstrates the internal consistency of the data set, highlights the role of magmatism in the mechanics of crustal stretching and reveals a complex post-rifting process including magma injection, fault creep and regional stretching. ISSN: 0012-821X.

Caulet J.P., Debrabant, P. and Fieux, M. 1988. "Dynamique Des Masses d'Eaux Oceaniques Et Sedimentation Quaternaire Sûr La Marge De l'Afrique De l'Est Et Dans Le Bassin De Somalie. Resultats Preliminaires De La Mission MD 44-INDUSOM Du Marion- Dufresne." Comptes Rendus- Academie Des Sciences, Serie II. Volume 307, Issue 3, Pages 281-288. Descriptors: Geographical Abstracts: Physical Geography- volume 71; Sediments and sedimentary processes- transport; Somali current; upwelling; Pleistocene sedimentation; Quaternary; Pleistocene. Notes: Indian Ocean- East African Margin Indian Ocean- Somalia Basin. Abstract: Sediments deposited under the Somalian eddies show high contents in biosiliceous components with high rates of accumulation. Between 3000 to 1000m, black levels and high contents in organic materials are related to the low oxygen intermediate water. Sedimentary hiatuses in Plio-Pleistocene cores from the bottom of the slope reflect the effects of the Deep Western Boundary Current between 4000 and 3000m. Clay minerals composition shows latitudinal changes related to the oceanic circulation pattern.

Caulet, J.P., Venec, Peyre M.T., Vergnaud, Grazzini C.; Nigrini, C., Summerhayes, C.P et al. 1992. "Variation of South Somalian Upwelling during the Last 160 Ka: Radiolarian and Foraminifera Records in Core MD 85674." Geological Society, London; Special Publication, 64. Descriptors: 379-389; LANDFORMS; Palaeoclimatology; variation; upwelling; radiolaria; Foraminifera; core MD 85674. Abstract: Indicators of upwelling activity and surface-water productivity for the last 160 ka have been studied in the "Marion Dufresne' core MD 85674 taken off Somalia. Our data indicate that, under the south Somalian gyre, upwelling activity was maximal during transition between isotope stages 6 and 5, isotope stage 3, and transition between isotope stages 2 and 1 (respectively at about 130 ka, 65 to 25 ka, and 15 to 10 ka). These data also suggest that, at least during the last 60 ka, periods of increased activity in the Somalian, Arabian and Peruvian upwelling systems were synchronous. Notes: In: Upwelling systems: evolution since the early Miocene (1992) p. 379-389; Geographic: Indian Ocean Somalia; Update: A 01 JAN 1993. OCLC: Accession Number: 0966858.

Cavazza, S. 1985. "Le Risorse Idriche Come Elemento Di Sviluppo Della Somalia." Translated title: "Water Resources as an Element of Development in Somalia." Somalia: Universita Nazionale Somala, Mogadishu, Somalia. Descriptors: Africa; development; East Africa; Somali Republic; water resources. Database: GeoRef. Accession Number: 1998-015613.

Cavell, S. 1984. Dam Camp, Water Supplies. Somalia: OXFAM, Mogadishu, Somalia. Descriptors: Africa; Dam Camp; dams; East Africa; northern Somali Republic; Somali Republic; water resources; water supply; Woqooyi Galbeed Somali Republic. Database: GeoRef. Accession Number: 1998-015614.

Cavendish, H. S. H. 1898. "Through Somaliland and Around and South of Lake Rudolf." Geogr. J. Volume 11, Pages 372-396. Descriptors: Africa; areal geology; East Africa; East African Lakes; Lake Turkana; Somali Republic. ISSN: 0016-7398.

Cecchi, A. 1886. "Da Zeila Alle Frontiere Del Caffa; Societa Geografica Italiana; Spedizione Italiana Nell Africa Equatoriale." Tranlsated title: "rom Zeila to the Caffa Border; Societa Geografica Italiana; Italian Expedition in Equatorial Africa." Italy: Loescher, Roma, Italy. Descriptors: Africa; areal geology; Caffa; East Africa; Somali Republic; Zeila Somalia. Notes: Part I. Database: GeoRef. Accession Number: 1996-038503.

Cecchi, Antonio. 1887. "Da Zeila Alle Frontiere Del Caffa; Societa Geografica Italiana; Spedizione Italiana Nell Africa Equatoriale."Translated title: "From Zeila to the Caffa Border; Societa Geografica Italiana; Italian Expedition in Equatorial Africa." Italy: Loescher, Rome, Italy. Descriptors: Africa; areal geology; Caffa Somali Republic; East Africa; Ethiopia; field trips; geologic maps; maps; Somali Republic; Zeila. Notes: Part III; illustrations incl. 4 plates. Database: GeoRef. Accession Number: 1998-011342.

Cecchi, Antonio. 1885. "Da Zeila Alle Frontiere Del Caffa; Societa Geografica Italiana; Spedizione Italiana Nell Africa Equatoriale." Translated title: "From Zeila to the Caffa Border; Societa Geografica Italiana." Italian Expedition in Equatorial Africa. Italy: Loescher, Rome, Italy. Descriptors: Africa; areal geology; Caffa Somali Republic; East Africa; equatorial region; Ethiopia; geography; Somali Republic; Zeila. Notes: Part II; illustrations incl. 2 plates. Database: GeoRef. Accession Number: 1998-011341.

Cecioni, G. 1940. "I Bur Della Somalia. Bur of Somalia." Rivista Coloniale. Volume 14/10, Descriptors: Africa; areal geology; Bay; Bur Somali Republic; East Africa; Somali Republic. Database: GeoRef.

Cecioni, Giovanni. 1940. "I Cedimenti Come Nuovo Aspetto Carsico." Translated title: "Sediments with new aspects of karst." Soc.Toscana Sci.Nat., Atti, Mem. Volume 48, Pages 102-110. Descriptors: Italian Somaliland; karst; physiographic geology; tabular hills. Notes: illustrations; Abstract: Describes a special type of karst form developed on cliff slopes of tabular hills of Italian Somaliland. It is considered the product of solution of gypsum beds underlying the compact sedimentary formations capping the hills. ISSN: 0365-7477.

Centers for Disease Control (CDC). 1989. "Nutritional Status of Somali Refugees- Eastern Ethiopia, September 1988-may 1989." MMWR Morbidity and Mortality Weekly. Rep. Jul 7. Volume 38, Issue 26, Pages 455-6, 461-3. Descriptors: Anthropometry; Body Weight; Child, Preschool; Ethiopia; Humans; Nutrition Disorders/diagnosis/prevention & control; Nutritional Status; Population Surveillance; Refugees; Scurvy/prevention & control; Somalia/ethnology; Time Factors; Water Supply. ISSN: 0149-2195.

Cerulli, Enrico; Lewis, I. M. and Burton, Richard Francis, et al. 1996. Somali MO04. New Haven, Conn: Human Relations Area Files. Descriptors: Somalis; Ethnology- Somalia; Muslims- Somalia; Somalia- Civilization, Modern. Abstract: The Muslim Somalis of the Horn of Africa speak the Somali language and live primarily in Somalia. This file consists of 32 documents, 10 of which are translations from the original Italian, two from French, and one from German. They cover a time span from the 1600s to about the mid 1980s. The majority of these works concentrate on the nomadic Somali of the Djibouti region of southeastern Ethiopia in what is known (in 1996) as the Somali Democratic Republic, composed of the former protectorate of British Somaliland, the former Italian U.N. Trusteeship for Somali, and the French territory of the Afars and the Issas. First footsteps in East Africa, Richard F. Burton; The shaping of Somali society, Lee V. Cassanelli; How a Hawiye tribe used to live; New notes on Islam in Somalia; New notes on the astronomical ideas of the Somalis; Observations on the Moslem movement in Somaliland; Personal names in Somali; Somali songs and little texts; Texts of the consuetudinary law of the Marrehân Somali; The Somali tribe; The consuetudinary law of northern Somalia (Mijirtein); The dancing of the Somali; The lunar stations in the astronomical ideas of the Somalis and the Danaki; The origin of the lower castes of Somalia, Enrico Cerulli; British Somaliland; Ralph Evelyn Drake-Brockman; The terminology and practice of Somali weather lore, astronomy, and astrology, Muusa H.I. Galaal; The slaughtered camel, Bernhard Helander; The Yibirs and Midgàns of Somaliland; J.W.C. Kirk; French Somaliland, André Leroi-Gourhan; A pastoral democracy; Clanship and contract in northern Somaliland; Dualism in Somali notions of power; Marriage and the family in northern Somaliland; Modern political movements in Somaliland, I & II; Peoples of the Horn of Africa; Sufism in Somaliland; The Somali lineage system and the total genealogy; The names of God in northern Somali; I.M. Lewis; Somali games; G. Marin; Contributions to the ethnography and anthropology of the Somali, Galla, and Harari, Philipp Viktor Paulitschke; Anthropology and ethnography of the peoples of Somalia, Nello Puccioni; Seventeen trips through

Somaliland and a visit to Abyssinia, H.G.C. Swayne. Worldcat Accession Number: 51249677.

Chaper, Maurice. 1888. Note sûr les Prétendus Combustibles Minéraux du Territoire d'Obokh. Paris. Extrait du Bulletin de la Société Géologique de France, 3e série, tome XVI, séance du 18 Juin 1888. Descriptors: Coal; mines and mineral resources. 819 pages. USGS Library.

Chaubey A.K, Srinivas K, Yatheesh V, Dyment J, Bhattacharya G.C and Royer J.-Y. 2002. "Paleogene Magnetic Isochrons and Palaeo-Propagators in the Arabian and Eastern Somali Basins, NW Indian Ocean." Geological Society Special Publication. Issue 195, Pages 71-85. Descriptors: Structural geology and tectonics; Magnetism; seafloor spreading; paleogeography; Paleogene; magnetic anomaly; geochronology; tectonic evolution Species Term: Somalia. Notes: Additional Info: United Kingdom; References: Number: 46; Geographic: Arabian Sea Indian Ocean- Arabian Basin Indian Ocean- Somali Basin. Abstract: We present a revised magnetic isochron map of the conjugate Arabian and Eastern Somali basins based on an up-to-date compilation of Indian, French, and other available sea- surface magnetic data. We have used the magnetic anomaly and the modulus of the analytical signal computed from the magnetic anomaly to identify and precisely locate the young and old edges of magnetic chrons in both basins. In addition to the major, well-defined anomalies, we have also used correlatable second-order features of the magnetic anomalies, the 'tiny wiggles', to strengthen the interpretation. The resulting isochrons and tectonic elements have been validated using the stochastic method of palaeogeographical reconstruction. The magnetic anomaly pattern in both basins depicts clear oblique offsets, characteristics of pseudofaults associated with propagating ridge segments. Our tectonic interpretation of the area revealed: (1) a complex pattern of ridge propagation between Chrons 28n (c. 63 Ma) and 25n (c. 56 Ma), with dominant eastward propagation between Chrons 26n (c. 58 Ma) and 25n (2) numerous, systematic westward propagations between Chrons 24n (c. 53 Ma) and 20n (c. 43 Ma); (3) asymmetric crustal accretion (caused by ridge propagation and asymmetric sea-floor spreading) in the conjugate basins during the whole period; (4) a slowing of India-Somalia motion after c. 52 Ma. ISSN: 0305-8719.

Chauhan, O. S. 1996. "Aeolian Deposition of Arabia and Somalia Sediments on the Southwestern Continental Margin of India." Curr. Sci. Volume 71, Issue 3, Pages 233. ISSN: 0011-3891.

Chelzzi, G., Deneubourg, J.L. and Focardi, S. 1984. "Cooperative Interactions and Environmental Control in the Intertidal Clustering of Nerita Textilis (Gastropoda; Prosobranchia)." Behaviour. Volume 90, Pages 150-166. Descriptors: Somalia; intertidal gastropod. Abstract: Somalian populations of Nerita textilis show a vertical separation between resting (mid-upper eulittoral) and feeding (lower eulittoral) zone. During high tides and mostly diurnal low tides some snails remain scattered (SF); these form aggregations (AF) at lower levels of the rocky shore. During the mostly nocturnal low tides both SF and AF migrate downward to feed on microalgae. Lower and upper aggregations are mostly frequented around neap (NT) and spring tides (ST), respectively. Clustering significantly reduces the exposure to waves during high tide, and probably overheating and dehydration during diurnal low tide. The periodical (NT) increase in density at the lower rest-zone (tide-pools belt), long-lasting chemical marking of collective rest sites and releasing mucus trails during feeding excursions cause the clustering. According to the hypothesis, a primer group stops at each aggregation site at NT, mostly returning to it by self-trailing after each feeding migration, while the progressive recruitment of snails from SF is caused by the interindividual trail-following during the return migration from the feeding ground. ISSN: 0005-7959.

Cherchi, A. and Schroeder, R. 1999. "Late Barremian Orbitolinid Foraminifera from Northern Somalia." Bollettino Della Societa Paleontologica Italiana. Volume 38, Issue 1, Pages 3-13. Descriptors: Micropalaeontology; Cretaceous; foraminifera; phylogenetics Species Term: Foraminifera; Somalia; Orbitolinidae; Palorbitolina lenticularis. Notes: Additional Info: Italy; References: Number: 28; Geographic: Somalia. Abstract: The lower part of the Early Cretaceous shallow marine deposits in the eastern Ahl Medo Range (Bosaso, northern Somalia) is characterized by three associations of orbitolinid foraminifers (from bottom to top): (1) Valserina transiens n.sp., (2) Palorbitolina lenticularis (Blumenbach), (3) Dictyoconus arabicus Henson and Palorbitolina lenticularis. This sequence of associations indicating a Late Barremian age is identical with that one in the central Oman Mountains. Valserina transiens n.sp. is regarded as the phylogenetic link between the genus Valserina and the Palorbitolina lenticularis group. ISSN: 0375-7633.

Cherchi, Antonietta and Schroeder, Rolf. 1999. "Late Barremian Orbitolinid Foraminifera from Northern Somalia." Bollettino Della Societa Paleontologica Italiana. Volume 38, Issue 1, Pages 3-13. Descriptors: Africa; Ahl Medo Range; Arabian Peninsula; Asia; assemblages; Barremian; biostratigraphy; biozones; Cretaceous; depositional environment; Dictyoconus arabicus; East Africa; Foraminifera; Invertebrata; lithostratigraphy; Lituolacea; Lower Cretaceous; marine environment; Mesozoic; microfossils; morphology; Mustahil Formation; new taxa; Oman; Oman Mountains; Orbitolinidae; Palorbitolina lenticularis; Protista; shallow-water environment; Somali Republic; taxonomy; Textulariina; Valserina transiens. Notes: References: 28; illustrations incl. strat. color, 3 plates, geol. sketch map. Abstract: The lower part of the Early Cretaceous shallow marine deposits in the eastern Ahl Medo Range (Bosaso, northern Somalia) is characterized by three associations of orbitolinid foraminifers (from bottom to top): (1) Valserina transiens n.sp., (2) Palorbitolina lenticularis (Blumenbach), (3) Dictyoconus arabicus Henson and Palorbitolina lenticularis. This sequence of associations indicating a Late Barremian age is identical with that one in the central Oman Mountains. Valserina transiens n.sp. is regarded as the phylogenetic link between the genus Valserina and the Palorbitolina lenticularis group. ISSN: 0375-7633.

Chetwynd, Archer S. R. 1954. Report of Hargeisa Water Supply. Somalia: Publisher unknown, Somalia. Descriptors: Africa; East Africa; Hargeysa Somali Republic; Somali Republic; water resources; water supply; Woqooyi Galbeed Somali Republic. Database: GeoRef. Accession Number: 1998-017911.

Chiesi, G. 1906. "L'Acqua Nel Benadir." Translated title: "Water in Benadir." Rivista Coloniale. Volume 1, Descriptors: Africa; Benadir Somali Republic; East Africa; Somali Republic; southern Somali Republic; water resources. Database: GeoRef.

China Water Prospecting, Beijing, China. 1971. Reports of Prospecting and Drilling of Wells for Water Supply to Hargeisa City in the Somali Democratic Republic. China: China Water Prospec., Beijing, China. Descriptors: Africa; drilling; East Africa; exploration; Hargeysa Somali Republic; Somali Republic; water resources; water supply;

water wells; wells; Woqooyi Galbeed Somali Republic. Database: GeoRef. Accession Number: 1998-017913.

China Well Drilling Team, Hargeysa, Somalia. 1983. Report on Geological Exploration of Exploratory Boreholes and Test Wells in the Northwest Region of the Somali Democratic Republic. Somalia: China Well Dril. Team, Hargeysa, Somalia. Descriptors: Africa; boreholes; East Africa; ground water; northwestern Somali Republic; observation wells; Somali Republic; water resources; water wells; wells. Database: GeoRef. Accession Number: 1998-017914.

"Chisimaio Port Facilities: Investigation of Retaining Wall Repair and Wave Action Damage Moles 1 and 2." 1966. Livorno, Italy: US Army Engineer Division, Mediterranean. Volume: TechRpt TC 423 .M43 1966b EC c. 001; TechRpt TC 423 .M43 1966b 2 EC c. 002, Descriptors: Harbors-Somalia-Design and construction; Retaining walls; Water waves; Somalia. Notes: iii, 20, 34 leaves, 1 folded leaf of plates: ill.; 27 cm.

"Chisimaio, Somali Democratic Republic: Final Narrative Report, Phase I Port Facilities." 1966. Livorno, Italy: U.S. Army Engineer Division Mediterranean. Volume: "DA-91-211-ENG.260.", pages: 157. Descriptors: Harbors-Somalia-Design and construction; Somalia. Abstract: With: Review of Alpina report and other data concerning proposed port at Kisimaio, Somalia, prepared by U.S. Army Engineer Division Mediterranean. Leghorn, Italy: The Division, Jan. 1960 and supplement to report dated Feb. 1961. Notes: 157 leaves in various foliations: ill., maps (some folded); 27 cm. DTIC.

Chorowicz, J. 2005. "The East African Rift System." Journal of African Earth Sciences. Volume 43, Issue 1-3, Pages 379-410. Descriptors: Regional structure and tectonics; rift zone; tectonic evolution. Notes: Additional Info: United Kingdom; References: Number: 212; Geographic: East African Rift Sub-Saharan Africa Africa. Abstract: This overview paper considers the East African rift system (EARS) as an intra-continental ridge system, comprising an axial rift. It describes the structural organization in three branches, the overall morphology, lithospheric cross-sections, the morpho-tectonics, the main tectonic features- with emphasis on the tension fractures- and volcanism in its relationships with the tectonics. The most characteristic features in the EARS are narrow elongate zones of thinned continental lithosphere related to asthenospheric intrusions in the upper mantle. This hidden part of the rift structure is expressed on the surface by thermal uplift of the rift shoulders. The graben valleys and basins are organized over a major failure in the lithospheric mantle, and in the crust comprise a major border fault, linked in depth to a low angle detachment fault, inducing asymmetric roll-over pattern, eventually accompanied by smaller normal faulting and tilted blocks. Considering the kinematics, divergent movements caused the continent to split along lines of preexisting lithospheric weaknesses marked by ancient tectonic patterns that focus the extensional strain. The hypothesis favored here is SE-ward relative divergent drifting of a not yet well individualized Somalian plate, a model in agreement with the existence of NW-striking transform and transfer zones. The East African rift system comprises a unique succession of graben basins linked and segmented by intracontinental transform, transfer and accommodation zones. In an attempt to make a point on the rift system evolution through time and space, it is clear that the role of plume impacts is determinant. The main phenomenon is formation of domes related to plume effect, weakening the lithosphere and, long after, failure inducing focused upper mantle thinning, asthenospheric intrusion and related thermal uplift of shoulders. The plume that had formed first at around 30 Ma was not in the Afar but likely in Lake Tana region (Ethiopia), its almost 1000 km diameter panache weakening the lithosphere and preparing the later first rifting episode along a preexisting weak zone, a Pan-African suture zone bordering the future Afar region. From the Afar, the rift propagated afterward from north to south on the whole, with steps of local lithospheric failure nucleations along preexisting weak zones. These predisposed lines are mainly suture zones, in which partial activation of low angle detachment faults reworked former thrust faults verging in opposite directions, belonging to double verging ancient belts. This is responsible for eventual reversal in rift asymmetry from one basin to the next. Supposing the plume migrated southward, or other plumes emplaced, the rift could propagate following former weaknesses, even outside areas influenced by plumes. This view of rift formation reconciles the classical models: active plume effect triggered the first ruptures; passive propagations of failure along lithospheric scale weak zones were responsible for the onset of the main rift segments. Various other aspects are shortly considered, such as tectonics and sedimentation, and relationships of the 'cradle of Mankind' with human evolution. By its size, structure and occurrence of oceanic lithosphere in the Afar, the EARS can be taken as a model of the prelude of oceanic opening inside a continent. ISSN: 0899-5362E; 1464-343X.

Chorowicz, J. and Mukonki, Mwamba na Bantu. 1980. "Lineaments Anciens, Zones Transformantes Recentes Et Geotectonique Des Fosses De l'Est Africain, d'Apres La Teledetection Et La Microtectonique. » Translated title : « Ancient Lineaments, Recent Transform Zones, and East African Geotectonic Trenches According to Remote Sensing and Microtectonics." Rapport Annuel- Musee Royal De l'Afrique Centrale.Departement De Geologie Et De Mineralogie. Volume 1979, Pages 143-167. Descriptors: Africa; African Plate; Assoua Lineament; Cenozoic; Central Africa; composition; Congo Democratic Republic; displacements; East Africa; East African Rift; faults; geophysical surveys; interpretation; Landsat; lineaments; mineralization; Neogene; orientation; plate tectonics; Precambrian; Quaternary; remote sensing; Shaba Congo Democratic Republic; Shaba Zaire; Somalian Plate; strike-slip faults; structural analysis; structural geology; surveys; Tanganyika-Malawi Lineament; tectonics; tension; Tertiary; transform faults; Zaire; Zambezi Lineament; Zambia. Notes: References: 7; illustrations incl. sketch maps. ISSN: 0304-5285.

Chorowicz, J., Mukonki, N. B. and Pottier, Y. 1979. "Mise En Evidence d'Une Compression Horizontale Liee a l'Ouverture Des Fosses Est-Africains (Branche Occidentale), Dans Le Seuil Entre Les Lacs Kivu Et Tanganyika. » Translated title : « Evidence of a Horizontal Compression Related to the Opening of the East African Rift (Western Branch) in the Ridge between Lakes Kivu and Tanganyika." Compte Rendu Sommaire Des Seances De La Société Géologique de France. Volume 21, Issue 5-6, Pages 231-234. Descriptors: Africa; African Plate; Burundi; Central Africa; Congo Democratic Republic; East African Lakes; East African Rift; faults; fractures; joints; Lake Kivu; Lake Tanganyika; microtectonics; neotectonics; plate tectonics; rift zones; Rwanda; Somali Plate; structural geology; systems; tectonics; tectonophysics; Zaire. Notes: References: 3; illustrations incl. geol. sketch map. ISSN: 0037-9417.

Christodoulias, J. C.; Giannaros, H. C. and Stamatopoulos, A. C. 1997. "Comparison of engineering properties of swelling soils from Cyprus, Greece and

Somalia." Proceedings of the International Conference on Soil Mechanics and Foundation Engineering- International Society for Soil Mechanics and Foundation Engineering 1997; Issue 14; Vol 1, p.61, 4 p. Balkema Press. ISSN: 0534-882X.

Chu, D. and Gordon, R.G. 1999. "Evidence for Motion between Nubia and Somalia Along the Southwest Indian Ridge." Nature. 04 MAR. Volume 398, Issue 6722, Pages 64-67. Descriptors: Regional structure and tectonics; plate motion; rifting Species Term: Somalia; Nubian. Notes: Additional Info: United Kingdom; References: Number: 27; Geographic: Africa- East African Rift. Abstract: The East African rift marks the northern boundary of the Nubian (West African) and Somalian (East African) plates, and has formed by horizontal stretching due to the separation of these plates. South of ~20°S, any expression of deformation or seismicity due to the relative motion of these two distinct plates vanishes, although the boundary must continue until it intersects another plate boundary. The nearest such boundary is that of the Antarctic plate, marked by the Southwest Indian ridge. But previous analyses of plate-motion data have indicated no significant difference between Nubia- Antarctica and Somalia-Antarctica motion. Here we show, using a large compilation of plate-motion data, that Nubia-Antarctica motion does differ from Somalia-Antarctica motion, and we determine a relative angular velocity of the two plates that has compact confidence limits. Our analysis places the pole of rotation near to the southern limit of African seismicity, implying that the southern part of the Nubian-Somalian plate boundary is a diffuse zone of convergence (up to ~2 mm yr-1), whereas up to ~6 mm yr-1 of separation is accommodated across the East African rift- about half the separation rate of the slowest mid-ocean ridge. ISSN: 0028-0836.

Chu, Dezhi. 1995. Studies in Plate Kinematics: Plate Motions in East Africa and a Comparison of Paleomagnetic Methods. Ph. D. dissertation. Illinois: Northwestern University. Descriptors: Nubia; Arabia; Somalia; Gulf of Aden; Red Sea Rift; Gulf of Suez. Abstract: Based on a new and systematic examination of all available magnetic profiles, improved estimates of the angular velocities between Arabia and Nubia and between Arabia and Somalia have been determined. Published inconsistencies between the interpretation of spreading rates in adjacent segments of the Red Sea have been reconciled. A new estimate of the angular velocity between Nubia and Somalia is determined solely from data in the Gulf of Aden and Red Sea. Uncertainties in the Nubia-Arabia motion are quantified using singular-value decomposition to find the eigenvectors of the covariance matrix. Through forward modeling and the use of synthetic data, I also test the robustness of estimates of Nubia-Somalia angular velocity that assume the existence and location of a hypothetical Nubia-Somalia-Antarctica triple junction. I find that this type of assumption can give both erroneous results and misleadingly small confidence limits. Fault azimuth and slip rate data from the Dead Sea Rift are used to estimate the motion of the Sinai block relative to Arabia and, indirectly, relative to Nubia. This permits the placement of bounds on the velocity across the Gulf of Suez. Preliminary results give an upper bound of 4 mm/yr of divergence perpendicular to the Gulf but no useful lower bound. Through statistical calculations and random error simulations, I have shown that increasing the window length can not always improve the accuracy of the running average. There always is a window length that corresponds to the minimum error. This window length has an exact relation with the data error, taken herein to be the root-mean-squared difference between true pole locations and estimated pole locations in an apparent polar wander path. With this relation, an optimal window for the running average can be chosen that does not smooth more than is needed. It is shown that the 95% confidence region calculated from Fisher statistics needs to be revised when the real dispersion of paleomagnetic poles is elliptical rather than circular. The advantages and disadvantages of the running average and circle fitting methods have been compared through various simplified models. Notes: M3: 9614720; M1: Ph.D.

Chu, Dezhi and Gordon, Richard G. 1999. "Evidence for Motion between Nubia and Somalia Along the Southwest Indian Ridge." Nature. March 4. Volume 398, Issue 6722, Pages 64-67. Descriptors: Rifts (Geology); Geology/East Africa; Continental drift; General Science; Biological & Agricultural. Notes: Bibliography; Illustration. Abstract: Movement between the Nubian (West African) and Somalian (East African) plates along the Southwest Indian ridge is reported. Analysis of a large compilation of plate-motion data revealed that Nubia-Antarctica motion differs from Somalia-Antarctica motion and facilitated the quantification of a relative angular velocity of the 2 plates with compact confidence limits. The results place the pole of rotation close to the southern limit of African seismicity, thus indicating that the southern portion of the Nubia-Somalia plate boundary is a diffuse zone of convergence that moves at a rate of up to around 2 mm per year. In contrast, as much as approximately 6 mm per year of divergence, around half the separation rate of the slowest mid-ocean ridge, is accommodated across the East African rift. ISSN: 0028-0836.

Chung, Y. 1987. "(Super 226) Ra in the Western Indian Ocean." Earth & Planetary Science Letters. Sep. Volume 85, Issue 1-3, Pages 11-27. Descriptors: alkaline earth metals; Amirante Passage; geochemistry; Indian Ocean; isotopes; Madagascar Basin; Mascarene Basin; metals; Ra-226; radioactive isotopes; radium; salinity; sea water; Somali Basin; western Indian Ocean. Notes: References: 18; illustrations incl. 1 table. ISSN: 0012-821X.

Church, J., Pople, D. and Komu, S. 2003. "High Hopes and Red Tides." Samaki News. Jul. Volume 2, Issue 1, Pages 13-14. Descriptors: Red tides; Algal blooms; Biological poisons; Phytoplankton; Poisonous organisms; Toxicity; Endotoxins; Lethal effects; Fish kill; Fishery resources; Toxicology; Kenya, Coast; Somalia; Marine. TR: KE0400017. Abstract: In the last week of January 2002, thousands of fish died along the north coast of Kenya and southern Somalia. Marine scientists believe it was caused by a 'Red Tide' or Harmful Algal Bloom (HAB). This article report on the impact of the event on the Kiunga Marine National Reserve (KMNR) and the local fishing community. Database: Environmental Sciences and Pollution Mgmt. LCCN: 2003408724.

Ciampi G. "Recenti Spostamenti Di Popolazione Su Base Etno-Politica in Africa Orientale." Istituto Interfacoltà di Geografia dell'Università di Firenze, 1990. Descriptors: Population; Water; nomads; demographic data; population change; government initiative; Bajuni settlement; sedentarisation programme; developing country; migration rate; sedentarisation; migration. Notes: Source: (1990) p. 33p; Geographic: Africa- (East) Somalia Kenya Tanzania. Abstract: Migrations of non-Somalian peoples living in Somalia, following initiatives which were taken by the Somalian government during the 1970s are examined. These government initiatives were part of extensive programs for the sedentarisation of nomads and the creation of agricultural and fishing cooperatives. Smaller initiatives had military causes. Information is also given on the demographic data and the distribution of the Bajuni settlements in Somalia, Kenya and Tanzania.

Ciani, M. and Diriye, F. U. 1995. "Presence of Rhizobia in Soils in Somalia." World J. Microbiol. Biotechnol. Volume 11, Issue 6, Pages 615-617. Descriptors: soil microorganisms; agricultural practices; Bradyrhizobium japonicum; Somalia. Abstract: Bradyrhizobium japonicum was absent and only low numbers of Rhizobium were present in nodulating forage legumes in 13 soil samples from the Jubba and Shabelle rivers of Somalia. ISSN: 0959-3993.

Citaco, Mogadishu, Somalia. 1975. Proposal for Sablale Agricultural Resettlement; an Irrigation Schema. Somalia: Citaco, Mogadishu, Somalia. Descriptors: Africa; agriculture; East Africa; irrigation; Sablale; Shebeellaha Hoose; Somali Republic; water resources; water supply. Database: GeoRef. Accession Number: 1998-011903.

Citaco, Rome, Italy. 1974. Progetto Esecutivo Di Una Piantagione Di Pompelmi; Parte II, Vol. 7, Studio Idrogeologico, Rapporto. Translated title: Executive Report of a Plantation of Grape Fruits; Part II, Vol. 7, Hydrogeologic Study, Report. Italy: Citaco, Rome, Italy. Descriptors: Africa; agriculture; East Africa; Genale Somali Republic; irrigation; Shebeellaha Hoose; Somali Republic; water resources; water supply. Database: GeoRef. Accession Number: 1998-011902.

Citerni, C. 1913. "Ai Confini Meridionali Dell'Ethiopia. Note Di Un Viaggio Attraverso l'Etiopia Ed i Paesi Galla e Somali." Translated title: "Southern Border of Ethiopia; Trip through Ethiopia, Galla and Somali Cities." Italy: Hoepli, Milan, Italy. Descriptors: Africa; areal geology; East Africa; Ethiopia; expeditions; Galla; Somali Republic; southern Ethiopia. Database: GeoRef. Accession Number: 1996-038296.

Clarke, W. S. 1992. "Somalia Background Information for Operation Restore Hope 1992-93. Special Rept." Army War Coll. Strategic Studies Inst., Carlisle Barracks, PA. Dec. pages: 45. Descriptors: Foreign aid; Somalia; Peace keeping; Restore Hope Operation; Behavior and society; Behavior and society International relations. Abstract: No abstract available. DTIC.

Clarke, W. S. 1992. "SOMALIA: Background Information for Operation Restore Hope. 1992-1993." Carlisle Barracks, PA; United States: Army War Coll. Strategic Studies Inst. 31 Dec. Volume: ADA2599595, page(s): 48. Descriptors: Somalia; Africans; Conflict; Culture; History; Societies; Strategic areas; Geographic areas; East Africa; Government(Foreign); Strategic intelligence; Foreign aid; Republic of Somalia; Clan; Subclan; RESTORE HOPE Operation; Ethnic conflict; Said Barre; Jihad. Abstract: The author contends that knowledge of a country's history and culture is essential to opening doors and establishing and maintaining friendships and contacts. Perhaps more that most African societies, the Somalis have a deep awareness of their history, culture and past achievements. The author attempts to stimulate some appreciation of Somali history and culture. For this reason, details are given of recent history so that readers will have some name and event recognition which outline the circumstances which led to the present situation in Somalia.... Republic of Somalia, Clan, Sub-clan, Operation Restore Hope, Ethnic conflict, Siad Barre, Political/clan organizations, Jihad. DTIC Accession Number: ADA2597771XSP; NTIS Accession Number: ADA2599595.

Clemens, Steven Curtis. 1990. Quaternary Variability of Indian Ocean Monsoon Winds and Climate. United States: Brown University, Providence, RI, United States. Descriptors: aerosols; Africa; Arabian Peninsula; Arabian Sea; Asia; Cenozoic; East

Africa; glacial geology; grain size; Indian Ocean; mineral composition; monsoons; Owen Ridge; paleoclimatology; Quaternary; satellite methods; seasonal variations; sediments; Somali Republic; stratigraphy; upwelling; wind transport. Database: GeoRef. Accession Number: 1991-013955.

Clin, Michel and Pouchan, Pierre. 1968. "Sûr un Reseau ue Failles ue Cisaillement Synvolcaniques dans le Territoire Francais des Afars et des Issas. » Translated title: "A Network of Synvolcanic Shear Faults in the French Territory of the Afars and Issas." Comptes Rendus Hebdomadaires Des Seances De l'Academie Des Sciences, Serie D: Sciences Naturelles. Volume 267, Issue 20, Pages 1553-1554. Descriptors: Afars; Africa; East Africa; faults; Issas; lava flows; shear zones; Somali Republic; structural geology; tectonics. Abstract: Basaltic flows, Somalia. ISSN: 0567-655X.

Cochran, J. R. 1988. "Somali Basin, Chain Ridge, and Origin of the Northern Somali Basin Gravity and Geoid Low." Journal of Geophysical Research. Solid Earth and Planets. 10 Oct. Volume 93, Issue B10, Pages 11985-12008. Descriptors: East Africa; Basins Geographic; Boundaries; Coastal regions; Continental drift; Discontinuities; Faults Geology; Geoids; Gradients; Gravity; Madagascar; Low density; Magnetic anomalies; Magnetic fields; Mozambique; North Direction; Reprints; Seafloor spreading; Ocean bottom topography; Somalia; Gravity anomalies; Ocean ridges; Ocean basins; Somali Basin; Chain Ridge; Ocean technology and engineering Marine geophysics and geology. Notes: Performer: Lamont-Doherty Geological Observatory, Palisades, NY. Abstract: The Northern Somali Basin, located between Chain Ridge and the Horn of Africa north of 4 deg N, is characterized by a distinct 5-m geoid low and by large negative gravity anomalies. Boundaries of the basin are marked by steep gradients in both gravity and geoid. Basement in Northern Somali Basin is 1-2 km deeper than on the Carlsberg Ridge flank to the southeast or on the Sheba Ridge flank to the north with a sharp discontinuity across the boundary of the basin. The Western Somali Basin, to the south, was created in the Late-Jr and Early Kr by the movement of Madagascar away from Africa. Reinterpretation of magnetic anomalies in the Western Somali Basin shows that they record both limbs of a mid-ocean ridge that was active by M22 time (Kimmeridgian) and died soon after MO (Aptian). Magnetic and gravity data allow the relict ridge crest to be traced from Davie Ridge near the African coast to the Dhow-VLCC-ARS fracture zone complex at 50 deg E. Davie ridge is a transform fault connecting the Western Somali Basin spreading center with a similar age spreading center in the Mozambique Basin. OCLC: 11994697.

Coffin, Millard F. and Rabinowitz, Philip D. 1983. "Geologic Evolution of the Western Somali Basin and East African Continental Margin; American Geophysical Union; 1983 Spring Meeting." EOS Trans. Am. Geophys. Union. 03 May. Volume 64, Issue 18, Pages 238-239. Descriptors: Africa; continental drift; continental shelf; controls; Cretaceous; Deep Sea Drilling Project; DSDP Site 240; DSDP Site 241; DSDP Site 242; East Africa; faults; Indian Ocean Islands; Jurassic; Kenya; Leg 25; Madagascar; marine geology; Mesozoic; oceanography; passive margins; plate tectonics; reconstruction; regional; rifting; sedimentation; Somali Republic; strike-slip faults; structural controls; Tanzania; tectonophysics; transform faults; volcanism; Western Somali Basin. Notes: DSDP, Deep Sea Drilling Project. ISSN: 0096-3941.

Coffin, Millard F. and Rabinowitz, Philip D. 1982. "A Multichannel Seismic Transect of the Somalian Continental Margin." Proceedings- Offshore Technology Conference, no.14, Vol.2. Pages 421-430. Descriptors: Africa; buried channels; Cenozoic; chemically precipitated rocks; continental rise; Deep Sea Drilling Project; diapirism; DSDP Site 241; East Africa; evaporites; fluvial features; geophysical methods; geophysical surveys; Indian Ocean; Jurassic; Kenya; Leg 25; marine transport; Mesozoic; Miocene; Neogene; rifting; sedimentary rocks; sedimentation; seismic methods; Somali Basin; Somali Republic; structural geology; surveys; tectonics; Tertiary; transport; Vema Cruise 3618; Vema Cruise 3619. Notes: DSDP, Deep Sea Drilling Project; References: 34; illustrations incl. sketch maps; Latitude: N010000,S040000 Longitude: E0510000, E0400000. ISSN: 0160-3663.

Cogne, J. P. and Humler, E. 2006. "Trends and Rhythms in Global Seafloor Generation Rate." Geochemistry, Geophysics and Geosystems. MAR 28. Volume 7, Pages Q03011. Abstract: The primary purpose of this paper is to investigate the spreading and production rates of oceanic ridges for the last 180 Myr, based on the detailed analysis of eight oceanic units (North, Central, and South Atlantic basins, Southwest, Central, and Southeast Indian Ridge systems, Somalia basin, and the Pacific plate) and using the most recent timescale for oceanic isochrons. The global study of oceanic ridges presented here shows that (1) the average rate of spreading, which we computed by weighting the rates obtained at each basin by the relevant ridge lengths, is constant since similar to 125 Ma at 53.4 +/- 5.9 mm yr(-1) (full rate), (2) the average surface production rate is 2.7 +/- 0.2 km 2 yr(-1), and (3) the minimum oceanic crust production in volume, or flux, is 18.7 +/- 2.9 km(3) yr(-1). These estimations are in close agreement (within +/- 10%) with other studies. However, the new results emerging from this analysis are the following: (1) The Cretaceous flux rates (in volume) might be only 10% higher than today over a short period of time (125-100 Myr). (2) The "pulse" of ocean crustal production (120-80 Ma) in the world total is predominantly the result of contributions from mantle temperature and oceanic plateaus but is not linked to the global spreading rate of oceanic ridges, as generally accepted. (3) The rates presented here differ from previously published models for the Cenozoic and show a general increasing trend in the last 50 Myr. (4) We finally suggest a possible similar to 25 Myr pseudo-periodicity of the oceanic production rate (in surface and in volume) at least during the last 75-80 Myr. These data could have a profound impact on a vast number of models including sea-level changes and more generally on the chemical mass balance between ocean and continent, which is known to be a key parameter in the history of the Earth's climate and ocean chemistry. ISSN: 1525-2027.

Coleman, Jr, Sterling. 2000. "Librarianship and Information Science in Islamic East Africa 1966-1999: An Annotated Bibliography." The International Information & Library Review. Volume 32, Issue 2, Pages 149-211. Abstract: This work is an annotated bibliography that consists of articles, books, conference papers, dissertations, and reports, etc. published in various library and information science forums on the subject of librarianship in Islamic East Africa. The goal of preparing this work is to provide a list of citations with abstracts that librarians, library students, and library scholars can use to perform research within this subject area and further the body of knowledge. The research methodology that was used to find these citations involved searching the database versions of ERIC, Dissertations Abstracts Online, and Library Literature within the online public access catalog of the Auburn University library system. It also involved searching the online databases of Library and Information Science Abstracts (LISA), British Education Index, and Education Abstracts within the DIALOG database as well as the respective print copies of these resources. While this work is by no means an exhaustive analysis of the entire East African library literature, it does strive to be comprehensive in terms of its country-by-country breakdown of librarianship within the region. Islamic East Africa for the purposes of this work incorporates the nations of Comoros, Djibouti, Eritrea, Ethiopia, Somalia, Sudan, and Tanzania. In providing a citation for a non-English language work, the English equivalent of the title of that work will be given next to the non-English title.

Collins, A.S and Pisarevsky, S.A. 2005. "Amalgamating Eastern Gondwana: The Evolution of the Circum-Indian Orogens." Earth-Sci. Rev. Volume 71, Issue 3-4, Pages 229-270. Descriptors: Plate tectonics; Gondwana Species Term: Somalia. References: Number: 341. Abstract: The Neoproterozoic global reorganisation that saw the demise of Rodinia and the amalgamation of Gondwana took place during an incredibly dynamic period of Earth evolution. To better understand the palaeogeography of these times, and hence help quantify the interrelations between tectonics and other Earth systems, we here integrate Neoproterozoic palaeomagnetic solutions from the various blocks that made up eastern Gondwana, with the large amount of recent geological data available from the orogenic belts that formed as eastern Gondwana amalgamated. From this study, we have: (1) identified large regions of pre-Neoproterozoic crust within late Neoproterozoic-Cambrian orogenic belts that significantly modify the geometry and number of continental blocks present in the Neoproterozoic world; (2) suggested that one of these blocks, Azania, which consists of Archaean and Palaeoproterozoic crust within the East African Orogen of Madagascar, Somalia, Ethiopia and Arabia, collided with the Congo/Tanzania/Bangweulu Block at 650-630 Ma to form the East African Orogeny; (3) postulated that India did not amalgamate with any of the Gondwana blocks until the latest Neoproterozoic/Cambrian forming the Kuunga Orogeny between it and Australia/Mawson and coeval orogenesis between India and the previously amalgamated Congo/Tanzania/Bangweulu-Azania Block (we suggest the name 'Malagasy Orogeny' for this event); and, (4) produced a palaeomagnetically and geologically permissive model for Neoproterozoic palaeogeography between 750 and 530 Ma, from the detritus of Rodinia to an amalgamated Gondwana. ISSN: 0012-8252.

Collins, Allan S. and Windley, Brian F. 2002. "The Tectonic Evolution of Central and Northern Madagascar and its Place in the Final Assembly of Gondwana." J. Geol. May. Volume 110, Issue 3, Pages 325-339. Descriptors: Geology/Madagascar; Gondwana; Plate tectonics/Africa; Orogeny; Geological models; General Science; Applied Science & Technology. Notes: PD: Bibliography; Illustration; Map; Table. Abstract: Recent work in central and northern Madagascar has identified five tectonic units of the East African Orogen (EAO), a large collisional zone fundamental to the amalgamation of Gondwana. These five units are the Antongil block, the Antananarivo block, the Tsaratanana sheet, the Itremo sheet, and the Bemarivo belt. Geochronological, lithological, metamorphic, and geochemical characteristics of these units and their relationships to each other are used as a type area to compare and contrast with surrounding regions of Gondwana. The Antananarivo block of central Madagascar, part of a broad band of pre-1000-Ma continental crust that stretches from Yemen through

Somalia and eastern Ethiopia into Madagascar, is sandwiched between two suture zones we interpret as marking strands of the Neoproterozoic Mozambique Ocean. The eastern suture connects the Al-Mukalla terrane (Yemen), the Maydh greenstone belt (northern Somalia), the Betsimisaraka suture (east Madagascar), and the Palghat-Cauvery shear zone system (south India). The western suture projects the Al-Bayda terrane (Yemen) through a change in crustal age in Ethiopia to the region west of Madagascar. Our new framework for the central EAO links the Mozambique belt with the Arabian/Nubian Shield and highlights the power of tectonic analysis in unraveling the complex tectonic collage of the EAO. Reprinted by permission of the publisher. ISSN: 0022-1376.

Collins-Longman Atlases. 1973. "Atlas for Somalia 2." United Kingdom: Collins-Longman for the Ministry of Education. Descriptors: Atlases, British; Atlas (atl). Notes: Description: 1 atlas (32 p.); color maps; 27 cm. Geographic: Somalia- Maps. Africa-Maps. Scales differ. Cartgrph Code: Category of scale: a; Notes: Cover title. "Printed and bound in Scotland by Wm. Collins Sons & Co. Ltd."--P. [1]. Other Titles: Atlas for Somalia two. ISBN: 0003601854; 0582001099; LCCN: 84-675380.

Collins-Longman Atlases. 1973. "Atlas for Somalia 1." United Kingdom: Collins-Longman for the Ministry of Education. Descriptors: Atlases, British; Atlas (atl). Notes: Description: 1 atlas (16 p.); color maps; 27 cm. Geographic: Somalia- Maps. Scales differ. Cartgrph Code: Category of scale: a; Notes: Cover title. "Printed and bound in Scotland by Wm. Collins Sons & Co. Ltd."--P. [1]. Other Titles: Atlas for Somalia one. ISBN: 0003601846 (Collins); ISBN: 0582001080 (Longman); LCCN: 84-675379.

Collinson, Alan S., Davies, Michael J. and Ibbotson, Peter. 1987. "Geological and Mineral Deposit Maps of the Arab World." Geology Today. Aug. Volume 3, Issue 4, Pages 133-135. Descriptors: Africa; Arabian Peninsula; Asia; East Africa; economic geology; economic geology maps; Indian Ocean; Iraq; maps; Middle East; mineral resources; Morocco; North Africa; Oman; Red Sea; Saudi Arabia; Somali Republic; Syria. Notes: illustrations incl. sketch map. ISSN: 0266-6979.

Colombini, I., Aloia, A., Fallaci, M., Pezzoli, G. and Chelazzi, L. 1998. "Spatial use of an Equatorial Coastal System (East Africa) by an Arthropod Community in Relation to Periodically Varying Environmental Conditions." Estuarine, Coastal and Shelf Science. Volume 47, Issue 5, Pages 633-647. Descriptors: zonation patterns; behavioural strategies; arthropods; sandy beach; East Africa. Abstract: During the dry season (1971) and the wet season (1973) two field studies were conducted at Sar Uanle, a locality along the coast of Somalia (East Africa). The study analysed the beach-dune system considering ecological aspects in a dynamic way. During the two periods of study and for an entire synodic period, a system of traps (directional pitfall traps) was used. At the same time the climatic factors were registered. For the most abundant arthropod species mean zonations were calculated during surface activity and considered in relation to the season, diel, synodic and tidal phases. The variations in mean zonation were then correlated with the main environmental parameters. During the wet season, for some taxa orientation indices were calculated. The results indicate that two distinct subcommunities are present at Sar Uanle: a more dynamic one on the beach and another on the dune. Arthropod species living in this beach-dune system have evolved behavioural strategies in relation to cyclical environmental changes according to their physiological needs, locomotory capacities and zonation.

Compagnon D. 1995. "Somalie: Les Limites De l'Ingerence "Humanitaire'. l'Echec Politique De l'ONU." CEAN, Bordeaux; distributed Karthala, Paris. Descriptors: pages 193-202; Water; humanitarian intervention; military intervention; UN peace keeping. Abstract: The severe famine of 1992 in Somalia was the official justification of the international military intervention, firstly by civil peace-keeping forces (UNITAF) under American command, then under the aegis of UN Operation in Somalia. The argument for quick intervention was specious, on the one hand because of the UN's and the great powers' passivity facing violent anarchy which had followed the collapse of Muhamed Siyaad Barre at the end of January 1991; on the other hand, because most of those who were in a more vulnerable position had already disappeared when the UNITAF appeared in December 1992. Notes: In: L'Afrique politique 1995 (1995) p. 193-202; Geographic: Somalia; Update: A 01 JAN 1996. OCLC Accession Number: 1171341.

Comprehensive Groundwater Development Project; End of Project Report, Draft Copy; Vol. 2, General Activities. 1986. Somalia: LBI/MMWR/WDA, Mogadishu, Somalia. Descriptors: Africa; development; East Africa; ground water; Somali Republic; water resources. Database: GeoRef. Accession Number: 1999-004726.

Comprehensive Ground Water Development Project; End of Project Report, Draft Copy; Vol. 4, Appendices to Vol. 2. 1986. Somalia: LBI/MMWR/WDA, Mogadishu, Somalia. Descriptors: Africa; development; East Africa; ground water; Somali Republic; water resources. Database: GeoRef. Accession Number: 1999-004727.

Comprehensive Groundwater Development Project; End of Project Report, Draft Copy; Maps. 1986. Somalia: LBI/MMWR/WDA, Mogadishu, Somalia. Descriptors: Africa; development; East Africa; ground water; maps; Somali Republic. Database: GeoRef. Accession Number: 1999-004728.

Comprehensive Groundwater Development Project; End of Project Report, Draft Copy; Vol. 3, Hydrogeology. 1986. Somalia: LBI/MMWR/WDA, Mogadishu, Somalia. Descriptors: Africa; development; East Africa; ground water; hydrology; Somali Republic; water resources. Database: GeoRef. Accession Number: 1999-004731.

Comprehensive Groundwater Development Project, Final Report; Vol.1, General Activities. 1985. Somalia: LBI/MMWR/WDA, Mogadishu, Somalia. Descriptors: Africa; development; East Africa; ground water; Somali Republic; water resources. Database: GeoRef. Accession Number: 1999-004717.

Comprehensive Groundwater Development Project, Final Report; Vol.2, Hydrology. 1985. Somalia: LBI/MMWR/WDA, Mogadishu, Somalia. Descriptors: Africa; development; East Africa; ground water; hydrology; Somali Republic; water resources. Database: GeoRef. Accession Number: 1999-004718.

Comprehensive Groundwater Development Project, Final Report; Vol.3, Appendix. 1985. Somalia: LBI/MMWR/WDA, Mogadishu, Somalia. Descriptors: Africa; development; East Africa; ground water; Somali Republic; water resources. Database: GeoRef. Accession Number: 1999-004719.

Comprehensive Groundwater Development Project; Interim Report. 1985. Somalia: LBI/MMWR/WDA, Mogadishu, Somalia. Descriptors: Africa; development; East Africa; ground water; Somali Republic; water resources. Database: GeoRef. Accession Number: 1999-004720.

Comprehensive Groundwater Development Project; Specification for Civil Works. 1985. Somalia: LBI/MMWR/WDA, Mogadishu, Somalia. Descriptors: Africa;

development; East Africa; ground water; Somali Republic; water resources; water supply. Database: GeoRef. Accession Number: 1999-004721.

Comprehensive Groundwater Development Project; Structure and Applications of a Water Industry System Model for Somalia; Vol. II. 1985. Somalia: LBI/MMWR/WDA, Mogadishu, Somalia. Descriptors: Africa; development; East Africa; ground water; Somali Republic; water resources. Database: GeoRef. Accession Number: 1999-004722.

Comprehensive Groundwater Development Project; Structure and Applications of a Water Industry System Model for Somalia; Vol. I. 1985. Somalia: LBI/MMWR/WDA, Mogadishu, Somalia. Descriptors: Africa; development; East Africa; ground water; models; Somali Republic; water resources. Database: GeoRef. Accession Number: 1999-004723.

Comprehensive Groundwater Development Project, Exploratory Report for the Bay Region, (Main Report + Annex). 1983. Somalia: LBI/MMWR/WDA, Mogadishu, Somalia. Descriptors: Africa; Bay Somali Republic; development; East Africa; ground water; Somali Republic; southern Somali Republic; water resources. Database: GeoRef. Accession Number: 1999-004716.

Comprehensive Groundwater Development Project, Economic Evaluation of the Comprehensive Groundwater Development Project. 1985. Somalia: LBI/MMWR/WDA, Mogadishu, Somalia. Descriptors: Africa; development; East Africa; evaluation; ground water; Somali Republic; water resources. Database: GeoRef. Accession Number: 1999-004724.

Conan, S. M.H., and Brummer, G. J. A. 2000. "Fluxes of Planktic Foraminifera in Response to Monsoonal Upwelling on the Somalia Basin Margin." Deep-Sea Research, Part 2: Topical Studies in Oceanography. Volume 47, Issue 9, Pages 2207-2227. Abstract: Sediment trap samples collected off Somalia (bi)weekly from early June 1992 through mid February 1993 show large seasonal variations in the shell flux and species composition of planktic foraminifera. These variations mirror the monsoon-driven circulation, which resulted in massive upwelling and offshore eddy transport in summer during the SW Monsoon (June-September) and accounted for nearly 90% of the estimated annual shell flux of planktic foraminifera. G. bulloides dominated the SW Monsoon along with G. glutinata, T. quinqueloba, N. dutertrei and T. iota in a fauna-rich species but with a low diversity and equitability. During the autumn intermonsoon (November-December) the ocean became stratified and the nutrient exhausted, whereas during the NE Monsoon (January-March) a warm surface mixed layer developed with the nutrients entrained by deep wind mixing. It was dominated by G. ruber in association with G. tenella, G. aequilateralis, G. trilobus/G. sacculifer and G. menardii in a fauna poorer in species but with a higher diversity and equitability. Consequently, the ratio in the abundance of G. bulloides and G. ruber follows the monsoonal cycle closely, as both species persist throughout the year. Only the rare G. theyeri seems to be restricted to the SW Monsoon. In October at the end of the SW Monsoon, the trap at 265 m above the bottom intercepted massive amounts of fine-grained carbonate (<125 μm) resuspended from the shelf and upper slope. It contained extreme fluxes of bioclastic fragments, benthic foraminifera and benthic ostracods as well as enhanced fluxes of small-sized planktic foraminifera <125 μm, notably G. rubescens. Integrated over the year, the species composition intercepted by the trap shows very good match with the shell fauna in a core top sediment from the same site. Apparently, the year-estimated monsoonal record of planktic foraminifera, which is strongly dominated by the SW Monsoon production, survives burial intact. Consequently, the species composition of planktic foraminifera and measures such as the bulloides/ruber ratio can be applied as a proxy for past changes in the intensity of the SW Monsoon off Somalia compared to the modern condition. ISSN/ISBN: 0967-0645.

Conan, S. M. -H, Ivanova, E. M. and Brummer, G. -J A. 2002. "Quantifying Carbonate Dissolution and Calibration of Foraminiferal Dissolution Indices in the Somali Basin." Marine Geology. Volume 182, Issue 3-4, Pages 325-349. Descriptors: Dissolution; Calibration; Fluxes; Rubber; Carbonates. Abstract: Two sediment traps moored off Somalia in 1992-1993 collected similar settling fluxes of carbonate and siliceous shells formed by various plankton groups. Planktic foraminifera showed large seasonal variations, with more than 74% of the total planktic foraminifera flux collected during the SW monsoon (summer upwelling), when Globigerina bulloides was dominating along with Globigerinita glutinata and Neogloboquadrina dutertrei. The intermonsoon and NE monsoon assemblages were dominated by Globigerinoides ruber. We used the trap records as our 'no dissolution' reference for comparison with three boxcore recorders in order to quantify the carbonate dissolution along a depth transect. Dissolution increases downslope, from Station 905 to 907 and 915 at depths of 1567 m, 2807 m and 4035 m, respectively. The carbonate fraction of the sediment at Station 915, which is located near the CCD, is the most affected by dissolution, with more than 97% of the planktic foraminifera dissolved. Here, the planktic foraminifera assemblage is strongly modified, with thick walled species such as N. dutertrei, Globorotalia tumida and Pulleniatina obliquiloculata as the most resistant. It is not representative of the settling assemblage. The planktic foraminiferal assemblages of the sediment surface at Stations 905 and 907 remain similar to the trap assemblages and the foraminifera are well preserved, although only 25% of foraminifera are apparently preserved at Station 905 and 8% at Station 907. Those numbers are surprisingly low and infer that only a small fraction of the foraminiferal carbonate production is buried and removed from the carbon cycle. This discrepancy between the export and buried flux is partly be due to (bio)mechanical destruction by benthic processes and to supralysoclinal dissolution, due to metabolic CO_2 generated by the benthic organisms. Another important factor is the interannual variability of the productivity, well known in the Arabian Sea. The calibration of commonly used foraminiferal dissolution indices (percentage of foraminiferal fragments, percentage of resistant species, foraminiferal dissolution index (FDX)) to our data only shows reliable results for high dissolution levels (> 97%). Off Somalia the most accurate of the proxies is the percentage of foraminiferal fragments compared to the other methods tested, i.e., FDX, planktic foraminiferal loss (L), percentage of radiolarians and diatoms, percentage of benthic foraminifera. The species assemblage appears to be not significantly modified by dissolution unless the estimated shell loss is high, > 92% of the arriving shells in our samples. This level is expressed in the percentage of fragmentation, the percentage of radiolarians and diatoms, the percentage of resistant foraminifera species, and FDX as 80%, 35%, 25% and 1.8, respectively. The relative abundance of Globigerina bulloides is a valid SW monsoon/upwelling proxy only when dissolution is moderate (less than or equal 92%). Globigerina bulloides and Globigerinoides ruber have similar burial efficiencies and susceptibilities to dissolution. Thus, the ratio G. bulloides/G. tuber is a valid proxy for past changes in the intensity of the SW monsoon even in the strongly dissolved samples. In our sediment record the ratio G. bulloides/G. ruber indicates that the SW monsoon was stronger in the recent past than in 1992-93. ISSN: 0025-3227.

Conedera, C. 1970. Photogeology of the Hol, God and Cursi Area. Somalia: Hammar Petroleum Comp., Mogadishu, Somalia. Descriptors: Africa; areal geology; cartography; Cursi Somali Republic; East Africa; Gedo Somali Republic; geologic maps; God Somali Republic; Hol Somali Republic; maps; photogeology; remote sensing; Somali Republic. Database: GeoRef. Accession Number: 1998-017974.

Conforti, E. 1955. "Le Acque Dello Uebi Scebeli." Translated title: "The Water of the Shebelle River." Somalia: AFIS, Mogadishu, Somalia. Descriptors: Africa; East Africa; hydrology; Shebelle River; Somali Republic; southern Somali Republic; surface water. Database: GeoRef. Accession Number: 1998-017926.

Conforti, E. 1954. "Studi Per La Valorizzazione Delle Risorse Idriche Del Basso e Medio Scebeli." Translated title: "Study about the Exploitation of Water Resources of the Lower and Middle Shebelle." Somalia: AFIS, Mogadishu, Somalia. Descriptors: Africa; East Africa; exploitation; exploration; hydrology; Shebelle River; Somali Republic; southern Somali Republic; water resources. Database: GeoRef. Accession Number: 1998-011908.

Conforti, E. 1954. "Studi Per Opere Di Risanamento e Valorizzazione Dei Descek Del Giuba." Translated title: "Study on Reclamation and Exploitation Works of the Descek of Juba." Somalia: AFIS, Mogadishu, Somalia. Descriptors: Africa; Descek Swamp; East Africa; exploitation; hydrology; Juba River; reclamation; Somali Republic; southern Somali Republic; water resources. Database: GeoRef. Accession Number: 1998-017924.

Conforti, E. 1954. Studio Sulle Possibilita' Agricole Della Regione Del Basso Giuba. Study on the Agricultural Possibility of the Lower Juba Region. Somalia: AFIS, Mogadishu, Somalia. Descriptors: Africa; agriculture; East Africa; hydrology; irrigation; Juba River; Somali Republic; southern Somali Republic; surface water. Database: GeoRef. Accession Number: 1998-017925.

Conforti, E. 1953. "Aspetti Del Problema Idrico Nella Agricoltura Somala." Translated title: "Aspects of the Water Problem in Somali Agriculture." Riv. Agric. Subtrop. Trop. Volume 47, Pages 7-9. Descriptors: Africa; agriculture; East Africa; irrigation; Somali Republic; water resources; water supply. ISSN: 0035-6026.

Conover, Helen Field, compiler. 1960. "Official Publications of Somaliland, 1941-1959; A Guide." Card Division. General Reference and Bibliography Division. Library of Congress. 41 pages. "Divided into British, French and Italian Somaliland, subdivided by issuing agency and with subject index to the whole. Includes LC Call Numbers or location symbols for other libraries."[13] Library of Congress.

Conover, Helen Field. 1964. Africa South of the Sahara. Library of Congress. African Section. Washington, General Reference and Bibliography Division, Reference Dept., Library of Congress, 1963 [i.e. 1964]. Africa, Sub-Saharan- Bibliography. LCCN: 63060087.

---

[13] Anglemyer, Mary. 1970. Natural Resources: A Selection of Bibliographies." Second edition. Washington, DC: Engineer Agency for Resource Inventories. EARI Development Research Series. Report #3. Page 81.

Conover, Helen Field. 1942-1943. The British Empire in Africa: selected references. Compiled by Helen F. Conover under the direction of Florence S. Hellman, chief bibliographer. Library of Congress. Division of Bibliography. [Washington] 1942-43. 4 volumes; 27 cm. Contents: I. General.--[II] British West Africa.--[III] British East and Central Africa.--IV. The Union of South Africa. Notes: Reproduced from type-written copy. Title of vol. IV varies slightly; vols. II and III have special t.p. only. Africa--Bibliography. Great Britain- Colonies- Africa- Bibliography. LCCN: 43051626; OCLC:3825749.

Consiglio nazionale delle ricerche (Italy). AGIP mineraria. 1957-1960. "Carta Geologica Della Somalia e Dell' Ogaden." AGIP Mineraria, Consiglio Nazionale delle Ricerche; Somalia. Descriptors: Geology- Somalia- Maps; Geology- Ethiopia- Ogaden-Maps. Notes: Description: 8 maps +; legend. Geographic: Somalia- Maps. Ogaden (Ethiopia)- Maps. Scale 1:500,000; Lambert equal area proj. Cartgrph Code: Category of scale: a Constant ratio linear horizontal scale: 500000; Notes: Shows: international boundaries, roads, tracks, geological information. General Info: Coverage: Somalia and Ogaden area, Ethiopia. OCLC Accession Number: 48599903.

Continental Oil of Somalia, Mogadishu, Somalia. 1975. Geology Study of the Somali Democratic Republic. Somalia: Continental Oil of Somalia, Somalia. Descriptors: Africa; areal geology; East Africa; petroleum; petroleum exploration; Somali Republic. Database: GeoRef. Accession Number: 1998-015623.

Cook, C. W. and C.W. Cook Mapping Service. 1935. "Map of Ethiopia and Adjacent Countries." Huntington Park, Calif: C.W. Cook Mapping Service. Notes: Description: 1 map; color, mounted on linen; 86 x 112 cm. Geographic: Africa, Northeast- Maps. Ethiopia- Maps. Somalia- Maps. Djibouti- Maps. Scale [ca. 1:2,534,400]. Cartgrph Code: Category of scale: a Constant ratio linear horizontal scale: 2534400; Notes: Grommets in upper corners for hanging. Insets: Area of probable naval conflicts. Scale [ca. 1:8,448,000]- Relationship of Ethiopia to European nations. Scale [ca. 1:25,344,000]. Responsibility: by C.W. Cook, licensed surveyor. OCLC Accession Number: 17893154.

Coppo, A., Colombo, M. and Pazzani, C., et al. 1995. "Vibrio Cholerae in the Horn of Africa: Epidemiology, Plasmids, Tetracycline Resistance Gene Amplification, and Comparison between O1 and Non-O1 Strains." Am. J. Trop. Med. Hyg. Oct. Volume 53, Issue 4, Pages 351-359. Descriptors: Adolescent; Adult; Age Distribution; Aged; Aged, 80 and over; Animals; Case-Control Studies; Child; Child, Preschool; Cholera/epidemiology/microbiology; Comparative Study; DNA, Bacterial/analysis; Disease Outbreaks; Female; Humans; Infant; Male; Middle Aged; Prevalence; R Factors; Research Support, Non-U.S. Gov't; Somalia/epidemiology; Tetracycline Resistance-genetics; Vibrio cholerae/drug effects/genetics; Water Microbiology. Abstract: The prevalence of Vibrio cholerae O1 and non-O1 has been investigated in numerous Somali regions of the Horn of Africa from 1983 to 1990. From January 1983 to January 1985 and between December 1986 and December 1990, no strains of V. cholerae O1 and 226 strains (5.3%) of V. cholerae non-O1 were isolated from 4,295 diarrhea cases. During a cholera epidemic in 1985 and 1986, the overall case-fatality rate was 13% and the attack rate was 3-3.5 per 1,000 population. Matched case-control studies identified a waterborne route of transmission. A drug-susceptible Ogawa strain from Ethiopia caused the introduction of the disease into northern Somalia. There were two major resistant derivatives of the original strain, and the one resistant to ampicillin, kanamycin, streptomycin, sulfonamide, and tetracycline (TC) predominated in the spreading disease. In 1986, susceptible Ogawa strains quickly displaced this resistant strain. The two incompatibility group C plasmids responsible for the resistance patterns had complex and scattered differences in their structures. Physical analysis of the plasmid DNA region coding for TC resistance demonstrated its genetic amplification in highly resistant variants of Ogawa strains. ISSN: 0002-9637.

Cornacchia, M. and Dars, R. 1983. "Un Trait Structural Majeur du Continent Africain; Les Lineaments Centrafricains du Cameroun au Golfe d'Aden. » Translated title : « A Major Structural Feature of the African Continent; the Central African Lineaments from Cameroon to the Gulf of Aden." Bull. Soc. Geol. Fr. Volume 25, Issue 1, Pages 101-109. Descriptors: Africa; African Plate; Cameroon; Central Africa; Congo; Congo Democratic Republic; cratons; East Africa; Ethiopia; faults; geophysical methods; geophysical surveys; intraplate tectonics; lineaments; magnetic anomalies; magnetic methods; metamorphic rocks; mylonites; plate tectonics; Somali Republic; structural geology; Sudan; surveys; tectonics; tectonophysics; West Africa; wrench faults; Zaire. Notes: References: 2 p. ISSN: 0037-9409.

Corni, G. 1937. "Somalia Italiana; Laghetti, Stagni e Paludi; Acque Sotterranee, Sorgenti e Pozzi." Translated title: "Italian Somaliland; Lakes, Ponds and Marshes; Ground Water, Springs and Wells." Italy: Edit. Arte e Storia, Milan, Italy. Descriptors: Africa; East Africa; ground water; hydrology; lakes; marshes; mires; ponds; Somali Republic; springs; surface water; water resources; water wells; wells. Database: GeoRef. Accession Number: 1998-017975.

Coronaro, R. 1914. "Il Giuba." Translated title: "Juba River." Italy: De Agostina Editore, Novara, Italy. Descriptors: Africa; East Africa; hydrographs; hydrology; Juba River; Somali Republic; surface water. Database: GeoRef. Accession Number: 1998-015625.

Cortes-Burns, H., Schrire, B. D., Pennington, R. T. and Miller, A. G. 2004. "A Taxonomic Revision of Socotran Indigofereae (Leguminosae Papilionoideae) with Insights into the Phytogeographical Links of the Socotran Archipelago." Nordic Journal of Botany. Volume 22, Issue 6, Pages 693-711. Abstract: This revision of Socotran Indigofereae (Leguminosae) treats two genera, Microcharis and Indigofera, and 16 species, one with two varieties. Indigofera coerulea var. coerulea and I. nugalensis are new records for the archipelago. Indigofera nephrocarpoides, L marmorata (lectotypified here) and I. socotrana are endemic, whilst Microcharis disjuncta var. fallax and Indigofera nugalensis are near-endemic to the Socotran islands. This study indicates that the closest relatives of the Socotran Indigofereae are to be found in African lineages. The strongest affinities are between Socotra and extreme NE Somalia (Indigofera pseudointricata, L nugalensis) and, to a lesser extent, with SW Oman (Microcharis disjuncta var. fallax), SW Pakistan (I. nephrocarpa) and Southern Yemen. Socotran Indigofereae are mainly derived from Tertiary African palaeotropical and drought-adapted lineages. We also suggest that following the separation of the islands from mainland Africa some Indigofereae would have reached the archipelago as a result of recent, long-distance dispersal events. ISSN: 0107-055X.

Coulie, E., Quidelleur, X., Gillot, P. Y., Courtillot, V., Lefevre, J. C. and Chiesa, S. 2003. "Comparative K-Ar and Ar/Ar Dating of Ethiopian and Yemenite Oligocene

Volcanism: Implications for Timing and Duration of the Ethiopian Traps." Earth and Planetary Science Letters. FEB 15. Volume 206, Issue 3-4, Pages 477-492. Abstract: It is now generally accepted that continental flood basalt (CFB) volcanism bears a strong relationship with continental breakup. The Ethiopian Afar plume has been linked to the opening of the Afar depression. Propagation of the Red Sea and Gulf of Aden rifts within the depression, still an ongoing process, has rifted away the Ethiopian and Yemenite trap sequences. They are in some locations more than 2 kin, thick and comprise a wide range of volcanic products, from tholeiitic basalts, in the lower part, to more acidic material in the upper part. Recent studies have established that the bulk of trap volcanism erupted about 30 Ma ago over a period of I Myr in the Ethiopian sections, while ages obtained on the Yemenite sections seem more distributed through time. Here, for the first time in a single study, we present geochronological results obtained for basalts and more evolved products for both Ethiopian and Yemenite traps. This approach eliminates inter-laboratory biases and discrepancies in the ages of standards, and imposes better constraints on the eruptive chronology of this CFB province. In addition, both the K-Ar and Ar-40/Ar-39 techniques have been applied simultaneously, in order to demonstrate that similar ages are indeed obtained for undisturbed samples. The two dating techniques used here yield concordant ages for most samples. On both sides of the Afar depression, our results support that the onset of basaltic volcanism is coeval, with undistinguishable ages of 30.6 +/- 0.4 and 30.2 +/- 0.4 Ma obtained from Ethiopia and Yemen, respectively. Most of the basaltic lava pile has been erupted in less than I Myr, but acidic volcanism seems more spread out through time. It is coeval with basalts in northern Ethiopia but extends to about 26 Ma in Yemen, as already recognized. A younger rhyolitic episode, probably related to the major 20 Ma phase of opening of the Red Sea and Gulf of Aden, as expressed in the Afar depression, is also observed in Yemen and central Ethiopia. ISSN: 0012-821X.

Cowiconsult, Copenhagen, Denmark. 1981. Rehabilitation of Rural Water Reservoir; Feasibility Study Report. Denmark: Cowiconsult, Copenhagen, Denmark. Descriptors: Africa; East Africa; feasibility studies; reservoirs; rural environment; Somali Republic; water resources. Database: GeoRef. Accession Number: 1998-015628.

Craster, S. L. 1903. "Military Railway in Somaliland 2'6" Gauge." Major S.L. Craster; Somalia. Descriptors: Railroads- Somalia- Maps; Manuscript (mss). Notes: Description: 2 maps. Geographic: Somalia- Maps, Manuscript. Notes: Manuscript on linen. Relief shown as form lines. Shows: line of proposed railways with distances in miles. Scale 1:126,720. Cartgrph Code: Category of scale: a Constant ratio linear horizontal scale: 126720; General Info: Coverage: Somalia. OCLC Accession Number: 48599528.

Crawford, Mark. 1986. "Rhinos Pushed to the Brink for Trinkets and Medicines." Science. October 10. Volume 234, Pages 147. Descriptors: Wildlife conservation; Rhinoceroses; General Science; Readers' Guide (Current Events); Applied Science & Technology. Notes: Illustrations. Abstract: At a House science and technology subcommittee hearing, geographer Esmond Bradley Martin testified that the world's rhinos are threatened with extinction. Fewer than 11,500 rhinoceroses survive worldwide, and these may be destroyed within a decade by poachers, who hunt the animals for their horns. Black rhinos are in particular danger; in the past six years they have been wiped out in Angola, Chad, Ethiopia, Mozambique, Rwanda, Somalia, Sudan, Uganda, and

Zaire. Martin believes that control of trade in rhino products--conducted primarily through North Yemen and Singapore--is the key to stopping poaching. The U.S. government is attempting to increase the pressures on these two countries to take rapid action. Even if such measures prove effective, however, protected reserves may be needed to ensure the survival of rhino populations. ISSN: 0036-8075.

Crema, C. 1923. "Osservazioni Sulla Geologia Del Medio Scebeli (Somalia Italiana) in Base Ai Materiali Raccolti Da S.A.R. Il Principe Luigi Di Savoia, Duca Degli Abruzzi (1919-1920)." Translated title: "Observations on the Geology of the Middle Shebelle, Italian Somaliland, Based on the Materials Collected by S.A.R. in the District of Luigi Di Savoia, Duke of Abruzzi, 1919-1920." Atti della Accademia Nazionale Dei Lincei, Rendiconti, Classe Di Scienze Fisiche, Matematiche e Naturali. Volume 5, no.32, Descriptors: Africa; areal geology; East Africa; Giohar Somali Republic; regional; Shabeellaha Dhexe Somali Republic; Shebelle River; Somali Republic. ISSN: 0392-7881.

Crowe, Sam. 1997/8/9. "Gedo Six Years of a Delicate Balancing Act in Somalia." The Lancet. Volume 350, Issue 9075, Pages 419-419.

Curtis, John D., Lersten, Nels R. And Lewis, Gwilym P. 1996. "Leaf Anatomy, Emphasizing Unusual 'Concertina' Mesophyll Cells, of Two East African Legumes (Caesalpinieae, Caesalpinioideae, Leguminosae)." Annals of Botany. Volume 78, Issue 1, Pages 55-59. Descriptors: Leguminosae; Caesalpinioideae; Cordeauxia; Stuhlmannia; concertina; mesophyll cells; desert adaptation; hollow glandular trichomes; leaf anatomy; wall thickenings. Abstract: Cordeauxia edulis (Somalia and Ethiopia), and Stuhlmanniamoavii (Tanzania, Kenya and Madagascar) are evergreen shrubs or small trees of dry areas. They have similar leaf anatomy as revealed by resin sectioning and scanning electron microscopy. The cuticle is extremely thick and all vascular bundles lack bundle sheath extensions. The most unusual feature is the mesophyll, three to seven layers consisting entirely of cylindrical palisade cells with lateral walls capable of changing vertical length by folding in a concertina-like manner. The matching outward folds of two adjacent cells always remain attached by means of a row of wall thickenings ('pegs'). The pegs can elongate, especially so between the widely separated mesophyll cells that occupy the substomatal chamber area. The unattached flexible inward wall folds enable these 'concertina' cells to shorten or lengthen vertically without disrupting cell interconnections in the interior of each relatively long-lived leaf as it periodically loses and gains water. Concertina cells may be an anatomical adaptation allowing these leaves to remain evergreen and survive extended periods of drought and yet to store water quickly when it becomes available. ISSN: 0305-7364.

D

Dacque, E. 1905. "Beitraege zur Geologie des Somalilandes; Untere Kreide, II Oberer Jura. » Translated title : « Contributions to the Geology of Somaliland; Upper Cretaceous, Lower Jurassic." Beitraege zur Palaeontologie und Geologie Oesterreich-Ungarns und des Orients. Volume 17, Descriptors: Africa; areal geology; Cretaceous; East Africa; Jurassic; Lower Jurassic; Mesozoic; Somali Republic; Upper Cretaceous. ISSN: 1019-892X.

d'Acremont, E., Fournier, M., Robin C., et al. 2005. "Structure and Evolution of the Eastern Gulf of Aden Conjugate Margins from Seismic Reflection Data."

Geophysical Journal International. Volume 160, Issue 3, Pages 869-890. Descriptors: Applied seismology; Structural geology and tectonics; continental margin; ocean-continent transition; rifting; seafloor spreading; seismic reflection Species Term: Somalia; Socotra. Notes: Additional Info: United Kingdom; References: Number: 73; Geographic: Gulf of Aden Arabian Sea Indian Ocean oceanic regions World. Abstract: The Gulf of Aden is a young and narrow oceanic basin formed in Oligo-Miocene time between the rifted margins of the Arabian and Somalian plates. Its mean orientation, strikes obliquely to the opening direction. The western conjugate margins are masked by Oligo-Miocene lavas from the Afar Plume. This paper concerns the eastern margins, where the 19-35 Ma breakup structures are well exposed onshore and within the sediment-starved marine shelf. Those passive margins, about 200 km distant, are non-volcanic. Offshore, during the Encens-Sheba cruise we gathered swath bathymetry, single-channel seismic reflection, gravity and magnetism data, in order to compare the structure of the two conjugate margins and to reconstruct the evolution of the thinned continental crust from rifting to the onset of oceanic spreading. Between the Alula-Fartak and Socotra major fracture zones, two accommodation zones trending separate the margins into three-trending segments. The margins are asymmetric: offshore, the northern margin is narrower and steeper than the southern one. Including the onshore domain, the southern rifted margin is about twice the breadth of the northern one. We relate this asymmetry to inherited Jurassic/Cretaceous rifts. The rifting obliquity also influenced the syn-rift structural pattern responsible for the normal faults trending from. The fault pattern could be explained by the decrease of the influence of rift obliquity towards the central rift, and/or by structural inheritance. The transition between the thinned continental crust and the oceanic crust is characterized by a 40 km wide zone. Our data suggest that its basement is made up of thinned continental crust along the southern margin and of thinned continental crust or exhumed mantle, more or less intruded by magmatic rocks, along the northern margin. ISSN: 0956-540X.

Dainelli, Giotto. 1943. Geologia Dell'Africa Orientale. Vol. I, Il Progresso Delle Conoscenze; Vol. II, L'Imbasamento Cristallino e La Serie Sedimentaria Mesozoica; Vol. III, La Successione Terziaria e i Fenomeni Del Quaternario. Geology of East Africa; Vol. I, Evolution of Knowledge; Vol. II, Crystalline Basement and Mesozoic Sedimentary Sequences; Vol. III, Tertiary Deposits and Quaternary Phenomena. Italy: R. Accad. Italia, Centro Studi per l'Africa Orientale Italiana, Rome, Italy. Descriptors: Abyssinia-Eritrea-Italian Somaliland; Africa; areal geology; East Africa; Eritrea, geology; Ethiopia; geologic history; geologic maps; geology; Italian Somaliland, geology; maps; mineral resources; Somali Republic. Abstract: A monographic study of the geology of Ethiopia, Eritrea, and Italian Somaliland, eastern Africa. Volume 1 deals with known data on the geology and mineral resources of the region; an extensive bibliography is included. Volume 2 deals with the crystalline basement and the Mesozoic sedimentary series and volume 3 with the Tertiary stratigraphic succession and Quaternary land forms. Volume 4 is composed of geologic and other maps. Notes: 4 Volumes, Vol. I, 464 pages; Vol. II, 704 pages and Vol. III, 748 pages; illustrations incl. 10 plates, tables. LCCN: 72224072.

Dal Piaz, G. V., Ibrahim, H. A., Martin, S. and Piccardo, G. B. 1993. Pan-African Metabasalts from the Maydh Area, Northeastern Somalia. Firenze; Istituto Agronomico L'Oltremar; 1993. Relazioni E Monografie Agrarie Subtropicali E Tropicali Nuova Serie113 A, pages: 41-58. International Meeting; 1st- 1987 Nov: Mogadishu. Notes:

Geology of Somalia and surrounding regions: Geology and mineral resources of Somalia and surrounding regions. OCLC: 34175029.

Dal Pra', A. 1985. "Contenuto Salino Delle Acque Superficiali e Sotterranee Nel Bacino Idrologico Del Fiume Shebelli." Translated title: "Saline Content of the Surface and Ground Water in the Hydrologic Basin of the Shebelle River." Somalia: Universita Nazionale Somala, Fac. Agr., Mogadishu, Somalia. Descriptors: Africa; East Africa; ground water; hydrology; saline composition; salinity; Shebelle River; Somali Republic; southern Somali Republic; surface water; water resources. Notes: I problemi connessi alla salinit dei suoli e delle acque conparticolare riferimento all'agricoltura Somala. Database: GeoRef. Accession Number: 1998-015640.

Dal Pra', A. 1985. Contenuto Salino Delle Acque Superficiali e Sotterranee Nel Bacino Del Fiume Shebeli. Saline Content of the Surface and Underground Water in Shebelle River Basin. Somalia: Universita Nazionale Somala, Mogadishu, Somalia. Descriptors: Africa; East Africa; ground water; hydrology; saline composition; Shebelle River; Somali Republic; southern Somali Republic; surface water. Notes: I problemi connessi alla salinit dei suoli e delle acque conparticolare riferimento all'agricoltura somala. Database: GeoRef. Accession Number: 1998-015641.

Dal Pra', A., Benvenuti, G., Omar, Shire Y., Osman, Mohamed A., Mumin, Mohamed G. and Ahmed, Yusuf I. 1983. "Indagine Idrogeologica Nel Territorio Circostante La Citta'Di Qorioley Sul Fiume Shabelle (Somalia) Per La Ricerca Di Acque Sotterranee Ad Uso Potabile. » Translated title : « Hydrogeologic Investigation in the Surrounding Territory of Qorioley City on Shebelle River, Somalia, about the Research of Ground Water as the use of Drinking Water." Quaderni Di Geologia Della Somalia (Mogadiscio). Volume 7, Descriptors: Africa; drinking water; East Africa; ground water; Qorioley Somali Republic; Shebelle River; Somali Republic; water resources. LCCN: 93093377.

Dal Pra', A., de Florentiis, N., Mumin, Mohamed and Hussen, Salad Mohamed. 1983. "Oscillazioni Della Superficie Piezometrica Della Falda Costiera Provocate Dalle Escursioni Di Marea Lungo Il Litorale Di Mogadiscio; Somalia. » Translated title : « Oscillations of the Piezometric Surface of the Coastal Aquifer Caused by Tides Along the Coast of Mogadishu, Somalia." Memorie Di Scienze Geologiche. Volume 36, Pages 371-375. Descriptors: Africa; aquifers; carbonate rocks; East Africa; fractures; ground water; hydrogeology; Mogadishu; porosity; sedimentary rocks; Somali Republic; surveys; tides; water table. Notes: References: 9; illustrations. ISSN: 0391-8602.

Dal Pra', A., Hussein Salad, M. and Mumin, Mohamed G. 1983. "Situazione Idrogeologica Della Zona Di Balcad in Relazione Al Rifornimento Idrico Dell'Azienda Agricola Della Universita' Nazionale Somala." Translated title: "Hydrogeologic Situation of the Balcad Zone in Relation to the Water Supply of the Somali National University Farm." Quaderni Di Geologia Della Somalia (Mogadiscio). Volume 7, Descriptors: Africa; agriculture; Balcad Somali Republic; East Africa; hydrology; irrigation; Mogadishu Somali Republic; Shebelle River region; Somali Republic; water resources; water supply. LCCN: 93093377.

Dal Pra', Antonio, De Florentiis, Nicola and Hussen, S., et al. 1986. "Ricerche Idrogeologiche Sulla Falda Costiera Della Somalia Centrale Tra Merka e Uarscek (Mogadiscio)." Translated title: "Hydrogeological Research on Coastal Aquifer of Central Somalia between Merka and Uarscek, Mogadishu." Memorie Di Scienze

Geologiche. Volume 38, Pages 91-110. Descriptors: Africa; aquifers; calcium carbonate; carbonate rocks; coastal environment; drinking water; East Africa; electrical conductivity; fissures; fresh water; ground water; hydraulic conductivity; hydrodynamics; limestone; Merka Somali Republic; Mogadishu Somali Republic; ocean circulation; salt-water intrusion; saturation; sea-level changes; sedimentary rocks; Somali Republic; temperature; tides; Uarscek Somali Republic; water quality; water table. Notes: References: 34; illustrations incl. 9 tables, sketch map. ISSN: 0391-8602.

Dal Pra', Antonio and Salad, Hussen. 1987. "Hydrogeological Research on the Coastal Aquifer of Central Somalia." GeoSom 87; International Meeting; Geology of Somalia and Surrounding Regions; Abstracts. Somalia: Somali Natl. Univ., Mogadishu, Somalia. GeoSom 87; Geology of Somalia and Surrounding Regions, Mogadishu. Somalia Conference: Nov. 23-30, 1987. Descriptors: Africa; aquifers; central Somali Republic; coastal environment; drinking water; East Africa; fluoride ion; fluorine; ground water; halogens; hydrogeology; salinity; salt-water intrusion; Somali Republic; surveys; water quality; water table; water wells; wells. OCLC Accession Number: 34175029.

Dal Pra', Antonio and Salad, Hussen. 1986. "Ricerche Sperimentali Sui Rapporti Tra Acque Dolci Di Falda e Acque Salate Di Intrusione Marina Lungo La Costa Della Somalia Centrale Nella Zona Di Jasira (Mogadiscio). » Translated title : « Experimental Studies on the Relationship between Aquifer Fresh Water and Marine Salt Water Intrusion Along the Central Somalia Coast in the Jasira Area, Mogadishu." Memorie Di Scienze Geologiche. Volume 38, Pages 169-186. Descriptors: Africa; aquifers; coastal environment; East Africa; electrical conductivity; experimental studies; fluctuations; fresh water; ground water; hydrodynamics; Jasira Somali Republic; Mogadishu Somali Republic; salinity; salt-water intrusion; sea-level changes; Somali Republic; temperature; water quality; water table. References: 47; illustrations incl. 5 tables, 2 plates. ISSN: 0391-8602.

Dalla Vedova, G. 1895. "Introduzione Ai Risultati Geologici Della Missione Bottego Al Giuba." Translated title: "Introduction to the Geologic Results of the Bottego Mission at Juba." Annali Del Museo Civico Di Storia Naturali Giacomo Doria. Volume 15, Descriptors: Africa; areal geology; Bottego Mission; East Africa; expeditions; Juba River; Somali Republic. ISSN: 0365-4389.

Danfulani, S.A. 1999. "Regional Security and Conflict Resolution in the Horn of Africa: Somalian Reconstruction After the Cold War." International Studies. Volume 36, Issue 1, Pages 35-61. Descriptors: Political; International Relations, International Financial Institutions; conflict management; geopolitics; regional security; territorial dispute. Notes: Additional Info: India; Geographic: Somalia. Abstract: The Horn here represents a microcosm of most of the problems in African politics, where territorial disputes and the issue of self-determination, among the different states in the region, have clashed with efforts directed at upholding the principle of Utis-possidetis Juris. In this article I shall analyze the geopolitical picture of the Horn, examining the composition of the original Somali Nation, relating this to its regional and extra-regional aspects with a view to understanding the conflict. Finally, I shall study the different efforts at containing and managing the conflicts, emphasizing the importance, problems, and prospects for regional security. The scope of this article will, however, exclude the United Nations military intervention, christened 'Operation Restore Hope' by the Americans who led the operation, on which significant research work has been undertaken. Succinctly, my effort, while using Somalia as a case study, is to give a geopolitical and geo-strategic interpretation of events with a conclusion, in the light of the situation in the Horn and the great lakes states of Central Africa, on the need to de-emphasize ethnicism, clanism, and fanaticism in the management and resolution of inter-African conflicts as an important parameter for peace, security, and development. Specifically, the article pleads for the reconstruction of Somalia through the laid down conflict resolution mechanisms of the OAU and through assistance by the international community. ISSN: 0020-8817.

Daniels, John L. 1970. Gebile. United Kingdom: Directorate Overseas Surveys, United Kingdom. Descriptors: Africa; areal geology; East Africa; Gebile; geologic maps; maps; Somali Republic. Notes: Latitude: N093000, N100000 Longitude: E0440000, E0433000. Database: GeoRef. Accession Number: 1989-007772.

Daniels, John L. 1961. "An Introduction to the Photogeological Study of the Bur Region, Southern Somali Republic." Somalia: Volume: JLD/23, Descriptors: Africa; areal geology; Bur; East Africa; photogeology; relief; remote sensing; Somali Republic; southern Somali Republic. Report of the Geological Survey, Somaliland Protectorate, Report: JLD/23, 1961. Database: GeoRef. Accession Number: 1998-018217.

Daniels, John L. 1960. "Geology of Hargeisa and Borama Districts." Somalia: Descriptors: Africa; areal geology; Borama Somali Republic; East Africa; Hargeisa Somali Republic; northern Somali Republic; Somali Republic; Woqooyi Galbeed Somali Republic. Notes: Rep. JDL. Database: GeoRef. Accession Number: 1998-015647.

Daniels, John L. 1960. Minerals and Rocks of Hargeisa and Borama Districts. Hargeisa: Ministry of Natural Resources. Mineral Resources Pamphlet Number 3. 25 pages; folded maps. Bibliography pages 21-22. Descriptors: Mines and mineral resources; petrology. USGS Library.

Daniels, John L. 1957. "Preliminary Report on the Geology of the Eastern Part of D.O.S. Sheet n. 33 "Gebile", Hargeisa District, with Notes and Observations made in the Western District." Somalia: 1957. Descriptors: Africa; areal geology; East Africa; Gebile Somali Republic; Hargeisa Somali Republic; Somali Republic; Woqooyi Galbeed Somali Republic. Database: GeoRef. Accession Number: 1998-015645.

Daniels, John Leonard, Skiba, W. J. and Sutton, J. 1965. "The Deformation of some Banded Gabbros in the Northern Somalia Fold-Belt." Quarterly Journal of the Geological Society of London. Volume 121, Part 2, Issue 482, Pages 111-142. Descriptors: Africa; banded; banded gabbro; East Africa; fold belt; gabbros; geologic maps; Hamar-Mora district; igneous rocks; intrusions; maps; petrology; plutonic rocks; Somali Republic; tectonics. Notes: With discussion; illustrations (incl. geol. map); Abstract: Among the many layered basic intrusions known to occur in the late Precambrian-lower Paleozoic fold belt in northeast Africa, several are found in the area of northern Somalia bordering on the Gulf of Aden in the vicinity of Berbera and Hargeisa. They are olivine gabbro mostly, with some pyroxenite, peridotite, troctolite, and anorthositic leucogabbro. They were intruded into rocks metamorphosed by an earlier period of deformation. Subsequent deformation and granite intrusion have resulted in fold structures, steeply dipping layering, and development of secondary structures affecting both the intrusions and country rock. Other mineral assemblages have also appeared. The intrusions are used to demonstrate three stages in the deformation process, ranging from those consisting largely of unaltered rock and with the thermal aureole preserved (as in the Hamar mass) to those where the intrusions are now concordant metagabbro lenses with intensely folded banding. The olivine and pyroxene in the gabbros are replaced by a blue-green amphibole, while plagioclase remains fresh. ISSN: 0370-291X.

"Dante" 1942. EAF 707. Scale: 1:100 000. [Nairobi]: [Survey Directorate, East Africa Command], 1942. 2 maps. Subjects: Somalia- Maps and charts 1942. Added Name: Great Britain. War Office. General Staff. Geographical Section. 2 sheets. British Library: Maps MOD EAF 707.

Darch, Colin. 1980. A Soviet View of Africa: An Annotated Bibliography on Ethiopia, Somalia, and Djibouti. Boston, Mass: G.K. Hall. Descriptors: 200. Notes: Ethiopia- Bibliography. Somalia- Bibliography. Djibouti- Bibliography. Éthiopie-Bibliographie. Somalie, République démocratique de- Bibliographie. Djibouti-Bibliographie. Notes: Includes indexes. Library of Congress. ISBN: 0816183651; LCCN: 79-28116.

"Data on Foreign Regions Where Uranium Resources are Developed, 1. Asia and Africa." 1982. Tokyo (Japan); Japan: Power Reactor and Nuclear Fuel Development Corp. Jan. Volume: PNCN4428112PT1; DE85700372, pages: 241. Descriptors: Africa; Asia; Afghanistan; Algeria; Angola; Botswana; Burma; Cameroon; Central African Republic; Chad; China; Congo Peoples Republic; Egyptian Arab Republic; Ethiopia; Exploration; Gabon; Geological Surveys; Ghana; India; Indonesia; Iran; Iraq; Kenya; Laos; Libya; Madagascar; Malawi; Malaysia; Mali; Mining; Morocco; Mozambique; Niger; Nigeria; North Korea; Oman; Pakistan; Philippines; Republic of Korea; Saudi Arabia; Senegal; Somalia; South Africa; Sri Lanka; Sudan; Tanzania; Thailand; Togo; Tunisia; Turkey; Foreign technology; ERDA/050100; ERDA/050200. Abstract: This book was published in July 1976. But thereafter, the information increased enormously, therefore the revised edition is to be published this time. The Asian regions are divided into Asia and Middle and Near East. The African regions are divided into northern, eastern, central, western and southern Africa. The general situation, the policy of uranium mining, the history of uranium ore exploration, the geological features, the uranium deposits and indications, the promising regions, the room Japan can step in, and the drawing showing the outline of geological features in respective countries are shown. In Niger, Japanese companies have taken part in the development of mines, and in Mali, the Power Reactor and Nuclear Fuel Development Corp. has carried out the exploration work. Also in Zambia, it has participated in the exploration project. The Japanese cooperation with the People's Republic of China and Thailand in uranium exploration seems to be promising. (Atomindex citation 16:001820). Notes: In Japanese. Database: NTIS. Accession Number: DE85700372.

Davies, P.A. 2005. "Wave-Powered Desalination: Resource Assessment and Review of Technology." Desalination. 30 DEC. Volume 186, Issue 1-3, Pages 97-109. Notes: Additional Info: Netherlands; References: Number: 52. Abstract: The growing scarcity of freshwater is driving the implementation of desalination on an increasingly large scale. However, the energy required to run desalination plants remains a drawback. The idea of using renewable energy sources is fundamentally attractive and many studies have been done in this area, mostly relating to solar or wind energy. In contrast, this study focuses on the potential to link ocean-wave energy to desalination. The extent of the resource is assessed, with an emphasis on the scenario of wave energy being massively exploited to supply irrigation in arid regions. Technologies of wave-powered desalination are reviewed and it is concluded that relatively little work has been done in this area. Along arid, sunny coastlines, an efficient wave-powered desalination plant could provide water to irrigate a strip of land 0.8 km wide if the waves are 1 m high, increasing to 5 km with waves 2 m high. Wave energy availabilities are compared to water shortages for a number of arid nations for which statistics are available. It is concluded that the maximum potential to correct these shortages varies from 16% for Morocco to 100% for Somalia and many islands. However, wave energy is mainly out-of-phase with evapotranspiration demand leading to capacity ratios of 3-9, representing the ratios of land areas that could be irrigated with and without seasonal storage. In the absence of storage, a device intended for widespread application should be optimized for summer wave heights of about 1 m. If storage is available, it should be optimized for winter wave heights of 2-2.5 m. ISSN: 0011-9164.

Davis, Lieut. 1903. "Sketch of Nogal district." IDWO 1800. Scale: 1:506,880. [London]: [War Office, Intelligence Division], 1903. 1 map. Note(s): Hand-lettered map of area between Burao and Ehil (Eyl). Subjects: Somalia- Maps and charts; Ethiopia- Maps and charts. Added Name: Great Britain. War Office. Intelligence Division. Added Name: Great Britain. War Office. General Staff. Geographical Section. British Library: Maps MOD IDWO 1800.

Davis, Dawne M. 1997. "Operational Logistics in MOOTW: What Your CINC Needs to Know." Naval War College Newport RI: Naval War College Newport RI. February 7, 1997. Volume: ADA325110. 17 Pages(s). Descriptors: *Military Operations; *Military Planning; *Logistics Planning; Military Reserves; Warfare; Iraq; Nations; Lessons Learned; Government(Foreign); Tools; Water; Operation; Wake; Logistics; California; Somalia; North(Direction); Feeding; Communism; Democracy; Haiti; Elections. Abstract: Much has been made of the changing role of the military since the threat from communism has all but been eliminated with the fall of the Soviet Union. In fact, without an 'evil empire' to provide a focus for our military plans the military has been forced to redefine itself in the wake of these startling changes. We must do more with less, downsize our forces, shift many logistical assets to the reserve components to save money, reduce our forward presence while at the same time maintaining capabilities to fight two major regional contingencies and to conduct military operations other than war (MOOTW). MOOTW have become increasingly predominant in the roles and missions of our armed forces. Since the fall of the iron curtain United States armed forces have been involved in Operation Provide Comfort to feed the Kurds in Northern Iraq, Operation Restore Hope to feed the starving masses in Somalia, Operation Uphold Democracy to restore a democratic government in Haiti, Joint Task Force Los Angeles during the riots in California and Operation Support Hope to stop the dying in Rwanda. These military operations demonstrate a gradual shift in the use of military forces from simply winning our nation's wars to providing the tools to feed the hungry, provide water to the thirsty and uphold foreign government elections all in support of our national interests. Notes: Final report. Database: DTIC. URL: http://handle.dtic.mil/100.2/ADA325110

Davis, B. G. and Haworth, H. F. 1963. Water Resources Development Project. Somalia: U. S. Agency for International Development, Mogadishu, Somalia. Descriptors: Africa; East Africa; exploration; Somali Republic; water resources. Database: GeoRef. Accession Number: 1998-015652.

Davis, Bruce G. and Haworth, Howard F. 1963. Review of USAID Water Resources Development Project, Somali Republic. Mogadiscio: Somalia. Descriptors: 91; Water resources development- Somalia; Water and water use- Somalia; Reclamation- Somalia. Notes: Responsibility: by Bruce G. Davis [and] Howard F. Haworth. LCCN: 65-62299.

Davis, Robert L. and Feierstein, Mark D. 1994. "Return to Somalia: The construction of Victory Base." Engineer. April 1994, Vol. 24 Issue 2, page 8 et seq., 7 pages, 1 diagram, 4bw photos. Subject Terms: Military Bases, American; United States- Armed Forces. Geographic Terms: Somalia; Mogadishu (Somalia). Abstract: Focuses on the second deployment of the United States 43rd Engineer Combat Battalion in Mogadishu, Somalia to construct the Victory Base, a 1,700-soldier base camp. Preparation for the deployment; Four phases of the Base construction; Lessons learned during the stay in Somalia. ISSN: 0046-1989.

De Chabalier, Jean-Bernard and Avouac, Jean-Philippe. 1994. "Kinematics of the Asal Rift (Djibouti) Determined from the Deformation of Fieale Volcano." Science. September 16. Volume 265, Pages 1677-1681. Descriptors: Volcanoes/Djibouti; Rifts (Geology); General Science; Readers' Guide (Current Events); Applied Science & Technology. Notes: Bibliography; Illustration; Map. Abstract: The present topography of the Asal rift in East Africa has been created by the dismemberment of the Fieale volcanic edifice, which formed astride the rift zone between 300,000 and 100,000 years ago. Topographic and geologic analyses indicate that spreading toward the northeast, at a rate of 17 to 29 millimeters per year, accounts for most of the separation between Arabia and Somalia. The low rate of topographic subsidence relative to extension suggests that crustal thinning has been balanced by injection and underplating of magmatic material that is similar in density to the crust. ISSN: 0036-8075.

De Groot, H., Wanke, S. and Neinhuis, C. 2006. "Revision of the Genus Aristolochia (Aristolochiaceae) in Africa, Madagascar and Adjacent Islands." Botanical Journal of the Linnean Society. June. Volume 151, Issue 2, Pages 219-238. Abstract: A taxonomic revision of the genus Aristolochia in Africa, Madagascar, the West African islands, the Comores and the Mascarenes is presented. In total, 11 indigenous species are accepted: A. albida, baetica, bracteolata, embergeri, heppi, hockii, fontanesii, paucinervis, pistolochia, rigida and sempervirens. Descriptions, distribution maps and a complete taxonomic synonymy are presented. The investigation revealed three clades: (1) A group from northern Africa, including the West African islands (A. baetica, A. sempervirens, A. fontanesii, A. pistolochia, A. paucinervis). Those taxa are characterized by unilabiate flowers with a sessile utricle. (2) A group from central and East Africa as well as Madagascar and adjacent islands. Those plants also have unilabiate flowers, but with a stiped utricle (A. bracteolata, A. albida, A. embergeri, A. heppii and A. hockii). The latter four species are endemic to the region. (3) A single species restricted to Somalia (A. rigida). Its flower characters do not match those of the remaining African species. The flower is trumpet-like and bilabiate, resembling species from the Near East. These relationships are also supported by the molecular analysis of the trnL-F region. ISSN: 0024-4074.

De Luca, R., Balestrieri, A. and Dinle, Y. 1983. "Measurement of Cutaneous Evaporation. 6. Cutaneous Water Loss in the People of Somalia." Boll. Soc. Ital. Biol. Sper. Oct 30. Volume 59, Issue 10, Pages 1499-1501. Descriptors: Adult; Body

Water/metabolism; English Abstract; Female; Humans; Male; Methods; Reference Values; Somalia; Volatilization; Water Loss, Insensible. Abstract: The purpose of this research is the measurement of the cutaneous water loss in the 29 young and healthy Somalian people by the Evaporimeter Ep I in order to compare it with the values of the European people. We did not find any significant difference. ISSN: 0037-8771.

De Luca, R., Balestrieri, A., Dinle, Y. and Costa, M. 1984. "Cutaneous Evaporimetry. VIII. Cutaneous Water Loss in Somalian Newborn Infants." Boll. Soc. Ital. Biol. Sper. Jan 30. Volume 60, Issue 1, Pages 161-164. Descriptors: English Abstract; Female; Humans; Infant, Newborn; Male; Somalia; Water Loss, Insensible. Abstract: The cutaneous water loss (CWL) were investigated in 21 new-born infants by Evaporimeter Ep-1. The mean value of the CWL resulted 13,3 +/- 7,5 g/h. This result is much greater compared with the value of European new-born infants equal to 6,4 +/- 1,8 g/h. ISSN: 0037-8771.

De Luca, R., Balestrieri, A., Dinle, Y. and De Lipsis, E. 1984. "Cutaneous Evaporimetry. VII. Cutaneous Water Loss in Basedow's Disease." Boll. Soc. Ital. Biol. Sper. Jan 30. Volume 60, Issue 1, Pages 157-159. Descriptors: Adolescent; Adult; Aged; English Abstract; Female; Graves Disease/physiopathology; Humans; Middle Aged; Water Loss, Insensible. Abstract: The cutaneous water loss was investigated in 12 patients affected with Basedow's disease by Evaporimeter Ep-1. The mean value of the CWL resulted 49,8 +/- 18,2 g/h. This result is significantly higher compared with the value of normal Somali people equal to 25,24 +/- 14,6 g/h. ISSN: 0037-8771.

De Marco, Giovanni; Fagotto, Flavio; Università di Roma and Centro di cartografia ecologica applicata. 1978. "Geobotanical and Ecofaunistic Map of the Lower Scebeli River (Somalia)." Roma: Print: Borgia. Descriptors: Biogeography- Somalia- Scebeli River Region- Maps; Phytogeography- Somalia- Shebelle River Region- Maps; Zoogeography- Somalia- Scebeli River Region- Maps. Notes: Description: 1 map; color; 41 x 90 cm. Scale 1:130,000. Cartographic Code: Category of scale: a Constant ratio linear horizontal scale: 130000; Notes: "Directed by V. Giacomini." "Surveyed in 1974-75." Animal species shown by pictographs. Includes key map. Includes cross-section and graphs: Relation between Somali seasons, monsoon winds and European calendar, Median monthly total flow of Scebeli River at Audegle (1952-1965), Pluviothermic diagram of Genale (1962-1966). OCLC Accession Number: 25820024.

De Sanctis, G. 1941. « Il Bacino Idrico Nel Descek Uamo; Opera Di Romana Grandezza Ai Confini Tra La Somalia Ed Il Chenia. » Translated title : « Water Basin in Descek Uamo; Work of Roman Greatness at the Borders between Somalia and Kenya." Somalia: L'Italia d'Oltremare, Rome, Somalia. Descriptors: Africa; Descek Uamo; East Africa; hydrology; irrigation; Kenya; Somali Republic; water resources. Database: GeoRef. Accession Number: 1998-015660.

De Waal, A. 1993. "Somalia: The Shadow Economy." Africa Report. Volume 38, Issue 2, Pages 24-28. Descriptors: National Trade and Consumption; developing country; economic assessment; hidden economy; national economy; socio-political problem; parallel economy; shadow economy; black market; smuggling; corruption. Notes: Somalia. Abstract: Somalia is now a universal symbol of an African basket case. Despite official statistics showing the country to be in a state of chronic disaster throughout the 1980s, the economy was thriving. But it was not the easily measurable economy. Instead, it was a hidden economy- comprising black marketeering, smuggling, corruption, stolen food aid, most of it in the hands of various elites- and it is prospering today as the political and social structures of Somalia continue to disintegrate. ISSN: 0001-9836.

Dede, Christian. 1996. Beiträge zur Hydrogeologie Nordost-Somalias. Berlin: Oberhofer. Descriptors: 112, 7; Hydrogeology- Somalia; Water-supply- Somalia. Notes: Four maps folded in pocket. "Dissertation ... Berlin, 1996"--T.p. verso. Includes bibliographical references. ISBN: 3925410244.

Dede, Christian, Hesse, Karl-Heinz and Schnaecker, Eckhard. 1990. "Hydrological and Hydrogeological Investigations in North-East Somalia; Research in Sudan, Somalia, Egypt and Kenya; Results of the Special Research Project "Geoscientific Problems in Arid and Semiarid Areas" (Sonderforshungsbereich 69); Period 1987-1990." Berliner Geowissenschaftliche Abhandlungen, Reihe A: Geologie Und Palaeontologie. Volume 120, Issue 2, Pages 719-753. Descriptors: Africa; anisotropy; aquifers; Cenozoic; Cretaceous; discharge; East Africa; Eocene; geochemistry; ground water; hydrochemistry; hydrodynamics; hydrogeologic maps; lithofacies; maps; Mesozoic; northeastern Somali Republic; Paleogene; permeability; recharge; Somali Republic; Tertiary; water resources; water table. Notes: References: 38; illustrations incl. 7 tables, sects. ISBN: 0172-8784.

DeMets, C., Gordon, R. G. and Royer, J. Y. 2005. "Motion between the Indian, Capricorn and Somalian Plates since 20 Ma: Implications for the Timing and Magnitude of Distributed Lithospheric Deformation in the Equatorial Indian Ocean." Geophysical Journal International. MAY. Volume 161, Issue 2, Pages 445-468. Abstract: Approximately 2200 magnetic anomaly crossings and 800 fracture zone crossings flanking the Carlsberg ridge and Central Indian ridge are used to estimate the rotations of the Indian and Capricorn plates relative to the Somalian Plate for 20 distinct points in time since 20 Ma. The data are further used to place limits on the locations of the northern edge of the rigid Capricorn Plate and of the southern edge of the rigid Indian Plate along the Central Indian ridge. Data south of and including fracture zone N (the fracture zone immediately south of the Vema fracture zone), which intersects the Central Indian ridge near 10°S, are well fit assuming rigid Capricorn and Somalian plates, while data north of fracture zone N are not, in agreement with prior results. Data north of fracture zone H, which intersects the Central Indian ridge near 3.2°S, are well fit assuming rigid Indian and Somalian plates, while data south of and including fracture zone H are not, resulting in a smaller rigid Indian Plate and a wider diffuse oceanic plate boundary than found before. The data are consistent with Capricorn- Somalia motion about a fixed pole since approximate to 8 Ma, but require rotation about a pole 15 degrees farther away from the Central Indian ridge from 20 to approximate to 8 Ma. The post-8-Ma pole also indicates Capricorn Somalia displacement directions that are 7 degrees clockwise of those indicated by the pre-8-Ma stage pole. In contrast, India- Somalia anomaly and fracture crossings are well fit by a single fixed pole of rotation for the past 20 Ma. India- Somalia motion has changed little during the past 20 Myr. Nonetheless, astronomically calibrated ages for reversals younger than 12.9 Ma allow resolution of the following small but significant changes in spreading rate: India- Somalia spreading slowed from 31 to 28 mm yr(-1) near 7.9 Ma and later sped up to 31 mm yr(-1) near 3.6 Ma; Capricorn- Somalia spreading slowed from 40 to 36 mm yr(-1) near 11.0 Ma, later sped up to 38 mm yr(-1) near 5.1 Ma and further sped up to 40 mm yr(-1) near 2.6 Ma. The motion between the Indian and Capricorn plates is estimated by differencing

India- Somalia and Capricorn- Somalia rotations, which differ significantly for all 20 pairs of reconstructions. India has rotated relative to the Capricorn Plate since at least approximate to 20 Ma. If about a pole located near 4° S, 75° E, the rate of rotation was slow, 0.11 degrees +/- 0.01° Myr(-1) (95 per cent confidence limits), from 20 to 8 Ma, but increased to 0.28° +/- 0.01° Myr(-1) (95 per cent confidence limits) at approximate to 8 Ma. The onset of more rapid rotation coincides, within uncertainty, with the inferred onset at 7- 8 Ma of widespread thrust faulting in the Central Indian basin, and with the hypothesized attainment of maximum elevation and initiation of collapse of the Tibetan plateau at approximate to 8 Ma. The plate kinematic data are consistent with steady India- Capricorn motion since 8 Ma and provide no evidence for previously hypothesized episodic motions during that interval. The convergence since 8 Ma between the Indian and Capricorn plates significantly exceeds (by 13 to 20 km) the convergence estimated from three north- south marine seismic profiles in the Central Indian basin. Where and how the additional convergence was accommodated is unclear. ISSN: 0956-540X.

d'Erasmo, Geremia. 1960. "Nuovi Avanzi Ittiolitici Della "Serie Di Lugh" in Somalia Conservati Nel Museo Geologico Di Firenze." Palaeontographia Italica. Volume 55, Pages 1-23. Descriptors: Africa; Chordata; collections; Cretaceous; East Africa; Geological Museum of Florence; ichthyoliths; Jurassic; Lugh series; Mesozoic; morphology; paleontology; Pisces; preservation; skeletons; Somali Republic; Vertebrata. Notes: illustrations Abstract: Describes remains of fishes (mainly isolated teeth) from the Lugh series in southern Somalia, now in the collection of the geologic museum of the university in Florence (Italy). Priohybodus arambourgi n.g. n.sp. is represented among the forms described. The fauna exhibits some affinity with middle-upper Jurassic faunas and marked affinities with Wealden faunas. ISSN: 0373-0972.

Derry, D. R. 1978. "Sources of Uranium; Present and Future; Australia's Mineral Energy Resources, Assessment and Potential." Occasional Publication- Earth Resources Foundation. Jul. Issue 1, Pages 71-83. Descriptors: Africa; Arctic region; Asia; Australasia; Australia; Canada; clastic rocks; deposits; East Africa; economic geology; Europe; France; global; Greenland; India; Indian Peninsula; metal ores; mining; Niger; ore deposits; production; reserves; sandstone deposits; sedimentary rocks; shale; Somali Republic; United States; uranium ores; West Africa; Western Europe. Notes: illustrations incl. tables. ISSN: 0156-9066.

Deshmukh, I. 1992. "Estimation of Wood Biomass in the Jubba Valley, Southern Somalia, and its Application to East African Rangelands." Africa J. Ecol. Volume 30, Issue 2, Pages 127-136. Descriptors: rangeland; biomass; bush land; woodland Species Term: Somalia. Notes: Somalia- Jubba Valley. Abstract: A comparative analysis was made of regression equations relating wood biomass (above-ground) of individual trees to their canopy dimensions in semi-arid rangelands of East Africa. "Best estimates' of wood biomass of bush lands and woodlands in the Jubba Valley were derived using these relationships. Bush lands carry 1-30 kg ha-1 dry weight of wood, woodlands 2-75 kg ha-1. Subdivision of bush land and woodland categories by density of wood canopy gives finer subdivisions of biomass categories. ISSN: 0141-6707.

Deutsche Texaco Akt, Federal Republic of Germany. 1973. Somalia; Coriole Basin Interpretation. Federal Republic of Germany: Deutsche Texaco Akt., Federal Republic of Germany. Descriptors: Africa; areal geology; Coriole Basin; East Africa;

Shebeellaha Hoose; Somali Republic. Database: GeoRef. Accession Number: 1998-015661.

Development Consultants Association, Cairo, Egypt. 1969. Feasibility Study for Lower Juba Project. Egypt: Development Consultants Association, Cairo, Egypt. Descriptors: Africa; East Africa; feasibility studies; hydrology; Juba River; Lower Juba Project; Somali Republic; surface water. Database: GeoRef. Accession Number: 1998-015662.

Dhanani, S. and FAO Representative's Office in Somalia. 1988. Forestry and Range: A Sector Review. Mogadishu: FAO Representative's Office in Somalia. Descriptors: 60, 3; Forests and forestry- Somalia; Rangelands- Somalia; Forest management- Somalia; Range management- Somalia. Notes: Cover title. "May 1988." Includes bibliographical references ([p. 1-3]). Responsibility: S. Dhanani. LCCN: 89-981129. OCLC Accession Number: 26160024.

Di Paola, G. M. Seife-Michael, B. and Arno, V. 1993. The Kella Horst: Its Origin and Significance in Crustal Attenuation and Magmatic Processes in the Ethiopian Rift Valley. Firenze; Istituto Agronomico L'Oltremar; 1993. Relazioni E Monografie Agrarie Subtropicali E Tropicali Nuova Serie113 A, pages: 323-338. International Meeting; 1st-1987 Nov: Mogadishu. Notes: Geology of Somalia and surrounding regions: Geology and mineral resources of Somalia and surrounding regions. OCLC: 34175029.

Di Savoia Aosta, L. A. 1932. La Esplorazione Dello Uabi-Uebi-Scebeli Dalla Sue Sorgenti Nell'Etiopia Meridionale Alla Somalia Italiana (1928-1929). Translated title: Shebelle River Exploration from its Sources in Southern Ethiopia to Italian Somalia, 1928-1929. Somalia: Mondadori, Milan, Somalia. Descriptors: Africa; East Africa; Ethiopia; expeditions; fluvial features; geography; rivers; Shebelle River; Somali Republic; southern Ethiopia. Database: GeoRef. Accession Number: 1998-014313.

"Diagram of Principal Road Communications, Italian East Africa." 1941. EAF 487. First ed. Scale: 1:3 000 000. [Nairobi]: [Survey Directorate, East Africa Command], 1941. 2 maps. Note(s): Map in 2 parts. Copy of Italian map dated 1940. Subjects: Somalia- Maps and charts; Ethiopia- Maps and charts; Eritrea- Maps and charts. Added Name: Great Britain. War Office. General Staff. Geographical Section. 2 sheets. British Library: Maps MOD EAF 487.

Dijon, W. 1971. Water Problems in the Somali Democratic Republic; Report. Somalia: Publisher unknown, Somalia. Descriptors: Africa; East Africa; report; Somali Republic; water resources; water supply. Database: GeoRef. Accession Number: 1998-015666.

Dilli, K. 1987. Clay Mineral Studies in the Somali Basin; Prospects for Hydrocarbons. GeoSom 87; International Meeting; Geology of Somalia and Surrounding Regions; Abstracts. Somalia: Somali Natl. Univ., Mogadishu, Somalia. GeoSom 87; Geology of Somalia and Surrounding Regions, Mogadishu. Somalia Conference: Nov. 23-30, 1987. Descriptors: areal studies; basins; clastic sediments; clay mineralogy; clay minerals; Deep Sea Drilling Project; diagenesis; economic geology; Indian Ocean; marine sedimentation; ooze; petroleum; petroleum exploration; sedimentary basins; sedimentation; sediments; sheet silicates; silicates; smectite; Somali Basin; terrigenous materials; transformations; upwelling. Notes: DSDP, deep sea drilling project. OCLC Accession Number: 34175029.

Dirie, Mohamed F., Wallbanks, K. R., Aden, Abdi A., Bornstein, S. and Ibrahim, M. D. 1989. "Camel Trypanosomiasis and its Vectors in Somalia." Veterinary Parasitology. Volume 32, Issue 4, Pages 285-291. Abstract: Blood samples from 3000 Somali camels (Camelus dromedarius) were examined for trypanosome infection. Of these, 160 (5.33%) were infected with Trypanosoma evansi, one (0.03%) with T. congolense and one (0.03%) with T. brucei. Camel trypanosomiasis occured in most areas of tabanid infestation throughout the country. The tabanids Philoliche zonata and P. magretti are incriminated as the major vectors of the disease. LCCN: sv 87009650.

Dolan, Robert. 1983. "Marine Science of the Northwest Indian Ocean." European Science Notes. Dec. Volume 37, Issue 12, Pages 466-570. Descriptors: Africa; Arabian Sea; areal geology; Egypt; expedition; Indian Ocean; John Murray Expedition; Mabahiss; marine geology; Mediterranean Sea; monsoons; Nile River; North Africa; ocean circulation; oceanography; petroleum; pollution; sea-floor spreading; Somali Current; Suez Canal; symposia. Database: GeoRef.

Dominco, E. 1967. Geology of El Wak-Mandera Basin, Southern Somalia; Vedi; Geology of the Southern Flank of the Bur Uplift. Somalia: Hammar Petroleum Company, Mogadishu, Somalia. Descriptors: Africa; areal geology; Bur Uplift; East Africa; El Wak-Mandera Basin; Gedo Somali Republic; Somali Republic; southern Somali Republic. Database: GeoRef. Accession Number: 1998-016783.

Dominco, E. 1967. Geology of the Garba Harre Area. Somalia: Hammar Oil Company, Mogadishu, Somalia. Descriptors: Africa; areal geology; East Africa; Garba Harre Somali Republic; Gedo Somali Republic; Somali Republic. Database: GeoRef. Accession Number: 1998-016784.

Dominco, E. 1967. Geology of the Northern Flank of the Bur Uplift; Final Report. Somalia: Hammar Oil Company, Mogadishu, Somalia. Descriptors: Africa; areal geology; Bur Uplift; East Africa; northern Somali Republic; Somali Republic; uplifts. Database: GeoRef. Accession Number: 1998-016785.

Dominco, E. 1966. Geology of the Baidoa-Oddur Area. Somalia: Hommar Petroleum Company, Mogadishu, Somalia. Descriptors: Africa; areal geology; Baidoa Somali Republic; Bakool Somali Republic; Bay Somali Republic; East Africa; Oddur Somali Republic; Somali Republic; southern Somali Republic. Database: GeoRef. Accession Number: 1998-016782.

Donaldson, S. A. D. 1896. "Expedition through Somaliland to Lake Rudolf; Galla Tribes." Geogr. J. Volume 3, Descriptors: Africa; areal geology; East Africa; East African Lakes; expeditions; Galla Tribes; Lake Turkana; Somali Republic. ISSN: 0016-7398.

Doorenbos, J. Smith, M. and United Nations, Food and Agriculture Organization, Rome, Italy. 1977. "Water use in Irrigated Agriculture." Italy: FAO, Rome, Italy. Descriptors: Africa; agriculture; East Africa; hydrology; irrigation; Somali Republic; water supply; water use. Database: GeoRef. Accession Number: 1998-016787.

Doornbos M and Markakis J. Stone J.C. 1991. "The Crisis of Pastoralism and the Role of the State: Trends and Issues." Aberdeen University African Studies Group. Descriptors: pages 270-278; Conflict, protest, human rights; state role; political marginalization; pastoralists; political violence. Abstract: This paper aims at a preliminary exploration of the political dimensions of the crisis of African pastoralism, with particular reference to the Horn. It attempts a better understanding of the political ramifications and repercussions of the processes of marginalization to which many pastoralists are subjected- taking the form of political violence and the appearance of new political movements- while drawing attention to some of the longer term political implications of current policies vis-à-vis the pastoralist sector. The second part of the paper reflects upon one example of processes of political marginalization of African pastoralists, namely the shift in the locus of political control from the pastoralist domain to the urban-based state in the Somali context. Notes: In: Pastoral economies in Africa and long term responses to drought. Proc. colloquium, Aberdeen, 1990 (1991) p. 270-278; Geographic: Africa- Horn of Africa Somalia. OCLC Accession Number: 0890481.

Dorre, A. S. and Rapolla, A. 1993. Regional Gravity Study of Somalia. Firenze; Istituto Agronomico L'Oltremar; 1993. Relazioni E Monografie Agrarie Subtropicali E Tropicali Nuova Serie113 A, pages: 397-440. International Meeting; 1st- 1987 Nov: Mogadishu. Notes: Geology of Somalia and surrounding regions: Geology and mineral resources of Somalia and surrounding regions. OCLC: 34175029.

Dorre, Abdulkadir S., Radina, B. and Rapolla, A. A. 1984. "Breve Nota Su Di Un Programma Di Indagini Geofisiche e Geologiche Applicate Ad Un Progetto Di Ricerca Di Falde Idriche Sotterranee e Di Sbarramenti Subalvei Di Alcuni Torrenti Nei Pressi Di Beled W. » Translated title : « Brief Note on the Geophysical and Applied Geology Investigation Program of a Research Project on a Ground Water Aquifer and on Blocking River Beds of some Streams Around Beled W." Cilmi Iyo Farsamo; Wargeyska Jaamacadda Ummadda Soomaaliyeed. Volume 10, Descriptors: Africa; aquifers; Beled Weyne Somali Republic; East Africa; geophysical surveys; ground water; Hiraan Somali Republic; Somali Republic; surface water; surveys. Database: GeoRef.

Douthwaite, R. J. 1987. "Lowland Forest Resources and their Conservation in Southern Somalia." Environ. Conserv. Volume 14, Issue 1, Pages 29-35. Descriptors: resource availability; forests; nature conservation; resource management; forestry; Somalia. Abstract: The survival of primary forest on the coastal lowlands of eastern Africa is now seriously in doubt despite widespread recognition of its importance for Nature conservation. Inroads by farming and logging, and locally by mining and water-resource development, reflect the growing pressure of human numbers on diminishing resources. Previous concern has been directed towards the remnant forests of Kenya and Tanzania, but there is little information on the current status and importance of lowland forests in Somalia. This paper reviews and extends such information as there is, and makes a case for conservation in Somalia before it is too late. ISSN: 0376-8929.

Douthwaite, R. J. 1986. "Effects of Drift Sprays of Endosulfan, Applied for Tsetse-Fly Control, on Breeding Little Bee-Eaters in Somalia." Environmental Pollution Series A, Ecological and Biological. Volume 41, Issue 1, Pages 11-22. Abstract: Feeding behaviour and breeding success in an insectivorous bird, the little bee-eater Merops pusillus, were monitored in an area of southern Somalia treated five times with drift sprays of the insecticide, endosulfan, applied at 14-41 g ha-1. Birds fed mainly on bees and wasps but flies and beetles were also commonly eaten. After the heaviest spray application day-flying insects virtually disappeared from one area for 24h; the feeding rate fell and breeding failure at three nests followed. However, at the end of the spraying operation, fledging success in the sprayed area was generally the same as outside. Disrupted laying and incubation, and co-operative breeding amongst little bee-eaters in

Somalia, may indicate a poor food supply and a population particularly vulnerable to insecticidal treatments. ISSN: 0143-1471.

Doxiades, Ionides A. 1961. Preliminary Study on the Waters of Somalia. United Kingdom: D.I.A., London, United Kingdom. Descriptors: Africa; East Africa; Somali Republic; water resources. Database: GeoRef. Accession Number: 1998-016790.

Drechsel, P. and Zech, W. 1988. "Site Conditions and Nutrient Status of Cordeauxia Edulis (Caesalpiniaceae) in its Natural Habitat in Central Somalia." Econ. Bot. Volume 42, Issue 2, Pages 242-249. Descriptors: environmental conditions; nutrient status; Somalia; Cordeauxia edulis. Abstract: Cordeauxia edulis, the yeheb (yicib, ye-eb), has attracted much interest as a potential multi-purpose shrub well adapted to (semi)arid conditions. Since there exist only limited information and analyses, mainly of the seeds, the authors try to answer some questions arising about site conditions, soil requirements, and, by leaf analyses, the state of nutrition. Finally they determined the fodder-value of the foliage, as the bush is grazed too. The results underline the low requirements of the yeheb for water and soil fertility. ISSN: 0013-0001.

Drechsel, P., Zech, W. and Kaupenjohann, M. 1989. "Soils and Reforestation in the Central Rangelands of Somalia." Arid Soil Resarch and Rehabilitation. Volume 3, Issue 1, Pages 41-64. Descriptors: reforestation; rangelands; Somalia; soil. Abstract: Representative soils of the Central Rangelands of Somalia were described between Bulo Burti and Galcayo. Limitations in land use are caused by aridity, salinity, shallowness, erosion, and alkalinity, as well as by the low nutrient status. Therefore, only less intensive reforestation is possible. Suitable methods are rainwater harvesting and irrigation with river water. Fuelwood and fodder species for community (fuelwood and fodder) and environmental reforestation are recommended. The protection of the natural regeneration is another important part of forestry and range management. Initially established plantations in the shebelli floodplain show that introduced species of Prosopis chilensis and Parkinsonia aculeata , as well as the native Acacia nilotica and Balanites aegyptiaca, are well adapted for reforestation in this area, whereas Azadirachta indica can be established only with difficulty and is not considered suitable. Database: Environmental Sciences and Pollution Mgmt. ISSN: 0890-3069.

Drexler, N. 1985. "Dynamics of Foreign Involvement in the Horn of Africa. Final Rept." Library of Congress, Washington, DC. Federal Research Div. Aug. pages: 32. Descriptors: Ussr; National security; Military assistance; Middle east; Dynamics; Somalia; Military planning; Instability; International relations; Catalytic conflict Warfare; Oil fields; Ports Facilities; Strategic areas; East west relations; Ethiopia; Sudan; Foreign policy; East Africa; Horn of Africa; Behavior and society International relations. Abstract: Describes and analyzes the complex interplay of forces affecting foreign involvement in the Horn of Africa and attempts to discern future trends and their implications for the West. Emphasis is placed on Ethiopia, Somalia, and Sudan. DTIC Accession Number: ADA3034196XSP.

Dreyfuss, M. 1932. « Études De Geologie Et De Geographie Physique Sûr La Côte Française Des Somalis. » translated title : « Geology and Physical Geography of the Somalian French Coast." France: Universite de Paris, Paris, France. Descriptors: Africa; areal geology; basalts; Djibouti; East Africa; igneous rocks; rhyolites; Somali Republic; volcanic rocks. Database: GeoRef. Accession Number: 1998-010503.

Drolia, R. K. and DeMets, C. 2005. "Deformation in the Diffuse India-Capricorn-Somalia Triple Junction from a Multibeam and Magnetic Survey of the Northern Central Indian Ridge, 3°S-10° S." Geochemistry Geophysics Geosystems. SEP 9. Volume 6, Pages Q09009. Abstract: [1] We use new multi-beam and magnetic observations from 3°S-10° S along the Central Indian ridge to describe morphotectonic features and spreading fabrics within the study area and quantify deformation across the diffuse India-Capricorn-Somalia triple junction. A megamullion adjacent to the Vityaz transform fault is the first reported for the northern Central Indian ridge. The principal transform displacement zone (PTDZ) of the Vema transform fault consists of several closely spaced faults that can be traced for more than 150 km. Serpentinite or volcanic intrusions that are offset by the Vema PDTZ have inferred ages of 30,000- 45,000 years, testifying to the youth of active features on the valley floor. Abyssal hill orientations throughout the study area have remained constant since at least 4 Ma, implying little or no change in the opening direction during this period. Abyssal hills from opposite sides of the ridge have the same orientations within our 1 degrees-3 degrees errors, implying that slip along reactivated fracture zones in the diffuse triple junction neither significantly rotates nor shears the intervening seafloor. Transform fault azimuths in the survey area agree with the NUVEL-1 Africa-Australia slip direction but are 5 degrees-9 degrees clockwise from the Africa-India direction. In contrast, seafloor spreading rates in the study area agree better with the NUVEL-1 Africa-India predictions. These seemingly conflicting observations are reconciled by a model in which seafloor within the diffuse triple junction rotates about the India-Capricorn pole at angular rotation rates that increase monotonically from zero to the full rotation rate across the deforming zone. ISSN: 1525-2027.

Droubi, A. 1998. For a Sustainable use of Groundwater in Agriculture in the Arab Region; Agricultural Threats to Groundwater Quality; Workshop Proceedings. France: Agricultural Threats to Groundwater Quality, Saragossa, Spain. Conference: Oct. 27-30, 1996. Descriptors: Africa; agriculture; aquifer vulnerability; aquifers; Arabian Peninsula; arid environment; Asia; brackish water; drawdown; East Africa; Egypt; fertilizers; ground water; international cooperation; Jordan; leaching; legislation; Mauritania; Middle East; Morocco; North Africa; policy; pollution; preventive measures; salinization; salt-water intrusion; Saudi Arabia; soils; Somali Republic; Sudan; sustainability; Syria; terrestrial environment; Tunisia; water quality; water resources; water scarcity; water supply; water use; West Africa; Yemen. Abstract: Arab countries are facing critical problems of environmental degradation. Population increase and economic growth have spurred higher demands for the limited water resources. Groundwater is often the best and sometimes the only source of water supply. The region faces a future of increasingly acute water scarcity. Intensification of agriculture, often by means of irrigation, has produced groundwater quality problems, particularly with respect to nitrate pollution, pesticides, fertilizers, salinisation, logging and over exploitation of aquifers. Water quality degradation is quickly joining water scarcity as a major issue in the region. A sustainable use of groundwater in agriculture require's preventive measures to deal with groundwater quality degradation. It also requires informing the public of environmental risk, increased awareness of environmental problems. An integral management policy taking into account cost-benefit issues is also required based on the estimation of the real cost of the water allocated, fees to be paid in the case of pollution, and monitoring.

Strengthening the capacity of the environmental institutions is also one of the priorities. Notes: References: 23; illustrations incl. 3 tables. ISBN: 8484979563.

Dualeh, A.H.A, Reuther, C.-D and Scheck, P. 1990. "Basement Structure and Sedimentary Cover of Somalia." Berliner Geowissenschaftliche Abhandlungen, Reihe A. Volume 120, Issue 2, Pages 505-518. Descriptors: Plate tectonics; basement; rifting; continental breakup; sedimentation. Notes: Somalia. Abstract: The structural and sedimentary patterns of Somalia reflect the Permo-Triassic to recent geodynamic events during the break-up of East-Gondwana, the initiation of Indian Ocean formation and the formation of the Gulf of Aden. Present day Somalia is characterized by three main structural highs: the Bur-Uplift in southern Somalia, the Nogal-High in northern Somalia and the Northern Uplift which parallels the Gulf of Aden. These present highs separated five major Jurassic basins: Berbera-Borama basin, Al Mado-Darror basin, Somali Embayment, Somali coastal basin and Lugh-Mandera basin. East-Gondwana underwent intracratonic extension and continental rifting during Permo-Triassic times, leading to the formation of a rift system. The grabens hosted mainly continental sediments (Karoo) and evaporites during the Permo-Triassic time. This Permo-Triassic rifting facilitated the drift of Madagascar away from East-Africa. At this time all the five Somali major basins were flooded and probably connected. This event marks the first of three major geodynamic-eustatic episodes which affected the Somali basins. The second episode occurred during the Aptian-Maastrichtian and the third during the Paleocene-Eocene. ISSN: 0172-8784.

Dualeh, A. H. A. and Nairn, A. E. M. 1993. The Continuing Story of the Fragmentation of Gondwana: A Contribution from Somalia. Firenze; Istituto Agronomico L'Oltremar; 1993. Relazioni E Monografie Agrarie Subtropicali E Tropicali Nuova Serie113 A, pages: 345-356. International Meeting; 1st- 1987 Nov: Mogadishu. Notes: Geology of Somalia and surrounding regions: Geology and mineral resources of Somalia and surrounding regions. OCLC: 34175029.

Dualeh, Abdirahman H. Ali. 1986. Geological and Stratigraphic Evolution of the Northeastern Somalia Continental Margin and Adjacent Areas. Descriptors: 129 leaves; Geology- Somalia; Geology, Stratigraphic- Cenozoic; Geology, Stratigraphic- Mesozoic. Notes: Typescript. Includes bibliographical references (leaves 127-129). Dissertation: Thesis (MS), University of South Carolina, 1986. OCLC Accession Number: 17216499.

Ducreux, C., Mathurin, G. and Latreille, M. 1991. "Red Sea- Gulf of Aden Source Rock Geochemical Evaluation; AAPG International Conference; Abstracts." AAPG Bull. Aug. Volume 75, Issue 8, Pages 1409-1410. Descriptors: Africa; Arabian Peninsula; Arabian Sea; Asia; carbon; Cenozoic; chemically precipitated rocks; correlation; drilling; East Africa; economic geology; evaporites; genesis; geochemistry; Gulf of Aden; Indian Ocean; maturity; Neogene; organic carbon; organic compounds; organic materials; petroleum; possibilities; Red Sea; sedimentary rocks; simulation; Somali Republic; source rocks; Tertiary; thermal history; Yemen. ISSN: 0149-1423.

Dueing, W. 1977. "Large-Scale Eddies in the Somali Current." Geophys. Res. Letters. Volume 4, Issue 4, Pages 155-158. Descriptors: Monsoons; Water circulation; Somali Current; Somalia; Marine. Notes: Records keyed from 1977 ASFA printed journals. TR: 1977. Abstract: Velocity records from a subsurface mooring in the Somali Current exhibit direction reversals during a time when the SW Monsoon is still steadily blowing. The records are interpreted in terms of large-scale eddies which seem to be generated near the equator and propagate slowly northward along the coast of Somalia.

The Somali Current constitutes the most important branch of the complex East African Current System. It is generated by the Summer Monsoon and flows along the coast of East Africa, from the latitude of Mombassa to the island of Socotra, corresponding to a distance of {approx} 2,200 km. Translated to the path of the Gulf Stream, this corresponds to the distance between Yucatan and Cape Hatteras. Despite the face that this current has been immensely important to seafarers for {approx} 2,000 yrs, it is one of the least explored boundary current systems in the Northern Hemisphere. In contrast to the Gulf stream, which shows only insignificant seasonal variations, the Somali Current reverses its direction following the seasonal change of the monsoon winds. A 1st long-term current meter record from the region indicates that the classical picture of a straightforward oceanic response to the forcing of the wind stress may be greatly oversimplified. It is shown that some of the key features of the record are not related to fluctuations of the wind field, but more likely to large-scale eddies propagating northward along the coast. Database: ASFA: Aquatic Sciences and Fisheries Abstracts.

Duignan, Peter, editor. 1971. Guide to research and reference works on Sub-Saharan Africa. Compiled by Helen F. Conover and Peter Duignan. With the assistance of Evelyn Boyce, Liselotte Hofmann and Karen Fung. Stanford, Calif., Hoover Institution Press, Stanford University [1971 or 2]. 1102 pages. 29 cm. Hoover Institution bibliographical series; 46. Subjects: Africa, Sub-Saharan- Bibliography. ISBN: 0817924612.

Duing, W. and Schott, F. 1978. "Measurements in the Source Region of the Somali Current during the Monsoon Reversal." J.Phys.Oceanogr., 8(2), 278-289. (1978). Descriptors: water temperature; current measurements; monsoons; thermocline; water currents; Somali Current; Equatorial Counter Current; winds; Marine. Notes: TR: IR7812444. Abstract: Temperature and current records were obtained from four subsurface moorings deployed in the source region of the Somali Current from mid-January to mid-July 1976. The first part of the records from January until early April showed that the array straddled the convergence zone of the northward-flowing East African Coast Current and the southward-flowing Somali Current. During this time, except for the southernmost location off Mombasa, the mean flow at all locations was weak and variable. The predominant variability had a time-scale of 4-5 days. Around 20 April the wind shifted to the southeast and three days later the flow in the upper 80 m turned northward and intensified. Development of strong northward flows below the thermocline took several more weeks. The observations imply that a switching mechanism took place at that time. The initially eastward-flowing Equatorial Counter Current is shifted rapidly 45 degree to the left to run northward along the coast. This mechanism may, in part, be responsible for the impulsive beginning of the Somali Current during the early stages of the monsoon onset south of the equator. Database: ASFA: Aquatic Sciences and Fisheries Abstracts.

Duing, Walter and Szekielda, Karl-Heinz. 1971. "Monsoonal Response in the Western Indian Ocean." Journal of Geophysical Research, Vol 76, no 18, P 4181-4187, June 20. 15 REF. ONR Contract: N0014-67-A-0201. Volume 1 TAB, Project Number: 083-06017-16-70(481). Descriptors: Ocean Currents; Indian Ocean; Monsoons; Synoptic Analysis; Winds; Advection; Currents (Water); Convection; Temperature; Heat Balance; Seasonal; Climatology; Somali Current. Abstract: infrared observations from spacecraft were used to investigate the response of the Somali current to the onset of the southwest monsoon. Selected satellite observations from three years were available for this study. The time-dependent development of horizontal temperature gradients at the sea surface serves as an indicator for the formation of the baroclinic structure of the Somali current. A comparison is made with the simultaneous development of the southwest component of the monsoon wind. The investigation reveals that the temperature gradient during the early formation stage in all years are directly proportional to the wind speed. The phase lag between the development of wind and temperature gradient during the buildup of the boundary current has a mean value of twelve days. During the decay period in late summer and fall, the lag increases continuously up to forty days. The observations suggest that two phenomena of different spatial scales play an important role during the formation of the Somali current; in the early stage (may, June), local wind-induced upwelling seems to be the more important source of baroclinicity; in the latter stage of the buildup (July), large-scale geostrophic effects seem to be dominating. Database: Environmental Sciences and Pollution Mgmt.

Dumont, H. J. and Maas, S. 1987. "Cladocera and Copepoda (Crustacea) from Somalia." Monit.Zool Ital., Suppl. Volume 22, Issue 8, Pages 87-99. Descriptors: new records; endemic species; biogeography; geographical distribution; Cladocera; Crustacea; Somali Dem. Rep. Halicyclops thermophilus; Marine; Brackish. Abstract: Two species of Cladocera and eight species of Copepoda (Crustacea) are reported from Somalia; six are first records for that country. No species are new to science. Besides a few that appear to be endemic to North-East Africa, several are interesting range extensions. Halicyclops thermophilus, a brackish-water species fringing the coasts of the Indian Ocean, is figured in detail. Database: ASFA: Aquatic Sciences and Fisheries Abstracts.

Dundas, F.G., Commander. 1893. "River Juba." Scale: 1:292 187. [London]: [General Staff, Geographical Section], 1893. 1 map. Note(s): Covers river from its mouth to Bardera (Baardheere). Subjects: Somalia- Maps and charts. Added Name: Great Britain. War Office. Intelligence Division. Added Name: Great Britain. War Office. General Staff. Geographical Section. British Library: Maps MOD IDWO 937.

Durrill, Wayne K. 1986. "Atrocious Misery: The African Origins of Famine in Northern Somalia, 1839-1884." American Historical Review; April 1986, Vol. 91 Issue 2, p287, 20p. Subject Terms: Famines; Herders; Sultanate. Geographic Terms: Somalia. Abstract: Discusses the role of Majeerteen sultanate, a group of European sailors, in the famine of Northern Somalia in 1839-1884. Participation of sultanate in overseas commodities market; Division of Majeerteen into three lineages; Impact of trade in livestock on Majeerteen; Effect of the 1868 drought on herders. ISSN: 0002-8762.

E

Eagles, Graeme, Gloaguen, Richard and Ebinger, Cynthia. 2002. "Kinematics of the Danakil Microplate." Earth & Planetary Science Letters. 30 Oct. Volume 203, Issue 2, Pages 607-620. Descriptors: accretion; Afar; Afar Depression; Africa; Bouguer anomalies; Danakil Microplate; East Africa; faults; gravity anomalies; horsts; Indian Ocean; kinematics; microcontinents; microplates; movement; oblique-slip faults; plate boundaries; plate rotation; plate tectonics; reconstruction; Red Sea; Red Sea region; remote sensing; rifting; shear zones; Somalia Plate; strike-slip faults; systems. References: 44; sketch maps; Abstract: A refinement and extrapolation of recent motion estimates for the Danakil microplate, based on ancient kinematic indicators in the Afar region, describes the evolution of a microplate in the continental realm. The Danakil horst is an elevated part of this microplate, exposing a Precambrian basement within the Afar depression, the site of the Nubia-Somalia-Arabia triple junction. We compare evidence for strike- or oblique-slip faults in data from the Afar depression and southern Red Sea to small circles about published poles of rotation for the Danakil microplate with respect to Nubia. A reconstruction about the preferred pole reunites lengths of a Precambrian shear zone on the Nubia and Danakil sides and preserves a uniform basement fabric strike through Nubia, Danakil and Yemen. Since at least magnetic chron C5 (approximately 11 Ma) Danakil rotated about a different pole with respect to Nubia than either Somalia or Arabia, but between chrons C5 and C2A Nubia-Danakil motion was a close approximation to Nubia-Somalia motion. Since C2A relative motions of the Danakil microplate have been independent of movements on any of the neighbouring plate boundaries. We relate this to the onset of oceanic-type accretion within Afar. The resulting eastwards acceleration of Danakil was accommodated by westwards propagation of the Gulf of Aden rift that became the new, discrete, plate boundary between the Danakil microplate and the Somalia plate. Present-day activity suggests that the Red Sea and Aden rifts will link through Afar, thereby isolating the Danakil horst as a microcontinent on the Arabian margin. ISSN: 0012-821X.

Eagleton, G.E, Mohamed, A.A, Odowa, A.A and Muse H.A. 1991. "A Comparison of Moisture-Conserving Practices for the Traditional Sorghum-Based Cropping System of the Bay Region, in Somalia." Agric., Ecosyst. Environ. Volume 36, Issue 1-2, Pages 87-99. Descriptors: Crop and livestock production- technical aspects; cropping system; intercropping; sorghum. Notes: References: Number: 10; Geographic: Somalia- Bay Region; Notes: Special Feature: 1 graph, 1 map, 10 references, 8 tables. Abstract: Four practices for conserving and better utilising soil moisture were examined in the principal sorghum-growing regions of Somalia. Even in an above-average wet season, the practice of bunding raised the total dry matter yield of the local sorghum cultivar from 4412 to 6007 kg ha-1 and its grain yield from 898 to 1103 kg ha-1. Another traditional practice, intercropping sorghum with a low density of cowpea, increased the land equivalent ratio by 10-30% despite marginal reductions in the yields of the component crops. Incorporating sorghum stover into the soil at the end of one season raised the next season's seed yield of sorghum from 403 to 551 kg ha-1 and of mungbean from 247 to 343 kg ha-1. When compared with continuous sorghum cropping, a clean fallow break in one season- storing moisture for the next- did not increase the subsequent yield of local sorghum, but increased peanut nut-in-shell yields from 333 to 604 kg ha-1 at an unfertilised farm site and from 562 to 973 kg ha-1 when phosphorus was applied. New systems that increase the availability of moisture and of the principal limiting nutrient, phosphorus, would lead to substantial human benefits. ISSN: 0167-8809.

"Economic and Political Turmoil Plague this Continent." 1994. World Oil. August. Volume 215, Pages 87 et seq. Descriptors: Petroleum industry- Africa; Applied Science & Technology. Abstract: Part of a special issue on the outlooks for the world's oil- and gas-producing regions. Economic and political turmoil are hampering drilling activity in Africa. However, new hydrocarbon laws with favorable terms and conditions and opportunities in European markets will enable companies to profit from oil and gas in northern Africa, while deepwater prospects appear promising in the long term in western Africa. Prospects for the oil and gas industries in Egypt, Libya, Tunisia, Algeria, the

Ivory Coast, Nigeria, Cameroon, Gabon, Congo, Angola, Namibia, South Africa, Benin, Chad, Equatorial Guinea, Ethiopia, Ghana, Guinea Bissau, Mauritania, Morocco, Mozambique, Niger, Senegal, Seychelles, Somalia, Sudan, Uganda, and Zaire are discussed. ISSN: 0043-8790.

El Gabaly, M. M. 1977. Problems and Effects of Irrigation in the Near East Region; Arid Land Irrigation in Developing Countries; Environmental Problems and Effects. United Kingdom: Pergamon Press, Oxford, United Kingdom. International Symposium on Arid Land Irrigation in Developing Countries, Alexandria. Egypt Conference: Feb. 16-21, 1976. Descriptors: Afghanistan; Africa; agriculture; Arabian Peninsula; arid environment; Asia; Bahrain; Cyprus; developing countries; East Africa; Egypt; environmental geology; field crops; ground water; Indian Peninsula; Iran; Iraq; irrigation; Jordan; Kuwait; land use; Lebanon; levels; Libya; Middle East; North Africa; Oman; Pakistan; Qatar; salinity; Saudi Arabia; seepage; soils; Somali Republic; Sudan; Syria; United Arab Emirates; water quality; Yemen; yields. Notes: References: 8; tables. Database: GeoRef. Accession Number: 1978-005860.

El-Baz, Farouk. 1990. Mapping from Space Photographs for the Monitoring and Exploitation of Water Resources; Sand Transport and Desertification in Arid Lands. Singapore: World Sci., Singapore. First International Workshop on Sand Transport and Desertification in Arid Lands, Khartoum. Sudan Conference: Nov. 17-26, 1985. Descriptors: Africa; applications; cartography; case studies; changes of level; East Africa; economic geology; economic geology maps; Egypt; Ethiopia; geomorphology; high-resolution methods; hydrology; Lake Nasser; land use; levels; maps; monitoring; Nile Valley; North Africa; radar methods; remote sensing; Somali Republic; space photography; surveys; water resources. Notes: References: 10; illustrations ISBN: 9971508583.

Elder, D. and United Nations Environment Programme, Nairobi, Kenya. 1987. "Sectoral Report on Marine Protected Areas and Reserves; Coastal and Marine Environmental Problems of Somalia; Annexes." International (III): 1987. Volume: 84, pages: 111-130. Descriptors: Africa; concepts; conservation; East Africa; ecosystems; environment; environmental geology; geomorphology; marine environment; protection; research; shore features; shorelines; Somali Republic. Notes: References: 17; illustrations incl. sketch maps; ISSN: 1014-8647. Database: GeoRef. Accession Number: 1992-030545.

Elewa, Ashraf M. T., Luger, Peter and Bassiouni, Mohamed A. 2001. "Middle Eocene Ostracoda from Northern Somalia; Paleoenvironmental Appraisal." Revue De Micropaleontologie. Dec. Volume 44, Issue 4, Pages 279-289. Descriptors: Africa; Arthropoda; assemblages; biostratigraphy; biozones; Cenozoic; cluster analysis; correlation coefficient; Crustacea; dissolved materials; dissolved oxygen; East Africa; Eocene; Invertebrata; Mandibulata; microfossils; middle Eocene; multivariate analysis; northern Somalia; Ostracoda; oxygen; paleoclimatology; paleoenvironment; Paleogene; reconstruction; Somali Republic; statistical analysis; Tertiary. References: 10; illus. incl. strat. cols., 2 tables, sketch map. Abstract: The present study is focused on the reconstruction and interpretation of the paleoenvironmental conditions that prevailed during the deposition of the Middle Eocene deposits of Northern Somalia by means of the ostracod assemblages distinguished. Quantitative analyses of 75 ostracod species, recognized in 36 ostracod-bearing samples from the study area, has revealed six types of assemblages (a, b1, b2, b3, b+c, c), obtained by applying Q-mode cluster analysis based on the Jaccard coefficient of similarity (the complete linkage method). These types represent, from older to younger, the Thracella indica Assemblage Zone (type b1; middle Lutetian), the Paijenborchellina reticulata-Paijenborchellina pustulata Assemblage Zone (type b2; middle Lutetian), the Hornibrookella moosae Assemblage Zone (type b3; middle Lutetian), the Paragrenocythere dhofarensis somaliensis Assemblage Zone (type a; late Lutetian), the Asymmetricythere tricostata Assemblage Zone (type b+c; late Lutetian to Bartonian) and the Neocyprideis rotundata Assemblage Zone (type c; late Lutetian to Bartonian), respectively. Q-mode analysis based on the extraction of latent roots and vectors of a non-parametric matrix produced by the Spearman correlation coefficient, led to the conclusion that the most effective environmental factors in the study area are: salinity (1st latent vector), substrate (2nd latent vector), dissolved oxygen of water (3rd latent vector), turbidity currents (4th latent vector). The general paleoenvironmental conditions prevailing during the deposition of the Middle Eocene strata in Northern Somalia are discussed in details. ISSN: 0035-1598.

Elmi, A.A. 1991. "Livestock Production in Somalia with Special Emphasis on Camels." Nomadic Peoples. Volume 29, Pages 87-103. Descriptors: Agriculture; Crop and livestock production- general; camel population; range animal; pastoral economy; livestock production; developing country; agricultural practice; pastoralism; meat production; camels; milk production. Notes: Somalia. Abstract: The Somali pastoral economy is mainly based on range animal resources. Over 70% of the Somali human population subsist in pastoralism. Pastoralists contribute more than 60% of the GDP and 80% of national export. Camel, cattle and small ruminants are widely distributed in all ecological zones of the country. Sheep and goats are dominant animals, but camel population exceeds cattle in number. A large number of camels and small ruminants are found in the drier northern and central zones. Cattle dominate in the wetter southern and Trans-Juba regions. Small ruminants make up 70% of the livestock marketed producing 60% of total meat production. Camels and goats provide 75% of the total milk supply. In the pastoral subsistence economy camels are primarily raised for milk production and small ruminants for generating cash income for the family. ISSN: 0822-7942.

Elmi, Abdullahi S. 1983/8. "The Chewing of Khat in Somalia." Journal of Ethnopharmacology. Volume 8, Issue 2, Pages 163-176. Abstract: Khat (Catha edulis Forsk.), known in Somalia as "qaad" or "jaad", is a plant whose leaves and stem tips are chewed for their stimulating effect. From the Harar area, khat has been introduced at different times into the present day territories of Somalia, Djibouti, South and North Yemen, Kenya, Madagascar, Tanzania and down to south eastern Africa. The plant, which belongs to the Celestraceae family, grows wild at altitudes of 1500-2000 m above sea level. Among the various compounds present in the plant (more than forty alkaloids, glycosides, tannins, terpenoids, etc.), two phenyl-alkylamines, namely cathine ((+)-norpseudoephedrine) and cathinone ((-)S-o-aminopropiophenone) seem to account mostly for the effect. The consumers get a feeling of well-being, mental alertness and excitement. The after effects are usually insomnia, numbness and lack of concentration. The excessive use of khat may create considerable problems of social, health and economic nature. These problems have been summarily reviewed. Khat chewing started at different times in different parts of Somalia. Since World War II, the prevalence of the practice has continuously increased and no social group is excluded. An epidemiological research to compare Northern and Southern regions of Somalia and to obtain a rough estimate of prevalence, definition of social characteristics of the groups of consumers, specification of the motivations, patterns of use and effects during and after consumption has been conducted. Consumers and non-consumers (7485 people) were randomly interviewed in the two regions. Khat consumption in relation to sex, age, occupation and grade of education is presented. LCCN: sv88033077.

Elmi, Abdullahi S. 1980/3. "Present State of Knowledge and Research on the Plants used in Traditional Medicine in Somalia." Journal of Ethnopharmacology. Volume 2, Issue 1, Pages 23-27. Abstract: Plants used in traditional medicine in Somalia have not so far been studied completely, and there are many difficulties owing to the secrecy maintained by traditional healers. Present research is being developed in a centre established by the Ministry of Health and in the Somali National University. A Somali traditional pharmacopoeia is envisaged. Pharmacological screening, chemical, and clinical studies are in progress on a number of plants. An herbarium with 2000 plants has been established. LCCN: sv88033077.

Elmi, A.A, Thurow T.L and Box T.W. 1992. "Composition of Camel Diets in Central Somalia." Nomadic Peoples. Volume 31, Pages 51-63. Descriptors: Crop and livestock production- technical aspects; camel diet; camel husbandry. Notes: Somalia. Abstract: The composition of camel diets was studied in Ceeldheer District in central Somalia in 1986/1987. Milking and non-milking camel diets were determined in both dry and wet seasons. Both types of camels consumed almost the same kinds of plants in any given seasons. Even though milking camels seemed more selective than dry camels, the animals were extremely flexible and opportunistic in diet selection and foraging behaviour. ISSN: 0822-7942.

"Encampments of the Dispossessed." 1981. National Geographic. June 1981, Vol. 159 Issue 6, page 756 et seq., 20 pages. Subject Terms: Epidemics; Refugee Camps; Refugees. Geographic Terms: Somalia. Abstract: Focuses on rise in number of refugees in Somalia, during 1980s. Population of refugees in the country; Epidemic diseases prevailing in the refugee camps; Difficulties faced by refugees in the refugee camps of the country; Impact of rise of refugees on economic conditions of the country. ISSN: 0027-9358.

"Encouraging the Private Sector in Somalia." 1982. Berg (Elliot) Associates, Alexandria, VA. Sep. Volume: AIDPNAAL378, pages: 185. Descriptors: Somalia; Agriculture; Fishing; Growth; Marketing; Construction; Economic policy; Private organizations; Developing country application; Strategies; Laws; Private land; International Monetary Fund; Behavioral and social sciences Economics; Agriculture Agricultural economics; Agriculture and food Agricultural economics; Business and economics Foreign industry development and economics. Abstract: Slow growth and an overburdened beaucracy were precursors to an economic crisis that led the Government of Somalia (GOS) to abandon its 1970's experiment with 'scientific socialism'. This report, intended to help the GOS define strategies to stimulate growth in the private sector, examines poor past economic performance; the general environment for the private sector; and recent liberalization and privitization efforts and their impact on agriculture, settlements, and on agreements with the International Monetary Fund. Notes: Funder: Agency for International Development, Washington, DC. NTIS Accession Number: PB85134567XSP.

Esenkov, O.E, Olson, D.B and Bleck R. 2003. "A Study of the Circulation and Salinity Budget of the Arabian Sea with an Isopycnic Coordinate Ocean Model." Deep-Sea Research Part II: Topical Studies in Oceanography. Volume 50, Issue 12-13, Pages 2091-2110. Descriptors: Water masses and fronts; wind forcing; mesoscale eddy; mixed layer; salinity; oceanic circulation. Somalia; Socotra. References: Number: 33; Geographic: Arabian Sea. Abstract: The evolution of surface circulation and salinity budget are studied with the open-boundary version of the Miami Isopycnic Coordinate Ocean Model (MICOM) that uses a global MICOM simulation as a boundary condition. Under climatological wind and thermodynamic forcing, the model develops solutions that are in good agreement with the climatologically forced global MICOM results and with observations. When the observed winds force the model, interannual variability of the surface fields increases significantly. However, coalescence of the two large eddies off Somalia in the end of the summer monsoon suggested in earlier observations does not occur in the model. To identify what processes facilitate or restrict the merger, a series of experiments was performed with modified model parameters and forcing fields. The eddies coalesced when half-slip, rather than no-slip, boundary conditions were used. In this case, less positive vorticity was produced at the coast, resulting in reduced blocking effect on the propagation of the southern eddy. The Socotra Island, which is submerged in the standard model, hinders a northward movement of the Great Whirl, leading to a stronger interaction between the eddies, which results in their subsequent merging. A more realistic coalescence occurs in an experiment where winds are held constant after reaching the peak summer value. Freshwater fluxes from the east and south are important for the salinity budget in the Arabian Sea, where evaporation exceeds precipitation. The only significant cross-equatorial transport of low-salinity water occurs in the upper 400m in the model. Most of this water is advected below the surface mixed layer at the western boundary. The strongest interaction between the mixed layer and the oceanic interior occurs during the summer in the coastal upwelling regions off Somalia. Almost half of all upwelled water comes from depths between 100 and 200m, thus signifying the importance of mid-depth circulation and water mass distribution for the surface processes. ISSN: 0967-0645.

"Establishing the Structure of an IGADD National and Sub-Regional Early Warning System." 1989. Sioux Falls, SD. Intergovernmental Authority for Drought and Development, Djibouti; United States: EROS Data Center. Dec. Volume: AIDPNABE469; PB90259771, pages: 187. Descriptors: Droughts; Early warning systems; Information systems; Models; User needs; Remote sensing; Djibouti; Ethiopia; Kenya; Somalia; Sudan; Uganda; Disaster planning; Famine; Disaster recovery; Case studies; Intergovernmental Authority for Drought and Development; Agency for International Development. Abstract: The Final Report on Establishing the Structure of an Intergovernmental Authority for Drought and Development (IGADD) National and Sub-Regional Early Warning System consists of three volumes. Collectively, they provide a complete treatment of the findings, conclusions, and recommendations suggested by the U.S. Geological Survey team. The three volumes are: Volume 1-Executive Summary. Contents include a brief summary of a strategic plan for an IGADD sub-regional early warning system (EWS). Topics include (1) a suggested EWS model, (2) IGADD's role in early warning, (3) priorities and actions, and (4) critical issues; Volume 2- Main Report. The volume provides a detailed discussion of the overall findings, conclusions and recommendations for an IGADD sub-regional EWS; Volume 3- Country Reports. Contents include country reports detailing the status and needs for early warning systems in the six IGADD member states. The country reports are the working documents prepared by the USGS team to develop the overall conclusions and recommendations. This document is Volume 1. Notes: Prepared in cooperation with Intergovernmental Authority for Drought and Development, Djibouti. Sponsored by Agency for International Development, Nairobi (Kenya). Contract AID/AFR-0510-P-IC-7022-22. Database: NTIS. Accession Number: PB90259771.

"Ethiopia and the Somalilands. Section 23. Weather and Climate. Revision." 1965. Central Intelligence Agency, Washington DC. April 1965. 80 Pages. Descriptors: *Ethiopia; *Weather; *Climate; Military geography; Monsoons; High temperature; Dust storms. Abstract: Revision of report dated May 50. Includes envelope with map. Original contains color plates: All DTIC and NTIS reproductions will be in black and white. Notes: Approved for Public Release. DTIC: ADA950065.

Eurospace. 1975-1979. "Somalia." Paris: Eurospace. Notes: Description: 1 map; photocopy; 40 x 28 cm. Geographic: Somalia- Maps. Scale [ca. 1:5,000,000]. Cartgrph Code: Category of scale: a Constant ratio linear horizontal scale: 5000000; Notes: Shows towns with more than 2500 inhabitants, power stations, radio links, airports, and populated areas. One source dated December 1974. Includes list of sources. Responsibility: Eurospace. OCLC Accession Number: 30746366.

Evans, Robert H. and Brown, Otis B. 1981. "Propagation of Thermal Fronts in the Somali Current System." Deep-Sea Research, Oxford. Volume 28, Issue 5A, Pages 521-527. Descriptors: Oceanic fronts; Upwelling water temperatures; Somali Current; Coastal Waters of East Africa. Notes: May, 1981. Refs. Abstract: Surface ship-of-opportunity and satellite remotely sensed sea surface temperature (SST) data were obtained during the mature phase of the Southwest Monsoon in the northwestern Indian Ocean for the years 1976-1979. Large, wedgelike areas of upwelled water are observed at 5° and 10°N after Somali Current spinup, indicative of a two-gyre circulation in the Somali Current system. Several months later, the southern separation region is observed to translate poleward (5-75 cm/sec) in three of the four years. This translation occurs in several distinct phases over a 1-mo interval, with the southern wedge eventually coalescing with the northern wedge between 8° and 10°N (1976, 1979). (In 1978, the northern wedge translated past Socotra and dissipated in the Arabian Sea.) The thermal signature of the new northern wedge becomes indistint 2-3 mo later. Database: Meteorological & Geoastrophysical Abstracts.

Eyow, Abdullahi Hassan. 1993. An Evaluation of Hydrocarbon Potential of Bihindule Basin, Northern Somalia. Descriptors: 146 leaves; Petroleum- Geology-Somalia; Geology- Somalia. Notes: Typescript. Some illustrations on folded leaves. Includes bibliographical references (leaves 143-146). Dissertation: Thesis (master's)-University of South Carolina, 1993. Responsibility: by Abdullahi Hassan Eyow. OCLC Accession Number: 31101266.

F

Fagotto, F. 1987. "Sand-Dune Fixation in Somalia." Environ. Conserv. Volume 14, Issue 2, Pages 157-163. Descriptors: dunes; environment management; desertification; erosion control; dune stabilization; Somalia; Somali Dem. Dep.; Marine.

Abstract: Sand-dune fixation is an important step in the struggle against desertification, the latter is gaining strength in Somalia. The objective is not only to fight erosion and to stabilize the dunes, but also to raise some revenue for residents through reafforestation and agricultural schemes. Education and training must be directed towards a steady prolification throughout the country of all sandfixing activities, though the experience acquired so far suggests that efforts should be mainly focused on a few successful fixing species, particularly Mesquite on stabilized sands and Pes-capræ on the shifting dunes. ISSN: 0376-8929.

Fagotto, F. 1985. "The Lion in Somalia." Mammalia. Volume 49, Issue 4, Pages 587-588. Descriptors: habitat preferences; Panthera leo; Somalia. Abstract: A specific research for Lions was carried out in 1984: all the information reported here concerning their status and distribution refers to this year. The description of the environment is limited to the southern territories of Somalia where lions are still not rare. The lion's habitat includes different types of plant communities typical of Southern Somalia, open woodlands, dense woodlands and hygrophilous savannas. ISSN: 0025-1461.

Faillace, C. 1998. "Brief Note on the Occurrence of High Fluoride Content in Groundwater of Somalia." Geologica Romana. Issue 34, Pages 51-57. Descriptors: Water Resources: Economic; Groundwater quality: phreatic zone; Water resources; Water supply and sanitation; fluoride; groundwater; health risk; water chemistry; water resource. Notes: Italy; References: 15; Geographic: Somalia. Abstract: Excess of fluoride in drinking water in Central Somalia and along the coastal belt of Shabeelle Dhexe and Benadir in Southern Somalia is the cause, together with high concentration of salts in water, of several illnesses. Information on areal distribution of fluoride in excess of WHO's standard in the country is not known. The available information consists of about 80 chemical analyses showing a fluoride content higher than 1 mg/l. This short note is meant to stimulate the interest for a future country-wide investigation on fluoride distribution and problem of fluorosis in Somalia and to suggest controlling its effects on health. ISSN: 0435-3927.

Faillace, C. 1993. "Results of a Country-Wide Ground-Water Quality Study in Somalia; Geology and Mineral Resources of Somalia and Surrounding Regions (with a Geological Map of Somalia 1:1,500,000)." Relazioni e Monografie Agrarie Subtropicali e Tropicali. Nuova Serie. Volume 113B, Pages 615-632. Descriptors: Africa; East Africa; ground water; imagery; remote sensing; satellite methods; Somali Republic; surface water; water quality; water resources; water use. Notes: sketch maps. Database: GeoRef.

Faillace, C. 1993. "The Need to use Simple and Appropriate Technology to Develop the Water Resources in Somalia; Geology and Mineral Resources of Somalia and Surrounding Regions (with a Geological Map of Somalia 1:1,500,000)." Relazioni e Monografie Agrarie Subtropicali e Tropicali. Nuova Serie. Volume 113B, Pages 633-648. Descriptors: Africa; dams; deep-water environment; design; East Africa; ground water; Juba River; Shabelle River; Somali Republic; surface water; technology; water resources; water wells; wells. Notes: illustrations, incl. sketch maps. Notes: Geology of Somalia and surrounding regions: Geology and mineral resources of Somalia and surrounding regions. OCLC Accession Number: 34175029.

Faillace, C. 1993. "Hydrogeological Importance of the Sub-Surface Basalts in the Mudug-Galgadud Plateau; Geology and Mineral Resources of Somalia and Surrounding Regions (with a Geological Map of Somalia 1:1,500,000)." Relazioni e Monografie

Agrarie Subtropicali e Tropicali. Nuova Serie. Volume 113B, Pages 649-664. Descriptors: Africa; aquifers; artesian waters; basalts; central Somalia; East Africa; ground water; igneous rocks; Mudug-Galgadud Plateau; recharge; Somali Republic; volcanic rocks; water wells; wells. References: 5; 3 tables, sects., sketch maps. Notes: Geology of Somalia and surrounding regions: Geology and mineral resources of Somalia and surrounding regions. OCLC Accession Number: 34175029.

Faillace, C. 1993. Results of a Country-Wide Ground-Water Quality Study in Somalia. Firenze; Istituto Agronomico L'Oltremar; 1993. Relazioni E Monografie Agrarie Subtropicali E Tropicali, Nuova Serie113 B, pages: 615-632. International Meeting; 1st- 1987 Nov: Mogadishu. Notes: Geology of Somalia and surrounding regions: Geology and mineral resources of Somalia and surrounding regions. OCLC Accession Number: 34175029.

Faillace, C. 1987. "The Need to use Simple Low-Cost Technology to Develop Water Resources in Somalia." GeoSom 87; Geology of Somalia and Surrounding Regions; Late Abstracts. Somalia: Somali Natl. Univ., Dep. Geol., Mogadishu, Somalia. GeoSom 87; Geology of Somalia and Surrounding Regions, Mogadishu. Somalia Conference: Nov. 24-Dec. 1, 1987. Descriptors: Africa; dams; East Africa; economic geology; economics; ground water; Juba River; resources; Shabeelle River; Somali Republic; surveys; technology; water quality; water resources; water supply; water wells; wells. OCLC Accession Number: 34175029.

Faillace, C. 1987. "Hydrogeological Importance of the Sub-Surface Basalt in the Mudug-Galgadud Plateau." GeoSom 87; International Meeting; Geology of Somalia and Surrounding Regions; Abstracts. Somalia: Somali Natl. Univ., Mogadishu, Somalia. GeoSom 87; Geology of Somalia and Surrounding Regions, Mogadishu. Somalia Conference: Nov. 23-30, 1987. Descriptors: Africa; aquifers; artesian waters; basalts; confined aquifers; East Africa; exploration; Galgudud; ground water; hydrogeology; igneous rocks; lava flows; Mudug; Ogaden; resources; Somali Republic; surveys; volcanic rocks; water resources. OCLC Accession Number: 34175029.

Faillace, C. 1987. "Results of a Country-Wide Groundwater Quality Study in Somalia." GeoSom 87; International Meeting; Geology of Somalia and Surrounding Regions; Abstracts. Somalia: Somali Natl. Univ., Mogadishu, Somalia. GeoSom 87; Geology of Somalia and Surrounding Regions, Mogadishu. Somalia Conference: Nov. 23-30, 1987. Descriptors: Africa; East Africa; economic geology; exploration; ground water; Somali Republic; surveys; water quality; water resources. OCLC Accession Number: 34175029.

Faillace, C. 1984. "Water Quality of Schebelli and Juba Valley." Quaderni Di Geologia Della Somalia (Mogadiscio). Volume 8, Descriptors: Africa; East Africa; geochemistry; hydrochemistry; Juba River; Shebelle River; Somali Republic; surface water; water quality. LCCN: 93093377.

Faillace, C. 1984. Development of Resources in Gedo Region by Appropriate Technology. Somalia: Publisher unknown, Mogadishu, Somalia. Descriptors: Africa; development; East Africa; Gedo Somali Republic; Somali Republic; technology; water resources. Database: GeoRef. Accession Number: 1998-016824.

Faillace, C. 1983. "A Brief Review of the Surface and Groundwater Resources of the North-West Region of Somalia." Quaderni Di Geologia Della Somalia (Mogadiscio).

Volume 7, Descriptors: Africa; East Africa; ground water; northwestern Somali Republic; Somali Republic; surface water; water resources. LCCN: 93093377.

Faillace, C. 1983. "Appropriate Technology for the Development of Water Resources in Somalia." Quaderni Di Geologia Della Somalia (Mogadiscio). Volume 9, Descriptors: Africa; development; East Africa; Somali Republic; technology; water resources. LCCN: 93093377.

Faillace, C. 1983. "Possibilities of Developing Water Resources for Irrigated Agriculture in Erigavo Area." Quaderni Di Geologia Della Somalia (Mogadiscio). Volume 7, Descriptors: Africa; agriculture; development; East Africa; Erigavo Somali Republic; irrigation; Sanaag Somali Republic; Somali Republic; water resources. LCCN: 93093377.

Faillace, C. 1983. Groundwater Condition of Beletweyn Area and Potential Water Resources for the Township Water Supply. Somalia: WDA, Mogadishu, Somalia. Descriptors: Africa; Belet Weyne Somali Republic; East Africa; ground water; Hiraan Somali Republic; Somali Republic; water resources; water supply. Database: GeoRef. Accession Number: 1998-016821.

Faillace, C. 1973. Review of Surface and Groundwater Resources of Northern Region of Somalia. Somalia: WDA, Mogadishu, Somalia. Descriptors: Africa; East Africa; ground water; northern Somali Republic; Somali Republic; surface water; water resources. Database: GeoRef. Accession Number: 1998-016818.

Faillace, C. 1964. "Le Risorse Idriche Sotterranee Dei Comprensori Agricoli Di Afgoi e Di Genale. » Translated title : « Ground Water Resources of the Agricultural Districts of Afgoi and Genale." Riv. Agric. Subtrop. Trop. Volume 58, no.10-12, Descriptors: Afgoi Somali Republic; Africa; agriculture; Benadir Somali Republic; East Africa; Genale Somali Republic; ground water; Shabeelaha Hoose; Somali Republic; water resources. ISSN: 0035-6026.

Faillace, C. 1964. "Surface and Underground Water Resources of the Shebeli Valley." Somalia: Descriptors: Africa; East Africa; ground water; Shebelle River; Somali Republic; southern Somali Republic; surface water; water resources. Database: GeoRef. Accession Number: 1998-016815.

Faillace, C. 1962. "Linee Programmatiche Per La Valorizzazione Delle Risorse Idriche in Somalia (Esclusa La Regione Nord)." Translated title: "Utilization of Water Resources in Somalia, Exclusive of the Northern Region." Somalia: Min. Lav. Pubbl., Mogadishu, Somalia. Descriptors: Africa; central Somali Republic; East Africa; Somali Republic; southern Somali Republic; water resources; water use. Database: GeoRef. Accession Number: 1998-018232.

Faillace, C. and Faillace, E. R. 1987. Water Quality Data Book of Somalia. Somalia: WDA, Mogadishu, Somalia. Descriptors: Africa; data; East Africa; geochemistry; hydrochemistry; Somali Republic; water quality; water resources. Notes: GTZ Project no. 80.2193.3-09.112. Database: GeoRef. Accession Number: 1998-016826.

Faillace, C. and Granata, P. 1993. A Contribution to the Stratigraphic Knowledge of Central Somalia with Special Reference to the Area of Gaalkacyo. Firenze; Istituto Agronomico L'Oltremar; 1993. Relazioni e Monografie Agrarie Subtropicali e Tropicali, Nuova Serie113 A, pages: 197-210. International Meeting; 1st- 1987 Nov: Mogadishu. Notes: Geology of Somalia and surrounding regions: Geology and mineral resources of Somalia and surrounding regions. OCLC Accession Number: 34175029.

Faillace, Costantino. 1984. Development of Water Resources in Gedo Region by Appropriate Technology. Mogadishu: Somali Democratic Republic, Ministry of Mineral & Water Res., Water Development Agency. Descriptors: 125 leaves; Water resources development- Somalia- Gedo; Water-supply- Somalia- Gedo; Appropriate technology- Somalia- Gedo. Notes: "March 1984." Includes bibliographical references. LCCN: 90-981812.

Faillace, Costantino; Faillace, E. R.; Somalia and Wakaaladda Horuumarinta Biyaha. 1987. Water Quality Data Book of Somalia: General Report. Eschborn, Federal Republic of Germany: Deutsche Gesselschaft für technische Zusammanarbeit; Somalia; Rossdorf, Federal Republic of Germany: Water Development Agency; Distribution, TZ-Verlagsgesellschaft. Descriptors: 141; Water quality- Somalia; Water-supply- Somalia; Water resources development- Somalia. Notes: At head of title: Water Development Agency Somalia. "GTZ project no. 80.2193.3-09.112." "May 1987." "1981-1990." Library of Congress; LCCN: 90-981361.

Faillace, Costantino; Faillace, E. R. Somalia and Wakaaladda Horuumarinta Biyaha. 1987. Water Quality Data Book of Somalia: General Report: 1981-1990. Eschborn: Deutsche Gesellschaft für Technische Zusammenarbeit (GTZ). Descriptors: 141; Water quality- Somalia; Water-supply- Somalia; Water resources development- Somalia. Notes: At head of title: Water Development Agency (WDA) Somalia. "GTZ project no. 80.2193.3-09.112." "May 1987." OCLC Accession Number: 20742567.

Faillace, Costantino; Somalia and Wakaaladda Horuumarinta Biyaha. 1983. Possibilities to Develop Water Resources for Irrigated Agriculture in Erigavo Zone: (Technical Report and Project Proposal). Mogadishu: Somali Democratic Republic, Ministry of Mineral & Water Resources, Water Development Agency. Descriptors: 18 leaves; Irrigation farming- Somalia- Erigavo Region; Agricultural resources- Somalia- Erigavo Region; Water resources development- Somalia- Erigavo Region. Notes: Responsibility: by Costantino Faillace. Library of Congress; LCCN: 84-981154.

Faillace, E. R.; Somalia and Wakaaladda Horuumarinta Biyaha. 1987. Water Quality Data Book of Somalia: Computer User's Manual and Technical Description. Eschborn, Federal Republic of Germany: Deutsche Gesselschaft für Technische Zusammanarbeit; Somalia; Rossdorf, Federal Republic of Germany: Water Development Agency; Distribution, TZ-Verlagsgesellschaft. Descriptors: 74 pages; Water quality- Somalia- Data processing; Water-supply- Somalia- Data processing. Notes: At head of title: Water Development Agency (WDA) Somalia. "GTZ project no. 80.2193.3-09.112." "May 1987." "1981-1990." Library of Congress; LCCN: 93-227341; OCLC Accession Number: 20742582.

Falciai, M. and Calamini, G. 1981. "Analisi Frequenziale Dei Deficit Agricoli in Alcune Stazioni Somale." Translated title: "Frequent Analysis of the Agricultural Deficit at some Somali Areas." Riv. Agric. Subtrop. Trop. Volume 70, no. 4, Descriptors: Africa; agriculture; East Africa; hydrology; irrigation; Somali Republic; water supply. ISSN: 0035-6026.

Fano, R. 1911. "Del Regime Delle Acque Nella Somalia Italiana." Translated title: "Water Regimes in Italian Somalia." Atti Dei Convegni Lincei, Accademia Nazionale Dei Lincei. Volume 2, Descriptors: Africa; East Africa; hydrology; Italian Somaliland; Somali Republic; southern Somali Republic; water regimes; water resources. ISSN: 0391-805X.

Fantozzi, P.L. 1996. "Transition from Continental to Oceanic Rifting in the Gulf of Aden: Structural Evidence from Field Mapping in Somalia." Tectonophysics. Volume 259, Issue 4, Pages 285-311. Descriptors: Plate tectonics; rifting; continental rifting; oceanic rifting; tectonic evolution. Geographic: Gulf of Aden Somalia. Abstract: The analysis of the spatial/temporal relationships between the tectonic structures of the continental margins bordering the Gulf of Aden and the adjacent oceanic structures provides a key for understanding the transition from continental to oceanic rifting. On the basis of structural and stratigraphical data from northeastern Somalia and southeastern Yemen it is possible to reconstruct the pre-Oligocene tectonic setting for the Gulf of Aden and its later rifting evolution. This study brings new evidence to demonstrate that: 1) During the Early-Middle Oligocene the Afro-Arabian Plate underwent a phase of intense faulting which led to the formation of small syn-rift basins elongated in a WNW-ESE direction, separated by transfer zones and structural highs. 2) Progressive crustal extension led to formation of spreading centers of oceanic crust in correspondence with the syn-rift basins. 3) The progressive separation of the Somalian and Yemenite continental margins led to the formation of a complete oceanic basin. ISSN: 0040-1951.

Fantozzi P.L and Ali Kassim M. 2002. "Geological Mapping in Northeastern Somalia (Midjiurtinia Region): Field Evidence of the Structural and Paleogeographic Evolution of the Northern Margin of the Somalian Plate." Journal of African Earth Sciences. Volume 34, Issue 1-2, Pages 21-55. Descriptors: Regional Geology; plate boundary; continental margin; stratigraphy; geological mapping. Notes: Additional Info: United Kingdom; References: Number: 95; Geographic: Somalia- Midjiurtinia. Abstract: A detailed geological investigation of the continental margins of northeastern Somalia, in land areas contiguous to major oceanic structures (e.g. the Alula-Fartaq fracture zone), was carried out from 1988 to 1991, and the results are given in the enclosed 1:200,000 geological map, derived from photo-interpretation and fieldwork at the scale 1:50,000. With respect to the rift of the Gulf of Aden, the stratigraphic sequences illustrated in the map have been distinguished as pre- and syn- and post-rift sediments. The pre-rift sediments rest on the low-grade pre-Palaeozoic phyllites and pan-African granitic intrusions and consist of sedimentary cover of continental, lagoon and marine facies ranging in age from Dogger to Eocene. Cretaceous-Eocene sediments pass eastward into clastic and marly deposits of the pelagic domain of the Indian Ocean. The syn- and post-rift sequences ranging in age from Oligocene to Miocene crop out only in narrow "en-echelon" basins striking WNW-ESE along the coast of the Gulf of Aden. The deposits belonging to the syn- and post-rift sequences are discordant and transgressive over the Meso-Cenozoic substratum and are composed by organogenic neritic marine, lagoonal and continental deposits. The tectonic setting of the area is characterized by half-graben bordered by faults with displacements of several kilometers striking WNW-ESE. Faults bordering contiguous half-graben systems often dip in opposite directions, so that in the transition area between different ones a "transfer" or "accommodation" zone develops; the zones are on the landward projection of the oceanic fracture zones, whilst the syn-rift basins are on the landward projection of the oceanic ridge. Based on our fieldwork the following reconstruction is put forward: 1. During the Early Oligocene the Afro-Arabian plate underwent a phase of intense faulting which led to the formation of small syn-rift basins elongated in the WNW-ESE direction, separated by transfer zones and structural highs. 2. Progressive crustal extension led to the formation of spreading centers of oceanic crust corresponding to earlier syn-rift basins. The fracture zones linking spreading centers formed in alignment with continental transfer zones. ISSN: 0899-5362.

FAO Representative's Office in Somalia. 1990. Summary- FAO Executed Projects, Somalia. Mogadishu: FAO Representative's Office. Descriptors: 8 leaves; Agricultural development projects- Somalia. Library of Congress; LCCN: 91-981433.

FAO Representative's Office in Somalia. 1987. Summary of FAO Executed Projects in Somalia. Mogadishu: FAO Representative's Office in Somalia. Descriptors: 17 leaves; Agricultural development projects- Somalia. Notes: "November 1987." Library of Congress; LCCN: 89-980072.

Farah, Ahmed Yusuf and WSP Transition Programme. 2000. "Opportunities for the Improvement of Essential Services: Primary Education, Health and Water." Nairobi: WSP Somali Programme. Descriptors: 64 pages; Human services- Somalia; Medical care- Somalia; Education- Somalia; Water-supply, Rural- Somalia; Animal health-Somalia. Notes: At head of title: WSP Transition Programme, Dib-u-dhiska Beelaha Degaalku Burburiyey, Somali Programme. At head of title on cover: WSP Somali Programme in Puntland. "This paper is a draft only. The final version will be published as part of a single volume in late 2000." "Draft, 2000" taken from the cover. Includes bibliographical references (p. 64). Responsibility: principal author: Ahmed Yusuf Farah; researchers: Adam J. Bihi, Abdulghaffar H. Mohamud Abdulle. OCLC Accession Number: 51960088.

Farax, Ibrahim Maxamed and Sommavilla, E. 1987. Morphogenetic Data on the Middle Shabeli Valley (Somalia). Somalia: Somali Natl. Univ., Mogadishu, Somalia. GeoSom 87; Geology of Somalia and Surrounding Regions, Mogadishu. Somalia Conference: Nov. 23-30, 1987. Descriptors: Africa; deposition; East Africa; erosion; fluvial features; geomorphology; ground water; high-energy environment; hydrogeology; landform description; sedimentation; Shebelle Valley; Somali Republic; water management; water resources. OCLC Accession Number: 34175029.

Farax, Ibrahim Maxamed; Sommavilla, E.; Maxamed, Cali Qassim; Marchesi, S.; Dhorre, Cabdulqadir Salad and Mase, G. 1987. Hydrogeology in Lower Shabeelle Valley through Geological, Geomorphological and Geophysical Models. Somalia: Somali Natl. Univ., Mogadishu, Somalia. GeoSom 87; Geology of Somalia and Surrounding Regions, Mogadishu. Somalia Conference: Nov. 23-30, 1987. Descriptors: Africa; aquifers; boreholes; Bur; drainage; East Africa; economic geology; electrical methods; environmental geology; fluvial features; geophysical methods; geophysical surveys; ground water; hydrogeology; land use; landform description; Landsat; models; paleohydrology; remote sensing; resistivity; Shebelle Valley; soils; Somali Republic; surveys; water resources. OCLC Accession Number: 34175029.

Farina, L., Vitale, G., Barletti, L. and Bobba, S. 1995. Rehabilitation of Existing Water Points in Somalia (Nugal Region). Food and Agriculture Development Centre, German Foundation for International Development; 1996. Livestock Production and Diseases in the Tropics. 8th, volume 2, pages: 478. International Conference; 8th. 1995 Sep: Berlin. Notes: Association of institutions of tropical veterinary medicine; livestock production and diseases in the tropics livestock production and human welfare; sponsor: Association of institutions of tropical veterinary medicine. ISBN: 3931227030.

Farquharson, Robert Alexander. 1924. "First Report on the Geology and Mineral Resources of British Somaliland (with map)." Alternate title: "Geology and Mineral

Resources of British Somaliland." London: Waterlow and Sons, Ltd. 53 pages, maps. Descriptors: geology; mines and mineral resources. Notes: Map on folded leaf in pocket. No more published. Includes bibliographical references. Responsibility: by R.A. Farquharson, Goverment Geologist. OCLC Accession Number: 27880778; USGS Library.

Farzin, Y.H. 1991. "Food Aid: Positive or Negative Economic Effects in Somalia?" Journal of Developing Areas. Volume 25, Issue 2, Pages 261-282. Descriptors: National Trade and Consumption; International Development Abstracts; foreign aid; international relations; developing country; government policy; policy formulation; national economy; food import; empirical evidence; import dependence; food import dependence; food aid; food strategy. Abstract: Somalia presents a particularly interesting case because despite having a predominantly agricultural/pastoral economy and being virtually self-reliant in food grain until the early 1970s, it became increasingly and to a striking degree dependent on food imports from the mid-1970s to the mid-1980s. The present paper investigates the possible causes of this development and provides empirical evidence to show that ill-formulated food aid programs in concert with unsound domestic economic policies were central to the emergence of Somalia's marked dependence on food imports during this period. The first section of the paper examines the magnitude, composition, and trend of food import dependence in Somalia, while the following section concentrates on the role of food aid in it, and the third section looks at the effect of the government's unwise policies. The concluding section highlights the key elements to be considered in formulating a food strategy in Somalia and suggests measures for effective use of food aid within that strategy. ISSN: 0022-037X.

Farzin, Y. H. 1988. "Food Import Dependence in Somalia: Magnitude, Causes, and Policy Options. Discussion Paper." International Bank for Reconstruction and Development, Washington, DC. Volume: WORLDBANKDP23, pages: 44. Descriptors: Developing countries; Foreign aid; Agricultural economics; Grains Food; Dependence; Food supply; International trade; Somalia; Foreign trade; Developing country application; Imports; Business and economics International commerce marketing and economics; Agriculture and food Agricultural economics. Abstract: The paper has three main objectives (a) to examine in detail the trends in food import, its composition and the degree of food import dependency in Somalia, (b) to analyze various micro and macro economic channels through which both foreign food aid as well as imprudent domestic economic policies have resulted in Somalia's excessive reliance on imported food, and (c) to highlight main policy options that both Somali decision makers and her foreign aid donors should consider in order to arrest Somalia's accelerating trend in food import dependency. LCCN: 87-34672; NTIS Accession Number: PB88164637XSP; ISBN: 0821310240.

Favilla, G. 1960. "Captazione e Sfruttamento Di Acque Superficiali Ad Uso Agricolo e Pastorale in Somalia." Translated title: "Utilization and Exploitation of Surface Water in Agriculture and Pastures in Somalia.' Riv. Agric. Subtrop. Trop. Volume 54, no.4-6, pages 7-9, Descriptors: Africa; agriculture; East Africa; exploitation; irrigation; Somali Republic; surface water; utilization; water resources; water use. ISSN: 0035-6026.

Favilla, G. 1955. "Studio Per l'Adduzione Delle Acque Del Giuba Al Descek Uamo." Translated title: "Study about the Adduction of Water of the Juba to the Descek

Uamo." Somalia: AFIS, Mogadishu, Somalia. Descriptors: Africa; construction; Descek Uamo; East Africa; Juba River; Somali Republic; surface water; water resources. Database: GeoRef. Accession Number: 1998-016863.

Federici, G. 1983. "La Ricerca Finalizzata Sull'Idrologia Dei Principali Corsi d'Acqua Della Somalia." Translated title: "Final Research on the Hydrology of the Principal Water Course of Somalia." Cilmi Iyo Farsamo; Wargeyska Jaamacadda Ummadda Soomaaliyeed. Volume 10, Descriptors: Africa; East Africa; hydrographs; hydrology; Somali Republic; surface water. Database: GeoRef.

Federici, G. 1980. "Le Risorse Idriche Della Somalia; Guida Bibliografica." Translated title: "Water Resources of Somalia; Bibliography Guide." Somalia: Universita Nazionale Somala, Dipart. Idraulica, Mogadishu, Somalia. Descriptors: Africa; bibliography; East Africa; Somali Republic; water resources. Database: GeoRef. Accession Number: 1998-016868.

Federici, Giorgio and Vallario, Antonio. 1989. « Guida Bibliografica Della Somalia; Scienze Geologiche, Idraulica, Risorse Idriche. » Translated title : « Bibliographic Guide to Somalia; Geological Science, Hydraulics and Water Resources." Somalia: Universita Nazionale Somala, Divisione di Geologia, Facolta di Ingegneria, Mogadishu, Somalia. Descriptors: Africa; areal geology; bibliography; East Africa; hydrogeology; Somali Republic; water resources. Notes: Extracted from Quaderni di geologia della Somalia, Vol. 10, 1988. Bibl. Nazionale Centrale di Firenze. Library of Congress.

Fernandes, R. M. S., Ambrosius, B. A. C. and Noomen, R., et al. 2004. "Angular Velocities of Nubia and Somalia from Continuous GPS Data: Implications on Present-Day Relative Kinematics." Earth & Planetary Science Letters. May. Volume 222, Issue 1, Pages 197-208. Descriptors: Plates (tectonics); Africa; Time series analysis; Earth rotation; Angular velocity; Global Positioning System; Mathematical models; Geological surveys; Lakes; Somalia. Abstract: This study focuses on the break-up of the African tectonic plate into separate Nubian and Somalian blocks, based on recent Global Positioning System (GPS) data. A new, unique velocity field has been obtained by processing all available observations of permanent GPS stations on Africa since 1996. The quantity and distribution of the stations and the length of the time-series of observations exceed that of previous studies by a considerable margin, allowing one to derive a reliable estimate of the differential motion between the Nubia and Somalian plates, which are considered as a single (African) block in the prevailing global tectonic plate models. The estimated relative pole of rotation of Somalia with respect to Nubia is located at 54.8 S; 37.0 E with magnitude- 0.069 /Ma, implying distinct opening in the Ethiopian Rift of magnitude ~7 mm/year and azimuth ~N94 E, whereas in southeastern South Africa this value is reduced to ~2 mm/year in almost the same direction. This is in accordance with some of the independent geological and geophysical tectonic models of the Nubia-Somalia plate boundary region. However, the spatial density of the current tracking network is still not optimal to establish the exact location of the entire Somalia-Nubia plate boundary; in particular, the possible branch east of Lake Victoria and heading towards the Mozambique Channel is impossible to confirm or reject at this moment. ISSN: 0012-821X.

Fernandes, R. M. S., Ambrosius, B. A. C. and Noomen, R., et al. 2003. "The Relative Motion between Africa and Eurasia as Derived from ITRF2000 and GPS Data."

Geophysical Research Letters. August 16, 2003. Volume 30, Issue 16, Pages 1828. Abstract: Studies of intra- and inter-plate deformation typically need a model describing the motions of the stable part of the tectonic plates for reference purposes. We have developed DEOS2k, a model for the current motion of seven major tectonic plates derived from space-geodetic observations. This paper focuses on relative motion between Africa and Eurasia. In the past, this motion has been poorly established because of poor data coverage for Africa. DEOS2k is based on ITRF2000 and new African GPS observations. It is an improvement over the NUVEL-1A model for predicting the present-day relative motions of these two plates. DEOS2k predicts in northeastern Africa that Africa-Eurasia relative motion is about 40% smaller in magnitude than NUVEL-1A and trends more to the northwest. This is consistent with independent local geodetic observations. A similar shift in orientation, clockwise, is observed at the western tip of the plate boundary. ISSN: 0197-7482.

Ferrandi, Ugo. 1896. « La Seconda Spedizione Bottego Nella Somalia Australe. » Translated title: "The Second Bottego Expedition in Southern Somalia." Italy: Societ Geografica Italian, Rome, Italy. Descriptors: Africa; areal geology; East Africa; expeditions; Somali Republic; southern Somali Republic. Database: GeoRef. Accession Number: 1998-016871.

Ferrandi, Ugo. 1891. "Notizie Sulla Spedizione in Somalia." Translated title: "News on the Expedition in Somalia." Italy: B.S. Esplorazione Commerciale in Africa, Rome, Italy. Descriptors: Africa; areal geology; East Africa; expeditions; Somali Republic. Database: GeoRef. Accession Number: 1998-016870.

Ferrari, M.C, Sgavetti, M and Chiari, R. 1996. "Multi-Spectral Facies in Prevalent Carbonate Strata of an Area of Migiurtinia (Northern Somalia): Analysis and Interpretation." International Journal of Remote Sensing. Volume 17, Issue 1, Pages 111-130. Descriptors: carbonate; multispectral imagery; Thematic Mapper; weathering product; reflectance spectra; multispectral image facies; weathering; rock expression. Geographic: Somalia- Migiurtinia. Abstract: Rock expression in a multi-spectral image is referred to as multi-spectral image facies. Thematic Mapper (TM) multi-spectral facies of sedimentary rock bodies were interpreted by: (1) comparison with laboratory reflectance spectra of weathered surfaces of rock samples from the study area; (2) considering the correlation between spectra of weathered and unweathered surfaces; and (3) petrographic analysis. The rocks studied comprise mainly carbonate and evaporitic strata. TM facies appeared to be related to rock composition, presence and nature of weathering processes, scale of the lithological variability, stratal pattern in the erosional profile and outcrop morphology. Specific weathering products are, in their turn apparently related to properties of the host rocks, such as composition and texture. ISSN: 0143-1161.

Ferrari, M. C. 1992. "Improved Decorrelation Stretching of TM Data for Geological Applications: First Results in Northern Somalia." International Journal of Remote Sensing. Volume 13, Issue 5, Pages 841-851. Descriptors: Thematic Mapper; image processing; Somalia; geology. Abstract: An effort was made to produce the best Thematic Mapper color image for detailed geological visual interpretation. Northern Somalia sedimentary rocks, ranging in age from Jurassic to Quaternary, were chosen as test-object. Bands 7 and 1 were selected as giving the best color composite; decorrelation stretching with spatial filtering of principal components was chosen as the most suitable enhancement technique. Improved image characteristics facilitate recognition and mapping of lithologic units which can be used to frame ongoing stratigraphical studies for the area under investigation. ISSN: 0143-1161.

Ferrari, M. C., Sgavetti, M. and Chiari, R. 1996. "Multi-Spectral Facies in Prevalent Carbonate Strata of an Area of Migiurtinia (Northern Somalia): Analysis and Interpretation." International Journal of Remote Sensing. Volume 17, Issue 1, Pages 111. ISSN: 0143-1161.

Ferraris, G. 1993. "Arfvedsonite from a Peralkaline Rhyolite (Mojo, Ethiopia)." Firenze; Istituto Agronomico L'Oltremar; 1993. Relazioni E Monografie Agrarie Subtropicali E Tropicali, Nuova Serie113 A, pages: 339-344. International Meeting; 1st-1987 Nov: Mogadishu. Name: Geology of Somalia and surrounding regions: Geology and mineral resources of Somalia and surrounding regions. OCLC Accession Number: 34175029.

Fici, S. 1991. "Floristic Relations between Eastern Africa and the Mediterranean Region with Special References to Northern Somalia." Flora Mediterranea. Volume 1, Pages 175-185. Descriptors: Palaeoecology and palaeobiogeography; General Microbial Ecology; palaeobotany; vegetation; migration; palaeobiogeography; palaeoecology; phytogeography; palaeogeography Species Terms: Buxus hildebrandtii; Juniperus procera. Abstract: The paleogeography, geology and climate of the Horn of Africa are presented in outline, and its floristic affinities with the Mediterranean region are discussed. These relations are particularly strong in northern Somalia, where 40 percent of the genera are in common with the Mediterranean region. The Buxus hildebrandtii evergreen shrub lands and the Juniperus procera forests of the Somali highland are characterized by the presence of a noteworthy element of holarctic, particularly Mediterranean, affinity. Migration in opposite directions must have occurred since late Tertiary times. ISSN: 1120-4052.

Finch, J. W. and Somalia, National Water Centre, Mogadishu, Somalia. 1987. "Data Processing and Management using [a] Personal Computer." National Water Centre Consultancy Report. Somalia: Descriptors: Africa; computers; data bases; data processing; East Africa; microcomputers; National Water Centre; Somali Republic; water resources. Notes: Pieces: 3. Database: GeoRef. Accession Number: 1998-016873.

Finizio A. 1990. "Somalia: Aspetti Demografici e Socio-Economici." Translated title: "Somalia: Demographic aspects and associated economics." Universo. Volume 70, Issue 6, Pages 734-759. Descriptors: demographic aspect; developing country; socio-economic aspect. Notes: Somalia. Abstract: Gives a general description of Somalia: climate, population, birth and death rates, medical provision, education, the economy (cattle rearing, crops, fisheries), and commerce. Two maps of Mogadishu are provided; Italy is the main trading partner. ISSN: 0042-0409.

Finn, Daniel P. and Byrne, John V. 1982. "Soil Loss in Developing Countries and its Relationship to Marine Resources; Examples from East Africa." Oceans 82 Conference Record. Oceans (New York). September. Volume 1982, Pages 942-949. Descriptors: Africa; coastal environment; conservation; continental shelf; controls; dams; deforestation; developing countries; East Africa; environment; environmental geology; erosion; human ecology; hydrology; Indian Ocean Islands; Kenya; Madagascar; marine transport; Mozambique; oceanography; sedimentation; siltation; soil surveys; soils; Somali Republic; stream transport; Tanzania. Sketch maps. ISSN: 0197-7385.

Fiori, M. Garbarino, C. Grillo, S. M. and Solomon, T. 1993. The Primary Gold Deposit of Lega Dembi (Sidamo): Ore Mineral Association and Genetic Significance. Preliminary Report. Firenze; Istituto Agronomico L'Oltremar; 1993. Relazioni E Monografie Agrarie Subtropicali E Tropicali Nuova Serie113 B, pages: 579-594. International Meeting; 1st- 1987 Nov: Mogadishu. Name: Geology of Somalia and surrounding regions: Geology and mineral resources of Somalia and surrounding regions. OCLC Accession Number: 34175029.

Fitzgerald, E. C. 1979. The Predrift Fit of Continents Around Southern and Eastern Africa and the Timing of Breakup Related to Petroleum Exploration. London: Phillips Petroleum Co. Europe-Africa. Descriptors: 24 leaves (+ attachments); Geology, Structural; Continental drift; Plate tectonics; Geology- Africa, Southern; Geology- Africa, East; Geology- Kenya; Geology- Mozambique Channel; Geology- Madagascar; Geology- Somalia; Geology- Tanzania; Geology- South Africa; Petroleum- Geology- Africa, Southern; Petroleum- Geology- Africa, East; Petroleum- Geology- Africa, Eastern; Petroleum- Geology- Kenya; Petroleum- Geology- Mozambique Channel; Petroleum- Geology- Madagascar; Petroleum- Geology- Somalia; Petroleum- Geology- Tanzania; Petroleum- Geology- South Africa. OCLC Accession Number: 14769599.

Flater, M. E. (Rhett). 1993. "Dispatch from Somalia." Vertiflite. Volume 39, Issue 1, Pages 6. ISSN: 0042-4455.

Flater, M. E. (Rhett). 1993. "Helicopter Aviation Supports Somalia Relief Mission." Vertiflite. Volume 39, Issue 2, Pages 38. ISSN: 0042-4455.

Flater, M. E. (Rhett). 1993. "Commentary. Somalia and Beyond. the U.S. Defines a New Post-Cold War Role." Vertiflite. Volume 39, Issue 1, Pages 4. ISSN: 0042-4455.

Flatz, C. 2001. "ICT for Africa-Ein Neues Entwicklungspolitisches Paradigma." Translated title: "A New Paradigm of Development Policy." Asien Afrika Lateinamerika. Volume 29, Issue 6, Pages 585-608. Descriptors: Transport and Communicatons; Internet; telecommunications; information technology; technological development; communication. References: 29; Geographic: Senegal, Botswana, Somalia. Abstract: Reviewing the development of the African information infrastructure the article tries to show the shortcomings of the information and communication structure in Africa and the cleavage between the bright promises of "cyber-development"; and the reality of information society in Africa. The paper explains why information and communication technologies won't be able to solve the African developmental problems and why telelearning, telemedicine and cyber-economy will just be another "chimera"; in the history of development in Africa. In the first part the article reviews the state of information technology in Africa and analyzes the communication infrastructure on the African continent. Case studies of Somalia, Botswana and Senegal show the inner-African differences in technological development. In the second part the paper tries to value the Okinawa Charter on Global Information Society in the context of these insights. ISSN: 0323-3790.

Fobair, I. W. 1986. "End of Tour Report." Descriptors: 1 v. (various pagings); Soil conservation- Somalia; Water conservation- Somalia. Notes: Named Corp: Mashruuca Daaqa Gobollada Dhexe (Somalia); Notes: Caption title. On cover: CRDP soil and water conservation, technical report. "July 1986." Other Titles: CRDP Soil and water conservation, technical report. C.R.D.P. soil and water conservation, technical report. OCLC Accession Number: 28827056.

Fobair, I. W. and Mashruuca Daaqa Gobollada Dhexe (Somalia). 1986. End of Tour Report. Mogadishu: Central Rangelands Development Project, Somali Democratic Republic. Descriptors: 100, 5; Soil conservation- Somalia; Water conservation- Somalia. Notes: "July 1986." "No. 649-0108." Library of Congress; LCCN: 90-980206.

Food and Agriculture Organization of the United Nations. 1900. Project for the Water Control and Management of the Shebelli River, Somalia. United Nations Development Program (Special Fund). Descriptors: Water resources development-Shebeli River Watershed (Ethiopia and Somalia). Abstract: v. 2A. The Balad Flood Irrigation Project, feasibility study, Technical annex. Notes: "November 1969." Includes bibliographical references. Responsibility: Food and Agriculture Organisation of the United Nations. Library of Congress; LCCN: 85-208339.

Food and Agriculture Organization of the United Nations; United Nations Development Programme and Somalia. 1967-1968. Agricultural and Water Surveys: Somalia; Final Report. Rome: United Nations Development Program. Descriptors: 6 volumes, illustrations, 29 cm. Agriculture- Somalia; Water resources development-Somalia; Physical geography- Somalia. Abstract: v. 1. General.--v. 2. Water resources.--v. 3. Landforms and soils.--v. 4. Livestock and crop production.--v. 5. Engineering aspects of development.--v. 6. Social and economic aspects of development. Notes: "FAO/SF:36/SOM." "Report prepared for the Government of Somalia by the Food and Agriculture Organization of the United Nations acting as executing agency for the United Nations Development Program." Vols. 2-6 prepared for the FAO by Lockwood Survey Corporation, Ltd. Includes bibliographical references. Library of Congress; LCCN: 73-156978.

Food and Agriculture Organization. 2001. "Special Operations Service -Somalia." 2001-03-02. Descriptors: Business; Commerce; Economy; AMED; Somalia; Somaliland. OCLC Accession Number: 49884192.

"Food Relief by Air, by Sea to Somalia." 1993. Modern Material Handling. May. Volume 48, Pages 12-13. Descriptors: Food relief; Somalia- History, 1991- Civil War; Relief work; Applied Science & Technology. Abstract: The efforts undertaken by the 7th Transportation Group in the recent Operation Restore Hope in Somalia are described. As part of the objective to deliver food aid, the unit was responsible for setting up port and airfield operations. To smooth the flow of equipment and supplies, analysts created schedules using the Airlift Deployment Analysis System. Two computer systems and automatic data collection technology helped keep the supply chain intact. ISSN: 0026-8038.

Forman, H. D. 1963. A Photogeological Interpretation of the Bur Region, Somali Republic (Northern Region). Somalia: Publisher unknown, Mogadishu, Somalia. Descriptors: Africa; areal geology; Bari Somali Republic; Bur Somali Republic; East Africa; geologic maps; maps; northern Somali Republic; photogeology; remote sensing; Somali Republic. Database: GeoRef. Accession Number: 1998-016875.

Fornari, G. 1929. "L'Uebi Scebeli Dalle Prime Esplorazioni a Quella Del Duca Degli Abruzzi." Translated title: "First Exploration of Shebelle River by the Duke of Abruzzi." Rassegna Economica delle Colonie. Volume 23, Descriptors: Africa; East Africa; exploration; Shebelle River; Somali Republic; streams; surface water. ISSN/ISBN: 0370-4319.

Fornari, G. 1906. "L'Uebi Scebeli e La Sua Regione." Translated title: "The Uebi Shebelle River Region." Bollettino della Societa Africana d'Italia. Volume 1906, Descriptors: Africa; East Africa; hydrographs; hydrology; Shebelle River; Somali Republic; surface water. ISSN: 0392-1468.

Forti, P., Francavilla, F., Gunay, G. et al. 1993. "The Hydrogeology of some Coastal Paleodunes in an Equatorial Area and their Karst Related Morphologies: The Case of Gesira (Somalia)." IAHS; Publication: 207. Descriptors: Hydrochemistry; palaeodune; karst; carbonate cementation; hydrochemistry; water mixing. Abstract: Close to Gesira village paleodunes nearly parallel to the seashore consist of cemented calcareous sand. The sea has partially eroded the paleodunes creating an abrasion platform and a cliff, in which several marine karst microforms are still now actively forming. Several caves have developed along the cliff where major cementation took place. At low tide inside the largest of these cavities, water constantly flows outward and in the first 5-10 m of the platform, peculiar small canyon-shaped structures up to 1 m deep and with a smooth rounded bottom have developed. The genesis of the largest caves and the canyon-shaped structures is related to the presence inside the paleodunes of two different waters, which mingle together: one, prevailing during high tides, is characterized by sea water, and the second, prevalent at low tides, is a fresh meteoric or partially mixed water. Hydrochemical data on both types of water are discussed and a model is presented to explain the genesis of both the largest caves and the canyon-shaped structures. Notes: In: Hydrogeological processes in karst terranes. Proc. international symposium and field seminar, Antalya, Turkey, 1990 (1993) pages 133-138; Geographic: Somalia- Gesira. OCLC Accession Number: 0995578.

Fournier, M., Bellahsen, N., Fabbri, O. and Gunnell, Y. 2004. "Oblique Rifting and Segmentation of the NE Gulf of Aden Passive Margin." Geochemistry, Geophysics, Geosystems. November 6. Volume 5, Pages Q11005. Abstract: The Gulf of Aden is a young, obliquely opening, oceanic basin where tectonic structures can easily be followed and correlated from the passive margins to the active mid-oceanic ridge. It is an ideal laboratory for studies of continental lithosphere breakup from rifting to spreading. The northeastern margin of the Gulf of Aden offers the opportunity to study on land the deformation associated with oblique rifting over a wide area encompassing two segments of the passive margin, on either side of the Socotra fracture zone, exhibiting distinct morphologic, stratigraphic, and structural features. The western segment is characterized by an elevated rift shoulder and large grabens filled with thick synrift series, whereas the eastern segment exhibits low elevation and is devoid of major extensional structures and typical synrift deposits. Though the morphostructural features of the margin segments are different, the stress field analysis provides coherent results all along the margin. Four directions of extension have been recognized and are considered to be representative of two tensional stress fields with permutations of the horizontal principal stresses sigma(2) and sigma(3). The two dominant directions of extension, N150°E and N20° E, are perpendicular to the mean trend of the Gulf of Aden (N75° E) and parallel to its opening direction (N20° E-N30° E), respectively. Unlike another study in the western part of the gulf, our data suggest that the N150° E extension stage is older than the N20° E extension stage. These conflicting chronologies, which are nowhere unambiguously established, suggest that the two extensions coexisted during the rifting. On-land data are compared with offshore data and are interpreted with reference to oblique rifting. The passive margin segmentation represents a local accommodation of the extensional deformation in a homogeneous regional stress field, which reveals the asymmetry of the rifting process. The first-order segmentation of the Sheba Ridge is inherited from the prior segmentation of the passive margin. ISSN: 1525-2027.

France, Ministere de la Recherche et Technologie, Service Information et Communication. 1990. "Observatoire du Sahara et du Sahel. Colloque De l'Observatoire Du Sahara Et Du Sahel." Translated title: "Sahara and Sahel Observatory; Colloquium of the Sahara and Sahel Observatory." Recherche Technologie (1985). Volume 64-65, Pages 17. Descriptors: Africa; Algeria; Burkina Faso; Chad; climate; desertification; drought; East Africa; ecology; Gambia; geologic hazards; Guinea-Bissau; Kenya; Libya; Mali; Mauritania; Morocco; Niger; North Africa; observatories; Sahara; Sahel; Senegal; Somali Republic; Sudan; Tunisia; Uganda; West Africa. Notes: illustrations. ISSN: 0765-0779.

Franceschetti, B. and Abdulkadir, Salad D. 1983. "Indagine Preliminare Sulla Potenzialita' Idrica Dei Bacini Torrentizi Situati Sulla Sinistra Dello Uebi Scebeli, Tra Halgen e Il Pozzo Di Ceel Gal, e Sulle Possibilita' Di Realizzare in Essi." Translated title: "Preliminary Investigation on the Potential of the Torrential Basins Situated on the Left of Webi Shabelle, between Halgen and Ceel Gal Well, and the Implementation Possibility." Quaderni Di Geologia Della Somalia (Mogadiscio). Volume 7, Descriptors: Africa; Belet Weyne Somali Republic; East Africa; Hiran Somali Republic; hydrology; Somali Republic; streams; water resources; water wells; Webi Shebelle; wells. LCCN: 93093377.

Frizzo, P. 1993. "Ore Geology of the Crystalline Basement of Somalia." Firenze; Istituto Agronomico L'Oltremar; 1993. Relazioni E Monografie Agrarie Subtropicali E Tropicali Nuova Serie113 B, pages: 517-540. International Meeting; 1st. 1987 Nov: Mogadishu. Notes: Geology of Somalia and surrounding regions: Geology and mineral resources of Somalia and surrounding regions. OCLC Accession Number: 34175029.

Frizzo, P. 1987. "Tematiche Per Ricerche Geominerarie in Somalia." Translated title: "Themes for Geomining Research in Somalia." Quaderni Di Geologia Della Somalia (Mogadiscio). Volume 9, Pages 1. Descriptors: Africa; East Africa; mining; mining geology; research; Somali Republic. LCCN: 93093377.

Frizzo, P. 1987. "Metallogenic Sketch of Somalia." GeoSom 87; International Meeting; Geology of Somalia and Surrounding Regions; Abstracts. Somalia: Somali Natl. Univ., Mogadishu, Somalia. GeoSom 87; Geology of Somalia and Surrounding Regions, Mogadishu. Somalia Conference: Nov. 23-30, 1987. Descriptors: Africa; basement; Bur; Cenozoic; chemically precipitated rocks; crystalline rocks; East Africa; economic geology; gabbros; granites; igneous rocks; iron formations; marbles; Mesozoic; metal ores; metamorphic rocks; mineral exploration; pegmatite; placers; plutonic rocks; Precambrian; Quaternary; sedimentary rocks; Somali Republic; veins. OCLC Accession Number: 34175029.

Frizzo, P. 1987. "Subjects for Prospecting Research in Somalia." GeoSom 87; International Meeting; Geology of Somalia and Surrounding Regions; Abstracts. Somalia: Somali Natl. Univ., Mogadishu, Somalia. GeoSom 87; Geology of Somalia and Surrounding Regions, Mogadishu. Somalia Conference: Nov. 23-30, 1987. Descriptors: Africa; clay minerals; clays; East Africa; economic geology; geochemical methods; gold ores; granites; heavy mineral deposits; igneous rocks; lead-zinc deposits; metal ores;

mineral exploration; mineral resources; pegmatite; platinum ores; plutonic rocks; scheelite; sepiolite; sheet silicates; silicates; Somali Republic; tungstates; tungsten ores. OCLC Accession Number: 34175029.

Frizzo, P. and Abdulcadir, Hagi Hassan. 1983. "Scheelite in Stream Sediments in the Crystalline Basement of Northern Somalia (Area NNE of Hargheisa)." Bollettino Della Societa Geologica Italiana. Volume 102, Issue 4, Pages 385-390. Descriptors: Africa; basement; Borama Mora; clastic rocks; East Africa; economic geology; Hargheisa; igneous rocks; metabasite; metagabbro; metaigneous rocks; metal ores; metamorphic rocks; mineral exploration; northern Somali Republic; Paleozoic; plutonic rocks; scheelite; sedimentary rocks; Somali Republic; stream sediments; syenites; tungstates; tungsten ores; upper Paleozoic; volcaniclastics. Notes: References: 1 p. illustrations incl. geol. sketch map. ISSN: 0037-8763.

Fulghum, David A. 1992. "Restrictions Shape Somalia Airlift." Aviation Week & Space Technology. December 14-21. Volume 137, Pages 24. Descriptors: Airplanes in relief work; Somalia- History, 1991- (Civil War)- Relief work; United States Air Force, Air Mobility Command; Applied Science & Technology; Readers' Guide (Current Events); Business; Aircraft; Public order and safety, National security; Aircraft Manufacturing. Abstract: The U.S. Air Force's Air Mobility Command (AMC) has created an air bridge of tankers and emergency refueling sites extending from the west coast of America to the east coast of Africa, in order to overcome limited facilities and fuel shortages in Somalia. The AMC's C-5 fleet and a select force of 100 C-141s will soon begin flying into Mogadishu, Somalia, at a target rate of between 10 and 11 landings a day. After landing in Mogadishu or at another hub in Mombasa, Kenya, relief cargo will be transferred to C-130s, which are better equipped for operations in isolated regions of Somalia. ISSN: 0005-2175.

Funaioli, U. 1960. "Bonifiche Lungo l'Uebi Scebeli." Translated title: "Reclamation of Shebelle River." Riv. Agric. Subtrop. Trop. Volume 54, Pages 4-6, 7-9. Descriptors: Africa; agriculture; East Africa; hydrology; reclamation; Shebelle River; Somali Republic; surface water; water resources. ISSN: 0035-6026.

G

Gabbani, G. 1978. "L'Uso Del Computer Nella Cartografia Geologica; Le Isopache Del Giurassice e Del Cretaceo Nella Somalia Settentrionale." Translated Title: "Use of the Computer for Geological Cartography; Jurassic and Cretaceous Isopachs in Northern Somalia." Bollettino Della Societa Geologica Italiana. Volume 97, Issue 4, Pages 485-493. Descriptors: Africa; areal geology; cartography; computer languages; Cretaceous; data processing; East Africa; Fortran; isopachs; Jurassic; maps; Mesozoic; northern Somali Republic; Somali Republic. Notes: References: 7; illustrations incl. sketch maps. ISSN: 0037-8763.

Gaciri, S. J. and Mailu, G. M. 1993. "Groundwater Investigations, Development and use in the Athi Basin, Kenya." Firenze; Istituto Agronomico L'Oltremar; 1993. Relazioni E Monografie Agrarie Subtropicali E Tropicali Nuova Serie113 B, pages: 711-733. International Meeting; 1st- 1987 Nov: Mogadishu. Notes: Geology of Somalia and surrounding regions: Geology and mineral resources of Somalia and surrounding regions. OCLC Accession Number: 34175029.

Gadomski, Christopher and Zilz, Hillary Sara. 1988. "Challenging Diesel's Domination in Somalia, Africa." Alternative Sources of Energy. September. Pages 53-54. Descriptors: Electric industry- Somalia; Wind power; Applied Science & Technology. Notes: Illustration; Map. ISSN: 0146-1001.

Galaal, Muusa H. I. 1992. "Stars, Seasons and Weather in Somali Pastoral Tradition." Niamey, Niger: CELHTO. Descriptors: 90; Weather- Folklore. Notes: Somalia- Folklore. OCLC Accession Number: 29707199.

Galaal, Muusa H. I. 1970. Stars, Seasons and Weather in Somali Pastoral Traditions. S.l: Descriptors: 96 leaves; Folklore- Somalia; Stars- Folklore; Astronomy- Folklore- Somalia; Astrology, Somali; Weather- Folklore- Somalia. Notes: Cover title. Includes bibliographical references. Responsibility: by Muusa H.I. Galaal. OCLC Accession Number: 32040738.

Galaal, Muusa H. I. 1968. The Terminology and Practice of Somali Weather Lore, Astronomy, and Astrology. Mogadishu: Somalia. Descriptors: 77 pages; map. 33 cm. Weather lore- Somalia. Notes: Includes bibliographical references. Responsibility: by Muusa H.I. Galaal. Library of Congress; LCCN: 74-390772.

Gamsakhurdiya, G. R., Meshchanov, S. L. and Shapiro, G. K. 1991. "Seasonal Variations in the Distribution of Red Sea Waters in the Northwestern Indian Ocean." Oceanology of the Academy of Sciences of the USSR. Volume 31, Issue 1, Pages 32-37. Descriptors: Sea water; Seasonal variations; Oceans; Somalia; Circulation; Coastal environments; Discharge; Fluctuation; Gulfs. Abstract: Hydrologic data for 1935-1981 are used to investigate seasonal variations in the distribution of Red Sea waters in the region between 0 degree and 20 degree N and between 43 degree and 70 degree E (other than the Gulf of Aden). The area occupied by Red Sea waters increases by about 9% from winter to spring. A maximum of 1% of this increase is caused by seasonal fluctuations in the discharge of Red Sea waters through the Strait of Bab-el-Mandeb. Most of the contribution is made by the seasonal rearrangement of the ocean circulation at intermediate depths (600-800 m). In winter, large amounts of Red Sea waters move southward along the Somali coast, while in summer, eastward transport intensifies. Database: CSA Technology Research Database.

Ganz H. 1990. "Facies Evolution and Hydrocarbon Potential of Upper Cretaceous Organic-Rich Sediments from North Somalia." Berliner Geowissenschaftliche Abhandlungen, Reihe A. Volume 120, Issue 2, Pages 621-632. Descriptors: Economic Geology; organic matter; hydrocarbon potential; source rock potential; Cretaceous. Notes: Somalia. Abstract: The AGIP No.1 Darin and No.1 Sagaleh wells prove the existence of nearly 3000 m sediments in the area. Data of Rock Eval pyrolysis and IR-kerogen-analysis indicate that the oil window is already reached at 1500 m (VRE = 0.7%), thus hydrocarbons could have been expelled. However, TOC contents of most horizons are below 0.5%. BGR core samples from the Hodmo Hed-Hed area about 400 km further west, probably cover an equivalent organic-rich layer. The Upper Cretaceous organic-rich sediments of Somalia which were previously regarded as coals, are unusually rich in lipids and algal material, which classifies them as coals transitional to lacustrine oil shales. The thickness of these horizons varies over short lateral distances. Generally the complete series never exceeds 15-20 m. ISSN: 0172-8784.

Gates M.A. 1990. "Morphological Drift Accompanying Nascent Population Differentiation in the Ciliate Euplotes Vannus." J. Protozool. Volume 37, Issue 2, Pages

78-86. Descriptors: Global Ecology; protist; ciliate; morphological drift Species Term: Euplotes vannus; Ciliophora; Somalia; Protista; Euplotes. Notes: Somalia. Abstract: Protists of the genus Euplotes can undergo genetic recombination by the normal outbreeding process of conjugation following mild starvation. Occasionally, the dominant mutation for the autogamy trait arises. Individuals possessing the trait show obligate self-fertilization upon mild starvation. This yields, after normal asexual division, a population of individuals that are reproductively isolated from the parental outbreeding strain. A morphometric analysis of sympatric autogamous and non-autogamous populations of Euplotes vannus from Somalia demonstrates that there has been morphological drift in gross body proportions in the autogamous populations. ISSN: 0022-3921.

Gatti, V. 1960. "Il Basso Corso Dell'Uebi Scebeli." Translated Title: "Lower Water Course of the Shebelle River." Riv. Agric. Subtrop. Trop. Volume 54, no.4-6, 7-9, Descriptors: Africa; East Africa; hydrographs; hydrology; movement; Shebelle River; Somali Republic; streams; surface water. ISSN: 0035-6026.

Gatti, V. and Conforti, E. 1960. "Contributo Ad Una Bibliografia Italiana Su Eritrea e Somalia Fino al 1952." Translated title: "Italian Bibliography on Eritrea and Somalia at the End of 1952." Italy: Istituto Agronomy, Florence, Italy. Descriptors: Africa; areal geology; bibliography; East Africa; Eritrea; Somali Republic. Database: GeoRef. Accession Number: 1998-016890.

Gaulon R, Roult G, Chorowicz J, Vidal G and Romanowicz B. 1992. "Regional Geodynamic Implications of the May-July 1990 Earthquake Sequence in Southern Sudan." Tectonophysics. Volume 209, Issue 1-4, Pages 87-103. Descriptors: Earthquakes; seismicity; Sudan earthquake 1990; strike slip faulting; lineament; source parameters; centroid moment tensor; block rotation. Notes: Sudan; Somalia. Abstract: Several large earthquakes occurred in the south of the Sudan in May and July 1990. The focal mechanism solution of the main shock shows left-lateral strike-slip faulting. The data suggest the existence of a wide, diffuse fault zone which is an active intra-continental transform area linking the two main branches of the rift system. If one considers this displacement and the slip directions obtained from the centroid moment tensors solutions as characterizing the motion between the Nubian and Somalian plates, new constraints on the regional geodynamics can be inferred; the Somalian block appears to be moving southeastwards relative to the African block. The rupture occurred in at least three steps, affecting a zone roughly 50 km long. ISSN: 0040-1951.

Gellatly, D. C. 1963. The Geology of the Darkainle Nepheline Syenite Complex, Borama District, Somali Republic. United Kingdom: University of Leeds, Leeds, United Kingdom. Descriptors: Africa; areal geology; Borama Somalia; Darkainle Complex; East Africa; igneous rocks; nepheline syenite; plutonic rocks; Somali Republic; syenites. Database: GeoRef. Accession Number: 1996-040136.

Gellatly, D. C. 1961. "A Geological Reconnaissance of the Ras Hantara Basement Inlier." Report of the Geological Survey, Somaliland Protectorate. Volume DCG/13, Descriptors: Africa; areal geology; basement; East Africa; northern Somali Republic; Ras Hantara; Somali Republic. Database: GeoRef.

Gellatly, D. C. 1961. "Further Notes on the Geology of the Las Dureh Area." Report of the Geological Survey, Somaliland Protectorate. Volume DCG/15, Descriptors: Africa; areal geology; Burao Somali Republic; East Africa; Las Dureh Somali Republic; northern Somali Republic; Somali Republic. Database: GeoRef.

Gellatly, D. C. 1961. "Interim Report on the Darkainle Alkaline Complex, Borama District." Report of the Geological Survey, Somaliland Protectorate. Volume DCG/18, Descriptors: Africa; alkalic composition; areal geology; basement; Borama Somali Republic; crystalline rocks; Darkainle alkaline complex; East Africa; northern Somali Republic; Somali Republic; Waqooyi Galbeed. Database: GeoRef.

Gellatly, D. C. 1961. "The Geology of the Area Around Dalan, Near Elayu, Erigavo District." Report of the Geological Survey, Somaliland Protectorate. Volume DCG/7, Descriptors: Africa; areal geology; Dalan Somali Republic; East Africa; Elayu Somali Republic; Erigavo Somali Republic; Sanaag Somali Republic; Somali Republic. Database: GeoRef.

Gellatly, D. C. 1961. "The Geology of the Area Around Manja Yihan, Bosaso District." Report of the Geological Survey, Somaliland Protectorate. Volume DCG/8, Descriptors: Africa; areal geology; Bari Somali Republic; Bosaso Somali Republic; East Africa; Manja Yihan; Somali Republic. Database: GeoRef.

Gellatly, D. C. 1960. "Report on the Geology of the Las Dureh Area." Report of the Geological Survey, Somaliland Protectorate. Volume 6, Pages 117-180. Descriptors: Africa; areal geology; East Africa; Las Dureh Somali Republic; Somali Republic. Database: GeoRef.

Gellatly, D. C. 1960. "Report on the Geology of the Las Dureh Area, Burao District, Quarter Degree Sheet Nos. 25, 26, 37, 38." Somalia: Report of the Geological Survey, Somaliland Protectorate. Volume: 6, Descriptors: Africa; areal geology; Burao Somali Republic; East Africa; geologic maps; Las Dureh Somali Republic; maps; northern Somali Republic; Somali Republic; Waqooyi Galbeed. Database: GeoRef. Accession Number: 1998-016896.

Gellatly, D. C. 1960. A Short Note on the Feldspar Reserves of the Pegmatite Around Mandera, Berbera District. Somalia: Somali Republic Geol. Survey, Hargeisa, Somalia. Descriptors: Africa; Berbera Somali Republic; East Africa; feldspar deposits; geologic maps; granites; igneous rocks; Mandera Somali Republic; maps; pegmatite; plutonic rocks; reserves; Somali Republic; Waqooyi Galbeed. Database: GeoRef. Accession Number: 1998-016895.

Gellatly, D. C. 1959. "The Geology of the Hudiso-Dabder Area, Berbera District." Report of the Geological Survey, Somaliland Protectorate. Volume DCG/3, Descriptors: Africa; areal geology; Berbera Somali Republic; East Africa; Hudiso-Dabder; northern Somali Republic; Somali Republic; Waqooyi Galbeed. Database: GeoRef.

Gellatly, D. C. and Stewart, J. A. B. 1970. Burao. United Kingdom: Directorate Overseas Survey, United Kingdom. Descriptors: Africa; areal geology; Burao; East Africa; geologic maps; maps; Somali Republic. Notes: Latitude: N093000, N100000 Longitude: E0460000, E0453000. Database: GeoRef. Accession Number: 1989-007774.

Gellatly, D. C. and Stewart, J. A. B. 1970. El Dur Elan. United Kingdom: Directorate Overseas Survey, United Kingdom. Descriptors: Africa; areal geology; East Africa; El Dur Elan; geologic maps; maps; Somali Republic. Notes: Latitude: N100000, N103000 Longitude: E0463000, E0460000. Database: GeoRef. Accession Number: 1989-007767.

Gellatly, D. C. and Stewart, J. A. B. 1970. Las Dureh. United Kingdom: Directorate Overseas Survey, United Kingdom. Descriptors: Africa; areal geology; East

Africa; geologic maps; Las Dureh; maps; Somali Republic. Notes: Latitude: N100000, N103000 Longitude: E0460000, E0453000. Database: GeoRef. Accession Number: 1989-007768.

Gellatly, D. C. and Stewart, J. A. B. 1970. Habaji. United Kingdom: Directorate Overseas Survey, United Kingdom. Descriptors: Africa; areal geology; East Africa; geologic maps; Habaji; maps; Somali Republic. Notes: Latitude: N093000, N100000 Longitude: E0463000, E0460000. Database: GeoRef. Accession Number: 1989-007775.

Gemmel, B. A. P. 1982. Hydrological Data Collection and Upgrading of the National Hydrometric Network on the Juba and Shebelle Rivers; Vol. III, Juba Hydrometric Data. United Kingdom: FAO, Harrow, United Kingdom. Descriptors: Africa; East Africa; hydrology; Juba River; Shebelle River; Somali Republic; streams; surface water. Database: GeoRef. Accession Number: 1998-016911.

Gemmel, B. A. P. 1982. Hydrological Data Collection and Upgrading of the National Hydrometric Network on the Juba and Shebelle Rivers; Vol. II, Juba Hydrometric Data. United Kingdom: FAO, Harrow, United Kingdom. Descriptors: Africa; East Africa; hydrology; Juba River; Shebelle River; Somali Republic; streams; surface water. Database: GeoRef. Accession Number: 1998-016912.

Gemmel, B. A. P. 1982. Hydrological Data Collection and Upgrading of the National Hydrometric Network on the Juba and Shebelle Rivers; Vol. IV, Working Drawings and Data Sheets. United Kingdom: FAO, Harrow, United Kingdom. Descriptors: Africa; East Africa; hydrology; Juba River; Shebelle River; Somali Republic; streams; surface water. Database: GeoRef. Accession Number: 1998-016913.

Gemmel, B. A. P. 1982. Hydrological Data Collection and Upgrading of the National Hydrometric Network on the Juba and Shebelle Rivers; Vol. I, Main Report. United Kingdom: FAO, Harrow, United Kingdom. Descriptors: Africa; East Africa; hydrology; Juba River; Shebelle River; Somali Republic; streams; surface water. Database: GeoRef. Accession Number: 1998-017930.

Gensior A and Alaily F. 1990. "Physical Soil Properties of an Alluvial Fan in the Northern Part of the Darror Valley (North-East Somalia)." Berliner Geowissenschaftliche Abhandlungen, Reihe A. Volume 120, Issue 2, Pages 705-712. Descriptors: Groundwater processes: vadose zone; soil depth; clay content; stone content; groundwater recharge; soil property; alluvial fan. Notes: Somalia- Darror Valley Somalia- Cal Miscaat. Abstract: In the north of the Darror Valley and southern Cal Miscaat (northeast Somalia) the parent material of the investigated soils consists of Tertiary conglomerates, Quaternary playa sediments, fluvial gravel and aeolian sand sheets. Part in the Darror Valley is considered to be an important groundwater recharge area. This is reflected by the low electrical conductivity of the investigated soils, as well as the increase of values with depth. The water needed for groundwater recharge is mainly coming from the mountain area where precipitation is seven times higher than in the Darror Valley. According to our investigations the amount of groundwater recharge must decrease with the increasing soil depth, clay content and with decreasing stone content. ISSN: 0172-8784.

Geology and Mineral Resources of Somalia and Surrounding Regions (with a Geological Map of Somalia 1:1,500,000). Relazioni e monografie agrarie subtropicali e tropicali nuova serie; no 113, A-B. Firenze: Istituto Agronomico L'Oltremar, 1993. 2 pts. Includes geological map in back pocket of pt B. 1993. Italy: Descriptors: Africa; areal geology; East Africa; geologic maps; maps; mineral resources; Somali Republic. Database: GeoRef. Accession Number: 1996-039694; British Library.

George Philip & Son. 1940. "Abyssinia (Ethiopia) with Eritrea and British, Italian & French Somaliland." London. Notes: Description: 1 map. color 22 x 29 in. Geographic: Ethiopia- Maps. Eritrea- Maps. Somalia- Maps. OCLC Accession Number: 68915919.

George Philip & Son. 1936. "Abyssinia (Ethiopia) with Eritrea and British, Italian and French Somaliland." London: George Philip & Son; Africa; Northeastern. Notes: Description: 1 map. Geographic: Africa, Northeast- Maps. Ethiopia- Maps. Eritrea- Maps. Somalia- Maps. Scale 1:3,300,000; Mollweide projection. Category of scale: a Constant ratio linear horizontal scale: 3300000; Notes: Relief shown as hachures and heights in feet. Shows: international boundaries, railways, roads, shipping routes. Coverage: Horn of Africa. OCLC Accession Number: 48603366.

George Philip & Son and London Geographical Institute. 1930s. "Abyssinia (Ethiopia): With Eritrea and British, Italian & French Somaliland." London: George Philip & Son. Notes: Description: 1 map; color; 57 x 74 cm., on sheet 63 x 81 cm. Geographic: Ethiopia- Maps. Eritrea- Maps. Somalia- Maps. Djibouti- Maps. Africa, East- Maps. Scale 1:3,300,000. 52 mi. = 1 in. Category of scale: a Constant ratio linear horizontal scale: 3300000; Notes: Relief shown by shading and spot heights. Shows international boundaries, railways, roads, and shipping routes. In lower left margin: The London Geographical Institute. OCLC Accession Number: 55093250.

George, U. 1978. "Geburt Eines Ozeans. Birth of an Ocean." Geo. Jul. Issue 7, Pages 50-80. Descriptors: Afar Depression; Africa; continental drift; Djibouti; East Africa; East African Rift; Ethiopia; Indian Ocean; Kenya; Natron Lake; photogeology; plate tectonics; popular geology; Red Sea; rift zones; Somalian; Tanzania; tectonophysics. Notes: illustrations; Latitude: N300000, E0520000. ISSN: 0342-8311.

GeoSom 87; Geology of Somalia and Surrounding Regions. 1987. Somalia: Somali National University, Department of Geology, Mogadishu, Somalia. Descriptors: Africa; areal geology; East Africa; regional; Somali Republic; symposia; West Africa. Notes: Individual abstracts are cited separately. OCLC Accession Number: 34175029.

German Ag. Tec. Coop., Federal Republic of Germany. 1975. Draft Programme for Water Resources Development Project in Somalia. Federal Republic of Germany: German Ag. Tec. Coop., Federal Republic of Germany. Descriptors: Africa; development; East Africa; Somali Republic; water resources. Database: GeoRef. Accession Number: 1998-017935.

Gestro, R. 1904. "Collezioni Geologiche Del Ten; Citerni." Translated Title: "Geologic Collection of Ten." Bollettino Della Societa Geografica Italiana. Volume 5, Descriptors: Africa; areal geology; East Africa; Somali Republic; Ten. ISSN: 0037-8755.

Getahun, Abebe. 2000. Systematic Studies of the African Species of the Genus Garra (Pisces; Cyprinidae). United States: City College, New York, NY. Descriptors: Abbay Basin; Actinopterygii; Africa; Angola; Arabian Peninsula; Asia; Cenozoic; Central Africa; Chordata; cladistics; Congo Democratic Republic; Cyprinidae; Cypriniformes; drainage basins; East Africa; Egypt; Eritrea; Ethiopia; fresh-water environment; Garra; Garra aethiopica; Garra blanfordii; Garra dembecha; Garra duobarbis; Garra geba; Garra hindii; Garra microstoma; Garra ornata; Garra quadrimaculata; Garra tana; Garra tibanica; Garra trewavasae; Gondwana; Guinea; Labeinae; lower Paleocene; morphology; morphometry; North Africa; Osteichthyes;

Paleocene; Paleogene; Pisces; Saudi Arabia; Somali Republic; Tanzania; taxonomy; Tekezze Basin; Teleostei; Tertiary; Vertebrata; vicariance; West Africa; Yemen. Database: GeoRef. Accession Number: 2002-007095.

Getaneh, A. Pretti, S. and Valera, R. 1993. An Outline of the Metallogenic History of Ethiopia. Firenze; Istituto Agronomico L'Oltremar; 1993. Relazioni E Monografie Agrarie Subtropicali E Tropicali Nuova Serie113 B, pages: 569-578. International Meeting; 1st- 1987 Nov: Mogadishu. Notes: Geology of Somalia and surrounding regions: Geology and mineral resources of Somalia and surrounding regions. OCLC Accession Number: 34175029.

Gherardi, F. and Vannini, M. 1989. "Field Observations on Activity and Clustering in Two Intertidal Hermit Crabs, Clibanarius Virescens and Calcinus Laevimanus (Decapoda, Anomura)." Mar. Behav. Physiol. Volume 14, Issue 3, Pages 145-159. Descriptors: aggregation behavior; intertidal environment; tidal rhythm; activity patterns; organism aggregations; tidal cycles; Anomura; Decapoda; Crustacea; Clibanarius virescens; Calcinus laevimanus; Somalia; factors affecting; Marine. Abstract: Activity and clustering an two intertidal species of hermit crabs, Clibanarius virescens and Calcinus laevimanus, were studied during a semilunar tide cycle in Somalia. Three groups were marked and examined throughout 16 low waters, and 12 samples of clustered and scattered crabs were collected from similar habitats. During high water, the animals took refuge in mass in holes of the cliff; they began to reappear (especially at night) and feed during ebb-tide (feeding phase). When exposed to the air, the crabs entered a resting phase with some aggregating in clusters, then when the tide rose became active again before going to their refuges. Clusters composed of up to 95 quiescent individuals (which were even more crowded during nocturnal and spring low tides) occurred at every low tide and at approximately the same place. Database: Environmental Sciences and Pollution Mgmt.

Giannitrapani, L. 1926. "Il Giuba. Juba." L' Universo. Volume 9, Descriptors: Africa; East Africa; geography; hydrographs; hydrology; Juba River; Somali Republic; streams. ISSN: 0042-0409.

Gibb & Partners, Kenya. 1984. Mogadishu Water Supply Expansion, Second Water Resources Investigation, Draft Report; Vol. II, Appendix B. Somalia: Water Agency, Mogadishu, Somalia. Descriptors: Africa; Benadir Somali Republic; East Africa; exploration; Mogadishu Somali Republic; report; Somali Republic; water resources; water supply. Database: GeoRef. Accession Number: 1998-017945.

Gibb & Partners, Kenya. 1980. Source Investigation for Mogadishu Water Supply Expansion; Final Report. Kenya: Gibb & Partners, Nairobi, Kenya. Descriptors: Africa; Benadir Somali Republic; East Africa; exploration; Mogadishu Somali Republic; Somali Republic; water resources; water supply. Notes: 2 Volumes. Database: GeoRef. Accession Number: 1998-017942.

Gibb & Partners, Kenya. 1977. Feasibility Study for Mogadisho [sic] Water Supply Expansion; Proposal for Hydrogeological Investigations for a New Water Resources. Kenya: Gibb & Partners, Nairobi, Kenya. Descriptors: Africa; Benadir Somali Republic; East Africa; exploration; feasibility studies; Mogadishu Somali Republic; Somali Republic; water resources; water supply. Database: GeoRef. Accession Number: 1998-017940.

Gibb & Partners, Kenya. 1977. Feasibility Study for Mogadishu Water Supply Expansion; Preliminary Report. Kenya: Gibb & Partners, Nairobi, Kenya. Descriptors: Africa; Benadir Somali Republic; East Africa; feasibility studies; Mogadishu Somali Republic; report; Somali Republic; water resources; water supply. Notes: In two volumes. Database: GeoRef. Accession Number: 1998-017941.

Gibb & Partners, Kenya. 1976. Feasibility Study for Mogadisho [sic] Water Supply Expansion. Kenya: Gibb & Partners, Nairobi, Kenya. Descriptors: Africa; East Africa; feasibility studies; Somali Republic; water resources; water supply; Mogadishu Somali Republic. Database: GeoRef. Accession Number: 1998-017939.

Gigli, G. 1930. "Sull'Uebi Scebelli." Translated title: "The Shebelle River." Rivista Delle Colonie Italiane. Volume 4, Descriptors: Africa; East Africa; geography; hydrographs; hydrology; Shebelle River; Somali Republic; surface water. ISSN: 0394-4026.

Gilbert, M. G. and Thulin, M. 2005. "Caralluma Lamellosa (Apocynaceae), a Remarkable New Species from Somalia." Nord. J. Bot. Volume 23, Issue 5, Pages 523-525. Abstract: The new species Caralluma lamellosa, from a rocky limestone slope about 200 m a.s.l. in north-eastern Somalia is described and illustrated. The plant is reminiscent of C. furta in habit, but its outer corona, with undivided linear-oblanceolate lobes opposite to the anthers and at first examination looking like lamellae on the corolla tube, is unparalleled among stapeliads.

Giovanni, C. 1940. "I Bur Della Somalia." Translated title: "The Bur of Somalia." Rivista Delle Colonie. Volume 14, no.10, Descriptors: Africa; areal geology; Bur Somali Republic; East Africa; Somali Republic. ISSN: 0394-4107.

GKW Consult. Engin., Mannheim, Federal Republic of Germany. 1984. Water Supply for Towns in Southern Somalia (Water Supply III); Xuddur Vol. VI, Hydrogeological Report. Federal Republic of Germany: GKW, Mannheim, Federal Republic of Germany. Descriptors: Africa; Bakool Somali Republic; East Africa; report; Somali Republic; southern Somali Republic; water resources; water supply; Xuddur Somali Republic. Database: GeoRef. Accession Number: 1998-017948.

GKW Consult. Engin., Mannheim, Federal Republic of Germany. 1984. Water Supply for Towns in Southern Somalia (Water Supply III); Baraawe, Vol. II, Hydrogeological Report. Federal Republic of Germany: GKW, Mannheim, Federal Republic of Germany. Descriptors: Africa; Baraawe Somali Republic; East Africa; report; Shabeelaha Hoose; Somali Republic; water resources; water supply. Database: GeoRef. Accession Number: 1998-017949.

GKW Consult. Engin., Mannheim, Federal Republic of Germany. 1984. Water Supply for Towns in Southern Somalia (Water Supply III); Bardheere, Vol. IV, Hydrogeological Report. Federal Republic of Germany: GKW, Mannheim, Federal Republic of Germany. Descriptors: Africa; Bardheere, Somali Republic; East Africa; Gedo, Somali Republic; report; Somali Republic; southern Somali Republic; water resources; water supply. Database: GeoRef. Accession Number: 1998-017950.

GKW Consult. Engin., Mannheim, Federal Republic of Germany. 1984. Water Supply for Towns in Southern Somalia (Water Supply III); Bardheere, Vol. V, Feasibility Study. Federal Republic of Germany: GKW, Mannheim, Federal Republic of Germany. Descriptors: Africa; Bardheere Somali Republic; East Africa; feasibility studies; Gedo

Somali Republic; Somali Republic; water resources; water supply. Database: GeoRef. Accession Number: 1998-017951.

GKW Consult. Engin., Mannheim, Federal Republic of Germany. 1984. Water Supply for Towns in Southern Somalia (Water Supply III); Luuq, Vol. I, Hydrogeological Report. Federal Republic of Germany: GKW, Mannheim, Federal Republic of Germany. Descriptors: Africa; East Africa; Gedo Somali Republic; Luuq Somali Republic; report; Somali Republic; southern Somali Republic; water resources; water supply. Database: GeoRef. Accession Number: 1998-017952.

GKW Consult. Engin., Mannheim, Federal Republic of Germany. 1984. Water Supply for Towns in Southern Somalia (Water Supply III); Vol. VIII, Hydrogeological Report. Federal Republic of Germany: GKW, Mannheim, Federal Republic of Germany. Descriptors: Africa; East Africa; report; Somali Republic; southern Somali Republic; water resources; water supply. Database: GeoRef. Accession Number: 1998-017953.

GKW Consult. Engin., Mannheim, Federal Republic of Germany. 1976. Water Resources Development in Somalia; Feasibility Study; Jowhar, Balad, Afgooye, Marka. Federal Republic of Germany: GKW, Mannheim, Federal Republic of Germany. Descriptors: Afgooye Somali Republic; Balad Somali Republic; Benadir Somali Republic; development; feasibility studies; Jowhar Somali Republic; Marka Somali Republic; Shabeelaha Dhexe; Shabellaha Hoose; water resources. Database: GeoRef. Accession Number: 1998-017946.

GKW Consult. Engin., Mannheim, Federal Republic of Germany. 1976. Water Resources Development in Somalia, Interim Report; Cerigeralo, Qardho, Garowe, Gallaeyo, Dhuresxa, Mareab. Federal Republic of Germany: GKM, Mannheim, Federal Republic of Germany. Descriptors: Africa; Cerigeralo Somali Republic; development; Dhuresxa Somali Republic; East Africa; Gallaeyo Somali Republic; Garowe Somali Republic; Mareab Somali Republic; Qardho Somali Republic; report; Somali Republic; water resources; Nugaal Somali Republic. Database: GeoRef. Accession Number: 1998-017947.

Glantz M.H. 1980. "El Nino: Lessons for Coastal Fisheries in Africa?" Oceanus. Volume 23, Issue 2, Pages 9-17. Descriptors: Exchange And Development; Industry And Services; Peru; fishing industry; anchoveta fishery; Mauritania; Somalia; Africa. Notes: Special Feature: 2 figs, 10 photos, 10 refs. Abstract: The major features of El Nino and of Peru's fishing industry during the last 15 years are summarized. The problem in using a forecast of El Nino as an effective management tool for the exploitation of Peruvian anchoveta fishery are discussed. The exploitation of coastal fisheries in African upwelling regions off Mauritania and Somalia by local and foreign vessels is described, and the importance of avoiding the over-capitalization of a fishing industry which in Peru led to its collapse is stressed. ISSN: 0029-8182.

Glassborow, Katy. 2005. "Somali Pirates Change Tack." Jane's Navy International. Issue V, Pages 39-40. Descriptors: Naval vessels; Ammunition; Guns (armament); Boats. Abstract: An apparent shift in the tactics of Somali pirates has led the International Maritime Bureau's (IMB's) Piracy Reporting Centre to push its recommendation that vessels keep at least 50 n miles from the coast out to beyond 200 n miles. The warning followed three attacks well off the coast, suggesting several boats are working in tandem to sustain themselves at greater distances from shore. On 5 November, 2005, a small boat with six heavily armed pirated pursued and fired rocket-propelled grenades at the cruise liner Seabourn Spirit, 70 n miles off the east coast of Somalia. On 6 November, 2005 pirates armed with RPGs and machine guns fired on another ship 110 n miles off the coast, damaging the bridge windows. ISSN: 1358-3719.

God, Mumin Mohamud. 1995. "Climatic Conditions Deduced from Late Pleistocene Deposits at Karin Gap (NE Somalia)." Berliner Geowissenschaftliche Abhandlungen, Reihe A: Geologie Und Palaeontologie. Volume 178, Pages 145. Descriptors: absolute age; Africa; Arthropoda; C-14; calcrete; carbon; carbonate rocks; carbonates; Cenozoic; clastic rocks; clastic sediments; clay; conglomerate; Crustacea; dates; East Africa; framework silicates; Gastropoda; gypsum; Holocene; Invertebrata; isotopes; Karin Gap; lacustrine environment; lower Holocene; Mandibulata; marl; microfossils; Mollusca; northeastern Somali Republic; Ostracoda; paleoclimatology; Pleistocene; quartz; Quaternary; radioactive isotopes; sand; sedimentary rocks; sedimentation; sediments; shallow-water environment; silica minerals; silicates; Somali Republic; spectra; sulfates; travertine; upper Pleistocene; X-ray diffraction data; X-ray fluorescence spectra. Notes: illustrations incl. stratigraphic columns, sects., 20 tables, geol. sketch maps; Reference includes data from Geoline, Bundesanstalt fur Geowissenschaften und Rohstoffe, Hanover, Germany. ISBN: 3895820172.

God, Mumin Mohamud. 1995. Climatic Conditions Deduced from Late Pleistocene Deposits at Karin Gap (NE-Somalia). Berlin: Selbstverlag Fachbereich Geowissenschaften, FU Berlin. Descriptors: 145; Geology- Somalia; Paleoclimatology- Somalia; Geology, Stratigraphic- Pleistocene. Notes: Includes bibliographical references (p. 114-136). ISBN: 3895820172 Series 0172-8784.

Goracci, C., Marci, F. and Negri, P. L. 1983. "Relation of the Amount of Fluoride in the Drinking Water of Somalia and in Dental Enamel." Odontostomatol. Trop. Sep. Volume 6, Issue 3, Pages 139-143. Descriptors: Adult; Comparative Study; Dental Enamel- analysis; English Abstract; Fluorides/administration & dosage/analysis; Humans; Italy; Middle Aged; Somalia; Water Supply- analysis. ISSN: 0251-172X.

Goracci, G., Marci, F., Negri, P. L. and Treccani, A. 1983. "Aspects of Dental Fluorosis in Subjects from Regions with Water Rich in Fluorine and their Classification." Minerva Stomatol. Nov-Dec. Volume 32, Issue 6, Pages 795-802. Descriptors: Adolescent; Adult; Child; Disease Reservoirs; English Abstract; Fluorides- analysis; Fluorosis, Dental classification- epidemiology; Humans; Somalia; Water Supply; water analysis. ISSN: 0026-4970.

Governo de Somalia Italiana, Mogadishu, Somalia. 1938. "Pozzi e Abbeveratoi Della Somalia Italiana." Translated title: "Wells and Watering Places of Italian Somaliland." Somalia: Descriptors: Africa; drinking water; East Africa; Italian Somaliland; Somali Republic; southern Somali Republic; water resources; water wells; wells. Database: GeoRef. Accession Number: 1998-017956.

Governo de Somalia Italiana, Mogadishu, Somalia. 1927. "Monografie Della Regione Somala, Il Giuba, La Migiurtinia Ed Il Territorio Di Nogal." Translated Title: "Monograph of the Somali Region, Juba, Migiurtania and Nogal Territory." Italy: De Agostini Ed., Novara, Italy. Descriptors: Africa; East Africa; geography; Juba Somali Republic; Migiurtania Somali Republic; Nogal Somali Republic; Somali Republic. Notes: In 3 volumes. Database: GeoRef. Accession Number: 1998-017955.

Gozes, R. and Torrent, H. 1985. National Ground Water Resources Assessment; Proposed Methodology and Specifications. France: BRGM, Orleans, France. Descriptors:

Africa; East Africa; ground water; Somali Republic; water resources. Database: GeoRef. Accession Number: 1998-017957.

Graham, N. J. D. and Hartung, H. 1988. "Package Water Treatment Facilities for Refugee Communities." Proceedings of the Institution of Civil Engineers PCIEAT Vol. 84. June 1988. Pages 8. Descriptors: Water treatment facilities; Developing countries; Refugee camps; Sand filters; Potable water; Somalia. Abstract: Package water supply systems have been developed by Oxfam and Imperial College for rapid deployment in emergency situations such as refugee settlements. A component of these is a water treatment unit for treating polluted surface waters by the process of slow sand filtration. Many such units have been deployed at refugee camps in Southern Somalia and have performed well. Database: Environmental Sciences and Pollution Mgmt.

Granath, J. W. Buchholz, P. E. Soofi, K. A. and Smith, R. S. U. 1993. Neogene Reactivation of Older Structural Features in the Northern Somalia: Inferences from Landsat Interpretation. Firenze; Istituto Agronomico L'Oltremar; 1993. Relazioni E Monografie Agrarie Subtropicali E Tropicali Nuova Serie113 A, pages: 357-368. International Meeting; 1st- 1987 Nov: Mogadishu. Notes: Geology of Somalia and surrounding regions: Geology and mineral resources of Somalia and surrounding regions. OCLC Accession Number: 34175029.

Granath, J. W. Buchholz, P. E. Soofi, Khalid A. and Smith, R. S. U. 1987. "Landsat Imagery and Gulf of Aden Aged Reactivation of Older Structural Features in Northern Somalia." GeoSom 87; International Meeting; Geology of Somalia and Surrounding Regions; Abstracts. Somalia: Somali Natl. Univ., Mogadishu, Somalia. GeoSom 87; Geology of Somalia and Surrounding Regions, Mogadishu. Somalia Conference: Nov. 23-30, 1987. Descriptors: Africa; Arabian Sea; Cenozoic; Darror Valley; data processing; East Africa; extension tectonics; geophysical surveys; Guban; Gulf of Aden; imagery; Indian Ocean; Jurassic; Mesozoic; mosaics; Neogene; Nogal Valley; northern Somali Republic; reactivation; remote sensing; Somali Republic; structural geology; surveys; tectonics; Tertiary. OCLC Accession Number: 34175029.

Graphi-Ogre and Maps.com. 1998. "Somalia." Santa Barbara, Calif: http://www.maps.com Notes: Description: 1 map; color; 98 x 69 cm. Geographic: Somalia- Administrative and political divisions- Maps. Scale [ca. 1:2,320,000]. Category of scale: a Constant ratio linear horizontal scale: 2320000; Notes: Wall map. "Geoatlas." Includes color ill. of flag. Library of Congress; LCCN: 2006-629319.

Gray F.A, Kolp B.J and Mohamed M.A. 1990. "A Disease Survey of Crops Grown in the Bay Region of Somalia, East Africa." FAO Plant Prot. Bull. Volume 38, Issue 1, Pages 39-47. Descriptors: Heavy metals; disease; sorghum; charcoal rot; covered kernel smut; sooty stripe; ashy stem blight; powdery mildew; leaf spot; pod rot; mung bean; groundnut; fungi; cowpea Species Term: Vigna unguiculata; Vigna radiata; Sphacelotheca sorghi; Ramulispora sorghi; Macrophomina phaseolina; Xanthomonas; Erysiphe polygoni; Isariopsis griseola; Rhizoctonia solani; Aspergillus niger; Cercospora arachidicola. Notes: Somalia- Bay Region. Abstract: The most common sorghum diseases were charcoal rot Macrophomina phaseoline, covered kernel smut Sphacelotheca sorghi and sooty stripe Ramulispora sorghi. Cowpea diseases were ashy stem blight Macrophomina phaseolina, bacterial blight Xanthomonas sp. and powdery mildew Erysiphe polygoni. Mung bean diseases were Macrophomina phaseolina, Xanthomonas sp. and angular leaf spot Isariopsis griseola. Groundnut diseases included Rhizoctonia root and pod rot Rhizoctonia solani, Aspergillus crown rot Aspergillus niger and early leaf spot Cercospora arachidicola. M. phaseolina caused disease in sorghum, cowpeas and mung beans and appears to be the most important single fungal pathogen in the Bay Region. FAO Library; ISSN: 0254-9727.

Grazzini C.V, Caulet J.P. and Venec, Peyre M.T. 1995. "Index De Fertilite et Mousson dans le Bassin de Somalie. Evolution au Quaternaire Superieur." Bulletin-Société Géologique de France. Volume 166, Issue 3, Pages 259-270. Descriptors: Oceans; Palaeoclimatology; Quaternary (late); fertility indicator; monsoon; upwelling. Notes: Indian Ocean Somalia. Abstract: Time series of geochemical and biological markers in pelagic sediments, such as N. dutertrei and URI (Upwelling Radiolarian Index), have been used to reconstruct upwelling changes in the Somalian upwelling area (NW Indian Ocean). Other fertility indicators, such as G. bulloides, SiO_2 and Ba fluxes, have been used to monitor monsoon changes in the Arabian Sea. Here we compare time series of upwelling proxies from two cores located under the Equator (Core MD 85668) and under the Somalian gyre (Core MD 85674) to data previously published for the Arabian Sea. To quantify relationships between fertility and climatic indicators, cross spectral comparisons with ETP (eccentricity + tilt + precession composite signal) have been used to estimate coherences in conjunction with phase relationships. The results demonstrate that fertility or upwelling tracers cannot be used to reconstruct palaeomonsoon changes in the Somalian area. As it has been already observed in the Arabian Sea, control on the nutrient supply may not necessarily be related only to monsoon intensity. Variations of geochemical and biological markers in the Somalian basin appear to be mostly related to water mass reorganization induced by global climate changes and in some cases to the distribution of solar radiation. There is an abridged English version. ISSN: 0037-9409.

Great Britain. Army. East Africa Command. East African Survey Group. 1943. "Index map to East Africa 1:50,000 Coastal Strips and 1:25,000 Coastal sheets." GP 1911. Scale: 1:2 000 000. [Nairobi]: [Survey Directorate, East Africa Command], 1943. maps: color Note(s): Graphic index. Subjects: Somalia- Maps and charts; Kenya- Maps and charts; Tanganyika- Maps and charts 1943. Added Name: Great Britain. War Office. General Staff. Geographical Section. 1 sheet. British Library: Maps MOD GP 1911.

Great Britain. Army. East Africa Command. East African Survey Group. 1942. "British Somaliland." [EAF] Misc 28. Scale: 1:2 000 000. [Nairobi]: [Survey Directorate, East Africa Command], 1942. Maps. Subjects: Somalia- Maps and charts; Ethiopia- Maps and charts 1942. Added Name: Great Britain. War Office. General Staff. Geographical Section. 1 sheet. British Library: Maps MOD [EAF] Misc 28.

Great Britain. Army. East Africa Command. East African Survey Group. 1941-1944. "Africa 1:2000000." Series EAF AA 2M. Scale 1:2 000 000. Nairobi: [Survey Directorate, East Africa Command], 1941-1944. 7 maps: color Note: 7 maps in series. Subject: Ethiopia- Maps and charts 1941-1944; Sudan- Maps and charts 1941-1944; Somalia- Maps and charts 1941-1944; Kenya- Maps and charts 1941-1944; Uganda-Maps and charts 1941-1944; Tanganyika- Maps and charts 1941-1944; Seychelles- Maps and charts 1941-1944. Added name: Great Britain. War Office. General Staff. Geographical Section. 12 sheets. British Library.

Great Britain. Army. East Africa Command. East African Survey Group. 1941. "Afgoi." EAF 229. First ed. Scale 1:8250 and 1:25 000. Nairobi: [Survey Directorate,

East Africa Command], 1941. Maps. Note: Inset of ammunitions storage area. Subject: Somalia- Maps and charts. Added name: Great Britain. War Office. General Staff. Geographical Section. 1 sheet. British Library.

Great Britain. Army. East Africa Command. East African Survey Group. "Brava." EAF 188. First ed. Scale: 1:100 000. [Nairobi]: [Survey Directorate, East Africa Command], 1941. Maps. Note(s): Copy of Italian map dated 1911. Subjects: Somalia- Maps and charts 1941. Added Name: Great Britain. War Office. General Staff. Geographical Section. 1 sheet. British Library: Maps MOD EAF 188.

Great Britain. Army. East Africa Command. East African Survey Group. 1941. "Budana & environs." EAF 164. First ed. Scale: 1:14 600. [Nairobi]: [Survey Directorate, East Africa Command], 1941. Maps. Note(s): Subjects: Somalia- Maps and charts 1941. Added Name: Great Britain. War Office. General Staff. Geographical Section. 1 sheet. British Library: Maps MOD EAF 164

Great Britain. Army. East Africa Command. East African Survey Group. 1941. "Chisimaio (Kismayo)." EAF 215. Second ed. Scale: 1:50 000. [Nairobi]: [Survey Directorate, East Africa Command], 1941. Maps. Subjects: Somalia- Maps and charts 1941. Added Name: Great Britain. War Office. General Staff. Geographical Section. 1 sheet. British Library: Maps MOD EAF 215.

Great Britain. Army. East Africa Command. East African Survey Group. 1941? "Chisimaio (Kismayo)." EAF Chisimaio 25k. Ed.1. Scale: 1:25 000. [Nairobi]: E A Survey Group, [19--]. 3 maps. Note(s): Sheets 1 to 3. These maps do not have EAF numbers. Subjects: Somalia- Maps and charts 19--. Added Name: Great Britain. War Office. General Staff. Geographical Section. 3 sheets. British Library: Maps MOD EAF Chisimaio 25k

Great Britain. Army. East Africa Command. East African Survey Group. "East Africa 1:1 000 000." Series EAF EA 1M. Scale: 1:1 000 000. [Nairobi]: [Survey Directorate, East Africa Command], 1941. 5 maps: color Subjects: Kenya- Maps and charts; Sudan- Maps and charts; Ethiopia- Maps and charts; Somalia- Maps and charts; Tanganyika- Maps and charts 1941. Added Name: Great Britain. War Office. General Staff. Geographical Section. 5 sheets. British Library: Maps MOD EAF EA 1M.

Great Britain. Army. East Africa Command. East African Survey Group. 1941. "Ferfer- Gorrahei Road" EAF 334 to EAF 336. First ed. Scale: 1:32 100 and 1:39 145. [Nairobi]: [Survey Directorate, East Africa Command], 1941. 3 maps. Subjects: Somalia- Maps and charts 1941. Added Name: Great Britain. War Office. General Staff. Geographical Section. 3 sheets. British Library: Maps MOD EAF 334 to EAF 336.

Great Britain. Army. East Africa Command. East African Survey Group. 1941. "Gelib (Jelib)- Bardera Road." EAF No 236 to EAF 239. First ed. Scale: 1:25 000. [Nairobi]: [Survey Directorate, East Africa Command], 1941. 4 maps. Note(s): Sheets 1 to 4. Subjects: Somalia- Maps and charts 1941. Added Name: Great Britain. War Office. General Staff. Geographical Section. 4 sheets. British Library: Maps MOD EAF 236 to EAF 239.

Great Britain. Army. East Africa Command. East African Survey Group. 1941. "Kolbio- Chisimaio (Kismayo) Road." EAF 148 & EAF 149. First ed. Scale: 1:50 000. [Nairobi]: [Survey Directorate, East Africa Command], 1941. 2 maps. Note(s): Sheets 1 & 2. Subjects: Somalia- Maps and charts 1941. Added Name: Great Britain. War Office.

General Staff. Geographical Section. 2 sheets. British Library: Maps MOD EAF 148 and EAF 149.

Great Britain. Army. East Africa Command. East African Survey Group. 1941. Merca, EAF 190. First ed. Scale: 1:50 000. [Nairobi]: [Survey Directorate, East Africa Command], 1941. Maps. Note(s): Copy of Italian map dated 1911. Subjects: Somalia-Maps and charts 1941. Added Name: Great Britain. War Office. General Staff. Geographical Section. 1 sheet. British Library: Maps MOD EAF 190.

Great Britain. Army. East Africa Command. East African Survey Group. 1941. "Mogadiscio (Mogadishu)." EAF 230. First ed. Scale: 1:30 000. [Nairobi]: [Survey Directorate, East Africa Command], 1941. Maps. Subjects: Somalia- Maps and charts 1941. Added Name: Great Britain. War Office. General Staff. Geographical Section. 1 sheet. British Library: Maps MOD EAF 230.

Great Britain. Army. East Africa Command. East African Survey Group. 1941. "River Giuba (Juba)." EAF 232 to 235 & EAF 279 to 288. First ed. Scale: 1:50 000. [Nairobi]: [Survey Directorate, East Africa Command], 1941. 14 maps. Note(s): Sheets 1 to 13. Subjects: Somalia- Maps and charts 1941. Added Name: Great Britain. War Office. General Staff. Geographical Section. 14 sheets. British Library: Maps MOD EAF 232 to 235 and EAF 279 to 288.

Great Britain. Army. East Africa Command. East African Survey Group. 1941. "Somalia 1:250 000." Series EAF Soma 250k. Scale: 1:250 000. [Nairobi]: [Survey Directorate, East Africa Command], 1941. 41 maps: color Subjects: Somalia- Maps and charts. Added Name: Great Britain. War Office. General Staff. Geographical Section. 41 sheets. British Library: Maps MOD EAF Soma 250k.

Great Britain. Army. East African Engineers. East African and Southern Rhodesian Base Survey Company, 157th. 1945. "Horn of Africa- British military administrations, East Africa Command- Static administration map." GP 2524. Scale: 1:3 000 000. [Nairobi]: [Survey Directorate, East Africa Command], 1945. maps: color Subjects: Ethiopia- Maps and charts; Somalia- Maps and charts 1945. Added Name: Great Britain. War Office. General Staff. Geographical Section. 1 sheet. British Library: Maps MOD GP 2524.

Great Britain. Army. East African Engineers. East African and Southern Rhodesian Base Survey Company, 157th. 1944. "British and Italian Somaliland, British Military Administrations East Africa Command." GP 2121. Scale: 1:3 000 000. [Nairobi]: [Survey Directorate, East Africa Command], 1944. maps: color Subjects: Somalia- Maps and charts 1944. Added Name: Great Britain. War Office. General Staff. Geographical Section. 1 sheet. British Library: Maps MOD GP 2121.

Great Britain. Army. East African Engineers. East African and Southern Rhodesian Base Survey Company, 157th. 1943. "British and Italian Somaliland." GP 2085. Scale: 1:2 000 000. [Nairobi]: [Survey Directorate, East Africa Command], 1943. Maps. Subjects: Somalia- Maps and charts 1943. Added Name: Great Britain. War Office. General Staff. Geographical Section. 1 sheet. British Library: Maps MOD GP 2085.

Great Britain. Army. East Africa Force. Survey Group. 1941. "A collection of E.A.F. maps of single roads in Somalia/Somaliland." 1st ed. Scales differ. Nairobi: E.A. Survey Group, 1941. Physical description: maps. Notes: Scales range from ca. 1:25 000 to ca. 1:50 000. Subject: Somaliland- Roads- maps (collections)- 1941. British Library.

Great Britain. Army. Royal Engineers. Field Survey Company, 19th. 1947. "Somalia [cultivated areas]." MDR Misc 10990. Scale: 1:4 000 000. [Cairo]: [Survey Directorate, Middle East], 1947. maps: color Note(s): Drawn by JIBME GHQ MELF Aug 1947. Shows areas where seven different crops are cultivated. Subjects: Somalia- Maps and charts. Added Name: Great Britain. War Office. General Staff. Geographical Section. 1 sheet. British Library: Maps MOD MDR Misc 10990.

Great Britain. Army. Royal Engineers. Base Survey Company, 157. 1945. "Geological Survey; Italian, British, French Somaliland." 157 Base Svy. Co. R.E.; Somalia. Descriptors: Geology- Somalia- Maps. Notes: Description: 2 maps. Geographic: Somalia- Maps. Scale 1:2,000,000. Category of scale: a Constant ratio linear horizontal scale: 2000000; Notes: Date of information 1933. Reproduced from Italian map by G. Stefanini. Relief shown as heights in metres. Shows: international boundaries, geological information. OCLC Accession Number: 48600821.

Great Britain. Army. Royal Engineers. Base Survey Company, 157. 1945. "Tribal Map of Somalia and British Somaliland (1 January 1945)." 157 (E.A. & S.R.) Base Survey. Co.; Somalia. Descriptors: Tribes- Somalia- Maps. Notes: Description: 1 map. Geographic: Somalia- Maps. Scale 1:2,500,000. Category of scale: a Constant ratio linear horizontal scale: 2500000; Notes: international boundaries, boundaries of Anglo-Ethiopian agreements, tribal boundaries and land, tribal migrations. "GP 2571." OCLC Accession Number: 48602044.

Great Britain. Army. Royal Engineers. Field Survey Company, 512th. 1940. "Belet Uen." DR 32. Scale: 1:1 000 000. [Cairo]: [Survey Directorate, Middle East], 1940. Maps. Subjects: Somalia- Maps and charts 1940. Added Name: Great Britain. War Office. General Staff. Geographical Section. 1 sheet. British Library: Maps MOD DR 32.

Great Britain. Army. Royal Engineers. Field Survey Company, 512th. 1940. "I-E-A Somaliland Frontier." DR 39. Scale: 1:1 000 000. [Cairo]: [Survey Directorate, Middle East], 1940. maps: color Subjects: Somalia- Maps and charts 1940. Added Name: Great Britain. War Office. General Staff. Geographical Section. 1 sheet. British Library: Maps MOD DR 39.

Great Britain. Army. Royal Engineers. Field Survey Company, 512th. "Somaliland 1:250,000." Series DR 76. Scale: 1:250 000. [Cairo]: [Survey Directorate, Middle East], 1940. 8 maps: color Note(s): Copied from a War Office map dated 1934. Subjects: Somalia- Maps and charts 1940. Added Name: Great Britain. War Office. General Staff. Geographical Section. 8 sheets. British Library: Maps MOD DR 76.

Great Britain. Army. Royal Engineers. Field Survey Company, 514th. 1941. "Italian East Africa." R 7. Scale: 1:6 000 000. [Cairo]: Survey Directorate Middle East Command], 1941. maps: color Subjects: Ethiopia- Maps and charts; Somalia- Maps and charts; Eritrea- Maps and charts 1941. Added Name: Great Britain. War Office. General Staff. Geographical Section. 1 sheet. British Library: Maps MOD R 7.

Great Britain. Army. Royal Engineers. Survey Engineers Regiment, 42nd. 1954. "Hargeisa Township." MDR Misc 11920. Scale: 1:5 000. [S.l.]: [Survey Directorate, Middle East], 1954. Maps. Subjects: Somalia- Maps and charts 1954. Added Name: Great Britain. War Office. General Staff. Geographical Section. 1 sheet. British Library: Maps MOD MDR Misc 11920.

Great Britain. Army. Royal Engineers. Survey Engineers Regiment, 42nd. 1949. "North-east Africa." MDR Misc 11369. Scale: not given. [Cairo]: [Survey Directorate,

Middle East], 1949. Maps. Note(s): Shows historical development of various admin regions. Subjects: Ethiopia- Maps and charts; Somalia- Maps and charts; Eritrea- Maps and charts; Djibouti- Maps and charts 1949. Added Name: Great Britain. War Office. General Staff. Geographical Section. 1 sheet. British Library: Maps MOD MDR Misc 11369.

Great Britain. Directorate of Colonial Surveys. 1957. "Somaliland Protectorate 1:125,000: Geological Survey." London: Directorate of Colonial Surveys. Descriptors: Geology- British Somaliland- Maps; Geology- Somalia- Maps. Notes: Description: maps; 44 x 44 cm. Scale 1:125,000. Category of scale: a Constant ratio linear horizontal scale: 125000; Notes: Later sheets titled Somali Democratic Republic. OCLC Accession Number: 54321296.

Great Britain. Directorate of Colonial Surveys. 1952. "Somaliland Protectorate." Teddington, England: The Directorate. Descriptors: Government publication; National government publication. Notes: Description: maps; 56 x 55 cm. Geographic: Somalia- Maps. Scale ca. 1:50,000; Transverse Mercator projection. Category of scale: a Constant ratio linear horizontal scale: 50000; Notes: Relief shown by hachures. "Preliminary plot." Includes index to adjoining sheets. OCLC Accession Number: 38741423.

Great Britain. Directorate of Military Survey. 2000. "Town Plans, Somali Republic. Hargeisa." Washington, D.C. NIMA. Descriptors: Government publication; National government publication. Notes: Description: 1 map; color; 66 x 94 cm. Geographic: Hargeisa (Somalia)- Maps. Scale 1:5,000. 12.672 in. to 1 mile. Category of scale: a Constant ratio linear horizontal scale: 5000; Notes: Standard map series designation: Series Y921. "Compiled from miscellaneous information up to January, 1959. Edition 2 revised from information supplied by HQ BFAP dated 1960. Published by D. Survey, Ministry of Defence, United Kingdom 1970. Reprinted by NIMA 8-00." "Universal Transverse Mercator grid, zone 38." Other Titles: Hargeisa; Series Y921; OCLC Accession Number: 49951272.

Great Britain. Directorate of Military Survey. 1970. "Town Plans, Somali Republic. Hargeisa." London; D. Survey. Descriptors: Government publication; National government publication. Notes: Description: color map; 66 x 94 cm. Scale 1:5,000; 12.672 in. to 1 mile. Geographic: Hargeisa (Somalia)- Maps. Notes: Standard map series designation: Series Y921. "Compiled from miscellaneous information up to January, 1959. Edition 2 revised from information supplied by HQ BFAP dated 1960." Projection: Transverse Mercator. "Universal Transverse Mercator grid, zone 38." Other Titles: Hargeisa. Series Y921. Library of Congress; LCCN: 73-691099.

Great Britain. Directorate of Military Survey. 1968. "Bender Beila." London: D. Survey, Ministry of Defence; Printed by the Survey Production Centre, RE. Descriptors: Government publication; National government publication. Notes: Description: 1 map; color; 45 x 66 cm. Geographic: Somalia- Maps, Topographic. Scale 1:500,000; Lambert conical orthomorphic projection. standard parallels 8040' and 11020'; (E 490--E 520/N 100--N 80). Category of scale: a Constant ratio linear horizontal scale: 500000 Coordinates--westernmost longitude: E0490000 Coordinates--easternmost longitude: E0520000; Notes: Standard map series designation: Series 1404. Relief indicated by contours, altitude tints and spot heights. "Heights in feet." Shows boundaries, highways and roads, railways, airports, rivers and water features and other details. Map overprinted with 10,000 metre Universal transverse Mercator grid. Includes glossary and index to adjoining sheets. "Sheet 791 D, series 1404." "Crown copyright 1968." Other Titles: Series 1404. OCLC Accession Number: 32420409.

Great Britain. Directorate of Military Survey. 1968. "Dante." Tolworth: D Survey; U.S. Army Topographic Command. Descriptors: Government publication; National government publication. Notes: Description: 1 map; color; 45 x 66 cm. Geographic: Dante Region (Somalia)- Maps, Topographic. Scale 1:500,000; Lambert conical orthomorphic projection. Notes: Relief indicated by color, contours, and spot heights. Includes key map. Responsibility: compiled by No. 1 SPC. RE in 1959. OCLC Accession Number: 27954011.

Great Britain. Directorate of Military Survey. 1968. "Town Plans: Somali Republic." London: Directorate of Military Survey, Ministry of Defence. Descriptors: Cities and towns- Somalia- Maps. Notes: Description: maps; 75 x 63 cm. Geographic: Somalia- Maps. Scale 1:5,000. Notes: "Series Y921." OCLC Accession Number: 54328936.

Great Britain. Directorate of Military Survey. 1964. "Tabda." London: Printed by SPC, RE. Descriptors: Government publication; National government publication. Notes: Description: 1 map; color; 45 x 68 cm. Geographic: Kenya- Maps, Topographic. Somalia- Maps, Topographic. Scale 1:250,000; Transverse Mercator projection. (E 40030'--E 42000'/N 1000'00"--N 0000'00"). Category of scale: a Constant ratio linear horizontal scale: 250000 Coordinates--westernmost longitude: E0403000 Coordinates--easternmost longitude: E0420000; Notes: "Heights in feet." Relief shown pictorially, and by altitude tints, and spot heights. Shows boundaries, roads, trails, water features, and other details. Map overprinted with the 10,000 meter UTM grid zone 37, Clarke 1880 spheroid. Includes adjoining sheets diagram, compilation diagram, and boundaries diagram. Standard map series designation: Series Y503. Other Titles: Series Y503. OCLC Accession Number: 41028186.

Great Britain. Directorate of Military Survey. 1964. "El Wak." London: Printed by SPC, RE. Descriptors: Government publication; National government publication. Notes: Description: 1 map;color; 45 x 67 cm. Geographic: Kenya- Maps, Topographic. Somalia- Maps, Topographic. Scale 1:250,000; Transverse Mercator projection. (E 40030'--E 42000'/N 3000'00"--N 2000'00"). Category of scale: a Constant ratio linear horizontal scale: 250000 Coordinates--westernmost longitude: E0403000 Coordinates--easternmost longitude: E0420000; Notes: "Heights in feet." Relief shown pictorially, and by contours, form lines, altitude tints, and spot heights. Shows boundaries, roads, trails, water features, and other details. Map overprinted with the 10,000 meter universal transverse Mercator grid, zone 37, Clarke 1880 spheroid. Includes adjoining sheets diagram, compilation diagram, and boundaries diagram. "British Crown copyright reserved." Standard map series designation: Series Y503. "Stock No. Y503XNA378." Other Titles: El Wak; Series Y503; Responsibility: published by D. Survey, War Office and Air Ministry; compiled by SPC, RE. OCLC Accession Number: 41021401.

Great Britain. Directorate of Military Survey. 1964. "Gherille." London: Printed by No. 1, SPC, RE. Descriptors: Government publication; National government publication. Notes: Description: 1 map;color; 45 x 67 cm. Geographic: Kenya- Maps, Topographic. Somalia- Maps, Topographic. Scale 1:250,000; Transverse Mercator projection. (E 40030'--E 42000'/N 2000'00"--N 1000'00"). Category of scale: a Constant ratio linear horizontal scale: 250000 Coordinates--westernmost longitude: E0403000

Coordinates--easternmost longitude: E0420000; Notes: "Heights in feet; spot heights are approximate." Relief shown pictorially, and by altitude tints, and spot heights. Shows boundaries, roads, trails, water features, and other details. Map overprinted with the 10,000 meter UTM grid, zone 37, Clarke 1880 spheroid. Includes adjoining sheets, compilation, and boundaries diagrams. "British Crown copyright reserved." Standard map series designation: Series Y503. "Stock No. Y503XNA3712." Other Titles: Series Y503; Responsibility: published by D. Survey, War Office and Air Ministry; compiled SPC, RE. OCLC Accession Number: 41021720.

Great Britain. Directorate of Military Survey. 1963. "Jara Jila." London: Printed by SPC RE. Descriptors: Government publication; National government publication. Notes: Description: 1 map;color; 45 x 67 cm. Geographic: Kenya- Maps, Topographic. Somalia- Maps, Topographic. Scale 1:250,000; Transverse Mercator projection. (E 40030'--E 42000'/S 0000'00"--S 1000'00"). Coordinates--westernmost longitude: E0403000 Coordinates--easternmost longitude: E0420000; Notes: "Heights in feet; local relief indicated by pecked form lines." Relief shown by form lines. Shows boundaries, roads, trails, water features, and other details. Map overprinted with the 10,000 meter universal transverse Mercator grid zone 36, Clarke 1880 spheroid. Includes adjoining sheets diagram, compilation diagram, and boundaries diagram. Standard map series designation. Other Titles: Series Y503. OCLC Accession Number: 38925953.

Great Britain. Directorate of Military Survey. 1963. "Town Plans, Somali Republic: Hargeisa." London; D. Survey. Descriptors: Government publication; National government publication. Notes: Description:color map; 65 x 93 cm. Scale 1:5,000; 12.672 in. to 1 mile. Geographic: Hargeisa (Somalia)- Maps. Notes: Standard map series designation: Series Y921. "Projection: Transverse Mercator." "Grid: East Africa-belt J." "Compiled from miscellaneous information up to March 1959. Edition 2 revised from information supplied by HQ BFAP dated 1960." Other Titles: Hargeisa. Series Y921. Library of Congress; LCCN: 73-696190.

Great Britain. Directorate of Military Survey. 1962. "Town Plans, Somali Republic. Burao." London: D. Survey. Descriptors: Government publication; National government publication. Notes: Description: 1 map; color; 75 x 63 cm. Geographic: Burao (Somalia)- Maps. Scale 1:5,000; 12.672 in. to 1 mile. Category of scale: a Constant ratio linear horizontal scale: 5000; Notes: "Projection: Transverse Mercator." "Grid: East African grid-belt K." "Compiled from miscellaneous information up to January 1959. Edition 2 revised from information supplied by HQ BFAP dated 1960." Other Titles: Burao; Standard map series designation: Y921. OCLC Accession Number: 55715337.

Great Britain. Directorate of Military Survey and Aerospace Center (U.S.). 1993. "Tactical Pilotage Chart, TPC. K-6C, Somali Democratic Republic, People's Democratic Republic of Yemen." London; Riverdale, MD: The Survey; NOAA Distribution Branch (N/ACC3), National Ocean Service, distributor. Descriptors: Aeronautical charts- Somalia; Aeronautical charts- Yemen; Government publication; National government publication. Notes: Description: 1 map; color; on sheet 106 x 146 cm. folded to 27 x 37 cm. Scale 1:500,000; Lambert conformal conic projection, standard parallels 9020' and 14040'; (E 490--E 550/N 120--N 80). Category of scale: a Constant ratio linear horizontal scale: 500000 Coordinates--westernmost longitude: E0490000 Coordinates--easternmost longitude: E0550000; Notes: Relief shown by contours, shading, tints, and spot heights.

Produced by Military Survey for DMA Aerospace Center to issue as series TPC K-6C. "Air information current at 11 August 1983." Shipping list no. 99-2027-S. "Compiled 1983." "Copyright HMSO London 1983." "Lithographed by DMAAC 3-93." Includes elevation tints and interchart relationship diagrams. "39150/10/83/821432M." Other Titles: TPC, Somali Democratic Republic, People's Democratic Republic of Yemen; Somali Democratic Republic, People's Democratic Republic of Yemen; Responsibility: produced under the direction of the Director of Military Survey, Ministry of Defence, United Kingdom. OCLC Accession Number: 42442027.

Great Britain. Directorate of Military Survey and Aerospace Center (U.S.). 1992. "Tactical Pilotage Chart, TPC. L-6D, Ethiopia, Somalia." London: The Survey; St. Louis, Mo; Denver, CO: The Center; USGS Branch of Distribution, Box 25286, distributor. Descriptors: Aeronautical charts- Ethiopia; Aeronautical charts- Somalia; Government publication; National government publication. Notes: Description: 1 map; color; on sheet 105 x 146 cm., folded to 27 x 37 cm. Scale 1:500,000; Lambert conformal conic projection, standard parallels 1020' and 6040'; (E 420--E 480/N 40--N 00). Category of scale: a Constant ratio linear horizontal scale: 500000 Coordinates--westernmost longitude: E0420000 Coordinates--easternmost longitude: E0480000; Notes: Relief shown by shading, tints, and spot heights. "Air information current at 28 December 1992." "Compiled 1985, revised December 1992." "Lithographed by DMAAC 12-92." "Copyright HMSO London 1986." Includes elevation tints and interchart relationship diagrams. Other Titles: Ethiopia, Somalia. Tactical pilotage chart. TPC. Responsibility: produced under the direction of the Director General of Military Survey, Ministry of Defence, United Kingdom; U.S. Defense Mapping Agency Aerospace Center. OCLC Accession Number: 27773194.

Great Britain. Directorate of Military Survey and Aerospace Center (U.S.). 1992. "Tactical Pilotage Chart, TPC. M-5B, Kenya, Somalia, Tanzania." London: The Survey; St. Louis, Mo; Denver, CO: The Center; USGS Branch of Distribution, Box 25286, distributor. Descriptors: Aeronautical charts- Kenya; Aeronautical charts- Somalia; Aeronautical charts- Tanzania; Government publication; National government publication. Notes: Description: 1 map; color; on sheet 105 x 146 cm., folded to 27 x 37 cm. Scale 1:500,000; Lambert conformal conic projection, standard parallels 1020' and 6040'; (E 370--E 430/S 00--S 40). Category of scale: a Constant ratio linear horizontal scale: 500000 Coordinates--westernmost longitude: E0370000 Coordinates--easternmost longitude: E0430000; Notes: Relief shown by shading, tints, and spot heights. "Air information current at 24 September 1991." "Revised 1991." "Lithographed by DMAAC 12-92." "Crown copyright 1991." Includes elevation tints and interchart relationship diagrams. Other Titles: Kenya, Somalia, Tanzania. Tactical pilotage chart. TPC. Responsibility: produced under the direction of the Director General of Military Survey, Ministry of Defence, United Kingdom; [DMA Aerospace Centre]. OCLC Accession Number: 27773324.

Great Britain. Directorate of Military Survey and Aerospace Center (U.S.). 1987. "Tactical Pilotage Chart, TPC. L-6B, Somalia." London; Riverdale, MD: The Survey; NOAA Distribution Branch (N/ACC3), National Ocean Service, distributor. Descriptors: Aeronautical charts- Somalia; Government publication; National government publication. Notes: Description: 1 map; color; on sheet 106 x 146 cm. folded to 27 x 37 cm. Scale 1:500,000; Lambert conformal conic projection, standard parallels 1020' and 6040'; (E

480--E 540/N 80--N 40). Category of scale: a Constant ratio linear horizontal scale: 500000 Coordinates--westernmost longitude: E0480000 Coordinates--easternmost longitude: E0540000; Notes: Relief shown by contours, shading, tints, and spot heights. Produced by Military Survey for DMA Aerospace Center to issue as series TPC L-6B. "Air information current at 9 December 1986." Shipping list no. 99-2027-S. "Compiled 1986." "Copyright HMSO London 1987." "Printed by MCE RE." Includes elevation tints and interchart relationship diagrams. "16500/2/87/861678M." Other Titles: TPC, Somalia; Somalia; Responsibility: produced under the direction of the Director of Military Survey, Ministry of Defence, United Kingdom. OCLC Accession Number: 42442067.

Great Britain. Directorate of Military Survey; Great Britain; Army and Royal Engineers. 1968. "Town Plans, Somali Republic. Burao." London: D. Survey, Ministry of Defence. Descriptors: Government publication; National government publication. Notes: Description: 1 map; color; 75 x 63 cm. Geographic: Burao (Somalia)- Maps. Scale 1:5,000; 12.672 in. to 1 mile. Notes: "Compiled from miscellaneous information up to January, 1959. Ed. 2 rev. from information supplied by HQ BFAP, dated 1960. UTM Grid added by SPC, RE, 1968." "Zone 38, Clarke 1880 Spheroid." Other Titles: Standard map series designation: Y921; OCLC Accession Number: 55715319.

Great Britain. Directorate of Military Survey; Great Britain; Army; Royal Engineers; United States and National Imagery and Mapping Agency. 2002. "Town Plans, Somali Republic. Berbera." Washington, D.C. NIMA. Descriptors: Government publication; National government publication. Notes: Description: 1 map; color; 75 x 63 cm., on sheet 88 x 66 cm. Geographic: Berbera (Somalia)- Maps. Scale 1:5,000. 12.672 in. = 1 mile. Category of scale: a Constant ratio linear horizontal scale: 5000; Notes: "Compiled from miscellaneous information up to February, 1959. Ed. 2 rev. from information supplied by HQ BFAP, dated 1960. UTM grid added by SPC, RE, 1968. Published by D. Survey, Ministry of Defence, United Kingdom, 1968. Reprinted by NIMA 6/02." "NIMA ref. no. Y921XBERBERA." Other Titles: Berbera; Standard map series designation: Series Y921; OCLC Accession Number: 56907914.

Great Britain. Directorate of Military Survey; United States and Army Map Service. 1965. "Erigavo." Washington, D.C. Army Map Service, Corps of Engineers. Descriptors: Government publication; National government publication. Notes: Description: 1 map; color; 45 x 66 cm. Geographic: Somalia- Maps, Topographic. Scale 1:500,000; Lambert conical orthomorphic projection. standard parallels 8040' and 11020'; (E 460--E 490/N 120--N 100). Category of scale: a Constant ratio linear horizontal scale: 500000 Coordinates--westernmost longitude: E0460000 Coordinates--easternmost longitude: E0490000; Notes: Standard map series designation: Series 1404. Relief indicated by contours, altitude tints and spot heights. "Heights in feet." Shows boundaries, highways and roads, railways, airports, rivers and water features and other details. Map overprinted with 10,000 metre East Africa grid. Includes glossary and index to adjoining sheets. Originally published: [London]: D Survey, War Office and Air Ministry, 1959. "Sheet 790 B, series 1404." "British Crown copyright reserved." Other Titles: Series 1404. Responsibility: published by D. Survey, War Office and Air Ministry; printed by [U.S.] Army Map Service, Corps of Engineers. OCLC Accession Number: 32405773.

Great Britain. Directorate of Military Survey; United States and Army Map Service. 1965. "Djibouti." Washington, D.C. Army Map Service, Corps of Engineers. Descriptors: Government publication; National government publication. Notes: Description: 1 map; color; 45 x 66 cm. Geographic: Djibouti- Maps, Topographic. Somalia- Maps, Topographic. Scale 1:500,000; Lambert conical orthomorphic projection. standard parallels 8040' and 11020'; (E 430--E 460/N 120--N 100). Category of scale: a Constant ratio linear horizontal scale: 500000 Coordinates--westernmost longitude: E0430000 Coordinates--easternmost longitude: E0460000; Notes: Standard map series designation: Series 1404. Relief indicated by contours, altitude tints and spot heights. "Heights in feet." Shows boundaries, highways and roads, railways, airports, rivers and water features and other details. Map overprinted with 10,000 metre universal transverse Mercator and East Africa grids. Includes glossary and index to adjoining sheets. Originally published: [London]: D Survey, War Office and Air Ministry, 1959. "Sheet 790 A, series 1404." "British Crown copyright reserved." Other Titles: Series 1404. Responsibility: published by D. Survey, War Office and Air Ministry; printed by [U.S.] Army Map Service, Corps of Engineers. OCLC Accession Number: 32405817.

Great Britain. Directorate of Military Survey; United States and Army Map Service. 1965. "Bender Beila." Washington, D.C. Army Map Service, Corps of Engineers. Descriptors: Government publication; National government publication. Notes: Description: 1 map; color; 45 x 66 cm. Geographic: Somalia- Maps, Topographic. Scale 1:500,000; Lambert conical orthomorphic projection. standard parallels 8040' and 11020'; (E 490--E 520/N 100--N 80). Category of scale: a Constant ratio linear horizontal scale: 500000 Coordinates--westernmost longitude: E0490000 Coordinates--easternmost longitude: E0520000; Notes: Standard map series designation: Series 1404. Relief indicated by contours, altitude tints and spot heights. "Heights in feet." Shows boundaries, highways and roads, railways, airports, rivers and water features and other details. Map overprinted with 10,000 metre East Africa grid. Includes glossary and index to adjoining sheets. Originally published: [London]: D Survey, War Office and Air Ministry, 1959. "Sheet 791 D, series 1404." "British Crown copyright reserved." Other Titles: Series 1404. Responsibility: published by D. Survey, War Office and Air Ministry; printed by [U.S.] Army Map Service, Corps of Engineers. OCLC Accession Number: 32404600.

Great Britain. Directorate of Military Survey; United States and Army Map Service. 1960. "Hargeisa." Washington, D.C. Army Map Service, Corps of Engineers. Descriptors: Government publication; National government publication. Notes: Description: 1 map; color; 45 x 66 cm. Geographic: Ethiopia- Maps, Topographic. Somalia- Maps, Topographic. Scale 1:500,000; Lambert conformal conical orthomorphic projection. standard parallels 8040' and 11020'; (E 430--E 460/N 100--N 80). Category of scale: a Constant ratio linear horizontal scale: 500000 Coordinates--westernmost longitude: E0430000 Coordinates--easternmost longitude: E0460000; Notes: Standard map series designation: Series 1404. Relief indicated by contours, altitude tints and spot heights. "Heights in feet." Shows boundaries, highways and roads, railways, airports, rivers and water features and other details. Map overprinted with 10,000 metre East Africa grid. Includes glossary and index to adjoining sheets. Originally published: [London]: D Survey, War Office and Air Ministry, 1959. "Sheet 790 D, series 1404." "British Crown copyright reserved." Other Titles: Series 1404. Responsibility: published by D. Survey, War Office and Air Ministry; printed by [U.S.] Army Map Service, Corps of Engineers. OCLC Accession Number: 32404373.

Great Britain. Directorate of Military Survey; United States and Army Map Service. 1960. "Las Anod." Washington, D.C. Army Map Service, Corps of Engineers. Descriptors: Government publication; National government publication. Notes: Description: 1 map; color; 45 x 66 cm. Geographic: Ethiopia- Maps, Topographic. Somalia- Maps, Topographic. Scale 1:500,000; Lambert conformal conical projection. standard parallels 8040' and 11020'; (E 460--E 490/N 100--N 80). Category of scale: a Constant ratio linear horizontal scale: 500000 Coordinates--westernmost longitude: E0460000 Coordinates--easternmost longitude: E0490000; Notes: Standard map series designation: Series 1404. Relief indicated by contours, altitude tints and spot heights. "Heights in feet." Shows boundaries, highways and roads, railways, airports, rivers and water features and other details. Map overprinted with 10,000 metre East Africa grid. Includes glossary and index to adjoining sheets. Originally published: [London]: D Survey, War Office and Air Ministry, 1959. "Sheet 790 C, series 1404." "British Crown copyright reserved." Other Titles: Series 1404. Responsibility: published by D. Survey, War Office and Air Ministry; printed by [U.S.] Army Map Service, Corps of Engineers. OCLC Accession Number: 32405297.

Great Britain. Directorate of Military Survey; United States and Army Topographic Command. 1970. "Aiscia." Washington, D.C. Printed by the U.S. Army Topographic Command. Descriptors: Government publication; National government publication. Notes: Description: 1 map; color; 45 x 66 cm. Geographic: Djibouti- Maps, Topographic. Ethiopia- Maps, Topographic. Somalia- Maps, Topographic. Scale 1:500,000; Lambert conical orthomorphic projection. standard parallels 8040' and 11020'; (E 400--E 430/N 120--N 100). Category of scale: a Constant ratio linear horizontal scale: 500000 Coordinates--westernmost longitude: E0400000 Coordinates--easternmost longitude: E0430000; Notes: Standard map series designation: Series 1404. Relief indicated by contours, altitude tints and spot heights. "Heights in feet." Shows boundaries, highways and roads, railways, airports, rivers and water features and other details. Map overprinted with 10,000 metre universal transverse Mercator and East Africa grids. Includes glossary and index to adjoining sheets. "Sheet 789 B, series 1404." "British Crown copyright reserved." Originally published: [London]: D Survey, Ministry of Defence; printed by 42 Survey Engineer Regiment, 1968. Other Titles: Series 1404. Responsibility: published by D. Survey, Ministry of Defence; printed by the U.S. Army Topographic Command. OCLC Accession Number: 32423670.

Great Britain. Directorate of Military Survey; United States and Army Topographic Command. 1970. "Dante." Washington, D.C. U.S. Army Topographic Command. Descriptors: Government publication; National government publication. Notes: Description: 1 map; color; 45 x 68 cm. Geographic: Somalia- Maps, Topographic. Scale 1:500,000; Lambert conical orthomorphic projection. standard parallels 8040' and 11020'; (E 490--E 520/N 120--N 100). Category of scale: a Constant ratio linear horizontal scale: 500000 Coordinates--westernmost longitude: E0490000 Coordinates--easternmost longitude: E0520000; Notes: Standard map series designation: Series 1404. Relief indicated by contours, altitude tints and spot heights. "Heights in feet." Shows boundaries, highways and roads, railways, airports, rivers and water features and other details. Map overprinted with 10,000 metre universal transverse Mercator grid. Includes glossary and index to adjoining sheets. "Sheet 791 A, series 1404." Originally published: [London]: D. Survey, Ministry of Defence, 1968. "British Crown copyright reserved." Other Titles: Series 1404. Responsibility: published by D. Survey, Ministry of Defence; printed by the U.S. Army Topographic Command. OCLC Accession Number: 32407276.

Great Britain. Directorate of Military Survey; United States and National Imagery and Mapping Agency. 2000. "Town Plans, Somali Republic. Hargeisa." Bethesda, MD: NIMA. Descriptors: Government publication; National government publication. Notes: Description: 1 map; color; 66 x 94 cm. Geographic: Hargeisa (Somalia)- Maps. Scale 1:5,000. 12.672 in. to 1 mile; Transverse Mercator. Category of scale: a Constant ratio linear horizontal scale: 5000; Notes: Reprint. Originally published: London: D. Survey, 1970. "Reprinted by NIMA 8-00." Standard map series designation: Series Y921. "Compiled from miscellaneous information up to January, 1959. Edition 2 revised from information supplied by HQ BFAP dated 1960." "Universal Transverse Mercator grid, zone 38." Other Titles: Hargeisa; Series Y921. Responsibility: D. Survey, Ministry of Defence, United Kingdom. OCLC Accession Number: 49605776; 56374364.

Great Britain. Directorate of Military Survey; United States and National Imagery and Mapping Agency. 1996. "Tactical Pilotage Chart, TPC. K-6D, Djibouti, Ethiopia, Somali Democratic Republic." London: The Survey; Bethesda, Md. Riverdale, MD: National Imagery and Mapping Agency; NOAA Distribution Branch (N/CG33), National Ocean Service, distributor. Descriptors: Aeronautical charts- Djibouti; Aeronautical charts- Ethiopia; Aeronautical charts- Somalia; Government publication; National government publication. Notes: Description: 1 map; color; on sheet 106 x 146 cm. folded to 27 x 37 cm. Scale 1:500,000; Lambert conformal conic projection, standard parallels 9020' and 14040'; (E 430--E 490/N 120--N 80). Category of scale: a Constant ratio linear horizontal scale: 500000 Coordinates--westernmost longitude: E0430000 Coordinates--easternmost longitude: E0490000; Notes: Relief shown by contours, shading, tints, and spot heights. Depths shown by isolines. "Air information current at 29 August 1983." Shipping list no. 99-2009-S. Produced by the Military Survey for NIMA to issue as series TPC K-6D. "Compiled 1983." "Copyright HMSO London 1983." "Printed by DMA 08-96." Includes elevation tints and interchart relationship diagrams. Other Titles: Djibouti, Ethiopia, Somali Democratic Republic; TPC, Djibouti, Ethiopia, Somali Democratic Republic; Responsibility: produced under the direction of the Director of Military Survey, Ministry of Defence, United Kingdom. OCLC Accession Number: 42135039.

Great Britain. Directorate of Overseas Surveys. 1973. "Qabri Bahar." Tolworth, England: published by Directorate of Overseas Surveys for the Somali Democratic Republic under arrangements sponsored by the Institute of Geological Sciences London. Descriptors: Geology- Somalia- Qabri Bahar- Maps; Geology- Somalia- Maps; Government publication; National government publication. Notes: Description: 1 map; color; 44 x 44 cm. Geographic: Somalia- Maps. Qabri Bahar (Somalia)- Maps. Scale 1:125,000; Category of scale: a Constant ratio linear horizontal scale: 125000; Notes: "Base map derived from 1:125,000 series, D.C.S. 39. 1/73/7722622 S." Includes index to adjoining sheets. Responsibility: prepared by the Directorate of Overseas Surveys 1972. OCLC Accession Number: 59136640.

Great Britain. Directorate of Overseas Surveys. 1971. "1:125,000 Somali Democratic Republic: Geological Survey." Tolworth, England: The Directorate. Descriptors: Geology- Somalia- Maps; Government publication; National government publication. Notes: Description: maps; color; 59 x 44 cm. or smaller on sheets 82 x 54 cm. orsmaller. Geographic: Somalia- Maps. Scale 1:125,000; (E 410--E 520/N 120--S 10). Category of scale: a Constant ratio linear horizontal scale: 125000 Coordinates-- westernmost longitude: E0410000 Coordinates--easternmost longitude: E0520000; Notes: Includes index to adjoining sheets. Other Titles: Somali Democratic Republic. Responsibility: prepared by the Directorate of Overseas Surveys. OCLC Accession Number: 19210442.

Great Britain. Directorate of Overseas Surveys. 1959. "Somaliland Protectorate: 1:25,000." Tolworth, Surrey: The Directorate. Descriptors: Government publication; National government publication. Notes: Description: maps; color; 66 x 55 cm. or smaller on sheets 95 x 89 cm. or smaller. Geographic: British Somaliland- Maps, Topographic. Somalia- Maps, Topographic. Scale 1:25,000; Transverse Mercator projection. Clarke 1880 (Modified) Spheroid; (E 430--E 450/N 10030'--N 90). Category of scale: a Constant ratio linear horizontal scale: 25000 Coordinates--westernmost longitude: E0430000 Coordinates--easternmost longitude: E0450000; Notes: Includes sheet location diagram, reliability diagram, magnetic variation diagram, and index to adjoining sheets. Standard map series designation: Series Y823. Relief shown by contours, hachures, spot heights and landforms. Other Titles: Series Y823. Responsibility: constructed, drawn and photographed by Directorate of Overseas Surveys. OCLC Accession Number: 16407945.

Great Britain. Foreign Office. Research Dept. 1966. "Boundary between Ethiopia and the Former British Somaliland Protectorate." London: F.O. Descriptors: Government publication; National government publication. Notes: Description: 1 map; 16 x 23 cm. Geographic: Ethiopia- Boundaries- Somalia- Maps. Somalia- Boundaries- Ethiopia- Maps. Scale [ca. 1:5,000,000]; (E 420--E 500/N 130--N 70). Category of scale: a Constant ratio linear horizontal scale: 5000000 Coordinates--westernmost longitude: E0420000 Coordinates--easternmost longitude: E0500000; Notes: "Map 6." "Appendix E." "D.33501. R.C.1646." Responsibility: Research Dept., F.O., Feb. 1966. Library of Congress; LCCN: 2001-628350.

Great Britain. General Staff. Geographical Section. 1926. "Somaliland [with routes]." GSGS 2924(b). Scale: 1:3 000 000. [London]: [GSGS, War Office], 1926. 1 map: color Note(s): To accompany Military Report on British Somaliland 1926. Routes described in Vol. II, Chapters VI and VII. Subjects: Somalia- Maps and charts 1926; Ethiopia- Maps and charts. Added Name: Great Britain. War Office. General Staff. Geographical Section. British Library: Maps MOD GSGS 2924(b).

Great Britain. General Staff. Geographical Section. 1926. "British Somaliland [showing routes]." GSGS 2991(b). Scale: 1:1 000 000. [London]: [GSGS, War Office], 1926. 1 map: color Note(s): To accompany Military Report on British Somaliland 1926. Subjects: Somalia- Maps and charts 1926. Added Name: Great Britain. War Office. General Staff. Geographical Section. British Library: Maps MOD GSGS 2991(b).

Great Britain. General Staff. Geographical Section. 1925. "Africa: Italian possessions." GSGS 2710d. Scale 1:20 000 000. London: [GSGS, War Office], 1925. 1 map: color Note: Base map of Africa with Italian possessions (Libya, Eritrea, Somaliland) overprinted in green. Subjects: Eritrea- Maps and charts 1925; Libya- Maps and charts 1925; Somalia- Maps and charts 1925. Added name: Great Britain. War Office. General Staff. Geographical Section. British Library.

Great Britain. General Staff. Geographical Section. 1922. "British Somaliland." GSGS 2991. Scale: 1:1 000 000. [London]: [GSGS, War Office], 1922. maps: color Note(s): Extended and revised from TSGS 1781. Subjects: Somalia- Maps and charts 1922. Added Name: Great Britain. War Office. General Staff. Geographical Section. 2 sheets. British Library: Maps MOD GSGS 2991.

Great Britain. General Staff. Geographical Section. 1922. "Sketch map of Kenya Colony to illustrate negotiations in the cession of Jubaland to Italy." GSGS 2989. Scale: 1:3 000 000. [London]: [GSGS, War Office], 1922. 1 map. Note(s): Shows territory claimed by Italy in July 1922. Subjects: Kenya- Maps and charts; Somalia- Maps and charts 1922. Added Name: Great Britain. War Office. General Staff. Geographical Section. British Library: Maps MOD GSGS 2989

Great Britain. General Staff. Geographical Section. 1919. "Somaliland." GSGS 2924. Scale: 1:3 000 000. [London]: [GSGS, War Office], 1919. 1 map: color Note(s): Two versions held of same date; one with hypsometric tints; other is monochrome. Appears to be reprint of an earlier map dated 1907. See GSGS 3730 for index. Subjects: Somalia- Maps and charts; Ethiopia- Maps and charts. Added Name: Great Britain. War Office. General Staff. Geographical Section. 2 sheets. British Library: Maps MOD GSGS 2924.

Great Britain. General Staff. Geographical Section. 1918. "Sketch map of Somaliland [showing tribal names and districts]." GSGS 1675b. Scale: 1:3 000 000. [London]: [General Staff, Geographical Section], [1918]. 1 map: color Note(s): With inset of Sokotra [sic]. This version shows boundary between Abyssinia and Italian Somaliland. See also IDWO 1875A. Subjects: Somalia- Maps and charts; Ethiopia- Maps and charts. Added Name: Great Britain. War Office. General Staff. Geographical Section. British Library: Maps MOD GSGS 1675b.

Great Britain. General Staff. Geographical Section. 1915-1924. "Jubaland: Map to accompany draft of Anglo-Italian Convention 1924." GSGS 2966. Scale: 1:1 000 000. [London]: [GSGS, War Office], 1915-1924. 1 map: color Note(s): Revised 1924. 1915 version not held. Subjects: Somalia- Maps and charts; Kenya- Maps and charts 1915-1924. Added Name: Great Britain. War Office. General Staff. Geographical Section. British Library: Maps MOD GSGS 2966.

Great Britain. General Staff. Geographical Section. 1909. "Sketch map of Somaliland [showing Line of Illalo Scouts, Mullah's grazing ground in 1907, and tribal names and districts]." GSGS 1675a. Scale: 1:3 000 000. [London]: [General Staff, Geographical Section], 1909. 1 map: color Subjects: Somalia- Maps and charts; Ethiopia- Maps and charts 1909. Added Name: Great Britain. War Office. General Staff. Geographical Section. British Library: Maps MOD GSGS 1675a.

Great Britain. General Staff. Geographical Section. 1908. "Jubaland." GSGS 2318. Scale: 1:1 000 000. [London]: [GSGS, War Office], 1908. 1 map: color Note(s): Extract from Africa 1:1 million, sheets 87, 88, 94 & 95 overprinted with grazing and tsetse fly areas. Subjects: Kenya- Maps and charts; Somalia- Maps and charts 1908. Added Name: Great Britain. War Office. General Staff. Geographical Section. British Library: Maps MOD GSGS 2318.

Great Britain. General Staff. Geographical Section. 1908. "Route diagram to accompany Vol. II, Military report on Somaliland." GSGS 2405. Scale: 1:2 000 000. [London]: [GSGS, War Office], 1908. 1 map: color Note(s): Shows road numbers.

Subjects: Somalia- Maps and charts; Ethiopia- Maps and charts 1908. Added Name: Great Britain. War Office. General Staff. Geographical Section. British Library: Maps MOD GSGS 2405.

Great Britain. General Staff. Topographical Section. 1907. "Sketch map of Somaliland." TSGS 2262. Scale: 1:3 000 000. [London]: [TSGS, War Office], 1907. 1 map: color Subjects: Somalia- Maps and charts. Added Name: Great Britain. War Office. General Staff. Geographical Section. British Library: Maps MOD TSGS 2262.

Great Britain. General Staff. Topographical Section. 1906. "Map of a portion of Somaliland to illustrate Operations of 1901-04." TSGS 1781(a). Scale: 1:1 000 000. [London]: [General Staff, Topographical Section], 1906. 1 map: color Note(s): Identical to IDWO 1781, but with numerous locations underlined in British Somaliland, Italian Somaliland and Ogaden. Subjects: Somalia- Maps and charts; Ethiopia- Maps and charts 1906. Added Name: Great Britain. War Office. General Staff. Geographical Section. British Library: Maps MOD TSGS 1781(a).

Great Britain. MI9. 1944. "Somaliland; Kenya Colony & Juba River." London: MI9. Descriptors: Maps, Military; Government publication; National government publication. Notes: Description: 2 maps on 1 sheet; both sides, color, rayon; 44 x 54 cm. and 33 x 56 cm., sheet 48 x 59 cm. Geographic: Somalia- Maps. Kenya- Maps. Juba River Valley (Ethiopia and Somalia)- Maps. Scale 1:2,000,000. 1 in. to 31.56 miles. 1.014 in. to 32 miles; (E 420--E 520/N 120--N 40). Scale 1:2,000,000; (E 370--E 480/N 40--S 20). Category of scale: a Constant ratio linear horizontal scale: 2000000 Coordinates--westernmost longitude: E0420000 Coordinates--easternmost longitude: E0520000 Category of scale: a Constant ratio linear horizontal scale: 2000000 Coordinates--westernmost longitude: E0370000 Coordinates--easternmost longitude: E0480000; Notes: Relief shown by spot heights. Publisher's designations: P- Q. Cloth map. Other Titles: Kenya Colony & Juba River. Kenya Colony and Juba River. Library of Congress; LCCN: 86-694630.

Great Britain. War Office. 1957. "British Somaliland 1:100,000." Series GSGS 4868. Scale: 1:100 000. [London]: D. Survey, War Office & Air Ministry, 1957. 69 maps. Note(s): Series in 69 sheets. Later redesignated as Series Y623. Mixed monochrome and color. Subjects: Somalia- Maps and charts 1957. Added Name: Great Britain. War Office. General Staff. Geographical Section. 80 sheets. British Library: Maps MOD GSGS 4868.

Great Britain. War Office. 1937. "Ethiopia, Italian Somaliland, Eritrea." ADI (Maps) AM 204AB(W). Scale: 1:4 600 000. [S.l.]: Air Ministry, United Kingdom, 1937. maps: color Note(s): Based on OR 934. Declassified: originally classified Most Secret. Subjects: Ethiopia- Maps and charts; Somalia- Maps and charts; Eritrea- Maps and charts. Added Name: Great Britain. War Office. General Staff. Geographical Section. 1 sheet. British Library: Maps MOD ADI 204A(W).

Great Britain. War Office. 1936. "Anglo-Italian Somaliland Boundary." London: War Office; Somalia; Northern. Descriptors: Manuscript (mss). Notes: Description: 13 maps + index. Geographic: British Somaliland- Boundaries- Italian Somaliland- Maps. Italian Somaliland- Boundaries- British Somaliland- Maps. Somalia- Maps. Notes: Manuscript tracings of Italian originals. Includes index. Relief shown by contours and heights in metres. Shows: roads, tracks, villages, wells. Scale 1:100,000. Category of scale: a Constant ratio linear horizontal scale: 100000; General Info: Coverage: Northern Somalia. OCLC Accession Number: 48603069.

Great Britain. War Office. General Staff. Geographical Section. 1957. "British Somaliland." London: G.S.G.S. Somalia. Notes: Description: 68 maps. Geographic: Somalia- Maps, Topographic. Scale 1:100,000. Category of scale: a Constant ratio linear horizontal scale: 100000; Notes: Enlarged from D.C.S. 39. Date of information 1956- Relief shown as heights in feet. Shows: grid, international and administrative boundaries, railways (2 categories), roads (2 categories), tracks, woods. General Info: Coverage: Somalia. OCLC Accession Number: 48598757.

Great Britain. War Office. General Staff. Geographical Section. 1942. "Frontier of British Somaliland and the French Somali Coast." GSGS 4075. Scale: 1:50 000. [S.l.]: War Office, [1942]. 2 maps: color Note(s): Maps show detail 4km each side of frontier. Subjects: Djibouti- Maps and charts; Somalia- Maps and charts 1942. 2 sheets. British Library: Maps MOD GSGS 4075.

Great Britain. War Office. General Staff. Geographical Section. 1941. "Afmadu- Gelib Road." South Africa. Survey Company. EAF 187. First ed. Scale 1:50 000. Nairobi: [Survey Directorate, East Africa Command], 1941. Maps. Subject: Somalia- Maps and charts. 1 sheet. British Library.

Great Britain. War Office. General Staff. Geographical Section. 1941. "Afmadu- Soia Road." South Africa. Survey Company. EAF 158. First ed. Scale 1:50 000. Nairobi: [Survey Directorate, East Africa Command], 1941. Maps. Subject: Somalia- Maps and charts. 1 sheet. British Library.

Great Britain. War Office. General Staff. Geographical Section. 1940. "Road information, Italian East Africa. " GSGS 4045a. Scale 1:3 000 000. GSGS, War Office, [1940]. 1 map : color Notes: Overprint shows six categories of roads and tracks. Subject(s): Ethiopia- Maps and charts- 1940. Somalia- Maps and charts- 1940. 1 sheet. British Library.

Great Britain. War Office. General Staff. Geographical Section. 1934-1956. "Somaliland 1:250 000." Series GSGS 3927. Scale: 1:250 000. [S.l.]: War Office, 1934- 1956. 17 maps: color Note(s): Series in seventeen sheets covers former British Somaliland only. Sheetlines based on grid. Sheets numbered consecutively; 1 to 17. Subjects: Somalia- Maps and charts 1934-1956. 82 sheets. British Library: Maps MOD GSGS 3927.

Great Britain. War Office. General Staff. Geographical Section. 1940. "Italian East Africa." London: G.S.G.S. Somalia. Notes: Description: 1 map. Geographic: Italian East Africa- Maps. Somalia- Maps. Scale 1:3,000,000. Category of scale: a Constant ratio linear horizontal scale: 3000000; Notes: Shows: grid, international, intercolonial and administrative boundaries, principal towns, railways. General Info: Coverage: Somalia. OCLC Accession Number: 48602097.

Great Britain. War Office. General Staff. Geographical Section. 1939. "Commission de la delimitation de la frontier entre l'Ethiopie et le Somaliland Britannique: Carte de la frontier." Series GSGS 4082. Scale: 1:50 000. [London]: War Office, 1939. maps: color Note(s): GSGS number added by hand. Title and marginalia in French only. This GSGS series is a direct copy of a series which constituted Annexe 2 to agreement signed on 28 March 1935 between Abyssinia and the UK. Detail extends 4 km each side of frontier. Language: English and French. Subjects: Somalia- Maps and charts; Ethiopia- Maps and charts. 24 sheets. British Library: Maps MOD GSGS 4082.

Great Britain. War Office. General Staff. Geographical Section. 1934. "Somaliland." London: G.S.G.S.; Somalia. Notes: Description: 17 maps. Geographic: Somalia- Maps, Topographic. Scale 1:250,000. Category of scale: a Constant ratio linear horizontal scale: 250000; Notes: Relief shown as contours or form lines (interval 100 metres) and heights. Shows: international and administrative boundaries, railways, roads, tracks. General Info: Coverage: Somalia. OCLC Accession Number: 48599597.

Great Britain. War Office. General Staff. Geographical Section. 1934. "Map illustrating the Anglo-Italian exchange of notes of 22nd Novr. 1933, relating to the boundary between Kenya & Italian Somaliland." GSGS 3934. Scale: 1:1 000 000. [S.l.]: War Office, 1934. 1 map: color Note(s): Detail limited to area within 20 or 30 miles of frontier. Subjects: Kenya- Maps and charts; Somalia- Maps and charts 1934. 1 sheet. British Library: Maps MOD GSGS 3934.

Great Britain. War Office. General Staff. Geographical Section. 1932. "Anglo-Italian Boundary Somaliland. Appendix Iv to our Agreement of 1st June 1931." London: G.S.G.S.; Somalia. Notes: Description: 18 maps. Geographic: British Somaliland-Boundaries- Italian Somaliland- Maps. Italian Somaliland- Boundaries- British Somaliland- Maps. Somalia- Maps. Scale 1:50,000. Category of scale: a Constant ratio linear horizontal scale: 50000; Notes: Relief shown as contours (interval 25 metres). Shows: Anglo-Italian boundary, boundary pillars, roads, tracks, vegetation. General Info: Coverage: Somalia. OCLC Accession Number: 48598316.

Great Britain. War Office. General Staff. Geographical Section. 1931-1947. "Anglo-Italian boundary Somaliland ." "Confine fra le Somalie Italiana ed Inglese." Series GSGS 3918. Scale 1:50 000. [S.l.]: War Office, 1931-1947. 18 maps: color Note: "This map in 18 sheets, constitutes Appendix 4 to our agreement of 1st June 1931". Last four sheets show boundary between British Somaliland and Abyssinia (Ethiopia). Reprinted in 1947. Earlier versions not held. Language: English and Italian. Subject: Somalia- Maps and charts 1931-1947; Ethiopia- Maps and charts 1931-1947. 18 sheets. British Library.

Great Britain. War Office. General Staff. Geographical Section. 1925. "Somaliland." London: G.S.G.S.; Somalia. Notes: Description: 1 map. Geographic: Somalia- Maps. Scale 1:3,000,000; conic projection. Category of scale: a Constant ratio linear horizontal scale: 3000000; Notes: Shows: international boundaries, caravan and main routes, tracks. General Info: Coverage: Somalia. OCLC Accession Number: 48602081.

Great Britain. War Office. General Staff. Geographical Section. 1925. "British Somaliland." London: G.S.G.S.; Somalia; Northern. Notes: Description: 1 map. Geographic: Somalia- Maps. Scale 1:1,000,000; conic projection. Category of scale: a Constant ratio linear horizontal scale: 1000000; Notes: Relief shown as form lines and heights. Shows: international boundaries, railways, caravan and main routes, tracks. General Info: Coverage: Northern Somalia. OCLC Accession Number: 48603778.

Great Britain. War Office. General Staff. Geographical Section. 1924. "Jubaland, Map to Accompany Anglo-Italian Convention, 1924." London: Gen. Staff, Geog. Section. Descriptors: Government publication; National government publication. Notes:

Description: 1 map; color; 71 x 40 cm. Geographic: Juba River Valley (Ethiopia and Somalia)- Maps. Scale 1:1,000,000. OCLC Accession Number: 40541401.

Great Britain. War Office. General Staff. Geographical Section. 1919. "Somaliland." London: Geographical Section, General Staff, War Office. Notes: Description: 1 map; 48 x 38 cm. Geographic: Somalia- Maps. British Somaliland- Maps. Italian Somaliland- Maps. Scale 1:3,000,000. OCLC Accession Number: 54334440.

Great Britain. War Office. General Staff. Geographical Section. 1909. "Map of the Southern Frontier of Abyssinia." Southampton, England: Ordnance Survey Office. Descriptors: Government publication; National government publication. Notes: Description: 1 map on 3 sheets; sheets 60 x 76 cm. or smaller. Geographic: Ethiopia- Boundaries- Somalia- Maps. Somalia- Boundaries- Ethiopia- Maps. Ethiopia- Boundaries- Kenya- Maps. Kenya- Boundaries- Ethiopia- Maps. Scale 1:500,000. 1.014 in. = 8 mi. Category of scale: a Constant ratio linear horizontal scale: 500000; Notes: Relief shown by contours and spot heights. "This map is based on a triangulation executed by Capt. Maud, R.E. (1904), extended by Capt. Waller, R.E., 1908." "The topography is mainly from plane-table sketches executed by Khan Sahib Sher Jung, 1904, and by Major Gwynn, R.E., 1909." Other Titles: Southern frontier of Abyssinia. OCLC Accession Number: 53509283.

Great Britain. War Office. General Staff. Geographical Section. 1908. "Jubaland." London: Geog. Section, Gen. Staff. Descriptors: Government publication; National government publication. Notes: Description: 1 map; color; 78 x 56 cm. Geographic: Juba River Valley (Ethiopia and Somalia)- Maps. Scale 1:1,000,000. Category of scale: a Constant ratio linear horizontal scale: 1000000; Notes: "Parts of sheets 87, 88, 94 & 95." Other Titles: Africa 1:1,000,000 Jubaland. OCLC Accession Number: 40541400.

Great Britain. War Office. General Staff. Topographical Section. 1907. "Somaliland." London: T.S.G.S.; Somalia. Notes: Description: 6 maps. Geographic: Somalia- Maps, Topographic. Scale 1:250,000. Category of scale: a Constant ratio linear horizontal scale: 250000; Notes: Relief shown as form lines. Shows: international boundaries, railways, tracks, telegraphs. Coverage: Somalia. Accession Number: 48599572.

Great Britain. War Office. General Staff. Topographical Section. 1906. "Map of a Portion of Somaliland." London: T.S.G.S.; Somalia. Notes: Description: 1 map. Geographic: Somalia- Maps. Scale 1:1,000,000; conic projection. Category of scale: a Constant ratio linear horizontal scale: 1000000; Notes: Relief shown as form lines and heights. Shows: international boundaries, tracks. General Info: Coverage: Somalia. OCLC Accession Number: 48600558.

Great Britain. War Office. Intelligence Branch. 1887. "Map of the Somali country." IB 663. Scale: 1:1 267 200. [London]: [War Office, Intelligence Branch], 1887-1893. maps: color Note(s): Later version hand annotated with the route of Major Hood and Lieut. Ffinch [sic] of the North Staffordshire Regiment, 1893. Covers area north of 6° N. only. Subjects: Somalia- Maps and charts; Ethiopia- Maps and charts; Djibouti- Maps and charts 1887-1893. Added Name: Great Britain. War Office. General Staff. Geographical Section. 3 sheets. British Library: Maps MOD IB 663.

Great Britain. War Office. Intelligence Branch. 1887. "Map of eastern tropical Africa, colored to illustrate the agreement between Great Britain & Germany relative to their respective spheres of influence and to the limits of the Zanzibar Dominions, October

1886." IB 630. Scale: approx [1:4 377 600]. [London]: [War Office, Intelligence Branch], 1887. 1 map: color Note(s): Map highlights areas "over which Great Britain reigns and Germany agrees not to extend her influence" and coastal areas ruled by Zanzibar. Subjects: Tanganyika- Maps and charts; Kenya- Maps and charts; Uganda- Maps and charts; Somalia- Maps and charts; Comoro Islands- Maps and charts 1887. Added Name: Great Britain. War Office. General Staff. Geographical Section. British Library: Maps MOD IB 630.

Great Britain. War Office. Intelligence Branch. 1885. "Map of eastern tropical Africa showing the limits of the Zanzibar Dominions, including the recent discoveries of Mr Joseph Thomson." IB 403. Scale: [1:4 377 600]. [London]: [War Office, Intelligence Branch], 1885. 1 map: color Note(s): Shows Zanzibar Dominions, which stretch from Mogdisho [sic] to Cape Delgado and inland to Ujiji. Mr Thomson travelled from Mombasa to Lake Victoria via Kilimanjaro and southern Kenya. Subjects: Tanganyika- Maps and charts; Kenya- Maps and charts; Somalia- Maps and charts; Uganda- Maps and charts 1885. Added Name: Great Britain. War Office. General Staff. Geographical Section. British Library: Maps MOD IB 403.

Great Britain. War Office. Intelligence Branch. 1885. "Map showing the boundaries of the Eesa and Gadabursi territories." IB 497. Scale: approx [1:1 267 200]. [London]: [War Office, Intelligence Branch], 1885. 1 map: color Note(s): Shows tribal boundaries in area bounded by Berbera, Harar and Gulf of Tajura. Subjects: Somalia- Maps and charts; Djibouti- Maps and charts 1885. Added Name: Great Britain. War Office. General Staff. Geographical Section. British Library: Maps MOD IB 497.

Great Britain. War Office. Intelligence Division. 1905. "Sketch map to illustrate Jubaland Expedition 1901." IDWO 1690. Scale: 1:506 880. [London]: [War Office, Intelligence Division], 1905. 1 map. Note(s): Shows route taken by Ogaden Punitive Force 12 January to 6 June 1901. Subjects: Somalia- Maps and charts. Added Name: Great Britain. War Office. General Staff. Geographical Section. British Library: Maps MOD IDWO 1690.

Great Britain. War Office. Intelligence Division. 1903. "Somaliland: Reconnaissance survey of road Berbera to Bohodleh." IDWO 1771. Scale: 1:126 720. [London]: [War Office, Intelligence Division], 1903. Maps. Note(s): Map in four parts with title etc. only on sheet 1. Sheets numbered 1, [1-b], 2 and [2-b]. Bohodleh (now spelt Buuhoodle) is on the border with Ethiopia. Subjects: Somalia- Maps and charts 1903. Added Name: Great Britain. War Office. General Staff. Geographical Section. 4 sheets. British Library: Maps MOD IDWO 1771.

Great Britain. War Office. Intelligence Division. 1903. "Berbera- Harrar railway: Berbera harbour." IDWO 1842. Scale: 1:10 560. London: [War Office, Intelligence Division], 1903. 1 map. Note: Town plan shows proposed railway line, station and jetty. See also IDWO 1838, 1840 and 1841. Subjects: Somalia- Maps and charts 1903. Added Name: Great Britain. War Office. General Staff. Geographical Section. Brisith Library: Maps MOD IDWO 1842

Great Britain. War Office. Intelligence Division. 1903. "Berbera- Harrar railway." IDWO 1838. Scale: 1:1 000 000. [London]: [War Office, Intelligence Division], [1903]. Maps: color Note: Overprinted on sheet 68 of Africa 1:1 million series (IDWO 1539), this map shows proposed railways from Berbera to Harrar via Hargeisa and Jig-Jiga, and also from Berbera to Burao and Bohodle. None of these lines were ever built. Subjects:

Somalia- Maps and charts 1903; Ethiopia- Maps and charts 1903. Added Name: Great Britain. War Office. General Staff. Geographical Section. British Library: Maps MOD IDWO 1838.

Great Britain. War Office. Intelligence Division. 1903. "Berbera- Harrar railway: Index section of Burao branch." IDWO 1841. Scale: 1:1 013 760. [London]: [War Office, Intelligence Division], 1903. 1 map. Note: Not a map, but a cross-section showing distances and altitudes of proposed railway. See also IDWO 1838, 1840 and 1842. Subjects: Somalia- Maps and charts 1903. Added Name: Great Britain. War Office. General Staff. Geographical Section. British Library: Maps MOD IDWO 1841.

Great Britain. War Office. Intelligence Division. 1903. "Berbera- Harrar railway: Index section of main line." IDWO 1840. Scale: 1:1 013 760. [London]: [War Office, Intelligence Division], 1903. 1 map. Note: Not a map, but a cross-section showing distances and altitudes of proposed railway. See also IDWO 1838, 1841 and 1842. Subjects: Somalia- Maps and charts 1903; Ethiopia- Maps and charts 1903. Added Name: Great Britain. War Office. General Staff. Geographical Section. British Library: Maps MOD IDWO 1840

Great Britain. War Office. Intelligence Division. 1933. "Berbera- Harrar railway reconnaissance." IDWO 1933. Scale: 1:250 000. [London]: [War Office, Intelligence Division], 1904. 1 map. Subjects: Somalia- Maps and charts 1904. Added Name: Great Britain. War Office. General Staff. Geographical Section. British Library.

Great Britain. War Office. Intelligence Division. 1904. "Sketch map of Somaliland showing tribal boundaries." IDWO 1873A. Scale: 1:3 000 000. [London]: [War Office, Intelligence Division], [1904]. 1 map: color Note(s): Overprinted on IDWO 1675, this map shows tribal boundaries and area occupied by treaty tribes (i.e. in British Protectorate and nearby parts of the Ogaden). Subjects: Somalia- Maps and charts; Ethiopia- Maps and charts 1904. Added Name: Great Britain. War Office. General Staff. Geographical Section. British Library: Maps MOD IDWO 1873A.

Great Britain. War Office. Intelligence Division. 1903. "Sketch Map of Somaliland." London: Intell. Div., War Office. Notes: Description: 1 map; 48 x 37 cm. Geographic: Somalia- Maps. Scale 1:3,000,000. OCLC Accession Number: 62723067.

Great Britain. War Office. Intelligence Division. 1902. "Map of a Portion of Somaliland." London: I.D.W.O.; Somalia. Notes: Description: 1 map. Geographic: Somalia- Maps. Scale 1:1,000,000; conic projection. Category of scale: a Constant ratio linear horizontal scale: 1000000; Notes: Relief shown as shading, form lines and heights. Shows: international boundaries, tracks. OCLC Accession Number: 48600513.

Great Britain. War Office. Intelligence Division. 1902. "Map to accompany the Report on the Somaliland Operations." IDWO 1664. Scale: 1:1 000 000. [London]: [War Office, Intelligence Division], 1902. 1 map. Note(s): A general map which does not show troop movements etc. Subjects: Somalia- Maps and charts; Ethiopia- Maps and charts 1902. Added Name: Great Britain. War Office. General Staff. Geographical Section. British Library: Maps MOD IDWO 1664.

Great Britain. War Office. Intelligence Division. 1895. "General map of the Italian Colony of Eritrea and adjoining countries." IDWO 1083. Scale: 1:1 584 000. London: Edward Stanford, 1895. 1 map: color Note(s): See IB 397 for 1886 version with different title. Subjects: Eritrea- Maps and charts; Ethiopia- Maps and charts; Sudan- Maps and charts; Yemen- Maps and charts; Somalia- Maps and charts; Djibouti- Maps and charts 1895. Added Name: Great Britain. War Office. General Staff. Geographical Section. British Library: Maps MOD IDWO 1083

Great Britain. War Office. Intelligence Division. 1895. "The Italian sphere of influence in Africa." IDWO 1895. Scale: 1:4 000 000. [London]: [War Office, Intelligence Division], 1895. 1 map. Note(s): Second version (November 1895) has additional place-names but no changes to boundaries. Subjects: Ethiopia- Maps and charts; Eritrea- Maps and charts; Somalia- Maps and charts; Sudan- Maps and charts 1895. Added Name: Great Britain. War Office. General Staff. Geographical Section. 2 sheets. British Library: Maps MOD IDWO 1085

Great Britain. War Office. Intelligence Division. 1903. "Map of a portion of Somaliland." IDWO 1676. Scale: 1:1 000 000. [London]: [War Office, Intelligence Division], 1903. maps: color Note(s): IDWO 1676 and 1676 provide partial coverage of British and Italian Somaliland in two considerably overlapping sheets. Detail taken from five sheets of the Africa 1:1 million Series IDWO 1539. Subjects: Ethiopia- Maps and charts; Somalia- Maps and charts 1903. Added Name: Great Britain. War Office. General Staff. Geographical Section. 5 sheets. British Library: Maps MOD IDWO 1676

Great Britain. War Office. Intelligence Division. 1903-1904. "Map of a portion of Somaliland." IDWO 1781. Scale: 1:1 000 000. [London]: [War Office, Intelligence Division], 1903-1906. maps: color Note(s): Covers area approx. 5° to 11°N. and 44 to 51 degrees East. Detail extracted from four sheets of IDWO 1539. See also IDWO 1817. Subjects: Somalia- Maps and charts; Ethiopia- Maps and charts 1903-1906. Added Name: Great Britain. War Office. General Staff. Geographical Section. 3 sheets. British Library: Maps MOD IDWO 1781

Great Britain. War Office. Intelligence Division. 1900. "Map of Capt. M S Wellby's routes in Somali-land." IDWO 1508. Scale: 1:1 267 200. [London]: [War Office, Intelligence Division], 1900. 1 map. Note(s): Shows route from Berbera south through Ogaden and back. Subjects: Somalia- Maps and charts; Ethiopia- Maps and charts 1900. Added Name: Great Britain. War Office. General Staff. Geographical Section. British Library: Maps MOD IDWO 1508.

Great Britain. War Office. Intelligence Division. 1897. "Map of the basin of the Juba River and of the countries to the north and west." IDWO 1249. Scale: 1:5 977 382. [London]: [War Office, Intelligence Division], 1897. 1 map: color Note(s): See IDWO 1250 for sheet to the south of same specification. Subjects: Ethiopia- Maps and charts; Somalia- Maps and charts; Kenya- Maps and charts; Sudan- Maps and charts 1897. Added Name: Great Britain. War Office. General Staff. Geographical Section. British Library: Maps MOD IDWO 1249.

Great Britain. War Office. Intelligence Division. 1898. "Map of the countries bordering on the R. Nile from its source to Assuan 1898." IDWO 1300. Scale: 1:11 222 688. [London]: [War Office, Intelligence Division], 1898. 1 map: color Note(s): Map shows a variety of boundaries; agreed, claimed, spheres of influence etc. Subjects: Sudan- Maps and charts; Ethiopia- Maps and charts; Eritrea- Maps and charts; Somalia- Maps and charts 1898. Added Name: Great Britain. War Office. General Staff. Geographical Section. British Library: Maps MOD IDWO 1300.

Great Britain. War Office. Intelligence Division. 1898. "Map of the river Juba below the Le Hele rapids and of the Aff Madu district." IDWO 1336. Scale: 1:506 880. [London]: [War Office, Intelligence Division], 1898. 1 map. Note(s): See also IDWO

1414. Subjects: Somalia- Maps and charts 1898. Added Name: Great Britain. War Office. General Staff. Geographical Section. British Library: Maps MOD IDWO 1336.

Great Britain. War Office. Intelligence Division. 1897. "Map of Zanzibar and parts of British & German East Africa." IDWO 1250. Scale: 1:5 977 382. [London]: [War Office, Intelligence Division], 1897. 1 map: color Note(s): See IDWO 1249 for sheet to the north of same specification. Subjects: Tanganyika- Maps and charts; Kenya- Maps and charts; Somalia- Maps and charts 1897. Added Name: Great Britain. War Office. General Staff. Geographical Section. British Library: Maps MOD IDWO 1250.

Great Britain. War Office. Intelligence Division. 1899. "British East Africa: Map of part of the Province of Jubaland." IDWO 1414. Scale: 1:506 880. [London]: [War Office, Intelligence Division], 1899. 1 map. Note(s): Detail only on right (west) bank of River Juba. The other side was an Italian possession. A revision of IDWO 1336. Subjects: Somalia- Maps and charts 1899. Added Name: Great Britain. War Office. General Staff. Geographical Section. British Library: Maps MOD IDWO 1414.

Great Britain. War Office. Intelligence Division. 1897-1898. "Map showing the limits of the East Africa Protectorate." IDWO 1205. Scale: 1:1 584 000. [London]: [War Office, Intelligence Division], 1897-1898. 1 map: color Note(s): Shows province and district boundaries. Subjects: Kenya- Maps and charts; Somalia- Maps and charts; Tanganyika- Maps and charts 1897-1898. Added Name: Great Britain. War Office. General Staff. Geographical Section. British Library: Maps MOD IDWO 1205

Great Britain. War Office. Intelligence Division. 1897. "Plan of the roads between Zeila and Antotto." IDWO 1238. Scale 1:1 000 000. London: [War Office, Intelligence Division], 1897. 1 map. Note: Copied from Italian map "Carta dimostrativa dell'Etiopia". Zeila now spelt Saylac. Antotto is near Addis Ababa. Subject(s): Ethiopia -- Maps and charts- 1897. Somalia -- Maps and charts- 1897. Added name: Great Britain. War Office. General Staff. Geographical Section. British Library.

Great Britain. War Office. Intelligence Division. 1897. "Skeleton map of Abyssinia & Somaliland to illustrate the Report of the Mission to Abyssinia in 1897." IDWO 1269. Scale: 1:4 000 000. [London]: [War Office, Intelligence Division], 1897. 1 map: color Note(s): Another version, apparently of the same date, has title "Abyssinian boundaries" but does not show Abyssinian province boundaries, province names or names of governors. Both maps are overprints on IDWO 1085. Subjects: Ethiopia- Maps and charts; Eritrea- Maps and charts; Somalia- Maps and charts; Djibouti- Maps and charts; Sudan- Maps and charts 1897. Added Name: Great Britain. War Office. General Staff. Geographical Section. 2 sheets. British Library: Maps MOD IDWO 1269.

Great Britain. War Office. Intelligence Division. 1897. "Skeleton map of part of British East Africa." IDWO 1160. Scale: 1:253 440. [London]: [War Office, Intelligence Division], 1897. Maps. Note(s): Series on irregular sheet-lines. Sheet has title "Sketch showing the proposed boundary of the Sultanate of Witu". Subjects: Kenya- Maps and charts; Somalia- Maps and charts; Tanganyika- Maps and charts 1897. Added Name: Great Britain. War Office. General Staff. Geographical Section. 4 sheets. British Library: Maps MOD IDWO 1160a-d

Great Britain. War Office. Intelligence Division. 1894. "Sketch of the routes in Somali Land travelled by Lts. H. C. Lowther & C. F. S. Vandeleur, Scots Guards in 1894." IDWO 1047. Scale: 1:443 520. [London]: [War Office, Intelligence Division], 1894. 1 map. Note(s): Route goes inland from Berbera to Hargeisa and Jijiga. Subjects:

Somalia- Maps and charts; Ethiopia- Maps and charts. Added Name: Great Britain. War Office. General Staff. Geographical Section. British Library: Maps MOD IDWO 1047.

Great Britain. War Office. Intelligence Division. 1892. "The course of the River Juba from Bardera to its mouth." ID 882. Scale: 1:500 000. [London]: [War Office, Intelligence Division], 1892. 1 map. Note(s): "From the travels of Baron C C v.d. Decken, 1862-5". Bardera now (2000) spelt Baardheere. Subjects: Somalia- Maps and charts. Added Name: Great Britain. War Office. General Staff. Geographical Section. British Library: Maps MOD ID 882.

Great Britain. War Office. Intelligence Division. 1890? "Africa 1:1,000,000." Series IDWO 1539. Scale 1:1 000 000. London: [War Office, Intelligence Division], [18- -] maps: color. Note: 56 sheets held out of the 132 required to cover all of mainland Africa (excludes Madagascar). Coverage does not equate with British possessions but is limited to, roughly, Morocco, coastal Algeria, Libya and Egypt, Nigeria and coast of West Africa, Sudan, Ethiopia, Somalia, Namibia and southern Angola. 4 by 6 degree sheetlines identical to IMW but with different numbering system, also used by IDWO 1479 and 1489. Revised editions in this series also designated TSGS 1539 or GSGS 1539. Four sheets of Sudan colored to show provincial boundaries. See IDWO 1476(a) for original index. Subjects: Africa -- Maps and charts. Added name: Great Britain. War Office. General Staff. Geographical Section. 62 sheets total. British Library.

Great Britain; War Office; General Staff and Geographical Section. 1919. "Somaliland." London: War Office. Descriptors: Government publication; National government publication. Notes: Description: 1 map; mounted on linen, color; 52 x 46 cm. Scale 1:3,000,000. Geographic: Somalia- Maps. Notes: Relief shown by gradient tints. Includes inset. Original: Compiled by the War Office. OCLC Accession Number: 38586630.

Green R.H. 2000. "Rehabilitation: Strategic, Proactive, Flexible, Risky?" Disasters. Volume 24, Issue 4, Pages 343-362. Descriptors: International Aid and Investment; post-war; aid policy; poverty alleviation; disaster management. Additional Info: United Kingdom; References: 12; Geographic: Mozambique; Somalia; Uganda. Abstract: Rehabilitation after armed conflict is a direct intellectual descendant of thinking about rehabilitation after natural calamity. It is related, generally, to poverty reduction and, operationally, to associated action at the micro level. This history has limited its strategic conceptualization and, in particular, its links with reconciliation and state re-legitimation and also with macro-economic stabilization and renewed growth. In post-war- or more generally, a lull in conflict with the potential to become permanent- a country's rapid, focused, prioritized action within a strategic framework is urgent. It is not risk-free for political and natural disasters as well as for those with economic and social bases. Flexibility, learning from initial experience and asking intended household beneficiaries about their needs in advance can reduce risk as can pre-positioning of contingency or standby resources to avert post-war calamities (for example, drought, flood) and catastrophes (renewed violence) from delaying and discrediting rehabilitation efforts. ISSN: 0361-3666.

Green R.H. 1999. "Khatt and the Realities of Somalis: Historic, Social, Household, Political and Economic." Rev. Afr. Polit. Econ. Issue 79, Pages 33-49. Descriptors: National Trade and Consumption; drugs trade; income; informal sector; national economy. Notes: Additional Info: United Kingdom; Geographic: Somalia.

Abstract: This article presents a brief review of khatt; a macro analysis of its roles in the Republic of Somaliland; briefer sketches of divergences of roles in other Somali heartland territories and Kenya concluding with a speculative section on what might be desirable (for Somalis) ways forward. All of these topics are bedeviled by limited prior research, the special pleading strands in much writing, the difficulty in knowing of or securing much of what is known and written and the intensely emotional context of most discourse. The last is not inherently a bad thing- khatt matters. Half of urban household absolute poverty, of farmer cash income and of female instituted divorces are not matters particularly appropriate for mild, disassociated academic curiosity. However, emotion leading to a rush to unanalyzed action, and to inventing 'symbolic truths', which have meaning but not analytical veridicality, can be the enemy of, just as much as the catalyst toward, the possible. ISSN: 0305-6244.

Greenwood, E. 1980. "Staff Appraisal Report; Somalia, Second Mogadishu Water Supply Project." United States: Volume: 3252-SO, Descriptors: Africa; Benadir Somali Republic; East Africa; Mogadishu Somali Republic; Mogadishu Water Supply Project; Somali Republic; water resources; water supply. Database: GeoRef. Accession Number: 1998-017958.

Greenwood, J. E. G. W. 1960. "Report on the Geology of the Las Khoreh-Elayu Area, Erigavo District Quarter Degree Sheets Nos. 5 and 6." Somaliland, Geological Survey, Rept. Volume 3, Pages 36. Descriptors: Africa; Cenozoic; Cretaceous; East Africa; geologic history; geologic maps; geology; Jurassic; Las Khoreh-Elayu area; Las Khoreh-Elayu area (parts); maps; Mesozoic; metals; metamorphic rocks; Somali Republic; Tertiary; tin. Notes: illustrations (incl. colored geol. maps); Abstract: Explanatory text for the Elayu and Las Khoreh sheets of the 1:125,000 geologic map of Somaliland; includes an appendix, by J. A. B. Stewart, dealing with the cassiterite deposits of the Dalan area. The mapped area is underlain principally by metasediments of the Inda Ad series. Details of their stratigraphic, structural, and metamorphic relationships are summarized. Jurassic to Tertiary sedimentary formations are also represented in the region. Database: GeoRef.

Greenwood, J. E. G. W. 1960. "Report on Geological Investigations in Somaliland Protectorate with Particular Reference to the North-Eastern Area." Report of the Geological Survey, Somaliland Protectorate. Volume JEG/1, Descriptors: Africa; areal geology; East Africa; northeastern Somali Republic; northern Somali Republic; Somali Republic; surveys. Database: GeoRef.

Greenwood, J. E. G. W. 1960. Elayu. United Kingdom: Directorate Overseas Survey, United Kingdom. Descriptors: Africa; areal geology; East Africa; Elayu; geologic maps; maps; Somali Republic. Notes: Latitude: N110000, N113000 Longitude: E0490000, E0483000. Database: GeoRef. Accession Number: 1989-007759.

Greenwood, J. E. G. W. 1960. Las Khoreh. United Kingdom: Directorate Overseas Survey, United Kingdom. Descriptors: Africa; areal geology; East Africa; geologic maps; Las Khoreh; maps; Somali Republic. Notes: Latitude: N110000, N113000 Longitude: E0483000, E0480000. Database: GeoRef. Accession Number: 1989-007760.

Greenwood, J. E. G. W. 1960. Report on the Geology of the Las Khoreh-Elayu Area, Erigavo District. with an Appendix on the Cassiterite Deposits of the Dalan Area. Hargeisa: Somaliland Stationery Office. Descriptors: 36 pages illustrations, maps. Notes:

6 maps in end pocket. Responsibility: by J.A.B. Stewart. OCLC Accession Number: 34363005.

Greenwood, J. E. G. W. and Stewart, J. A. B. 1960. Report of the Geology of the Las Khoreh-Elayu Area: Erigavo District: Quarter Degree Sheets Nos 5 and 6 (D.O.S. 1:125,000). Hargeisa: Stationary Office. Descriptors: VIII, 36. Notes: Responsibility: J.E.G.W. Greenwood; with an appendix on the Cassiterite deposits of the Dalan Area by J.A.B. Stewart. Accession Number: 67822761.

Greenwood, W. R. 1982. "A Preliminary Evaluation of the Nonfuel Mineral Potential of Somalia." United States: 1982. Volume: OF 82-0788, pages: 46. Descriptors: Africa; East Africa; economic geology; evaluation; geologic maps; index maps; maps; mineral exploration; mineral resources; Somali Republic; USGS. Notes: USGS Open-file Reports; Publications of the U. S. Geological Survey; References: 77; illustrations; ISSN: 0196-1497.

Gregory, J. W. 1925. "The Geology of Somaliland and its Relations to the Great Rift Valley." Monographs of the Geological Department of the Hunterian Museum. Volume 1, Descriptors: Africa; areal geology; East Africa; East African Rift; Great Rift Valley; Somali Republic. Database: GeoRef.

Gregory, J. W. 1900. "Geology and Fossil Corals and Echinoids of Somaliland." Quarterly Journal of the Geological Society of London. Volume 56, Pages 26-45. Descriptors: Africa; Anthozoa; areal geology; biostratigraphy; Coelenterata; East Africa; Echinodermata; Echinoidea; Echinozoa; fossils; Invertebrata; Somali Republic. Notes: 1 plate. ISSN: 0370-291X.

Gregory, J. W. 1896. "Note on the Geology of Somaliland." Geol. Mag. Volume 4, Pages 289-294. Descriptors: Africa; areal geology; East Africa; Somali Republic. ISSN: 0016-7568.

Gregory, J. W. 1896. "A Note on the Geology of Somaliland Based on the Collections made by Mrs. E. Lort Phillips, Miss Edith Cole and Mr. G. P. V. Aylmer." Geol. Mag. Volume 33, Descriptors: Africa; areal geology; collections; East Africa; northern Somali Republic; Somali Republic. ISSN: 0016-7568.

Griffiths, C. L. 2005. "Coastal Marine Biodiversity in East Africa." Indian J. Mar. Sci. MAR. Volume 34, Issue 1, Pages 35-41. Abstract: The Indian Ocean coastline of mainland Africa is over 9 500 km long and comprises the tropical coasts of Somalia, Kenya, Tanzania and Mozambique, plus the subtropical and warm-temperate Indian Ocean coastline of South Africa. The regional marine fisheries catch (Indian Ocean catch only for South Africa) is about 200 000 t, more than 80% of which is taken from Mozambique and Tanzania. Regional fisheries are dominated by intense artisanal and subsistence sectors, although there are also several commercially important industrial fisheries, mostly targeting lobsters, prawns and squid. No reliable species lists exist for individual countries of tropical Africa, but 11257 marine species are recorded from the western Indian Ocean region (island states included) and 11980 species from South Africa (including Atlantic coast). Comparing these lists by taxonomic group, and with similar. lists for Europe (29713 known species) reveals great disparities in taxonomic coverage and large gaps in the data, especially for smaller sized organisms. It is concluded that less than half of species actually present in East Africa have been described. Existing data are also based largely on shallow-water surveys and the benthic invertebrates of deeper waters, especially those of the continental slope and abyssal zone, remain virtually unexplored. ISSN: 0379-5136.

Griffiths, John F. 1972. Climates of Africa; Amsterdam: New York, Elsevier Pub. Co. Descriptors: 604; Climatologie; Klimaat. Abstract: "The Horn of Africa" is found on pages 133-165. Covers Somalia, Jubuti [sic], and the Eritrean coast of Ethiopia. a narrative summary with maps, charts and bibliography. Notes: Africa- Climate. Afrique-Climat. 6. ISBN: 0444408932; LCCN: 72-135485.

Grimes, Annie E. 1968. Bibliography of Climatic Maps for Ethiopia, Eritrea, and the Somalilands. Washington, D.C: U.S. Dept. of Commerce, Weather Bureau. Notes: Ethiopia- Climate- Maps- Bibliography. Somalia- Climate- Maps- Bibliography. Notes: Photocopy. OCLC Accession Number: 65852096.

Grinyer, A. W. and Jones, S. A. 1977. "Somalia. Reconnaissance Survey of the East Coast from Brava to Hafun for Establishment of Fish Landing Centres. A Report Prepared for the Fisheries Development Project." FAO, Rome (Italy). pages: FAO, Rome (Italy). 31. Descriptors: Echo surveys; Water currents; Coastal zone; Fishing harbors; Somali Democratic Republic; Marine. Abstract: The consultants were appointed to carry out surveys by regions of the coastline of Somalia, as part of the UNDP/FAO Fisheries Development Project in Somalia, to locate possible fishing ports and landing places. Following discussions with the Ministry of Fisheries and Marine Transport, the extent of the survey work was limited to the east coast from Brava to Bargaal. Selected locations of potential importance to fishing industry development were examined. Measurements of water depths were obtained by means of a portable echo sounder and further information was obtained by field survey, local inquiry and observation. The findings indicate a somewhat featureless coast, exposed to a seasonal reversal of strong monsoon winds blowing parallel to the coast, with accompanying strong currents and littoral drift. Natural protection is limited and is often accompanied by minimal depth of water. Proposals have been prepared for the possible development of fish landing facilities at the 3 most suitable locations, namely Brava, Bender Beila and Hafun. Notes: 5 photographs; 14 maps. Records keyed from 1977 ASFA printed journals. NU: FAO/FI/DP/SOM--73/005/3; TR: 1977. Database: ASFA: Aquatic Sciences and Fisheries Abstracts. Accession Number: 5262469.

Grolee, J. 1961. "Le Probleme de l'Eau en Côte Française del Somalis." Translated title: "Water Problems on the French Somali Coast." France: BCEOM, Paris, France. Descriptors: Africa; Djibouti; East Africa; French Somaliland; water resources. Database: GeoRef. Accession Number: 1998-017963.

Grubb P. 2002. "Types, Type Locality and Subspecies of the Gerenuk Litocranius Walleri (Artiodactyla: Bovidae)." J. Zool. Volume 257, Issue 4, Pages 539-543. Descriptors: Animal ecology: mammals: population ecology- 73.4.7.5.3; subspecies; geographical distribution; type locality; ungulate Species Term: Litocranius walleri; Bovidae; Artiodactyla; Ungulata; Vertebrata. Notes: Additional Info: United Kingdom; References: Number: 26; Geographic: Somalia- Chisimaio Ethiopia Tanzania Kenya. Abstract: From what is known about the discovery of the gerenuk Litocranius walleri, the type locality is here restricted to the vicinity of Chisimaio, Somalia. Two discrete subspecies can be recognized. The smaller L. w. walleri occurs in north-east Tanzania, Kenya, southern Somalia and southern Ethiopia. The larger L. w. sclateri is restricted to

Djibouti, northern Somalia and adjacent parts of northern Ethiopia and is present in very few protected areas. ISSN: 0952-8369.

GTZ, Mogadishu, Somalia. 1984. Proposed Programme for Implementation in Somalia Dem. Rep. Final Draft. Somalia: GTZ, Mogadishu, Somalia. Descriptors: Africa; East Africa; programs; Somali Republic; water resources. Database: GeoRef. Accession Number: 1998-017969.

GTZ, Mogadishu, Somalia. 1984. Water Resources Development Project in Somalia; Water Supply II, Burao Technical Report Extension of Well Field. Somalia: GTZ/WDA, Mogadishu, Somalia. Descriptors: Africa; Burao, Somali Republic; development; East Africa; Somali Republic; Togdheer Somali Republic; water resources; water supply; water wells; wells. Database: GeoRef. Accession Number: 1998-017970.

GTZ, Mogadishu, Somalia. 1981. Water Resources Development Project in Somalia; Water Supply II, Dusamareeb Technical Report. Somalia: GTZ/WDA, Mogadishu, Somalia. Descriptors: Africa; development; Dusamareeb Somali Republic; East Africa; Galguduud Somali Republic; Somali Republic; water resources; water supply. Database: GeoRef. Accession Number: 1998-017968.

GTZ, Mogadishu, Somalia. 1976. Water Resources Development Project in Somalia; Interim Report; Vol. I. Somalia: GTZ, Mogadishu, Somalia. Descriptors: Africa; development; East Africa; Somali Republic; water resources. Database: GeoRef. Accession Number: 1998-017967.

Guébourg J.-L. 2002. "Le Somaliland, un Jeune État; en Quête de Reconnaissance." Mappemonde. Volume 2002, Issue 3, Pages 39-42. Descriptors: International Trade and Economic Association; competition (economics); regional economy; regional trade. Notes: Additional Info: France; Geographic: Somalia-Somaliland. Abstract: Somaliland, a former British colony, composed of the Issak clan and three million people strong, unilaterally seceded from Somalia in 1991. The restored sea link between Berbera, the only harbor, and Addis Abeba, and the separation between Ethiopia and Eritrea have boosted Somaliland's trade and enable it to compete modestly with Djibouti. ISSN: 0764-3470.

H

Haakonsen, M. J. 1985. Report on the December Rains and Flood and their Effect in the North-Western Regions. Somalia: UNICEF, Hargeisa, Somalia. Descriptors: Africa; atmospheric precipitation; climate; East Africa; floods; geologic hazards; hydrology; northern Somali Republic; rain; Somali Republic. Database: GeoRef. Accession Number: 1998-017972.

Haakonsen, M. J. 1984. A Brief Description of some Water Sources and Nomadic Settlements in Berbera District. Somalia: UNICEF, Hargeisa, Somalia. Descriptors: Africa; Berbera Somali Republic; East Africa; Somali Republic; Waqooyi Galbeed; water resources. Database: GeoRef. Accession Number: 1998-017971.

Haaland G and Keddeman W. 1984. "Poverty Analysis: The Case of Rural Somalia." Economic Development & Cultural Change. Volume 32, Issue 4, Pages 843-860. Descriptors: International Development Abstracts; Poverty analysis; Agro pastoral systems of production; Somalia; Socioeconomic monitoring. Abstract: Deals with different approaches in poverty analysis. Methodological problems are discussed in relation to concrete material from Somalia, where more than half the population are livestock producers. A specific methodological problem in this context is comparison of the poverty level of pastoralists and subsistence farmers. Attempts to throw light on this problem through a combination of a case study anthropological approach and larger statistical surveys. ISSN: 0013-0079.

Haas, Otto; Miller, A. K. MacFadyen, William A. and Hunt, John A. 1952. Eocene Nautiloids of British Somaliland. New York: American Museum of Natural History. Descriptors: p. 317-354, 11; Nautiloidea, Fossil- Somalia; Paleontology- Eocene; Paleontology- Somalia; Geology- Somalia. Notes: "With a note on the geology of the Daban area and the localities of the described nautiloids by W.A. MacFadyen." "Issued August 5, 1952"--T.p. verso. "Some 20 years ago, W.A. Macfadyen and J.A. Hunt collected well over a hundred Tertiary nautiloids from British Somaliland (which had previously yielded only two specimens), and these constitute the basis for this study"--P. 319 Includes bibliographical references (p. 350-354). Other Titles: Note on the geology of the Daban area and the localities of the described nautiloids. OCLC Accession Number: 18189808.

Habashi, F. and Bassyouni, F. A. 1982. Mineral Resources of the Arab Countries. United Kingdom: Chemecon, London, United Kingdom. Descriptors: Africa; Arabian Peninsula; Asia; East Africa; economic geology; industrial minerals; Mauritania; metal ores; Middle East; mineral resources; monographs; North Africa; Somali Republic; Sudan; West Africa. Abstract: An inventory of mineral resources and industrial minerals by country with tables of reserves and production. Notes: Edition: 2; References: 3 p. illustrations incl. geol. sketch map. ISBN: 0902777548.

Habbane, A. Y. and McVeigh, J. C. 1986. "Energy Trends in Somalia: The Place for the Renewable Resources." Int J Ambient Energy. Volume 7, Issue 2, Pages 59-68. Descriptors: Energy Utilization- Somalia; Economics; Fuel Economy; Solar Energy; Biomass. Abstract: The first section of this paper examines the relationship between energy consumption and economic performance in Somalia. This is followed by a broad description of the major energy demand and supply patterns. The respective roles of traditional and commercial fuels are outlined and the relative importance of the traditional fuels is emphasized. The environmental impacts, such as deforestation resulting from an increasing fuelwood consumption; and the economic burdens of rising oil imports, are also discussed. Finally, the renewable resources and their applications are presented, followed by an appraisal of the national energy strategy. ISSN: 0143-0750.

Habbane, A. Y. and McVeigh, J. C. 1986. Solar Radiation Model for Somalia. Montreal, Que, Can: Pergamon Press, New York, NY, USA. Volume: 4, pages: 2523-2527. Intersol 85, Proceedings of the Ninth Biennial Congress of the International Solar Energy Society. Descriptors: Solar Radiation; Meteorology. Abstract: While the industrialized countries have well-established solar radiation networks based on many years of detailed observations, solar irradiance data is still scarce in most of the tropical developing countries. However, some stations record sunshine hours and there have been a number of attempts to establish a relationship between sunshine hours and solar irradiance. This paper shows how one particular formula, the Barbaro et al model, has been modified to determine solar irradiance from sunshine hours in Somalia. ISBN: 0-08-033177-7.

Haider, A. 1989. "Geologie De La Formation Ferrifere Precambrienne Et Du Complexe Granulitique Encaissant De Buur (Sud De La Somalie); Implications Sûr l'Evolution Crustale Du Socle De Buur." Translated Title: "Geology of the Precambrian Iron Formation and Surrounding Granulitic Complex of Buur, Southern Somali Republic; Implications for Crustal Evolution of the Buur Basement." France: I.N.P.L., Nancy, France. Descriptors: Africa; amphibolite facies; basement; Buur; chemically precipitated rocks; East Africa; economic geology; evolution; facies; geochemistry; geologic barometry; geologic thermometry; granulite facies; iron formations; iron ores; major elements; metaigneous rocks; metal ores; metals; metamorphic rocks; metamorphism; petrology; Precambrian; rare earths; retrograde metamorphism; sedimentary petrology; sedimentary rocks; Somali Republic; southern Somali Republic; trace elements. Notes: illustrations incl. 300 anals., 7 plates, 16 tables, geol. sketch maps. Iron ores- Geology- Somalia- Mogadishu Region; Geology, Stratigraphic- Precambrian; Granite- Somalia- Mogadishu Region. Notes: Somalia- Geology- Mogadishu Region. Notes: Explanations on verso of plates. "Thèse de l'Institut national polytechnique de Lorraine soutenue le 22 Mars 1989." Includes bibliographical references (p. 141-160). ISBN: 2905532300; ISSN: 0755-978X.

Halcrow & Partners, United Kingdom. 1982. North-West Somalia Refugee Water Supply Consultancy; Interim Report. United Kingdom: Halcrow, London, United Kingdom. Descriptors: Africa; East Africa; northern Somali Republic; northwestern Somali Republic; Somali Republic; water resources; water supply. Database: GeoRef. Accession Number: 1998-028510.

Hall, John B. 1992/7/15. "Ecology of a Key African Multipurpose Tree Species, Balanites Aegyptiaca (Balanitaceae): The State-of-Knowledge." Forest Ecology and Management. Volume 50, Issue 1-2, Pages 1-30. Abstract: Published information on the distribution of Balanites aegyptiaca (Linnaeus) Del. was assembled and summarised as a distribution map interpreted using soil and vegetation maps of Africa and the Middle East, meteorological data and information in the ecological literature. Balanites aegyptiaca is reported from most African countries and also from the Arabian peninsula and adjacent parts of the Middle East. Within Africa the range is from Mauritania in the west to Somalia in the east and from Egypt southwards to Zimbabwe.Occurrence south of the Sahara desert is generally related to rain-fed situations with a mean annual rainfall of 400-800 mm, but there has been spread to disturbed sites in wetter climates. Further north, Balanites aegyptiaca is present in numerous localities where sub-optimal rainfall is augmented with water from other sources. Other conditions associated with the species are a mean annual temperature 20°C and freedom from frost. Typical sites are level and low-lying, with relatively fertile loamy or clayey soil of low salinity but, particularly in the northern Sahel, Balanites aegyptiaca is also widely present on lighter, better-drained soil.Balanites aegyptiaca is characteristic of vegetation in which woody species are widely spaced, with full crown exposure. Associated woody species vary through the range but listings frequently include the families Caesalpiniaceae, Capparaceae, Combretaceae (especially Combretum) and Mimosaceae (especially Acacia). When present in abundance, the number of individuals more than 5 cm diameter at breast height may exceed 25 ha-1, representing 15% of woody individuals present.Attention is drawn to numerous relationships with animals (including Man and domestic livestock) and the significance of these in the conservation and management context. Practicable action in terms of provenance trials, which would lead to improved understanding of the potential of the species as a resource, is outlined and suggestions are made for positive management measures and planting initiatives. ISSN: 0378-1127.

Halloway, A. E. 1961. Potable Water Supply for Mogadiscio; Preliminary Report. Somalia: International Cooperation Administration, Mogadishu, Somalia. Descriptors: Africa; Benadir Somali Republic; drinking water; East Africa; Mogadishu Somali Republic; Somali Republic; water resources; water supply. Database: GeoRef. Accession Number: 1998-028511.

Harms & Brady and Earth Satellite Corporation. 1989. "Somali Democratic Republic." Chevy Chase, Md. The Corporation. Descriptors: Geology- Somalia- Maps; Geology, Structural- Somalia- Maps; Oil wells- Somalia- Maps; Gas wells- Somalia- Maps. Notes: Description: 2 maps on 4 sheets; photocopies, color; 83 x 104 cm. and 83 x 104 cm., sheets 102 x 77 cm. Geographic: Somalia- Remote-sensing maps. Scale 1:1,000,000; (E 400--E 510/N 120--S 020). Notes: Shows geology and fault lines on photomap base. Based on Landsat satellite imagery. Glossy photographs on Kodak paper. Reproductions of printed maps. Includes text, cross section, tectonic synthesis, geologic and rock columns, index to oil and gas wells and dry holes, Landsat location diagram, and index of map sources. Abstract: Northern sheet- Southern sheet. LCCN: 90-684290.

Harrison & Sons Lith and Irish Academic Press. 1874, 1978. "Sketch of Northern Dominions of the Sultan of Zanzibar, Visited by Vice Consul F. Holmwood during the Months of October & November. 1874." Dublin: Irish Academic Press. Descriptors: Coasts- Somalia- Maps; Coasts- Kenya- Maps; Maps- Facsimiles. Notes: Description: 1 map; color; 22 x 21 cm. Geographic: Somalia- Maps. Kenya- Maps. Scale [ca. 1:3,600,000]. Notes: Relief shown by hachures and spot heights. Facsimile. "(1150.8/76.1305) (F.O. 475)." Publisher's no. from its catalog of nineteenth century maps of Africa from the British Parliamentary papers: 35. OCLC Accession Number: 54640123.

Harrison, J. J. 1901. "A Journey from Zeila to Lake Rudolf." Geogr. J. Volume 18, Pages 258-275. Descriptors: Africa; areal geology; East Africa; East African Lakes; Ethiopia; expeditions; Lake Turkana; Somali Republic; Zeila Somalia. ISSN: 0016-7398.

Hartnady C.J.H. 2002. "Earthquake Hazard in Africa: Perspectives on the Nubia-Somalia Boundary." S. Afr. J. Sci. Volume 98, Issue 9-10, Pages 425-428. Descriptors: Earthquakes; Natural Hazards; Hazards and Disaster Planning; earthquake; natural disaster; disaster management Species Term: Somalia. Notes: Additional Info: South Africa; References: 28; Geographic: Africa. Abstract: A wide plate boundary zone between the Nubia and Somalia plates extends through eastern and southern Africa, from the Red Sea-Gulf of Aden region to the mid-oceanic Southwest Indian Ridge. The observed pattern of earthquake activity divides it into seismic belts surrounding relatively stable aseismic blocks. In eastern Africa, the Ukerewe Nyanza plate and the Rovuma plate are separately distinguishable, but in southern Africa and the adjacent Southwest Indian Ocean, the separation of the Transgariep and Lwandle blocks remains to be demonstrated. Because of the slow rates of plate motion along the wide Nubia-Somalia plate boundary and the correspondingly long recurrence times of major events, the quantitative assessment of earthquake hazard requires a new method of estimating maximum magnitudes in the seismic belts, based on the principle of seismic moment conservation. Application of this method requires that the rates and directions of motion of the major plates and the boundary zone blocks be known with sub-mm/yr accuracy. A proposed new project to extend the network of space-geodetic observatory sites in Africa and establish a unified continental reference frame would determine these motions and thus contribute to a long-term African international strategy for natural disaster reduction. ISSN: 0038-2353.

Harvey P. 1998. "Rehabilitation in Complex Political Emergencies: Is Rebuilding Civil Society the Answer?" Disasters. Volume 22, Issue 3, Pages 200-217. Descriptors: political conflict; international aid; civil society. Notes: United Kingdom; References: Number: 16; Geographic: Somalia. Abstract: The paper examines the challenge of rehabilitation from complex political emergencies (CPEs) and identifies a strategy that is characterized as a civil society rebuilding approach. It focuses on Somalia and a case study of a CARE project that aims to build the capacity of local NGOs. The paper argues that civil society in CPEs is simultaneously being undermined and contested by warring parties and emerging after state collapse. The scope of this paper is limited to one case and that case study examines only a single aspect of civil society: national and international NGOs. The paper therefore presents tentative and preliminary results based on limited research. However, in reviewing the literature and presenting a way of approaching the subject, it aims to suggest a starting point for developing a theoretical framework for such research. The paper finds that international agencies have tended to focus on civil society institutions simply as conduits for aid money and that this had tended to creat organizations which lack downward accountability, are dependent on donors and are not addressing the wider roles for civil society envisaged in the approach. Rebuilding civil society does hold out for the promise of giving non-military interests a stronger voice and starting a process of changing the aid delivery culture. Achieving these objectives, however, will be a slow and largely indigenous process and there is a need for lowered expectations about what outside assistance can achieve. ISSN: 0361-3666.

Harvey, R. B. L. 1920. "Somaliland." Major R.B.L. Harvey; Somalia; Northern. Descriptors: Manuscript. Notes: Description: 1 map. Geographic: Somalia- Maps, Manuscript. Notes: Manuscript. Relief shown as form lines. Shows: tracks. Scale 1:125,000. Category of scale: a Constant ratio linear horizontal scale: 125000; General Info: Coverage: Northern Somalia. OCLC Accession Number: 48603132.

Hassan, A. D. Nur, A. M. Omar Shire Yusuf and Ahmed, O. M. 1993. Some Geotechnical Characteristics of the Superficial Alluvial Soils of Shabelle River between Afgoye and Jannale (Southern Somalia). Firenze; Istituto Agronomico L'Oltremar; 1993. Relazioni E Monografie Agrarie Subtropicali E Tropicali Nuova Serie113 B, pages: 551-568. International Meeting; 1st- 1987 Nov: Mogadishu. Notes: Geology of Somalia and surrounding regions: Geology and mineral resources of Somalia and surrounding regions. OCLC Accession Number: 34175029.

Hassan, Abdirahman Dahir; Nur, Abdirisak Mohamed; Yusuf, Omar Shire and Ahmed, Osman Mohamed. 1987. Some Geotechnical Properties of the Fine Alluvial Deposits of Shabelle River from Afgoye to Jannale (Southern Somalia); GeoSom 87; International Meeting; Geology of Somalia and Surrounding Regions; Abstracts. Somalia: Somali Natl. Univ., Mogadishu, Somalia. GeoSom 87; Geology of Somalia and Surrounding Regions, Mogadishu. Somalia Conference: Nov. 23-30, 1987. Descriptors: Afgoye; Africa; alluvium; Atterberg limits; clastic sediments; clay; compressibility; East Africa; engineering geology; Jannale; materials, properties; sediments; Shebelle River;

soil mechanics; Somali Republic; southern Somali Republic; statistical analysis; strength. OCLC Accession Number: 34175029.

Hassan, Abucar Moallim. 1987. Sedimentological and Micropaleontological Data on the El Bur and Hamur Sepiolite (Galgudud, Central Somalia); GeoSom 87; International Meeting; Geology of Somalia and Surrounding Regions; Abstracts. Somalia: Somali Natl. Univ., Mogadishu, Somalia. GeoSom 87; Geology of Somalia and Surrounding Regions, Mogadishu. Somalia Conference: Nov. 23-30, 1987. Descriptors: Africa; argillite; Arthropoda; Bithynia; brackish-water environment; Bur; Carophyta; Cenozoic; central Somali Republic; clastic rocks; clastic sediments; clay minerals; clays; Crustacea; East Africa; economic geology; Foraminifera; fresh-water environment; Galguduud; Hamur; Invertebrata; lacustrine environment; lithostratigraphy; Mandibulata; microfossils; Miocene; Neogene; organic compounds; organic materials; Ostracoda; oxides; properties; Protista; reworking; secondary minerals; sedimentary rocks; sediments; sepiolite; sheet silicates; silicates; Somali Republic; stratigraphy; Tertiary; upper Miocene. OCLC Accession Number: 34175029.

Hassan, Hassan Moham. 1990. Cost-Benefit Analysis of the Baardheere Dam Project and Proposed Water Management in the Juba Valley, Somalia. Descriptors: 140 leaves; Water resources development- Somalia- Cost effectiveness. Notes: Named Corp: Baardheere Dam Project (Somalia); Geographic: Juba River Valley (Ethiopia and Somalia); Notes: Typescript. Includes bibliographical references (leaves 138-140). Dissertation: Thesis (Master of Earth Resources Management)--University of South Carolina, 1990. OCLC Accession Number: 24534447.

Hayder, A. 1993. Preliminary Data on the Magmatic Origin and Emplacement Context of the Precambrian Amphibolites of the Buur Region (Southern Somalia). Firenze; Istituto Agronomico L'Oltremar; 1993. Relazioni E Monografie Agrarie Subtropicali E Tropicali Nuova Serie113 A, pages: 129-142. International Meeting; 1st-1987 Nov: Mogadishu. Notes: Geology of Somalia and surrounding regions: Geology and mineral resources of Somalia and surrounding regions. OCLC Accession Number: 34175029.

Heare, S.F., Rohrer, W. and Humphrey, A.M. 1986. "Wreck of the MV Ariadne--Hazardous Substance Emergency Response in a Third World Country." Conference Title: National Conference on Control of Hazardous Material Spills. Date Held: May 5-8, 1986, St. Louis, Missouri. Publication Date: Not Provided. Description: 7 pages. Abstract: This case history describes the grounding of the containership MV Ariadne in the harbor of Mogadishu, Somalia, in August 1985, and the subsequent shipboard fire and release of hazardous materials into the harbor area. A U.S. Coast Guard/Environmental Protection Agency team was sent to Mogadishu to advise the Somali government on how best to cope with the situation. This paper discusses the team's experiences in Somalia and identifies some of the unique problems noted. The authors indicate that the MV Ariadne incident exemplifies the kinds of problems associated with hazardous emergency response in a third world country and points up the need for greater attention to the problems inherent in the handling and shipment of hazardous substances in these countries. Supplemental Information: Conference paper. Descriptors: Case studies; Developing countries; Disasters and emergency operations; Marine safety; Maritime accidents; Oil spills. Other Terms: Ariadne (Vessel); Emergency response systems; Groundings; Hazardous material spills; Marine accidents; Shipboard fires; Toxic vapors.

Document Source: Maritime Technical Information Facility. Database: TRIS Online: Accession No: 00654867.

Heitman, Helmoed-Romer. 2005. "Peacekeeping Force for Somalia Proposed." Jane's Defence Weekly. Issue MAR, Pages 1. Descriptors: Military operations; Logistics; Personnel; Decision making; Military aircraft; Aerospace industry; Ships. Abstract: East African defence officials met in Entebbe, Uganda, on 7 and 8 March to discuss the proposed deployment of an African Union (AU) peacekeeping force in Somalia. The meeting was focused on practical aspects of force levels, means of deployment and logistics support. A meeting of the chiefs of defence staff of the countries of the Inter-Governmental Agency for Development (IGAD), which brokered the settlement, is to follow before the ministers of defence meet to take final decisions. The IGAD countries-Djibouti, Eritrea, Ethiopia, Kenya, Sudan and Uganda (Somalia is also a member, albeit only pro-forma at present) have been authorised by the AU to deploy a peacekeeping force in Somalia, and several of the countries have agreed to provide forces. Several of the groups currently in de facto control of Somalia, as well as members of the new government have, however, objected to the deployment of any troops from Ethiopia or Djibouti, as both of those countries are seen to have been too closely involved with other groups in the country. US officials have also warned against using troops from any of the neighboring countries lest their neutrality be in doubt. ISSN: 0265-3818.

Heitman, Helmoed-Romer. 2005. "IGAD to Deploy in Somalia." Jane's Defence Weekly. Issue MAR, Pages 1. Descriptors: Military operations; Logistics; Personnel; Decision making; Military aircraft; Aerospace industry; Ships. Abstract: The Inter-Governmental Authority on Development (IGAD), which comprises most of the governments of eastern Africa, will deploy a peacekeeping force to Somalia in April. The deployment is intended to protect the new Transitional National Government of Somalia while it establishes itself in Mogadishu and will take place despite opposition by several Somali factions. The decision was announced by President Yoweri Museveni of Uganda in Entebbe on 14 March, following an IGAD defence ministers' meeting. Museveni stressed that "we are going to deploy, with or without the support of the warlords". Several factions in Somalia have objected specifically to Ethiopian troops being included in the force. The IGAD deployment has been authorised by the African Union. The troops for the force, dubbed IGASOM, will be provided by the IGAD member states. The proposal is for a force of some 10,000 troops- half the number that the Somali government had hoped for- but it remains to be seen whether the IGAD countries can muster sufficient troops and the funding for a deployment of that strength. The initial cost estimate has been put at $500 million. ISSN: 0265-3818.

Heitman, Helmoed-Romer. 2005. "AU Accepts Peace Mission for Somalia." Jane's Defence Weekly. Issue JA, Pages 3. Descriptors: Military operations; Social aspects; Public policy; Laws and legislation; Installation; Personnel training; Societies and institutions. Abstract: The Peace and Security Council (PSC) of the African Union (AU) accepted in principle the deployment of an AU Peace Support Mission in Somalia. The council also approved the establishment of an AU Advance Mission in Nairobi to liaise with the Transitional Federal Government (TFG), the Inter-Governmental Agency for Development (IGAD) and other parties. The AU Peace Support Mission will deploy in support of the efforts of the Somali Transitional government in the security sector. The stabilization of the security situation in Somalia will require that these fractions hand over control of key installations to the AU Mission on behalf of the TFG and are disarmed and demobilized. ISSN: 0265-3818.

Helander B. 1990. "Getting the most Out of it: Nomadic Health Care Seeking and the State in Southern Somalia." Nomadic Peoples. Volume 25-27, Pages 122-132. Descriptors: International Development Abstracts; health seeking behavior; pastoralists; nomads; state; health care. Notes: Somalia- Bay Region. Abstract: The author attempts to maintain a dual perspective: to see local events in the light of national patterns of health care, and to use rural patterns of health seeking to shed light on problems in the organization of national health. Case material from southern Somalia, and in particular the agro-pastoral population of Bay region, is presented. More specifically, discusses some of the factors that affect the character of health services for nomadic pastoralists, eg the spatial dispersion of the population, seasonal mobility, and the special difficulty of establishing workable reach-out mechanisms for a continuously shifting population. While emphasis throughout is on grass-root events and the micro-politics of health care, the discussion of the local level is framed by a consideration of the national environment of health policy guidelines and the political and economic constraints within the public health sector. ISSN: 0822-7942.

Helms, P. B., Fisher, Robert L., Smith, Warren and Jantsch, Marie Z. 1974. "Surveys of Four Sites in the Tropical Western Indian Ocean as Preparation for Deep Sea Drilling Project Leg 24." Initial Reports of the Deep Sea Drilling Project. Oct. Volume 24, Djibouti, F.T.A.I. to Port Louis, Mauritius, May-June 1972, Pages 637-650. Descriptors: acoustical methods; ANTIPODE Expedition; Argo Fracture Zone; Chagos-Laccadive Ridge; Deep Sea Drilling Project; geophysical methods; geophysical surveys; Indian Ocean; Leg 24; magnetic methods; marine; marine geology; Mascarene Plateau; Melville; photography; seismic methods; Somali Basin; structural geology; surveys; topography; west. Notes: SP: DSDP, Deep Sea Drilling Project; illustrations incl. sketch maps. ISSN: 0080-8334.

Hemrich G. 2005. "Matching Food Security Analysis to Context: The Experience of the Somalia Food Security Assessment Unit." Disasters 29, SUPPL. Pages S67-S91. Descriptors: International Aid and Investment; food security; humanitarian aid; Somalia. Notes: United Kingdom; References: Number: 36; Geographic: Somalia; East Africa; Sub-Saharan Africa; Africa; Eastern Hemisphere. Abstract: This case study reviews the experience of the Somalia Food Security Assessment Unit (FSAU) of operating a food security information system in the context of a complex emergency. In particular, it explores the linkages between selected features of the protracted crisis environment in Somalia and conceptual and operational aspects of food security information work. The paper specifically examines the implications of context characteristics for the establishment and operations of the FSAU field monitoring component and for the interface with information users and their diverse information needs. It also analyses the scope for linking food security and nutrition analysis and looks at the role of conflict and gender analysis in food security assessment work. Background data on the food security situation in Somalia and an overview of some key features of the FSAU set the scene for the case study. The paper is targeted at those involved in designing, operating and funding food security information activities. ISSN: 0361-3666.

Hendrikson, K. H. 1973. A Programme for the Allocation of Deep Wells to the Rural Areas. Somalia: German Planning and Economic Advisory Group, Mogadishu,

Somalia. Descriptors: Africa; East Africa; rural environment; Somali Republic; water resources; water supply; water wells; wells. Database: GeoRef. Accession Number: 1998-028521.

Hendrikson, K. H. 1965. Water Supply in the Region of Hargeisa-Burao, Northern Somalia. Federal Republic of Germany: German Planning and Economic Advisory Group, Frankfurt, Federal Republic of Germany. Descriptors: Africa; Burao Somali Republic; East Africa; Hargeisa Somali Republic; Somali Republic; Togdheer Somali Republic; Waqooyi Galbeed; water resources; water supply. Database: GeoRef. Accession Number: 1998-028520.

Henry, J. C. 1979. Present and Future Irrigated Agriculture in Shebelle and Juba River Basins. Italy: FAO, Rome, Italy. Descriptors: Africa; agriculture; East Africa; hydrology; irrigation; Juba River; Shebelle River; Somali Republic; surface water. Database: GeoRef. Accession Number: 1998-028522.

Herren U.J. 1992. "Cash from Camel Milk: The Impact of Commercial Camel Milk Sales on Garre and Gaaljacel Camel Pastoralism in Southern Somalia." Nomadic Peoples. Volume 30, Pages 97-113. Descriptors: Social, economic and political aspects of rural change; Garre; Gaaljacel; milk production; camel milk; pastoralism; household access; commercial impact; developing country; household characteristics; agricultural commercialization; camel husbandry; pastoral production system; milk sales; Garre people; Gaaljacel people. Notes: Somalia. Abstract: The production system of the two agropastoral societies in this study is based on camel husbandry but also includes important cattle and crop components. The author looks at the access of households to the basic productive assets, livestock herds and agricultural plots, and at their distribution between households. The study examines the camel milk off take which is possible within the traditional pastoral production system. The incidence and frequency of milk sales according to season is discussed, and an attempt is made to assess the amounts which are consumed and sold by sample households of different wealth strata. The impact of commercial marketing on the local societies is discussed. ISSN: 0822-7942.

Herisson, Eustache. 1829. "Northeast Africa and the Near East" (942K). A portion of "Carte Generale De L'Afrique" by Eustache Herisson, 1829. See: http://www.lib.utexas.edu/maps/historical/africa_ne_1829.jpg

Hesse, K. H. Dede, C. Schnaecker, E. and Yussuf Abdi Salah. 1990. "Teilprojekt F2. Hydrogeologische Und Hydrologische Untersuchung NordSomalias." Federal Republic of Germany: Techn. Univ., Berlin, Federal Republic of Germany. Descriptors: Africa; carbonate rocks; chemically precipitated rocks; current directions; East Africa; evaporites; ground water; ground-water provinces; hydrodynamics; hydrogeology; hydrology; karst hydrology; recharge; sedimentary rocks; Somali Republic; watersheds. Database: GeoRef In Process. Accession Number: 192820-13.

Hitchcock G.L, L. Key E and Masters J. 2000. "The Fate of Upwelled Waters in the Great Whirl, August 1995." Deep-Sea Research Part II: Topical Studies in Oceanography. Volume 47, Issue 7-8, Pages 1605-1621. Descriptors: Circulation; gyre; monsoon; oceanic circulation; upwelling. Notes: Additional Info: United Kingdom; References: Number: 41; Geographic: Arabian Sea Somalia. Abstract: The Great Whirl is a large, anticyclonic gyre that develops off the northern Somali coast during the Southwest Monsoon. In August 1995 the NOAA Ship Malcolm Baldrige surveyed the seaward edge of the upwelling zone associated with this gyre. The fate of recently upwelled water was followed by mapping surface property distributions along a cool surface feature that extended seaward along the northern edge of the Great Whirl. Surface properties (T, S, and chlorophyll a), surface velocity (ADCP), and XBT and CTD casts were interpreted in relation to the trajectories of three instrumented surface drifters deployed in the feature. Cool surface waters correspond in space to the shoaling of the upper thermocline and offshore advection from the coast. Surface chlorophyll a concentrations decreased from 2 to 3g l-1 in the Upwelling zone to 0.5-1.5g l-1 in the surface feature and contiguous waters. Maximum surface velocities in the Great Whirl were 250 cm s-1 with velocities > 100 cm s-1 along the northern perimeter of the gyre. Decorrelation time-scales for u and v velocity components, and chlorophyll a fluorescence, from the drifters were on the order of 4 to 7 days. These times are comparable to those over which the drifters were ejected from the Great Whirl into the Socotra Gyre. Decorrelation times for sea-surface temperature were somewhat longer (10 days). All three platforms passed between the Somali coast and Socotra within a week of their deployment and then traveled east into the northern Arabian Sea. ISSN: 0967-0645.

Hitchcock, R. K. and Hussein, H. 1987. "Agricultural and Non-Agricultural Settlements for Drought-Affected Pastoralists in Somalia." Disasters. Volume 11, Issue 1, Pages 30-39. Descriptors: droughts; agriculture; sociology; economics; Somalia. Abstract: The severe drought of 1973-1975 in Somalia had major impacts on pastoral populations, many of whom moved into specially established camps where food, water and medical assistance were provided by the government and international agencies. At the end of the drought it was decided to settle the remaining 120,000 people in six settlements, three of which were agricultural and three oriented towards fishing. This paper analyzes these settlements, with particular attention paid to agricultural systems, organization and socioeconomic characteristics. Some of the problems facing the settlements included their location, administrative structure and relative overemphasis on social services as opposed to development of production and income generating activities. It is shown that the socioeconomic viability of settlements would be enhanced if careful studies were undertaken beforehand and if a diversified development strategy were employed. ISSN: 0361-3666.

Hjertson, Mats L. 1995/12. "Taxonomy, Phylogeny and Biogeography of Lindenbergia (Scrophulariaceae)." Botanical Journal of the Linnean Society. Volume 119, Issue 4, Pages 265-321. Descriptors: cytology; generic position; morphology; revision; sexual systems. Abstract: The Afro-Asiatic genus Lindenbergia Lehm. (Scrophulariaceae) is revised and 12 species are recognized. The position of the genus within the Scrophulariaceae is analysed, and it is concluded that Lindenbergia belongs in the tribe Gratioleae. Morphological aspects of the genus are discussed and a cladistic analysis including all species is presented. The somatic chromosome number 2n = 32 is reported for three species. Biogeographical matters are briefly discussed and distribution maps for all species are presented. Two new species are described: L. awashensis Hjertson from central Ethiopia and L. serpyllifolia Hjertson from north-eastern Somalia. A key and descriptions are provided for all species. Several lecto- and neotypes are selected and several names are placed in synonymy.

Hjort Af Ornas A. 1993. "The Multi-Purpose Camel: Interdisciplinary Studies on Pastoral Production in Somalia." Uppsala University, Research Programme on Environmental Policy & Society, EPOS. Descriptors: 246p; Livestock; research cooperation; social organization; camel production; pastoralism; developing country; management issue. Abstract: This volume reports on the activities of the Somali Camel Research Project, which started in 1982 and which emphasized sustained local research above short-term expatriate inputs. The papers are arranged in five sections. The first provides an overview of bilateral Somali-Swedish research cooperation, outlining the project design, context and life-cycle, and the political circumstances of its collapse. The second section covers the social organization of camel production systems, with two major themes: milk production and camel marketing. Husbandry, management and production are examined in the following section, which is concerned largely with the physical aspects of production systems based on camel rearing. Section IV has two chapters on camel health and disease, while the final section looks at bilateral research project methods. In: The multi-purpose camel: interdisciplinary studies on pastoral production in Somalia (1993) p. 246p; Geographic: Somalia; Update: 01 JAN 1995. ISBN: 9150610090.

Hjort Af Ornas A, Hussein M.A and Krokfors C. 1991. "Food Production and Dryland Management: A Somali Camel Research Agenda." Nomadic Peoples. Volume 29, Pages 113-124. Descriptors: Agricultural research and extension; dryland management; camel research; food production. Notes: Somalia. Abstract: This essay is about a project and its life as an interdisciplinary bilateral undertaking in the zone between research and training, until interrupted by a violent death caused by the current Somali civil war. Research content is summarized as an indicator of the interdisciplinary approach. Significant attention is also given to organizational experiences from bilateral cooperation, from combining training and research, and from seeking to introduce research findings into policy making. ISSN: 0822-7942.

Hoag, R. B., Jr and Ingari, J. C. 1990. Modern Approach to Ground Water Exploration in Arid and Semi-Arid Lands; Sand Transport and Desertification in Arid Lands. Singapore: World Sci., Singapore, Singapore. First International Workshop on Sand Transport and Desertification in Arid Lands, Khartoum. Sudan Conference: Nov. 17-26, 1985. Descriptors: Africa; aquifers; arid environment; case studies; East Africa; exploration; ground water; Mesa Programme; Mineral Exploration System Analysis; naturally fractured reservoirs; programs; recharge; remote sensing; resources; semi-arid environment; Somali Republic; water resources; Woqooyi Galbeed. Notes: illustrations ISBN: 9971508583.

Hoag, Roland B., Jr and Ingari, Joseph C. 1988. "Successful Application of an Advanced Groundwater Exploration Program in the High Desert of the Northwest Somalia; Proceedings of the Sahel Forum on the State-of-the-Art of Hydrology and Hydrogeology in the Arid and Semi-Arid Areas of Africa." Proceedings of the Sahel Forum. Volume 1988, Pages 348. Descriptors: Africa; arid environment; East Africa; economic geology; exploration; ground water; northwestern Somalia; Somali Republic; water resources. Database: GeoRef.

Hoben, Allan. 1985. Resource Tenure Issues in Somalia. Boston: Boston University, African Studies Center. Descriptors: 46 leaves; Land tenure- Somalia; Water rights- Somalia; Natural resources, Communal- Somalia. Notes: Cover title. "January 1985." Includes bibliographical references (leaves 44-46). LCCN: 90-981418.

Hoering U. 1994. "Neue Weltordnung: Trial and Error. Nach-Gedanken Zu Einer Intervention." Peripherie. Volume 55-56, Pages 56-67. Descriptors: Geographical

Abstracts: Human Geography; Water; peacekeeping; interventionism; new world order; American intervention; UN strategy; geopolitical studies; UN peacekeeping; humanitarian intervention; military intervention. Notes: Somalia. Abstract: Examining the reasons and interests of the US government for the US-led military intervention in Somalia the article argues that the main rational for "Operation Restore Hope' was to test the new instrument of UN-sanctioned "robust peacekeeping operations'. But instead of improving the peace keeping capacity of the UN due to contradictions between military considerations and humanitarian, political and economic objectives the intervention worsened the preconditions for a long term overall solution of the conflict in Somalia. And while the failure of the military intervention proved the need for preventive political and economic measures, the same negligence of prevention and mediation was repeated in the case of Rwanda. So the main lesson of the test case Somalia has been a considerably weakened enthusiasm for military peacekeeping operations. -English summary. ISSN: 0173-184X.

Hofstetter, R. and Beyth, M. 2003. "The Afar Depression: Interpretation of the 1960-2000 Earthquakes." Geophysical Journal International. NOV. Volume 155, Issue 2, Pages 715-732. Abstract: We studied the seismic activity of the Afar Depression (AD) and adjacent regions during the period 1960-2000. We define seven distinct seismogenic regions using geological, tectonic and seismological data. Based on the frequency-magnitude relationships we obtain b-values of about 1 for the different regions. The pattern of the distribution of the location of epicenters fits with the known active fault zone in the AD and the axial volcanic ridges. The Bab el Mandab area and the Danakil-Aysha'a blocks are less active. For 125 intermediate to strong earthquakes the seismic moment and source parameters were calculated. The results of the fault plane solutions for the Afar Depression indicate mainly strike-slip and normal sense of movement originating from fault planes striking NW-SE. These results indicate a clockwise block rotation described previously as a bookshelf model in central AD. There are a few right-lateral faults east of Massawa with E-W-striking fault planes. At the southern Red Sea, north of the Danakil block, the mixed focal mechanisms, with axial plane striking NW-SE, comprise several reverse faulting, strike-slip motion and normal faulting. Right-lateral movement was also calculated for a cluster of seismic events between the Manda Hararo and Alyata volcanic ridges along NW-SE-striking faults. Along the N-S-striking faults in the escarpment, at the western Afar margins, there are two distinct clusters of epicenters. The strong earthquakes at the southern cluster exhibit normal or strike-slip motions. The intermediate to small earthquakes in the northern cluster exhibit reverse and strike-slip motions. Mainly normal faults were calculated along NE-SW-striking faults of the Ethiopian East African Rift. Estimates of the seismic efficiency suggest that the maximal values are about 50 per cent or less, implying that most of the motion is taken aseismically. OCLC: 20060716.

Holden, Constance. 1989. "Somalia Pledges Human Rights Reforms." Science. February 10. Volume 243, Pages 734. Descriptors: Civil rights; Somalia/Social conditions; General Science; Readers' Guide; geography; Applied Science & Technology. Abstract: In response to pressure from international human rights groups, the U.S. Congress, and the National Academy of Sciences, the government of Somalia has announced its intention to release all political prisoners. The prisoners include 12 scientists and engineers. Congress is withholding $19 million in foreign assistance appropriations for fiscal 1988 pending assurance that the government has fulfilled its promises to release political prisoners and to correct other lapses in human rights. ISSN: 0036-8075.

Holden, Constance. 1988. "NAS Delegation Asks Somalis to Free 13." Science. January 29. Volume 239, Pages 458. Descriptors: Scientists/Somalia; Civil rights; National Academy of Sciences (U.S.); General Science; Applied Science & Technology. ISSN: 0036-8075.

Hollister, D. T. Birch, L. E. Bruce, J. B. and Wright, M. M. 1981. "Investigation of Alternative Roofing Systems Available for use in the Kurtunwaare Pilot Housing Project." Florida Agricultural and Mechanical Univ., Tallahassee. Experimental Low-Cost Construction Unit. Performer: Berger (Louis), Inc., East Orange, NJ. Jun. Volume: PUB5, AIDPNAAM638, pages: 77. Descriptors: Roofs; Evaluation; Roofing materials; Houses; Somalia; Developing country application; Mechanical industrial civil and marine engineering Construction equipment materials and supplies; Mechanical industrial civil and marine engineering Structural engineering; Building industry technology. Abstract: Results of a study on roof systems considered for use in a pilot housing project for displaced nomads in Kurtunwaare, Somalia are presented in this report. The study begins with a description of the pilot project and a list of all roofing systems and building materials available for it. Full evaluation is made of makuti roofing systems--including traditional makuti, makuti modified with tar paper, and preservative and fire retardant treated makuti--as well as of roofing systems employing local fired clay tiles, corrugated fiber and asbestos cement tiles, corrugated aluminum and galvanized iron sheets, sulphur impregnated corrugated cardboard shingles, and reinforced concrete slab. Structural framing systems for both light and heavy roofs are examined as well. Comparative charts summarize information on the physical structure (including construction procedures and thermal effects), state of development, local environmental and economic impacts, and short-and long-term costs of each roofing type. In conclusion, corrugated fiber cement tiles, traditional makuti, corrugated galvanized iron sheets, reinforced concrete slab, and/or preservative treated makuti are recommended; local fired clay tiles and sulphur impregnated corrugated cardboard are recommended with reservations. The other types of roofs are not recommended. Notes: Funder: Agency for International Development, Washington, DC. NT: Prepared in cooperation with Berger (Louis), Inc., East Orange, NJ. NTIS Accession Number: PB85150191XSP.

Holmes, Ralph Jerome. 1954. "A Reconnaissance Survey of the Mineral Deposits of Somalia (Former Italian Somaliland), a Report of the Technical Assistance Mission. New York: World Mining Consultants. 83 pages, 3 folded maps, 5 mounted maps, photographs, tables. Bibliography pages 82-83. At head of Title: United States of America. Foreign Operations Administration. Special Mission to Italy for Economic Cooperation. Rome, Italy. TA 45-125. escriptors: economic geology; geologic maps; iron; Italian Somaliland; maps; metals; mineral resources. Abstract: Ferruginous quartzites of the Bur region are considered the most significant potential economic resource of Somalia. The iron is believed to have been originally deposited in the form of sedimentary limonite or hematite in sandstone, which was converted to quartzite with intergrowths of specular hematite and magnetite during intense metamorphism. USGS Library.

Holtzman, J. S. 1982. "Market for Livestock and Meat in Saudi Arabia: Implications for Somalia." Michigan State Univ., East Lansing. Jun. Volume: AIDPNAAN488, pages: 63. Descriptors: Trends; Imports; Prices; Demand Economics; Supply Economics; Exports; Somalia; Assessments; Beef; Sheep; Tables Data; Marketing; Livestock; Meat; International trade; Saudia Arabia; Developing country application; Frozen foods; Behavioral and social sciences Economics; Business and economics International commerce marketing and economics. Abstract: Trends in Saudi Arabia's importation of livestock and meat and the resulting implications for Somalia and other countries in the region are assessed. Discussion is given in turn to: (1) trends in the composition of live animal and chilled/frozen meat imports to Saudi Arabia and the estimated consumption of red and white meat; (2) livestock prices and consumer preferences for different types of meat; (3) the structure of Saudi Arabia's livestock trade with Near Eastern/East African suppliers and Australia; and (4) the projected demand and supply of meat in Saudi Arabia, with reference to earlier projections. Notes: Funder: Agency for International Development, Washington, DC. NTIS Accession Number: PB85191351XSP.

Hopkins, Stephen T. and Jones, Douglas E. 1983. Research Guide to the Arid Lands of the World. Phoenix, AZ: Oryx Press. Descriptors: 391; Arid regions- Bibliography; Arid regions- Research- Bibliography; Régions arides- Bibliographie; Régions arides- Recherche- Bibliographie; Aride gebieden; Onderzoek. Abstract: The section on the dry land areas of Somalia is found on pages 125-126. Notes: Includes indexes. Includes bibliographical references (p. 367-368). ISBN: 0897740661; LCCN: 83-42500.

"Horn of Africa." 2002. Journal of African Earth Sciences (1994). Feb. Volume 34, Issue 1-2, Pages 93. Descriptors: Africa; areal geology; East Africa; Ethiopia; Horn of Africa; Somali Republic. Individual papers are cited separately. ISSN: 1464-343X.

Horner-Johnson B.C, Gordon R.G, Cowles S.M and Argus D.F. 2005. "The Angular Velocity of Nubia Relative to Somalia and the Location of the Nubia-Somalia-Antarctica Triple Junction." Geophysical Journal International. Volume 162, Issue 1, Pages 221-238. Descriptors: Plate tectonics; triple junction; plate tectonics Species Term: Somalia; Nubian. Notes: United Kingdom; References: Number: 28. Abstract: A new analysis of geologically current plate motion across the Southwest Indian ridge (SWIR) and of the current location of the Nubia-Antarctica-Somalia triple junction is presented. Spreading rates averaged over the past 3.2 Myr are estimated from 103 well-distributed, nearly ridge-perpendicular profiles that cross the SWIR. All available bathymetric data are evaluated to estimate the azimuths and uncertainties of transform faults; six are estimated from multibeam data and 12 from precision depth recorder (PDR) data. If both the Nubian and Somalian component plates are internally rigid near the SWIR and if the Nubia-Somalia boundary is narrow where it intersects the SWIR. Thus, the boundary is either along the spreading ridge segment just west of the Andrew Bain transform fault complex (ABTFC) or along some of the transform fault complex itself. These limits are narrower than and contained within limits previously found by Lemaux et al. from an analysis of the locations of magnetic anomaly 5. The data are consistent with a narrow boundary, but also consistent with a diffuse boundary as wide as 700 km. The new Nubia Somalia pole of rotation lies north of the Bouvet triple junction, which places it far to the southwest of southern Africa. The new angular velocity determined only from data along the SWIR indicates displacement rates of Somalia relative to Nubia of 3.6 ± 0.5 mm yr-1 (95 per cent confidence limits) between Somalia and Nubia near the SWIR, and of 8.3 ± 1.9 mm yr-1 (95 per cent confidence limits) near Afar. The new Nubia-Somalia angular velocity differs significantly from the Nubia-Somalia angular velocity estimated from Gulf of Aden and Red sea data. This significant difference has three main alternative explanations: (i) that the plate motion data have substantial un modeled systematic errors, (ii) that the Nubian component plate is not a single rigid plate, or (iii) that the Somalian component plate is not a single rigid plate. We tentatively prefer the third explanation given the geographical distribution of earthquakes within the African composite plate relative to the inferred location of the Nubia-Somalia boundary along the SWIR. ISSN: 0956-540X.

Hosier R.H and Milukas M.V. 1992. "Two African Woodfuel Markets: Urban Demand, Resource Depletion, and Environmental Degradation." Biomass & Bioenergy. Volume 3, Issue 1, Pages 9-24. Descriptors: Water; urban fuel wood demand; charcoal market; environmental degradation; wood fuel market. Notes: Somalia Rwanda. Abstract: This paper examines charcoal markets in two African cities: Mogadishu, Somalia and Kigali, Rwanda. Although Rwanda and Somalia represent drastically different physical environments, both are considered to be wood-scarce. But neither market has demonstrated straightforward depletion effects. In Mogadishu, the price first rose and then fell in reaction to shifts in the substructure of the charcoal market, relaxed regulations and economic contraction. In Rwanda, the price began rising only after the closing of the Bugasera Region to charcoal producers. Charcoal must be increasingly produced from private farmland. These two case studies highlight the importance of agricultural land clearance, conflicting government regulations, and shifts in market structure in determining whether or not charcoal prices will demonstrate depletion effects, and whether or not charcoal production will lead to local environmental degradation. ISSN: 0961-9534.

Hotchkiss, Patricia F. 1989. Somali Refugee Programme (Ethiopia): Evaluation Report. Descriptors: 20, 6; Refugees- Somalia; Refugees- Ethiopia; Refugee camps-Ethiopia- Evaluation; Somalis; Emergency relief; Refugee camps; Water supply; Housing; NGOs; Evaluation. Notes: Named Corp: Oxfam. Notes: "12 April 1989". OCLC Accession Number: 70139823.

Howard Humphreys & Sons. 1960. "Hargesia Plan of Existing Mains and Proposed Reticulation. Jan. 1960." London. Descriptors: Water-supply- Somalia-Hargesia- Maps; Water- Storage- Somalia- Hargesia- Maps. Notes: Description: map; 30 x 49 cm. Scale 1:10,000. Notes: Blueprint. Relief shown by contours. "Reproduced from maps prepared by Messrs. Hunting Aerosurvey Ltd." "AH53 drawing no. P3." Indexed for points of interest. LCCN: 73-691270; WorldCat Accession Number: 5403330.

Hrbek, T. and Meyer, A. 2003. "Closing of the Tethys Sea and the Phylogeny of Eurasian Killifishes (Cyprinodontiformes: Cyprinodontidae)." J. Evol. Biol. Jan. Volume 16, Issue 1, Pages 17-36. Descriptors: Animals; Asia; Base Sequence; Comparative Study; DNA, Mitochondrial genetics; Europe; Evolution, Molecular; Geography; Geology; Killifishes- genetics; Likelihood Functions; Molecular Sequence Data; Phylogeny; Research Support, Non-U.S. Gov't; Research Support, U.S. Gov't, Non-P.H.S. Abstract: To test vicariant speciation hypotheses derived from geological evidence of the closing of the Tethys Sea, we reconstruct phylogenetic relationships of the predominantly fresh-water killifish genus Aphanius using 3263 aligned base pairs of mitochondrial DNA from samples representing 49 populations of 13 species. We use additional 11 cyprinodontid species as outgroup taxa. Genes analysed include those encoding the partial 12S and 16S ribosomal RNAs; transfer RNAs for valine, leucine, isoleucine, glutamine, methionine, tryptophan, alanine, asparagine, cysteine and tyrosine; and complete nicotinamide adenine dinucleotide dehydrogenase subunit I and II. Molecular substitution rate for this DNA region is estimated at of 8.6 +/- 0.1 x 10(-9) substitutions base pair(-1) year(-1), and is derived from a well dated transgression of the Red Sea into the Wadi Sirhan of Jordan 13 million years ago; an alternate substitution rate of 1.1 +/- 0.2 x 10(-8) substitutions base pair(-1) year(-1) is estimated from fossil evidence. Aphanius forms two major clades which correspond to the former eastern and western Tethys Sea. Within the eastern clade Oligocene divergence into a fresh-water clade inhabiting the Arabian Peninsula and an euhaline clade inhabiting coastal area from Pakistan to Somalia is observed. Within the western Tethys Sea clade we observe a middle Oligocene divergence into Iberian Peninsula and Atlas Mountains, and Turkey and Iran sections. Within Turkey we observe a large amount of genetic differentiation correlated with late Miocene orogenic events. Based on concordance of patterns of phylogenetic relationships and area relationships derived from geological and fossil data, as well as temporal congruence of these patterns, we support a predominantly vicariant-based speciation hypothesis for the genus Aphanius. An exception to this pattern forms the main clade of A. fasciatus, an euhaline circum-Mediterranean species, which shows little genetic differentiation or population structuring, thus providing no support for the hypothesis of vicariant differentiation associated with the Messinian Salinity Crisis. The two phylogenetically deepest events were also likely driven by ecological changes associated with the closing of the Tethys Sea. ISSN: 1010-061X.

Hsu, S. A. 1982. "Wind-Wave Interactions Under the Influence of the Somali Low-Level Jet." Louisiana State Univ Baton Rouge Coastal Studies Inst: Louisiana State Univ Baton Rouge Coastal Studies Inst. 1982. 7 pages. Descriptors: *Ocean Waves; *Wind; Stresses; Somalia; Evaporation; Air Water Interactions; Heat Flux; Low Altitude; Indian Ocean. Abstract: Since the speed of the Somali low-level jet may reach 40 m/s in certain regions over the Indian Ocean during the summer, a constant value for the drag coefficient, C sub 10, has been found inadequate for many air-sea interaction studies such as heat flux and evaporation computations. During MONEX 1979 (May and June), high-resolution rawinsonde stations were set up at Mogadishu and at Gardo, in Somalia, to measure the structure of this jet. Simultaneous observations of surface winds and waves downwind from the jet were made by research and merchant ships. It is the purpose of this paper to study the significant heights and periods of sea waves under the influence of the Somali jet. In order to improve input to the study of atmosphere-ocean systems, such as heat budget, wave and current forecasting, and numerical modeling of the marine boundary layer, a recently developed wind stress formulation which incorporates wave-breaking characteristics is also investigated. Notes: Contract Number: N00014-75-C-0192; TR-355. DTIC Number: ADA120222.

Hsu, S. A. 1985. Improved Formulas for Estimating Offshore Winds. Houston, TX, USA: ASCE, New York, NY, USA. Volume: 3, pages: 2220-2231. Nineteenth Coastal Engineering Conference, Proceedings of the International Conference. Descriptors: Meteorology; Mathematical Models; Mechanical Variables Measurement-

Velocity. Abstract: On the basis of many pairs of simultaneous measurements of wind speed onshore, U(LAND), and offshore, U(SEA), in areas ranging from Somalia, near the equator, to the Gulf of Alaska, and under conditions ranging from breeze to hurricane, it was found that for operational use U(SEA) equals 3. 93 U** one-half (LAND) for U(LAND) less than 10 m s** minus **1 (or 20 kt); and U(SEA) equals 1. 24 U(LAND) for U(LAND** greater than equivalent to 10 m s** minus **1. These formulas were developed mainly from theoretical considerations and were verified by field measurements. ISSN: 0-87262-438-2.

Hsu, S. A. 1982. "Interactions between Mesoscale and Synoptic-Scale Wind Systems in Somalia. Technical Rept." Louisiana State Univ., Baton Rouge. Coastal Studies Inst. Volume: TR354, pages: 7. Descriptors: Wind; Indian Ocean; Somalia; Low altitude; Diurnal variations; Atmospheric temperature; Humidity; Barometric pressure; Reprints; Somali Jet; Atmospheric sciences Meteorology; Atmospheric sciences Dynamic meteorology. Abstract: No abstract available. Notes: In Extended Abstracts, International Conference on the Scientific Results of the Monsoon Experiment, p3-4-3-50, 26-30 Oct 81 . Sponsored in part by Grant NSF-ATM80-13644. DTIC Accession Number: ADA1202423XSP.

HSU, S. A. and Louisiana State Univ., Baton Rouge. Inst. of Coastal Studies. 1981. Diurnal Variation of the Low-Level Jet. WMO Intern. Conf. on Early Results of FGGE and Large-Scale Aspects of its Monsoon Expt. 5 p (SEE N82-23851 14-47); International Organization; 1981; WMO Intern. Conf. on Early Results of FGGE and Large-Scale Aspects of its Monsoon Expt. 5 p (SEE N82-23851 14-47); International Organization. Descriptors: Diurnal Variations; Jet Streams (Meteorology); Somalia; Air Land Interactions; Air Water Interactions; Arabian Sea; Atmospheric Pressure; Atmospheric Temperature; Global Atmospheric Research Program; Pressure Gradients; Radiosondes; Sea Breeze; Wind Velocity. Abstract: Diurnal variation of the Somalia jet were investigated by collecting simultaneous measurements of wind, temperature, and humidity from radiosondes launched four times daily. Diurnal variation of the pressure gradient is insufficient to explain the diurnal variation of the jet. The speed of the jet increases from 10 to 15 mps during the day to 20 to 23 mps during the night. The cooler temperature associated with the Somali current offshore, originally produced by this jet, feeds back to the atmosphere by enhancing the temperature gradient across the coastal zone. These interactions in turn affect the velocity of the land/sea breeze system along the Somali coast. This diurnal and pulsational phenomenon is further augmented by the existence of the Gulf of Aden and the mountain and valley winds associated with the regional geography. Notes: Contract: NSF ATM-78-13388; NSF ATM-80-13644; sponsored in cooperation with Office of Naval Research, Arlington, Virginia; available from MF E06; HC WMO. Database: CSA Technology Research Database. Accession Number: N82-23943 (AH).

Hubbard M, Merlo N, Maxwell S and Caputo E. 1992. "Regional Food Security Strategies: The Case of IGADD in the Horn of Africa." Food Policy. Volume 17, Issue 1, Pages 7-22. Descriptors: Water; regional cooperation; food security Species Term: Somalia. Notes: Africa- Horn of Africa. Abstract: This article examines the work of the Intergovernmental Authority on Drought and Development (IGADD), formed by the governments of Djibouti, Ethiopia, Kenya, Somalia, Sudan and Uganda, in striving to improve food security through regional cooperation. The potential benefits of cooperation are noted, alongside the obstacles to their realization. IGADD's strategy is described in detail, and the steps being taken to implement it are discussed. ISSN: 0306-9192.

Huchon P, Al Khirbash S, Gafaneh A, Jestin F, Cantagrel J.M and Gaulier J.M. 1991. "Extensional Deformations in Yemen since Oligocene and the Africa- Arabia- Somalia Triple Junction." Annales Tectonicae. Volume 5, Issue 2, Pages 141-163. Descriptors: Plate tectonics; African plate; Arabian plate; Somalia plate; extensional deformation; triple junction. Notes: Yemen. Abstract: The geometry and chronology of the successive extensional tectonic events which affected part of the triple junction between the Africa, Arabia and Somalia plates are described. E-W extension prevailed during the Oligocene and early Miocene and was associated with the emplacement of the Yemen Volcanics. From 22 to 18 Ma ago, the uppermost part of the Yemen Volcanics was emplaced under a N-S trending extension related to the increasing influence of the proto-Gulf of Aden continental rift. Finally, Yemen was submitted to NE-SW extension between 18 and 10 Ma, around the time of acceleration of rifting and subsequent oceanization of the Red Sea and Gulf of Aden. This phase of extension is sealed in the southernmost part of Yemen by late Miocene basaltic series. The relationships between deformation and kinematics are discussed. ISSN: 0394-5596.

Huchon, P. and Khanbari, K. 2003. "Rotation of the Syn-Rift Stress Field of the Northern Gulf of Aden Margin, Yemen." Tectonophysics. APR 10. Volume 364, Issue 3-4, Pages 147-166. Abstract: Remote sensing and field studies of several extensional basins along the northern margin of the Gulf of Aden in Yemen show that Oligocene-Miocene syn-rift extension trends N20°E on average, in agreement with the E-W to N120°E strike of main rift-related normal faults, but oblique to the main trend of the Gulf (N70°E). These faults show a systematic reactivation under a 160°E extensional stress that we interpret also as syn-rift. The occurrence of these two successive phases of extension over more than 1000 kin along the continental margin suggests a common origin linked to the rifting process. After discussing other possible mechanisms such as a change in plate motion, far-field effects of Arabia-Eurasia collision, and stress rotations in transfer zones, we present a working hypothesis that relates the 160°E extension to the westward propagation since about 20 Ma of the N70°E-trending, obliquely spreading, Gulf of Aden oceanic rift. The late 160°E extension, perpendicular to the direction of rift propagation, could result from crack-induced extension associated with the strain localization that characterizes the rift-to-drift transition. ISSN: 0040-1951.

Hughes, G. W. and Beydoun, Z. R. 1992. "Red Sea- Gulf of Aden: Biostratigraphy, Lithostratigraphy and Palaeoenvironments." J. Pet. Geol. Volume 15, Issue 2, Pages 135-156. Descriptors: Petroleum geology; Stratigraphy. Abstract: Sediments of Palaeozoic, Mesozoic and early Palaeogene ages experienced a similar geological history in Ethiopia, Yemen and Somalia. During the late Eocene, however, uplift and differential erosion took place, prior to rift development in the middle Oligocene, when the proto-Gulf of Aden became established. To a certain extent, a similar sequence of events had also taken place in those regions of Egypt, Sudan, Ethiopia, Yemen and Saudi Arabia which border the Red Sea, but post-Eocene subsidence is now believed to have commenced during the late Oligocene in the southern Red Sea and progressed later, during the early Miocene, in the northern Red Sea and Gulf of Suez. Timing of this progressive development of the Gulf of Aden rift complex through the Red Sea and Gulf of Suez is well constrained by biostratigraphy, and provides a new approach to the understanding of lithological variations within the region. These lithological variations have, until now, only been considered on a country-specific basis, thus hindering establishment of a regional history of sedimentation. The well-understood and well-documented lithostratigraphic nomenclature of the Miocene succession of the Gulf of Suez has been used as a reference, or type, with which the lateral facies variations within the Red Sea have been compared. By this method only has it been possible to produce palaeoenvironmental maps for the entire study region, and for each formation and member. Lateral equivalents of the Gulf of Suez Nukhul, Rudeis, Kareem, Belayim, South Gharib, Zeit, Wardan and Shagara Formations have been identified within the Red Sea syn-rift and post-rift episodes. ISSN: 0141-6421.

Huliaras A. 2002. "The Viability of Somaliland: Internal Constraints and Regional Geopolitics." Journal of Contemporary African Studies. Volume 20, Issue 2, Pages 157-182. Descriptors: Economy; Political; International Relations, International Financial Institutions; regional politics; independence; political relations; boundary delimitation; geopolitics. Notes: Additional Info: United Kingdom; References: Number: 117; Geographic: Somalia. Abstract: Throughout the Cold War, secession was taboo in the state-centric international system. However, the breakup of the Soviet Union, Yugoslavia and Czechoslovakia, the independence of Eritrea and East Timor as well as recent developments in Kosovo seemed to weaken the principle of inviolable state boundaries. From one point of view, these events may have far-reaching repercussions for Africa where borders are generally considered to be more arbitrary than elsewhere (Herbst 1989). And no other area in Africa is closer to secession than the northern region of Somalia- an area whose boundaries largely correspond to the former British Protectorate of Somaliland. (The British and Italian colonies of Somaliland became the Republic of Somalia in 1960.) In 1991, the northern region of Somalia declared its independence from a state that was collapsing into chaos. Thus, a regional administration (retaining the name 'Somaliland') has in the last decade overseen the creation of a modest state structure, a safe environment and a revitalised commercial economy. This contrasts sharply with the anarchy that characterizes the south. However, Somaliland has failed to achieve international recognition as a separate state. This paper examines the viability of Somaliland as an independent entity by analyzing three important factors influencing the prospects for its continuing existence and also attempts to define the success of its efforts for international recognition: first, economic viability; secondly, the viability of its political institutions; and thirdly, the international environment, including the prospects for reconstitution of political order in the southern part of Somalia. ISSN: 0258-9001.

Hunt, John Anthony. 1960. Report on the Geology of the Berbera-Sheikh Area: Berbera and Burao Districts: Quarter Degree Sheets, Nos. 24 and 26 (D.O.S. 1:25.000). Hargeisa: Stationary Office. Descriptors: VII, 27. Notes: Met bibliography. OCLC Accession Number: 67822767.

Hunt, John Anthony. 1960. Report on the Geology of the Berbera-Sheikh Area, Berbera and Burao Districts. Hargeisa: Somaliland Stationery Office. Descriptors: 27 pages maps. Notes: Maps in end pocket. OCLC Accession Number: 34426443.

Hunt, John Anthony. 1958. Report on the Geology of the Adadleh Area, Hargeisa and Berbera Districts, Quarter Degree Sheet no. 35 (D.O.S. 1:25,000). Hargeisa: Somaliland Protectorate, Geological Survey, Report Number 2. Descriptors: Adadleh Somali Republic; Africa; Cenozoic; Cretaceous; East Africa; geologic history; geologic maps; geology; maps; Mesozoic; mineral resources; Precambrian; Somali Republic; Tertiary. Abstract: The rocks of the area are Basement System (Archean?) metamorphics and gabbros, for which six series have been tentatively outlined, later dikes, sills, and pegmatites, Cretaceous Nubian sandstone and Dubato clays, lower Eocene limestone, Tertiary clays and volcanics, and Pleistocene sediments. A tentative tectonic history is suggested. Beryl, columbite, and muscovite are found in the pegmatites, and some small amounts have been recovered. Building stone and road metal for local use are abundant. Notes: Somaliland, Geol. Surv., Rept. illustrations (incl. colored geol. map). 16 pages, two maps, 1 color; bibliography page 15. Database: GeoRef. Accession Number: 1959-002450.

Hunt, John Anthony. 1956. "Somaliland Protectorate; Annual Report of Geological Survey, 1955 (April, 1955-March, 1956)." Descriptors: Africa; British Somaliland, Geological Survey, report; East Africa; Geological Survey; Somali Republic; Surveys, reports. Notes: (processed), illustrations (incl. sketch map), Hargeisa, [n.d. Abstract: Includes summarized data on the results of reconnaissance mapping and notes on mineral occurrences and water supply. Database: GeoRef.

Hunt, John Anthony. 1954. "Gypsum anhydrite." Hargeisa: Geological Survey. Mineral Resources Pamphlet Number 1. 4 pages; references on page 4. USGS Library.

Hunt, John Anthony. 1953. "Somaliland Protectorate; Annual Report of Geological Survey, 1952 (April, 1952-March, 1953)." 13 Pages 0 ?0. Descriptors: Africa; East Africa; Geological Survey; report; Somali Republic; Surveys, reports. Notes: 1953 (April, 1953-March, 1954), 11 pages, sk. maps, Hargeisa, [n.d., 1954?]. (processed); Abstract: Includes reports on the results of reconnaissance mapping, mainly in potentially mineralized areas of the Precambrian basement system. Database: GeoRef.

Hunt, John Anthony. 1951. A General Survey of the Somaliland Protectorate, 1944-1950. Final Report on "An Economic Survey and Reconnaissance of the British Somaliland Protectorate 1944-1950," Colonial Development and Welfare Scheme D. 484. Hargeisa: Somaliland Protectorate. Pages: 203. Descriptors: Africa; East Africa; geologic history; geologic maps; geology; maps; mineral resources; soils; Somali Republic; water supply. Abstract: Includes a section on geology (p. 95-106), with data on the composition of the terrain, mineral occurrences, water supply, and soils. The region is composed of a pre-Triassic (probably Precambrian) basement, Triassic to Tertiary formations, Quaternary sediments and volcanic rocks, and younger dune sands, sandy clay, and secondary limestones. Notes: Somalia- Description and travel. Somalia-Gazetteers. Notes: Bibliography: p. 180-201; illustrations (incl. geol. map), Hargeisa. LCCN: 54-42945; OCLC Accession Number: 3011788.

Hunt, John Anthony. 1951. "Note on the Geology of the Somaliland Protectorate." Colonial Geology and Mineral Resources. Volume 2, Issue 1, Pages 29-30. Descriptors: Africa; East Africa; Geology and mineral resources; mineral resources; Somali Republic. Abstract: Brief historical review of geologic reconnaissance, mineral exploration, and water-supply investigations which have been carried out in British Somaliland. Database: GeoRef.

Hunt, John Anthony. 1939. "Report II on Geology and Water Supply of Part of Zeila District for Colonial Development Fund Scheme." John A. Hunt; Somalia; Saylac. Descriptors: Geology- Somalia- Maps. Notes: Description: 5 maps +; 1 appendix sheet. Geographic: Saylac (Somalia)- Maps. Scales vary. Category of scale: a Constant ratio linear horizontal scale: various; Notes: Includes illustration 1. General geology at 1:250,000- II. Detail map of Taqusha at 1:2,500- III. Detail map of Zeila at 1:25,000- IV. Detail map of Ashado at 1:25,000- v. Detail map of Sillil Hemal at 1:25,000. General Info: Coverage: Saylac district, northwestern Somalia. OCLC Accession Number: 49653265.

Hunting Technical Services Ltd; Sir M. MacDonald & Partners; United Nations Development Programme and Food and Agriculture Organization of the United Nations. 1970s. Project for the Water Control and Management of the Shebelli River Somalia. Descriptors: 5 v. in 8; Water resources development- Shebelli River watershed, Somalia. Notes: Shebelli River Watershed (Somalia); Notes: "Executing agency: Food and Agriculture Organisation of the United Nations." Reproduction: Microfiche. [s. l. s. n., 197-?]. 37 microfiches. OCLC Accession Number: 17001709.

Hussein, A. H. 1984. Chemical and Nutritional Evaluation of River Shebelle Water. Somalia: Universita Nazionale Somala, Fac. Scienze, Mogadishu, Somalia. Descriptors: Africa; East Africa; geochemistry; hydrochemistry; hydrology; nutrition; Shebelle River; Somali Republic; surface water. Database: GeoRef. Accession Number: 1999-004622.

Hussein, M. A. 1983. "Ricerche Sulla Determinazione Delle Caratteristiche Idrologiche Dei Terreni Di Mordinle (Somalia)." Translated title: "Research on the Determination of the Hydrologic Characteristics of the Mordinle Terrains, Somalia." Riv. Agric. Subtrop. Trop. Volume 77, no.2, Descriptors: Afgoi Somalia; Africa; Benadir; East Africa; hydrology; irrigation; Mordinle; Somali Republic; terrains; water supply. ISSN: 0035-6026.

Hussein, M. A., Rassim, A. M., Ahmed, A. I. and Benini, G. 1988. "Aspetti e Problemi Irrigui Degli Agrumi in Somalia. Aspects and Problems of Citrus Irrigation in Somalia." Irrigazione e Drenaggio. Volume 35, no.2, Descriptors: Africa; agriculture; drainage basins; East Africa; hydrology; irrigation; Somali Republic; water supply; water use. ISSN: 0394-9338.

Hutchings, L., Pitcher, G. C., Probyn, T. A. and Bailey, G. W. 1995. "The Chemical and Biological Consequences of Coastal Upwelling; Upwelling in the Ocean; Modern Processes and Ancient Records." Environmental Sciences Research Report. Volume 18, Pages 65-81. Descriptors: Atlantic Ocean; atmosphere; bathymetry; Benguela Current; biology; bottom water; California Current; Canary Current; carbon; Chordata; climate change; climate-induced circulation; coastal environment; concentration; currents; feeding; geochemistry; Humboldt Current; Indian Ocean; mixing; nutrients; ocean circulation; ocean currents; Pacific Ocean; Pisces; plankton; productivity; sedimentation; sediments; Somali Current; stratification; surface water; temperature; upwelling; Vertebrata; winds. References: 60; illustrations. ISBN: 0471960411.

Hutchinson, P. 1992. "The Southern Oscillation and Prediction of ``Der" Season Rainfall in Somalia." Journal of Climate, Boston, MA. Volume 5, Issue 5, Pages 525-531. Descriptors: Southern Oscillation-rainfall relationships; Seasonal rainfall forecasting; Somalia. May 1992. Refs., figs., tables. Abstract: Somalia survives in semiarid to arid conditions, with annual rainfall totals rarely exceeding 700 mm, which are divided between two seasons. Many areas are arid, with negligible precipitation. Seasonal totals are highly variable. Thus, any seasonal rainfall forecast would be of significant importance to both the agricultural and animal husbandry communities. An investigation was carried out to determine whether there is a relationship between the Southern Oscillation and seasonal rainfall. No relationship exists between the Southern Oscillation and rainfall during the midyear ``Gu" season, but it is shown that the year-end ``Der" season precipitation is affected by the Southern Oscillation in southern and central areas of Somalia. Three techniques were used: correlation, regression, and simple contingency tables. Correlations between the SOI (Southern Oscillation index) and seasonal rainfall vary from zero up to about -0.8, with higher correlations in the south, both for individual stations and for area-averaged rainfall. Regression provides some predictive capacity, but the ``explanation" of the variation in rainfall is not particularly high. The contingency tables revealed that there were very few occasions of both high SOI and high seasonal rainfall, although there was a wide scatter of seasonal rainfall associated with a low SOI. It is concluded that the SOI would be useful for planners, governments, and agencies as one tool in food /famine early warning but that the relationships are not strong enough for the average farmer to place much reliance on forecasts produced solely using the SOI. ISSN: 0894-8755.

Hutchinson, P. 1990. "Frequency Distributions of Daily, Monthly, Seasonal and Annual Rainfalls in Somalia, and their use in the Generation of Rainfall Distributions in Data-Deficient Areas." WMO Tropical Meteorological Research Programme, Extended Abstracts of Papers Presented at the Third WMO Symposium on Meteorological Aspects of Tropical Droughts with Emphasis on Long-Range Forecasting (Niamey, Niger, 30 April-4 may 1990). Pages 239-245. Descriptors: Rainfall regime; Rainfall probability estimation; Somalia. Refs., figs., tables. (World Meteorological Organization. WMO Tropical Meteorology Research Programme Report Series, No. 36). (WMO/TD-No. 353). Abstract: The rainfall regime in Somalia, which possesses inadequate precipitation records, both temporally and spacially, has been modeled by simplifying the fitting of non-stationary Markov chains to the basic probability of rainfall occurrence without taking into account what happened in previous events, thereby calculating the probability of rainfall occurrence within an interval. The gamma distributions with parameters, which vary with the time of the year, were then fitted to the rainfall amounts. The results for annual and seasonal data, for monthly data and for daily data are discussed on the basis of analysis of the correlation between the shape parameter of the gamma distribution and the mean rainfall for year, season, month and day. For monthly rainfall, the relationships between the parameters of the gamma distribution are weaker and it is also necessary to consider dry months. In the case of daily rainfall, the proportion of dry and wet days becomes important. While the mean of the fitted gamma distribution varies with the season, it does not appear to vary very much geographically; the shape factor varies neither with season nor geography. Database: Meteorological & Geoastrophysical Abstracts.

Hutchinson, Peter; Polishchouk, O. Somalia and Food Early Warning System Dept. 1989. The Agroclimatology of Somalia. Mogadishu: Somali Democratic Republic, Ministry of Agriculture, Food Early Warning Dept. Descriptors: 186 leaves; Weather-Somalia. Notes: Cover title. Includes bibliographical references (leaves 182-186). Reproduction: Photocopy. OCLC Accession Number: 20185801.

Hyde, W. L. 1984. Preparatory Assistance for the Establishment of a National Water Centre. Somalia: MMWR/FAO, Mogadishu, Somalia. Descriptors: Africa; East

Africa; government agencies; National Water Centre; Somali Republic; water resources. Database: GeoRef. Accession Number: 1999-004629.

Hydrotechnic Corporation; Somalia; United States and Agency for International Development. 1965. Ground Water Investigation for Mogadiscio Water Supply. New York: Hydrotechnic Corp. Descriptors: 79 leaves in various foliations, 39 leaves of plates; Groundwater- Somalia- Mogadiscio; Water-supply- Somalia- Mogadishu. Notes: At head of title: Republic of Somalia and Agency for International Development. Library of Congress; LCCN: 76-379820.

Hydrotecnic Company, New York, NY, United States. 1963. Groundwater Investigation for Mogadiscio. Somalia: Hydrotecnic Company, Mogadishu, Somalia. Descriptors: Africa; Benadir; East Africa; ground water; Mogadishu Somali Republic; Somali Republic; water resources. Database: GeoRef. Accession Number: 1999-004631.

Hydrotecnic Company, New York, NY, United States. 1963. "Feasibility Report on Mogadiscio Water Supply System." United States: Descriptors: Africa; Benadir; development; East Africa; feasibility studies; Mogadishu Somali Republic; Somali Republic; water resources; water supply. Database: GeoRef. Accession Number: 1999-004630.

I

I codici e le leggi civili della Somalia: 1a, 2a e 3a Carta della rivoluzione della Repubblica democratica somala: appendice delle leggi complementari: indice sommario ed analitico-alfabetico: aggiornato al 30 giugno 1978. A cura di Hassan Scek Ibrahim (Hassey). Translated title: "The codes and the civil laws of the Somalia: 1a, 2a and 3a Papers of the revolution of the Republic Democratic Somali: appendix to the complementary laws: highly summarized and analytical-alphabetical index: From the begining to the June 30, 1978. Edited by Hassan Scek Ibrahim (Hassey)." Mogadiscio: Universita nazionale della Somalia, 1978. (Padova): La garangola. 768 pages; 25 cm. Biblioteca Nazionale Centrale di Firenze: B.26.8. 7468.

Iannelli, Pierino. 1984. The Principles of Pasture Improvement and Range Management and their Application in Somalia. Rome: Food and Agriculture Organization of the United Nations. Descriptors: 213; Pastures- Somalia; Range management- Somalia. Notes: Bibliography: p. 164-169. Note: The Principles of Pasture Improvement and Range Management and Their Applications in Somalia. ISBN: 9251022232; LCCN: 86-165955.

Iano, R. 1911. "Del Regime Delle Acqua Nell Nostre Colonia." Translated Title: "The Water Regime of our Colony." Italy: Istituto Coloniale Italiano, Rome, Italy. Descriptors: Africa; East Africa; hydrology; Somali Republic; water resources. Notes: 1, section 8, theme 9. Database: GeoRef. Accession Number: 2000-016563.

Ibrahim, Hersi A. and Sassi, F. P. 1982. "Outline of the Somalia Basement." Quaderni Di Geologia Della Somalia (Mogadiscio). Volume 6, Descriptors: Africa; areal geology; basement; East Africa; Somali Republic. LCCN: 93093377.

Ibrahim, Maxamed F. 1987. "Osservazioni Geologiche Nell'Area Compresa Tra Bur Weyn, Buulo Burti e Buqda Aqable (Somalia Centro-Meridionale)." Translated Title: "Geologic Observation in the Area between Bur Weyn, Buulo Burti and Buqda Aqable, South-Central Somalia." Quaderni Di Geologia Della Somalia (Mogadiscio). Volume 9, Descriptors: Africa; areal geology; Buqda Aqable Somali Republic; Bur Weyn

Somali Republic; Buulo Burte Somali Republic; East Africa; Hiraan; regional; Somali Republic; south-central Somali Republic. LCCN: 93093377.

Ighe, H. H. 1982. Zur Klarung Der Grundwasserbeschaffenhieit in Nord Somalia." Translated Title: "The Clarification of Ground Water Conditions in Northern Somalia." RuhrWestfalische Technische Hochschule, Aachen, Federal Republic of Germany. Descriptors: Africa; East Africa; ground water; northern Somali Republic; Somali Republic; water resources. Database: GeoRef. Accession Number: 1999-004647.

Ilyin, A. 1976. Geology of the Bur Area. Somalia: UNDP, Mogadishu, Somalia. Descriptors: Africa; areal geology; basement; Bur Somali Republic; East Africa; Somali Republic; southern Somali Republic. Database: GeoRef. Accession Number: 1999-004641.

Ilyin, A. 1967. Geological Map of the Area Bur, Somali Republic. Somalia: UNDP, Mogadishu, Somalia. Descriptors: Africa; Bur Somali Republic; East Africa; geologic maps; maps; Somali Republic; southern Somali Republic. Database: GeoRef. Accession Number: 1999-004640.

Ilyin, A. V. 1978. "Late Precambrian-Early Cambrian Phosphorites in Africa." Newsletter of the International Geological Correlation Program, Project 156: Phosphorites. Apr. Issue 2, Pages 8-9. Descriptors: Africa; Benin; Burkina Faso; Cambrian; chemically precipitated rocks; East Africa; economic geology; Lower Cambrian; Mauritania; Niger; Nigeria; Paleozoic; phosphate deposits; phosphate rocks; Precambrian; Proterozoic; reserves; resources; sedimentary rocks; Somali Republic; upper Precambrian; West Africa. Notes: IGCP, International Geological Correlation Programme; IGCP Project No. 156; References: 2. Database: GeoRef.

Inoue, Toshiro. 1997. "Contrast of 87/88 Indian Summer Monsoon Observed by Split Window Measurements." Advances in Space Research. Volume 19, Issue 3, Pages 447-455. Abstract: The split window on board NOAA-9 was used to see the contrast of convective activity, sea surface temperature (SST) and water vapor amount associated with the 1987 and 1988 Indian summer monsoons. Convective activity estimated by the split window over southern India was depressed in 1987, especially during July. SST over the Arabian Sea was a little colder in 1988 than in 1987. On the other hand, water vapor amount over the Arabian Sea was higher in 1988 than in 1987. The enhancement (suppression) of convection over India during July corresponded to larger (smaller) water vapor amount and to colder (warmer) SST over the Arabian Sea. Water vapor amount over the Arabian Sea tends to increase about 30% in comparison to east of Somalia in both years. SST to east of Somalia decrease (increase) with the increase (decrease) of low-level Somali jet. However, the magnitude of SST decrease over the area does not directly correspond to the magnitude of low-level Somali jet. COSPAR. ISSN: 0273-1177.

Institute for Development Anthropology (Binghamton, N.Y.). 1983. USAID/Somalia: Juba Development Analytical Studies (649-0134): Project Paper. Descriptors: 1 v. (various paging); Applied anthropology- Juba River Valley (Ethiopia and Somalia); Soils- Juba River Valley (Ethiopia and Somalia); Soils- Analysis; Land use- Juba River Valley (Ethiopia and Somalia). Notes: Juba River Valley, Ethiopia and Somalia- Environmental conditions. Juba River Valley, Ethiopia and Somalia- Social conditions. Juba River Valley, Ethiopia and Somalia- Economic conditions. Notes: Title from caption. Other Titles: Juba development analytical studies. OCLC Accession Number: 61851559.

Istituto Geographico Militare (Italy). 1920. "Somalia Italiana 1:10,000." Scale: 1:10,000. [S.l.]: Istituto geografico militare, 1920. 5 maps. Subjects: Somaliland- Maps and charts (collections) 1920. Missing sheet 1. British Library: Maps X.4035.

Italian East Africa: The principal known land and water aerodromes. 1939. MBAM 3368. Scale: 1:4 600 000. [S.l.]: Air Ministry, United Kingdom, 1939. Maps. Note(s): Based on OR 934A. Subjects: Ethiopia- Maps and charts; Eritrea- Maps and charts; Somalia- Maps and charts. Added Name: Great Britain. War Office. General Staff. Geographical Section. 1 sheet. British Library: Maps MOD MBAM 3368.

[Italian hydrographic charts of Somalia]. 1942. East Africa Force 762, 768, 769 & 772. Scale: 1:300 000. [Nairobi]: [Survey Directorate, East Africa Command], 1942. 8 maps. Note(s): Copies of four Italian hydrographic charts, produced by Istituto Idrografico della R. Marina. Each one in two parts (North and South). EAF 762: Da Obbia a El meghet. 768: Da Eil a Obbia. 769: Baia del Negro. 772: Da Ras Hafun a Eil. Subjects: Somalia- Maps and charts. Added Name: Great Britain. War Office. General Staff. Geographical Section. 8 sheets. British Library: Maps MOD EAF 762, 768, 769 & 772.

International Cooperation Administration, Washington, DC, United States. 1961. Inter River Economic Exploitation; the Somali Republic. United States: Inter. Coop. Administr., Washington, DC, United States. Descriptors: Africa; East Africa; exploration; Somali Republic; water resources. Database: GeoRef. Accession Number: 1999-004645.

International Development Association. 1978. Development Credit Agreement (Mogadishu Water Supply Project) between Somali Democratic Republic and International Development Association. Washington, D.C: International Development Association. Descriptors: 18 pages; Loans, Foreign- Somalia; Debts, External- Somalia; Water-supply- Somalia- Finance; Municipal water supply- Somalia- Mogadishu-Finance; Economic assistance- Somalia. Notes: Named Corp: International Development Association- Finance. Notes: "Conformed copy"--Cover. Cover title. Other Titles: Treaties, etc. Somalia, 1978 June 30; Mogadishu water supply project. OCLC Accession Number: 48880251.

International Research and Training Institute for the Advancement of Women. 1988. Women, Water Supply, and Sanitation: A National Training Seminar, Mogadiscio, Somalia, 13-18 February 1988. Santo Domingo, Dominican Republic: INSTRAW. Descriptors: 98 pages; Water-supply, Rural- Somalia- Management- Citizen participation- Congresses; Sanitation, Rural- Somalia- Citizen participation- Congresses; Women in rural development- Somalia- Congresses; International Drinking Water Supply and Sanitation Decade, 1981-1990; Eau- Approvisionnement rural- Somalie-Gestion- Participation des citoyens- Congrès; Femmes dans le développement rural-Somalie- Congrès; Assainissement rural- Somalie- Participation des citoyens- Congrès. Notes: Named Conf: Décennie internationale de l'eau potable et de l'assainissement (1981-1990). Notes: Includes bibliographical references. LCCN: 89-165689; OCLC Accession Number: 20721827.

International Travel Maps. 2000. "International Travel Maps, Somalia, Scale 1:1,117,000." Vancouver, B.C: International Travel Maps. Descriptors: Roads- Somalia-Maps; Routes- Somalie- Cartes. Notes: Description: 1 map; color; 69 x 89 cm., folded to

25 x 11 cm. Geographic: Somalia- Maps, Tourist. Somalie- Cartes touristiques. Scale 1:1,117,000; Universal Transverse Mercator proj. (E 410--E 510/N 130--N 10). Category of scale: a Constant ratio linear horizontal scale: 1117000 Coordinates--westernmost longitude: E0410000 Coordinates--easternmost longitude: E0510000; Notes: Relief shown by gradient tints and spot heights. Panel title. Includes text, index, 2 insets, location map, and color ill. ISBN: 0921463685; LCCN: 2001-620706.

Italconsult, Rome, Italy. 1968. Lower Juba Region Development Plan. Italy: Italconsult, Rome, Italy. Descriptors: Africa; development; East Africa; hydrology; Jubbada Hoose; Somali Republic; water resources. Database: GeoRef. Accession Number: 1999-004652.

Italy, Commission Generale de la Colonie, Rome, Italy. 1927. Oltre Giuba. Upper Juba. Italy: Sindacato Naz. Arti Grafiche, Rome, Italy. Descriptors: Africa; East Africa; fluvial features; hydrology; Juba River; rivers; Somali Republic; southern Somali Republic; surface water. Database: GeoRef. Accession Number: 1998-011906.

Italy, Istituto Coloniale Italiano, Italy. 1909. Notizie Ed Informazioni Sul Benadir. Notice and Information on the Benadir. Italy: Istituto Coloniale Italiano, Italy. Descriptors: Africa; Benadir; East Africa; geography; hydrology; Somali Republic. Database: GeoRef. Accession Number: 1999-004648.

Italy, Istituto de Geografia Militare, Italy. 1912. "Relazione Sui Lavori Compiuti in Somalia Dal Giugno 1910 Al Giugno 1912." Translated Title: "Report on the Work Completed in Somalia from June 1910 to June 1912." Italy: Ministero delle Colonie, G. Bertero, Rome, Italy. Descriptors: Africa; areal geology; East Africa; Somali Republic. Database: GeoRef. Accession Number: 1999-004649.

Italy, Istituto de Idrografia Regia Marina, Italy. 1936. "Relazione Sulla Spedizione Idrografica in Somalia Nell'Anno 1935." Translated Title: "Report on the Hydrographic Expedition in Somalia in the Year of 1935." Italy: Tipogr. Ist. Idrogr. Regia Marina, Rome, Italy. Descriptors: Africa; East Africa; expeditions; hydrographs; Indian Ocean; marine geology; Somali Republic. Database: GeoRef. Accession Number: 1999-004650.

Italy, Servizio Geologico d'Italia, Rome, Italy. 1936. "Bibliografia Geologica Italiana Per Gli Anni 1915-1933, Africa Orientale Italiana." Translated Title: "Italian Geologic Bibliography for the Years 1915-1933, Italian East Africa." Bollettino Dell'Ufficio Geologico d'Italia. Volume 61, Descriptors: Africa; areal geology; bibliography; East Africa; Italian East Africa; Somali Republic. ISSN: 0366-2314.

Italy, Studio Ingegneri Architetti Specializzati, Italy. 1966. "Rapporto Giustificativo Del Programmma Di Attuazione Immediata Delle Sfruttamento e Delle Conservazione Delle Disponibilita' Idriche." Translated title: "Report to Justify the Program of Exploitation and Conservation of Water Resources." Italy: Descriptors: Africa; conservation; East Africa; Somali Republic; water resources; water use. Database: GeoRef. Accession Number: 2000-016540.

IUCN East Africa Regional Office; IUCN Eastern Africa Programme and Somali Natural Resources Management Programme. 2000. Towards Environmentally Sound Water Projects in Somalia: Introduction to EIA. Nairobi, Kenya: IUCN Eastern Africa Regional Office. Descriptors: 20; Water resources development- Environmental aspects-Somalia; Environmental impact analysis- Somalia. Notes: Cover title. At head of title: IUCN Eastern Africa Programme; Somali Natural Resources Management Programme.

"May 2000." "Project no. 6/SO-82/95+6/SO-83/04"--P. 1. Includes bibliographical references (p. 20). Library of Congress; LCCN: 2002-375345; OCLC Accession Number: 51060471.

IUCN Eastern Africa Programme and Somali Natural Resources Management Programme. 2000. Towards Environmentally Sound Water Projects in Somalia: Introduction to EIA. Nairobi: IUCN-EARO. Descriptors: 20 pages; Environmental impact analysis- Somalia; Water- Environmental aspects- Somalia. Notes: EIA-Environmental Impact Assessment. Project no. 6/SO-82/95 +6/SO-83/04. OCLC Accession Number: 49331145.

IUREP Orientation Phase Mission; Summary Report; Somalia. 1984. France: Descriptors: Africa; East Africa; economic geology; International Uranium Resources Evaluation Project; metal ores; mineral exploration; programs; resources; Somali Republic; uranium ores. Notes: International Uranium Resources Evaluation Project; illustrations incl. geol. sketch map; Latitude: N120000, E0513000 Longitude. Database: GeoRef. Accession Number: 1985-070726.

Ivanik, M. M., Kozak, S. O. and Orlovs'kiy, G. M. 1987. "Paleoklimat i Poshirennya Fe-Mn Mikrokonkretsiy v Osadkakh Somaliys'Koyi Kotlovini." Translated Title: "Paleoclimate and Distribution of Fe-Mn Micronodules in the Sediments of the Somali Basin." Dopovidi Akademiyi Nauk Ukrayins'Koyi RSR, Seriya B: Geologichni, Khimichni Ta Biologichni Nauki. Volume 1987, Issue 3, Pages 17-19. Descriptors: Cenozoic; controls; economic geology; ferromanganese composition; genesis; Indian Ocean; mineral deposits, genesis; mineral resources; nodules; oceanography; paleoclimatic controls; paleoclimatology; Quaternary; sedimentary processes; Somali Basin. References: 4. ISSN: 0002-3523.

J

Jachowski, Robert A. and Lee, Charles E. 1967. "Evaluation of the Design and Construction of the Chisimaio Phase I Port Project, Somali Republic Prepared by Robert A. Jachowski and ." Washington, D.C. U.S. Army Corps of Engineers. February 1967. Volume: TechRpt UG 23 .C3 1967b, Descriptors: Harbors-Somalia-Design and construction; Somalia; Kismaayo (Somalia)-Harbor. Notes: 8, [2] leaves: plans; 27 cm. Cover title: Chisimaio, Somali Republic: evaluation of the design and construction of the Phase I Port Project. NTIS.

Jalludin, Mohamed and Razack, Moumtaz. 1997. "Modelisation d'Un Aquifere En Milieu Volcanique Fracture Souss Climat Aride (Republique De Djibouti)." Translated Title: "Modelling a Fractured Volcanic Aquifer Under an Arid Climate (Republic of Djibouti); Hard Rock Hydrosystems." IAHS-AISH Publication. Volume 241, Pages 91-101. Descriptors: Africa; aquifers; arid environment; basalts; Cenozoic; cinder cones; climate; concepts; Dalha Basalts; dikes; Djibouti; East Africa; effects; fractured materials; ground water; Gulf Basalts; Holocene; hydrodynamics; igneous rocks; intrusions; lithostratigraphy; numerical analysis; Pleistocene; potentiometric surface; Quaternary; recharge; simulation; Somalia Basalts; spatial distribution; transmissivity; volcanic features; volcanic rocks. References: 8; illustrations incl. 1 table, geol. sketch map. ISSN: 0144-7815.

Jama, Aden, A. and Frizzo, P. 1996. "Geochemistry and Origin of Low and High TiO2 Mafic Rocks in the Barkasan Complex: A Comparison with Common

Neoproterozoic Gabbros of Northern Somali Crystalline Basement." Journal of African Earth Sciences. Volume 22, Issue 1, Pages 43-54. Abstract: The Barkasan complex comprises numerous metagabbroic little bodies intruded along a Neoproterozoic shear zone in the crystalline basement of northern Somalia. The constituting rocks can be subdivided into a low TiO_2 (0.41-1.39 wt.%) group, (e.g. the lower portion of the Ago Marodi intrusion and the Wahmadhigato body) and a high TiO_2 (2.13-6.75 wt.%) group, typical of the upper portion of the Ago Marodi intrusion. The compositional and mineralogical changes result from fractionation and differentiation in a cooling tholentic magma. The low titanium rocks are related to a first stage of crystallization and the high titanium rocks represent a relatively later stage. By comparison, the Sheikh gabbro (one of the principal Neoproterozoic gabbroic intrusions in northern Somalia), despite its common contents in phases of the first stage of fractional crystallization (olivine and pyroxene), has lower nickel, chromium and Mg. Significant differences also exist in terms of Zr/Y, Zr/Rb and Zr/Nb ratios, reflecting a different parental magma composition. ISSN: 1464-343X.

Jamal, V. 1988. "Somalia: Survival in a "Doomed' Economy." Int. Labour Rev. Volume 127, Issue 6, Pages 783-812. Descriptors: agriculture; International Development Abstracts- volume 74; food production; consumption; nomads; farmers; urban area; remittance; free market rate; structural adjustment. Abstract: The article focuses on food production and consumption in the special circumstances of the Somali economy. It is shown that Somali nomads and farmers have enjoyed a much higher level of real welfare than hitherto assumed, while in the urban areas poverty has been averted despite falling wages because of remittances sent home from Somali workers abroad. Since most transactions, including remittances, livestock exports and internal trade, take place at essentially free market rates, the standard structural adjustment remedies applied in Somalia are peripheral to the real problems of adjustment. By creating the illusion that they provide the answer to all of the country's problems they may even be harming its economy. ISSN: 0020-7780.

Jamal, V. 1988. "Somalia: Understanding an Unconventional Economy." Development & Change. Volume 19, Issue 2, Pages 203-265. Descriptors: International Development Abstracts- volume 74; agriculture; unconventional economy; welfare; food production; consumption; structural adjustment; devaluation. Abstract: The basic objectives of this paper are to look at the real welfare situation of the Somali people, as depicted by food production and consumption statistics, to analyze the major impacts on the economy of external and internal factors and to examine the premises of the extant remedies for the economy. Section II focuses on an estimation of the food production and consumption situation in the last few years and its evolution since the 1970s. The urban sector's special position is also discussed here. In section III, the analysis of the unconventional economy is applied to assess the appropriateness of the current structural adjustment programmes in Somalia. Attention is particularly focused on the efficacy of devaluation as a remedy. ISSN: 0012-155X.

Janzen, Jorg. 1991. "Somalias Kustemfischerei. Gegenwartige Situation Und Zukunftige Entwicklungsmoglichkeiten." Erde. Volume 122, Issue 2, Pages 131-143. Descriptors: Fisheries; Fisheries and aquatic resources; developing country; fishery economics; coastal area; artisanal fishing; fish conservation Species Term: Somalia. Abstract: People living on coasts make fish an important part of their diet, and may export fish, the author describes the structure, economic importance and development problems of Somalia. Artisan fishing on the coast is an important source of income, earlier and current developments of fishing are shown. The fishermen need better housing, fishing gear and marketing, with a better conservation of fish its economic importance may increase. The author recommends private investment rather than state aid, cooperation to produce and sell is important. ISSN: 0013-9998.

Janzen, Jorge. 1990. "Staudamme Und Grossflachiger Bewasserungsfeldbau Contra Mobile Viehwirtschaft? Das Baardheere-Staudammprojekt in Sud-Somalia Und Mogliche Folgen Fur Die Mobile Viehwirtschaft." Translated title: "Back-up dam and large flat water ways versus nomadic cattle economics? The Baardheere dam project in South Somalia and possible consequences for cattle economics." Erde. Volume 121, Issue 1, Pages 25-38. Descriptors: Crop and livestock production- general; pastoralism; dam construction; irrigation Species Term: Somalia; Bos taurus; Animalia; Riparia. Notes: Somalia. Abstract: Agricultural development in the Sahelian region is concentrated in river valleys where the main population centers are also found. However, dam construction and cropping of riverside areas (formerly pastures) have had an unfavorable impact on the traditional pastoral economy of cattle-keeping tribes. Apart from the loss of dry-season fodder, access to watering points on the river banks is often severely restricted. ISSN: 0013-9998.

Janzen, Jorge. 1990. "Anthropogeographische Forschung in Den Bewasserungsregionen Sud- Somalias- Ziele, Methodik, Ergebnisse." Translated Title: "Anthropo-geographical research in irrigated regions South Somalia- goals, methodology, results." Geomethodica. Volume 15, Pages 71-106. Descriptors: Agriculture; Social, economic and political aspects of rural change; rural area; developing country; irrigation; agricultural practice; socio-economic conditions; geographical research; smallholders; structural change; ecological conditions; irrigated agriculture; rural development; smallholder sector. Notes: Somalia- Jubba Valley. Abstract: Describes a human geographical research project on rural development, started in Somalia in 1984, presenting partial results of this work. Spatially concentrated in the valleys of the Jubba and Shabeelle Rivers in southern Somalia, the explanation deals substantially with the irrigated agriculture of smallholders. An introduction to the framework of the conditions of irrigated agriculture in Somalia is followed by a presentation of the aims of the project and the methodical aspects of human geographical field research, and an overview about the different forms of irrigated agriculture in southern Somalia is given. A case study from the middle Jubba Valley shows and analyses structural changes of smallholder irrigated farming, as well as the effects on the socio-economic and ecological conditions. To close, some pragmatic recommendations for future rural development and irrigated agriculture in Somalia are put up for discussion. ISSN: 0171-1687.

Japan, International Cooperation Agency, Japan. 1985. The Basic Design Study Report on the Groundwater Development Project in Lower Shebelle Related to ICARA II. Japan (JPN): Japan Int. Coop. Agency, Japan. Descriptors: Africa; design; development; East Africa; ground water; programs; Shebelle River; Somali Republic; southern Somali Republic; water resources. Database: GeoRef. Accession Number: 1999-004655.

Jaworski, E. Daar, A. Reed, L. E. and Pershouse, J. 1982. A Land Use Survey of Northwest Somalia as Interpreted from Landsat Imagery. MI, Environmental Research Institute of Michigan: Ann Arbor. Volume 2, pages: 1089-1098. Remote Sensing of Arid and Semi-Arid Lands; Proceedings of the International Symposium on Remote Sensing of Environment, Cairo, Egypt; United States; 19-25 Jan. 1982; Remote Sensing of Arid and Semi-Arid Lands. Proceedings of the International Symposium on Remote Sensing of Environment, Cairo, Egypt; United States Conference: 19-25 Jan. 1982. Descriptors: Land Use; Photointerpretation; Photomapping; Remote Sensing; Satellite-Borne Photography; Somalia; Agriculture; Earth Observations (From Space); Geology; Grasslands; Landsat Satellites; Satellite Observation. Abstract: False color Landsat imagery was utilized to map land use changes and geologic lineaments in Northwest Somalia. Manual photograph interpretation and multi-temporal image comparison comprised the research techniques. Map products which were generated included maps of land use types, rangeland grazing intensity, and geologic lineaments. Lack of access to available aerial photography and existing large-scale maps precluded computer processing, except for a Band 7/Band 5 ratio image. This study demonstrates that manual interpretation of Landsat images can provide useful products at a cost of less than $0.2/square kilometer. Notes: NU: A83-24526 09-43; NR: 6. ISSN: 0275-5505.

Jenner, A. C. 1899. "Expeditions from Kismayu to Logh, on the Juba." Geogr. J. Volume 14, Descriptors: Africa; areal geology; East Africa; expeditions; Juba River; Kismayo Somali Republic; Luuq Somali Republic; Somali Republic. ISSN: 0016-7398.

Jensen, Tommy G. 1990. "A Numerical Study of the Seasonal Variability of the Somali Current." Florida State Univ Tallahassee Geophysical Fluid Dynamics Inst: Florida State Univ Tallahassee Geophysical Fluid Dynamics Inst. February 1990. pages: 129 Pages(s). Descriptors: Air water interactions; Computerized simulation. Abstract: A new numerical ocean model with multiple isopycnal layers is used to model the Indian Ocean. Normal vertical modes are used for initialization and in a new open boundary formulation. A 21 year integration with the Hellerman Rosenstein wind stress is made with a 3.5 layer and a 1.5 layer version of the model. The solution with three active layers reproduces the observed general circulation and variability of the Indian Ocean, e.g., the semiannual equatorial undercurrent and Yanai wave field in the west. Seasonal changes in the Somali Current system are studied in more detail. It is found that barotropic instability is likely to cause the generation of the Great Whirl in early June. We find good agreement between the observed undercurrents and the simulations in the model. Equatorial onshore flow below the thermocline in June is associated with the disappearance of the undercurrent below the Somali current. Return of this undercurrent in the fall is caused by instability of the Great Whirl. Experiments where the duration of the summer monsoon is extended show that the initial decrease in the magnitude of the Great Whirl is due to eastward and downward energy transfer rather that due to relaxation of the wind. The model solutions indicate that baroclinic instability plays an important role in the decay of the Great Whirl. The solution with a single active layer is essentially the same for the upper layer until the late summer monsoon, when the flow becomes unstable. Different decay patterns of the whirl and associated eddies leads to different flows during the winter monsoon. Notes: Contract Number: N00014-85-G-0240. Database: DTIC.

Jensen, T. G. 2001. "Arabian Sea and Bay of Bengal Exchange of Salt and Tracers in an Ocean Model." Geophysical Research Letters. Oct. Volume 28, Issue 20, Pages 3967-3970. Descriptors: Water exchange; Salt advection; Salinity data; Tracers; Ocean circulation; Ocean currents; Transport processes; Fresh water; Mass transport; Water mass tracers; Mass flux; Oceanic mixing tracers; Freshwater transport in oceans; Arabian Sea; Indian Ocean, Bengal Bay; India, Bengal Bay; Indian Ocean, Somali Current. Notes: TR: CS0221901. Abstract: Exchanges of mass and salt between the mixed layers in the Arabian Sea and the Bay of Bengal are examined in a model of the Indian Ocean using passive tracers as a tool to map the pathways. Inflow of high salinity water from the Arabian Sea into the Bay of Bengal is significant and occurs after the mature phase of the southwest monsoon. Freshwater transport out of the Bay of Bengal is southward throughout the year along the eastern boundary of the Indian Ocean. Low salinity water transport into the Arabian Sea occurs in the Somali Current during the southwest monsoon, closing a clockwise path of water mass transport. Only a small fraction of low salinity water is advected into the eastern Arabian Sea from the Bay of Bengal. ISSN: 0094-8276.

Jestin F, Huchon P and Gaulier J.M. 1994. "The Somalia Plate and the East African Rift System: Present-Day Kinematics." Geophysical Journal International. Volume 116, Issue 3, Pages 637-654. Descriptors: Plate tectonics; plate motion; Arabian plate; Antarctic plate; Somalia plate; inversion; triple junction. Notes: Africa- East African Rift. Abstract: The motion of the Somalia plate relative to the Nubia (Africa), Arabia and Antarctica plates is re-evaluated using a new inversion method based on a Monte Carlo technique and a least absolute value misfit criterion. The results confirm that the motion of Arabia with respect to Africa is significantly different from the motion relative to Somalia. It is further shown that the data along the SW Indian Ridge are compatible with a pole of relative motion between Africa and Somalia located close to the hypothetical diffuse triple junction between the ridge and the East African Rift. The resulting Africa-Somalia motion is then compatible with the geological structures and seismological data along the East African Rift system. ISSN: 0956-540X.

Jobin, W. R. 1986. "Designing Hydro Reservoirs to Prevent Tropical Diseases." International Water Power and Dam Construction IWPCDM Vol.38. November. Pages 9 ref. Descriptors: Water pollution effects; Pest control; Hydroelectric plants; Reservoir design; Reservoir operation; Tropic zone; Human diseases; Animal diseases; Biocontrol; Liver flukes; Cattle; Sheep; Malaria; Bilharzia; River blindness; Onchocerciasis; Mosquitoes; Snails; Africa; Central America; Orinoco River; South America; Tilapia; Fish; Somalia. Abstract: The characteristics of malaria, bilharzia, river blindness (onchocerciasis), and liver flukes of cattle and sheep are described and techniques for preventive design and operation of reservoirs to control these diseases are discussed. Mosquito and snail populations can be controlled by manipulating reservoir levels; proper design of spillways can limit the populations of the tropical blackfly, transmitter of river blindness in Africa and Central America; and design of reservoir shorelines can be focused on elimination of suitable habitats by including rapid drainage systems. Diseases of cattle and sheep can be limited by controlling access of animals to the water with fencing. Biological control of the bilharzia snail has been accomplished by the introduction of Marsia, a larger snail from the Orinoco River area South America. The large snail out competes the bilharzia snail in reservoirs. The Tilapia fish and two new species from Somalia have been used to control mosquito larvae. ISSN: 0306-400X.

Jobstraibizer, G. and Omar Shire, Y. 1977. "Il Quaternario Della Somalia." Translated Title: "The Quaternary of Somalia." Quaderni Di Geologia Della Somalia (Mogadiscio). Volume 1, Pages 51-59. Descriptors: Africa; areal geology; Cenozoic; East Africa; Quaternary; Somali Republic. LCCN: 93093377.

John Bartholomew and Son and the Edinburgh Geographical Institute. 1927-1935. "Ethiopia and Adjoining Territories." Edinburgh: John Bartholomew. Notes: Description: 1 map; color, mounted on linen; 62 x 59 cm., folded to 21 x 12 cm. Geographic: Africa, Northeast- Maps. Ethiopia- Maps. Somalia- Maps. Djibouti- Maps. Scale 1:4,000,000. Cartgrph Code: Category of scale: a Constant ratio linear horizontal scale: 4000000; Notes: Relief shown by color, form lines, and spot heights. Depths shown by color and form lines. Grommets in upper corners for hanging which obscure series title. Linen backing not original; folded title and accompanying information unreadable. Inset: Northeast Africa, scale 1:23,000,000. OCLC Accession Number: 17892916.

Johnson, Gregory C. 1990. "Near-Equatorial Deep Circulation in the Indian and Pacific Oceans." Woods Hole Oceanographic Institution Ma: Woods Hole Oceanographic Institution Ma. September 1990. Volume: ADA232635, pages: 161 Pages(s). Descriptors: Africa, Atlantic Ocean, Basins(Geographic), Bathymetry, Bottom, Boundaries, Circulation, Deep Oceans, Diffusivity, East Pacific Rise, Equatorial Regions, Estimates, Feeding, Geometry, Indian Ocean, Inertial Systems, Marshall Islands, Mass Transfer, Ocean Basins, Ocean Currents, Pacific Ocean, Slope, Somalia, Transport, Vertical Orientation, Water; Ocean bottom, Deep water, Ocean currents, Water masses, Mass transport, Equatorial regions, Theses, Ocean circulation. Abstract: Flow of deep and bottom water in the near-equatorial Indian Ocean, Atlantic Ocean, and Pacific Ocean is described through analysis of CTD data. Zero velocity surfaces are chosen using water-mass properties in conjunction with the thermal-wind field and used to obtain transport estimates. In the Somali Basin a deep western boundary current with an estimated transport of 4,000,000 cu m/s moves north on the continental rise of Africa at 3 deg S. Equatorial observations suggest that this current turns eastward at the equator. In the Pacific Ocean at 10 deg N two deep western boundary currents with estimated mass transport of 5,000,000 cu m/s and 8,100,000 cu m/s move north off the Caroline and Marshall Islands respectively. A current with an estimated transport of 4,700,000 cu m/s moves south over the western flank of the East Pacific Rise. This current feeds a westward jet at the equator and brings the net northward transport of bottom water at 10 deg N to 8,400,000 cu m/s. Effects of ocean basin geometry, bottom bathymetry and vertical diffusivity as well as a model meridional inertial current on a sloping bottom near the equator are discussed in conjunction with the observational results. Notes: Master's thesis. Database: DTIC. DTIC Accession Number: ADA2326353XSP.

Johnson, D. R. 1989. "Large Scale Sea Surface Topography Response to Wind Forcing in the GINSEA using GEOSAT Altimetry. (Abstract). (Reannouncement with New Availability Information)." Naval Oceanographic and Atmospheric Research Lab., Stennis Space Center, MS. Volume: NOARLAB89321091, pages: 1. Descriptors: Basins Geographic; Data bases; Flow; Height finding; Indian Ocean; North Atlantic Ocean; Rotation; Scale; Somalia; Synthesis; Reprints; Ocean currents; Wind stress; Ocean models; Scatterometers; SF298 Only; Air water interactions; Ocean surface; Ocean technology and engineering Dynamic oceanography; Atmospheric sciences Dynamic meterology; Natural resources and earth sciences Natural resource surveys. Abstract: Cross-over point difference data from the 16 month Pre-ERM GEOSAT mission were analyzed together with basin scale wind stress from the same period. Using complex empirical orthogonal functions (CEOF), the synthesized data set showed coherent patterns of variations indicating the locations of major currents and their comcomittant wind forcing. Applied to the northwestern Indian Ocean, the well-known Somali Current and Great Whirl system were clearly evident, giving validity to the analysis method. Applied to the GINSEA North Atlantic, the Irminger Current and North Atlantic inflow were evident, Increasing in strength with northward wind stress forcing. The Iceland Faeroes Front was not apparent, however. DTIC Accession Number: ADA2330785XSP; NTIS Accession Number: ADA2330785.

Johnson, J. H. 1978. "A Conceptual Review of Somali Groundwater Resources." Italy: Descriptors: Africa; East Africa; ground water; Somali Republic; theoretical models; water resources. Database: GeoRef. Accession Number: 1999-004657.

Johnson, John William. 1967. "Historical Atlas of the Horn of Africa." Mogadishu: U.S. Peace Corps. Descriptors: Atlas. Notes: Description: [3] l., 15 l. of maps. 33 cm. Geographic: Somaliland- Boundaries- Maps. Somalia- Boundaries- Maps. Somalia- Historical geography- Maps. Notes: Bibliographical references included in preface, 2d prelim. leaf. Library of Congress; LCCN: 71-653340; OCLC Accession Number: 26970.

Johnston, B. 1982. "Somalia Follow-Up." Journal of Family Health Training. Volume 1, Issue 2, Pages 15-77. "Dedicated to and written for family health trainers in Africa and the Middle East". Descriptors: Africa; Africa South of the Sahara; Africa, Eastern; Allied Health Personnel; Community Health Aides; Community Health Services; Consumer Participation; Curriculum; Delivery of Health Care; Demography; Developing Countries; Disease; Education; Emigration and Immigration; Health; Health Personnel; Health Planning; Health Services; Middle East; Midwifery; Organization and Administration; Population; Population Dynamics; Primary Health Care; Program Development; Public Health; Somalia; Teaching Materials; Arab Countries; Community Participation; Community Workers; Demographic Factors; Diseases; Eastern Africa; Educational Activities; Field Report; Field Workers; Midwives; Migration; Paramedical Personnel; Program Activities; Programs; Settlement And Resettlement; Training Programs. Notes: Also published as French title, Revue de Formation en Sante Familiale 1(2):15-7 1982. Abstract: This article is a sequel to 1 which described how plans were made to deliver primary health care in Somalian refugee camps, and provides information on the program which was developed for the training of health workers in the camps. A map is provided to show the location of the camps. The main problems to be dealt with were food, water, shelter, sanitation, drug supplies, staff, management, equipment, and facilities. Health problems included infectious diseases, malnutrition, parasites, and obstetrical and gynecological problems. A 2-week workshop was organized with health professionals, and the final program consisted of 5 modules, details of which are contained in the text. 5 teams went out, each having 5 members and an elected leader. The role of the community health worker was defined and the job description written. Detailed learning objectives were set for each module. The community health workers were selected from and by the camp population. The 1st module of training lasted 1 month and was followed by the other modules. At the end of the 1st training period, which involved 1,800 people, the teachers were called back for a 3-day evaluation period. The only change was a clarification of the immunization timetable. A country-wide evaluation was made at the end of 12 months, the result of which was a traditional birth attendant training program. OCLC: 08619242.

Johnstone, D. R., Cooper, J. F., Casci, F. and Dobson, H. M. 1990. "Interpretation of Spray Monitoring Data in Tsetse Control Operations using Insecticidal Aerosols Applied from Aircraft." Atmos. Environ. Volume 24A, Issue 1, Pages 53-61. Descriptors: Insect Control- Environmental Testing; Aerosols--Measurements; Insecticides--Applications; Computer Programs; Flow of Fluids--Monitoring. Abstract: Fresh advances in the interpretation of spray monitoring data obtained from rotary slide samplers are outlined. A predictive computer program permits volumetric spray measurements to be converted to the most probable insecticide dose received by tsetse flies resting adjacent to the samplers. Some applications to data collected during recent spray operations in Zimbabwe and Somalia are discussed. ISSN: 0004-6981.

Jolly, J. L. W., Petkof, B. and Singleton, R. H., et al. 1977. "The Mineral Industry of Other Areas of Africa." Minerals Yearbook. Volume 1974, Vol. 3, Area reports; international, Pages 1147-1210. Descriptors: 1974; Africa; Benin; Botswana; Burkina Faso; Burundi; Cameroon; Central Africa; Central African Republic; Chad; Congo; Djibouti; East Africa; economic geology; economics; Equatorial Guinea; Ethiopia; Gambia; gas; Guinea; Indian Ocean Islands; Ivory Coast; Lesotho; Madagascar; Malawi; Mali; Mauritania; Mauritius; metals; mineral resources; natural; natural gas; Niger; nonmetals; petroleum; production; regional; Rwanda; Sahara; Senegal; Somali Republic; Southern Africa; Sudan; Swaziland; Togo; West Africa; Zimbabwe. Notes: tables. ISSN: 0076-8952.

"Jubba Environmental and Socioeconomic Studies (JESS). Volume 1. Executive Report. Final Rept." 1989. Associates in Rural Development, Inc., Burlington, VT. pages: 141. Descriptors: Economic development; Foreign aid; Flood control; Reservoirs; Irrigation; Dams; Somalia; Developing countries; Jubba Valley; Environmental impacts; Socioeconomic factors; Baseline studies; Hydroelectric power plants; Baardheere Dam; Energy Electric power production; Civil engineering Civil engineering; Business and economics Foreign industry development and economics; Behavior and society International relations. Abstract: Jubba Environmental and Socioeconomic Studies (JESS) was carried out as a three-phase project to collect environmental and socioeconomic data in Somalia's Jubba Valley, the site of proposed development of a large hydroelectric dam. Complementary to construction of the dam, various plans are being prepared for subsequent development of irrigated agriculture in the middle and lower Jubba Valley. Numerous environmental and socioeconomic changes will occur with dam construction, filling of the reservoir, infrastructural enhancement, and intensification of agriculture. Volume I, the Executive Report, and reports based on JESS longer-term studies (TEBS and SEBS) represent the most comprehensive assessment of the overall JESS effort: these reports consider and, in most cases, summarize the findings of other investigations. Notes: Funder: Agency for International Development, Washington, DC. See also Volume 2, PB89-224141. Sponsored by the Agency for International Development, Washington, DC. NTIS Accession Number: PB89224133XSP.

"Jubba Environmental and Socioeconomic Studies (JESS). Volume 2. Environmental Studies. Final Rept." 1989. Associates in Rural Development, Inc., Burlington, VT. pages: 671. Descriptors: Flood control; Land use; Wildlife; Monitoring; Foreign aid; Dams; Somalia; Developing countries; Jubba Valley; Environmental impacts; Baseline studies; Resource management; Hydroelectric power plants; Baardheere Dam; Terrestrial ecosystems; Energy Electric power production; Energy Environmental studies; Civil engineering Civil engineering; Business and economics Foreign industry development and economics; Behavior and society International relations. Abstract: The Jubba Environmental and Socioeconomic Studies (JESS) investigated conditions in the Jubba Valley of southern Somalia. Projections from that baseline information were intended to elucidate changes likely to occur as a result of construction of a high dam near Baardheere and related developments. In particular, JESS was required to suggest ways of mitigating adverse impacts, enhancing potentially good impacts, and to draw up a program for future environmental and socioeconomic monitoring. The report contains an analysis of the Terrestrial Ecology Baseline Studies (TEBS) section of the JESS project. Human use of biological resources is examined from the perspectives of land use, forestry, rangelands, and biological conservation. TEBS activities are used as the basis for a future monitoring program of terrestrial ecology. Notes: Funder: Agency for International Development, Washington, DC. NT: See also Volume 1, PB89-224133 and Volume 3, PB89-224158. Sponsored by the Agency for International Development, Washington, DC. NTIS Accession Number: PB89224141XSP.

"Jubba Environmental and Socioeconomic Studies (JESS). Volume 4. Bibliography. Final Rept." 1989. Associates in Rural Development, Inc., Burlington, VT. pages: 122. Descriptors: Economic development; Foreign aid; Flood control; Reservoirs; Irrigation; Bibliographies; Dams; Somalia; Developing countries; Jubba Valley; Environmental impacts; Socioeconomic factors; Baseline studies; Hydroelectric power plants; Baardheere Dam; Energy Electric power production; Civil engineering Civil engineering; Business and economics Foreign industry development and economics; Behavior and society International relations; Library and information sciences Reference materials. Abstract: The bibliography is a revised version of the Jubba Environmental and Socioeconomic Study (JESS) bibliography that was published in 1986, relating specifically to social and environmental systems in Somalia as well as river basin assessment and planning, in general. Many new references have been added and the bibliography has been organized into 22 different sections, corresponding to the subject codes (BIBCODE) being used by the JESS team in Somalia. The bibliography includes selected monographs, conference papers, journal articles, book chapters, reports, JESS studies and dissertations. Users should recognize that the bibliography is neither definitive nor comprehensive, and that an inclusion of a particular reference does not imply that its scientific merit has been substantiated. Notes: Funder: Agency for International Development, Washington, DC. See also Volume 3, PB89-224158. Sponsored by the Agency for International Development, Washington, DC. NTIS Accession Number: PB89224166XSP.

Juch, D. and Schonfeld, M. 1970. "Report of Geological Investigations on the Somalian Escarpment, Ethiopia." Federal Republic of Germany: Descriptors: Africa; areal geology; East Africa; Ethiopia; faults; Harar Ethiopia; landforms; lithofacies;

paleomagnetism; report; sequence stratigraphy; Somalian Escarpment; tectonics; thickness; tilt; unconformities; volcanism. Notes: illustrations. Database: GeoRef. Accession Number: 1999-003232.

Jum'a, S. High Level WCARRD Follow-up and Lagos Plan of Action Strategy Review Mission (Somalia) and Food and Agriculture Organization of the United Nations. 1983. Review of Rural Development Strategy and Policies in Somalia Report of the High Level WCARRD Follow-Up and Lagos Plan of Action Strategy Review Mission, 1-12 may 1982. Rome: Food and Agriculture Organization of the United Nations. Descriptors: 121, 42; Rural development- Somalia. Notes: "WCARRD Follow-up Mission No. 8." "S. Jum'a ... mission leader"--Annex I. Includes bibliographical references (p. 118-121). Reproduction: Microfiche. Rome, Italy: David Lubin Memorial Library, FAO, 1983. 2 microfiches. FAO documentary unit no. 8333009. Other Titles: Report of the High Level WCARRD Follow-up and Lagos Plan of Action Strategy Review Mission, 1-12 May 1982. OCLC Accession Number: 21253367.

Jung S.J.A, Ganssen G.M and Davies G.R. 2001. "Multidecadal Variations in the Early Holocene Outflow of Red Sea Water into the Arabian Sea." Paleoceanography. Volume 16, Issue 6, Pages 658-668. Descriptors: Oceans; Depositional environments; Holocene; outflow; decadal variation; paleoclimate; climate forcing; oxygen isotope; evaporation; precipitation (climatology). Somalia. Notes: United States; References: Number: 50; Geographic: Red Sea Arabian Sea. Abstract: We present Holocene stable oxygen isotope data from the deep Arabian Sea off Somalia at a decadal time resolution as a proxy for the history of intermediate/upper deep water. These data show an overall reduction by 0.5 between 10 and ~6.5 kyr B.P. superimposed upon short-term variations at a decadal-centennial timescale… Changes in water temperature and salinity cause the outflowing Red Sea Water to settle roughly 800 m deeper than today. ISSN: 0883-8305.

Jung S.J.A, Ganssen G and Kroon D, et al. 2002. "Centennial-Millennial-Scale Monsoon Variations Off Somalia Over the Last 35 Ka." Geological Society Special Publication. Issue 195, Pages 341-352. Descriptors: Palaeoclimatology; Geochronology, stratigraphy and palaeontology; Palaeoclimatology; paleoproductivity; sediment core; paleoclimate; monsoon; Younger Dryas; Holocene Species Term: Somalia; Foraminifera. Notes: Additional Info: United Kingdom; References: Number: 42; Geographic: Arabian Sea; Somalia. Abstract: We present a multi-proxy study of sediment Core 905 from the Arabian Sea offshore Somalia to assess the validity of a number of proxies for productivity, temperature and wind strength, to reconstruct the monsoon history in the western Arabian Sea. The present-day seasonal variation in productivity in the modern Arabian Sea off Somalia reflects the change from the high-productivity SW monsoon to the low-productivity NE monsoon seasons. Annual productivity is therefore largely controlled by SW monsoon driven upwelling. The geochemical records of Core 905 document millennial-scale variations, for example, in Ba/Al and Corg content. The Younger Dryas and the time equivalent period to Heinrich event 1 show low annual productivity whereas the early Holocene periods are characterized by high productivity. The upwelling-productivity peaked during Early Holocene time and was followed by a decrease toward the modern values. The total flux of planktic foraminifera and the concentration of the planktic foraminifera G. bulloides are not always controlled by the total productivity. Variations in calcite dissolution, the advection of expatriate fauna or a seasonal decoupling of primary and secondary production appear to hamper

6666666

6666666666

straightforward interpretations of those foraminifera records. We conclude that at significantly changed climatic boundary conditions compared with the present day, bulk-sediment-related proxies of productivity more consistently record the local upwelling history than foraminifer-based productivity proxies. ISSN: 0305-8719.

Jung, S. J. A., Davies, G. R., Ganssen, G. M. and Kroon, D. 2004. "Synchronous Holocene Sea Surface Temperature and Rainfall Variations in the Asian Monsoon System." Quaternary Science Reviews. NOV. Volume 23, Issue 20-22, Pages 2207-2218. Abstract: An increasing number of high-resolution paleoclimate records show substantial natural variation during the Holocene. In order to improve climate projections on human lifetime, the processes that potentially control teleconnections between different parts of the climate system need to be understood. A highly suitable area to study these processes is the Asian monsoon system, as it is one of the most dynamic climate systems on Earth and largely controls climate in Asia and the Indo-W-Pacific realm. Here, we present a Holocene stable O-isotope record from the summer-dwelling planktic foraminifer G. bulloides in Core 905 off Somalia. Initially dated by the radiocarbon method the record was tuned to atmospheric C-14-variations without violating the radiocarbon dates. The O-isotope variations in Core 905 imply monsoon-controlled Holocene average summer sea surface temperature variations of up to 2-2.5°C within decades to centuries. Other proxy records from marine and continental settings representing key locations in the Asian monsoon system were compared with the record of Core 905. Within the resolution of the individual age models, Holocene monsoonal records vary in concert for the Arabian Sea, continental China and the South China Sea and imply simultaneous warm/wet and cool/dry alternations in the Asian monsoon within centuries to decades. Two potential scenarios are discussed to explain the coherent monsoonal records. Scenario one involves trade wind-induced temperature variations in the W-Pacific warm pool as the driving force. The second scenario advocates the controlling process to synchronous variations in trade wind strength simultaneously occurring over the Pacific and Indian Ocean. ISSN: 0277-3791.

Jung, S. J. A., Davies, G. R., Ganssen, G. M. and Kroon, D. 2004. "Stepwise Holocene Aridification in NE Africa Deduced from Dust-Borne Radiogenic Isotope Records." Earth & Planetary Science Letters. APR 30. Volume 221, Issue 1-4, Pages 27-37. Abstract: Transfer of tropical heat to higher latitudes is the major driving force of the Earth's climate. Consequently, sediments in regions to the north and south of the tropics potentially retain an archive of past major climate reconfigurations. The climate of one such region, around the Arabian Sea, sensitively depends on the coupled Asian and African monsoons that also control the dust transport. Here, we use the Sr-Nd isotope ratios of the dust fraction from Core 905 (Arabian Sea off Somalia), as a novel tool to deduce the Holocene weathering history of the Horn of Africa with emphasis on the climate transition that took place from a wet early to a dry late Holocene. The highly variable Sr isotope ratios are interpreted to reflect mainly changes in the evaporation-precipitation balance over NE Africa whilst the Nd isotope measurements record no significant variations and point to a prevailing NE African dust source. The Sr isotope record shows that the first aridification step occurred at 8.5 kyr BP followed by an unstable transitional period up to 6 kyr BP, characterized by decadal-scale high-amplitude variations in the evaporation/precipitation balance. A second aridification step began at 6 kyr BP and ceased at 3.8 kyr BP when modern-day dry climate was established. The combined Sr and Nd isotope records probably reflect north-south shifts of the Intertropical Convergence Zone controlling the evaporation/precipitation balance over NE Africa. ISSN: 0012-821X.

Jung, Simon J. A., Ivanova, Ekaterina and Reichart, Gert Jan, et al. 2002. "Centennial-Millennial-Scale Monsoon Variations off Somalia Over the Last 35 Ka; the Tectonic and Climatic Evolution of the Arabian Sea Region." Geological Society Special Publications. Volume 195, Pages 341-352. Descriptors: absolute age; Africa; Arabian Sea; C-14; carbon; Cenozoic; centennial-factor; climate change; climate effects; cores; dates; deep-water environment; East Africa; Foraminifera; glacial environment; Heinrich events; Indian Ocean; Invertebrata; isotope ratios; isotopes; marine environment; microfossils; millennial-scale; monsoons; O-18/O-16; oxygen; paleoclimatology; paleotemperature; planktonic taxa; Pleistocene; productivity; Protista; Quaternary; radioactive isotopes; reconstruction; scale factor; Somali Republic; stable isotopes; upwelling; winds. Notes: References: 42; illustrations incl. 1 table, geol. sketch maps; Latitude: N050000, N250000; Longitude: E0650000, E0350000. Abstract: We present a multi-proxy study of sediment Core 905 from the Arabian Sea offshore Somalia to assess the validity of a number of proxies for productivity, temperature and wind strength, to reconstruct the monsoon history in the western Arabian Sea. The present-day seasonal variation in productivity in the modern Arabian Sea off Somalia reflects the change from the high-productivity SW monsoon to the low-productivity NE monsoon seasons. Annual productivity is therefore largely controlled by SW monsoon driven upwelling. The geochemical records of Core 905 document millennial-scale variations, for example, in Ba/Al and C (sub org) content. The Younger Dryas and the time equivalent period to Heinrich event 1 show low annual productivity whereas the early Holocene and Bolling-Allerod periods are characterized by high productivity. The upwelling-productivity peaked during Early Holocene time and was followed by a decrease toward the modern values. The total flux of planktic foraminifera and the concentration of the planktic foraminifera G. bulloides are not always controlled by the total productivity. Variations in calcite dissolution, the advection of expatriate fauna or a seasonal decoupling of primary and secondary production appear to hamper straightforward interpretations of those foraminifera records. We conclude that at significantly changed climatic boundary conditions compared with the present day, bulk-sediment-related proxies of productivity more consistently record the local upwelling history than foraminifer-based productivity proxies. ISBN: 1862391114.

K

Kamen-Kaye, M. and Barnes, S. U. 1978. "Exploration Outlook for Somalia, Coastal Kenya, and Tanzania." Oil Gas J. 31 Jul. Volume 76, Issue 31, Pages 224-248. Descriptors: Africa; basement; Cenozoic; East Africa; economic geology; Kenya; Mesozoic; natural gas; petroleum; petroleum exploration; petrology; Somali Republic; Tanzania; Tertiary; traps. Notes: sects., geol. sketch map. ISSN: 0030-1388.

Karanov, N. 1981. "Kaolin in Bur Area, South Somalia." Spisanie Na Bulgarskogo Geologichesko Druzhestvo. Volume 42, Issue 1, Pages 118-129. Descriptors: Africa; alteration; areal studies; Bur region; clay mineralogy; clay minerals; East Africa; economic geology; granites; granulometry; igneous rocks; industrial minerals; kaolin deposits; mineral composition; plutonic rocks; sheet silicates; silicates;

Somali Republic; surveys. Notes: References: 2; illustrations incl. 2 tables, geol. sketch map, sect. ISSN: 0007-3938.

Kassim Mohamed, Ali, Carmignani, L., Conti, P. and Fantozzi, P. L. 2002. "Geology of the Mesozoic-Tertiary Sedimentary Basins in Southwestern Somalia; Horn of Africa." Journal of African Earth Sciences (1994). Feb. Volume 34, Issue 1-2, Pages 3-20. Descriptors: Africa; areal geology; basins; Cenozoic; East Africa; geologic maps; geomorphology; igneous rocks; maps; Mesozoic; sedimentary basins; sedimentary rocks; Somali Republic; stratigraphy; tectonics; Tertiary; volcanic rocks. Notes: References: 89; illustrations incl. sects., strat. cols. Abstract: Two main sedimentary basins can be recognized in southern Somalia, the NE-SW trending Mesozoic-Tertiary Somali coastal basin, and the NNE-SSW Mesozoic Luuq-Mandera Basin. The two basins are separated by the Bur region where the Proterozoic-early Paleozoic Metamorphic basement of southern Somalia outcrops. The investigated area covers part of the Metamorphic basement of southern Somalia and of the Luuq-Mandera Basin, although this basement is not described in details in this paper. In the Bur region the basement outcrops discontinuously near inselbergs and monadnocks, which stand out of a blanket of recent sediments. Because of this patchy distribution and the limited areal extent of the outcrops, the structure of the metamorphic basement is difficult to reconstruct. A NW-SE trend of structures prevails and two metamorphic complexes (the Olontole and Diinsor complexes) can be recognized. The Luuq-Mandera Basin is a wide NNE-SSW synclinorium, delimited to the SE by the basement high of the Bur region, and to the west by the crystalline basement high of NE Kenya (Northern Frontier District). The extreme thickness of Triassic sediments in the axial part of the basin, and the thinner and younger succession on both sides of the basin suggest that the Luuq-Mandera Basin was a subsiding elongated area that was invaded by the sea in the early Mesozoic, during the dismembering of Gondwana. The Jurassic-Cretaceous succession that followed comprises two main cycles of transgression and regression; the carbonate sediments that lie at the bottom pass up section into shales, evaporites and sandstone deposits. Since Late Cretaceous, continental condition prevailed, with a long phase of peneplanation, and then a general uplift, which brought about the creation of lake depressions and the capture of the Dawa River, with formation of the present Jubba Valley. The main tectonic events in the study area, and throughout SW Somalia, are represented by strike-slip movements along vertical faults in the Sengif and Garbahaarrey Belt. Deformation is localized within a narrow belt that extends for more than a 100 km in a NE-SW direction. The near parallelism between the fold axes and the regional orientation of faults indicates a right-lateral movements along faults. The structure of the Garbahaarrey Belt consists of an anastomosing fault system that delimits elongated folded blocks, arranged in anticline-syncline structures, with subvertical axial surfaces and fold axes parallel to the main wrench faults. The orientation of folds and the typical "positive flower structure" profile of the anticlines indicate that shortening was perpendicular to the strike of the wrench, i.e. in a SE-NW direction. In the Garbahaarrey Belt, strike-slip and shortening, therefore, occurred contemporaneously and led to a relative transpression between the NW and SE blocks. The observed parallelism between fold and fault orientation cannot be explained with a simple rotation of pre-existing fold axes during transpression, but can be regarded as an example of folding and strike-slip movements that occurred simultaneously but independently along frictionless faults. The faults delimiting the anticlines accommodated the strike-slip component of transpression only, whereas the compressive component led to the generation of fold axes parallel to the wrench zone. Results of the field work are summarized in two geological maps of the Gedo, Bakool, and Bay regions (1:250,000) which accompany this report (maps are attached with this issue). ISSN: 1464-343X.

Kassim, M. Ali, Carmignani, L. and Fantozzi, P., et al. 1992. "Flood Basalts of the Gedo Region (Southern Somalia). Geology, Petrology and Isotope Geochemistry." ATTI Della Accademia Nazionale Dei Lincei Rendiconti Lincei Scienze Fisiche E Naturali. Volume 3, Issue 9, Pages 311. ISSN: 1120-6349.

Katskov, A. I. 1967. Brief Results of the Mineral and Groundwater Survey Project Activities and Recommendations on Phase II. Somalia: UNDP, Mogadishu, Somalia. Descriptors: Africa; East Africa; ground water; mineral resources; programs; Somali Republic; water resources. Database: GeoRef. Accession Number: 1999-004665.

Katskov, A. I. 1967. Semiannual Reports by Mineral and Groundwater Survey Project, Covered the Period January 1, 1964 through August 31, 1967 (Reports 1-7). Somalia: UNDP, Mogadishu, Somalia. Descriptors: Africa; East Africa; ground water; mineral resources; report; Somali Republic; water resources. Database: GeoRef. Accession Number: 1999-004667.

Katskov, A. I. and Lartsev, V. S. 1968. Mineral and Groundwater Survey in Somalia. UNDP, Mogadishu, Somalia. Descriptors: Africa; East Africa; ground water; mineral resources; Somali Republic; surveys. Database: GeoRef. Accession Number: 1999-004668.

Kaz'min, V. G. 1975. "Strukturnaya Evolyutsiya Riftov Krasnomorsko-Adenskogo Regiona." Translated title: "Structural Evolution of the Rifts in the Red Sea-Aden Region." Byulleten' Moskovskogo Obshchestva Ispytateley Prirody, Otdel Geologicheskiy. Jun. Volume 50, Issue 3, Pages 116-128. Descriptors: Aden Rift; Afar Depression; Africa; Arabian Plate; Arabian Sea; East Africa; Ethiopia; Ethiopian Rift; evolution; Gulf of Aden; Indian Ocean; mechanism; Nubian Plate; Red Sea; Red Sea region; Red Sea Rift; rift zones; Somalian Plate; structural geology; structure; tectonics. Notes: References: 43; illustrations incl. sketch map. ISSN: 0366-1318.

Kaz'min, V. G., Zonenshayn, L. P., Verzhbitskaya, A. I. and Savostin, L. A. 1984. "Instantaneous Kinematics of the Afro-Arabian Rift System; Special Session of the International "Lithosphere" Programme." International Geological Congress, Abstracts = Congres Géologique International, Resumes. Volume 27, Issue Vol. IX, Part 1, Pages 68-69. Descriptors: Afar Plate; Africa; African Plate; Arabian Plate; distribution; East Africa; Ethiopia; Ethiopian Rift; extension fractures; faults; fractures; Indian Ocean; neotectonics; plate tectonics; Red Sea; Red Sea Rift; rifting; rotation; Somalian Plate; strike-slip faults; structural geology; tectonics; tectonophysics; transform faults; Wongi fault belt. Notes: SP: IGC, International Geological Congress; Tezisy; 27-y mezhdunarodnyy geologicheskiy kongress; illustrations incl. 1 table. Database: GeoRef.

Keijzer, L. 1986. "Salvaging a Cocktail of Dangerous Substances." Fairplay International Shipping Weekly Vol. 298 No. 5372, Pages 25-27. English Abstract: Salvage of Panamanian Containership Ariadne, Carrying Hazardous Material Including Tetraethyl Lead, After Grounding Near Port of Mogadiscio, Somalia. ISSN: 0307-0220.

Kent, P. E. 1982. "The Somali Ocean Basin and the Continental Margin of East Africa." United States: Plenum Press, New York, United States. Descriptors: Africa;

basins; Carboniferous; Cenozoic; continental margin; East Africa; Ethiopia; Indian Ocean; Kenya; marine geology; Mesozoic; oceanography; Paleozoic; Phanerozoic; sedimentary basins; Somali Basin; Somali Republic; stratigraphy; Tanzania; Tertiary. Notes: References: 35; sects., strat. cols., sketch maps. ISBN: 0306377764.

Kent, P. E. 1974. "Continental Margin of East Africa; a Region of Vertical Movements." United States: Springer-Verlag, New York, United States. Descriptors: Africa; basalts; Cenozoic; continental margin; east; East Africa; faults; igneous rocks; Indian Ocean; Indian Ocean Islands; Kenya; Madagascar; marine geology; Mesozoic; Mozambique; Mozambique Basin; Mozambique Channel; oceanography; sedimentation; Somali Basin; Somali Republic; structure; Tanzania; tectonics; volcanic rocks. Notes: illustrations incl. sketch maps. Database: GeoRef. Accession Number: 1975-024757.

Khaled, H. 1982. "Chromium and Manganese Ores in the Arab Countries." Arab Mining Journal. Volume 2, Issue 4, Pages 30-35. Descriptors: Africa; Algeria; Arab Mining Company; Arabian Peninsula; Asia; chromite ores; East Africa; economic geology; Egypt; inventory; Jordan; manganese ores; metal ores; Middle East; mineral shows; mines; Morocco; North Africa; Oman; Somali Republic; Sudan. Abstract: Inventory of deposits and indexes of chromite with their principal grade characteristics and tonnage in Jordan, Egypt, Sudan, Somalia, Morocco, Syria, Oman and Algeria. ISSN: 0250-9881.

Khalif M.H and Doornbos M. 2002. "The Somali Region in Ethiopia: A Neglected Human Rights Tragedy." Rev. Afr. Polit. Econ. Volume 29, Issue 91, Pages 73-94. Descriptors: Conflict, protest, human rights; violence; social conflict; famine; natural gas. References: Number: 48; Geographic: Somalia. Abstract: This article reviews Ethiopia's human rights record with a particular focus on the human rights situation in the Somali region. Attention is paid to the atrocities committed against civilians, specifically community and political leaders as well as members of the Somali State legislature. Furthermore, the 2000 famine is discussed as a human rights issue in the light of indications that this is famine was deliberately choreographed. The article also explores human rights violations inflicted upon the Somali region's population following the discovery of natural gas and the denial of benefits thereof to the local community. In conclusion some future scenarios are examined to ascertain to what extent they might possibly change the prospects for the people in the Somali region. ISSN: 0305-6244.

Khatib, A. B. 1970. Duty Trip to Somalia 30 Jan.-4 Feb. (on Land and Water Resources). Italy: FAO, Rome, Italy. Descriptors: Africa; East Africa; field trips; natural resources; Somali Republic; water resources. Database: GeoRef. Accession Number: 1999-004673.

Khatib, A. B. 1966. Visit of FAO Regional Water Resources Office to Somalia. Italy: FAO, Rome, Italy. Descriptors: Africa; East Africa; Food and Agriculture Organization; government agencies; Somali Republic; water resources. Database: GeoRef. Accession Number: 1999-004672.

Kidane, T., Courtillot, V. and Manighetti, I., et al. 2003. "New Paleomagnetic and Geochronologic Results from Ethiopian Afar: Block Rotations Linked to Rift Overlap and Propagation and Determination of a Similar to 2 Ma Reference Pole for Stable Africa." Journal of Geophysical Research-Solid Earth. FEB 18. Volume 108, Issue B2, Pages 2102. Abstract: [1] Joint French-Ethiopian field trips in 1995-1996 yield new geochronologic and paleomagnetic data, which significantly expand our knowledge of the recent magmatic and tectonic history of the Afar depression. Twenty-four new K-Ar ages range from 0.6 to 3.3 Ma. There is quite good agreement between magnetic polarities and Geomagnetic Polarity Timescale (GPTS). Eight age determinations with uncertainty less than 50 kyr can be used in future reassessments of the GPTS (upper and lower Olduvai Matuyama reversals and Reunion and Mammoth subchrons). Paleomagnetic analysis of 865 cores from 133 sites confirms that low-Ti magnetites are the main carrier of the Characteristic Remanent Magnetization (ChRM). A positive tilt test (based on two subgroups with 63 and 23 sites, respectively) confirms that this ChRM is likely the primary magnetization. The main paleomagnetic results can be summarized as follows. A similar to2 Ma reference pole for stable Africa is determined based on 26 sites located on either side of the northern termination of the East African rift…

…Combined with earlier data of Acton et al. [2000], our new data allow mean paleomagnetic field directions to be determined for five individual, fault-bounded blocks previously identified by tectonic analysis within central Afar. These all have suffered negligible rotations about vertical axes since emplacement of the lava. This contrasts with the significant rotations previously uncovered to the east in Djiboutian Afar for three major individual blocks. Taken altogether, the declination differences with respect to reference directions are 2 +/- 4 degrees for central Afar and 13 +/- 4 degrees for eastern Afar, consistent with the model of Manighetti et al. [2001a]. It appears that in the last similar to3 Ma the Afar depression was extensively floored by trap-like basalts, which were deformed by a single but complex physical (tectonic) process, combining diffuse extension, rift localization, propagation, jumps and overlap, and bookshelf faulting.

Kielmas, M. 2001. "East Africa's Elusive Elephants." Pet. Rev. Volume 55, Issue 652, Pages 18-19. Descriptors: Oil and gas industry; Oil and gas production; Oil and gas exploration; Oil and gas fields; Offshore operations; Deep water; Deep-sea drilling; Potential resources; Licensing; Kenya, Coast; Tanzania; Mozambique; Somali Dem. Rep. Marine. Abstract: Current industry trends in favor of deep water exploration could renew hopes of finding oil in East Africa. Seismic surveys over the deep offshore have identified huge structures that may provide the long awaited breakthrough in oil production. This article reports on recent deep water offshore activity in Tanzania, Mozambique, Kenya and Somalia. ISSN: 0020-3076.

Kigotho, Anderson Wachira. 1997/12/13. "Belet Weyne Somalia--the Country the World Wants to Forget." The Lancet. Volume 350, Issue 9093, Pages 1757-1757.

Kigotho, Anderson and McGregor, Alan. 1998. "Rift Valley Fever Confirmed in "Mystery" African-Disease Outbreak." Lancet (North American Edition). January 10. Volume 351, Issue 9096, Pages 115. Descriptors: Public health/Kenya; Rift Valley fever; Hemorrhagic fevers; General Science. Abstract: Experts remain unclear about the causes of a mysterious hemorrhagic fever-like disease reported to have killed up to 350 people in Kenya in about a month, as well as an undetermined number in Somalia. Rift Valley Fever virus has been isolated, but, as the populations affected are undernourished and have infections associated with the lack of clean water, the WHO is also considering the involvement of diseases such as shigella and malaria. In addition, the numbers of domestic and wild animals involved has meant that consideration is being given to anthrax. ISSN: 0099-5355.

Kindle, J. C. 1991. "Topographic Effects on the Seasonal Circulation of the Indian Ocean." Sep. Pages 12. Descriptors: Barometric pressure; Circulation; Somalia;

Ocean bottom; Coastal regions; Cycles; Gravity; Islands; Madagascar; North Atlantic Ocean; Numerical analysis; Seasonal variations; Signals; Wind stress; Climatology; Indian Ocean; Ocean currents; Ocean models; Topography; Somali current; Equatorial dynamics; Ocean forecasting; Ocean circulation; Air water interactions; Ocean technology and engineering Hydrography; Ocean technology and engineering Dynamic oceanography; Atmospheric sciences Meteorological data collection analysis and weather forecasting. Notes: Performer: Naval Oceanographic and Atmospheric Research Lab., Stennis Space Center, MS. Funder: Shared Bibliographic Input. Abstract: The relative effects of islands and topography on the seasonal circulation of the southern Indian Ocean are examined. The numerical simulations use a one-active layer reduced gravity model and a two-layer model with full bottom topography forced by the Hellerman and Rosenstein wind stress climatology. The reduced gravity experiments show that the absence of a seasonal cycle of transport off the north coast of Madagascar is not caused by the presence of islands to the east. The two-layer experiments indicate that the presence of abrupt topography associated with the Mascarene Ridge and the Providence Bank north of Madagascar decrease the annual barotropic signal of the North Madagascar Current and increase the seasonal variation of the Southeast Madagascar Current. Furthermore, the two-layer experiments predict the existence of a significant seasonal variation of transport north of the Saya de Malha Bank along the Mascarene Ridge; the signal (12-16 Sv in amplitude) is sufficiently large that the model results may be tested by future observations. Finally, the differences between seasonal variation of the Indian and North Atlantic oceans are discussed.

King P. 1986. An African Winter. Penguin Books. Descriptors: 249p; International Development Abstracts. drought; famine; Ethiopia; Sudan; war; Kenya; Eritrea; Somalia. Abstract: A non-academic text which first describes the 1980's drought and famine in Ethiopia, and how and in what way food aid was supplied. Compares the situation in Ethiopia with Sudan, noting superpower interests. The second section on war looks at three areas of conflict: NE Kenya, Eritrea, and Somalia and includes a chapter on Djibouti. Concludes with options for a way out of Africa's problems. Notes: In: An African winter. (1986) p. 249p; Notes: Special Features: index. ISBN: 0140523650.

Kintz, D. 1989. "Approche Comparative Du Pastoralisme Somalien." Translated title: "Comparative approaches to pastoralism in Somalia." Mondes En Developpement. Volume 17, Issue 66, Pages 77-81. Descriptors: Social, economic and political aspects of rural change; pastoralism; commercialization; migration. Notes: Africa Somalia. Abstract: Pastoral societies as a whole and African ones in particular have many similarities. But the same clichés are used for all of them: one of these describes pastoral producers as non-commercial, which is incorrect in Somalia as well as in other African countries. ISSN: 0302-3052.

Klaver, A. 1964. Final Report; Surface Geological Studies of the Somali Basin. Somalia: Mobil Petroleum Co., Mogadishu, Somalia. Descriptors: Africa; areal geology; Dusa Mareb Somali Republic; East Africa; Galguduud; Indian Ocean; Somali Basin; Somali Republic. Database: GeoRef. Accession Number: 1999-004675.

Klaver, A. 1964. Geological Map Showing Structural Detail, Hiran Area, Scale 1:500,000. Somalia: Mobil Petroleum Co., Mogadishu, Somalia. Descriptors: Africa; East Africa; geologic maps; Hiran; maps; Somali Republic. Database: GeoRef. Accession Number: 1999-004676.

Klaver, A. and Jenkins, A. 1964. Report on Dusa Mareb n.1 and 2 and Field Trips. Somalia: Mobil Petroleum Co., Mogadishu, Somalia. Descriptors: Africa; areal geology; Dusa Mareb Somali Republic; East Africa; field trips; Galgaduud; Somali Republic. Database: GeoRef. Accession Number: 1999-004677.

Kleyner, W. 1976. Trans-Juba Livestock Project-Somalia; Report on the Chemical Analysis of Water from Juba River. Somalia: Publisher unknown, Somalia. Descriptors: Africa; chemical analysis; East Africa; geochemistry; hydrochemistry; Juba River; Somali Republic; surface water; Trans-Juba Livestock Project. Database: GeoRef. Accession Number: 1999-004678.

Klitzsch, E. 1988. "Sonderforschungsbereich 69; Geowissenschaftliche Probleme in Ariden Gebieten; Entwicklung Und Potential Kratonaler Grossstrukturen." Translated title: "Special Research Area 69; Geoscientific Problems in Arid Zones; Development and Potential of Large Scale Cratonic Strutures." Jahresbericht- Deutsche Forschungsgemeinschaft, Band 2, Programme Und Projekte. Volume 1988, Pages 844-845. Descriptors: Africa; arid environment; deserts; East Africa; eolian features; ground water; mineral resources; programs; research; Sahara; soils; Somali Republic; Sudan; tectonics. Reference includes data from Geoline, Bundesanstalt fur Geowissenschaften und Rohstoffe, Hanover, Germany. ISSN: 0340-1359.

Klitzsch, Eberhard. 1995. "Aethiopien-Somalia-Nordkenia." Translated title: "Ethiopia, Somalia, Northern Kenya." Regional Petroleum Geology of the World; Part II, Africa, America, Australia and Antarctica--Regionale Erdoel- Und Erdgasgeologie Der Erde; Teil II, Afrika, Amerika, Australien Und Antarktis." Beitraege Zur Regionalen Geologie Der Erde. Volume 22, Part 2, Pages 265-266. Descriptors: Africa; areal geology; boreholes; coastal environment; East Africa; Ethiopia; history; Kenya; northern Kenya; petroleum; petroleum exploration; Somali Republic. Notes: References: 2; sketch map. ISBN: 3443110223.

Klitzsch, Eberhard; Schrank, Eckart; Deutsche Forschungsgemeinschaft and Sonderforschungsbereich 69. 1990. Research in Sudan, Somalia, Egypt and Kenya: Results of the Special Research Project "Geoscientific Problems in Arid and Semiarid Areas" (Sonderforschungsbereich 69): Period 1987-1990. Berlin: Selbstverlag Fachbereich Geowissenschaften, FU Berlin. Descriptors: 2 v. Geology- Sudan; Geology- Somalia; Geology- Egypt; Geology- Kenya. Notes: "This volume is dedicated to Rushdi Said." "This research project is sponsored by Deutsche Forschungsgemeinschaft ... [et al.]."--T.p. verso. Pref. in German; summaries also in German. Each vol. has 4 maps on folded leaves in pocket. Includes bibliographical references. Other Titles: Results of the Special Research Project "Geoscientific problems in arid and semiarid areas" (Sonderforschungsbereich 69). Geoscientific problems in arid and semiarid areas. ISBN: 392754115X (vol. 1); 3927541168 (vol. 2) ISSN: 0172-8784; LCCN: 91-228540.

Klocker, R., Ganssen, G., Jung, S.J.A., Kroon, D. and Henrich, R. 2006. "Late Quaternary millennial-scale variability in pelagic aragonite preservation off Somalia." Marine Micropaleontology. May 24, 2006, volume 59, issue 3-4, pages 171-183.

Knoch, K. 1914. "Bericht Über Die Fortschritte Der Geographischen Meteorologie [Im Horn Von Afrika]." Translated title: "Report on the Advances in Geographical Meteorology in the Horn of Africa." Geographisches Jahrbuch. Volume 36, Pages 204. Descriptors: Africa; East Africa; geography; Horn of Africa; meteorology;

Somali Republic. Notes: Continuing report with supplements in no. 39, 1919-1923,1924, p.138; no. 41, 1926, p.106-107; no. 44, 1929-1930, p.133. ISSN: 0072-095X.

Knox, Bergman, Shearer, Denver, CO, United States. 1963. Areal Geology and Geomorphic Analysis of Southwest Somalia. United States: Knox, Bergman, Shearer, Denver, CO, United States. Descriptors: Africa; areal geology; East Africa; geomorphologic maps; geomorphology; maps; Somali Republic; southwestern Somali Republic. Database: GeoRef. Accession Number: 1999-004679.

Kocks, F. H. 1972. Installation of Wells, Project n. 211.016.25; Final Report. Federal Republic of Germany: Consulting Engineers, Düsseldorf, Federal Republic of Germany. Descriptors: Africa; East Africa; ground water; programs; Somali Republic; water wells; wells. Database: GeoRef. Accession Number: 1999-004682.

Kocks, F. H. 1971. Installation of Wells. Federal Republic of Germany: MMWR/WDA, Düsseldorf, Federal Republic of Germany. Descriptors: Africa; East Africa; ground water; Somali Republic; water wells; wells. Database: GeoRef. Accession Number: 1999-004681.

Kocks, F. H. 1966. Installation of 70 Pumps; Basic Project. Federal Republic of Germany: Consulting Engineers, Düsseldorf, Federal Republic of Germany. Descriptors: Africa; East Africa; ground water; pumping; Somali Republic; water wells; wells. Database: GeoRef. Accession Number: 1999-004680.

Koettlitz, R. 1900. "A Journey through Somaliland and Southern Abyssinia to the Shangalla or Berta Country and Blue Nile and through the Sudan to Egypt." Journal of the Manchester Geographical Society. Volume 16, Descriptors: Africa; areal geology; Berta Country; Blue Nile; East Africa; Egypt; Ethiopia; expeditions; North Africa; Shangalla; Somali Republic; Sudan. Database: GeoRef.

Koning, E., Brummer, G. -J, Van Raaphorst, W., Van Bennekom, J., Helder, W. and Van Iperen, J. 1997. "Settling, Dissolution and Burial of Biogenic Silica in the Sediments Off Somalia (Northwestern Indian Ocean)." Deep-Sea Res. (Ii Top. Stud. Oceanogr.). Volume 44, Issue 6-7, Pages 1341-1360. Descriptors: Particulate flux; Sedimentation; Sediment-water interface; Pore water; Sediment chemistry; Biogenic deposits; Leaching; Dissolution; Burying; Silica; Sediment traps; Coastal upwelling; Somali Dem. Rep. Indian Ocean, Somali Basin; Marine. Notes: Special issue: Netherlands Indian Ocean Program 1992-1993: First results. TR: CS9810138. Abstract: Particle fluxes of biogenic silica through the water column, silica burial fluxes into the sediments, and the flux of dissolved silica across the sediment-water interface estimated from pore water profiles are used to assess the behavior of biogenic silica at two stations 80 and 270 km offshore along a transect off the Somali coast in the northwestern Indian Ocean. Particulate biogenic silica fluxes varied from 0.3 mmol m super(-2) day super(-1) in the non-upwelling season to 6 mmol m super(-2) day super(-1) during upwelling on the Somali slope. Fluxes were lower in the Somali Basin, from 0.2 to 2.3 mmol m super(-2) day super(-1). Evaluation of the dissolution curves derived by wet chemical leaching in sediment trap and sediment samples shows that the K sub(m) values, the apparent reactivity rates in alkaline medium, are higher for the shallow sediment traps than for deep trap and boxcore sediments. Modelling of pore water profiles shows that in the sediment most dissolution occurs in the top half-centimeter, and pore water effluxes are in close agreement with those from in situ benthic incubations. Our results show that less than 10% of the biogenic silica arriving on the Somali Margin is buried in the sediments, giving a burial efficiency lower than the approximately 20% reported from the open Arabian Sea. ISSN: 0967-0645.

Koning, E., van Iperen, J. M., van Raaphorst, W., Helder, W., Brummer, G. -J A. and van Weering, T. C. E. 2001. "Selective Preservation of Upwelling-Indicating Diatoms in Sediments Off Somalia, NW Indian Ocean." Deep-Sea Research. (I Oceanogr. Res. Pap.). Nov. Volume 48, Issue 11, Pages 2473-2495. Descriptors: Silica; Diatoms; Phytoplankton; Sediments; Sediment analysis; Sediment traps; Sediment-water interface; Pelagic sediments; Biogenic deposits; Coastal upwelling; Microorganisms; Chaetoceros; Thalassionema nitzschioides; Somalia; Indian Ocean; Indian Ocean; Somalia; Somali Basin. Notes: TR: CS0122556. Abstract: The diatom species composition of settling biogenic silica particles collected in sediment traps was compared with the underlying sediment to determine the preservation of the various diatom species and to investigate the potential of biogenic silica as an indicator for changes in paleo-upwelling intensity. During the Netherlands Indian Ocean Programme (NIOP), settling particles were collected at two sampling sites off Somalia (NW Indian Ocean) for 9 months, from June 1992 to February 1993. One sediment trap array was deployed on the Somali slope directly below one of the main upwelling gyres, and a second array, meant as a reference site to reflect pelagic sedimentation, was moored in the Somali Basin away from direct coastal upwelling influence. At both sites diatoms represented over 90% of the total opal microorganisms. On the Somali slope, total annual diatom flux was 12.6x10 super(9) valves m super(2), 76% of which was collected during the 112d of the southwest monsoon, with peak fluxes in October, the end of the upwelling season. In the Somali Basin, the total annual flux was lower, 4.8x10 super(9) valves m super(2), and only 39% was collected during the SW monsoon period (98d). At both sampling sites, a distinct seasonal diatom species succession of 'pre-upwellers', 'upwellers' and 'oceanic species' was apparent. Although only a small part of the diatom assemblage escaped dissolution at the sediment-water interface, two species, Thalassionema nitzschioides and Chaetoceros resting spores, were preserved in the sediment, indicating that they are resistant to dissolution at the sediment-water interface. Eighty one percent of the deposition of Thalassionema nitzschioides and 78% of the deposition of Chaetoceros occurred during the upwelling period. Since these two species are the dominant component of the diatom assemblage in the sediments, and thus determine the biogenic silica content, we conclude that this preserved biogenic silica reflects the upwelling in the surface layer of the water column. On the Somali Margin, variations in biogenic silica flux as inferred from sedimentary records can therefore be used as an indicator for changes in paleo-upwelling intensity. ISSN: 0967-0637.

Konoike Construction Company, Tokyo, Japan (JPN). 1987. Groundwater Development Project in Lower Shebelli Related to ICARA II (Phase 10; Final Inspection Record; Chapter 2, Production Wells). Japan: Konoike/MMWR, Tokyo, Japan. Descriptors: Africa; development; East Africa; ground water; hydrology; Shebelle River; Somali Republic; southern Somali Republic; surface water; water resources; water wells. Database: GeoRef. Accession Number: 1999-004683.

Kononov, Y. V. 1974. A Report on Geochemical Reconnaissance in the Northern Province, Somali Democratic Republic. Somalia: United Nations, Mogadishu, Somalia. Descriptors: Africa; East Africa; exploration; geochemistry; northern Somali Republic; Somali Republic; water resources. Database: GeoRef. Accession Number: 1999-004684.

Konoplynsev, A. 1966. Explanatory Remarks to Mineral and Groundwater Survey Project in the Somali Republic. Somalia: UNDP, Mogadishu, Somalia. Descriptors: Africa; East Africa; explanatory text; ground water; mineral resources; Somali Republic; water resources. Database: GeoRef. Accession Number: 1999-004685.

Kozerenco, V. N. 1970. Geological Map of Somalia (Scala 1:1,000,000). Somalia: UNDP, Mogadishu, Somalia. Descriptors: Africa; East Africa; geologic maps; maps; Somali Republic. Notes: Latitude: N120000, E0513000 Longitude: E0410000,. Database: GeoRef. Accession Number: 1999-004686.

Kozerenko, V. N. and Lartsev, V. S. 1981. "O Damarsko-Somaliyskoy Zone Glubinnykh Razlomov v Svyazi s Osobennostyami Geologicheskogo Stroeniya Yuzhnoy i Vostochnoy Afriki." Translated title: "The Damaro-Somalian Deep Fault Zone in Relation to the Geologic Structure of Southern and Eastern Africa." Byulleten' Moskovskogo Obshchestva Ispytateley Prirody, Otdel Geologicheskiy. Volume 56, Issue 5, Pages 19-34. Descriptors: Africa; Damaro-Somalian Zone; dynamics; East Africa; fault zones; faults; igneous activity; metallogeny; mobile belts; Phanerozoic; Precambrian; Southern Africa; stratigraphy; structural geology; tectonics. Notes: References: 1 p. illustrations incl. geol. sketch map, sketch map. ISSN: 0366-1318.

Kozerenko, V. N. and Lartsev, V. S. 1978. "Block Tectonics of the East African Activation Region." Geotectonics. Aug. Volume 12, Issue 1, Pages 34-40. Descriptors: Africa; basement; block structures; Cenozoic; East Africa; Ethiopia; faults; Kenya; lithofacies; Mesozoic; sedimentary cover; Somali Republic; structural geology; systems; tectonics; thickness; troughs. Notes: References: 10; sect., strat. cols., sketch map. ISSN: 0016-8521.

Kreger, V. J. 1968. Final Report on Water Engineering. Italy: FAO, Rome, Italy. Descriptors: Africa; East Africa; hydraulics; report; Somali Republic; water resources; waterways. Database: GeoRef. Accession Number: 1999-004687.

Krenkel, E. 1926. "Abessomalien, (Abessinien Und Somalien). Abessomalien; Abyssinia and Somalia." Handbuch Der Regionalen Geologie. Volume 7, no.8, Descriptors: Africa; areal geology; East Africa; Ethiopia; Somali Republic. Database: GeoRef.

Krenkel, Erich. 1925. Abessomalien. Abyssinia-Somalia. Federal Republic of Germany: Gebruder Borntrager, Berlin, Federal Republic of Germany. Descriptors: Africa; areal geology; East Africa; Ethiopia; Somali Republic. Database: GeoRef. Accession Number: 1998-014614.

Krishnamurti, T. N., Molinari, John and Pan, Hua Lu. 1976. "Numerical Simulation of the Somali Jet." J. Atmos. Sci. Volume 33, Issue 12, Pages 2350-2362. Abstract: It is shown that many of the observed features of the cross-equatorial low-level jet of the Arabian Sea, Indian Ocean and Somalia can be numerically simulated by including the east African and Madagascar mountains, the beta effect, and a lateral forcing from the east around 75 degree E. The numerical model presented here is a one-level primitive equation model with a detailed bottom topography and a one-degree latitude grid size. When all the three above-mentioned parameter are included, features such as strong winds just downstream from the Madagascar mountains, an equatorial relative speed minimum, an intense jet off the Somali coast and a split of the jet over the northern Arabian Sea are simulated from an initial state of rest.

Krishnasamy K. 2001. "Building a 'Partnership' for Peace in Intrastate Wars: India's Peacekeeping Style in Somalia (1993-1995)." Low Intensity Conflict and Law Enforcement. Volume 10, Issue 1, Pages 27-46. Descriptors: Geography; International Relations, International Financial Institutions; Political; peacekeeping; military intervention; strategic approach; partnership approach. Notes: Additional Info: United Kingdom; References: Number: 42; Geographic: India; Somalia. Abstract: The Indian military often relates to its experience as part of the UN peace operation in Somalia not only positively but also as being unique. The Indian military contingent had a very different agenda in terms of managing the conflict in Somalia. While the UN was pursuing greater use of force by coercing co-operation, the Indian army adopted a peacekeeping style that relied less on force and more on seeking and fostering local community support. The Indian military's community-oriented and people-centered conflict management approaches enhanced not only its presence in the field but also its popularity among key disputants and wider local community. This article aims to shed some light on the kinds of strategies employed by the Indian army during its deployment in Somalia and also to draw some inferences from its experience for the future conduct of intrastate peacekeeping operations. ISSN: 0966-2847.

Kronberg, P. 1974. "Mapping of Lithologic and Structural Units using Multispectral Imagery (Afar-Triangle/Ethiopia and Adjacent Areas (Ethiopian Plateau, Somali Plateau, and Parts of Yemen and Saudi Arabia)) [Final Report, Aug. 1973- Mar. 1974]." Descriptors: Plateaus; Somalia; Gulfs; Yemen; Geological mapping; Mapping; Surface structure; Saudi Arabia; Belts; Covering; Satellites (artificial). Notes: RP: E75-10403; NASA-CR-143396; REPT-3. Abstract: The author has identified the following significant results. ERTS-1 MSS imagery covering the Afar-Triangle/Ethiopia and adjacent regions (Ethiopian Plateau, Somali Plateau, and parts of Yemen and Saudi Arabia) was applied to the mapping of lithologic and structural units of the test area at a scale 1:1,000,000. Results of the geological evaluation of the ERTS-1 imagery of the Afar have proven the usefulness of this type of satellite data for regional geological mapping. Evaluation of the ERTS images also resulted in new aspects of the structural setting and tectonic development of the Afar-Triangle, where three large rift systems, the oceanic rifts of the Red Sea and Gulf of Aden and the continental East African rift system, seem to meet each other. Surface structures mapped by ERTS do not indicate that the oceanic rift of the Gulf of Aden (Sheba Ridge) continues into the area of continental crust west of the Gulf of Tadjura. ERTS data show that the Wonji fault belt of the African rift system does not enter or cut through the central Afar. The Aysha-Horst is not a Horst but an autochthonous spur of the Somali Plateau. Database: NTIS. Accession Number: E7510403.

Kronberg, P. Schoenfeld, M. and Guenther, R. 1975. Small Scale Mapping in Ethiopia, Somalia and Yemen using Field Data, Aerial Photographs and ERTS-1 Imagery; Geoscientific Studies and the Potential of the Natural Environment. Federal Republic of Germany: Deutsche UNESCO-Komm., Koeln, Verlag Dok. Muenchen, Munich, Federal Republic of Germany. Geoscientific Studies and the Potential of the Natural Environment, Hanover. Federal Republic of Germany Conference: April 28-May 23, 1975. Descriptors: aerial photography; Afar; Africa; airborne; applications; Arabian Peninsula; areal geology; Asia; cartography; East Africa; ERTS; Ethiopia; field studies; geologic; geophysical methods; ground; imagery; maps; remote sensing; satellite; Somali

Republic; Yemen. Notes: References: 5; geol. sketch maps. Database: GeoRef. Accession Number: 1977-037786.

Kronberg, P. and Technische Univ., Clausthal-Zellerfeld (Germany). 1974. "Mapping of Lithologic and Structural Units using Multispectral Imagery (Afar-Triangle/Ethiopia and Adjacent Areas (Ethiopian Plateau, Somali Plateau, and Parts of Yemen and Saudi Arabia)) [Final Report, Aug. 1973- Mar. 1974]." Descriptors: African Rift System; Ethiopia; Geological Faults; Gulfs; Plateaus; Saudi Arabia; Somalia; Tectonics; Volcanology; Yemen; Erep; Lithology; Maps; Multispectral Photography; Skylab Program. Abstract: The author has identified the following significant results. ERTS-1 MSS imagery covering the Afar-Triangle/Ethiopia and adjacent regions (Ethiopian Plateau, Somali Plateau, and parts of Yemen and Saudi Arabia) was applied to the mapping of lithologic and structural units of the test area at a scale 1:1,000,000. Results of the geological evaluation of the ERTS-1 imagery of the Afar have proven the usefulness of this type of satellite data for regional geological mapping. Evaluation of the ERTS images also resulted in new aspects of the structural setting and tectonic development of the Afar-Triangle, where three large rift systems, the oceanic rifts of the Red Sea and Gulf of Aden and the continental East African rift system, seem to meet each other. Surface structures mapped by ERTS do not indicate that the oceanic rift of the Gulf of Aden (Sheba Ridge) continues into the area of continental crust west of the Gulf of Tadjura. ERTS data show that the Wonji fault belt of the African rift system does not enter or cut through the central Afar. The Aysha-Horst is not a Horst but an autochthonous spur of the Somali Plateau. Database: CSA Technology Research Database.

Kroner A and Sassi F.P. 1996. "Evolution of the Northern Somali Basement: New Constraints from Zircon Ages." Journal of African Earth Sciences. Volume 22, Issue 1, Pages 1-15. Descriptors: Geochronology; age determination; basement; zircon; Proterozoic. Notes: Somalia. Abstract: The single zircon ages confirm that the western northern Somali crystalline basement (NSB) is not a juvenile Pan-African terrain, but a composite basement consisting of pre-Pan-African Proterozoic crust ranging in age between ~1400 and ~1820 Ma. This was affected by granitoid magmatism and associated tectono-thermal events at ~840, ~800-760 and ~720 Ma, respectively, at the same time when, farther north in the Arabian-Nubian shield, subduction-related magmatism led to the formation of intra-oceanic island arcs and marginal basins. Pre-Pan-African Archaean to Proterozoic rocks similar in age to the western NSB also occur in eastern Ethiopia and the two regions may be part of the same continental terrane, sandwiched in between juvenile domains farther west in Ethiopia and farther east in northern Somalia. ISSN: 0899-5362.

Krummes D.C. 1980. "Transportation in East Africa: A Bibliography." Vance Bibliographies, Public Administration Series: Bibliography. Issue P-595, Pages 27p. Descriptors: Ethiopia; Djibouti; Kenya; Uganda; Tanzania; Somalia; Zambia; transportation. Abstract: The bibliography is multi-modal in approach and focuses on Ethiopia, Djibouti, Kenya, Uganda, Tanzania, Somalia and Zambia, and was compiled in late 1979. LCCN: 80623747.

Ku, T. L. and Chung, Y. 1980. "Ra-226 in the Indian Ocean; GEOSECS Results; American Geophysical Union; 1980 Fall Meeting." EOS Trans. Am. Geophys. Union. 11 Nov. Volume 61, Issue 46, Pages 988. Descriptors: alkaline earth metals; Arabian Sea;

Asia; Bay of Bengal; Carlsberg Ridge; geochemistry; GEOSECS; Indian Ocean; isotopes; metals; Mid-Indian Ridge; oceanography; Ra-226; radioactive isotopes; radium; sea water; sediments; Somali Basin; South Australian Basin; Southeast Indian Ridge; Sri Lanka. ISSN: 0096-3941.

Kucera, Joshua. 2005. "Somalia Likely to Get Fewer Peacekeepers." Jane's Defence Weekly. Issue SEPT, Pages 2. Descriptors: Military operations; Societies and institutions; Law enforcement. Abstract: A proposed 10,000-troop peacekeeping mission to Somalia is being scaled back as the country's transitional government is taking steps to take more control of its own security. The officials from the African Union (AU) and Inter-Government Authority on Development (IGAD) believes that their mission will probably be smaller and will focus on demobilizing militias rather than peacekeeping. The AU is considering to fund the demobilization centers and the resettlement of refugees and internally displaced person (IDPS). The United Nations Development Program has trained 2,000 of a planned 5,000 Somali police, while Kenya has agreed to train a further 5,000 personnel. ISSN: 0265-3818.

Kucera, Joshua. 2005. "Djibouti: Us Foothold in Africa- African Foothold." Jane's Defence Weekly. Issue CT, Pages 9. Descriptors: Military operations; Operations research; Military bases; Public works; Civil defense; Project management; Military engineering. Abstract: US operations in the Horn of Africa are showing the way ahead for the US military's increasing role in bringing order to ungoverned areas. The US has maintained an African base in the Horn of Africa since year 2002. US troops have trained 2,800 soldiers from regional armies in that time. Public works projects in the region are all strategically close to Somalia. Officials intend to expand Combined Joint Task Force (CJTF)-Horn of Africa (HOA's) activities into Uganda and Tanzania. ISSN: 0265-3818.

Kucera, Joshua and Koch, Andrew. 2005. "US Navy to Take Over Horn of Africa Mission." Jane's Defence Weekly. Issue SEPT, Pages 2. Descriptors: Military operations; Geographical regions; International cooperation; Public policy. Abstract: The US Navy is preparing to assume command of the Combined Joint Task Force-Horn of Africa (CJTF-HOA) which aims to disrupt violent Islamist networks in the area, including Somalia and Yemen. The command staff of CJTF-HOA has been dominated by the US Marine Corps (USMC) but the service has become overstretched as a result of the continuing operations in Iraq and Afghanistan. Along with the handover to the USMC, the US Department of Defense (DoD) is looking at an alternative location for the base which is likely going to be either Ethiopia, Kenya, or Uganda. None of the choices are ideal as choosing Ethiopia may make it appear that the US is favoring it in its continuing dispute with Eritrea, while force protection is an issue with Kenya. ISSN: 0265-3818.

Kumar, M. R. Ramesh, Muraleedharan, P. M. and Sathe, P. V. 1999. "On the Role of Sea Surface Temperature Variability Over the Tropical Indian Ocean in Relation to Summer Monsoon using Satellite Data." Remote Sens. Environ. Volume 70, Issue 2, Pages 238-244. Descriptors: atmosphere; NOAA satellite; monsoon; satellite imagery; sea surface temperature; Somalia. Notes: Additional Info: References: 17; Geographic: Indian Ocean- (Tropical). Abstract: The sea surface temperature (SST) variability over the tropical Indian Ocean is studied for the period January 1988 to December 1992 using the multichannel sea surface temperature (MCSST) from the NOAA series of satellites. The MCSST values were found to agree reasonably well with the Reynolds and Smith (1994) data for the same period, but the anomalies were found to differ significantly. An interesting result is the influence of the Somalia coast zonal anomaly of SST (SCZASST) and central Indian Ocean zonal anomaly of SST (CIOZASST) with the monsoon rainfall over different meteorological subdivisions of the Indian subcontinent. The SCZASST is negatively and significantly correlated with the monsoon rainfall over the western and central parts of India, while CIOZASST is positively and significantly correlated. ISSN: 0034-4257.

Kunkle, H. S. and Abel, N. O. J. 1976. Water Resources Potentials for Wildlife in the Proposed Lac Badana National Park, Southern Somalia; First Reconnaissance. Italy: FAO, Rome, Italy. Descriptors: Africa; East Africa; Lac Badana National Park; public lands; Somali Republic; southern Somali Republic; surveys; water resources. Database: GeoRef. Accession Number: 1999-004690.

Küster, D., Vachette, M., Matheis, G., Jacob, K.H and Mohamed, F.H. 1990. "Geochemical Indications of Late Pan-African Rare-Metal Potentials in NE-Africa: Variations in Time and Style." Berliner Geowissenschaftliche Abhandlungen, Reihe A. Volume 120, Issue 2, Pages 551-570. Descriptors: Metals; mineralization; rare earth element; granite; Pan African orogeny. Notes: Somalia Sudan Egypt. Abstract: Pan-African rare-metal potential granitoid complexes of various types and geological settings were studied to evaluate their geotectonic significance and feasible resources. Target areas are the muscovite pegmatites of the Bayuda Desert, northern Sudan, the Berbera region, northwestern Somalia, the albite pegmatites and quartz veins around Bosaso, northeastern Somalia, and the apogranitic complex of Wadi Abu Rusheid, Eastern Desert/Egypt. The Bayuda Desert pegmatites are interpreted as anatectic products of regional metamorphism, which have been formed contemporaneously with syntectonic granitoids during the main tectonic phase of the Pan-African. The slow cooling of the northern Somalian crust after the emplacement of late Pan-African granites caused the circulation of fluid phases in a tectonically overprinted crustal segment and thus contributed to the formation of rare-metal bearing vein-type systems over a longer period. The rather extreme specialization at Wadi Abu Rusheid, both geochemically and geotectonically, may have been produced by continuous activation along the deep-seated Nugrus-Hafafit shear zone; the most intensive alteration phase is nearly synchronous with the final vein-accumulations in northern Somalia. ISSN: 0172-8784.

Küster, D., Haider, A., Utke, A., Leupolt, L. and Lenoir, J.L. 1990. "Pan-African Granitoid Magmatism in Northeastern and Southern Somalia." Berliner Geowissenschaftliche Abhandlungen, Reihe A. Volume 120, Issue 2, Pages 519-536. Descriptors: Igneous petrology and processes; partial melting; magmatism; granite; petrology; geochemistry; Pan African orogeny. Abstract: In northeastern Somalia the granitoid magmas were discordantly emplaced into low-grade metamorphic schists and sandstones of the Inda Ad Group. The composition of the magmatic association corresponds to the calc-alkaline "granodioritic' series. The existence of two different zircon generations (U/Pb ages of 626 Ma for the juvenile magmatic and > 1000 Ma for the inherited zircons) and the intermediate initial Sr-isotope ratio indicate that both mantle material and pre-Pan-African crustal components participated in the generation of the post-kinematic magmas of northeastern Somalia. In the Buur region of southern Somalia the granitoid magmas were discordantly emplaced into high grade metamorphic granulite to amphibolite facies rocks. Two Pan-African granitoid associations have been distinguished. Compared with the post-kinematic granitoids of northeastern Somalia the post-kinematic granitoids of the Buur area are enriched in LIL- and HFS-elements and have also an elevated initial Sr-isotope ratio. It is therefore suggested that these magmas derived from partial melting of the crust in connection with post-orogenic crustal uplift and relaxation. ISSN: 0172-8784.

Küster, D. and Harms, U. 1998. "Post-Collisional Potassic Granitoids from the Southern and Northwestern Parts of the Late Neoproterozoic East African Orogen: A Review." Lithos. Volume 45, Issue 1-4, Pages 177-195. Descriptors: Igneous petrology and processes; Pan African orogeny; granite; petrogenesis. Notes: Additional Info: Netherlands; References: Number: 81; Geographic: Ethiopia Madagascar Somalia Sudan. Abstract: Potassic metaluminous granitoids with enrichments of HFS elements constitute part of widespread post-collisional magmatism related to the Late Neoproterozoic Pan-African orogeny in northeastern Africa (Sudan, Ethiopia, Somalia) and Madagascar. The plutons were emplaced between 580 and 470 Ma and comprise both subsolvus and hypersolvus biotite-granite, biotite-hornblende-granite, quartz-monzonite and quartz-syenite. Pyroxene-bearing granitoids are subordinate. Basic dikes and enclaves of monzodioritic composition are locally associated with the granitoid plutons. … The association with basic magmas derived from subduction-modified enriched mantle sources strongly suggests that the granitoids represent hybrid magmas produced by interaction and mixing of mantle and crust derived melts in the lower crust. The most intense period of this potassic granitoid magmatism occurred between 585 and 540 Ma, largely coeval with HT granulite facies metamorphism in Madagascar and with amphibolite facies retrogression in southeastern Africa (Somalia, Sudan). Granitoid magmatism and high-grade metamorphism are probably both related to post-collisional lithospheric thinning, magmatic underplating and crustal relaxation. However, the emplacement of potassic granites continued until about 470 Ma and implies several magmatic pulses associated with different phases of crustal uplift and cooling. The potassic metaluminous granites are temporally and spatially associated with post-collisional high-K calc-alkaline granites with which they share many petrographical, geochemical and isotopical similarities, except the incompatible element enrichments. The resemblance indicates a strongly related petrogenesis of both granite associations. ISSN: 0024-4937.

L

Lacroix, A. 1930. "Études Géologiques En Ethiopie, Somalie Et Arabie Meridionale." Translated title: "Geologic Studies in Ethiopia, Somalia and Southern Arabia." Memoires De La Société Géologique de France. Volume 6, Descriptors: Africa; Arabian Peninsula; areal geology; Asia; East Africa; Ethiopia; Somali Republic; Yemen. ISSN: 0369-2027.

Lahmeyer International, International (III). 1987. Somalia, Baardheere Dam Project; Economic Evaluation. International (III): Lahmeyer Int., International (III). Descriptors: Africa; Baardheere Dam Project; Baardheere Somali Republic; dams; East Africa; economics; evaluation; Gedo Somali Republic; hydrology; Somali Republic; water resources; water supply. Database: GeoRef. Accession Number: 1999-004696.

Lahmeyer International, International (III). 1986. Shebelle Water Strategy Study; Vol. 1, Main Report. International (III): Descriptors: Africa; East Africa; Shebelle River;

Somali Republic; southern Somali Republic; surface water; water resources. Database: GeoRef. Accession Number: 1999-004693.

Lahmeyer International, International (III). 1986. Shebelle Water Strategy Study; Vol.2, Appendices A,B,D (without Annexes) and E. International (III): Descriptors: Africa; East Africa; Shebelle River; Somali Republic; southern Somali Republic; water resources. Database: GeoRef. Accession Number: 1999-004694.

Lahmeyer International, International (III). 1986. Shebelle Water Strategy Study; Vol. 3, Appendices F to I (Only Table of Contents, some Relevant Chapter). International (III): Descriptors: Africa; East Africa; Shebelle River; Somali Republic; southern Somali Republic; water resources. Database: GeoRef. Accession Number: 1999-004695.

Lambert, M. R. 1997. "Environmental Effects of Heavy Spillage from a Destroyed Pesticide Store Near Hargeisa (Somaliland) Assessed during the Dry Season, using Reptiles and Amphibians as Bioindicators." Arch. Environ. Contam. Toxicol. Jan. Volume 32, Issue 1, Pages 80-93. Descriptors: Accidents; Amphibia/metabolism; Analysis of Variance; Animals; Climate; Comparative Study; Hydrocarbons, Chlorinated; Insecticides- adverse effects- toxicity; Lizards; Organothiophosphorus Compounds; Pesticide Residues; Poisoning- mortality; Ranidae- metabolism; Research Support, Non-U.S. Gov't; Soil Pollutants; Somalia; Species Specificity; Water Pollutants, Chemical analysis and toxicity. Abstract: A pesticide store near Hargeisa (Somaliland) was damaged by bombing in May 1988, and subsequently looted by local people, who removed and drained drums of chemicals. Assessment of the effects of resultant spillage during the dry season, March/April 1993, established that pesticides, mainly organochlorines (dieldrin and products and BHC isomers) and organophosphates (fenitrothion and malathion) had contaminated 3700 sq mof soil at up to 3728.0 (geometric mean 149.0) ppm (5180.0 g/m3) total insecticides. Reptiles avoided contamination above 1 ppm, and were absent above 10 ppm. Experimental contact with highly contaminated soil caused death in lizards-Hemidactylus parkeri and Mabuyas. striata-after 26.5 and 33. 5 h, respectively (residue levels elevated over 2000-and 149-fold), and 100% mortality within 65 min in frogs Tomopterna cryptotis (geometric mean residue level elevated 168-fold). Sediment 350 m downstream of the spill contained dieldrin at 0.50 ppm (0.03-0.05 ppm after 1.6, to 9.0, km). Whole body residues of spillage vicinity lizards were up to 1.52 ppm wet weight (193.6 ppm lipid) total insecticides. Geometric mean of 0.36 ppm was elevated fivefold above mean background level of Hargeisa lizards in the valley below. Dieldrin and products was highest; the level of BHC isomers was also significantly higher than DDT. Geometric mean total insecticide level in Chalcides ragazzii and H. parkeri was four times higher than in surface-dwelling Pseuderemias smithi. Reptile species richness was habitat-influenced. Frogs without abnormalities present in mud of river-bed wells indicated uncontaminated ground water (organochlorine residues undetected). Low levels in frogs, and of M. s. striata in the vicinity of wells [geometric means 0.09 and 0.07 (ranges 0-0.48 and 0.01-0.31) ppm, respectively], implied that by 2.7 km downstream of the spill few residues were entering food chains. ISSN: 0090-4341.

Lassonczyk B. 1990. "Characteristics of a Fossil Soil Developed on the Basement Complex in Northeast Somalia." Berliner Geowissenschaftliche Abhandlungen, Reihe A. Volume 120, Issue 2, Pages 713-718. Descriptors: Regional Geology; paleosol; weathering; basement. Notes: Somalia. Abstract: A fossil soil developed from the basement complex was characterized. The phyllites of Late Precambrian or Early Cambrian age show an intensive chemical weathering and desilification related to warm humid climatic conditions. Complete destruction of mica and formation of kaolinite and sesquioxides are the main characteristics. While the lower part of the profile is a saprolitic formation only the upper part represents a remnant of an eroded soil. An increase of the groundwater table probably related to the sea transgression at the beginning of the Lower Jurassic affected an alteration of the soil by gleyic processes. Soil formation was finally stopped by the sedimentation of the Adigrat-Formation (Lower Jurassic). ISSN: 0172-8784.

Lauderdale P and Toggia P. 1999. "An Indigenous View of the New World Order: Somalia and the Ostensible Rule of Law." Journal of Asian and African Studies. Volume 34, Issue 2, Pages 157-177. Descriptors: Political; Conflict, protest, human rights; civil war; ethnic conflict; indigenous population; new world order. Notes: Additional Info: Netherlands; References: 58; Geographic: Somalia. Abstract: In the 'New World Order,' Somalia is characterized as a deviant society, especially by Western countries. This characterization is magnified by focusing upon armed conflicts among different groups in Somalia and is marked by a neglect of global forces and history, including indigenous perspectives. The benchmark for judging the nature and scale of such crises is the condition of statelessness, measured by the absence of a central political authority and the modern claim of an ostensible universal rule of law. However, the attempted replacement of sacred places and kinship identities of indigenous peoples with the identity of the New World Order that emphasizes self-interested and self-maximizing individuals, i.e., Western individualism, has led not to a melting pot, but a boiling pot. The Somalis, as with many other ethnic and indigenous groups throughout the world, do not find a meaningful sense of life by being defined as modern individuals via the state. Any viable alternative to disentangling Somalia and similar indigenous peoples from current and future crises might benefit from recognition and accommodation to their traditional ways of life and systems of governance. Moreover, future work should include explications of the impact of global hegemony, the increasing role of the United Nations in advancing foreign policy, military interventions under the facade of peacekeeping, and the acceleration of a market economy ostensibly directed by global forces such as the World Bank and the International Monetary Fund. ISSN: 0021-9096.

Leckie, Edwards R.G. 1906. "Geological map of British Somaliland." Scale: 1:2 000 000. [London]: [TSGS, War Office], 1906. 1 map: color Note(s): Supplied by Mr R G Edwards Leckie. Subjects: Somalia- Maps and charts. Added Name: Great Britain. General Staff. Topographical Section. Added Name: Great Britain. War Office. General Staff. Geographical Section. British Library: Maps MOD TSGS 2206.

Lelgemann, D., Reuther, C. D., Gehlich, U., Cui, C., Salahchourian, M. and Angermann, D. 1990. "Teilprojekt G 5. Neotektonik im Bereich der Nubischen, Arabischen und Somalischen Platte." Translated title: "Subproject G 5. Neotectonics within the range of the Nubian, Arab and Somali plates." Federal Republic of Germany: Techn. Univ., Berlin, Federal Republic of Germany. Descriptors: Africa; African Plate; Arabian Peninsula; Arabian Plate; Asia; block structures; East Africa; Egypt; faults; fracture zones; geodetic networks; ground methods; neotectonics; North Africa; plate movement; plate tectonics; progress report; Red Sea region; remote sensing; report; satellite methods; Saudi Arabia; Somali Republic; structural geology; Sudan; systems;

tectonics. Notes: References: 4; geol. sketch maps. Database: GeoRef In Process. Accession Number: 192820-1.

Lemaux II, J, Gordon, R.G. and Royer, J.-Y. 2002. "Location of the Nubia-Somalia Boundary Along the Southwest Indian Ridge." Geology. Volume 30, Issue 4, Pages 339-342. Descriptors: Plate tectonics; tectonic setting; plate motion; plate boundary; rift zone; seafloor spreading; African plate. Notes: Additional Info: United States; References: 18; Geographic: Somalia Africa- East African Rift. Abstract: The northern boundary between the Nubian (West African) and Somalian (East African) plates is marked by the existence of the East African Rift, which is the locus of roughly east-west stretching due to the separation of these plates. South of 20°S, any expression of deformation or seismicity due to the relative motion of these two presumably distinct plates vanishes or is extremely subtle, although the boundary must continue until it intersects another plate boundary. Until recently, there were no observational limits on the location of that intersection, and the recent limits were very broad, being thousands of kilometers wide. Here, from an analysis of the locations of the old edge of magnetic anomaly 5 (which is over 11 Ma seafloor), we show that the main locus of deformation between Nubia and Somalia near the Southwest Indian Ridge over the past 11 m.y. can be no wider than a few hundred kilometers, from 100 km west of the Du Toit Fracture Zone to 50 km east of the Andrew Bain Fracture Zone complex. Deformation is most likely concentrated along a closely spaced set of several fracture zones known collectively as the Andrew Bain Fracture Zone complex, which is only ∼100 km wide. The displacement along the Nubia-Somalia boundary over the past 11 m.y. is 23 km ± 6 km (95% confidence limits), indicating a displacement rate of 2 mm/yr, much slower than typical rates of seafloor spreading and subduction, but comparable to that along the Owen fracture zone and Dalrymple trough, which separate the Arabian plate from the Indian plate, and along the Azores-Gibraltar line, which separates the Eurasian plate from the Nubian plate. ISSN: 0091-7613.

Lenoir, J.L., Haider, A., Matheis, G., Kuster, D., Liegeois, J.P. and Utke, A. 1994. "Origin and Regional Significance of Late Precambrian and Early Palaeozoic Granitoids in the Pan-African Belt of Somalia." Geol. Rundsch. Volume 83, Issue 3, Pages 624-641. Descriptors: Igneous petrology and processes; granitoid; isotope geochemistry; Pan African orogeny; geochronology; Proterozoic; Palaeozoic. Notes: Somalia. Abstract: Granitoids within the Precambrian basement of north-eastern and southern Somalia are subdivided on the basis of geology, geochronology and petrology into three different assemblages. The late Proterozoic to early Palaeozoic granitoids in Somalia express contrasting regimes, characterized by strong juvenile input in the north, close to the Arabian-Nubian Shield, whereas intense crustal reworking with little addition of juvenile material prevailed in the south. Somalia was definitively not a cratonic area during the pan-African, but a zone of high crustal mobility. ISSN: 0016-7835.

Leon, Jean-Francois and Legrand, Michel. 2003. "Mineral Dust Sources in the Surroundings of the North Indian Ocean." Geophysical Research Letters. Volume 30, Issue 6, Pages 42-1. Descriptors: Dust; Geographical regions; Wind; Satellite observatories; Infrared imaging. Abstract: We have used one year (1999) of Meteosat observations to monitor the dust activity in the arid and semi-arid areas surrounding the North Indian Ocean. The method is based on the IDDI product derived from the infrared channel of the satellite. Dust sources are located according to a threshold on the IDDI and another on the surface wind speed. The global features of the dust activity in this area are in a good agreement with previous studies. Main dust sources are located in the Nubian desert, the Arabian peninsula, Iran, Pakistan and Afghanistan, and North West India. It is found that Somalia is also an area of dust emission with a maximum of activity from May to October. ISSN: 0094-8276.

Leroy, S., Gente, P. and Fournier, M., et al. 2004. "From Rifting to Spreading in the Eastern Gulf of Aden: A Geophysical Survey of a Young Oceanic Basin from Margin to Margin." Terra Nova. AUG. Volume 16, Issue 4, Pages 185-192. Abstract: A geophysical survey in the eastern Gulf of Aden, between the Alula-Fartak (52°E) and the Socotra (55°E) transform faults, was carried out during the Encens-Sheba cruise. The conjugate margins of the Gulf are steep, narrow and asymmetric. Asymmetry of the rifting process is highlighted by the conjugate margins (horst and graben in the north and deep basin in the south). Two transfer fault zones separate the margins into three segments, whereas the present-day Sheba Ridge is divided into two segments by a transform discontinuity. Therefore segmentation of the Sheba Ridge and that of the conjugate margins did coincide during the early stages of oceanic spreading. Extensive magma production is evidenced in the central part of the western segment. Anomaly 5d was identified in the northern and southern parts of the oceanic basin, thus confirming that seafloor spreading in this part of Gulf of Aden started at least 17.6 Ma ago. ISSN: 0954-4879.

Leser, H. 1990. "Das 15. "Basler Geomethodische Colloquium 15: Landschaftsbewertungen Und Landschaftsplanung in Afrikanischen Entwicklungslandern- Bespiel Somalia." Translated title: "Basler geomethodical Colloquium: 15th. Landscape evaluations and landscape planning in African development lands- Somalia." Geomethodica. Volume 15, Pages 5-26. Veröffentlichungen des 15. Basler Geomethodischen Colloquiums. Basel: Basler Afrika Bibliographien : Auslieferung, Basler Geomethodisches Colloquium, 1990. Descriptors: environmental management; ecological planning; international program; landscape classification; regional planning; developing country; global change; landscape ecology; landscape classification and evaluation. Notes: Somalia. Abstract: The author postulates, that problems of the developing countries (hunger, economic disasters, exploitation of resources) are also problems of the environment. These environmental problems effect global change because they concern large areas and herein also the fluxes of matter and energy of the geobiosphere. Landscape classification and landscape evaluation systems as well as landscape and regional planning must take into consideration these ecological connections and relations of the earth. The environmental problems in the developing countries have their effects of course in the natural systems of the earth, but they are triggered off in the anthroposystem. Social life, economy, political decisions, national and ethnic traditions operate as system regulators. That means practical solutions of landscape ecosystem problems must include all subsystems of the landscape. Global programs, for instance of UNESCO or UN, and solutions to the environmental problems are concerned with the total geobiosphere. They have to take into consideration the eco-functional "on-the-spot' effect of many ecological, economic and social processes. It is their worldwide area effect only that causes global change. This results that environmental protection and landscape planning must begin on local objects of landscape ecosystems. A useful approach for this is represented by the "concept of ecological planning". ISBN: 3905141507.

Leslie, A.D. 1991. "Agroforestry Practices in Somalia." Forest Ecology & Management. Volume 45, Issue 1-4, Pages 293-308. Descriptors: Agriculture; Evolution And Palaeoecology; Agroforestry and intercropping; agroforestry practice; agricultural practice; developing country; agroforestry; bush fallow; soil fertility; pastoralism; shifting cultivation Species Term: Dobera glabra. Notes: Somalia- Jubba River; Somalia-Shabeelle River; Notes: 2 diagrams, 11 references, 2 tables. Abstract: Traditional agrogorestry methods in Somalia and attempts to introduce new practices are described. Physical, social and political constraints are discussed and recommendations for future developments are made. Nomadic pastoralism with shifting cultivation is practiced over most of the country. Settled communities in these areas plant live fencing. Most agroforestry is found near the two main rivers, the Jubba and the Shabeelle. On rainfed land scattered trees, most frequently Dobera glabra, are retained. These provide limited dry season browse, fruit and poles but are mainly used as shade for the farmer and his livestock. A bush fallow is often used to maintain soil fertility. On irrigated land, agricultural crops are commonly grown alongside young fruit trees until shade becomes too great. Other practices include growing crops in mature coconut plantations and with date palms. Large banana plantations are protected by shelterbelts, predominantly of Casuarina equisetifolia. ISSN: 0378-1127.

Levich, Robert A. 2003. "The Effect of Peace Corps on a Geologist's Career." Geological Society of America, 2003 Annual Meeting. Abstracts with Programs-Geological Society of America. Nov. Volume 35, Issue 6, Pages 107. Descriptors: Africa; Brazil; East Africa; geologists; Ghana; government agencies; mineral exploration; Peace Corps; practice; public policy; Somali Republic; South America; Uganda; United States; West Africa. Abstract: Early, in 1963, the late Dr. William Donn, Research Geophysicist at Lamont Geological Observatory and Professor of Geology at Brooklyn College, CUNY stated his belief that my prospective Peace Corps service in Ghana would provide career opportunities that might otherwise be unavailable. 40 years later, his insight has proved true. At 21 years of age, I held a B.Sc. in geology and had little experience beyond geology field camp in Utah. My academic interests were uncertain, and my professional goals were secondary to residing in New York City, the Center of my Universe. My Peace Corps service in Ghana, 1963-65, did lead to many opportunities and Returned Peace Corps Volunteer (RPCV) geologists played a major role. A fellow PCV geologist convinced me to attend graduate school at the University of Texas, where I studied hard rock economic geology, and my MA thesis evaluated a silver mining district in the Sierra Madre of Durango, Mexico. I returned to the Ghana Geological Survey in 1969 to study gold deposits. In 1973, an RPCV's recommendation led the US Atomic Energy Commission (AEC) to hire me to evaluate domestic uranium resources. Another RPCV was Vice President of the mining company that hired me in 1981. In 1982, my AEC and Peace Corps experience led the International Atomic Energy Agency (IAEA) to select me to evaluate uranium resources in Uganda and Somalia, and in 1984, an RPCV aided my return to DOE. One year later, DOE chose me to develop a multi-national study of a Brazilian uranium deposit as an analogue to a nuclear waste repository. In 1989, DOE selected me to manage all international technical cooperative programs in deep geologic disposal, and I was later appointed as US Delegate to an OECD/Nuclear Energy

Agency Technical Advisory Group in Paris, where I have served for 14 years. I've lived and worked in the West for 30 of the past 40 years, spent eight overseas and only two in the eastern US. I studied mineral deposits for 20 years, and since 1984, my efforts have been devoted to the deep geologic disposal of nuclear waste. The Peace Corps unquestionably defined my career and paved my way to participate in scientific programs in 20 countries on five continents. ISSN: 0016-7592.

Lewis I.M. 2004. "Visible and Invisible Differences: The Somali Paradox." Africa. Volume 74, Issue 4, Pages 489-515. Descriptors: Geography; Political economy, class and state analysis; nationalism; state building; identity construction; ethnopolitics; Somalia. References: 53; Geographic: Somalia East Africa Sub-Saharan Africa Africa Eastern Hemisphere World. Abstract: In exploring the difficulties experienced by the traditionally politically uncentralised Somalis in establishing a stable and effective state, based on their ethnicity, this article compares ethnicity, nationalism and lineage identity. In this case, ethnicity and nationalism are local products, influenced but not created by the colonial experience. They have had to contend with the intractable force of segmentary lineage identity, which has proved extremely difficult to adapt and accommodate to the requirements of modern statehood. In its cultural context, agnation is all the more pervasive and powerful in constituting an 'invisible' bond, conceived by Somalis as a biologically based distinction like 'race'. Unlike race, it is almost infinitely elastic and divisible. Ethnic identity, which rests on external distinctions such as language, culture and religion, cannot be broken down into a series of formally equivalent segments, but is less binding as a social force. Today, after the collapse of the state of Somalia in 1991, following protracted grass-roots peace-making between clans, two parts of the nation- the former British Somaliland, and the north-eastern region of Somalia ('Puntland', based on the Majerteyn clan, and other closely related clans)- have developed separate local states. Although Somaliland claims complete independence, which Puntland does not, both polities incorporate parliamentary institutions that accommodate traditional, and modern political leaders and processes. The ex-Italian residue, Southern Somalia, still without any form of government, is in what appears to be the final throes of its long-running, fourteenth grandiose international 'peace' conference in Kenya. Thousands of delegates, in various configurations, have already spent over eighteen months in these talks. Although its embryonic constitution now recognises 'clans' as constituent political units, this attempt to re-establish Somalia is based on the usual 'top-down' approach, rather than on spontaneous local negotiations amongst 'stakeholders' on the ground, such as those on which Somaliland and Puntland are founded. With contingents of foreign 'experts', the whole process seeks to reinstate a familiar Eurocentric state model, unadapted to Somali conditions. ISSN: 0001-9720.

Liede-Schumann, S. and Mere, U. 2005. "Notes on succulent Cynanchum (Apocynaceae, Asclepiadoideae) in East Africa." Novon. Volume 15 (2), pages 320-323. Abstract: Changes in succulent Cynanchum I., species are needed for the Flora of Somalia account of Apocynaceae subfamily Aselepiadoideae. The following changes are proposed within the tribe Asclepiadeae, subtribe Cynanchinae, Cynanchum riminale subsp. crassicaule, another subspecific taxon in the C. riminale complex, is described as new, based on morphological, ecological, and molecular evidence... The new combination Cynanchum riminale subsp. stipitaceum is proposed, and Sarcostemma andongense Hiern is recognized as a new synonym of C. riminale subsp. stipitaceum.

Lino, E. 1957. "I Pozzi Aerei Della Somalia." Translated title: "Somalia Surface Wells." Somalia d'Oggi. Volume 2, Descriptors: Africa; aquifers; East Africa; ground water; hydrodynamics; Somali Republic; water wells. Database: GeoRef.

Lino, E. 1956. "Possibilita Minerarie Della Somalia; Rapporto Sulle Prospezioni Effettuate in Somalia Dalla Missione Mineraria Delle M.S.A. dal 1/7/53 al 5/9/53." Translated title" "Mining Possibility in Somalia; Report of the Research Prospects in Somalia by the Mining Mission of the M.S.A. from 1/7/53 to 5/9/53." Somalia: Publisher unknown, Mogadischu, Somalia. Descriptors: Africa; East Africa; mineral exploration; mining; mining geology; research; Somali Republic. Database: GeoRef. Accession Number: 1996-040233.

Lino, E. 1954. "L'Acqua Vita Della Somalia; Storia Del Pozzo Tubolare Nel Dafet." Translated title: "Water of Somalia; Story of Tubular Wells in Dafet." Corriere Della Somalia. Volume 48, Descriptors: Africa; Dafet Somalia; East Africa; ground water; Shabeellaha Hoose; Somali Republic; water resources; water wells; wells. Database: GeoRef.

Lino, E. 1954. "Guida Breve Geo-Mineraria Della Somalia." Translated title: "Brief Guide to Geo-Mining of Somalia." Somalia: AFIS, Mogadishu, Somalia. Descriptors: Africa; East Africa; field trips; mining geology; Somali Republic. Database: GeoRef. Accession Number: 1999-004704.

Lino, E. 1939. "Relazione all'Ufficio Minerario Del Governo Della Somalia Di Una Missione in Migiurtnia Compinta Dal 22 Marzo Al 29 Maggio 1939." Translated title: "Somalia Bureau of Mining Report of the Mission to Migiurtnia from March 22 to may 29, 1939." Somalia: Publisher unknown, Mogadischu, Somalia. Descriptors: Africa; areal geology; East Africa; government agencies; Migiurtinia Somalia; Somali Bureau Of Mining; Somali Republic. Database: GeoRef. Accession Number: 1996-040232.

Litografia artistica cartografica. 1960. "Carta Geologica Della Penisola Somala." Translated title: "Geologic map of the Somali peninsula." Litografia Artistica Cartografica; Somalia. Descriptors: Geology- Somalia- Maps. Notes: Description: 1 map. Geographic: Somalia- Maps. Scale 1:4,000,000. Notes: Depths shown by bathymetric isolines. Shows: international boundaries, geological information. General Info: Coverage: Somalia. OCLC Accession Number: 48602255.

Little, Otway Henry. 1925. The Geography and Geology of Makalla (South Arabia). With two appendices: 1, Descriptions of fossils from South Arabia and the British Somaliland; 2. Notes on some terrestrial Mollusca from the hinterland of Makalla. Cairo: Government Press. 250 pages, 3 folded maps, tables. At head of title: Ministry of Finance, Egypt. Geological Survey of Egypt. USGS Library.

Little, P.D, Mahmoud, H. and Coppock, D.L. 2001. "When Deserts Flood: Risk Management and Climatic Processes among East African Pastoralists." Climate Research. Volume 19, Issue 2, Pages 149-159. Descriptors: Nature- Society Relations; Climate change; Information systems, climatic and soil conditions; pastoralism; El Nino; drought; flood; risk assessment; global climate; Somalia. Notes: Germany; References: 37; Geographic: Africa- (East). Abstract: Pastoral populations of East Africa confront multiple risks associated with drought, food shortages and insecurity. In this arid region, drought is a 'normal' event and herders pursue strategies of mobility, livestock loaning and diversification to combat its effects. What is not a norm are prolonged floods when precipitation cycles become inverted and dry season rainfall greatly exceeds the average amount for a year. This article examines the events and responses to 'El Niño' in the rangeland areas of northern Kenya and southern Somalia during 1997/98. It suggests that these global climatic episodes need to be assessed against local factors and processes, which strongly shape their impacts on human populations. ISSN: 0936-577X.

Lockwood Survey Company, Toronto, ON, Canada. 1968. Agricultural and Water Surveys, Somalia; Final Report, Vol. 2, Water Resources. Italy: Descriptors: Africa; agriculture; East Africa; irrigation; Somali Republic; surveys; water resources; water supply. Notes: SF-36, Som. I. Database: GeoRef. Accession Number: 1999-004711.

Lockwood Survey Company, Toronto, ON, Canada. 1966. Agricultural and Water Survey; Somali Republic. International (III): Descriptors: Africa; agriculture; East Africa; irrigation; Somali Republic; surveys; water resources; water supply. Database: GeoRef. Accession Number: 1999-004710.

Lombard, Jean. 1933. "Nouvelles Géologiques Africaines." Translated title: "African Geological News." Chronique Des Mines Coloniales. Jun. Volume 15, Pages 284-296. Descriptors: Africa; Anglo-Egyptian Sudan; Angola; bibliography; central; Central Africa; Congo Democratic Republic; East Africa; French Equatorial Africa; French West Africa; historical geology; Malawi; Somali Republic; Tanganyika; Uganda; Zaire. Notes: (Bur. d'Études Geol. et Min. Coloniales), an. 2; A summary account of progress in geology in 1932; includes contributions on Afrique occidentale francaise (French West Africa) by R. Furon, p. 285-286; on Afrique equatoriale francaise (French Equatorial Africa) by Jean Lombard, p. 287-288; on Angola by F. Mouta, p. 288-289; on British colonies (Uganda, Tanganyika Territory, Nyasaland, Anglo-Egyptian Sudan, British Somaliland) by John Parkinson, p. 289-293; on Congo Belge (Belgian Congo) by A. Jamotte, p. 293-295; Bibliography, p. 295-297. ISSN: 0366-7324.

"The Long-Term Perspective Study of Sub-Saharan Africa. Volume 1. Country Perspectives." 1990. World Bank; Background Papers. Descriptors: 184p; Economic Development; International Development Abstracts; economic crisis; development path; developing country; development policy. Abstract: Part of a set of four volumes of background papers which were prepared in the process of producing the Bank's 1989 report Sub-Saharan Africa: from crisis to sustainable growth. Describes the physical, social, political, economic and resource conditions in eight African countries: Ethiopia, Malawi, Mali, Mauritania, Somalia, Sudan, Tanzania, and Zambia. The papers examine the similarities and differences faced by the countries in dealing with the region's economic crisis. Also included are papers on alternative development paths followed by some non-African countries: the FRG, the Nordic countries, and a number of East Asian countries. Eight of the papers are abstracted separately. Notes: In: The long-term perspective study of sub-Saharan Africa. Volume 1. Country perspectives (1990); 184 pages; Geographic: Africa- (sub-Sahara) Ethiopia; Malawi; Mali; Mauritania; Somalia; Sudan; Tanzania. Update: 01 JAN 1991. ISBN: 0821316028; OCLC Accession Number: 0848041.

Long, G. H. and Lee, M. L. 1973. "Report of the 1972 Field Party. Somalia." Burmah Oil Somalia, Mogadishu, Somalia. Descriptors: Africa; areal geology; East Africa; geophysical surveys; petroleum; petroleum exploration; Somali Republic; surveys. Database: GeoRef. Accession Number: 1999-004714.

Longley, C., Saide, M.A, Dominguez, C. and Leonardo, W.J. 2002. "Do Farmers Need Relief Seed? A Methodology for Assessing Seed Systems." Disasters. Volume 26,

Issue 4, Pages 343-355. Descriptors: International Aid and Investment; assessment method; aid flow; knowledge; food aid. Notes: United Kingdom; References: 28; Geographic: Mozambique Somalia. Abstract: This article outlines a methodology to help agencies better determine whether or not relief seed is needed by farmers affected by disaster. A brief review of current seed needs assessment procedures in southern Somalia and Mozambique illustrates problems of knowing which crops and households are affected, the importance of seed access (not just availability) and the need to plan interventions earlier than at present. The development of a Seed Systems Profile is proposed to understand better both the socio-economic and agro-ecological aspects of farmers' seed systems. A five-step framework for assessing seed systems in disaster situations is also presented. These tools are currently being tested and further refined in Mozambique. A better understanding of farmers' seed systems will allow for the development of relief and rehabilitation interventions that effectively enhance the resilience and reduce the vulnerability of these systems. ISSN: 0361-3666.

Lorenz, F.M. 1994. "Confronting thievery in Somalia." Military Review. August 1994, Vol. 74 Issue 8, page 46, 10 pages, 5 bw photos. Subject Terms: Military Assistance, American; Somalia- Social Conditions; United States. Marine Corps. Geographic Terms: Somalia. Abstract: Cites the challenges US Marine Corps encountered concerning rules of engagement, use of nonlethal force and detainment of civilians after landing in Somalia in December 1992. Practical solutions developed during Operation Restore Hope to legally resolve the issues involved; Food for thought for exercise planners and leaders preparing for future humanitarian relief operations. ISSN: 0026-4148.

Louis, A. J. M. F. F. 1932. "La Esplorazione Dello Uabi-Uebi Scebeli Dalle Sue Sorgenti Nella Etiopia Meridionale Alla Somalia Italiana, 1928-29." Translated title: "Shebelle River Exploration from its Sources in Southern Ethiopia to Italian Somalia, 1928-29." Italy: Publisher unknown, Milan, Italy. Descriptors: Africa; East Africa; Ethiopia; exploration; fluvial features; rivers; Shebelle River; Somali Republic; southern Ethiopia; water resources. Database: GeoRef. Accession Number: 1998-014618.

Lovgren, Stefan. 1997. "Floods could bring famine." U.S. News & World Report; 11/24/97, Vol. 123 Issue 20, page 53, 1/3 page, 1c. Subject Terms: Floods. Geographic Terms: Somalia. Abstract: Focuses on the floods from El Nino in Somalia. People being displaced from their homes in the Juba Valley area; Malaria being rampant; Probability of famine due to loss of harvest; The effect of no central government in Somalia on relief efforts; Insufficient aid from the European Union and the United States. ISSN: 0041-5537.

Lubbock, Robin. 1994. "Lack of fuel, water hinders Somalia recovery." *National Catholic Reporter.* Volume 30 (12), Jan 21, 1994: page 14(1). Abstract: Somali farmers face a number of obstacles in producing food. For example, livestock banditry is a problem and there is a severe water shortage. International aid is available but will not not come in time to save many crops. ISSN: 0027-8939.

Luger, P. et al. 1994. "Comparison of the Jurassic and Cretaceous Sedimentary Cycles of Somalia and Madagascar: Implications for the Gondwana Breakup." Geol. Rundsch. Volume 83, Issue 4, Pages 711-727. Descriptors: Mesozoic; Gondwana; Cretaceous; Jurassic; continental breakup; sedimentary history. Notes: Somalia Madagascar. Abstract: The Jurassic and Cretaceous sedimentary history of northern

Somalia and the Morondava Basin of south-western Madagascar have been studied. Both regions display an independent facial development; however, a comparison of the sequential evolution of the Mesozoic sedimentary successions in these two presently widely separated areas reveals a surprisingly high level of similarity, which probably reflects major events during the disintegration of Eastern Gondwana during the Jurassic and Cretaceous. Although in Jurassic times the onset of transgressions and regressions in both areas compares well with eustatic development, major deviations in combination with the tectonic activities of different degrees are observed in the Early and Late Cretaceous synchronously in both regions. ISSN: 0016-7835.

Luger, P., Kallenbach, H. and Mette, W., et al. 1990. "The Jurassic and the Cretaceous of Northern Somalia: Preliminary Results of the Sedimentologic and Stratigraphic Investigations." Berliner Geowissenschaftliche Abhandlungen, Reihe A. Volume 120, Issue 2, Pages 571-594. Descriptors: Mesozoic; Jurassic; Cretaceous; sedimentation; transgression-regression; sea level variation. Notes: Somalia. Abstract: The deposits of Jurassic age in N Somalia belong to the Adigrat Sandstones (pre-Toarcian) and the Ahl Medo Sequences (Early Toarcian to Late Kimmeridgian). Whereas the first sedimentary succession comprises two lithofacies, the latter is divided into 12 lithofacies. The unconformably overlying deposits of Cretaceous age are represented by the Tisje Formation (pre-Aptian, Aptian to Campanian), consist of a predominantly calcareous lower sequence and a mainly siliciclastic upper sequence. This transgressive-regressive-transgressive sedimentary succession upward grades into the Auradu Limestones (Campanian, Maastrichtian to Early Eocene). ISSN: 0172-8784.

Luger, P. 2003. "Paleobiogeography of Late Early Cretaceous to Early Paleocene Marine Ostracoda in Arabia and North to Equatorial Africa." Palaeogeography, Palaeoclimatology and Palaeoecology. AUG 1. Volume 196, Issue 3-4, Pages 319-342. Abstract: About 1270 marine ostracod species of Aptian to Early Paleocene age from Central, West, North and East Africa (inclusive Madagascar) as well as the Arabian Peninsula and Pakistan/Western India have been examined on their spatial and stratigraphical distribution. Their paleobiogeographical relations have been examined based on the Jaccard-index. The results of these comparative studies are demonstrated on nine geographical maps for each stage (Aptian through Early Paleocene) and four paleogeographical reconstructions including the assumed land/sea distribution for Cenomanian, Turonian, Maastrichtian, and Early Paleocene times. The major results are the recognition of an increasingly uniform pan-'South Tethyan ostracod province' (North and Northeast Africa and the Arabian Peninsula) from the Aptian to the Cenomanian. Due to the emergence of major parts of the Arabian Shield during the Turonian, this pan-South Tethyan bioprovince was split into a western 'Protobuntonia numidica ostracod province' (North Africa and Western Middle East) and an eastern 'Kaesleria ostracod province' (Eastern Arabia, South Iran, Somalia) since the Coniacian. While the proliferation of the phylogenetic lines of the ostracod genera and species of the 'South Tethyan ostracod province' towards the Protobuntonia numidica faunas into the Turonian-Coniacian was continuous, the newly developing genera and species of the Protobuntonia numidica and the Kaesleria provinces of the 'Senonian'-Early Palcocene had almost nothing in common. Although most of Northeast Africa and the Arabian Shield underwent strong subsidence during the Campanian-Maastrichtian combined with a widespread transgression, these strictly separated bioprovinces did not reunite or show clear faunal interrelationships during these times. The reasons for the described phenomena are probably largely to be sought in the paleoecological peculiarities of the regarded ostracod faunas (i.e. a relatively cooler water fauna in the Protobuntonia province (probably partially influence by upwelling) and a relatively warmer, tropical fauna in the Kaesleria province). However, the strict separation of these ostracod assemblages along a narrow N-S directed borderline running from Northeast Saudi Arabia across Jordan into Syria is thought to be highly unusual and to require further investigation, which could probably stimulate new paleo-oceanographical and paleogeographical considerations for the Late Cretaceous-Early Paleocene of Northeast Africa, the Middle East and Arabia.

Luger, P. and Kuss, J. 1993. "Late Cretaceous to Basal Tertiary Shallow-Water Sedimentation in Northern Somalia." Geoscientific Research in Northeast Africa; Proceedings of the International Conference. Netherlands: A.A. Balkema, Rotterdam, Netherlands. International Conference on Geoscientific Research in Northeast Africa, Berlin. Federal Republic of Germany Conference: June 17-19, 1993. Descriptors: Africa; carbonate rocks; Cenozoic; clastic rocks; correlation; Cretaceous; East Africa; Foraminifera; Invertebrata; lower Tertiary; marine sedimentation; Mesozoic; microfossils; northern Somali Republic; Protista; regression; sea-level changes; sedimentary rocks; sedimentation; shallow-water environment; Somali Republic; stratigraphic boundary; stratigraphic columns; Tertiary; transgression; Upper Cretaceous. Notes: References: 8; strat. color, sketch map. ISBN: 9054103299.

Luger, Peter. 1995. "Grossforaminiferen-Biostratigraphie Der Kreide Von Nord-Somalia." Translated title "Largely foraminiferous biostratigraphie of the chelk of North Somalia." 65. Jahrestagung Der Palaeontologischen Gesellschaft, Hildesheim, 25. Bis 30. September 1995; Abstracts und Poster." Terra Nostra (Bonn). Volume 4-95, Pages 39. Descriptors: Africa; biostratigraphy; carbonate rocks; Cretaceous; East Africa; emergent coastlines; Foraminifera; Invertebrata; Lituolacea; Mesozoic; microfossils; Orbitolinidae; Protista; sea-level changes; sedimentary cycles; sedimentary rocks; shallow-water environment; shore features; shorelines; Somali Republic; Textulariina; transgression. ISSN: 0946-8978.

Luger, Peter and Schudack, Michael. 2001. "On Early Cretaceous (Earliest Aptian) Freshwater Charophyta and Ostracoda from Northern Somalia." Neues Jahrbuch Für Geologie und Palaeontologie. Abhandlungen. Volume 220, Issue 2, Pages 245-266. Descriptors: Africa; algae; Aptian; Arthropoda; biostratigraphy; Charophyta; Chlorophyta; Clavator; Cretaceous; Crustacea; Cypridea; East Africa; fresh-water environment; Invertebrata; lithostratigraphy; lower Aptian; Lower Cretaceous; Mandibulata; Mesozoic; microfossils; northern Somalia; Ostracoda; Plantae; Somali Republic. Notes: illustrations incl. strat. cols., sketch map. ISSN: 0077-7749.

Lusini, G. 1949. "L'Attivita Mineraria Dell'Italia in Somalia." Translated title: "Mining Activity of Italy in Somalia." Africa Italiana. Volume 4, Descriptors: Africa; East Africa; mining; mining geology; Somali Republic. Database: GeoRef.

Lustig, Lawrence K. 1967. "Geomorphology and Surface Hydrology of Desert Environments." In W.G. Mcginnies, B.J. Goldman, and Patricia Pay lore, Eds., Deserts of the World. Tucson, AZ. Pages Refs. Descriptors: Geomorphology; Surface Drainage; Deserts; Drainage Patterns(Geologic); Topography; Arid Climates; Dunes; Semiarid Climates; Aeolian Soils; Surface Waters; Hydrology; Arid Lands; Channel Morphology;

Kalahari Desert; Namib Desert; Sahara Desert; Somali-Chalbi Desert; Arabian Desert; Iranian Desert; Thar Desert; Australian Deserts; Monte-Patagonia Desert; Atacama-Peruvian Desert; North American Deserts. Abstract: a survey on the state of knowledge of geomorphology and surface hydrology of the world's arid regions (excluding O.S.S.R. and China). Backgrounds of modern geomorphic concepts are discussed. The detailed review of information evaluates each of the world's deserts separately, indicating authorities and depositories. Political subdivisions (nations; USSR, states) are employed. Desert definition is considered in terms of basic climatic data, mean annual values, and expansion and contraction of deserts. Equilibrium of desert landforms (ie. Plains, topographic lows, topographic highs, alluvial fans and pediments) is considered as the central problem of arid lands geomorphology. The basic problems in the study of desert drainage and mapping are treated. Recommendations include establishment of basic data stations on transect lines perpendicular to presumed boundaries and extending from semiarid to extremely arid, treatment of precipitation data in terms of some parameter such as kurtosis (which relates frequency distribution), use of quantitative descriptive techniques of landforms, determination of current processes and rates, development of a quantitative scale of rock weathering, and a search for some physical characteristics of stream channels that will serve as accurate guides to the probable discharge and frequency of discharge of desert drainage. Includes comprehensive bibliography of approximately 2,500 refs. Database: Environmental Sciences and Pollution Mgmt.

Luther, M. E. and O'Brien, J. J. 1989. "Modeling the Variability in the Somali Current." Tallahassee. Mesoscale Air-Sea Interaction Group; United States: Florida State Univ. Volume: 274ONR; pages: 16. Descriptors: Air water interactions; Eddies(Fluid mechanics); Ocean models; Ocean currents; Wind; Atmosphere models; Boundaries; Climatology; Coastal regions; Flow fields; Gravity; Indian Ocean; Mathematical models; Mean; Monsoons; Nonlinear systems; Physics; Recirculation; Reduction; Reprints; Rotation; Scale; Seasonal variations; Shear properties; Ships; Computerized simulation; Somalia; Temperate regions; Time; Variables; Yemen; Western boundary currents; Ocean circulation; Great Whirls; Somali Current. Abstract: A numerical model of the wind driven circulation in the Indian Ocean is used to study the variability of the circulation on seasonal and interannual time scales. The model is a nonlinear reduced gravity model driven by observed winds. Model simulations use a monthly mean climatology of ships' winds as forcing and the 23 year long monthly mean Cadet and Diehl winds as forcing. The model is very successful in simulating the observed features of the regional circulation, such as formation and decay of the two gyre systems in the Somali Current during the southwest monsoon and formation of decay of the eddies off the Oman and Yemen coasts. Examination of model statistics from many years of simulation using climatological monthly mean winds shows that the model fields repeat exactly from one year to the next over most of the basin, even in the highly nonlinear eddies like the great whirl. Exceptions occur in the smaller scale eddies that form in the strong shear zones around the great whirl and in the southern gyre recirculation region, where the flow field exhibits a more chaotic nature, but even these features are nearly repeating from one year to the next. When observed, inter-annually varying winds are used to drive the model and the variability from year to year increases dramatically. This indicates that interannual variability in the model fields is due solely to variability in the winds and not due to inherent variability in the model physics, as is seen in mid-latitude models of oceanic general circulation. Western boundary currents. Reprints. (EDC). Notes: Original contains color plates: All DTIC and NTIS reproductions will be in black and white. DTIC Accession Number: ADA2113835.

Luther, M. E. O'Brien, J. J. and Meng, A. H. 1985. "Morphology of the Somali Current System during the Southwest Monsoon." Office of Naval Research, Arlington, VA. pages: 32. Descriptors: Stresses; Indian ocean; Somalia; Monsoons; Seasonal variations; Reprints; Ocean currents; Wind; Somali current; Wind stress; Gyres; Earth sciences and oceanography; Dynamic oceanography; Atmospheric sciences; Meteorology; Ocean technology and engineering; Dynamic oceanography; Atmospheric sciences; Dynamic meterology. Abstract: Results are presented from two cases of a model of the seasonal circulation in the Northwest Indian Ocean. The model is a nonlinear reduced gravity transport model, with realistic basin geometry and using observed winds as forcing. One case is forced by the monthly mean of the FGGE winds while the other is forced by the monthly mean climatological winds. The two cases are compared and inferences are made as to the importance of the different mechanisms at work in the generation and decay of the Somali Current system during the southwest monsoon. The Southern Hemisphere currents are shown to reverse before the local winds, due to the relaxation of the northeast monsoon winds. The southern gyre of the two-gyre system responds primarily to the Southern Hemisphere tradewinds, while the northern gyre forms north of 4 N in response to the development of the Findlater jet and its associated wind stress curl. The two-gyre system collapse is highly correlated with a decrease in the westerly component of the equatorial wind stress. The circulation patterns are strongly influenced by the gradient of the wind stress curl, as well as by the curl itself. The transition from southwest to northeast monsoon conditions depends on the remnants of the previous season's circulation patterns. DTIC Accession Number: ADA1589837XSP.

Lyons, P. and Dennison, A. 1960. Somalia; Geological-Geophysical Report. Somalia: Sinclair Somali Corporation, Mogadishu, Somalia. Descriptors: Africa; areal geology; East Africa; geophysical surveys; Somali Republic; surveys. Database: GeoRef. Accession Number: 1999-004734.

M

MacCulloch, C. R. 1970. Report to the Government of Somalia on Agricultural Development in the Northern Region. Rome: Food and Agriculture Organization of the United Nations: United Nations Development Programme. Descriptors: 16; Agricultural development- Somalia; Regional development- Somalia; Agricultural development projects- Somalia. OCLC Accession Number: 31638119.

MacDonald Partners, Cambridge, Somalia. 1978. Jowhar Sugar Estate; Drainage and Reclamation Study. United Kingdom: MMP, Cambridge, United Kingdom. Descriptors: Africa; agriculture; drainage; East Africa; irrigation; Jowhar sugar estate; reclamation; Shebellaha Dhexe; Somali Republic; water resources. Database: GeoRef. Accession Number: 1999-004747.

MacDonald Partners, Cambridge, United Kingdom. 1986. Hydrometry Report; Somalia, Mission Report, Stage 2; Feb.-Jun. 1986. United Kingdom: MMP, Cambridge, United Kingdom. Descriptors: Africa; East Africa; hydrology; Somali Republic; water resources. Database: GeoRef. Accession Number: 1999-004764.

MacDonald Partners, Cambridge, United Kingdom. 1986. Hydrometry Project; Somalia, Annual Summaries of Daily River Flow from the Primary Gauging Stations Operated on the Juba and Shebelle Rivers (1984-1986). United Kingdom: MMP, Cambridge, United Kingdom. Descriptors: Africa; East Africa; gauging; hydrology; Juba River; movement; Shebelle River; Somali Republic; stream flow; surface water. Database: GeoRef. Accession Number: 1999-004765.

MacDonald Partners, Cambridge, United Kingdom. 1986. Hydrometry Project; Somalia; Annual Summaries of Daily Flow of the Juba River at Baardheere. United Kingdom: MMP, Cambridge, United Kingdom. Descriptors: Africa; Baardheere Somali Republic; East Africa; Gedo Somali Republic; hydrology; Juba River; movement; Somali Republic; stream flow; surface water. Database: GeoRef. Accession Number: 1999-004766.

MacDonald Partners, Cambridge, United Kingdom. 1985. Hydrometry Project; Somalia; Annual Summaries of Daily River Flow for the Primary Gauging Stations Operated on the Juba and Shebelle Rivers. United Kingdom: MMP, Cambridge, United Kingdom. Descriptors: Africa; East Africa; gauging; hydrology; Juba River; movement; Shebelle River; Somali Republic; surface water. Database: GeoRef. Accession Number: 1999-004763.

MacDonald Partners, Cambridge, United Kingdom. 1984. Hydrometry Project; Somalia; Final Report; Stage 1. United Kingdom: MMP, Cambridge, United Kingdom. Descriptors: Africa; East Africa; hydrology; Somali Republic; water resources. Database: GeoRef. Accession Number: 1999-004760.

MacDonald Partners, Cambridge, United Kingdom. 1984. Hydrometry Project; Somalia; Progress Report. United Kingdom: MMP, Cambridge, United Kingdom. Descriptors: Africa; East Africa; hydrology; progress report; report; Somali Republic; water resources. Database: GeoRef. Accession Number: 1999-004761.

MacDonald Partners, Cambridge, United Kingdom. 1984. Hydrometry Project; Somalia; Water Resources of Wadis of Northern Somalia. United Kingdom: MMP, Cambridge, United Kingdom. Descriptors: Africa; East Africa; fluvial features; northern Somali Republic; Somali Republic; wadis; water resources. Database: GeoRef. Accession Number: 1999-004762.

MacDonald Partners, Cambridge, United Kingdom. 1983. Baardheere Reservoirs; Comparison with Alternative Schemes. United Kingdom: MMP, Cambridge, United Kingdom. Descriptors: Africa; Baardheere Somali Republic; East Africa; Gedo Somali Republic; reservoirs; Somali Republic; surface water; water resources. Database: GeoRef. Accession Number: 1999-004758.

MacDonald Partners, Cambridge, United Kingdom. 1983. Lower Shebelle Reconnaissance Survey Report. United Kingdom: MMP, Cambridge, United Kingdom. Descriptors: Africa; East Africa; hydrology; Shebelle River; Somali Republic; surface water; surveys. Database: GeoRef. Accession Number: 1999-004759.

MacDonald Partners, Cambridge, United Kingdom. 1981. Flood Damage Assessment Study; Main Report. Italy: FAO/MMP/HTS, Rome, Italy. Descriptors: Africa; agriculture; damage; East Africa; floods; geologic hazards; Somali Republic. Database: GeoRef. Accession Number: 1999-004757.

MacDonald Partners, Cambridge, United Kingdom. 1980. Pre-Feasibility Report for Gedo Region. Somalia: MMP, Cambridge, Somalia. Descriptors: Africa; East Africa;

feasibility studies; Gedo Somali Republic; Somali Republic; water resources. Database: GeoRef. Accession Number: 1999-004753.

MacDonald Partners, Cambridge, United Kingdom. 1980. Pre-Feasibility Report for Jalalaqsie (Jowhar Region). United Kingdom: MMP, Cambridge, United Kingdom. Descriptors: Africa; agriculture; East Africa; feasibility studies; Hiiraan; irrigation; Jalalaqsi Somali Republic; Jowhar Somali Republic; Somali Republic; water resources; water supply. Database: GeoRef. Accession Number: 1999-004754.

MacDonald Partners, Cambridge, United Kingdom. 1980. Reconnaissance and Pre-Feasibility Studies in Jalalaqsie (Jowhar District) and Gedo Region. United Kingdom: MMP, Cambridge, United Kingdom. Descriptors: Africa; agriculture; East Africa; feasibility studies; Gedo Somali Republic; Hiiraan; irrigation; Jalalaqsi Somali Republic; Jowhar Somali Republic; Somali Republic; water resources; water supply. Database: GeoRef. Accession Number: 1999-004755.

MacDonald Partners, Cambridge, United Kingdom. 1978. Flood Embankment Economic Appraisal Report to Juba Sugar Project. United Kingdom: MMP, Cambridge, United Kingdom. Descriptors: Africa; East Africa; embankments; hydrology; irrigation; Juba Sugar Project; Jubbada Dhexe; Somali Republic; water supply. Database: GeoRef. Accession Number: 1999-004746.

MacDonald Partners, Cambridge, United Kingdom. 1978. Juba Sugar Project-Kamsuma North Area and Labadad South Area, Irrigation and Drainage Works; Note on Flood Protection of Projects. United Kingdom: MMP, Cambridge, United Kingdom. Descriptors: Africa; agriculture; East Africa; floods; irrigation; Jubbada Hoose; Kamsuma Somali Republic; Labadad Somali Republic; protection; reservoirs; Somali Republic; water resources; water supply. Database: GeoRef. Accession Number: 1999-004748.

MacDonald Partners, Cambridge, United Kingdom. 1978. Note on November Flood in Juba River, Somalia; Internal Report. United Kingdom: MMP, Cambridge, United Kingdom. Descriptors: Africa; East Africa; floods; hydrology; Juba River; Somali Republic; surface water. Database: GeoRef. Accession Number: 1999-004749.

MacDonald Partners, Cambridge, United Kingdom. 1978. Genale-Bulo Mererta Project. United Kingdom: MMP, Cambridge, United Kingdom. Descriptors: Africa; East Africa; Genale-Bulo Mererta Project; Shabeellaha Hoose; Somali Republic; water resources; water supply. Database: GeoRef. Accession Number: 1999-004750.

MacDonald Partners, Cambridge, United Kingdom. 1978. Bardheere Reservoir; Review. United Kingdom: MMP, Cambridge, United Kingdom. Descriptors: Africa; Baardheere Reservoir; East Africa; Gedo Somali Republic; reservoirs; Somali Republic; surface water; water resources. Database: GeoRef. Accession Number: 1999-004751.

MacDonald Partners, Cambridge, United Kingdom. 1977. Genale-Bulo Mererta Project; Inception Report. United Kingdom: MMP, Cambridge, United Kingdom. Descriptors: Africa; agriculture; East Africa; Genale-Bulo Mererta Project; hydrology; irrigation; Shabeellaha Hoose; Somali Republic; water supply. Database: GeoRef. Accession Number: 1999-004740.

MacDonald Partners, Cambridge, United Kingdom. 1977. Jowhar Sugar Estate; Drainage and Reclamation Study; Report 1977. United Kingdom: MMP, Cambridge, United Kingdom. Descriptors: Africa; agriculture; drainage; East Africa; irrigation;

Jowhar sugar estate; reclamation; Shabeellaha Dhexe; Somali Republic; water resources; water supply. Database: GeoRef. Accession Number: 1999-004741.

MacDonald Partners, Cambridge, United Kingdom. 1977. Sakow Dam, Juba River; Site Investigation. United Kingdom: MMP, Cambridge, United Kingdom. Descriptors: Africa; dams; East Africa; Juba River; Jubbada Dhexe; Sakow Dam; site exploration; Somali Republic; water resources. Database: GeoRef. Accession Number: 1999-004743.

MacDonald Partners, Cambridge, United Kingdom. 1976. Jowhar Sugar Estate; Drainage and Reclamation Study. United Kingdom: MMP, Cambridge, United Kingdom. Descriptors: Africa; agriculture; drainage; East Africa; Jowhar Somali Republic; reclamation; Shabeellaha Dhexe; Somali Republic; water resources; water supply. Database: GeoRef. Accession Number: 1999-004738.

MacDonald Partners, Cambridge, United Kingdom. 1976. Juba Sugar Project; Design Note. United Kingdom: MMP, Cambridge, United Kingdom. Descriptors: Africa; agriculture; design; East Africa; irrigation; Jubbada Hoose; Somali Republic; water resources; water supply. Database: GeoRef. Accession Number: 1999-004739.

MacDonald Partners, Cambridge, United Kingdom. 1975. Duduble Off-Stream Storage Reservoirs; Pre-Design Survey and Study. United Kingdom: MMP, Cambridge, United Kingdom. Descriptors: Africa; East Africa; reservoirs; Shabeellaha Dhexe; Somali Republic; streams; water resources; water storage. Database: GeoRef. Accession Number: 1999-004736.

MacDonald Partners, Cambridge, United Kingdom. 1975. Flood Relief Channels and Jowhar Off-Stream Storage; Pre Design Phase Report. United Kingdom: MMP, Cambridge, United Kingdom. Descriptors: Africa; channels; East Africa; floods; geologic hazards; Jowhar Reservoir; Shabeellaha Dhexe; Somali Republic; streams; water resources; waterways. Database: GeoRef. Accession Number: 1999-004737.

MacDonald Partners, Cambridge, United Kingdom. 1973. Jowhar Off-Stream Storage Reservoir; Pre-Design Phase Report. United Kingdom: MMP, Cambridge, United Kingdom. Descriptors: Africa; East Africa; Jowhar Reservoir; reservoirs; Shebellaha Dhexe; Somali Republic; water resources; water storage. Database: GeoRef. Accession Number: 1999-004735.

MacFadyen, William A. 1935. The Geology and Paleontology of British Somaliland. United Kingdom: Gov. Brit. Somaliland, London, United Kingdom. Descriptors: Africa; areal geology; Somali Republic; East Africa; northern Somali Republic; paleontology. Database: GeoRef. Accession Number: 1999-004773.

MacFadyen, William A. 1933. The Geology of British Somaliland with a Geological Map 1:1,000,000. United Kingdom: Crown Agents, London, United Kingdom. Descriptors: Africa; areal geology; Somali Republic; East Africa; geologic maps; maps; northern Somali Republic. Page(s): 87. Descriptors: Geology- Somalia. Note(s): Map in pocket. "June, 1933." Includes bibliographical references (p. 39-41). LCCN: gs 33-399.

MacFadyen, William A. 1932. "Late Geological History of British Somaliland." Nature (London). Volume 130, Descriptors: Africa; areal geology; Somali Republic; East Africa; northern Somali Republic; regional. ISSN: 0028-0836.

MacFadyen, William A. 1931. "Geology of British Somaliland." Quarterly Journal of the Geological Society of London. Volume 87, Descriptors: Africa; areal geology; East Africa; northern Somali Republic; Somali Republic. ISSN: 0370-291X.

MacFadyen, William A. 1931. "Water Supply in Somaliland." Geogr. J. Volume 1931, Descriptors: Africa; East Africa; northern Somali Republic; Somali Republic; Somaliland; water resources; water supply. ISSN: 0016-7398.

MacFadyen, William Archibald. 1952. Water Supply and Geology of Parts of British Somaliland. Hargeisa: Govt. Printer. pages: 119. Descriptors: Water-supply-Somalia; Geology- Somalia. Notes: Four folded leaves of illustrations, tables and maps in pocket. Title from cover. Stamped at head of title: "Geological Survey, Hargeisa, Somaliland Protectorate." Mimeographed. Bibliography pages 114-119. OCLC Accession Number: 34428293; USGS Library.

Mackay, C. Williams, G. Gilbert, D. L. F. and Widner, H. 1954. A Geological Reconnaissance of the Sedimentary Deposits of the Protectorate of British Somaliland. United Kingdom: Crown Agents, London, United Kingdom. Descriptors: Africa; areal geology; East Africa; northern Somalia; sedimentary rocks; Somali Republic. Database: GeoRef. Accession Number: 1996-037973.

MacKenzie, Debora. 1992. "Toxic Waste Adds to Somalia's Woes." New Scientist. September 19. Volume 135, Pages 5. Descriptors: Hazardous waste disposal; Export trade/Laws and regulations; General Science; Applied Science & Technology. ISSN: 0262-4079.

Madany M.H. 1992. "Intercropping Fodder Trees: A Case Study from Somalia." Rangelands. Volume 14, Issue 4, Pages 208-213. Descriptors: Agroforestry and intercropping; legumes; agroforestry project; fodder trees; intercropping Species Term: Somalia; Bos taurus; Leucaena leucocephala; Animalia; Leucocephala; Gliricidia sepium. Notes: Somalia- Julib District- Homboy. Abstract: This paper reports on the feasibility of intercropping leguminous fodder trees in an agropastoral community in southern Somalia. The Homboy Agroforestry Project is small-scale and was designed to use informal on-farm trials backed by extension work. The primary objective was to provide high-protein cattle fodder during the dry-season by intercropping leguminous trees, such as Leucaena leucocephala and Gliricidia sepium. ISSN: 0190-0528.

Madgwick J. 1989. "Somalia's Threatened Forests." Oryx. Volume 23, Issue 2, Pages 94-101. Descriptors: floodplain; forest. Abstract: The floodplain forest in Somalia is the northernmost outlier of moist lowland forest in coastal East Africa. Recent surveys reveal that only a few hundred hectares remain intact in the whole of Somalia. Clearance for cultivation has been encouraged by irrigation and drainage schemes in the upper and lower reaches of the Jubba valley. Only the poorly accessible Middle Jubba (Bu'aale District) with predominantly saline, alkaline, impermeable soils, has retained significant areas of relict floodplain forest. ISSN: 0030-6053.

MAFFI, M. 1960. "Larval Focus of A. Gambiae and A. Squamosus in a Salt-Water Well at Busc Busc (Coastal Region of Lower Oltre Giuba). First Finding in Somaliland." Riv. Malariol. Jun. Volume 39, Pages 131-133. Descriptors: Anopheles. ISSN: 0370-565X.

Magellan Geographix. 1993. "Somalia." Washington, D.C: Magna & Carta; Williams & Heintz Corp.). Notes: Description: 1 map; color; 39 x 41 cm., folded to 22 x 14 cm. Geographic: Somalia- Maps. Scale 1:4,250,000. Cartgrph Code: Category of scale: a Constant ratio linear horizontal scale: 4250000; Notes: At head of title: Magna Carta maps. Includes ill. of national flag, brief national information, and location map. Insets: Somali ethnic groups- United States and Somalia size comparison- Mogadishu [shows selected buildings]. LCCN: 93-682806; OCLC Accession Number: 28025017.

Magellan Geographix and Magna Carte. 1993. "Somalia." Washington, D.C: Magna Carta. Descriptors: Ethnology- Somalia- Maps. Notes: Description: 1 map; color; 40 x 42 cm. Geographic: Somalia- Maps. Scale 1:4,250,000; (E 410--E 510/N 120--S 20). Notes: Includes text, ill. of flag, location map, and insets of "Somali ethnic groups," "United States and Somalia size comparison" and "Mogadishu." ISBN: 0783412665. OCLC Accession Number: 27820525.

Mahajan P.N, Mujumdar V.R and Ghanekar S.P. 1989. "Excitation of Low-Level Jet as seen by GOES (I-O) Satellite off the Somali Coast." Adv. Atmos. Sci. Volume 6, Issue 4, Pages 475-482. Descriptors: GOES; coast; monsoon. Geographic: Somalia. Abstract: Excitation of Low-level cross-equatorial flow in the western Indian ocean results from an interaction between extra-tropical perturbations moving eastward across the South African-Malgassy region of the Southern Hemisphere. This excitation occurs 2-3 days after the first appearance of a northward propagation cold front across the South African-Malgassy region. Intensification of cross-equatorial flow is followed by an increase in rainfall activity along the west coast of India after 3-4 days. The study reveals that this association can be used to forecast an increase in rainfall activity along the west coast of India 5-7 days in advance. ISSN: 0256-1530.

Mainguy, Maurice. 1984. "Petrole et Gaz Naturel en Afrique au Sud du Sahara. Translated title: "Oil and Natural Gas in Sub-Saharan Africa." Revue De l'Energie. Volume 35, Issue 360, Pages 1-10. Descriptors: Petroleum Prospecting; Natural Gas Deposits- Africa. Abstract: Intensive oil prospecting efforts are going on in Africa. The author of this article sets South Africa aside and studies three zones: the east coast, central Africa and the west coast- particularly the Gulf of Guinea. The vast Ethiopia-Kenya sedimentary zone has given disappointing results despite quite important prospecting efforts; nevertheless, there are some encouraging signs for oil and especially for gas in Ethiopia and in Somalia. In central Africa, two types of basins with oil potential can be distinguished: the 'grabens' and the 'interior' or 'intratonic' basins. It seems that the best results have been achieved in Sudan. The hydrocarbon producing region on the West African coast is limited to the Gulf of Guinea. By far the main producer is Nigeria. The first authentic discovery in sub Saharan Africa was made in Angola in 1955 by the Belgian Petrofina Company; this was the Benifica deposit. Gabon and Angola began producing at about the same time as Nigeria. Gabon's production reached a culminating point between 1975 and 1977 and it is only due to strictly technical causes that its production in 1982 attained the level of 1972. Elf Gabon is the main producer, but a dozen other oil companies are engaged in prospecting. Shell and Gulf are producers. ISSN: 0303-240X.

Maloiy, G. M. 1971. "Temperature Regulation in the Somali Donkey (Equus Asinus)." Comp. Biochem. Physiol. A. Jul 1. Volume 39, Issue 3, Pages 403-412. Descriptors: Animals; Body Temperature; Body Temperature Regulation; Body Weight; Camels/physiology; Circadian Rhythm; Cold; Electrocardiography; Environmental Exposure; Female; Heat; Male; Oxygen Consumption; Perissodactyla- physiology; Respiration; Stress; Sweating; Temperature; Water Deprivation. ISSN: 0300-9629.

Maloiy, G. M. 1970. "Water Economy of the Somali Donkey." Am. J. Physiol. Nov. Volume 219, Issue 5, Pages 1522-1527. Descriptors: Adaptation, Physiological; Animal Feed; Animals; Appetite; Body Temperature Regulation; Body Weight; Dehydration- physiopathology; Desert Climate; Feces analysis; Feeding Behavior; Female; Male; Osmolar Concentration; Perissodactyla- physiology; Stress-physiopathology; Urine analysis; Water Deprivation; Water-Electrolyte Balance. ISSN: 0002-9513.

Maloiy, G. M. and Boarer, C. D. 1971. "Response of the Somali Donkey to Dehydration: Hematological Changes." Am. J. Physiol. Jul. Volume 221, Issue 1, Pages 37-41. Descriptors: Adaptation, Physiological; Animals; Blood Proteins/analysis; Blood Volume; Body Fluids/analysis; Body Weight; Cattle; Chlorides/blood; Dehydration; Drinking Behavior; Erythrocyte Count; Female; Hematocrit; Hemoglobins/analysis; Male; Osmolar Concentration; Osmosis; Perissodactyla/physiology; Sodium/blood; Time Factors; Tritium; Tropical Climate; Urine/analysis; Water. ISSN: 0002-9513.

Mamboya, F. A. Pratap, H. B. and Bjork, M. 2003. Heavy Metal Contamination in the Western Indian Ocean (a Review). Grenoble, France: EDP Sciences. Volume: 107, II, pages: 1437. XII International Conference on Heavy Metals in the Environment, may 26-30 2003. Descriptors: Heavy metals; Contamination; Marine pollution; Ecosystems; Impurities; Health hazards; Sediments; Particles (particulate matter); Seawater. Abstract: Western Indian Ocean Coast has many potential marine ecosystems such as mangrove, seagrass meadows, macroalgae, and coral reefs. It is largely unspoiled environment however, tourism and population growth in coastal urban centers, industrialization, are presenting a risk of pollutants input to the marine environment of the Western Indian Ocean. Mining, shipping and agricultural activities also input contaminants into the marine environment via runoff, vessel operations and accidental spillage. Heavy metals are among the pollutants that are expected to increase in the marine environment of the Western Indian Ocean. The increase in heavy metal pollution can pose a serious health problem to marine organism and human through food chain. This paper reviews studies on heavy metal contamination in the Western Indian Ocean. It covers heavy metal studies in the sediments, biota, particulates and seawater collected in different sites. In comparison to other regions, only few studies have been conducted in the Western Indian Ocean and are localized in some certain areas. Most of these studies were conducted in Kenyan and Tanzanian coasts while few of them were conducted in Mauritius, Somalia and Reunion. No standard or common method has been reported for the analysis or monitoring of heavy metals in the Western Indian Ocean. ISSN: 1155-4339.

Manfredi, G. 1952. "La Regolazione Del Giuba. Translated title: Regulation of the Juba River." La Vie Del Mondo. Volume 3, Descriptors: Africa; East Africa; hydrology; Juba River; Somali Republic; surface water; water resources. ISSN: 0393-8948.

Manfredi, G. 1938. "La Regolazione Delle Acque Dello Uebi Scebeli e l'Avvenire Della Somalia. Translated title: Water Regulation of the Shebelle River and the Future of Somalia." Ingegnere. Volume 11, Descriptors: Africa; East Africa; hydrology; Shebelle River; Somali Republic; surface water; water resources. ISSN: 0020-0905.

Mangialardi, L. and Mantriota, G. 1994. "Continuously Variable Transmissions with Torque-Sensing Regulators in Waterpumping Windmills." Renewable Energy. Volume 4, Issue 7, Pages 807-823. Abstract: One widespread application of wind energy is in water-pumping. Usually windmills are coupled to water pumps, though piston pumps are difficult to match to wind turbines. In the first approximation, the piston pump requires a constant torque to the turbine at all rotational speeds. Conversely, if the turbine were to operate at maximum efficiency levels, it would supply a torque proportional to the square of the wind speed. Hence, the combination of windmill and piston pump would be maximally efficient only for a single wind speed.This paper suggests incorporating a continuously variable transmission (C.V.T.) between the pump and the turbine to enhance the efficiency of windpumpers at different wind speeds. The feasibility of adopting automatically regulated C.V.T.s, which adjust their transmission ratio to the torque without requiring sophisticated regulators or control systems, is also examined since such a solution does not undermine the cost-effectiveness, reliability or simplicity of the system.A mathematical model has been relied on to simulate the operation of a windpumper equipped with a C.V.T. with a torque sensing regulator. The advantages in terms of efficiency and rate of flow of such a solution are demonstrated assuming that a water-pumping system equipped with a C.V.T. is installed in two locations in Somalia. Finally, the paper examines the dynamic behaviour of the system proposed.

Manighetti, I., Tapponnier, P. and Gillot, P. Y., et al. 1998. "Propagation of Rifting Along the Arabia-Somalia Plate Boundary- into Afar." Journal of Geophysical Research. 10 Mar. Volume 103, Issue B3, Pages 4947-4974. Descriptors: Plates (Tectonics); Geochronology; Geological Faults; Volcanoes; Lithosphere; African Rift System; Deformation; Somalia. Notes: 70. Abstract: It is generally accepted that the Aden ridge has propagated westward from about 58 deg E to the western tip of the Gulf of Aden/Tadjoura, at the edge of Afar. Here, we use new tectonic and geochronological data to examine the geometry and kinematics of deformation related to the penetration of that ridge on dry land in the Republic of Djibouti. We show that it veers northward, forming a narrow zone of dense faulting along the northeastern edge of the Afar depression. The zone includes two volcanic rifts (Asal-Ghoubbet and Manda Inakir), connected to one another and to the submarine part of the ridge by transfer zones. Both rifts are composite, divided into two or three disconnected, parallel, NW-SE striking subrifts, all of which appear to have propagated northwestward. In Asal-Ghoubbet as in Manda Inakir, the subrifts appear to have formed in succession, through north directed jumps from subrifts more farther south. At present, the northernmost subrifts of the Manda Inakir rift form the current tip of the northward propagating Arabia-Somalia plate boundary in Afar. We account for most observations by a mechanical model similar to that previously inferred for the Gulf of Aden, in which propagation is governed by the intensity and direction of the minimum horizontal principal stress. We interpret the northward propagation on land, almost orthogonal to that in the gulf, to be related to necking of the Central Afar lithosphere where it is thinnest. ISSN: 0148-0227.

Manni, F. 1932. "Le Acque Dello Scebeli Da Genale Ad Hawai e l'Uebi Scebeli." Translated title: "Shebelle River Water from Genale at Hawai and the Lower Shebelle." L'Agricoltura Coloniale. Volume 5, Descriptors: Africa; East Africa; Genale Somali Republic; Shabeellaha Hoose; Shebelle River; Somali Republic; surface water; water resources. ISSN: 0394-2945.

Manoncourt, Serge, Doppler, Brigitte and Enten, Francois, et al. 1992. "Public Health Consequences of the Civil War in Somalia, April, 1992." The Lancet. Volume 340, Issue 8812, Pages 176-177.

Mansse, E. 1916. "Rocce Della Somalia Italiana Raccolte Dalla Missione Scientifica Stefanni-Paoli Del 1913. Rocks of the Italian Somaliland Collected by the Stefanni-Paoli Scientific Mission of 1913." Annales De l'Institut Oceanographique (Monaco). Volume 31, Descriptors: Africa; areal geology; East Africa; expeditions; Italian Somaliland; rocks; Somali Republic; southern Somali Republic; Stefanni-Paoli Scientific Mission. ISSN: 0078-9682.

Manzelli M, Benedettelli S and Vecchio V. 2005. "Agricultural Biodiversity in Northwest Somalia- an Assessment among Selected Somali Sorghum (Sorghum Bicolor (L.) Moench) Germplasm." Biodivers. Conserv. Volume 14, Issue 14, Pages 3381-3392. Descriptors: Agriculture; Crop and livestock production- general; Arable; agricultural ecosystem; biodiversity; food security; sorghum; genetic variation Species Term: Sorghum bicolor; Somalia; Animalia. Notes: References: 19; Geographic: Somalia East Africa Sub-Saharan Africa. Abstract: The seed sector situation in Northwest Somalia is critical. The availability of food has decreased and many people are at risk of hunger. Food security can be restored by enhancing the local genetic resources and creating an efficient seed sector. Sorghum is important as a food and fodder crop in this region. It is close to Ethiopia, which is considered as the probable origin and domestication of Sorghum. Twelve morphological and productive characteristics were chosen to assess the phenotypic variability of 16 accessions of sorghum from Northwest Somalia. Univariate (analysis of variance and G test) and multivariate (discriminant and cluster analysis) methods were used to assess the morphological variation within the accession and to group the 16 accessions into clusters based upon quantitative and qualitative characters. Elmi Jama Cas, Masego Cas, Masego Cad and Carabi clearly represent distinct landraces with specific features suitable for different purpose, such as grain and/or forage production. Each landrace tested is able to grow under harsh environmental conditions, thus ensuring a low, but stable production for small poor resources farmers. Knowledge and conservation of local landraces will provide a broad base of genetic variability from which improved sorghum varieties can be developed, thus aiding in the stabilization of a secure and sustainable food supply for farmers of Northwest Somalia. ISSN: 0960-3115.

"Map of a Portion of Somaliland [Including Part of British and Italian Somaliland, and Abyssinia]." 1907. London. Notes: Description: 1 map; 72 x 67 cm. Geographic: Somalia- Maps. British Somaliland- Maps. Italian Somaliland- Maps. Scale 1:1,000,000. Notes: To accompany: Official history of the operations in Somaliland, 1901-4. Other Titles: Official history of the operations in Somaliland. OCLC Accession Number: 54334098.

March, L. 1973. Field Procedure Report, Geography Survey, South Juba Basin. Somalia: Texaco Deutsche A.G. Somalia, Somalia. Descriptors: Africa; areal geology; East Africa; field studies; geography; Juba Basin; Juba River; Somali Republic; surveys. Database: GeoRef. Accession Number: 1999-004790.

Marchal R. 1993. "Les Mooryaan De Mogadiscio. Formes De La Violence Dans Un Espace Urbain En Guerre." Cahiers d'Études Africaines. Volume 130, Pages 295-320. Descriptors: Cultural; Conflict, protest, human rights; war situation; mooryaan; cultural relations; civil disturbance; urban violence; adolescents; violence; delinquency;

war; urban area. Notes: Somalia- Mogadisho Somalia- Mogadishu. Abstract: From December 1990 till December 1992, the Somalian capital was the scene of fighting and violence that had less and less to do with political strategies. Using data from the field, the author has investigated the actors in this fighting and inquired into their social trajectories so as to shed light on some issues in this crisis. A society in war is being restructured by warfare. Social relations and politics, as practised by competing factions, are becoming criminal. Brief comparisons show that the mooryaan cannot be reduced to forms of social violence such as the warenleh, shifta, or Becker's "outsiders'. Questions are raised about the future of this social group, which is rapidly evolving toward a more classical form of delinquent violence. ISSN: 0008-0055.

Marchesini, E. 1938. "Osservazione Geologica Dall'Aeroplano Durante La Campagna Di Ricerche in Africa Orientale Italiana Dell'AGIP, 1937-1938. Translated title: Geological Observation from an Airplane during the Research Mission in Italian East Africa by the AGIP, 1937-1938." Rivista Aeronautica. Volume 11, Descriptors: Africa; airborne methods; areal geology; East Africa; Italian East Africa; Somali Republic. ISSN: 0391-6162.

Marchuk, G.I., Diansky, N.A., Moshonkin, S.N., Rusakov, A.S. and Zalesny, V.B. 2006. "High-resolution simulation of monsoon variability of the Indian Ocean currents." Russian Journal of Numerical Analysis and Mathematical Modelling. Volume 21 (2), pages 153- 168.

Marci, F., Goracci, G. and Negri, P. L. 1982. "Fluoride Content of Enamel in an Area Where the Water is Richly Fluoridated." Riv. Ital. Stomatol. Volume 51, Issue 10, Pages 867-870. Descriptors: Adult; Dental Caries/metabolism; Dental Enamel/analysis; Dental Enamel Solubility; English Abstract; Fluorides/analysis; Humans; Middle Aged; Somalia; Water Supply/analysis. ISSN: 0035-6905.

Marconi, A. 1918. Somalia Settentrionale, Scala 1:2,000,000. Northern Somalia, Scale 1:2,000,000. Italy: Ministero delle Colonie, Rome, Italy. Descriptors: Africa; areal geology; East Africa; geologic maps; maps; northern Somali Republic; Somali Republic. Database: GeoRef. Accession Number: 1999-004796.

Marconi, A. 1918. Somalia e Paesi Limitrofi, Scala 1:4,000,000. Somalia and Borderlands, Scale 1:4,000,000. Italy: Ministero delle Colonie, Rome, Italy. Descriptors: Africa; areal geology; East Africa; geologic maps; maps; Somali Republic. Database: GeoRef. Accession Number: 1999-004797.

Marconi, A. 1912. Carta Dimostrativa Della Somalia Italiana Meridionale, Scala 1:1,000,000. Demonstration Map of Southern Italian Somalia, Scale 1:1,000,000. Italy: Ministero Affari Esteri, Rome, Italy. Descriptors: Africa; areal geology; East Africa; geologic maps; maps; Somali Republic; southern Somali Republic. Database: GeoRef. Accession Number: 1999-004795.

Marconi, A. 1911. Somalia Italiana Meridionale. Southern Italian Somalia. Italy: Comando di Corpo di Stato Maggiore, Rome, Italy. Descriptors: Africa; areal geology; East Africa; geologic maps; maps; Somali Republic; southern Somali Republic. Database: GeoRef. Accession Number: 1999-004794.

Marconi, A. 1910. Carte Della Somalia, Scala 1:200,000. Map of Somalia, Scale 1:200,000. Italy: Istituto Geografico Militare, Florence, Italy. Descriptors: Africa; areal geology; East Africa; maps; Somali Republic. Database: GeoRef. Accession Number: 1999-004793.

Marquardt, Mark A. Marquardt, Mark A. Design phase completion report, Land Registration Program, Shabelli Water Management Project; University of Wisconsin--Madison and Land Tenure Center. 1989. End of Contract Report: Phase I, Land Registration Program, Shabelli Water Management Project. Madison, Wis: Land Tenure Center, University of Wisconsin-Madison. Descriptors: 31; Land titles- Registration and transfer- Somalia- Shebeli River Valley; Land tenure- Somalia- Shebeli River Valley; Project appraisal- Somalia- Shebeli River Valley; Land administration- Somalia- Shebeli River Valley. Abstract: Appendix I. Design phase completion report, Land Registration Program, Shabelli Water Management Project by Mark A. Marquardt. Notes: Named Corp: Shabelli Water Management Project. Notes: "September 1989." Other Titles: Phase I, Land Registration Program, Shabelli Water Management Project; Land Registration Program, Shabelli Water Management Project. OCLC Accession Number: 36071851.

Marrocchi, Giuliano. 1973. "Note Minerarie Sulla Somalia; 28 Maggio-11 Giugno 1972; 10 Ottobre-19 Ottobre 1972. Translated title: A Mineralogical Note on Somalia; May-June-October 1972." Industria Mineraria. Volume 24, Issue 4, Pages 173-175. Descriptors: Africa; distribution; East Africa; economic geology; mineral resources; possibilities; Somali Republic. Notes: illustrations (incl. geol. sketch map). ISSN: 0019-7696.

Mascarelli, L. 1931. "Relazione Di Analisi Di due Campioni Di Acqua Prelevati in Somalia Dal Fiume Uebi Scebeli Presso Il Villaggio Ducca Degli Abruzzi. Translated title: Analysis of Two Water Samples from Shebelle River Near Ducca Degli Abruzzi Town." Somalia: Descriptors: Africa; East Africa; geochemistry; hydrochemistry; Shebelle River; Somali Republic; surface water; water quality; water resources. Database: GeoRef. Accession Number: 2000-010220.

Mascarelli, L. 1931. "Relazione Sulla Potabilita' o Non Potabilita' Di Alcuni Campioni Di Acqua Dei Pozzi Della SAIS. Translated title: Relation between Potability and Non-Potability of some Water Samples from SAIS Wells." Somalia: Descriptors: Africa; drinking water; East Africa; geochemistry; ground water; hydrochemistry; Somali Republic; water quality; water wells. Database: GeoRef. Accession Number: 2000-010221.

Mascarelli, L. 1930. "La Composizione Delle Acque Dell'Uebi Scebeli. Translated title: "Water Composition of the Shebelle River." Bolletino Della Societa Italiana Per Il Progresso Delle Scienze. Volume 25, no.40, Descriptors: Africa; East Africa; geochemistry; hydrochemistry; Shebelle River; Somali Republic; surface water; water quality; water resources. Database: GeoRef.

Mascini, M. 1983. Rapporto Sulla Caratterizzazione Analitica Delle Acque Della Somalia. Translated title: Report on the Analytical Characterization of the Water of Somalia. Somalia: Universita Nazionale Somala, Dip. Chimica, Mogadishu, Somalia. Descriptors: Africa; East Africa; geochemistry; hydrochemistry; Somali Republic; water. Database: GeoRef. Accession Number: 1999-004800.

Mascini, M. and Liberti, A. 1984. "Ion Selective Electrodes for Measurements in Fresh Waters." Science of Total Environment Vol.37. Volume July, Pages 6 Ref. Descriptors: Ion selective electrodes; Potentiometry; Somalia; Well water; Detection limits; Fluorides; Chemical analysis. Abstract: Ion selective electrodes (ISE), now commonly employed for measurements in fresh waters, are simple, sturdy, and reliable.

Samples need only a minor pretreatment before being appropriate for analysis. The detection limit of a potentiometric measurement is usually in the range of 10 to the -5 power M to 10 to the -6 power M. Several features make potentiometry very attractive as an analytical method in water analysis. One advantage is that measurements can be performed in field conditions with portable instruments, this feature is of considerable importance in environmental monitoring of water quality, especially when storage and transportation may be sources of errors in analytical determinations. Applications of the ISE in field measurements have been realized in a Somalia survey with simple equipment. The water wells of several desert districts were characterized with the aim of evaluating future residential and cattle raising areas. About 100 wells were analyzed in a desert area located far from any village. Through this survey, experimental evidence was obtained of endemic illness of nomadic populations due to the high fluoride content of some wells. Database: Environmental Sciences and Pollution Mgmt.

Mashruuca Daaqa Gobollada Dhexe (Somalia). 1985. Central Rangelands Development Project Non-Formal Education Service Pamphlet. Mogadishu: The Project. Descriptors: v. Range management- Somalia; Grazing- Somalia; Soils- Somalia; Livestock- Somalia- Development. Abstract: no. 1. Protection and improvement of Somalia rangelands- no. 2. A description of the proposals for the central Rangelands Development Project- no. 3. Introduction to the general ecology and the environment related to animal production in Somalia- no. 4. Practical field methods in range management- no. 7. An introduction to botany for the range lands- no. 8. the soil- no. 9. Systems of grazing- no. 10. Duties of the range educators. Notes: Named Corp: Mashruuca Daaqa Gobollada Dhexe (Somalia); Notes: Cover title. Vols. no. 3-4, have title: Central Rangelands Dev. Project non formal education service pamphlet; no. 9: Central Rangelands Development Project booklet; no. 10: Central Rangelands Dev. Project non formal education service staff training pamphlet. Other Titles: Non-formal education service pamphlet. Central Rangelands Dev. Project non formal education service pamphlet. Central Rangelands Development Project booklet. Central Rangelands Dev. Project non formal education service staff training pamphlet. Library of Congress. LCCN: 87-155442; OCLC Accession Number: 29599011.

Masiero-Novelli, Raniero and Ahmed-Hassan, A. 1980. "Su Alcune Mineralizzazioni Della Somalia Settentrionale." "Some Mineralization of Northern Somalia." L' Industria Mineraria (1950). Oct. Volume 1, Issue 5, Pages 25-30. Descriptors: Africa; East Africa; economic geology; metal ores; mineral deposits, genesis; mineralization; northern Somali Republic; occurrence; Somali Republic. References: 14; illustrations incl. sketch maps. ISSN: 0391-1586.

Mason, J. E. 1962. "Report on the Geology of the Area North of Hargeisa and Laferug, and Hargeisa and Berbera Districts." Report of the Geological Survey, Somaliland Protectorate. Volume 7, Pages 34. Descriptors: Africa; areal geology; Berbera Somali Republic; East Africa; Hageysa Somali Republic; Somali Republic. Database: GeoRef.

Mason, J. E. 1962. "Report on the Geology of the Area North of Hargeisa and Laferug, Hargeisa and Berbera Districts, Quarter Degree Sheets Nos. 22, 23, 34." Somalia: Volume: 7, Descriptors: Africa; areal geology; Berbera Somali Republic; East Africa; Hargeisa Somali Republic; Laferug Somali Republic; Somali Republic; Waqooyi Galbeed. Database: GeoRef. Accession Number: 1999-004804.

Mason, J. E. 1957. Note on the Geology of the Barkileh-Ida Kabaita Neighbourhood. Somalia: Geol. Surv. Somalil. Prot., Hargeisa, Somalia. Descriptors: Africa; areal geology; Barkileh Mountain; East Africa; Ida Kabaita; regional; Somali Republic; Togdheer Somali Republic. Database: GeoRef. Accession Number: 1999-004803.

Mason, J. E. 1954. Report on the Geology of the Basement Rocks of the Sheikh-Auradli-Wajire Area. Somalia: Geol. Surv. Somalil. Prot., Hargeisa, Somalia. Descriptors: Africa; areal geology; Auradli Somali Republic; basement; East Africa; regional; Sheikh Somali Republic; Somali Republic; Togdheer Somali Republic; Wajire Somali Republic. Database: GeoRef. Accession Number: 1999-004802.

Mason, J. E. and Warden, A. 1957. Report on the Geology of the Heis-Mait-Waqderia Area, Erigavo District. Hargeisa: Somaliland Stationery Office. Descriptors: 23 pages maps. Notes: Entry: 19960326; Update: 19960326. Accession Number: 34463121.

Mason, J. E. and Warden, A. J. 1971. Saba Wanak and Bulhar. United Kingdom: Directorate Overseas Surv., United Kingdom. Descriptors: Africa; areal geology; Bulhar; East Africa; geologic maps; maps; Saba Wanak; Somali Republic. Notes: Latitude: N100000, N104000 Longitude: E0443000, E0440000. Database: GeoRef. Accession Number: 1989-007766.

Mason, J. E. and Warden, A. J. 1971. Hargeisa. United Kingdom: Directorate Overseas Surv., United Kingdom. Descriptors: Africa; areal geology; East Africa; geologic maps; Hargeisa; maps; Somali Republic. Notes: Latitude: N093000, N100000 Longitude: E0443000, E0440000. Database: GeoRef. Accession Number: 1989-007773.

Mason, J. E. and Warden, A. J. 1959. Waqderia. United Kingdom: Directorate Overseas Surv., United Kingdom. Descriptors: Africa; areal geology; East Africa; geologic maps; maps; Somali Republic; Waqderia. Notes: Latitude: N110000, N113000 Longitude: E0480000, E0473000. Database: GeoRef. Accession Number: 1989-007761.

Mason, J. E. and Warden, A. J. 1959. Mait. United Kingdom: Directorate Overseas Surv., United Kingdom. Descriptors: Africa; areal geology; East Africa; geologic maps; Mait; maps; Somali Republic. Notes: Latitude: N110000, N113000 Longitude: E0473000, E0740000. Database: GeoRef. Accession Number: 1989-007762.

Mason, J. E. and Warden, A. J. 1959. Medishe. United Kingdom: Directorate Overseas Surv., United Kingdom. Descriptors: Africa; areal geology; East Africa; geologic maps; maps; Medishe; Somali Republic. Notes: Latitude: N103000, N110000 Longitude: E0480000, E0473000. Database: GeoRef. Accession Number: 1989-007763.

Mason, J. E. and Warden, A. J. 1959. Erigavo. United Kingdom: Directorate Overseas Surv., United Kingdom. Descriptors: Africa; areal geology; East Africa; Erigavo; geologic maps; maps; Somali Republic. Database: GeoRef. Accession Number: 1989-007764.

Mason, J. E. and Warden, A. J. 1957. "Report on the Geology of the Heis-Mait-Waqderia Area, Erigavo District Quarter Degree Sheets 3, 4, 14, 15 and 16." Somaliland, Geol. Surv., Rept. Volume 1, Pages 23. Descriptors: Africa; Cenozoic; Cretaceous; East Africa; geologic history; geology; Heis-Mait- Waqderia area; Heis-Mait-Waqderia area; igneous rocks; Jurassic; Mesozoic; metamorphic rocks; Precambrian; Somali Republic; Tertiary; Triassic. Abstract: Explanatory text for a series of map sheets covering part of the Makhir coast inland to the escarpment forming the major topographic and geologic feature of the region. Data on the rock types represented in the basement complex and the unconformably overlying Inda Ad series, the four kinds of granitic intrusions cutting both ancient complexes, the various members of the Mesozoic and Tertiary sedimentary sequences, and the geologic history of the region are reported. Database: GeoRef.

Mason, J. E. and Warden, A. J. 1956. The Geology of the Heis-Mait-Waqderi Area, Erigavo District. Somalia: Geol. Surv. Somalil. Prot., Hargeisa, Somalia. Descriptors: Africa; areal geology; East Africa; Erigavo Somali Republic; geologic maps; Heis Somali Republic; Mait Somali Republic; maps; regional; Sanaag Somali Republic; Somali Republic; Waqderia Somali Republic. Database: GeoRef. Accession Number: 1999-004805.

Mason, J. E. and Warden, A. J. 1956. Report on the Geology of the Heis-Mait-Waqdera Area: Erigavo District: Quarter Degree Sheets Nos. 3,4,14,15,and 16 (D.O.S. 1:125,000): Based on a Reconnaissance Survey of the Basement System and Contiguous Formations by J.E. Mason, A.J. Warden, and J.A.B. Stewart. Hargeisa: Stationary office. Descriptors: VI, 23. OCLC Accession Number: 67822765.

Mason, J. E. Warden, A. J. and Stewart, J. A. B. 1959. Heis Area. United Kingdom: Directorate Overseas Surv., United Kingdom. Descriptors: Africa; areal geology; East Africa; geologic maps; Heis region; maps; Somali Republic. Notes: Latitude: N103000, N110000 Longitude: E0470000, E0463000. Database: GeoRef. Accession Number: 1989-007765.

Massey G. 1989. "Agropastoralism and Agropastoral Society in South-Central Somalia." Journal of Developing Societies. Volume 5, Issue 2, Pages 157-174. Descriptors: Crop and livestock production- general; agropastoralism; pastoralism; Rahanweyn people. Abstract: The characteristics of an agropastoral mode of production and agropastoral society are examined, illustrated by the case of the Rahanweyn of the interriverine area of south-central Somalia. A discussion of the viability and utility of agropastoralism in situations of social change concludes the paper. ISSN: 0169-796X.

Massoli-Novelli, R. 1981. "Il Deposito Ferrifero Precambrico Di Burgalan." Translated title: "Precambrian Iron Deposit of Bur Galan." Bollettino Della Associazione Mineraria Subalpina. Volume 18, Issue 1-2, Pages 19-38. Descriptors: Africa; banded structures; Bur Galan; East Africa; economic geology; iron ores; major elements; metal ores; metamorphic rocks; petrology; Precambrian; quartzites; Somali Republic; southern Somali Republic; stratigraphy. Notes: References: 1 p. illustrations incl. 4 anal., geol. sketch map. ISSN: 0392-355X.

Mateer N.J and Et Al. 1992. "Correlation of Nonmarine Cretaceous Strata of Africa and the Middle East." Cretaceous Res. Volume 13, Issue 3, Pages 273-318. Descriptors: Palaeogeography and palaeoclimatology; Cretaceous; plate tectonics; stratigraphic correlation; convergence; Continentale Intercalaire; Nubian Sandstone; Nubian; Somalia; Africa; Middle East. Abstract: The Saharan region is dominated by nonmarine "Continentale Intercalaire' and "Nubian Sandstone' clastics of Early to mid-Cretaceous age, and in Egypt, locally Late Cretaceous. "Nubian Sandstone' can be correlated into the Arabian Peninsular, where it was deposited around the Arabian craton prior to inundation by Tethyan transgressions. The opening of the South Atlantic invoked tectonic stresses forming pull-apart basins in West Africa, and exploited the Pan-African shear zone cutting central Africa forming basins from Nigeria to southern Sudan. Intracratonic basins in northern Sudan and Egypt resulted from tectonic stresses associated with the opening of the Red Sea and the northward convergence of Africa into the European continent. Nonmarine Cretaceous deposits in central and inland south-eastern Africa outcrop sporadically and are associated with original extensional tectonics related to the separation of Madagascar. Coastal basins of Mozambique and South Africa containing marginal nonmarine Cretaceous facies appear to be associated with the separation of Madagascar and the Falkland Plateau from Africa. Nonmarine sedimentation in Africa and the Middle East is predominantly pre-Cenomanian, with some Upper Cretaceous rocks occurring in Egypt, Somalia, Yemen, Nigeria and Madagascar during regressive episodes. ISSN: 0195-6671.

Maurigi, C. 1910. "Clima Ed Agricoltura Della Gascia." Translated title: "Climate and Agriculture of Gascia." Riv. Agric. Subtrop. Trop. Volume 1910, Descriptors: Africa; agriculture; climate; East Africa; Gascia Somali Republic; hydrology; irrigation; Somali Republic; water resources. ISSN: 0035-6026.

Maxamed Adan, G. 1981. "Studio Idrogeologico Delle Sorgenti Termali Della Zona Di Berbera." Translated title: "Hydrogeologic Study of the Thermal Springs of the Berbera Zone." Somalia: Univ. Naz. Somala, Dipart. Geologia, Mogadishu, Somalia. Descriptors: Berbera Somali Republic; ground water; hot springs; springs; thermal waters; Waqooyi Galbeed. Database: GeoRef. Accession Number: 1999-004809.

Maxamed, Aweys Cadde; Axmed, Yusuf Mahad; Axmed, Cumar Maxamed and Axmed, Cabdisalaam Maxamed. 1987. The Liquid and Solid Discharge of Somali Rivers; a Methodological Approach; GeoSom 87; International Meeting; Geology of Somalia and Surrounding Regions; Abstracts. Somalia: Somali Natl. Univ., Mogadishu, Somalia. GeoSom 87; Geology of Somalia and Surrounding Regions, Mogadishu. Somalia Conference: Nov. 23-30, 1987. Descriptors: Africa; East Africa; ephemeral streams; field studies; fluvial features; Giuba; hydrogeology; hydrology; monitoring; perennial streams; sedimentation; Shebelle River; Somali Republic; stream transport; streams; surveys; transport; wadis; water quality. OCLC Accession Number: 34175029.

Maxamed, Cabdulqaadir Maxamuud. 1987. "Some Thoughts on the Tethyan Characters of the Somali Shallow Benthic Faunas." GeoSom 87; International Meeting; Geology of Somalia and Surrounding Regions; Abstracts. Somalia: Somali Natl. Univ., Mogadishu, Somalia. GeoSom 87; Geology of Somalia and Surrounding Regions, Mogadishu. Somalia Conference: Nov. 23-30, 1987. Descriptors: Africa; benthonic taxa; biogeography; Bivalvia; Brachiopoda; Cenozoic; Cretaceous; East Africa; Echinodermata; Echinoidea; Echinozoa; Europe; faunal studies; Foraminifera; Gastropoda; Indian Ocean; Invertebrata; Jurassic; Mediterranean region; Mesozoic; microfossils; middle Miocene; Miocene; Mollusca; Neogene; paleogeography; plate collision; Protista; shallow-water environment; Somali Republic; stratigraphy; Tertiary; Tethys; Western Europe. OCLC Accession Number: 34175029.

Mazov, Sergei. 2003. The USSR and the Former Italian Colonies, 1945-50. Cold War History. April 2003, Vol. 3 Issue 3, pages 49-78. Subject Terms: International Relations. Geographic Terms: Eritrea; Italy; Libya; Somalia; Soviet Union. Abstract: This article studies the motives determining Soviet leadership's actions concerning the former Italian colonies. Analyzing the body of documents from the Archive of Russian Foreign Policy (the Foreign Ministry Archive), the author came to the conclusion that Realpolitik imperatives clearly dominated ideological considerations on this issue. The architects of the Soviet foreign policy were interested in the geostrategic position of Libya, Somalia and Eritrea, rather than in the prospects for revolutionary development there. Soviet diplomacy showed remarkable persistence and flexibility (the official position was changed four times), trying to achieve a Soviet presence in the Mediterranean- a traditional priority region of Russian geopolitical aspirations. ISSN: 1468-2745. Database: Military & Government Collection. Persistent link to this record: http://search.epnet.com/login.aspx?direct=true&db=mth&an=11407113&login page=Login.asp&scope=site

Mazzocchi, Alemannin. 1927. "Il Nuovo Problema Del Giuba." Translated title: "The New Problem of the Juba." Italy: Istituto Agricolo Coloniale Italiano, Florence, Italy. Descriptors: Africa; East Africa; Juba River; Somali Republic; water resources. Database: GeoRef. Accession Number: 1999-004813.

Mazzocchi, Alemannin. 1919. "I Nostri Grandi Problemi Coloniali; Lo Sbarramento Del Giuba." Translated title: "Our Big Colonial Problems; Juba Dam." L'Agricoltura Coloniale. Volume 10, Descriptors: Africa; dams; East Africa; Juba Dam; Juba River; Somali Republic; water resources. ISSN: 0394-2945.

Mazzocchi, G. 1973. "Note Minerarie Sulla Somalia." Translated title: "Mining Note on Somalia." Industria Mineraria. Volume 1973, Descriptors: Africa; East Africa; mining; mining geology; Somali Republic. ISSN: 0019-7696.

McCarthy, J. W. Clapp-Wincek, C. Londner, S. and Thomas, A. 1985. "Soil and Water Conservation Project in Two Sites in Somalia: Seventeen Years Later." Washington, DC. Center for Development Information and Evaluation; United States: Agency for International Development. Aug. Volume: Aid Project Impact Evaluation 62; AIDPNAAL064; PB87206553, pages: 57. Descriptors: Soil conservation; Water conservation; Somalia; Developing countries; Irrigation; Agricultural engineering; Semi-arid land; Developing country application. Abstract: In the early 1960s, the Agency for International Development (AID) and the Somalia Ministry of Agriculture and Animal Husbandry carried out a soil and water conservation project, the bunding of a cultivated watershed in the northwest. The project was intended to deliver conservation of the eroding topsoil and farm irrigation by arresting and concentrating the rainwater that runs off the hills during spring storms. The project aimed to demonstrate an easily replicable technology, water gathering, which could substantially increase grain production in other similar areas during years of subnormal to good rainfall, although not during drought. An evaluation showed that the scale of the AID bunding project was not so great as to create major adverse impacts on local herding; it enhanced the dry-farming activity without seriously supplanting the area's economic mainstay. However, the policy most likely to lead to long-term food security is probably not a global drive to intensify and extend crop agriculture, but one that exploits each area's profile of productive strengths and weaknesses to meet its profile of nutritional needs. Database: NTIS. Accession Number: PB87206553.

McCarthy, John W. United States and Agency for International Development. 1985. A Soil and Water Conservation Project in Two Sites in Somalia Seventeen Years Later. Washington, D.C. U.S. Agency for International Development. Descriptors: 1 v. (various pagings); Soil conservation- Somalia; Water conservation- Somalia. Notes: "August 1985." "PN-AAL-064" on cover. Bibliography: last p. Reproduction: Microfiche. Washington. D.C.?: SUPTDOCS/GPO, [1986]. 1 microfiche: negative. OCLC Accession Number: 14756496; 14686797.

McCoy D.E. 2000. "American Post-Cold War Images and Foreign Policy Preferences Toward 'Dependent' States: A Case Study of Somalia." World Affairs.

Volume 163, Issue 1, Pages 39-47. Descriptors: Political; International Relations, International Financial Institutions; foreign policy; international relations; new world order; post-Cold War. Notes: Additional Info: United States; Geographic: Somalia United States. Abstract: The analytical tasks of this case study were twofold: to determine the operative image of Somalia and to test the impact of that image, if any, on America's subsequent policy. The results highlight the existence of both a positive American self-image and a dependent image of Somalia. I also found the anticipated strategic and tactical preferences that I hypothesized would be associated with such images. Undoubtedly, the new world order does not indicate a new world view. ISSN: 0043-8200.

McCreary Jr., Julian P., Kohler, Kevin E., Hood, Raleigh R. and Olson, Donald B. 1996. "A Four-Component Ecosystem Model of Biological Activity in the Arabian Sea." Prog. Oceanogr. Volume 37, Issue 3-4, Pages 193-240. Abstract: A coupled, physical-biological model is used to study the processes that determine the annual cycle of biological activity in the Arabian Sea. The physical model is a 21/2-layer system with a surface mixed layer imbedded in the upper layer, and fluid is allowed to move between layers via entrainment, detrainment and mixing processes. The biological model consists of a set of advective-diffusive equations in each layer that determine the nitrogen concentrations in four compartments: nutrients, phytoplankton, zooplankton and detritus. Coupling is provided by the horizontal-velocity, layer-thickness, entrainment and detrainment fields from the physical solution. Surface forcing fields (such as wind stress and photosynthetically active radiation) are derived from monthly climatological data, and the source of nitrogen for the system is upward diffusion of nutrients from the deep ocean into the lower layer. Our main-run solution compares favorably with observed physical and biological fields; in particular, it is able to simulate all the prominent phytoplankton blooms visible in the CZCS data. Three bloom types develop in response to the physical processes of upwelling, detrainment and entrainment. Upwelling blooms are strong, long-lasting events that continue as long as the upwelling persists. They occur during the Southwest Monsoon off Somalia, Oman and India as a result of coastal alongshore winds, and at the mouth of the Gulf of Aden through Ekman pumping. Detrainment blooms are intense, short-lived events that develop when the mixed layer thins abruptly, thereby quickly increasing the depth-averaged light intensity available for phytoplankton growth. They occur during the fall in the central Arabian Sea, and during the spring throughout most of the basin. In contrast to the other bloom types, entrainment blooms are weak because entrainment steadily thickens the mixed layer, which in turn decreases the depth-averaged light intensity. There is an entrainment bloom in the central Arabian Sea during June in the solution, but it is not apparent in the CZCS data. Bloom dynamics are isolated in a suite of diagnostic calculations and test solutions. Some results from these analyses are the following. Entrainment is the primary nutrient source for the offshore bloom in the central Arabian Sea, but advection and recycling also contribute. The ultimate cause for the decay of the solution's spring (and fall) blooms is nutrient deprivation, but their rapid initial decay results from grazing and self shading. Zooplankton grazing is always an essential process, limiting phytoplankton concentrations during both bloom and oligotrophic periods. Detrital remineralization is also important: in a test solution without remineralization, nutrient levels drop markedly in every layer of the model and all blooms are severely weakened. Senescence, however, has little effect: in a test solution without senescence, its lack is almost completely compensated for by increased grazing. Finally, the model's detrainment blooms are too brief and intense in comparison to the CZCS data; this difference cannot be removed by altering biological parameters, which suggests that phytoplankton growth in the model is more sensitive to mixed-layer thickness than it is in the real ocean. ISSN: 0079-6611.

McMullen R.K. 1993. "Somaliland: The Next Eritrea?" Low Intensity Conflict & Law Enforcement. Volume 2, Issue 3, Pages 421-433. Descriptors: Political; Political systems and political change; developing country; political geography; international acceptability; independence claim; state formation; secessionism; state collapse. Notes: Somalia- Somaliland. Abstract: Whether or not Somaliland will follow in Eritrea's footsteps depends at least in part on its ability to meet the emerging criteria of an internationally acceptable bid for statehood. In general, the relative merits of the secessionist claim against the authority of the mother state must be weighed against the potential negative consequences of the separatist cause. Positive arguments for secession might include the claim of unjust incorporation, drastic oppression and/or severe economic discrimination by the mother state, or the need to protect a threatened culture or people. Various potential negative consequences include the impoverishment or geographical discontinuity of the mother state, the creation of double minorities and the establishment of an illiberal ethnocracy. Somaliland's recent bid for statehood has been structured in part by the political dynamics of two seminal events- independence and unification in 1960 and the collapse of the state of Somalia three decades later. Following a brief analysis of the continuing political importance of these two critical junctures, Somaliland's potential for successfully challenging the authority of the Somali state and gaining international acceptability will be assessed using the criteria outlined above.

Mehmet, O. 1971. "Effectiveness of Foreign Aid-the Case of Somalia." Journal of Modern African Studies. Volume P 31-47, Pages 14 REF. Descriptors: Arid Lands; Foreign Lands; Water Resources Development; Economic Feasibility; Financial Feasibility; Political Aspects; Planning; Somalia; Developing Countries; International Aid. Abstract: the arid nation of somalia is one of the poorest nations in the world, with an estimated annual per capita income of 50 dollars. However, it is one of the world's largest recipients of foreign aid, receiving an annual average of 15 dollars/per capita in 1965-1969, which was more than 3 times the average annual aid to other developing countries during the same period. Transportation, agriculture and industrial development received the largest allocations, while education and water supplies received surprisingly small allocations. Drought, resulting in reduced grain production and heavy livestock losses, combined with border incidents with Ethiopia and the eventual takeover by a military dictatorship, tended to retard development. While foreign aid came from many sources and was designated for many different developmental projects, all indications are that it was not for completely altruistic purposes. Much of it reflects cold-war rivalry and the strategic importance of somalia dominating the entrance into the red sea and facing the Indian ocean. American aid was mostly for deep water ports and construction of electricity and water supplies. In most cases, funds for aid came tied to unfavorable financial conditions, and with the inability of the country to develop fast enough, such aid will burden the country with almost impossible debt servicing obligations. Database: Environmental Sciences and Pollution Mgmt.

Melbinger, E. 1987. Caritas Somalia Water Project; Final Report. Somalia: Caritas, Mogadishu, Somalia. Descriptors: Africa; Carita Somalia Water Project; East

Africa; Somali Republic; water resources. Database: GeoRef. Accession Number: 1999-004816.

Melbinger, Ernst; Caritas Somalia and Water Project. 1987-1988. Caritas Somalia Water Project: Final Report. Mogadishu: The Project. Descriptors: 2 v. Wells- Somalia; Drinking water- Somalia; Water-supply- Somalia. Abstract: [1. Without special title- 2] Phase II. LCCN: 90-980213; OCLC Accession Number: 30781888.

Mellor, W. J. 1995. "Operation Restore Hope: The Australian Experience." Carlisle Barracks, PA; United States: Army War College. Volume: ADA2947042, page(s): 41. Descriptors: Warfare; Army personnel; Pacific ocean islands; Symposia; Coastal regions; Water; Population; Availability; Wounds and injuries; Somalia; Agreements; Army; Australia. Abstract: In the early hours of 17 Feb 93, an Australian Army section, on patrol in the town of Baidoa in central southern Somalia, was fired upon as it approached the town main water point. In the ensuing firefight, three Somalis were wounded, one of whom subsequently died. This small action marked the first time since the end of the Vietnam War that Australian soldiers had engaged in an exchange of hostile fire, and it came, ironically, while engaged on a massive humanitarian mission to save Somali lives. I had the privilege of being the Commander of the Australian contingent on Operation Restore Hope, the UN sanctioned, US led effort to provide security for the provision of humanitarian relief for the population of Somalia. This paper will address some of the issues which arose during the planning and execution of the Australian contribution, which was known as Operation Solace. The observations made are, of course, entirely my own, but with the limited availability of Australian documentation in my present circumstances, they are also made largely from memory, and readers should bear that in mind at all times. Database: NTIS. Accession Number: ADA2947042.

Menenti, M. Azzali, S. Verhoef, W. and Vanswol, R. 1991. "Mapping Agro-ecological Zones and Time Lag in Vegetation Growth by Means of Fourier Analysis of Time Series of NDVI Images." Soil and Water Research, Wageningen (Netherlands); Netherlands: Winand Staring Centre for Integrated Land. Volume: BCRS9122; ETN9291611; N92297175, pages: 35. Descriptors: Fourier analysis; Mapping; Remote sensing; Vegetation growth; Agriculture; Normalizing (Statistics); Satellite imagery; Vegetative index; Classifications; Environmental monitoring; Maps; Qualitative analysis; Quantitative analysis; Somalia; Zambia; Foreign technology. Abstract: A method which gives a quantitative meaning to general concepts such as 'variability of vegetation growth', 'growth cycle', and 'early or late growing season' is presented. This was accomplished by applying Fourier analysis to time series of 10 day composites of Normalized Difference Vegetation Index (NDVI) images of Zambia and Somalia. Time series of NDVI images are deconvoluted in a set of harmonic functions having different periods (or frequency). The amplitude of each harmonic gives the contribution of NDVI variation in time of the actual observations; for each harmonic the time lag with respect to the origin in time on the observations is obtained. Since the technique was applied on a pixel by pixel basis, maps were obtained of amplitude and time lag for each period. Different techniques are compared to map zones wherein vegetation development is similar and which will be termed 'isogrowth' zones. The obtained results indicate that differences in vegetation development among map units of an existing agroclimatic map are not significant, while reliable differences are observed among the map units obtained by means of the described procedure. Notes: Sponsored by Beleidscommissie Remote Sensing. NTIS Accession Number: N92297175.

Menkhaus K. 2003. "State Collapse in Somalia: Second Thoughts." Rev. Afr. Polit. Econ. Volume 30, Issue 97, Pages 405-422. Descriptors: Political; Conflict, protest, human rights; elite politics; political instability; state role; crime. Notes: Additional Info: United Kingdom; References: 51; Geographic: Somalia East Africa Sub-Saharan Africa. Abstract: Somalia's protracted crisis of complete state collapse is unprecedented and has defied easy explanation. Disaggregating the Somali debacle into three distinct crises- collapse of central government, protracted armed conflict, and lawlessness- helps to produce more nuanced analysis. Significant changes have occurred in the nature and intensity of conflict and lawlessness in Somalia since the early 1990s, with conflicts becoming more localized and less bloody, and criminality more constrained by customary law and private security forces. These trends are linked to changing interests on the part of the political and economic elite, who now profit less from war and banditry and more from commerce and service business that require a predictable operating environment. The prolonged collapse of Somalia's central government cannot be explained as a reflection of local interests. The country's elite would profit greatly from the revival of a recognized but ineffective 'paper' state. The inability of Somalia's leaders to cobble together such a state is best explained as a product of risk aversion. Political and economic actors in collapsed states fear a change in the operating environment which, though far from ideal, is one in which they have learned to survive and profit. ISSN: 0305-6244.

Menkhaus K and Prendergast J. 1995. "Governance and Economic Survival in Post-Intervention Somalia." Trocaire Development Review. Volume 1995, Pages 47-61. Descriptors: International Development Abstracts; developing country; military intervention; geopolitical studies; political instability; aid role; economic survival; international aid; governance; aid policy; political context; sustainable development. Notes: Somalia. Abstract: The international community, the UN, governments and donor agencies have traditionally viewed Somalia's localized or clan politics as anarchic, and therefore as an obstacle to long-term peaceful and sustainable development. This article sets out the challenges and opportunities facing them in working within rather than against Somalia's "stateless' political reality. In choosing to work with local structures it is essential that the legitimacy of various types of structures is evaluated based on clear criteria such as their local support, performance and adherence to principles of "good government'. The authors argue that the future role for aid and diplomacy should centre around engaging with local structures, and supporting Somali initiatives to rebuild their country from the ground up.

Merkouriev, S. and DeMets, C. 2006. "Constraints on Indian Plate Motion since 20 Ma from Dense Russian Magnetic Data: Implications for Indian Plate Dynamics." Geochemistry Geophysics Geosystems. FEB 2. Volume 7, Pages Q02002. Abstract: [1] We use more than 230,000 km of Russian marine magnetic and bathymetric data from the Carlsberg and northern Central Indian ridges, comprising one of the most geographically extensive, dense shipboard surveys anywhere in the ocean basins, to describe in detail seafloor spreading since 20 Ma along the trailing edge of the Indian plate. India-Somalia plate rotations for similar to 1 Myr intervals over the past 20 Myr are derived from inversions of more than 6600 crossings of 20 magnetic reversals and similar to 1400 crossings of fracture zones that offset these two ridges. Statistical analysis of the numerous data indicates that outward displacement of reversal boundaries due to finite seafloor emplacement widths and correlated noise for anomaly crossings from individual spreading segments constitute two distinct sources of systematic bias in the locations of magnetic anomaly crossings, contrary to the often-made assumption that random, Gaussian-distributed noise dominates the error budget. Seafloor spreading rates slowed gradually by 30% from 20 Ma to 10 +/- 1 Ma about a relatively stationary pole of rotation. From 11 Ma to 9 Ma the rotation axis migrated several angular degrees toward the plate boundary, modestly increasing the spreading gradient along the plate boundary. India-Somalia kinematic data for times since similar to 9 Ma are consistent with remarkably steady motion, with no evidence for a change in either the rotation pole or rate of angular opening within the few percent precision of our data. The timing and nature of changes in India-Somalia motion since 20 Ma closely resemble those for the Capricorn-Somalia plate pair, indicating that India and Capricorn plate motions are strongly coupled. We speculate that the slowdown in seafloor spreading at the trailing edges of the Indo-Capricorn composite plate from 20 Ma to 10 +/- 1 Ma resulted from the increasing amount of work that was needed to build topography in the Himalayan collisional zone. The transition to stable India-Somalia and Capricorn-Somalia seafloor spreading at similar to 10- 9 Ma corresponds well with the onset at 8 Ma of folding and faulting across an equatorial plate boundary separating the Indian and Capricorn plates, suggesting that the latter may have played a fundamental role in restoring equilibrium between the torques that were driving and resisting the northward motions of the Indian and Capricorn plates.

Merla, Giovanni and ConsiglioNnazionale delle Ricerche (Italy). 1979. A Geological Map of Ethiopia and Somalia (1973): 1:2.000.000, and Comment, with a Map of Major Landforms. Firenze; Elmsford, N.Y: Centro Stampa; distributed by Pergamon Press. Descriptors: Geology- Ethiopia; Geology- Somalia. Notes: At head of title: Consiglio nazionale delle ricerche, Italy. Two maps on folded leaves in pocket. Bibliography: p. 89-95. colored Geological Map Measuring 120 x 80 Cm Accompanied by Explanatory Notes (95 Pp.) Including 36 color Pictures and a Four-Colour Map of the Same Area on Scale 1/3.000.000 Featuring Major Landforms." 1980. Earth-Science Reviews. Volume 16, Pages 388-388. ISBN: 0080240283; LCCN: 80-40420.

Merla, G., et al. 1979. "Ethiopian-Somalian Geology." A Geological Map of Ethiopia and Somalia (1973) at 1:2000 000. (Plus an Explanatory Text Including 22 Black and White Illustrations and 37 color Illustrations and a 1:3000 000 Map of Major Landforms) Congsiglio Nazionale Delle Ricerche, Italy. Distributed by Pergamon Press, Oxford. 117 Pp. 1980. Journal of Structural Geology. Volume 2, Issue 4, Pages 472-473.

Merla, Giovanni and Consiglio nazionale delle ricerche (Italy). 1973. "Geological Map of Ethiopia and Somalia." Firenze, Italy: Consiglio nazionale delle ricerche. Descriptors: Geology- Ethiopia- Maps; Geology- Ethiopia; Geology- Somalia- Maps; Geology- Somalia; Government publication; National government publication. Notes: Description: 1 map; color; 124 x 182 cm. Scale 1:2,000,000; (E 340-E 560/N 180-N 20). Accompanied by: A Geological map of Ethiopia and Somalia (1973), 1:2,000,000 and comment with a map of major landforms by Giovanni Merla ... [et al.]. 1979. (viii, 95 pages, [7] leaves of plates: ill., maps (1 folded in pocket)). Label mounted on verso of t.p. of accompanying text: Distributed by Pergamon Press, Elmsford, N.Y. Includes bibliographical references (p. 89-95 of text). Other Titles: Carta geologica dell'Etiopia e della Somalia. Geological map of Ethiopia and Somalia (1973), 1:2,000,000 and comment with a map of major landforms. ISBN: 0080240283.

Messana, Giuseppe and Chelazzi, Lorenzo. 1986. "The Fauna of the Subterranean Waters of East Africa, and Particularly of Somalia." Stygologia. Dec. Volume 2, Issue 4, Pages 339-351. Descriptors: Africa; East Africa; faunal studies; ground water; hydrogeology; Somali Republic; surveys. Notes: References: 61. ISSN: 0169-3662.

Meumann, Oscar. 1902. "From the Somali-Coast through Southern Ethiopia to the Sudan." Geogr. J. Volume 20, no. 4, Descriptors: Africa; East Africa; Ethiopia; expeditions; geography; Somali Republic; Sudan. ISSN: 0016-7398.

Metz, H. C. 1992. "Area Handbook Series. Somalia a Country Study." Washington, DC. Federal Research Div. Sponsor: Department of the Army, Washington, DC; United States: Library of Congress. May. Volume: ADA2649978, page(s): 316. Descriptors: Somalia; Case studies; Religion; Economics; Government; Military forces; Societies; History; Environments; Geography; Political science; Political parties; Political revolution; Civilian population. Abstract: No abstract available. Database: NTIS. Accession Number: ADA2649978.

Meyer, W. R. 1994. "Somalia Environmental Monitoring of Ammunition in Open Storage, Fall 1993. Final Rept." Army Defense Ammunition Center and School, Savanna, IL. Validation Engineering Div. Sep. Volume: USADACS9314, pages: 51. Descriptors: Corrosion; Monitoring; High temperature; Atmospheric temperature; Field tests; Long range Time; Reliability; Data acquisition; Solar radiation; Surface to surface missiles; Surveillance; Safing and arming Ordnance; Humidity; Stockpiles; Somalia; Storage; Antitank missiles; Cartridges; Tow 2 missiles; Tank ammunition; Missile technology; Ordnance Ammunition explosives and pyrotechnics. Abstract: The U.S. Army Defense Ammunition Center and School (USADACS), Validation Engineering Division (SMCAC-DEV), was tasked by US. Army Armament Research, Development and Engineering Center (ARDEC) to monitor ammunition in open storage in Somalia. This was an ongoing test of ARDEC, Predictive Technology Branch, surveillance program to ensure long-term reliability of ammunition in the field. This report contains test results of selected ammunition monitored during November 1993. DTIC Accession Number: ADA2947786XSP.

Meyers, R. J. and Murrell, K. L. 1986. "Divestiture of Somali Marine Products: Privatization in Somalia." Center for Privatization, Washington, DC. Funder: Agency for International Development, Washington, DC. 24 Sep. Volume: AIDPNAAV907, pages: 48. Descriptors: Socioeconomic factors; Investments; Financing; Decentralization Dispersal; Developing countries; Somalia; Fisheries; Economic development; Entrepreneurship; Commercial development; Developing country application; Privatization; Business and economics Foreign industry development and economics; Agriculture and food Fisheries and aquaculture. Abstract: The Center for Privatization (CFP) was awarded a contract through USAID's Private Investment Bureau (PRE) to survey the privatization potentials in the developing world. CFP sent a team of two to Somalia for twelve days to survey the sociopolitical, financial and resources-based potentials for privatization of fisheries. The report details their findings and recommendations. Funder: Analysis Group, Inc., Washington, DC. NT: Sponsored by

Agency for International Development, Washington, DC., and Analysis Group, Inc., Washington, DC. NTIS Accession Number: PB87213724XSP.

Michelsen A and Rosendahl S. 1990. "Propagule Density of VA-Mycorrhizal Fungi in Semi-Arid Bushland in Somalia." Agric., Ecosyst. Environ. Volume 29, Issue 1-4, Pages 295-301. Descriptors: bushland; vesicular-arbuscular mycorrhizae Species Term: Cucumis sativus; Acacia nilotica; Leucaena leucocephala. Abstract: VAM propagule densities of clay soils from bush land in various stages of degradation and from cultivated areas were determined. Use of Cucumis sativus, Acacia nilotica and Leucaena leucocephala as host plants in assays was compared, proving C. sativus to give the highest propagule counts. Associations of vegetation with a high degree of cover showed higher propagule densities than degraded bushland and cultivated areas. Subsoils showed very low VAM propagule counts.

Michelsen, Anders. 1992. "Mycorrhiza and Root Nodulation in Tree Seedlings from Five Nurseries in Ethiopia and Somalia." Forest Ecology and Management. April. Volume 48, Issue 3-4, Pages 335-344. Abstract: Mycorrhizal fungi and nodulating bacteria are important for tree seedling establishment in degraded areas in the tropics. In this survey the presence and abundance of mycorrhizal fungi and the presence of nodulating bacteria were investigated in roots of tree seedlings and in soils from five nurseries in Ethiopia and Somalia.Each of the 41 tree species, 11 indigenous and 30 exotic, were colonized by mycorrhizal fungi in at least one of the nurseries. In two of the five nurseries, however, the average mycorrhizal colonization level of the seedlings was very low, as was the case for the density of propagules of vesicular-arbuscular mycorrhizal (VAM) fungi in the soil of one of these two nurseries. In that nursery none of the leguminous trees were nodulated either.The importance of maintaining a large and infective population of mycorrhizal fungi and rhizobia in nursery soil is discussed, and simple nursery procedures which could promote the colonization of tree seedlings by indigenous rhizobia and mycorrhizal fungi are suggested. ISSN: 0378-1127.

Middle West Service Company, Chicago, IL, United States. 1971. Mogadiscio Water Agency. United States: Middle West Serv. Co., Chicago, IL, United States. Descriptors: Africa; Benadir; East Africa; Mogadishu Somali Republic; Mogadishu Water Agency; Somali Republic; water resources. Database: GeoRef. Accession Number: 1999-004822.

Mifsud, M. F. and Wilkinson, K. G. 1984. Assistence in Water Legislation; Somalia; Report on Proposed National Water Resources Law. Italy: FAO, Rome, Italy. Descriptors: Africa; East Africa; legislation; regulations; Somali Republic; water resources. Database: GeoRef. Accession Number: 1999-004823.

Migliorini, C. 1945. Geological Work Carried Out by AGIP in Italian Somaliland and Neighbouring Territories. Italy: Univ. Firenze, Institute di Geologia, Florence, Italy. Descriptors: Africa; AGIP; areal geology; East Africa; research; Somali Republic. Database: GeoRef. Accession Number: 1996-032853.

"Military communications map, Somaliland." 1941. Ed. 1. Scale: 1:2 000 000. Nairobi: East Africa Survey Group, 1941. 2 maps: color Note(s): These maps do not have EAF numbers. Subjects: Somalia- Maps and charts; Ethiopia- Maps and charts. Added Name: Great Britain. War Office. General Staff. Geographical Section. 2 sheets. British Library: Maps MOD EAF Somali mil comms.

Mineraria Somala, Mogadishu, Somalia. 1963. Relazione Sull'Attivita Svolta Negli Anni 1953-1963. Report on the Activity between the Years of 1953-1963. Somalia: Mineraria Somala, Mogadishu, Somalia. Descriptors: Africa; development; East Africa; mining; mining geology; Somali Republic. Database: GeoRef. Accession Number: 1999-067960.

Mister, Robert. 1985. Participation in Refugee Development Programmes: Contrasting Experience from Somalia. Oxford: Oxfam. Descriptors: 19; Refugees-Services for- Somalia; Community development- Somalia; Water-supply- Somalia; Economic development; Refugee aid and development; Durable solutions; Water supply; Agriculture; Self-help projects. Notes: Named Corp: UNHCR. Notes: "A paper presented to a workshop on the development approach to refugee assistance jointly organized by UNHCR and ICVA" "November, 1985". Accession Number: 68401579.

Mo, Yanqing. 1990. An Assessment of Petroleum Potential and Economic Feasibility in Northern Somalia. Descriptors: 209 leaves; Petroleum- Geology- Somalia. Notes: Typescript. Includes bibliographical references (leaves 205-209). Dissertation: Thesis (masters')--University of South Carolina, 1990. OCLC Accession Number: 24176385.

[Mogadishu] Mogadiscio Region (1MB) and map key and scale (354K) Portion of AMS Series 2201, Sheet 25 "Mogadiscio", Original scale 1:2,000,000, U.S. Army Map Service, 1968. See: http://www.lib.utexas.edu/maps/africa/mogadiscio_1968.jpg

[Mogadishu] Mogadiscio Region Portion of sheet NA-38-67, Original scale 1:200,000, U.S. Defense Mapping Agency, 1992 (586K). See: http://www.lib.utexas.edu/maps/africa/mogadishu_1992.jpg

Mohamed Ahmed, M. M. and Dairri, M. F. 1987. "Trypanosome Infection Rate of Glossina Pallidipes during Wet and Dry Seasons in Somalia." Trop. Anim. Health Prod. Feb. Volume 19, Issue 1, Pages 11-20. Descriptors: Age Factors; Animals; Female; Fertility; Male; Rain; Seasons; Sex Factors; Sex Ratio; Somalia; Trypanosoma- isolation & purification; Tsetse Flies- parasitology & physiology. Abstract: Using biconical traps the distribution, population composition, insemination rate, pregnancy, age structure and trypanosome infection rate of G. pallidipes were studied during the wet season 1982 and the dry season 1983 at Mareerrey Somalia. Flies restricted to the riverine gallery forest in the dry season become dispersed into approximately 1 km of the Acacia thickets in the wet season. There was no significant variation in population components of G. pallidipes between wet and dry seasons. The male:female ratio remained at 0.5:1 in both seasons. The insemination rate of females was significantly high during the wet season. Pregnancy stages were not statistically different between the two study periods. A significantly greater proportion of females that were below age category 4 were found in the wet season. The trypanosome infection rate was 2.6% and 1.5% in the wet and dry seasons respectively. Flies were found infected with only the vivax and congolense group trypanosomes. A linear positive correlation existed between the trypanosome infection rate and the physiological age of females. ISSN: 0049-4747.

Mohamed Ali Hussein; Hjort, Anders and Sweden. Beredningen för u-landsforskning. 1986. Survival in Arid Lands: The Somali Camel Research Project. Stockholm: Swedish Agency for Research Cooperation with Developing Countries. pages: 15. Descriptors: Somali Camel Research Project; Camels; Somalia; Arid regions Research. Notes: Mohamed Ali Hussein, Anders Hjort. Cover title. ISBN: 9186826026.

Mohamed J. 2004. "The Political Ecology of Colonial Somaliland." Africa. Volume 74, Issue 4, Pages 534-566. Descriptors: Social History; political change; environmental degradation; colonialism; nineteenth century; environmental history. References: Number: 59; Abstract: The social basis of ecological change in Somaliland during the colonial period was politics, especially imperial politics: the division of the Somali country into various colonial spheres, the loss of territory under the 1897 Anglo-Ethiopian Treaty, and the pacification wars. These events, as it were, reduced the land available for use by the pastoralists, which led to overgrazing, soil erosion and ecological degradation. Moreover, the income of the population declined throughout the colonial period. Even though during the late colonial period the 'nominal' price of pastoral goods increased, the 'real' price of pastoral commodities did not increase to cover the loss of income caused by inflation and the high cost of imported goods. These two processes- on the one hand ecological degradation and on the other the decline of income- could be understood if they were read contrapuntally. Such reading is possible only if we give full attention to political ecology: why ecology had changed, the politics of that change, and the impact it had on the income and everyday life of the population. ISSN: 0001-9720.

Mohamed, A. A. 2003. "Study of Larvivorous Fish for Malaria Vector Control in Somalia, 2002." East. Mediterr. Health J. Jul. Volume 9, Issue 4, Pages 618-626. Descriptors: Animals; Anopheles- parasitology and physiology; Attitude to Health; Breeding; Cichlids-physiology; Culex- parasitology and physiology; Humans; Insect Vectors- parasitology and physiology; Larva- parasitology & physiology; Malaria-epidemiology and parasitology- prevention & control transmission; Male; Mosquito Control methods; Pest Control, Biological methods; Seasons; Somalia-epidemiology; Water Supply. Abstract: An intervention study was conducted on the introduction of the larvivorous fish Oreochromis spilurus as a method of malaria vector control in Kalabeydh village, northern Somalia. This species is resistant to chlorine in water up to a concentration of 1.0 mg/L. Fish were introduced into 25 berkit (reservoirs). After 1 month the number of larvae in each berkit was reduced by between 16.5% and 78.6% (mean 52.8%). Community acceptance and participation was good. The introduction of larvivorous fish is a cheap method of malaria vector control, but its sustainability needs special consideration and education of the community is important, especially to cover the dry season, when most of the berkit dry up. ISSN: 1020-3397.

Mohamed, Abdi Mohamed. 1978. Studio Idrologico Nel Bacino Dello Uebi Scebeli. Hydrologic Study in the Shebelle River Basin. Somalia: Univ. Naz. Somala, Dip. Geologia, Mogadishu, Somalia. Descriptors: Africa; East Africa; hydrology; rivers and streams; Shebelle River; Somali Republic; surface water. Database: GeoRef. Accession Number: 1999-067975.

Mohamed, Ali A. 1986. Studio Idrochimico Delle Acque Del Distretto Di Sablaale. Hydrochemical Study of the Water of Sablaale District. Somalia: Univ. Naz. Somala, Dip. Geol., Mogadishu, Somalia. Descriptors: Africa; East Africa; geochemistry; hydrochemistry; hydrology; Jubbada Hoose; Sablaale Somali Republic; Somali Republic; surface water; water. Database: GeoRef. Accession Number: 1999-067976.

Mohamed, Mohamed Abdi. 1987. Toward a Proper use of Different Water Resources; Application to the Low Juba Region, Southern Somalia. Somalia: Somali Natl. Univ., Mogadishu, Somalia. GeoSom 87; Geology of Somalia and Surrounding Regions, Mogadishu. Somalia Conference: Nov. 23-30, 1987. Descriptors: Africa;

drainage patterns; East Africa; economic geology; ground water; hydrogeology; hydrology; Juba; Juba River; salinity; Somali Republic; southern Somali Republic; surface water; surveys; water management; water quality; water resources; water wells; wells. OCLC Accession Number: 34175029.

Mohr, P. A. 1963. Geologic Map of Horn of Africa, Scale 1:2,000,000. United Kingdom: Philip and Tacey, London, United Kingdom. Descriptors: Africa; East Africa; geologic maps; Horn of Africa; maps; Somali Republic. Database: GeoRef. Accession Number: 1999-067978.

Mohr, P. A. and Smithsonian Astrophysical Observatory, Cambridge, MA. 1974. "Mapping of the Major Structures of the African Rift System (Ethiopia, Somalia, and Yemen)." Volume: E75-10024; NASA-CR-140738; SAO-SPECIAL-REPT-361; SAO-403-004; Pagination 90P, Descriptors: African Rift System; Ethiopia; Photomapping; Somalia; Structural Properties (Geology); Yemen; Earth Resources Program; Geological Faults; Lithology; Photogeology; Precambrian Period; Tectonics. Abstract: The author has identified the following significant results. The ERTS-1 satellite imagery has facilitated a major advance in accurate mapping and better understanding of the African Rift valleys. A unified scheme of mapping of the whole rift system, from Malawi to Ethiopia, has been accomplished. The structures revealed by the imagery are discussed in the light of known ground truth for the northern half of the African Rift System. The ERTS-revealed structures confirm the likelihood of a drift of Arabia away from Africa and impose constraints on the drift vectors. Numerous features have been newly recognized from the imagery: the gently curvilinear plan of virtually all the African Rift valleys, the pervasive but not overriding degree of influence of Precambrian structures on rift faulting, the deep structural control of many rift volcanic centers, the occurrence of several unsuspected calderas both in the rifts and on the plateaus, and massive cauldron subsidence phenomena previously unknown to eastern Africa. Lithological mapping in eastern Africa can be greatly refined using the ERTS imagery, and new or revised subdivisions of the Precambrian appear possible and, in some places, necessary. Valleys excavated in the Ethiopian highlands by Pleistocene glaciers are newly recognized. Contract: NAS5-21748; ERTS. Database: CSA Technology Research Database. Accession Number: N75-13340 (AH).

Mokma, D. L. and Abdulla, O. H. D. 1984. "Classification of Five Pedons Near Afgoi, Somalia." Geoderma. Volume 33, Issue 3, Pages 219-226. Abstract: Five pedons representing the major soils on the Central Agricultural Research Station located on the Shabelle River floodplain near Afgoi (Somalia) were described and sampled. The samples were analyzed for particle-size distribution, pH, calcium carbonate equivalent, electrical conductivity and sodium adsorption ratio. According to Soil Taxonomy, three pedons were classified as Udic Chromusterts, very fine, isohyperthermic; one as a Udic Chromustert, fine isohyperthermic; and one as a Typic Ustifluvent, fine-loamy isohyperthermic. Four of the pedons were classified as Chromic Vertisols and the fifth as a Calearic Fluvisol according to legend of the FAO-Unesco soil map of the world.

Montanari, L. 1963. Orientamenti Per La Geologia Della Somalia. Translated title: Orientation on Somali Geology. Somalia: Univ. Naz. Somala, Mogadishu, Somalia. Descriptors: Africa; areal geology; East Africa; Somali Republic. Database: GeoRef. Accession Number: 1999-067979.

Moore, P. S., Marfin, A. A. and Quenemoen, L. E., et al. 1993. "Mortality Rates in Displaced and Resident Populations of Central Somalia during 1992 Famine." The Lancet. April 10, 1993. Volume 341, Issue 8850, Pages 935-938. Abstract: Famine and civil war have resulted in high mortality rates and large population displacements in Somalia. To assess mortality rates and risk factors for mortality, we carried out surveys in the central Somali towns of Afgoi and Baidoa in November and December, 1992. In Baidoa we surveyed displaced persons living in camps; the average daily crude mortality rate was 16.8 per 10,000 population during the 232 days before the survey. An estimated 74% of children under 5 years living in displaced persons camps died during this period. In Afgoi, where both displaced and resident populations were surveyed, the crude mortality rate was 4.7 deaths per 10,000 per day. Although mortality rates for all displaced persons were high, people living in temporary camps were at highest risk of death. As in other famine-related disasters, preventable infectious diseases such as measles and diarrhoea were the primary causes of death in both towns. These mortality rates are among the highest documented for a civilian population over a long period. Community-based public health interventions to prevent and control common infectious diseases are needed to reduce these exceptionally high mortality rates in Somalia.

"More Debris Removal than Construction in Somalia." 1993. ENR. February 22. Volume 230, Pages 14. Descriptors: Somalia; Contractors; US Army, Corps of Engineers- Contracts and procurement; Brown & Root, Inc.; Business; Heavy construction. Abstract: Thus far, the $33 million set aside to fund U.S. construction efforts in Somalia have not gone toward rebuilding the country's infrastructure as planned. Instead, engineers and contractors in Somalia to provide logistical support to the U.S. troops in Operation Restore Hope have been removing debris, installing latrines and showers, and renovating buildings. The situation arose because the cost of infrastructure repair and the conditions proved to be prohibitive. The difficulties of supplying 25,000 troops with water, dealing with the harsh environment, and ensuring personal safety in Somalia experienced by contractor Brown & Root of Houston and the Naval Facilities Engineering Command, Atlantic Division, are discussed briefly. ISSN: 0891-9526.

Morgan, M. J. 1972. Irrigation Water use Study; Shebelli River, Balad to Baire. Somalia: Ministry of Agriculture, Mogadishu, Somalia. Descriptors: Africa; Balad Somali Republic; East Africa; hydrology; irrigation; Shabeellaha Dhexe; Shebelle River; Somali Republic; water resources; water use. Notes: Communications to the Ministry of Agriculture. Database: GeoRef. Accession Number: 1999-067987.

Morgan, M. J. 1972. Report on the Surface Water Resources in West Lower Juba Area. Italy: FAO, Rome, Italy. Descriptors: Africa; East Africa; hydrology; Jubbada Hoose; Somali Republic; surface water; water resources. Database: GeoRef. Accession Number: 1999-067988.

Morgan, M. J. 1972. Surface Water Resources of West Lower Juba Region. Somalia: Ministry of Agriculture, Mogadishu, Somalia. Descriptors: Africa; East Africa; hydrology; Jubbada Hoose; Somali Republic; surface water; water resources. Notes: Communications to the Ministry of Agriculture. Database: GeoRef. Accession Number: 1999-067989.

Morgan, M. J. 1971. Lower Juba Flood Tuculle to Camsuma. Somalia: Ministry of Agriculture, Mogadishu, Somalia. Descriptors: Africa; Camsuma Somali Republic; East Africa; floods; geologic hazards; hydrology; irrigation; Jubbada Hoose; Somali

Republic; Tuculle Somali Republic. Notes: Communications to the Ministry of Agriculture. Database: GeoRef. Accession Number: 1999-067983.

Morgan, M. J. 1971. November 1971 Flooding of Main Gelib to Camsuma Road, Bod Bodley Flood Irrigation Control. Somalia: Ministry of Agriculture, Mogadishu, Somalia. Descriptors: Africa; Bod Bodley Somali Republic; Camsuma Somali Republic; East Africa; floods; Gelib Somali Republic; geologic hazards; hydrology; irrigation; Jubbada Dhexe; Jubbada Hoose; Somali Republic; water supply. Notes: Communications to the Ministry of Agriculture. Database: GeoRef. Accession Number: 1999-067984.

Morgan, M. J. 1971. Surface Water Resources; North-Western Region. Somalia: Ministry of Agriculture, Mogadishu, Somalia. Descriptors: Africa; East Africa; northwestern Somali Republic; Somali Republic; surface water; water resources. Notes: Communications to the Ministry of Agriculture. Database: GeoRef. Accession Number: 1999-067985.

Morgan, M. J. 1971. Surface Water Resources and Floods Problems in the Lower Schebelli (Brava District). Somalia: Ministry of Agriculture, Mogadishu, Somalia. Descriptors: Africa; Banadir Somali Republic; Brava Somali Republic; East Africa; floods; geologic hazards; hydrology; Shebelle River; Somali Republic; surface water; water resources. Notes: Communications to the Ministry of Agriculture. Database: GeoRef. Accession Number: 1999-067986.

Morgan, M. J. 1970. Flood Irrigation Potential for Wadis Iran, Nur Fahan, Libah and Wadi El Dere, Somalia: Ministry of Agriculture, Mogadishu, Somalia. Descriptors: Africa; central Somali Republic; East Africa; floods; hydrology; irrigation; Somali Republic; Wadi El Dere; Wadi Iran; Wadi Nur Fahan; water supply. Notes: Communications to the Ministry of Agriculture. Database: GeoRef. Accession Number: 1999-067981.

Morgan, M. J. Food and Agriculture Organization of the United Nations and United Nations Development Programme. 1973. Report to the Government of Somalia on Surface Water Resources. Rome: FAO. Descriptors: 12; Water resources development-Somalia. Notes: At head of title: United Nations Development Programme, SOM/69/006, no. TA 3190. Includes bibliographical references (p. 11). Other Titles: Surface water resources. National Agricultural Library. OCLC Accession Number: 6605623.

Morrocco, John D. 1992. "Somalia to Impact Debate on Reshaping U.S. Forces." Aviation Week & Space Technology. December 14-21. Volume 137, Pages 23-24. Descriptors: United States-Armed forces; Somalia-History-1991- (Civil War)/Relief work; Applied Science & Technology; Readers' Guide (Current Events); Business; National security. Abstract: U.S. defense analysts are predicting that the relief mission in Somalia will have an impact on the debate over the size, roles, and missions of the U.S. armed forces. In addition to influencing a study on roles and missions currently being conducted by the Joint Chiefs of Staff, Operation Restore Hope will also help to shape the definition and rules of the low-intensity military operations that appear to be the wave of the future. The U.S. military has had little in the way of training or preparation for peacekeeping or peacemaking operations. Recently, as the armed forces have slowly warmed to the idea of taking on nontraditional missions, theories about dealing with such operations have been bandied about at war colleges, and Somalia will probably be a proving ground for many of these theories. ISSN: 0005-2175.

Morrocco, John D. 1992. "U.S. to Spearhead Somalia Relief Force." Aviation Week & Space Technology. December 7. Volume 137, Pages 26. Descriptors: Food relief; Famines; Applied Science & Technology. Abstract: A U.S. military force of as many as 28,000 troops is being assembled for deployment to Somalia as part of a massive UN effort to restore order in the famine-stricken African nation and protect humanitarian relief operations. A U.S. Marine amphibious task unit of 1,800 personnel arrived off the Somalian coast on December 3. At the same time, Pentagon officials briefed President George Bush on plans to deploy additional Marine and Army units. It was expected that Lt. Gen. Robert B. Johnston, commander of the 1st MEF, would be named head of the U.S. force. A resolution authorizing member states to deploy combat troops "to establish a secure environment for humanitarian relief operations in Somalia" was approved by the UN Security Council on December 3. It is expected that the U.S., as the largest contributor of troops, will assume overall command and control of the international force. ISSN: 0005-2175.

"Motion of Nubia Relative to Antarctica since 11 Ma: Implications for Nubia-Somalia, Pacific-North America, and India-Eurasia Motion." 2006. Geology. June. Volume 34, Issue 6, Pages 501-504. Abstract: Recent estimates of the rotation between Nubia and Somalia have resulted in disparate poles of rotation for the motion since 3.16 Ma (southwest of South Africa) compared with that since 11.03 Ma (near the east tip of Brazil). Here we use magnetic anomaly profiles unavailable in prior Nubia-Antarctica motion studies to significantly revise the estimate of the rotation between Nubia and Antarctica since 11.03 Ma. We use this newly estimated rotation to construct revised estimates of Nubia-Somalia, Pacific-North America, and India-Eurasia motion. The new Nubia-Somalia rotation indicates substantial displacement of Somalia relative to Nubia over the past 11.03 m.y. 129± 62 km extensional, 90± 42 km right-lateral transtensional, and 52± 21 km right-lateral transtensional near the northern extremity of the East African Rift, the northern Mozambique Basin, and the Andrew Bain Fracture Zone complex, respectively. The substantial rotation between Nubia and Somalia implies that prior plate motion estimates based on a circuit through Africa are biased by 60-85 km at 11.03 Ma and perhaps by much more for earlier reconstructions. Our results imply that India-Eurasia motion since 11.03 Ma has been [similar]12% ([similar]5 mm yr-1) slower than, and [similar]20[degree] clockwise of, estimates that neglect Nubia-Somalia motion. Our results further imply that Pacific-North America displacement since 11.03 Ma has been 5[degree]-10[degree] clockwise of prior estimates and require 58-75 km less extensional displacement across the Basin and Range since 11.03 Ma than inferred before. ISSN: 0091-7613.

Mubarak, J.A. 2002. "A Case of Private Supply of Money in Stateless Somalia." Journal of African Economies. Volume 11, Issue 3, Pages 309-325. Descriptors: Capital and Investments; price dynamics; currency market; financial crisis; exchange rate. Notes: References: Number: 13. Abstract: Since the collapse of the state in 1991 and the subsequent crash of the domestic currency, the Somali economy starved for liquidity to facilitate economic recovery and means to replace ageing banknotes. In the long absence of effective state authority, forged banknotes from factions competed to fill the void and became accepted. In much the way that Milton Friedman predicted, the consequent unregulated private supply of money seeking to capture seignoirage income soon raised prices, destabilized domestic markets, eroded the market value of Somali shilling banknotes and reduced it to pure commodity money. However, the experience did not lead to an infinite price level simply because the money issuers were unwilling to add zeros to denominations in fear that the public will not accept them. Rather, as the marginal cost of importing and injecting new reprints on existing highest denomination neared its exchange value, the incentive to import more banknotes subsided, and domestic prices started to stabilize. Incredibly, at this very low exchange value, the Somali currency survives and retains its service value, and real cash balances have actually increased. ISSN: 0963-8024.

Mubarak, J.A. 1997. "The 'Hidden Hand' Behind the Resilience of the Stateless Economy of Somalia." World Development. Volume 25, Issue 12, Pages 2027-2041. Descriptors: Economic Development; informal economy; developing country; state collapse; national economy; economic development; private sector; market forces. Abstract: This paper provides a framework of analysis of economic markets in stateless Somalia. It argues that the informal market that sustained private sector activities under the repressive policies of Siad Barre's government has provided a functioning system to the economy. In small communities that achieved internal law and order, their economy boomed in an unprecedented free-market environment, and the no-government situation has proven to be far better than the repressive government institutions and policies of Barre's era. However, the economic expansion is chaotic and there is a strong new demand for an accountable and effective government that could provide essential public goods for sustainable economic development. ISSN: 0305-750X.

Mueller, German. 1966. "Grain Size, Carbonate Content, and Carbonate Mineralogy of Recent Sediments of the Indian Ocean off the Eastern Coast of Somalia." Naturwissenschaften. Volume 53, Issue 21, Pages 547-550. Descriptors: Africa; East Africa; East Coast; ecology; marine; marine geology; sediments; Somali Republic. Notes: sketch maps; Abstract: Recent Indian Ocean sediments off the eastern coast of Somalia decrease progressively in grain size and carbonate content away from the coast. They form zones of various sediment types based on different sand-silt-clay ratios and a different carbonate content. These zones are approximately parallel to the coast. The sediments' carbonate mineralogy is typical of the different depositional environments: deep-water sediments (globigerina ooze, deep-sea clay) are composed nearly exclusively of low-magnesian calcite, whereas carbonate sands in the shelf area (mainly derived from reef debris) are rich in high-magnesian calcite and aragonite. Dolomite was found by x-ray diffraction analyses in almost half of the samples. ISSN: 0028-1042.

Mukhtar M.H. 1996. "The Plight of the Agro-Pastoral Society of Somalia." Rev. Afr. Polit. Econ. Volume 70, Pages 543-553. Descriptors: Conflict, protest, human rights; Political; agropastoral society; clan system; civil war; developing country; tribal structure; political conflict; geopolitical study; agropastoralism; resource conflict. Notes: Somalia. Abstract: This article assesses some of the factors involved in the current civil war in Somalia, especially as they pertain to the inter-riverine region of the south. Particular emphasis is placed on the Dighil/Mirifle clan in that region. The article investigates the social causes of the worst civil war in the modern history of this country. The basic tenet of the 'single cause analysis' is an invented homogeneity, which presents Somalia as one of the few culturally homogeneous countries in Africa. In fact, Somalia has always been divided into southern agro-pastoral clans and northern nomadic clans which have distinctively different cultural, linguistic, and social structures. The myth of

Somali homogeneity played a major role in the rise of nomadic clans to political predominance, and the appropriation of resources from the less warlike and intensely religious agro-pastoral groups in and around the inter-riverine region. A major factor in the Somali conflict is the struggle among clans for control of limited and increasingly scarce resources, especially land and water. More specifically, it is a violent competition between the Darood and Hawiye clan families for political and economic dominance of the inter-riverine region. ISSN: 0305-6244.

Mukhtar, Mohamed Haji; Castagno, Margaret and Historical dictionary of Somalia. 2003. Historical Dictionary of Somalia. Lanham, Md: Scarecrow Press. Descriptors: 353. Notes: Somalia- History- Dictionaries. 6.420. Notes: Rev. ed. of: Historical dictionary of Somalia by Margaret Castagno. 1975. Includes bibliographical references (p. 287-352). ISBN: 0810843447; LCCN: 2002-10300.

Mullen J. 1989. "Training and Visit System in Somalia: Contradictions and Anomalies." Journal of International Development. Volume 1, Issue 1, Pages 145-167. Descriptors: Agriculture and Rural Development; agricultural extension; farm management; training and visit system. Notes: Somalia. Abstract: After a brief introduction to Somalian agriculture, the author considers the objectives of the current (1987-1991) Five Year Plan, and the Agricultural Extension and Farm Management Training Project (AFMET) with which it hopes to achieve them. The organizational structure, staffing, funding and activities of AFMET are discussed and assessed, generally favorably, although with some reservations concerning future policy co-ordination. ISSN: 0954-1748.

Munslow B, O'Keefe P, Pankhurst D and Philips P. 1983. "Energy and Development on the African East Coast: Somalia, Kenya, Tanzania and Mozambique." Ambio. Volume 12, Issue 6, Pages 332-337. Descriptors: International Development Abstracts; Species Term: Somalia. Notes: Special Feature: 3 photos, 6 tables, 16 refs. Abstract: These four countries, despite divergent political systems and cultures, have a number of things in common: soaring populations, undeveloped or under-developed industrial infrastructures, national programmes which drive for self-sufficiency in food production and a dual-prolonged energy crisis in the form of too much dependency on imported oil and a growing shortage of fuelwood, the poor man's energy source. Profiles the energy situation in each country vis-a-vis development, and makes recommendations for changes. In order to push development goals, the energy dilemma must be dealt with first. ISSN: 0044-7447.

Murray, S. P. and Hsu, S. A. 1982. "A Collection of Reprints." Louisiana State Univ., Baton Rouge. Coastal Studies Inst. Dec. Volume: TR343, TR354, pages: 79. Descriptors: Oil spills; Wind; Weather; Meteorological phenomena; Ocean currents; Coastal regions; Somalia; Interactions; Ocean waves; Reprints; Somali Jet; Earth sciences and oceanography Dynamic oceanography; Atmospheric sciences Meteorology; Atmospheric sciences Dynamic meterology; Ocean technology and engineering Dynamic oceanography; Environmental pollution and control Water pollution and control. Contents: The Effects of Weather Systems, Currents, and Coastal Processes on Major Oil Spills at Sea; Interactions between Mesoscale and Synoptic-Scale Wind Systems in Somalia; Wind-Wave Interactions under the Influence of the Somali Low-Level Jet. Notes: Published in various journals. Includes Rept. no. TR-355. DTIC Accession Number: ADA1240662XSP.

Mutch, Thembi. 2005. "Killer in the Fold." New Scientist. Volume 186, Issue 2497, Pages 42-45. Descriptors: Vaccines; Biodiversity; Ecosystems; Disease control; Viruses; Population statistics; Rural areas; Finance; Surveying. Abstract: The difficulties faced while carrying out an eradication campaign to prevent the spread of virus that causes rinderpest in cattle in arid land across the Kenya-Somalia border, known as the southern Somali pastoral ecosystem, are discussed. The drive to eliminate rinderpest is failing due to scientific disagreements, infighting, funding crises, and other logistical problems. The elimination of the disease requires a combination of surveillance and vaccination, which involves identifying areas where rinderpest is circulating and carrying out a mass vaccination campaign. It is also logistically impossible to vaccinate all the cattle in the southern Somali ecosystem, which leaves traces of the disease intact in cattle populations assumed disease-free. ISSN: 0262-4079.

Myers, Norman. 1987. "Linking Environment and Security." The Bulletin of the Atomic Scientists. June. Volume 43, Pages 46-47. Descriptors: Internal security; Natural resources; General Science. Abstract: The effect of environmental factors on military conflict, the topic of a three-day workshop convened by the Norwegian minister of defense in Oslo late last year, is receiving increasing attention. Depletion of natural resources such as soil, water, forests, grasslands, and fisheries is rarely the only cause of war but can be of critical importance. For example, if Ethiopia had used the money it spent on military activities between 1976 and 1980 on preserving the resource base in traditional farmlands, the migration toward the Ogaden might have been smaller. Somalia has cited that migration as one of its reasons for invading Ethiopia. ISSN: 0096-3402.

N

Nairobi Map Service. 2006. "Map of Somalia Showing Administrative Boundaries." Nairobi: Nairobi Map Service, Limited. Notes: Description: 1 map: color; 53 x 36 cm., on sheet 58 x 39 cm., folded to 29 x 13 cm. Geographic: Somalia- Maps. Scale 1:3,500,000. Cartgrph Code: Category of scale: a Constant ratio linear horizontal scale: 3500000; Notes: Includes 1 inset map: "Africa- Political Development (1963-2005)". OCLC Accession Number: 70218163.

Naleye, A. M. and Harms, J. C. 1993. "Petroleum Exploration in Somalia." Firenze; Istituto Agronomico L'Oltremar; 1993. Relazioni E Monografie Agrarie Subtropicali E Tropicali Nuova Serie113 B, pages: 417-428. International Meeting; 1st-1987 Nov: Mogadishu. Notes: Geology of Somalia and surrounding regions: Geology and mineral resources of Somalia and surrounding regions. OCLC Accession Number: 34175029.

Naleye, A. M. and Harms, J. C. 1987. "Petroleum Exploration in the Somali Democratic Republic." GeoSom 87; Geology of Somalia and Surrounding Regions; Late Abstracts. Somalia: Somali Natl. Univ., Dep. Geol., Mogadishu, Somalia. GeoSom 87; Geology of Somalia and Surrounding Regions, Mogadishu. Somalia Conference: Nov. 24-Dec. 1, 1987. Descriptors: Africa; drilling; East Africa; economic geology; petroleum; petroleum exploration; possibilities; Somali Republic; source rocks. OCLC Accession Number: 34175029.

Nathanail P; Nathanail J., Maund J.G. and Eddleston M. 1998. "Mitigating Geohazards Affecting Mountain Roads in Northeast Somaliland." Geological Society Publishing House; UK. Descriptors: pages 231-237; erosion control; hazard assessment;

mitigation; mountain region; road. Abstract: Roads linking the Gollis plateau with sea ports on the Gulf of Aden descend a steep escarpment. The roads are continually affected by erosion and degradation. These usually gentle processes constitute significant geohazards to the communities on the plateau which rely on the roads to export their agricultural produce as they attempt to rebuild the local economy following years of civil war. Engineering geological advice was provided to local communities involved in maintaining and repairing three roads, in areas occupied by different Somali clans. The absence of advice at an earlier stage had resulted in much wasted effort in inappropriate realignments and gradients. By tailoring the advice to the materials, tools and skills available locally, one road could be re-opened and two could be significantly improved, reducing travel times for laden lorries by up to half. In: Geohazards in Engineering Geology. Geological Society Special Publication No.15. (1998) p. 231-237; References: 7; Somalia; July 1, 1999. ISBN: 1862390126.

Neary, C. 1978. "Evolution and Mineralization of the Arabian-Nubian Shield." Mining Magazine (London). Jul. Volume 139, Issue 1, Pages 47-48. Descriptors: Africa; Arabian Peninsula; Arabian Shield; Arabian-Nubian Shield; Asia; East Africa; economic geology; Egypt; evolution; igneous rocks; island arcs; metamorphic rocks; mineral resources; mineralization; North Africa; Nubia; petrology; plate tectonics; report; Somali Republic; Sudan; symposia. ISSN: 0308-6631.

Nechayev, S. V. 1981. "Strukturno-Vozrastnyye Sootnosheniya Kompleksov Dokembriya Ukrainskogo Shchita; Cheshskogo Massiva, Nekotoryye Geotektonicheskiye i Metalogenicheskiye Problemy." Translated title: "The Structural-Geochronological Relationship of Precambrian Complexes in the Ukrainian Shield and the Bohemian Massif; some Geotectonic and Metallogeny Problems." Geologicheskiy Zhurnal (Kiev, 1968). Volume 41, Issue 2, Pages 38-50. Descriptors: Africa; age; areal geology; Bohemian Massif; Central Europe; Commonwealth of Independent States; correlation; Czechoslovakia; East Africa; economic geology; Europe; lineaments; metal ores; metallogeny; plate tectonics; Precambrian; regional; Russian Platform; Somali Republic; stratigraphy; structural geology; tectonics; tectonophysics; tin ores; Ukraine; Ukrainian Shield; USSR. References: 2 p. illustrations incl. 1 table, geol. sketch map. ISSN: 0367-4290.

Nehlig, P., Asfirane, F. and Braux, C., et al. 2000. "Reappraisal of the Mineral Potential of the Arabian Shield; a GIS Synthesis and Multi-criteria Cross-Analysis of Geologic and Metallogenic Data; Brazil 2000; 31st International Geological Congress; Abstracts Volume." International Geological Congress, Abstracts = Congres Géologique International, Resumes. Aug. Volume 31, Descriptors: Africa; Arabian Peninsula; Arabian Shield; Asia; East Africa; Egypt; epithermal processes; Eritrea; Ethiopia; gold; mathematical geology; exothermal processes; metallogeny; metals; mineral deposits, genesis; mineral exploration; North Africa; potential deposits; Precambrian; Proterozoic; Saudi Arabia; Somali Republic; statistical analysis; Sudan; upper Precambrian; Yemen. Notes: IGC, International Geological Congress; File G2401015.PDF; Latitude: N123000, N304000; Longitude: E0594000, E0343000. Database: GeoRef.

Membrane, P. G. and Conti, R. 1997. "Water for Mogadishu; Water Supply in a War Torn Town." Bulletin Für Angewandte Geologie = Bulletin Pour La Geologie Appliquee = Bollettino Per La Geologia Applicata = Bulletin for Applied Geology. Volume 2, Issue 2, Pages 151-165. Descriptors: Africa; East Africa; fresh water;

geographic information systems; ground water; information systems; mapping; monitoring; optimization; salt water; salt-water intrusion; Somali Republic; urbanization; water harnessing; water management; water resources; water supply; water wells. Notes: References: 10; illustrations incl. 1 table, sect. ISSN: 1420-6846.

Neumann, O. 1904. "From the Somali Coast through Ethiopia to the Sudan." Annual Report- Smithsonian Institution. Volume 1903, Pages 775-792. Descriptors: Africa; areal geology; East Africa; Ethiopia; Somali Republic; Sudan. ISSN: 0097-644X.

Neumann, O. 1902. "From the Somali Coast through Southern Ethiopia to the Sudan." Geography Journal. Volume 20, Pages 373-401. Descriptors: Africa; areal geology; East Africa; Ethiopia; Somali Republic; southern Ethiopia; Sudan. ISSN: 0016-7398.

New TransCentury Foundation and USAID/Somalia. 1983. Water Resource and Road Development for Refugee Settlement in Northwest Somalia: A Proposal to USAID/Somalia. Washington, D.C: The Foundation. Descriptors: 35 leaves; Water resources development- Somalia; Roads- Somalia- Design and construction; Refugees- Somalia. Notes: "December 15, 1983." LCCN: 91-216239; OCLC Accession Number: 29521344.

Nigh, Robert. 1993. "A Ground Water Consultant in Somalia." Water Well Journal. Feb. Volume 47, Issue 2, Pages 29-30. Descriptors: Africa; developing countries; drilling; East Africa; ground water; Somali Republic; water quality; water resources; water wells; wells. ISSN: 0043-1443.

Nimis P, Molin G and Visona D. 1996. "Crystal Chemistry of Danalite from Daba Shabeli Complex (Northern Somalia)." Mineralogical Magazine. Volume 60, Issue 2, Pages 375-379. Descriptors: crystal chemistry; danalite; helvite. Notes: Somalia- Daba Shabeli Complex. Abstract: Danalite is the Fe2+ end-member of the minerals of the helvite group. These minerals are relatively uncommon, although limited amounts are known at many localities round the world. Their typical host rocks are skarns, but occurrences in mineralized veins pegmatites, and altered alkali granites have also been reported. This work reports the first finding of danalite from the Daba Shabeli Complex, Somalia, and provide new crystal-chemical data on a Fe65Mn23Zn12 danalite sample from this locality. ISSN: 0026-461X.

Northern Cartographic, Inc. 1989. "Jubba Environmental and Socioeconomic Studies (JESS): Areas of Project Research and Potential Development Impact." Burlington, Vt: Northern Cartographic, Inc. Descriptors: Natural resources- Somalia- Maps; Natural resources- Juba River Valley (Ethiopia and Somalia)- Maps. Notes: Description: 1 map; color; 103 x 75 cm. Geographic: Somalia- Economic conditions- Maps. Juba River Valley (Ethiopia and Somalia)- Economic conditions- Maps. Scale 1:500,000; (E 41000'--E 44000'/N 4000'--S 0030'). Notes: Shows Juba River Valley in Somalia. "JESS ... implemented in southern Somalia by the U.S. Agency for International Development with the Ministry of National Planning and Jubba Valley Development. Associates in Rural Development, Inc. of Burlington, Vermont was contracted to provide technical assistance in research and preparation of the JESS final report." LCCN: 92-684537; OCLC Accession Number: 26210912.

Northern Somalia (1.75MB) and map key and scale (345K). Portion of AMS Series 2201, Sheet 21 "Djibouti", original scale 1:2,000,000. Published by the U.S. Army Map Service, 1968. See: http://www.lib.utexas.edu/maps/africa/djibouti_1968.jpg

Ntale, H. K., Gan, T. Y. and Mwale, D. 2003. "Prediction of East African Seasonal Rainfall using Simplex Canonical Correlation Analysis." J. Clim. June. Volume 16, Issue 12, Pages 2105-2112. Descriptors: Seasonal rainfall forecasting; Statistical-numerical long-range forecasting techniques; Correlation analysis; Algorithms; Pressure anomalies; Sea surface temperature anomalies; Principal component analysis; Probability forecasting evaluation; Rainfall; Coasts; Climates; Precipitation; Weather; Statistical models; Ocean-atmosphere system; Climate; Africa, East; Indian Ocean; Atlantic; Tanzania; Kenya; Somalia; Angola; Guinea Gulf; Africa. Notes: TR: CS0513260. Abstract: A linear statistical model, canonical correlation analysis (CCA), was driven by the Nelder-Mead simplex optimization algorithm (called CCA-NMS) to predict the standardized seasonal rainfall totals of East Africa at 3-month lead time using SLP and SST anomaly fields of the Indian and Atlantic Oceans combined together by 24 simplex optimized weights, and then 'reduced' by the principal component analysis. Applying the optimized weights to the predictor fields produced better March-April-May (MAM) and September-October-November (SON) seasonal rain forecasts than a direct application of the same, unweighted predictor fields to CCA at both calibration and validation stages. Northeastern Tanzania and south-central Kenya had the best SON prediction results with both validation correlation and Hanssen-Kuipers skill scores exceeding +0.3. The MAM season was better predicted in the western parts of East Africa. The CCA correlation maps showed that low SON rainfall in East Africa is associated with cold SSTs off the Somali coast and the Benguela (Angola) coast, and low MAM rainfall is associated with a buildup of low SSTs in the Indian Ocean adjacent to East Africa and the Gulf of Guinea. ISSN: 0894-8755.

Nur A. 2003. "A Preliminary Magnetic Study Over the Mandera-Lugh Basin of Southwest Somalia." Journal of Mining and Geology. Volume 39, Issue 2, Pages 117-120. Descriptors: Plate tectonics; Paleozoic; geophysical survey; sedimentary basin; tectonic evolution; rifting; Mesozoic; magnetic survey Species Term: Somalia. Notes: Additional Info: Nigeria; References: 9; Geographic: Somalia East Africa Sub-Saharan Africa. Abstract: A preliminary ground magnetic survey was carried out to determine the sediment thickness and the crustal structure of the Mandera-Lugh Basin of the Somali sector. The results obtained from the spectral analysis of two profiles show variations in sediment thickness reflecting on the geologic situation of the area. Profile AA' which connects between Garbaharry and Lugh, sediment thickness of about 6000 was obtained. On the other hand, sediment thickness of about 8000m was derived for the profile BB' which passes through Lugh town. From the analysis of Hilbert transformation, the structural pattern of the Manderal-Lugh was found to be mainly NE-SW direction. It is considered that the tectonic element and sediment deposition in the Manderal-Lugh Basin, are related to the geodynamic evolution of the East African rift system, which began in the late Paleozoic. ISSN: 1116-2775.

O

Olembo R. 1983. "The East African Region: A Collection of Diversity." Ambio. Volume 12, Issue 6, Pages 284-287. Descriptors: International Development Abstracts-74; Somalia; Kenya; Tanzania; Mozambique; environmental problems Species Term: Somalia. Notes: Special Feature: 5 figs, photo, 6 refs. Abstract: The mainland countries of the region- Somalia, Kenya, Tanzania, and Mozambique- though striving towards the collective goal of development, have approached it from very different paths. They are peopled by a staggering variety of ethnic groups and graced with some of the most spectacular scenery in the world. Despite a common heritage of colonialism and a wealth of human and natural resources, the region remains politically unstable and economically weak. But through the collective action of the East African Regional Seas Programme, these countries are attempting to cope with some of their most severe environmental problems. ISSN: 0044-7447.

Olembo, Reuben. 1983. "East African Region: A Collection of Diversity." Ambio. Volume 12, Issue 6, Pages 284-287. Descriptors: Environmental Protection; Ecology. Abstract: The mainland countries of the Region- Somalia, Kenya, Tanzania, and Mozambique- though striving toward the collective goal of development, have approached it from strikingly different paths. These countries are peopled by a staggering variety of ethnic groups and graced with some of the most spectacular scenery in the world. Despite a wealth of human and natural resources, the Region remains politically unstable and economically weak. Through collective efforts, these nations and the island states of Madagascar, Comoros, Mauritius, Reunion and Seychelles- which share the Southwest Indian Ocean- are attempting to come to grips with common environmental problems.

Olson, Steven F. 2000. "The Proterozoic Evolution of Africa." Information Circular- University of the Witwatersrand, Economic Geology Research Unit. Apr. Volume 343, Pages 61. Descriptors: absolute age; accretionary wedges; Africa; Archean; basins; Cambrian; Congo Craton; cratons; crust; dates; East African Craton; faults; greenstone belts; historical geology; intrusions; island arcs; K/Ar; lava fields; lithostratigraphy; magmatism; Mesoproterozoic; metamorphic belts; Neoproterozoic; Paleoproterozoic; Paleozoic; plate tectonics; Precambrian; Proterozoic; rifting; sedimentary basins; Somalia Craton; Southern African Craton; Rb/Sr; terranes; time scales; U/Pb; ultramafic composition; upper Precambrian; volcanic features. Notes: Includes appendices; References: 281; 11 tables, geol. sketch maps. ISSN: 0375-8087.

Omar, Haji D. 1973. The Shabelle River Quality. Somalia: CARS, Afgooye, Somalia. Descriptors: Africa; East Africa; geochemistry; hydrochemistry; hydrology; Shebelle River; Somali Republic; surface water; water quality. Database: GeoRef. Accession Number: 1999-068004.

Omar, Shire Y. 1987. "Attivita' Estrattiva Nella Regione Del Benadir (Sfruttamento Delle Cava e Difesa Dell'Ambiente)." Translated title: "Extractive Activity in the Benadir Region; Exploitation of the Quarries and Protection of the Environment." Quaderni Di Geologia Della Somalia (Mogadiscio). Volume 9, Descriptors: Africa; Benadir; East Africa; environmental effects; exploitation; land use; mining; mining geology; quarries; Somali Republic. LCCN: 93093377.

Omar, Shire Y. Piccoli, G. and Perissotto, A. 1978. "Prime Lauree in Geologia." Translated title: "First Degree in Geology." Somalia: Univ. Naz. Somala, Mogadishu, Somalia. Descriptors: academic institutions; geology; Somali University. Notes: Cilmi iyo Farsamo; 3. Database: GeoRef. Accession Number: 1999-068009.

Omar, Shire Y. and Vallario, A. 1988. "Aspetti Idrogeologici Della Pianificazione Delle Risorse Idriche; Un Approccio Ai Problemi Della Somalia." Translated title: "Hydrogeologic Aspects of the Planning of Water Resource; an Approach to the Problems of Somalia." Bollettino Della Societa Dei Naturalisti in Napoli. Volume 18,

Descriptors: Africa; East Africa; hydrology; planning; Somali Republic; water resources. ISSN: 0366-2047.

Omenetto, P. Binda, P. L. Frizzo, P. and Warden, A. J. 1987. Metallogenic Outlines of Somalia and Surrounding Regions. Somalia: Somali Natl. Univ., Mogadishu, Somalia. GeoSom 87; Geology of Somalia and Surrounding Regions, Mogadishu. Somalia Conference: Nov. 23-30, 1987. Descriptors: Africa; Arabian Peninsula; Asia; basement; clay minerals; clays; East Africa; economic geology; Egypt; lead-zinc deposits; metal ores; metallogenic provinces; metallogeny; metals; mineral exploration; North Africa; placers; precious metals; rare earth deposits; Saudi Arabia; sepiolite; sheet silicates; silicates; Somali Republic; titanium ores; Yemen; zirconium. OCLC Accession Number: 34175029.

Omer A and El Koury G. 2004. "Regulation and Supervision in a Vacuum: The Story of the Somali Remittance Sector." Small Enterprise Development. Volume 15, Issue 1, Pages 44-52. Descriptors: Capital and Investments; financial services; regulatory framework; banking; migrants remittance. Notes: Additional Info: United Kingdom; References: Number: 4; Geographic: Somalia East Africa Sub-Saharan Africa Africa. Abstract: This paper addresses the challenges that the Somali remittance sector is facing post September 11, 2001. In the absence of a central government and a banking system, the Somali remittance sector provides basic financial services, including the remittance of funds from the Somali diaspora. The lack of a central authority and a central bank poses challenges, such as absence of fiscal policies, lack of financial planning and investment guidelines, in addition to currency mismanagement and erratic fluctuation and pervasive insecurity, to remittance companies both in Somalia and abroad. In order to meet these challenges the remittance companies need to increase their current operating standards and commit to compliance with all host country rules and regulations. Under these circumstances the most effective option is self-regulation until such time that a formal banking sector is established in Somalia. In the mean time host country regulators along with the Somali remittance sector must collaborate to achieve compliance. ISSN: 0957-1329.

Omodei, D. 1937. "Notizie Sul Clima Della Somalia Italiana Meridionale e Sul Regime Idrometrico Dell'Uebi Scebelli." Translated title: "News on the Climate of Southern Italian Somalia and on the Hydrometric Regime of Shebelle River." Italy: Publisher unknown, Genoa, Italy. Descriptors: Africa; climate; East Africa; Shebelle River; Somali Republic; southern Somali Republic; surface water. Database: GeoRef. Accession Number: 1999-068015.

Omodei, D. 1926. "Notizie Sul Clima Della Somalia Italiana Meridionale e Sul Regime Dello Uebi Scebeli (Secondo Le Osservazioni Fatte Nelle Stazioni Istituite Dal Duca Degli Abruzzi)." Translated title: "News on the Climate of Southern Italian Somalia and on the Regime of the Shebelle River; the Observations made in the Established Station of the Duca Degli Abruzzi." Italy: Descriptors: Africa; climate; Duca degli Abruzzi; East Africa; fluvial features; hydrology; Jowhar Somali Republic; rivers; Shabeellaha Dhexe; Shebelle River; Somali Republic; surface water. Database: GeoRef. Accession Number: 1999-068011.

Onor, R. 1925. La Somalia Italiana. Translated title: Italian Somalia. Italy: Bocca Ed., Turin, Italy. Descriptors: Africa; areal geology; East Africa; Somali Republic. Database: GeoRef. Accession Number: 1999-068018.

Oriol, Ramon A. and Arnone, Robert A. 1990. "Seasonal Optical Properties Derived from Coastal Zone Color Scanner Satellite Data Along the Somali Coast and the Gulf of Aden." Stennis Space Center, Ms: Naval Ocean Research And Development Activity Stennis Space Center, Ms. April 1990. Volume: Ada226301, Pages: 39 Pages(S). Descriptors: Scientific Satellites; *Bathymetry; *Colors; Accuracy; Air Pollution; Airborne; Attenuation; Channels; Coastal Regions; Coefficients; Data Bases; Diffusion; Digital Systems; Aden Gulf; Image Processing; Images; Measurement; Monsoons; Oceans; Optical Properties; Optics; Planning; Scanners; Seasonal Variations; Ships; Signals; Somalia; Surface Properties; Surface Waters; Surveys; Upwelling; Variations; Visibility; Water; Wind; Ocean color; Ocean optics; Satellites; Multispectral. Abstract: Optical water properties of the world's oceans can be obtained from data collected with Coastal Zone Color Scanner (CZCS) aboard Nimbus-7. Our understanding of the spatial and temporal variability of surface optical properties is greatly improved from synoptic images from visible channel satellites. Current satellite data processing techniques can eliminate the atmospheric contamination that contributes 90% of the visible channel signal. The remaining signal, which constitutes the ocean color, is directly related to the diffuse attenuation coefficient (K) at 490 nanometers for the upper surface waters. Calculation and geographic registration of K can be performed on each sqm image-pixel of the CZCS data, and results show that the accuracy is within 92% of ship measurements. Regional costal optical atlases are required for planning bathymetric surveys using the Airborne Bathymetric System, CZCS data provide a method of deriving temporal and spatial variability of costal optical properties in regions where limited ship measurements are available. This report presents a demonstration of the capability of a regional optics data base generated using CZCS data. A series of CZCS images of the eastern Somali cost and the Gulf of Aden has been processed for the diffuse attenuation coefficient and have been used to define a regional optical database. This data base exists digital image form and clearly defines optical variability in response to continental winds, monsoon wind, and coastal upwelling. Notes: Prepared in collaboration with Planning Systems, Inc., Slidell, LA. Database: DTIC. Accession Number: ADA2263010XSP.

Oroda, Ambrose S. 2002. Application of Remote Sensing to Early Warning for Food Security and Environmental Monitoring in the Horn of Africa. International Society for Photogrammetry and Remote Sensing. Developments and Technology Transfer in Geomatics for Environmental and Resource Management. Dar es Salaam, Tanzania Conference: 25-28 March, 2002. Descriptors: Remote Sensing; Food Security; Early Warning; Famine; Eastern Africa. Abstract: The horn of Africa includes the countries of Burundi, Eritrea, Ethiopia, Kenya, Rwanda, Sudan, Tanzania, and Uganda with a land area of more than 6million km2 and a population in excess of 160 million people. Most of the countries in the sub-region are either semi-arid or arid with significant evidence of environmental degradation. The environmental and weather conditions in most of the countries are highly precarious, the sub-region experiencing frequent droughts and crop failures whose results are the ever recurring famines. The sub-region, therefore, requires a reliable and effective early warning system (EWS) for environmental monitoring and food security. The recent World Bank Workshop in Nairobi, Kenya, on Food Security situation in the Horn of Africa put emphasis on establishing reliable national and regional early warning systems for food security. The Regional Centre for Mapping of Resources for Development (RCMRD) has, over the years, implemented remote sensing based regional early warning systems for food security to supplement national early warning initiatives that are not so elaborately developed in the eastern Africa countries. Between 1995 and 2000, the RCMRD in collaboration with the Environmental Analysis and Remote Sensing (EARS) Consultants of Delft, The Netherlands with funding from the Netherlands Government carried out an early warning system project known as Regional Famine Early Warning System (REFEWS), which introduced new aspects in the early warning activities in the sub-region. The project developed a methodology for monitoring vegetation growth conditions as well as estimating end of season crop yields. The methodology estimates end of season crop yield forecasts half way the crop growing season and therefore if implemented gives a precise advance warning on the food situation before end of the season. Being a remote sensing based methodology, it has several advantages that include a wide and extensive coverage, regular data availability, timely data availability and dissemination as well as being highly cost-effective. This paper describes the early warning methodology developed for food security and environmental monitoring in the Horn of Africa. Notes: ISPRS Commission VI workshop. URL: http://www.photogrammetry.ethz.ch/general/persons/jana/daressalaam/papers/oroda.pdf.

Orr-Schelen-Mayeron & Associates. 1966. Procedures for the Establishment of the Mogadiscio Water Agency, Mogadiscio, Somali Republic. Minneapolis: Orr-Schelen-Mayeron. Descriptors: 68, 63A leaves; Water-supply- Somalia- Mogadishu. Notes: Named Corp: Mogadiscio Water Agency. Notes: Cover title. "Prepared for the Department of State, Agency for International Development, Washington, D.C." Other Titles: Procedures for the establishment of the Mogadiscio Water Agency ... Library of Congress; LCCN: 75-325944; OCLC Accession Number: 1858625.

Ortiz, S. 2003. "Two New Species of the Oxygonum Stuhlannii-Atriplicifolium Complex (Polygonaceae) from Somalia." Bot. J. Linn. Soc. JUL. Volume 142, Issue 3, Pages 341-345. Abstract: In the course of taxonomic research of the Polygonaceae for the Flora Zambesiaca and Conspectus Florae Angolensis projects, all the species of the genus Oxygonum Burch. ex Campd. in tropical Africa were studied. This resulted in the description of two new species: Oxygonum thulinianum and O. hastatum from Somalia. These species belong to the Oxygonum stuhlmannii-atriplicifolium complex, a group of species characterized by fruits with three prickles or teeth, and leaves always or sometimes sagittate to hastate. The Linnean Society of London, Botanical Journal of the Linnean Society, 2003, volume 142, 341-345.

Osman S.H. 1992. "Miocene Alluvial to Marine Facies Transition of the Mait Group in the Bosaso Basin Migiurtinia, Northern Somalia)." Giornale Di Geologia. Volume 54, Issue 1, Pages 67-76. Descriptors: Sediments and sedimentary processes-transport; alluvial fan; sea level; Miocene; Mait Group; fan delta; reef. Notes: Somalia-Migiurtinia- Bosaso Basin Indian Ocean- Gulf of Aden. Abstract: The Oligo-Miocene tectonic activity which caused the separation of the Arabian plate from the Somali plate, giving origin to the ocean basin of the Gulf of Aden, generated a dozen of sedimentary basins along the coastal belt of Northern Somalia. The Bosaso Basin is one of the largest basins developed during that period. Terrigenous materials, eroded from the uplifted Upper Precambrian-Lower Cambrian crystalline basement (Inda Ad Unit) and the overlying Jurassic-Eocene sedimentary sequences, were deposited together with shallow marine carbonate sediments (reefs, back-reef lagoonal strata and beach calcarenites). ISSN: 0017-0291.

Osman, A. S. Farag, H. A. and Abdi, M. S. 1976. Geology of Somalia. Somalia: Somali Democrat. Rep., Minist. Miner. and Water Resour., Mogadishu, Somalia. Descriptors: Africa; areal geology; Cenozoic; clastic rocks; complexes; crystalline rocks; East Africa; gneisses; lithostratigraphy; Mesozoic; metamorphic rocks; Precambrian; regional; sandstone; schists; sedimentary rocks; shale; siltstone; Somali Republic; stratigraphy. Notes: References: 20. Database: GeoRef. Accession Number: 1979-020757.

Ottaviani, M. 1994. "Geotechnical properties of the Somalia sepiolite." La Geotechnique des Sols Indures-Roches Tendres: Comptes Rendus d'un Symposium Internationale. 20-23 Septembre 1993. Page 723 et seq. Proceedings of the International Symposium Under the Auspices of the International Society for Soil Mechanics and Foundation Engineering (ISSMFE). Part 1 (of 2), Sep 20-23 1993. Athens, Greece. Descriptors: Sepiolite; El-Bur, Somalia; Geotechnical properties; Porosity; Strength; Hard soil; Triaxial tests. Abstract: A study of the physical and mechanical characteristics of the sepiolite deposit of El-Bur, Somalia has been carried out using a limited amount of samples. It has been found that porosity and strength properties are higher than any common hard soil but that the presence of water considerably decreases the strength values. Such behavior seems to depend primarily on the complex mineral structure of sepiolite. Notes: Geotechnical Engineering of Hard Soils-Soft Rocks: Proceedings of an International Symposium; Athens, Greece, 20-23 September 1993 (La Geotechnique des Sols Indures-Roches Tendres: Comptes Rendus d'un Symposium Internationale; Athenes, Grece, 20-23 Septembre 1993). Anagnostopoulos, A, et al, eds. A A Balkema, Rotterdam, 1993, volume 1, pages 723-728. ISBN: 90-5410-344-2; 90-5410-345-0.

P

Pallabazer, R. 1984. Water from Wind in Somalia; Annual Report 2. Somalia: Univ. Naz. Somala, Fac. Ing., Mogadishu, Somalia. Descriptors: Africa; East Africa; Somali Republic; water resources; wind energy; winds. Database: GeoRef. Accession Number: 1999-068026.

Pallabazer, R. 1983. Water from Wind in Somalia; Preliminary Investigation and Programme. Somalia: Univ. Naz. Somala, Fac. Ing., Mogadishu, Somalia. Descriptors: Africa; East Africa; Somali Republic; water resources; wind energy; winds. Database: GeoRef. Accession Number: 1999-068024.

Pallabazer, R. 1983. Water from Wind in Somalia; Progress Report Number 1. Somalia: Univ. Naz. Somala, Fac. Ing., Mogadishu, Somalia. Descriptors: Africa; East Africa; Somali Republic; water resources; wind energy; winds. Database: GeoRef. Accession Number: 1999-068025.

Pallabazzer, R and Gabow, Abdulkadir A. 1992. "Wind Generator Potentiality in Somalia." Renewable Energy. Volume 2, Issue 4-5, Pages 353-361. Descriptors: Wind power; Renewable energy resources; Electric power plants; Pumping plants; Climatology. Abstract: The wind data of several measurement sites in Somalia have been analysed in order to characterize the wind potentiality in relation to the type of wind generators; these have been defined by a simple model of the system output. The relation between machine and local frequency distribution as to energy extraction can be defined by a parameter ('site effectiveness'), which is maximized by a suitable combination of the rated and cut-in wind speed. On this basis it is shown that Somalia is characterized by wind frequency distributions that can be exploited in the best way by relatively slow rather than fast wind machines. ISSN: 0960-1481.

Pallabazzer, Rodolfo. 1986. Towards A Wind-Map of Somalia. Montreal, Que, Canada: Pergamon Press, New York, NY. Volume: 4, pages: 2577-2581. Intersol 85, Proceedings of the Ninth Biennial Congress of the International Solar Energy Society. Descriptors: Wind Power. Abstract: Somalia is on the Horn of Africa in the monsoon area of the West Indian ocean. Owing to this location it benefits from a regular resource of wind energy. An investigation on the wind territorial and time distributions is being carried out with the aim of realizing a wind intensity and wind energy map. First results are given. ISBN: 0-08-033177-7.

Pallabazzer, Rodolfo. 1986. Performance Model for Multibladed Water-Pumping Wind-Mills. Montreal, Que, Canada: Pergamon Press, New York, NY. Volume: 4, pages: 2117-2121. Intersol 85, Proceedings of the Ninth Biennial Congress of the International Solar Energy Society. Descriptors: Wind Power. Abstract: The steady and the dynamic equilibrium of a multibladed water pumping windmill has been studied under the assumption of a simple model. Good agreement has been found between theoretical and experimental results. The experimental research was performed in Somalia. ISBN: 0-08-033177-7.

Pallabazzer, Rodolfo. 1983. Water from Wind in Somalia. Muqdisho: National University of Somalia. Descriptors: v. Wind pumps- Somalia; Water-supply- Somalia. Abstract: [1] Preliminary investigation and program- [2] Progress report n. 4 R. Pallabazzer, C.A. Gabow, A.W. Yusuf- [3] Progress report n. 5 R. Pallabazzer, A.A. Gabow- [4] Synthesis report on wind data (1984-1988) A.A. Gabow, R. Pallabazzer, G. Pugliese. Notes: Cover title. Vol. 1: July 1983. Includes bibliographical references (v. [1], p. 12-13). LCCN: 87-980814; OCLC Accession Number: 18909512.

Pallabazzer, Rodolfo; Boccazzi, A. and Gabow, Abdulkadir A. 1990. "Program for Wind Resource Exploitation in Somalia." European Community Wind Energy Conference. Pages 10.

Pallabazzer, Rodolfo. 1995. "Evaluation of Wind-Generator Potentiality." Solar Energy. July. Volume 55, Pages 49-59. Descriptors: Direct energy conversion; Wind power/Mathematical models; Applied Science & Technology. Notes: Bibliography; Map; Diagram. Abstract: The paper presents a method to analyse wind data for wind-turbine siting. Two parameters are defined by means of which the siting can be optimised, the site effectiveness (ratio output/available energy) and the cut-in speed. It is shown that the site effectiveness achieves a maximum for a value of the cut-in speed which depends on the rated speed and on the site. The cut-in speed of a WECS should be chosen according to the rated speed of the turbine and the cubic mean wind speed of the site. The wind data of two regions (Somalia and Calabria, Italy) with very different climates are examined and compared, showing correlation with the cubic mean wind speed. It is also shown that the Weibull and Rayleigh wind-frequency distribution models are not reliable for evaluating the output energy of a WECS. ISSN: 0038-092X.

Pallabazzer, Rodolfo and Gabow, Abdulkadir A. 1991. "Wind Resources of Somalia." Solar Energy. Volume 46, Issue 5, Pages 313-322. Descriptors: Wind measurement; Wind power; Somalia/Climate; Applied Science & Technology. Notes: Bibliography; Map. ISSN: 0038-092X.

Pallister, John Weaver. 1958. "Somaliland Protectorate; Annual Report of the Geological Survey Department for Period April 1957-March 1958." Descriptors: Africa; East Africa; Geological Survey Department; report; Somali Republic; Surveys, reports. Notes: Hargeisa; Abstract: Includes summarized data on the results of geologic mapping, special studies of basic igneous masses, and mineral and water-supply investigations. Database: GeoRef.

Pallister, John Weaver. 1957. "Somaliland Protectorate; Annual Report of the Geological Survey for April 1956 to March 1957." Descriptors: Africa; East Africa; Geological Survey; report; Somali Republic; Surveys, reports. Notes: (processed), [Hargeisa]; Abstract: Includes summarized data on the results of reconnaissance mapping and notes on mineral occurrences and water supply. Database: GeoRef.

Pallister, John Weaver. 1960. "Somaliland Protectorate; Annual Report of the Geological Survey Department for Period April 1959-March 1960." Descriptors: Africa; British Somaliland, Geological Survey Department, report; East Africa; exploration; Geological Survey Department; mineral resources; Somali Republic; Surveys, reports. Notes: Hargeisa; Abstract: Includes summarized data on the results of geologic mapping and mineral and water-supply investigations. Database: GeoRef.

Pallister, John Weaver. 1959. "Somaliland Protectorate; Annual Report of the Geological Survey Department for Period April 1958-March 1959." Descriptors: Africa; East Africa; exploration; Geological Survey Department; mineral resources; report; Somali Republic; Surveys, reports. Notes: Hargeisa; Abstract: Includes summarized data on the results of geologic mapping and mineral and water-supply investigations. Database: GeoRef.

Pallister, John Weaver. 1959. Mineral Resources of [the] Somaliland Protectorate. London: HMSO. Mineral Resources Pamphlet Number 2. 165 pages; maps. Bibliography pages 164-165. Reprinted from Overseas Geology and Mineral Resources, Volume 7, number 2. Descriptors: Mines and mineral resources. USGS Library.

Pallister, John Weaver, Warden, A. J. and Allen, J. B. 1962. "The Suria Malableh Gypsum-Anhydrite Deposit of the Former Somaliland Protectorate." Overseas Geology and Mineral Resources. Volume 8, Issue 4, Pages 428-437. Descriptors: Africa; Cenozoic; chemically precipitated rocks; East Africa; evaporites; geologic maps; gypsum; gypsum-anhydrite; maps; sedimentary rocks; sediments; Somali; Somali Republic; sulfates; Suria Malableh area; Tertiary. Notes: illustrations (incl. geol. map); Abstract: Among the large deposits of anhydrite and gypsum in the former Somaliland protectorate, now part of the Somali Republic, the one best suited for development is at Suria Malableh, southeast of the port of Berbera on the Gulf of Aden. Sampling, coring, and analytical work indicate a total thickness of more than 160 feet of gypsum of 80 percent grade, in beds of five feet thickness or more. Anhydrite beds total 64 feet thickness with an average grade of more than 90 percent. Proven reserves are 10 million tons of workable ore, with, under the conditions of the area, almost limitless estimated reserves. The evaporite beds are lower Tertiary in age. ISSN: 0030-7467.

Panati, Andrea. 2005. "6OOCW-Somalia DXpedition." QST. May. Volume 89, Issue 5, Pages 81-82. Descriptors: DXpeditions; Amateur radio stations- Somalia; Applied Science & Technology. Notes: Illustration; Map. Abstract: The writer describes the 6O0CW DXpedition to Somalia. Using an FT-847, 500 W amplifier, and 30/40/80 m vertical antenna, he began operations on February 3, 2005. The effects of the civil war meant that assistance from locals, armed guards, and a generator were required. DX activity continued until February 17, when the DXpedtion returned to Italy. ISSN: 0033-4812.

Pape, M. B. 1982. Preliminary Economic Analysis of the Comprehensive Groundwater Development Project. Somalia: Descriptors: Africa; development; East Africa; economics; ground water; Somali Republic; water resources. Database: GeoRef. Accession Number: 1999-068031.

Parkinson, F. B. 1898. "Two Recent Journeys in Northern Somaliland, I." Geogr. J. Volume 11, Pages 15-34. Descriptors: Africa; areal geology; East Africa; field trips; northern Somali Republic; Somali Republic; topography. ISSN: 0016-7398.

Parkinson, J. 1932. "A Preliminary Note on the Borama Shists, British Somaliland." Geol. Mag. Volume 69, Descriptors: Africa; areal geology; Awdal Somali Republic; Borama Somali Republic; East Africa; metamorphic rocks; schists; Somali Republic; Woqooyi Galbeed. ISSN: 0016-7568.

Parkinson, J. 1920. "Report on the Geology and the Geography of the North Part of the East-Africa Protectorate, with a Note on the Gneisses and Schists of the District." Colonial Reports. Miscellaneous Series. Volume 91, Descriptors: Africa; areal geology; East Africa; East Africa Protectorate; gneisses; metamorphic rocks; northern Somali Republic; schists; Somali Republic. Database: GeoRef.

Parsons Company, Los Angeles, CA, United States. 1975. A Final Report, Production Well Construction, Phase II; Mogadiscio Water Supply. Somalia: Water Agency, Mogadishu, Somalia. Descriptors: Africa; Benadir; construction; East Africa; Mogadishu Somali Republic; Somali Republic; water resources; water supply; water wells; wells. Database: GeoRef. Accession Number: 1999-068041.

Parsons Company, Los Angeles, CA, United States. 1973. Feasibility Report on Mogadishu Water Supply. Somalia: Water Agency, Mogadishu, Somalia. Descriptors: Africa; Benadir; East Africa; feasibility studies; Mogadishu Somali Republic; Somali Republic; water resources; water supply. Database: GeoRef. Accession Number: 1999-068040.

Parsons Company, Los Angeles, CA, United States. 1970. Hydrogeological Studies; Mogadishu Water Supply Project. United States: Parsons Co., Los Angeles, CA, United States. Descriptors: Africa; Benadir; East Africa; hydrology; Mogadishu Somali Republic; Somali Republic; water resources; water supply. Database: GeoRef. Accession Number: 1999-068038.

Parsons Company, Los Angeles, CA, United States. 1970. Production Well Construction Phase III; Mogadishu Water Supply Project; Final Report. United States: Descriptors: Africa; Benadir; East Africa; ground water; Mogadishu Somali Republic; Mogadishu Water Supply Project; Somali Republic; water resources; water supply; water wells; wells. Database: GeoRef. Accession Number: 1999-068039.

Paulitschke, P. 1888. "Harar; Forschungsreise Nach Den Somal-Und Galla-Laendern Ost-Afrikas." Translated title: "Harar; Investigation in Somalia and Galla Regions of East Africa." Federal Republic of Germany: Brockhaus, Leipzig, Federal Republic of Germany. Descriptors: Africa; areal geology; East Africa; Ethiopia; Galla; Harar Ethiopia; Somali Republic. Database: GeoRef. Accession Number: 1999-067162.

Paulitschke, Philipages 1887. "Begleitworte Zur Geologische Routenkarte Fuer Die Strecke, Von Zejla Bis Bia Woraba." Translated title: "Description of the Geology of the Road Map for the Distance from Zeila to Bia Woraba." Mitteilungen Der Geographischen Gesellschaft Wien. Volume 30, Pages 212-219. Descriptors: Africa; areal geology; Bia Woraba; East Africa; field trips; geologic maps; maps; road log; Somali Republic; Zeila. Notes: table. ISSN: 1023-4713.

Pausewang S. 1999. "Humanitarian Assistance during Conflict in a State-Less Society: The Case of Somalia." Working Paper- Chr.Michelsen Institute. Issue 5, Pages 1-29. Descriptors: International Aid and Investment; Aid Flows and Policies; aid policy; humanitarian aid; political system. Notes: Additional Info: Norway; References: Number: 27; Geographic: Somalia. Abstract: This working paper was prepared for the workshop on 'Aid and Humanitarian Assistance in Africa' in Arusha, 27-29 June, 1998, with financial support from the Norwegian Ministry of Foreign Affairs. It is based on a series of interviews with aid agencies operating in Somalia after the crisis of 1992-1993. It tries to map out how different agencies reacted in a state-less society, which efforts they made to help re-build local administrative and political structures, which local partners they worked with, how they dealt with security issues, and how they assessed the prospects for the future of Somalia. The report first gives a critical overview over the problems and the different positions held on these issues, before giving a short overview over the responses of the different agencies. The central observation of the report is the dilemma faced by agencies which want to give humanitarian aid to the affected people without compromising the efforts of the local population to form new local structures, and without contributing to a continuation of internal warfare. It shows that agencies find different solutions to these problems, and act on different assumptions and hence with different objectives- though towards a common goal. ISSN: 0804-3639.

Payet, R. and Obura, D. 2004. "The Negative Impacts of Human Activities in the Eastern African Region: An International Waters Perspective." Ambio. Feb. Volume 33, Issue 1-2, Pages 24-33. Descriptors: Climate; Commerce; Conservation of Natural Resources; Ecosystem; Environment; International Cooperation; Policy Making; Poverty; Social Conditions; Water Pollution/prevention & control; Water Supply. Abstract: The complex interactions between human activities and the environment at the interface of land and water is analyzed with a focus on the Somali Current (East Africa), and Indian Ocean Island States, sub-regions of the Global International Waters Assessment (GIWA). These 2 sub-regions contain some of the world's richest ecosystems, including the high biodiversity forests of Madagascar and the diverse coastal habitats of the eastern African coast. These ecosystems support local communities and national and regional economies. Current and future degradation of these systems, from water basins to continental shelves, affects the livelihoods and sustainability of the countries in the region, and long-term efforts to reduce poverty. The assessments determined that pollution and climate change are the primary environmental and social concerns in the Islands of the Indian Ocean, while freshwater shortage and unsustainable exploitation of fisheries and other living resources are the primary environmental and social concerns in East Africa. The GIWA approach, through assessing root causes of environmental concerns, enables the development of policy approaches for mitigating environmental degradation. This paper explores policy frameworks for mitigating the impacts, and reducing the drivers, of 3 environmental concerns--freshwater shortage; solid waste pollution; and climate change--

addressing social and institutional causes and effects, and linking the sub-regions to broad international frameworks. The common theme in all 3 case studies is the need to develop integrated ecosystem and international waters policies, and mechanisms to manage conflicting interests and to limit threats to natural processes. ISSN: 0044-7447.

Pearce, Fred. 2002. "Female Intuition." New Scientist. July 6. Volume 175, Issue 2350, Pages 50-53. Descriptors: Environmental research; Scientists- Somalia; Jibrell, Fatima- Interviews; General Science; Applied Science & Technology. Notes: Portrait. Abstract: An interview with Fatima Jibrell, Somalia's leading environmentalist. Jibrell, who grew up as a nomad in the Somali desert, has spent most her adult life bringing up children and gaining academic qualifications in the U.S. Since her return to Somalia in the early 1990s, Jibrell has been campaigning to promote the role of women in politics, conflict resolution, and environmental issues. She received the Goldman Environmental Prize for grassroots environmentalists this year. ISSN: 0262-4079.

Pearce, Fred. 1993. "Somalia: An Ecopolitical Tragedy." Audubon. March/April. Volume 95, Pages 61. Descriptors: Soils- Somalia; General Science. Notes: Illustration. Abstract: The breakdown in law and order in Somalia has added to the famine in recent months, but the fundamental causes of the problem were misuse of the country's grazing lands and water resources. Soil degradation was aggravated by damming valleys, poor water management, and undermining the traditional systems of land management. ISSN: 0097-7136.

Pease, A. E. 1898. "Some Account of Somaliland; with Notes on Journeys through the Gadabursi and Western Ogaden Country." Scottish Geographical Magazine. Volume 14, Descriptors: Africa; areal geology; East Africa; Gadabursi; Ogaden Somali Republic; Somali Republic. ISSN: 0036-9225.

Perkins, C. L. 1993. "Somalia: Focal Point of a Revamped U.S. Regional Security Strategy for Southwest Asia. Final Rept." Air War Coll., Maxwell AFB, AL. 5 Apr. pages: 38. Descriptors: Military assistance; Strategic analysis; Balance of power; International relations; Military planning; Special forces; Foreign policy; International politics; Somalia; Peacekeeping; Southwest Asia; Behavior and society. Abstract: This paper asserts that the United States' humanitarian intervention in Somalia could provide a rare window of opportunity for improving America's strategic posture in Southwest Asis. United States Central Command's area of responsibility is largely deficient in the operational support facilities available to the other regional unified commands. Such facilities would help optimize the sustainability and responsiveness of USCENTCOM's warfighting forces. It calls for forward basing small-scale United States forces in Somalia to provide USCENTCOM the redundancy and flexibility to respond with a wide spectrum of war-fighting packages which would be suitable to adequately and decisively accomplish its regional security strategy--with less costly consequences. Peacemaking operations, using U.S. special operations forces, are also discussed along with the effect our presence will have on Somalia and other Arab states. Finally, the paper examines a recent news article about the exploration for potentially significant quantities of oil and gas in northern Somalia. DTIC Accession Number: ADA2830800XSP.

Peruzzo, L. 1993. The Pressure Character of the Inda Ad Metamorphism (Northern Somalia). Firenze; Istituto Agronomico L'Oltremar; 1993. Relazioni E Monografie Agrarie Subtropicali E Tropicali Nuova Serie113 A, pages: 119-128. International Meeting; 1st- 1987 Nov: Mogadishu. Notes: Geology of Somalia and surrounding regions: Geology and mineral resources of Somalia and surrounding regions. OCLC Accession Number: 34175029.

Petermann, A. 1860. "Somali Coast and Aden Gulf" (1MB). "Karte der Somali-Kuste und des Golf's von Aden" From Mittheilungen aus Justus Perthes' Geographischer Anstalt uber Wichtige Neue Erforschungen auf dem Gesammtgebiete der Geographie von Dr. A. Petermann. Volume 6, 1860. See:
http://www.lib.utexas.edu/maps/historical/aden_gulf_1860.jpg

Peterson, J. A. 1986. "Geology and Petroleum Resources of Central and East-Central Africa." Modern Geology. Volume 10, Issue 4, Pages 329-364. Descriptors: Africa; Central Africa; chemically precipitated rocks; clastic rocks; east-central Africa; economic geology; Ethiopian Plateau; evaporites; Indian Ocean; petroleum; petroleum exploration; Red Sea Basin; resources; rift zones; sedimentary rocks; Somali Basin; source rocks. Notes: Non-USGS publications with USGS authors; References: 60; illustrations incl. 2 tables, sects., sketch maps. ISSN: 0026-7775.

Peterson, J. A. 1983. "Assessment of Undiscovered Conventionally Recoverable Petroleum Resources of Northwestern, Central, and Northeastern Africa (Including Morocco, Northern and Western Algeria, Northwestern Tunisia, Mauritania, Mali, Niger, Eastern Nigeria, Chad, Central African Republic, Sudan, Ethiopia, Somalia, and Southeastern Egypt)." United States: 1983. Volume: Open-File Report 83-0598, pages: 28. Descriptors: Africa; Algeria; Central Africa; Central African Republic; Chad; East Africa; economic geology; Egypt; Ethiopia; evaluation; Mali; Mauritania; Morocco; Niger; Nigeria; North Africa; petroleum; resources; Somali Republic; Sudan; Tunisia; USGS; West Africa. Notes: Publications of the U. S. Geological Survey; 2 tables, charts, sketch map; ISSN: 0196-1497.

Peterson, James A. 1986. "Geology and Petroleum Resources of Central and East-Central Africa; AAPG Annual Convention with Divisions; SEPM/EMD/DPA; Technical Program and Abstracts." AAPG Bull. May. Volume 70, Issue 5, Pages 630-631. Descriptors: Africa; Benue Valley; Cenozoic; Central Africa; Chad Basin; Doba-Doseo Basin; East Africa; economic geology; energy sources; Ethiopian Plateau; Indian Ocean; Mesozoic; Niger Delta; Nigeria; reservoir rocks; rift zones; Somali Basin; source rocks; Upper Nile Basin; West Africa. Notes: Non-USGS publications with USGS authors. ISSN: 0149-1423.

Peterson, James A. 1985. "Geology and Petroleum Resources of Central and East-Central Africa." United States Geological Survey: 1985. Volume: OF 85-0589, pages: 48. Descriptors: Africa; Benue Valley; Central Africa; Chad Basin; East Africa; economic geology; energy sources; Indian Ocean; natural gas; Niger Valley; oil and gas fields; petroleum; petroleum exploration; petroleum geology; Red Sea region; resources; review; sections; Somali Basin; USGS; West Africa. Notes: USGS, Publications of the U. S. Geological Survey; References: 60; illustrations incl. 2 tables, sects., sketch maps; ISSN: 0196-1497.

Petroconsultants S.A. 1970s. Open Acreage Evaluation, Africa. Geneva: Petroconsultants S.A. Descriptors: 5 v. (loose-leaf); Geology- Mauritania; Geology-Senegal; Geology- Gambia; Geology- Guinea-Bissau; Geology- Guinea; Geology- Sierra Leone; Geology- Liberia; Geology- Côte d'Ivoire; Geology- Ghana; Geology- Togo; Geology- Benin; Geology- Nigeria; Geology- Cameroon; Geology- Equatorial Guinea; Geology- Gabon; Geology- Congo (Brazzaville); Geology- Congo (Democratic

Somalian Earth Sciences

Republic); Geology- Angola; Geology- South Africa; Geology- Botswana; Geology- Lesotho; Geology- Mozambique; Geology- Madagascar; Geology- Mauritius; Geology- Seychelles; Geology- Tanzania; Geology- Kenya; Geology- Uganda; Geology- Somalia; Geology- Sudan. OCLC Accession Number: 14223168.

Petroconsultants S.A and Integrated Exploration and Development Services Limited. 1995. "Somalia." Geneva, Switzerland: Petroconsultants. Descriptors: Oil and gas leases- Somalia- Maps; Petroleum- Somalia- Maps; Natural gas- Somalia- Maps; Petroleum in submerged lands- Somalia- Maps; Natural gas in submerged lands- Somalia- Maps. Notes: Description: maps; some photocopies, some color; 82 x 61 cm. or smaller. Scale 1:4,000,000; Lambert conic conformal proj. (E 360--E 560/N 180--S 40). Scale 1:2,000,000; universal transverse Mercator proj. Notes: Shows oil and gas leases, fields, and wells. Depths shown by contours. Some sheets produced jointly by Integrated Exploration and Development Services Limited and Petroconsultants. Includes text, list of rightholders, and location map. Other Titles: At head of title of some sheets: Foreign Scouting Service; At head of title of some sheets: Global exploration & production service. LCCN: 99-439948; OCLC Accession Number: 41660126.

Petroconsultants S.A; Integrated Exploration and Development Services Limited and IHS Energy. 1982. "Ethiopia." Geneva, Switzerland: Petroconsultants. Descriptors: Oil and gas leases- Ethiopia- Maps; Petroleum- Ethiopia- Maps; Natural gas- Ethiopia- Maps; Oil and gas leases- Eritrea- Maps; Petroleum- Eritrea- Maps; Natural gas- Eritrea- Maps; Petroleum in submerged lands- Eritrea- Maps; Natural gas in submerged lands- Eritrea- Maps; Oil and gas leases- Somalia- Maps; Petroleum- Somalia- Maps; Natural gas- Somalia- Maps; Petroleum in submerged lands- Somalia- Maps; Natural gas in submerged lands- Somalia- Maps. Notes: Description: maps; some photocopies, some color; 82 x 86 cm. or smaller. Scales differ; universal transverse Mercator; (E 360--E 560/N 180--S 40). Notes: Shows oil and gas leases, fields, and wells. Depths shown by contours. Some sheets produced jointly by Integrated Exploration and Development Services Limited and Petroconsultants. Later sheets produced by IHS Energy. Some maps do not have projection statement. Includes text, list of right holders, and location map. Some sheets include "Geological sketch map." Other Titles: Some maps titled: Ethiopia and Somali Republic; Some maps titled: Ethiopia and Eritrea; At head of title of some sheets: Foreign Scouting Service; At head of title of some sheets: Global exploration & production service. Library of Congress; LCCN: 83-693415; 81-692609; OCLC Accession Number: 9945187.

Philander, G. Roy-Delecluse, P. and Princeton Univ., NJ. Geophysical Fluid Dynamics Lab. 1981. The Somali Current. Nova Univ. Recent Progr. in Equatorial Oceanog. p 443-451 (See N83-24088 13-48); United States; 1981; Nova Univ. Recent Progr. in Equatorial Oceanog. p 443-451 (See N83-24088 13-48); United States. Descriptors: Air Water Interactions; Coastal Currents; Equatorial Regions; Indian Ocean; Somalia; Tropical Meteorology; Hydrography; Monsoons; Ocean Temperature; Surface Temperature; Upwelling Water; Wind Direction. Abstract: The response of an equatorial basin to winds that are parallel to the western coast, and that are spatially uniform so that the curl is zero is examined. Uniform cross-equatorial winds with zero curl drive an eastward jet centered on about 4 deg N and westward drift to the south. A western boundary current closes the circulation and is associated with intense upwelling where it turns offshore at about 4 deg N. The tendency of the coastal jet to overshoot is inhibited

Engineering Research and Development Center   284 by winds with a zonal component and by an inclined coast. There is, of course, a vorticity balance at all points in the basin, in particular the region where the jet separated from the coast. However, to attribute the separation of the jet to constraints imposed by the vorticity balance is to confuse cause and effect. The vorticity balance must be consistent with this constraint. Notes: Available from HC A20/MF A03. Database: CSA Technology Research Database. Accession Number: N83-24131 (AH).

Philippon, N., Camberlin, P. and Fauchereau, N. 2002. "Empirical Predictability Study of October-December East African Rainfall." Q. J. R. Meteorol. Soc. Oct. Volume 128, Issue 585, Pages 2239-2256. Descriptors: Rainfall-atmospheric circulation relationships; Tropical rainfall; Tropical rainfall variations; Atmospheric precipitations; Rainfall; Prediction; Tropical meteorology; Monsoons; Winds; Correlation analysis; Atmospheric motion; Sea surface; Surface temperature; Water temperature; Weather forecasting; Africa, East; Equatorial Indian Ocean; Somalia; South Africa; Tropical Atlantic; Equatorial Atlantic; Atlantic, Guinea Gulf; Marine. Notes: TR: CS0322880. Abstract: Relationships between September atmospheric dynamics over the global tropics and the East Africa short rains over the domain 4 degree S-4 degree N, 30-39 degree E were investigated to define a seasonal rainfall forecasting scheme. From total and partial correlation analyses over the period 1968-97, three major and useful signals appear in September. Precursor components of the October-December east-west circulation cell appear over the equatorial Indian Ocean (EIO). They were discovered through various indices of sea-surface temperature (SST), zonal wind, vertical velocity and moist static energy computed over the western, central and eastern parts of the EIO. Principal component analyses were applied to these indices to form two synthetic indices which discriminate ocean-atmosphere variability in the eastern and western parts of the basin. The Indian monsoon dynamics control the SSTs off the Somalia coast which are known to play an important role in the variability of the short rains. Ridge-trough systems develop off South Africa which, when persistent enough, drive SST anomalies in the eastern tropical and equatorial Atlantic; these in turn are associated with anomalies in the middle levels over an equatorial band ranging from the Guinea Gulf to the western Indian Ocean. The three September signals used as predictors in a multiple linear regression model explain 64% of the rainfall variance; a comparable efficiency is also achieved with a linear discriminant analysis model. ISSN: 0035-9009.

Piccoli, G. 1981. "Bibliografia Geologica Della Somalia, Aggiornamento 1981." Translated title: "Geologic Bibliography of Somalia, 1981 Addition." Quaderni Di Geologia Della Somalia (Mogadiscio). Volume 5, Descriptors: Africa; areal geology; bibliography; East Africa; Somali Republic. LCCN: 93093377.

Piccoli, G. 1980. "Bibliografia Geologica Della Somalia, Aggiornamento 1980." Translated title: "Geologic Bibliography of Somalia, 1980 Additions." Quaderni Di Geologia Della Somalia (Mogadiscio). Volume 4, Descriptors: Africa; areal geology; bibliography; East Africa; Somali Republic. LCCN: 93093377.

Piccoli, G. 1978. "Rassegna Bibliografica Geologica Della Somalia." Translated title: "Geologic Bibliographic Review of Somalia." Quaderni Di Geologia Della Somalia (Mogadiscio). Volume 2, Descriptors: Africa; areal geology; bibliography; East Africa; Somali Republic. LCCN: 93093377.

Piccoli, G. and Carrelli, Antonio (president). 1980. "Cretaceous Tectonic Movements in Central Somalia; Geodynamic Evolution of the Afro-Arabian Rift

System." Atti Dei Convegni Lincei, Accademia Nazionale Dei Lincei. Issue 47, Pages 187-191. Descriptors: Africa; central Somali Republic; Cretaceous; East Africa; East African Rift; eustacy; Indian Ocean; lithofacies; Mesozoic; plate tectonics; Red Sea; rift zones; Somali Republic; structural geology; tectonics; tectonophysics. Notes: References: 15; sketch map. ISSN: 0391-805X.

Piccoli, G. and Ibrahim, Hersi A. 1981. Dizionario Geologico Italiano-Somalo-Inglese, --Erayfuraha Cilmiga Dhulka Ee Afafka Talyaaniga-Soomaaliga-Ingiriiska, --Italian-Somali-English Geological Dictionary. Italy: Coop. Tip., Padova, Italy. Descriptors: Africa; dictionaries; East Africa; geology; Somali Republic. Database: GeoRef. Accession Number: 1999-068793.

Piccoli, G. and Savazzi, E. 1984. "Five Shallow Benthic Molluscs Faunas from the Upper Eocene (Baron, Priabona, Garoowe, Nanggulan, Takashima)." Bollettino Della Societa Paleontologica Italiana. Volume 22, Descriptors: Africa; benthonic taxa; Cenozoic; East Africa; Eocene; faunal studies; Garoowe Somali Republic; Invertebrata; Mollusca; Nugaal Somali Republic; Paleogene; shallow-water environment; Somali Republic; Tertiary; upper Eocene. ISSN: 0375-7633.

Pike, R. W. 1958. Occurrence of Ground Water in Central Somalia. Somalia: Sinclair Somali Corporation, Mogadishu, Somalia. Descriptors: Africa; central Somali Republic; East Africa; ground water; Somali Republic; water resources. Database: GeoRef. Accession Number: 1999-068797.

Pistolesi, A. 1966. Photogeology of Scebeli River Area, Somalia, (Carta Geol. 1:250.000). Somalia: Scebel Oil Co., Mogadishu, Somalia. Descriptors: Africa; areal geology; cartography; East Africa; geologic maps; maps; photogeology; Shebelle River; Somali Republic. Database: GeoRef. Accession Number: 1999-068798.

Pomies, Catherine, Davies, Gareth R. and Conan, Sandrine M. H. 2002. "Neodymium in Modern Foraminifera from the Indian Ocean; Implications for the use of Foraminiferal Nd Isotope Compositions in Paleo-Oceanography." Earth & Planetary Science Letters. 15 Nov. Volume 203, Issue 3-4, Pages 1031-1045. Descriptors: Africa; alkaline earth metals; biochemistry; calcium; Cenozoic; concentration; continental margin; East Africa; Foraminifera; geochemistry; Globigerina; Globigerina bulloides; Globigerinacea; Globigerinidae; Globigerinoides; Globigerinoides ruber; Holocene; Indian Ocean; Invertebrata; isotopes; marine sediments; metals; modern; modern analogs; neodymium; paleo-oceanography; partition coefficients; Protista; Quaternary; rare earths; Rotaliina; sample preparation; sea water; sediment traps; sediments; Somali Basin; Somali Republic; stable isotopes. Notes: References: 40; illustrations incl. 3 tables, sketch map. Abstract: The use of the Nd isotope composition of planktic foraminifera as a proxy for sea surface water is based on the assumption that Nd is incorporated during initial calcification and preserved after foraminifera death, subsequent transport through the water column and storage in the sediment pile. To test this assumption Nd concentrations are reported from foraminiferal tests collected from 20 to 3045 m depth (multinets and sediment traps) from offshore Somalia and in the Somali Basin of the Indian Ocean. Nd partition coefficients calculated for two species Globigerinoides ruber and Globigerina bulloides sampled in the uppermost water column (100 m) are 16-52. The measured Nd/Ca ratio of living foraminifera is approximately 8 nmol/mol, which is 8-190 times lower than reported in cleaned foraminifera from sediment cores (61-1500 nmol/mol). Moreover, the Nd content of foraminiferal tests varies markedly within the water column (8-220 nmol/mol Ca) recording a positive correlation with Mn content (r (super 2) = 0.91). The observed changes in the Nd concentration after foraminiferal death appear to reflect Nd addition from deeper in the water column and imply a major disturbance of the primary Nd isotope composition formed at the surface. The high Nd/Ca reported for cleaned foraminifera from sediment cores are interpreted to be a consequence of current cleaning techniques. Although cleaning successfully removes organics and secondary Mn coatings into solution any released rare earth elements (REE) are subsequently re-adsorbed onto the calcite. Cleaning removes elements with a 2+ valency state into solution (e.g. Cd and Mn) but more reactive elements with higher valency states (e.g. REE (super 3+) , Th (super 4+)) are re-absorbed onto residual carbonate. It is concluded that, although foraminifera clearly record temporal variations in Nd isotope composition, the relationship between these variations and changes in the sea surface Nd isotope composition has yet to be established unambiguously. Further work is required to rule out the possibility that the Nd in foraminifera represents a mixed signal derived from surface sediments and bottom and pore waters. ISSN: 0012-821X.

Pons J.M, Schroeder J.H, Hofling R and Moussavian E. 1992. "Upper Cretaceous Rudist Assemblages in Northern Somalia." Geologica Romana. Volume 28, Pages 219-241. Descriptors: Invertebrates; Cretaceous; bivalve; rudist; palaeobiogeography; Campanian; Maastrichtian. Abstract: Cretaceous rocks in the area of Bosaso, Somalia, are represented by the Tisje Fm. and the lower part of the Auradu Limestones. In the limestones of the upper Tisje Fm., of Campanian age, two carbonate microfacies have been differentiated (Coral-Biolithite Facies and Detritus Facies) and are described. A carbonate microfacies of Maastrichtian age is described from the lower part of the Auradu Limestones. The rudist fauna at each of these levels is described. Campanian-Maastrichtian Somalian rudist fauna results to have similarities to the one of the northern rim of the eastern and western Mediterranean Tethys and, during Campanian, also to the Caribbean. ISSN: 0435-3927.

Popov, A., Kidwai, A. L. and Karani, Said A. 1972. "Groundwater in the Somali Democratic Republic; the Role of Hydrology and Hydrometerology in the Economic Development of Africa." WHO Publication. Volume 301, Vol. 2, Pages 340-342. Descriptors: Africa; East Africa; ground water; Jubba River; Shabelle River; Somali Republic; surveys; United Nations; World Health Organization. Database: GeoRef.

Popov, A. P. and Kidwai, A. L. 1972. "Ground Water in Somalia Democratic Republic; Part I and Part II." United States: Descriptors: Africa; East Africa; ground water; hydrology; Somali Republic; water resources. Database: GeoRef. Accession Number: 1999-068801.

Popov, A. P. Kidwai, A. L. and Karani, A. 1972. "Ground Water in the Somali Republic." International (III): ICA-FAO, International (III). Descriptors: Africa; East Africa; ground water; Somali Republic; water resources. Database: GeoRef. Accession Number: 1999-068802.

Popov, A. P. Somalia; Ministry of Mining and Water Resources and United Nations Development Programme. 1970-1971. "Hydrogeological Map." Somali Democratic Republic, Ministry of Mining and Water Resources, United Nations Development Programme, Mineral and Groundwater Survey Project (Phase II). Descriptors: Groundwater- Somalia- Maps. Notes: Description: 1 map on 4 sheets;

photocopy; sheets 61 x 138 cm. Scale 1:1,000,000; (E 410--E 510/N 120--S 20). Notes: Black line print. Ancillary map showing distribution of total hardness in shallow groundwater. Responsibility: prepared by A.P. Popov, United Nations Senior Hydrogeologist. OCLC Accession Number: 15088800.

Popov, A. V. 1971. "Ground Water in Somali Democratic Republic." United States: Descriptors: Africa; East Africa; ground water; Somali Republic; water resources. Notes: Technical report. Database: GeoRef. Accession Number: 1999-068800.

Portis, A. and Robecchi-Brichetti. 1896. "Catalogo Della Collezione Geologica Raccolta Nel Viaggio Da Zeila all'Harrar." Translated title: "Catalogue of the Geological Collection in the Zeila and Harrar Trip." Italy: Publisher unknown, Milan, Italy. Descriptors: Africa; areal geology; catalogs; East Africa; Ethiopia; Harar Ethiopia; Somali Republic; Zeila Somalia. Database: GeoRef. Accession Number: 1996-038385.

Poveri, Fontana F. 1937. "Note Sul Comprensorio Di Bonifica Di Genale Nella Somalia Italiana." Translated title: "Note on the Drainage of Genale Disrict in Italian Somalia." Italy: IAO, Firenze, Italy. Descriptors: Africa; agriculture; drainage; East Africa; Genale Somali Republic; hydrology; Shabeellaha Hoose; Somali Republic; water resources; water supply. Database: GeoRef. Accession Number: 1999-068805.

Pozzi, R. and Benvenuti, G. 1979. "Studio Geologico Applicato e Geofisico Per Dighe Subalvee Nel Distretto Del Nogal (Somalia Settentrionale)." Translated title: "Geotechnical and Geophysical Study on Subsurface Dams in the Nogal District; Northern Somalia." Memorie Di Scienze Geologiche. Volume 32, 33 p, Descriptors: Africa; alluvium; Auradu Limestone; Cenozoic; clastic sediments; Cretaceous; East Africa; electrical methods; engineering geology; Garoe; geophysical methods; geophysical surveys; ground water; ground-water dams; hydrochemistry; hydrogeology; Karkar Formation; Lesomma Formation; lithostratigraphy; Mesozoic; Nogal Valley; sedimentary rocks; sediments; size distribution; Somali Republic; surveys; Taleh Evaporites; underground installations. Notes: With summary in Somali; References: 18; illustrations incl. tables, sect., geol. sketch map. ISSN: 0391-8602.

Pozzi, R., Benvenuti, G., Gatti, G. and Farah, Ibrahim Mahamed. 198?. "Water Supply and Agricultural use; a Proposal for the Adoption of Subsurface Dams in Somalia." Descriptors: Africa; dams; East Africa; ground water; ground-water dams; irrigation; Somali Republic; water supply. Notes: Pubblicazione, Dipartimento de Scienze della Terra dell'Universita degli Studi Milano. Sezione di Geologia e Paleontologia. Nuova Serie, vol. 449, 25 pages, 198? Reprinted from Proceedings of the Second international congress of Somali studies, edited by Labahn, Thomas, Vol. 3, Aspects of development, Helmut Buske Verlag, Hamburg; illustrations incl. sketch map. Database: GeoRef.

Pozzi, R., Benvenuti, G., Gatti, G. and Ibrahim, Maxamed F. 1983. "Water Supply and Agricultural use; a Proposal for the Adoption of Subsurface Dams in Somalia." Quaderni Di Geologia Della Somalia (Mogadiscio). Volume 7, Descriptors: Africa; agriculture; dams; East Africa; ground water; ground-water dams; Somali Republic; water resources; water supply. Notes: Proc. II Internat. Congr. Somali Studies, 3, Helmut Buske Verlag, Humburg. LCCN: 93093377.

Pozzi, R. Benvenuti, G. and Maxamed, Huseen Salaad. 1987. Influence of Paleogeography on Water Resources in Arid and Semi-Arid Lands; an Example from Central Somalia. Somalia: Somali Natl. Univ., Mogadishu, Somalia. GeoSom 87;

Geology of Somalia and Surrounding Regions, Mogadishu. Somalia Conference: Nov. 23-30, 1987. Descriptors: Africa; arid environment; carbonate rocks; Cenozoic; central Somali Republic; clastic rocks; East Africa; economic geology; exploration; Galgadud; ground water; Hiran; hydrogeology; lagoonal environment; marine environment; Miocene; Mudug Beds; Mudugh; Neogene; Nugal; Obbia; Oligocene; Paleogene; paleogeography; paleohydrology; resources; sedimentary rocks; semi-arid environment; shallow-water environment; Shebelle River; Somali Republic; surveys; Taleh Formation; terrestrial environment; Tertiary; Trap Series; water quality; water resources. OCLC Accession Number: 34175029.

Pozzi, R., Benvenuti, G., Mohamed, Cabdi Xaaji and Shuurije, Cabdi Iidle. 1982. "Groundwater Resources in Central Somalia." Memorie Di Scienze Geologiche. Volume 35, Pages 397-409. Descriptors: Africa; aquifers; central Somalia; East Africa; economic geology; ground water; hydrogeologic maps; hydrogeology; maps; Somali Republic; surveys; water resources; water wells; wells. Notes: References: 14; illustrations incl. 2 tables. ISSN: 0391-8602.

Pozzi, R. and Hussein, Salad M. 1984. "Ground Water Resources in Hobyo Area (Mudug Region-Central Somalia)." Memorie Di Scienze Geologiche. Volume 36, Pages 443-451. Descriptors: Africa; central Somali Republic; East Africa; ground water; Hobyo Somali Republic; Mudug Somali Republic; Somali Republic; water resources. ISSN: 0391-8602.

Pozzi, Renato, Benvenuti, Giovanni and Husseen, Salaad Maxamed. 1987. "Studio Idrogeologico e Geofisico Dell'Area Di Cadale (Somalia Centrale)." Translated title: "Hydrological and Geophysical Study of the Cadale Area, Central Somalia." Memorie Di Scienze Geologiche. Volume 39, Pages 245-256. Descriptors: Africa; aquifers; Cadale Somali Republic; discharge; East Africa; ground water; hydrodynamics; hydrology; pumping; recharge; salt-water intrusion; Somali Republic; water quality; water resources; water table; water use; well-logging. Notes: References: 18; illustrations incl. 1 table. ISSN: 0391-8602.

Pozzi, Renato, Benvenuti, Giovanni and Maxamed, Husseen Salaad. 1987. "Studio Idrogeologico e Geofisico Dell'Area Di Cadale (Somalia Centrale)." Translated title: "Hydrogeological and Geophysical Studies of the Cadale Area, Central Somalia." Pubblicazione, Dipartimento De Scienze Della Terra Dell'Universita Degli Studi Milano. Sezione Di Geologia e Paleontologia. Nuova Serie. Oct. Volume 523, Pages 12. Descriptors: Africa; aquifers; Cadale Somalia; East Africa; electrical methods; electrical sounding; exploitation; geophysical methods; geophysical surveys; ground water; salt-water intrusion; shallow aquifers; Somali Republic; surveys; water resources. Notes: Reprinted from Mem. Sci. Geol., Univ. Padova, Vol. 34, p. 245-256, 1987; References: 18; illustrations incl. 1 table, sketch map. Database: GeoRef.

Pozzi, Renato, Robba, Elio, Bernasconi, Maria Pia and Huseen, Salaad Maxamed. 1985. "Late Paleogene-Early Middle Miocene Formations in Obbia Area (Mudug Region, Central Somalia)." Memorie Di Scienze Geologiche. Volume 37, Pages 423-434. Descriptors: Africa; algae; biostratigraphy; Bivalvia; Cenozoic; East Africa; Foraminifera; Gastropoda; geologic maps; Invertebrata; lithofacies; maps; microfossils; middle Miocene; Miocene; Mollusca; Mudug Somali Republic; Neogene; Obbia Somali Republic; Paleogene; Plantae; Protista; sequence stratigraphy; Somali Republic;

stratigraphic units; Tertiary; thallophytes; unconformities; upper Paleogene. Notes: References: 27; illustrations incl. 1 plate. ISSN: 0391-8602.

"Preliminary Report on the Chemical Analysis of Fluoride Content in Water from Wells of Middle and Northern Somalia." 1980. Odontostomatol. Trop. Dec. Volume 3, Issue 4, Pages 155-157. Descriptors: Fluorides- analysis; Somalia; Water Supply and analysis. ISSN: 0251-172X.

Previatello, P. Radina, B. and Soranzo, M. 1987. A First Approach to the Study of the Cohesive Soil of the Shabelle Valley to be Utilized as River Embankments. Somalia: Somali Natl. Univ., Mogadishu, Somalia. GeoSom 87; Geology of Somalia and Surrounding Regions, Mogadishu. Somalia Conference: Nov. 23-30, 1987. Descriptors: Africa; Alluvial soils; clastic sediments; clay; cohesive materials; compaction; compression; East Africa; embankments; engineering geology; materials, properties; plasticity; sediments; shear strength; Shebelle Valley; slope stability; soil mechanics; soils; Somali Republic; triaxial tests. OCLC Accession Number: 34175029.

Previatello, Paolo, Radina, Bruno and Soranzo, Maurizio. 1990. "A Study of Cohesive Soils of the Shabelle Valley for River Embankments." Memorie Di Scienze Geologiche. Volume 42, Pages 35-40. Descriptors: Africa; clastic sediments; clay; compaction; construction; East Africa; embankments; experimental studies; floods; geologic hazards; mechanical properties; physical properties; plasticity; sediments; shear strength; Shebelle Valley; silt; soil mechanics; Somali Republic; testing. Notes: References: 14; illustrations incl. 2 tables, sketch map. ISSN: 0391-8602.

Prior J. 1994. "Pastoral Development Planning." Oxfam, Development Guidelines, number 9. Descriptors: 150 pages; Planning; project experience; pastoral development; rural planning; developing country. Abstract: Attempts to demystify pastoral development through an examination of the impacts of the more common forms of project experience. The first section examines the recent forces of change within the pastoral sector and appraises the development record in the light of the needs of pastoral peoples. The second looks in detail at the experiences and lessons of the Erigavo Erosion Control and Range Management Project in NW Somalia (Somaliland). The next section speculates on the future of pastoral development, discussing ways to reduce vulnerability and encourage social development through the strengthening institutional capacity of pastoral communities. Notes: In: Pastoral development planning (1994) 150 pages; Geographic: Somalia- Somaliland; Update: A 01 JAN 1995. ISBN: 0855982039.

Proposals for a Second Phase "Mineral and Groundwater Survey Project" of the Somali Government, to be Executed with Assistance of the United Nations Development Programme (Special Fund). 1967. Somalia: U.N.D.P., Mogadischu, Somalia. Descriptors: Africa; East Africa; ground water; mineral resources; Somali Republic; surveys; UNDP. Database: GeoRef. Accession Number: 1996-069192.

Purdy, John; Survey of Kenya; Robert Laurie and James Whittle; British Museum and Trustees. 1961. "Part of a Map of the Continent and Islands of Africa, John Purdy, 1809 A.D." Nairobi: Survey of Kenya. Descriptors: Maps- Facsimiles; Government publication; National government publication. Notes: Description: 1 map; 36 x 35 cm. Geographic: Ethiopia- Maps. Somalia- Maps. Kenya- Maps. Indian Coast (Africa)- Maps. Scale not given. Notes: Relief shown by hachures. Shows present-day area of eastern Ethiopia and coastal areas of Kenya and Somalia. "Reproduced by kind permission of the

Trustees of the British Museum, London." "44." Original: Purdy, John, 1773-1843. Atlas of Kenya. OCLC Accession Number: 39690191.

Puri, R. K. 1962. Bibliography Relating to Geology, Mineral Resources, Paleontology, etc., of Somali Republic. Somalia: Somali Republic Geol. Surv., Hargeisa, Somalia. Descriptors: Africa; areal geology; bibliography; East Africa; Somali Republic. Database: GeoRef. Accession Number: 2000-012044.

Puri, R. K. and Somalia Geological Survey (see above). 1961. Bibliography Relating to Geology, Mineral Resources, Palaeontology, etc., of Somali Republic. N.p. Descriptors: Geology- Somalia- Bibliography; Mines and mineral resources- Somalia- Bibliography; Paleontology- Somalia- Bibliography. Notes: At head of title: Somali Republic Geological Survey. OCLC Accession Number: 15537503.

R

R. Rumbold et Associés. 1980. Somalia on-Shore Areas: Summary of Stratigraphy, Tectonics and Hydrocarbon Potential. Saint Ambroix, France: R. Rumbold et Associés. Descriptors: 103, 20 leaves; Geology- Somalia; Geology, Stratigraphic- Somalia; Geology, Structural- Somalia; Petroleum- Geology- Somalia. "November, 1980." OCLC Accession Number: 14248212.

Rabinowitz, Philip D., Coffin, Millard F. and Falvey, David. 1982. "Salt Diapirs Bordering the Continental Margin of Northern Kenya and Southern Somalia." Science. 05 Feb. Volume 215, Issue 4533, Pages 663-665. Descriptors: Africa; chemically precipitated rocks; continental margin; continental shelf; continental slope; diapirs; East Africa; evaporites; geophysical methods; geophysical surveys; Indian Ocean; Jurassic; Kenya; Lower Jurassic; Mesozoic; oceanography; plate tectonics; rifting; salt tectonics; sedimentary rocks; seismic methods; Somali Republic; structural geology; surveys; tectonics; tectonophysics; West Indian Ocean. Notes: Lamont-Doherty Geol. Obs. Contrib. No. 3271; References: 19; illustrations incl. sketch map. ISSN: 0036-8075.

Ranganathan V. 1992. "Rural Electrification in Africa." Zed Books, in association with African Energy Policy Research Network, Gaborone. Descriptors: 182 pages; Fuel and Energy; Water; electricity supply; rural community; developing country; policy recommendation; economic development; electrification; rural development; energy supply. Abstract: This book, which brings together the findings of a research programme carried out under AFREPEN's auspices, provides an assessment of the extent to which African countries have succeeded in bringing electricity to rural communities. The case studies of Zambia, Ethiopia, Botswana, Lesotho and Somalia have a largely uniform framework, where three areas (one high-income electrified area, one low-income electrified area, and one unelectrified area) are compared and contrasted in order to explore the nexus between electrification and development. The second part of each of the country studies is the ex-ante study, carrying out the financial and economic analysis of the projected electrification of the third area in the survey. Some of the country studies have conducted the ex-ante appraisal for all three areas. A final chapter offers policy recommendations and conclusions. Notes: In: Rural electrification in Africa (1992) p. 182p; Geographic: Zambia Ethiopia Botswana Lesotho Somalia Africa; Notes: Special Features: index. ISBN: 1856491110.

Rao, G. V. and van de Boogaard, H. M. E. 1981. "Structure of the Somali Jet Deduced from Aerial Observations Taken during June-July, 1977." Joint IUTAM/IUGG

International Symposium on Monsoon Dynamics, New Delhi, Dec. 5-9, 1977, Monsoon Dynamics: Proceedings, N. Y. , Cambridge University Press, 1981. Pages 321-331. Descriptors: Jet stream studies; Water vapor flux; Coastal Waters of East Africa; Kenya; Somalia. Notes: Refs. Abstract: An aerial reconnaissance program was sponsored by the U.S. National Science Foundation in June and July 1977, under which NCAR's Electra was used in 13 flights to reconnoiter the Somali Jet by the east African coast. The base of the operations was Nairobi, Kenya. Cross sections along 2°S between 37 and 44° E disclosing the kinematic structure of the jet on four different days were constructed from observations taken on these flights. The major points of interest of these cross sections are the mesoscale variation of the jet, both in intensity and in its frequency of occurrence, and the existence of a secondary wind maximum over the Indian Ocean near the coast. The water vapor flux across 2° S for three different days (June 11, June 15, and July 5) and one night (June 29) was computed for the section between 1000 and 700 mb and 37 and 44°E. These fluxes revealed some day-to-day variations. The average flux was found to be 1.79 x 10 super(1) super(0) ton/day. This shows that the water vapor flux for Aug. 1964 across the Equator at 1000-450 mb between 420 and 75° E was previously underestimated. Database: Meteorological & Geoastrophysical Abstracts.

Rapetti, G. 1935. "L'Opera Della SAIS in Somalia." Translated title: "SAIS Operations in Somalia." Italy: Istituto di Agronomia Coloniale Italiana, Florence, Italy. Descriptors: Africa; East Africa; exploration; hydrology; Somali Republic; water resources. Database: GeoRef. Accession Number: 2000-018801.

Rapolla A, Cella F and Dorre A.S. 1995. "Gravity Study of the Crustal Structures of Somalia Along International Lithosphere Program Geotransects." Journal of African Earth Sciences. Volume 20, Issue 3-4, Pages 263-274. Descriptors: Gravity- 72.11.2; basement morphology; gravity survey; crustal structure. Notes: Somalia. Abstract: Gravity data have been used to examine the crystalline basement morphology along five geotransects in Somalia defined by the Global Geotransect Project. The gravity data were digitized from the 1:1000 000 Gravity Anomaly Map of Somalia produced by the African Gravity Project. After the removal of the non-crustal wavelength anomalies from the observed gravity field, the remaining gravity anomalies were interpreted in terms of 2.5D crustal models. The results of the 2.5D gravity modelling indicate that the basement beneath the southern Somali basins is partially or totally transformed to denser material and that, just a few hundred kilometres offshore from Somalia, the basement is of an oceanic nature. ISSN: 0899-5362.

Rapp, R. A., Alvarez, I. and Wendel, J. F. 2005. "Molecular Confirmation of the Position of Gossypium Trifurcatum Vollesen." Genet. Resour. Crop Evol. SEP. Volume 52, Issue 6, Pages 749-753. Abstract: Taxonomic understanding is a necessary prerequisite for intelligent germplasm maintenance and evaluation. Here, we use molecular evidence to address the generic position of the poorly known and morphologically unusual taxon Gossypium trifurcatum Vollesen. This species possesses dentate leaves, a feature not otherwise found in Gossypium L. but one that is common in Cienfuegosia Cav., a related genus in the small Malvaceous tribe Gossypieae. G. trifurcatum is a rare plant, restricted to deserts of Eastern Somalia and known from only two collections, the last in 1980. Using DNA extracted from an herbarium specimen, we amplified and sequenced the chloroplast gene ndhF. Phylogenetic analysis reveals G. trifurcatum to be cladistically nested within Gossypium. These data diagnose dentate leaves as an autapomorphy within a genetically diverse assemblage of African-Arabian species, which remain the least well-represented cottons in germplasm collections.

Rebmann, Georg Friedrich; Survey of Kenya; British Museum and Trustees. 1961. "Imperfect Sketch of a Map of East Africa, J. Rebmann, 1850 A.D." Nairobi: Survey of Kenya. Descriptors: Maps- Facsimiles; Government publication; National government publication. Notes: Description: 1 map; 29 x 36 cm. Geographic: Tanzania-Maps. Kenya- Maps. Somalia- Maps. Indian Coast (Africa)- Maps. Scale not given; (E 29000'--E 44000'/N 1030'--S 10030'). Notes: Relief shown by hachures. Shows present-day coastal areas of Tanzania, Kenya, and Somalia. "Reproduced by kind permission of the Trustees of the British Museum, London." Includes notes. "45." Original: Rebmann, Georg Friedrich, 1768-1824. Imperfect sketch of a map from 1 1/20 north to 10 1/20 south latitude and from 29 to 44 degrees east longitude. Atlas of Kenya. Other Titles: Original title; Imperfect sketch of a map from 1 1/20 north to 10 1/20 south latitude and from 29 to 44 degrees east longitude. OCLC Accession Number: 39686234.

"Recuperation of Uranium and Gold in Mineral Pulps by Adsorption." 1985. International Atomic Energy Agency, Vienna (Austria). Jun. Volume: INISBR577, CONF85093691, pages: 14. Descriptors: Ion Exchange; Ore Processing; Separation Processes; Slurries; Somalia; Gold; Gold Ores; Uranium; Uranium Ores; ERDA/400105; ERDA/050400; Foreign technology; Natural resources and earth sciences Mineral industries; Chemistry Industrial chemistry and chemical process engineering. Abstract: The technological routes for the treatment of the gold and uranium ores are presented. The results obtained during the continuous tests with the uraniferous Ores of Wabo in Somalia are presented. The utilization of 99% of the uranium content in the alkaline pulp is obtained. (Atomindex citation 18:012393). Notes: In Portuguese. Brazilian congress on mining, Brasilia, Brazil, 23 Sep 1985. NTIS Accession Number: DE87701446XSP.

Redding, Stephen; Sitnam, Paul; Wood, Graham and CARE International in Kenya. 1994. The Refugee Assistance Project of CARE-Kenya, 1991-1993. CARE-Kenya. Descriptors: 16; Refugees- Kenya; Refugees- Somalia; Refugee camps- Kenya; Humanitarian assistance- Kenya; Somalis; Refugee camps; Emergency relief; Refugee aid and development; Food supply; Sanitation; Water supply; Social services. Notes: Named Corp: CARE International in Kenya. OCLC Accession Number: 70147526.

Reddy S.R, Rao M.V, Easton A.K, Clarke S.R and Nath A.N. 1995. "Gyres Off the Somali Coast and Western Boundary Currents in the Bay of Bengal during the South-West Monsoon." Int. J. Remote Sens. Volume 16, Issue 9, Pages 1679-1684. Descriptors: Circulation; current system; monsoon; gyre; western boundary currents; monsoons; coastal circulation. Notes: Somalia Bay of Bengal. Abstract: Satellite remotely-sensed sea surface temperature (SST) data were obtained during the initial phase of the onset of the south-west monsoon in the north-western Indian Ocean for the years 1987 and 1988. Large wedge-like areas of upwelled water during 1988 were observed at 5°N and 10°N after the Somali current spinup, indicative of a 2-gyre circulation in the Somali current system. Satellite infrared observations of the Bay of Bengal during the early February 1990 revealed the existence of two bands of warm water that resembled a western boundary current (WBC) along the east coast of India. A warm core eddy, with a major axis of nearly 120km, appeared at 89°E and 19°N at the end of the axis of the current. The existence of the two gyre system off the Somali coast and the presence or absence of the

WBC in the Bay of Bengal in response to the onset of the summer monsoon has been discussed in detail. ISSN: 0143-1161.

Redfield, T. F., Wheeler, W. H. and Often, M. 2003. "A Kinematic Model for the Development of the Afar Depression and its Paleogeographic Implications." Earth & Planetary Science Letters. NOV 30. Volume 216, Issue 3, Pages 383-398. Abstract: The Afar Depression is a highly extended region of continental to transitional oceanic crust lying at the junction of the Red Sea, the Gulf of Aden and the Ethiopian rifts. We analyze the evolution of the Afar crust using plate kinematics and published crustal models to constrain the temporal and volumetric evolution of the rift basin. Our reconstruction constrains the regional-scale initial 3D geometry and subsequent extension and is well calibrated at the onset of rifting (similar to 20 Ma) and from the time of earliest documented sea-floor spreading anomalies (similar to6 Ma Red Sea; similar to10 Ma Gulf of Aden)... Syn-rift sedimentary and magmatic additions to the crust are taken from the literature. Our analysis reveals a discrepancy: either the base of the crust has not been properly imaged, or a (plume-related?) process has somehow caused bulk removal of crustal material since extension began. Inferring subsidence history from thermal modeling and flexural considerations, we conclude subsidence in Afar was virtually complete by Mid Pliocene time. Our analysis contradicts interpretations of late (post 3 Ma) large (similar to2 km) subsidence of the Hadar area near the Ethiopian Plateau, suggesting paleoclimatic data record regional, not local, climate change. Tectonic reconstruction (supported by paleontologic and isotopic data) suggests that a land bridge connected Africa and Arabia, via Danakil, up to the Early to Middle Pliocene. The temporal constraints on land bridge and escarpment morphology constrain Afar paleogeography, climate, and faunal migration routes. These constraints (particularly the development of geographic isolation) are fundamentally important for models evaluating and interpreting biologic evolution in the Afar, including speciation and human origins.

Rees, D. J., Omar, A. M. and Rodol, O. 1991. "Implications of the Rainfall Climate of Southern Somalia for Semi-Mechanized Rain-Fed Crop Production." Agricultural and Forest Meteorology. Volume 56, Issue 1-2, Pages 21-33. Descriptors: Rainfall effects on crops; Crop water economy; Somalia. Notes: July 1991. Refs., figs., tables. Abstract: Ten years of climate and yield data were used to evaluate semi-mechanized rain-fed farming practices on vertisols in southern Somalia. Median rainfall totals for the main rainy season varied between 384 and 432 mm, and maize production was reasonably successful, with a rainfall water use efficiency for grain of 9 kg ha super(-) super(1) mm super(-) super(1). Estimated probabilities of crop failure varied between 0.1 and 0.3, and of reasonable yields between 0.5 and 0.67. The local cowpea variety proved to have a low yield potential, with a water use efficiency of only 1 kg ha super(-) super(1) mm super(-) super(1) , but a low minimum water requirement, resulting in low yields but a low frequency of crop failure. Rainfall during the short rainy season was low and insufficient to justify cropping. Analysis of the relations of yield with current and preceding fallow rainfall indicated that fallow soil management failed to increase yields. Cumulative probabilities of planting rains did not reach 0.5 until the latter part of April which, together with the rapidly decreasing probabilities of rainfall during June, indicates that short-season maize varieties may be better adapted to the environment. The probabilities of too much rainfall hampering crop and soil management activities and causing crop waterlogging damage are sufficiently high to warrant investigations on methods of improved surface drainage. ISSN: 0168-1923.

Reeves, C. V., de Wit, M. J. and Sahu, B. K. 2004. "Tight Reassembly of Gondwana Exposes Phanerozoic Shears in Africa as Global Tectonic Players." Gondwana Research. JAN. Volume 7, Issue 1, Pages 7-19. Abstract: Aeromagnetic surveys help reveal the geometry of Precambrian terranes through extending the mapping of structures and lithologies from well-exposed areas into areas of younger cover. Continent-wide aeromagnetic compilations therefore help extend geological mapping beyond the scale of a single country and, in turn, help link regional geology with processes of global tectonics. In Africa, India and related smaller fragments of Gondwana, the margins of Precambrian crustal blocks that have escaped (or successfully resisted) fracture or extension in Phanerozoic time can often be identified from their aeromagnetic expression. We differentiate between these rigid pieces of Precambrian crust and the intervening lithosphere that has been subjected to deformation (usually a combination of extension and strike-slip) in one or more of three rifting episodes affecting Africa during the Phanerozoic: Karoo, Early Cretaceous and (post-) Miocene, Modest relative movements between adjacent fragments in the African mosaic, commensurate with the observed rifting and transcurrent faulting, lead to small adjustments in the position of sub-Saharan Africa with respect to North Africa and Arabia. The tight reassembly of Precambrian sub-Saharan Africa with Madagascar, India, Sri Lanka and Antarctica (see animation in http://kartoweb.itc.nl/gondwana) can then be extended north between NW India and Somalia once the Early Cretaceous movements in North Africa have been undone. The Seychelles and smaller continental fragments that stayed with India may be accommodated north of Madagascar. The reassembly includes an attempt to undo strike-slip on the Southern Trans-Africa Shear System. This cryptic tectonic transcontinental corridor, which first formed as a Pan-African shear belt 700-500 Ma, also displays demonstrable dextral and sinistral movement between 300 and 200 Ma, not only evident in the alignment of the unsuccessful Karoo rifts now mapped from Tanzania to Namibia but also having an effect on many of the eventually successful rifts between Africa-Arabia and East Gondwana. We postulate its continuation into the Tethys Ocean as a major transform or megashear, allowing minor independence of movements between West Gondwana (partnered across the Tethys Ocean with Europe) and East Gondwana (partnered with Asia), Europe and Asia being independent before the similar to250 Ma consolidation of the Urals suture. The relative importance of primary driving forces, such as subduction 'pull', and 'jostling' forces experienced between adjacent rigid fragments could be related to plate size, the larger plates being relatively closely-coupled to the convecting mantle in the global scheme while the smaller ones may experience a preponderance of 'jostling' forces from their rigid neighbours.

Resource Management & Research. 1983. "Somali Democratic Republic. Southern Rangelands Survey: Census Results, Maps." London: Resource Management & Research. Descriptors: Rangelands- Somalia- Maps. Notes: Description: maps; both sides; 76 x 67 cm., sheets 46 x 74 cm. Scale 1:1,000,000; (E 410--E 470/N 50--S 10). Cartgrph Code: Category of scale: a Constant ratio linear horizontal scale: 1000000 Coordinates--westernmost longitude: E0410000 Coordinates--easternmost longitude: E0470000; Notes: Relief of some sheets shown by contours. Originally bounded and numbered separately, e.g. Volume 1, part. 2. Each sheet individually numbered and titled, e.g. Somalia, southern rangelands, Figure 1.01: principal geographic features of the southern rangelands of Somalia. Each map printed on 2 sheets. Acommpanied by: Profiles. sheets (some on verso of map); 60 cm.- Texts R.M. Watson, J.M. Nimmo. vols. maps, index; 30 cm. LCCN: 94-685746; 87-675503. OCLC Accession Number: 31817004.

Resource Management & Research. 1981. "Somali Democratic Republic. Northern Rangelands Survey: Census Results, Maps." London: Resource Management & Research. Descriptors: Rangelands- Somalia- Maps. Notes: Description: maps; 55 x 50 cm. Scale 1:1,000,000; (E 430--E 510/N 120--N 80). Cartgrph Code: Category of scale: a Constant ratio linear horizontal scale: 1000000 Coordinates--westernmost longitude: E0430000 Coordinates--easternmost longitude: E0510000; Notes: Cover title. Published in book form and numbered separately, e.g. Volume 1, part. 2. Each sheet individually titled, e.g. Somalia, northern rangelands (western section): principal geographic features of the northern rangelands of Somalia (Western section). LCCN: 94-685744; OCLC Accession Number: 31817001.

Resource Management & Research. 1980-9999. "Somali Democratic Republic. Central Rangelands Survey: Static Range Resources, Maps." London: Resource Management & Research. Descriptors: Rangelands- Somalia- Maps. Notes: Description: maps; 55 x 50 cm. Scale 1:1,000,000; (E 450--E 490/N 80--N 30). Cartgrph Code: Category of scale: a Constant ratio linear horizontal scale: 1000000 Coordinates--westernmost longitude: E0450000 Coordinates--easternmost longitude: E0490000; Notes: Cover title. Originally bounded and numbered separately, e.g. Volume 1, part. 2. Each sheet individually numbered and titled, e.g. Somali, central rangelands, figure 1.01: principal geographic features of the three regions. Based on Somali Survey Department 1:100,000 map series. Acommpanied by profiles. 1 sheet; 60 cm. LCCN: 94-685745; OCLC Accession Number: 31817002.

Reznikov M, Niemi T.M, Ben-Avraham Z and Hartnady C. 2005. "Structure of the Transkei Basin and Natal Valley, Southwest Indian Ocean, from Seismic Reflection and Potential Field Data." Tectonophysics. 03 MAR. Volume 397, Issue 1-2, SPEC. ISS., Pages 127-141. Descriptors: Regional structure and tectonics; Structural geology and tectonics; oceanic crust; basin evolution; structural geology; seismic survey; plate boundary; magnetic anomaly Species Term: Somalia. Notes: Additional Info: Netherlands; References: 41; Geographic: Indian Ocean (Southwest) Indian Ocean oceanic regions World. Abstract: Marine geophysical data from the southern Natal Valley and northern Transkei Basin, offshore southeast Africa, were used to study the structure of the crust and sedimentary cover in the area. The data includes seismic reflection, gravity and magnetics and provides information on the acoustic basement geometry (where available), features of the sedimentary cover and the basin's development. Previously mapped Mesozoic magnetic anomalies over a part of the basin are now recognized over wider areas of the basin. The ability to extend the correlation to the southeast within the Natal Valley further confirms an oceanic origin for this region and provides an opportunity to amplify the existing plate boundary reconstructions. The stratigraphic structure of the southern Natal Valley and the northern Transkei Basin reflects processes of the ocean crust formation and subsequent evolution. The highly variable relief of the acoustic basement may relate to the crust formation in the immediate vicinity of the continental transform margin. Renewed submarine seismicity and neotectonic activity in the area is probably related to the diffuse boundary between the Nubia and Somalia plates. 2.5-D crustal models show that a 1.7-3.2-km-thick sediment sequence overlies a 6.3±1.2-km-thick normal oceanic crust in the deep southern Natal Valley and Transkei Basin. The oceanic crust in the study area is heterogeneous, made up of blocks of laterally varying remanent magnetization (0.5-3.5 A/m) and density (2850-2900 kg/m3). Strong modifications of accretionary processes near ridge/fracture zone intersections may be a reason of such heterogeneity. ISSN: 0040-1951.

Richardson, Jacques G. 1993. Decision Making Analysis and Simulation: The Novel Concept of `Humanitarian Intervention' in International Relations. Le Touquet, Fr: Publ by IEEE, Piscataway, NJ, USA. Volume: 2, pages: 438-443. Proceedings of the IEEE International Conference on Systems, Man and Cybernetics. Part 2 (of 5), Oct 17-20 1993. Descriptors: Decision theory; Social sciences; International law; International cooperation; Large scale systems; Social aspects; Simulation. Abstract: Until the 1990s it was canon law in foreign relations that a sovereign state did not interfere in another's internal affairs. After the Second World War, events saw a deterioration in this canon, both in fact and in law. The cases of Somalia and Bosnia-Herzegovina are examined here. In systemic terminology, dimensionality and both hard and soft variables raised the level of complexity of the decision processes involved. By the early 1990s, ethnic strife in Bosnia-Herzegovina and anarchy and famine in Somalia, together with major violations of human rights in both these areas, caused a radical change in attitude among governments of the major democracies as well as in both the Security Council and General Assembly of the United Nations. The decisional procedure involved needed to take into account a host of problems: historical, political and geopolitical, social and cultural, military, economic and environmental. These factors added dimensions of complexity sometimes bordering on chaos, portraying much contemporary decision-making in the search for justice and equitable resolution of conflict. ISBN: 0884-3627; 0-7803-0911-1.

Riddell J.C. 1989. "Rural Development and the Cadastre: Issues and Examples from Somalia." Land Reform, Land Settlement & Cooperatives. Volume 1-2, Pages 79-96. Descriptors: Social, economic and political aspects of rural change; dam construction; land tenure policy; cadastre. Notes: Somalia- Jubba River. Abstract: In all agrarian societies there are locally recognized means for acquiring, exchanging and transferring land even though these are often organized along traditional patterns and outside the national institutional structures. A development oriented land tenure policy to be effective will have to take into account the nature of these transfers, the actual rights exchanged and the security involved. This paper discusses the role a well designed multi-purpose cadastre can play in implementing an effective, development oriented land tenure policy without sacrificing equity and participatory goals. Examples are drawn from recent field studies in Somalia, where there is a plan to dam the Jubba River for energy production and agricultural development. The paper analyses the kinds of land tenure problems that exist and which will be exacerbated by the increase in demand for land following the construction of the dam, and the role of the cadastre in such circumstances.

Rirash M.A. 1992. "Somali Oral Poetry as a Vehicle for Understanding Disequilibrium and Conflicts in a Pastoral Society." Nomadic Peoples. Volume 30, Pages 114-121. Descriptors: pastoralism; social change; poetry; water. Notes: Somalia.

Abstract: Somali oral poetry is a vital aspect of Somali pastoral society and the medium through which the Somalis depict their history and express their feelings towards both friends and foes. This paper aims to explain how Somali oral poetry is transmitted from one generation to another and highlights the pastoralists' reaction to and rationalization of challenges and conflicts, internal and external. ISSN: 0822-7942.

Roark, Philip. 1985. "Development of Groundwater in Karst Zones of Somalia; Symposium on Tropical Hydrology and 2nd Caribbean Islands Water Resources Congress." American Water Resources Association Technical Publication Series TPS. Volume 85-1, Pages 3-6. Descriptors: Africa; Anole Formation; East Africa; economic geology; ground water; hydrogeology; Iscia Baidoa Formation; karst; Somali Republic; south-central Somali Republic; surveys; water management; water quality; water resources. References: 4; ISSN: 0731-9789.

Robba, E. Angelucci, A. Boccaletti, M. Piccoli, G. Arush, M. A. and Cabdulqaadir, M. M. 1987. Geologic Evolution of Central and Southern Somalia from the Triassic to Recent. Somalia: Somali Natl. Univ., Mogadishu, Somalia. GeoSom 87; Geology of Somalia and Surrounding Regions, Mogadishu. Somalia Conference: Nov. 23-30, 1987. Descriptors: Africa; Arabian Sea; areal geology; Cenozoic; central Somali Republic; chemically precipitated rocks; East Africa; evaporites; Gulf of Aden; Holocene; Indian Ocean; Indian Ocean Islands; Jurassic; Madagascar; Main Gypsum Formation; marine environment; Mesozoic; Neogene; Quaternary; regional; sedimentary rocks; Somali Basin; Somali Republic; southern Somali Republic; Tertiary; Triassic; Yesomma Formation. OCLC Accession Number: 34175029.

Robecchi Bricchetti, L. 1899. "Somalia e Benadir." Translated title: "Somalia and Benadir." Italy: Aliprandi, Milan, Italy. Descriptors: Africa; Benadir Somali Republic; East Africa; geography; Somali Republic. Database: GeoRef. Accession Number: 2000-012048.

Roberge D. 2005. "Après Le Tsunami." Geomatica. Volume 59, Issue 4, Pages 445-450. Descriptors: Natural Hazards; tsunami; natural hazard Species Term: Somalia. Abstract: In the early hours of the day after Christmas 2004, a major earthquake measuring 9.0 on the Richter scale struck the coast of many South-East Asian countries. The quake triggered a powerful tsunami, reaching ten to twenty metres in height moving through the Indian Ocean at over 500 kilometres an hour. The tsunami flooded coastal areas in India, Indonesia, Sri Lanka, Thailand, Maldives, Myanmar, Seychelles, and Somalia, wiping away homes and lives. This cataclysm made us realize how vulnerable humanity is in the face of nature's strength. The weeks that followed the tragedy were focused on rescue efforts and tending to the survivors; trying to meet the latter's primary needs by providing them care, food and water, and emergency shelters were obvious priorities. Reconstruction of infrastructures such as roads, bridges, sanitary networks, etc. would have to follow. How can we resettle communities while respecting land rights that prevailed before the deadly wave that wiped out all landmarks? During the reconstruction phase, the international community of land surveyors will have to be present in order to promote the importance of land rights issues and tenure. If the land related issues are well managed in the resettlement plan, it can translate into an opportunity to improve the land rights situation. If not, it can aggravate an already problematic situation. The International Federation of Surveyors (FIG) must take the lead along with the United Nations (UN) agencies to promote land tenure issues and ensure that they are taken into consideration in the reconstruction and resettlement plans for the communities affected by such natural disasters. ISSN: 1195-1036.

Roberts, D. G. 1969. "Structural Evolution of the Rift Zones in the Middle East." Nature (London). Volume 223, Issue 5201, Pages 55-57. Descriptors: Africa; Arabian Peninsula; Asia; East Africa; evolution; Middle East; Nubia; rift zones; sea-floor spreading; Somali Republic; structural geology; tectonics. Notes: sketch maps; Abstract: Plate tectonics, Nubia-Somalia-Arabia blocks, separation attributed to differing rates-directions of sea-floor spreading, postulated four-plate system developed since late Miocene-early Pliocene, Danakil horst block rotation. ISSN: 0028-0836.

Robertson Research International, compiler. 1987. Muqdisho. Mogadishu. Morocco: Arab Organisation for Mineral Resources, Rabat, Morocco. Descriptors: Africa; Arabian Peninsula; Asia; East Africa; economic geology maps; geochemistry; host rocks; maps; mineral deposits, genesis; mineral resources; mines; Mogadishu map sheet; morphology; Muqdisho map sheet; production; Somali Republic; Yemen. Scale: 1:2,500,000. Type: colored economic geology map. Database: GeoRef. Accession Number: 2005-008130.

Robinson A.P. 1988. "Charcoal-Making in Somalia: A Look at the Bay Method." Unasylva. Volume 40, Issue 159, Pages 42-49. Descriptors: tree regeneration; Bay Method; charcoal production; cooperatives; resource depletion Species Term: Somalia. Notes: Somalia. Abstract: The Bay Region of Somalia is the main charcoal production area serving the national capital and largest city of Mogadishu, some 300-350 km away. Charcoal production areas are licensed by the National Range Agency to charcoal cooperatives and operated by cooperative members living in temporary camps. An unusually efficient charcoal production method developed in the region forms the subject of this article. Unfortunately, however, nothing is being done to regenerate the trees used to make the charcoal, so that efficient production, in this case, is leading to steady resource depletion. ISSN: 0041-6436.

Roche, J. 1983. "Ornithological Research in the Republic of Somalia. Results of the 1973 Italian Expedition in Northern Somalia." Monit. Zool. Ital., Suppl. Volume 18, Pages 95-100. Descriptors: ecological distribution; Aves; Somalia. Abstract: Report of birds collected during an ecological survey in northern Somalia, particularly in the mountanous area of Galgala oasis (11° 00N-49°03'E), by the 1973 Italian expedition organized by the Centro di Studio per la Faunistica ed Ecologia Tropicali of the Consiglio Nazionale delle Ricerche (Florence). Results are presented by habitats and in a systematic order.

Ross, S. L. and United Nations Environment Program, Nairobi, Kenya. 1987. "Sectoral Report on Marine Oil and Chemical Spills; Coastal and Marine Environmental Problems of Somalia; Annexes." International (III): 1987. Volume: 84, pages: 51-86. Descriptors: Africa; controls; East Africa; environmental geology; marine environment; oil spills; pollution; preventive measures; Somali Republic. Notes: Includes appendix; References: 15; illustrations incl. 2 tables, sketch maps; ISSN: 1014-8647.

Roth, Michael J. Roth, Michael J. Somalia land policies and tenure impacts: the case of the Lower Shabelle; University of Wisconsin- Madison and Land Tenure Center. 1988. Land Titling Issues and Policy Recommendations, Shabelli Water Management Project: Proposed Scope of Work for an Advisor to Assist the MOA with Planning and Implementing a Land Registration Program. Descriptors: Land titles- Somalia; Project proposals; Land tenure; Agrarian structure; Policy evaluation. Abstract: Somalia land policies and tenure impacts: the case of the Lower Shabelle by Michael Roth. Notes: Named Corp: Shabelli Water Management Project. Notes: "May 1988." Includes bibliographical references. OCLC Accession Number: 29694889.

Royer, Jean-Yves, Richard G. Gordon and Benjamin C. Horner-Johnson. 2006. "Motion of Nubia relative to Antarctica since 11 Ma: implications for Nubia-Somalia, Pacific-North America, and India-Eurasia motion." *Geology*. Volume 34 (6), June 2006: page 501(4). Abstract: Recent estimates of the rotation between Nubia and Somalia have resulted in disparate poles of rotation for the motion since 3.16 Ma (southwest of South Africa) compared with that since 11.03 Ma (near the east tip of Brazil). Here we use magnetic anomaly profiles unavailable in prior Nubia-Antarctica motion studies to significantly revise the estimate of the rotation between Nubia and Antarctica since 11.03 Ma. We use this newly estimated rotation to construct revised estimates of Nubia-Somalia, Pacific-North America, and India-Eurasia motion. The new Nubia-Somalia rotation indicates substantial displacement of Somalia relative to Nubia over the past 11.03 m.y.: 129 [+ or -] 62 km extensional, 90 [+ or -]42 km right-lateral transtensional, and 52 [+ or -] 21 km right-lateral transtensional near the northern extremity of the East African Rift, the northern Mozambique Basin, and the Andrew Bain Fracture Zone complex, respectively. The substantial rotation between Nubia and Somalia implies that prior plate motion estimates based on a circuit through Africa are biased by 60-85 km at 11.03 Ma and perhaps by much more for earlier reconstructions. Our results imply that India-Eurasia motion since 11.03 Ma has been ~12% (~5 mm [yr.sup.-1]) slower than, and ~20° clockwise of, estimates that neglect Nubia-Somalia motion. Our results further imply that Pacific-North America displacement since 11.03 Ma has been 5°-10° clockwise of prior estimates and require 58-75 km less extensional displacement across the Basin and Range since 11.03 Ma than inferred before. Keywords: Nubia-Somalia motion, Southwest Indian Ridge, Pacific-North America motion, India-Eurasia motion.

Royer, Jean-Yves, Srinivas K. and Yatheesh V., et al. 2002. "Paleogene Plate Tectonic Evolution of the Arabian and Eastern Somali Basins." Geological Society Special Publication. Issue 195, Pages 7-23. Descriptors: Structural geology and tectonics; Regional structure and tectonics; gravity survey; Indian plate; Arabian plate; geochronology; magnetic anomaly; Paleogene; tectonic evolution; ocean basin; spreading center Species Term: Somalia. References: 53; Geographic: Indian Ocean- Somali Basin Indian Ocean- Arabian Basin Arabian Sea. Abstract: We review previous models for the Paleogene tectonic evolution of the Arabian and Eastern Somali basins and present a model based on a new compilation of magnetic and gravity data. Using plate reconstructions, we derive a self-consistent set of isochrons for Chron 27 to Chron 21 (61-46 Ma). The new isochrons account for the development of successive ridge propagation events along the Carlsberg Ridge, leading to an important spreading asymmetry between the conjugate basins. Our model predicts the growth of the outer and inner pseudo-faults associated with the ridge propagation events. The location of outer pseudo-faults appears to remain very stable despite a drastic change in the direction of ridge propagation before Chron 24 (c. 54 Ma). The motion of the Indian plate relative to the Somalian plate is stable in direction through Paleogene time; spreading velocities decrease from 6 to 3 cm a-1. Our reconstructions also confirm that the Arabia-India plate boundary was located west of the Owen Ridge along the Oman margin during Paleogene time; some compression is predicted at about Chron 21 (47 Ma) between the Indian and Arabian plates. ISSN: 0305-8719.

Russo, A., Bosellini, F.R, Mohamed, C.M and Yusuf, S.M. 1990. "Paleoenvironmental Analysis and Cyclicity of the Mustahil Formation (Cretaceous of Central Somalia)." Rivista Italiana Di Paleontologia e Stratigrafia. Volume 96, Issue 4, Pages 487-500. Descriptors: Mesozoic; palaeoenvironment; cyclic sedimentation; Cretaceous; rudist reef; sea level variation Species Term: Somalia; Albia; Eoradiolites. Notes: Somalia- Ogaden- Fafan Valley; Somalia- Bur Bitthale. Abstract: A Mustahil section, measured at Bur Bitthale near Belet Uen (Central Somalia), is here described. The succession, dated as Late Aptian to Early-Middle Albian age on the basis of good faunal evidence consists of two well developed thickening-coarsening sequences, where four different facies have been recognized. The cap of both sequences is represented by a rudistid framework dominated by Eoradiolites lyratus. We interpret these cycles as shoaling up parasequences, which are the result of two depositional regressions produced by the progradation of broad shallow-water carbonate systems over the adjacent ramp and deep shelf. ISSN: 0035-6883.

Russo, E. 1914. "Pozzi Esistenti Nel Territorio Galgial; Scala 1:100,000." Translated title: "Existing Wells in Galgial Territory; Scale 1:100,000." Italy: Ministero delle Colonie, Ufficio Cartografico, Rome, Italy. Descriptors: Africa; East Africa; Galgial Territory; geologic maps; maps; Somali Republic. Database: GeoRef. Accession Number: 2000-012059.

Ryding, O. 2005. "Plectranthus Igniarioides (Lamiaceae), a New Species from Somalia." Novon. Volume 15, Issue 2, Pages 361-363. Abstract: Plectranthus igniarioides Ryding, a new species from northern Somalia, is described and illustrated. Comparison is made with P. argentifolius Ryding, P.cuneatus (E.G.Baker) Ryding, P.gillettii J.K. Morton, P. igniarius (Schweinfurth) Agnew, and P. puberulentus, J.K. Morton. The circumscription of Plectranthus gillettii, which has been reported to occur in Somalia, is refined to include only the material known from Ethiopia and Kenya.

Ryding, O. 2004. "A New Species of Otostegia (Lamiaceae) from Somalia." Nord. J. Bot. Volume 23, Issue 3, Pages 265-267. Abstract: A new species from the gypsum plains in northern Somalia, Otostegia ericoidea Ryding is described and illustrated.

S

Saarberg Interplan, International (III). 1987. Saudi Arabian Programme for Rural Water Supply in Somalia; Summary of Technical Data. Somalia: SI/GTZ/MMWR, Mogadishu, Somalia. Descriptors: Africa; data; East Africa; hydrology; rural environment; Somali Republic; water resources; water supply; water wells; wells. Database: GeoRef. Accession Number: 1998-014291.

Saarberg Interplan, International (III). 1987. "Well n. MU 69, Jowle II, Mudug." Somalia: SI/GTZ/MMWR, Mogadishu, Somalia. Descriptors: Africa; East Africa; hydrology; Jowle Somalia; Mudug; rural environment; Somali Republic; water resources; water supply; water wells; wells. Database: GeoRef. Accession Number: 1998-014292.

Saarberg Interplan, International (III). 1986. "Well n. GA 12, Godon, Galgaduud." Somalia: SI/GTZ/MMWR, Mogadishu, Somalia. Descriptors: Africa; East Africa; Galgaduud; Godon Somalia; hydrology; rural environment; Somali Republic;

water resources; water supply; water wells; wells. Database: GeoRef. Accession Number: 1998-014270.

Saarberg Interplan, International (III). 1986. "Well n. GA 13 A, Ceel Dheere, Galgaduud." Somalia: SI/GTZ/MMWR, Mogadishu, Somalia. Descriptors: Africa; Ceel Dheere Somali Republic; East Africa; Galgaduud; hydrology; rural environment; Somali Republic; water resources; water supply; water wells; wells. Database: GeoRef. Accession Number: 1998-014271.

Saarberg Interplan, International (III). 1986. "Well n. GA 13, Ceel Dheere, Galgaduud." Somalia: SI/GTZ/MMWR, Mogadishu, Somalia. Descriptors: Africa; Ceel Dheere Somali Republic; East Africa; Galgaduud; hydrology; rural environment; Somali Republic; water resources; water supply; water wells; wells. Database: GeoRef. Accession Number: 1998-014272.

Saarberg Interplan, International (III). 1986. "Well n. MU 23, Bale Busle, Mudug." Somalia: SI/GTZ/MMWR, Mogadishu, Somalia. Descriptors: Africa; Bale Busle Somali Republic; East Africa; hydrology; Mudug; rural environment; Somali Republic; water resources; water supply; water wells; wells. Database: GeoRef. Accession Number: 1998-014273.

Saarberg Interplan, International (III). 1986. "Well n. MU 58, Jowle I, Mudug." Somalia: SI/GTZ/MMWR, Mogadishu, Somalia. Descriptors: Africa; East Africa; hydrology; Mudug; rural environment; Somali Republic; water resources; water supply; water wells; wells. Database: GeoRef. Accession Number: 1998-014274.

Saarberg Interplan, International (III). 1986. "Well n. GA 55, Nooleeye, Galgaduud." Somalia: SI/GTZ/MMWR, Mogadishu, Somalia. Descriptors: Africa; East Africa; Galgaduud; hydrology; Nooleeye Somalia; rural environment; Somali Republic; water resources; water supply; water wells; wells. Database: GeoRef. Accession Number: 1998-014275.

Saarberg Interplan, International (III). 1986. "Well n. MU 61, Mayle, Mudug." Somalia: SI/GTZ/MMWR, Mogadishu, Somalia. Descriptors: Africa; East Africa; hydrology; Mayle Somalia; Mudug; rural environment; Somali Republic; water resources; water supply; water wells; wells. Database: GeoRef. Accession Number: 1998-014276.

Saarberg Interplan, International (III). 1986. "Well n MU 17, Dhalwo, Mudug." Somalia: SI/GTZ/MMWR, Mogadishu, Somalia. Descriptors: Africa; Dhalwo Somali Republic; East Africa; hydrology; Mudug; rural environment; Somali Republic; water resources; water supply; water wells; wells. Database: GeoRef. Accession Number: 1998-014277.

Saarberg Interplan, International (III). 1986. "Well n. MU 27, Jerriban, Mudug." Somalia: SI/GTZ/MMWR, Mogadishu, Somalia. Descriptors: Africa; East Africa; hydrology; Jerriban Somalia; Mudug; rural environment; Somali Republic; water resources; water supply; water wells; wells. Database: GeoRef. Accession Number: 1998-014278.

Saarberg Interplan, International (III). 1986. "Well n. MU 21, Bitaale, Mudug." Somalia: SI/GTZ/MMWR, Mogadishu, Somalia. Descriptors: Africa; Bitaale Somali Republic; East Africa; hydrology; Mudug; rural environment; Somali Republic; water resources; water supply; water wells; wells. Database: GeoRef. Accession Number: 1998-014279.

Saarberg Interplan, International (III). 1986. "Well n. GA 57, Abudwag, Galgaduud." Somalia: SI/GTZ/MMWR, Mogadishu, Somalia. Descriptors: Abudwaq Somali Republic; Africa; East Africa; Galgaduud; hydrology; rural environment; Somali Republic; water resources; water supply; water wells; wells. Database: GeoRef. Accession Number: 1998-014280.

Saarberg Interplan, International (III). 1986. "Well n. MU 64, Jerriban II, Mudug." Somalia: SI/GTZ/MMWR, Mogadishu, Somalia. Descriptors: Africa; East Africa; hydrology; Jerriban Somalia; Mudug; rural environment; Somali Republic; water resources; water supply; water wells; wells. Database: GeoRef. Accession Number: 1998-014281.

Saarberg Interplan, International (III). 1986. "Well n. GA 67, Cadaado, Galgaduud." Somalia: SI/GTZ/MMWR, Mogadishu, Somalia. Descriptors: Africa; Cadaado Somali Republic; East Africa; Galgaduud; hydrology; rural environment; Somali Republic; water resources; water supply; water wells; wells. Database: GeoRef. Accession Number: 1998-014282.

Saarberg Interplan, International (III). 1986. "Well n. MU 65, Harfo II, Mudug." Somalia: SI/GTZ/MMWR, Mogadishu, Somalia. Descriptors: Africa; East Africa; Harfo Somalia; hydrology; Mudug; rural environment; Somali Republic; water resources; water supply; water wells; wells. Database: GeoRef. Accession Number: 1998-014283.

Saarberg Interplan, International (III). 1986. "Well n. MU 62, Xingood, Mudug." Somalia: SI/GTZ/MMWR, Mogadishu, Somalia. Descriptors: Africa; East Africa; hydrology; Mudug; rural environment; Somali Republic; water resources; water supply; water wells; wells; Xingood Somali Republic. Database: GeoRef. Accession Number: 1998-014284.

Saarberg Interplan, International (III). 1986. "Well n. GA 15, Docoley, Galgaduud." Somalia: SI/GTZ/MMWR, Mogadishu, Somalia. Descriptors: Africa; Docoley Somali Republic; East Africa; Galgaduud; hydrology; rural environment; Somali Republic; water resources; water supply; water wells; wells. Database: GeoRef. Accession Number: 1998-014285.

Saarberg Interplan, International (III). 1986. "Well n. GA 56, Xaawo, Galgaduud." Somalia: SI/GTZ/MMWR, Mogadishu, Somalia. Descriptors: Africa; East Africa; Galgaduud; hydrology; rural environment; Somali Republic; water resources; water supply; water wells; wells; Xaawo Somali Republic. Database: GeoRef. Accession Number: 1998-014286.

Saarberg Interplan, International (III). 1986. "Well n. MU 18, Xarardheere, Mudug." Somalia: SI/GTZ/MMWR, Mogadishu, Somalia. Descriptors: Africa; East Africa; hydrology; Mudug; rural environment; Somali Republic; water resources; water supply; water wells; wells; Xarardheere Somali Republic. Database: GeoRef. Accession Number: 1998-014287.

Saarberg Interplan, International (III). 1986. "Well n. MU 63, Semade, Mudug." Somalia: SI/GTZ/MMWR, Mogadishu, Somalia. Descriptors: Africa; East Africa; hydrology; Mudug; Semade Somalia; Somali Republic; water resources; water supply; water wells; wells. Database: GeoRef. Accession Number: 1998-014288.

Saarberg Interplan, International (III). 1986. "Well n. MU 24, Malasle, Mudug." Somalia: SI/GTZ/MMWR, Mogadishu, Somalia. Descriptors: Africa; East Africa; hydrology; Malasle Somalia; Mudug; rural environment; Somali Republic; water resources; water supply; water wells; wells. Database: GeoRef. Accession Number: 1998-014289.

Saarberg Interplan, International (III). 1986. "Well n. MU 60, Haarfo, Mudug." Somalia: SI/GTZ/MMWR, Mogadishu, Somalia. Descriptors: Africa; East Africa; Haarfo Somalia; hydrology; Mudug; rural environment; Somali Republic; water resources; water supply; water wells; wells. Database: GeoRef. Accession Number: 1998-014290.

Saarberg Interplan, International (III). 1985. "Well n. BA 35, Bakool." Somalia: SI/GTZ/MMWR, Mogadishu, Somalia. Descriptors: Africa; Bakool; East Africa; hydrology; rural environment; Somali Republic; water supply; water wells; wells. Database: GeoRef. Accession Number: 1998-014247.

Saarberg Interplan, International (III). 1985. "Well n. HI 7A, Moko Kori, Hiran." Somalia: SI/GTZ/MMWR, Mogadishu, Somalia. Descriptors: Africa; East Africa; Hiran Somali Republic; hydrology; Moko Kori Somalia; rural environment; Somali Republic; water supply; water wells; wells. Database: GeoRef. Accession Number: 1998-014248.

Saarberg Interplan, International (III). 1985. "Well n. BA 36, Kurtoon, Bakool." Somalia: SI/GTZ/MMWR, Mogadishu, Somalia. Descriptors: Africa; Bakool; East Africa; hydrology; Kurtoon Somalia; rural environment; Somali Republic; water resources; water supply; water wells; wells. Database: GeoRef. Accession Number: 1998-014249.

Saarberg Interplan, International (III). 1985. "Well n. GE 48, Bura, Gedo." Somalia: SI/GTZ/MMWR, Mogadishu, Somalia. Descriptors: Africa; Bura Somali Republic; East Africa; Gedo; hydrology; rural environment; Somali Republic; water resources; water supply; water wells; wells. Database: GeoRef. Accession Number: 1998-014250.

Saarberg Interplan, International (III). 1985. "Well n. GA 10, Garabla, Galgaduud." Somalia: SI/GTZ/MMWR, Mogadishu, Somalia. Descriptors: Africa; East Africa; Galgaduud; Garabla Somali Republic; hydrology; rural environment; Somali Republic; water resources; water supply; water wells; wells. Database: GeoRef. Accession Number: 1998-014251.

Saarberg Interplan, International (III). 1985. "Well n. HI 7, Moko Kori, Hiran." Somalia: SI/GTZ/MMWR, Mogadishu, Somalia. Descriptors: Africa; East Africa; Hiran Somali Republic; hydrology; Moko Kori Somalia; rural environment; Somali Republic; water resources; water supply; water wells; wells. Database: GeoRef. Accession Number: 1998-014252.

Saarberg Interplan, International (III). 1985. "Well n. HI 4, Halgan, Hiran." Somalia: SI/GTZ/MMWR, Mogadishu, Somalia. Descriptors: Africa; East Africa; Halgan Somalia; Hiran Somali Republic; hydrology; rural environment; Somali Republic; water resources; water supply; water wells; wells. Database: GeoRef. Accession Number: 1998-014253.

Saarberg Interplan, International (III). 1985. "Well n. GE 52, Tuulo Barwaago, Gedo." Somalia: SI/GTZ/MMWR, Mogadishu, Somalia. Descriptors: Africa; East Africa; Gedo; hydrology; rural environment; Somali Republic; Tuulo Barwaago Somali Republic; water resources; water supply; water wells; wells. Database: GeoRef. Accession Number: 1998-014254.

Saarberg Interplan, International (III). 1985. "Well n. HI 54, Aborey, Hiran." Somalia: SI/GTZ/MMWR, Mogadishu, Somalia. Descriptors: Aborey Somali Republic;

Africa; East Africa; Hiran Somali Republic; hydrology; rural environment; Somali Republic; water resources; water supply; water wells; wells. Database: GeoRef. Accession Number: 1998-014255.

Saarberg Interplan, International (III). 1985. "Well n. GA 8, Bargan, Galgaduud." Somalia: SI/GTZ/MMWR, Mogadishu, Somalia. Descriptors: Africa; Bargan Somali Republic; East Africa; Galgaduud; hydrology; rural environment; Somali Republic; water resources; water supply; water wells; wells. Database: GeoRef. Accession Number: 1998-014256.

Saarberg Interplan, International (III). 1985. "Well n. HI 5, Teedan, Hiran." Somalia: SI/GTZ/MMWR, Mogadishu, Somalia. Descriptors: Africa; East Africa; Hiran Somali Republic; hydrology; Somali Republic; water resources; water supply; water wells; wells. Database: GeoRef. Accession Number: 1998-014257.

Saarberg Interplan, International (III). 1985. "Well n. GE 39, Maikaraaby, Gedo." Somalia: SI/GTZ/MMWR, Mogadishu, Somalia. Descriptors: Africa; East Africa; Gedo; hydrology; Maikaraabey Somalia; rural environment; Somali Republic; water resources; water supply; water wells; wells. Database: GeoRef. Accession Number: 1998-014258.

Saarberg Interplan, International (III). 1985. "Well n. HI 6, Maxaas, Hiran." Somalia: SI/GTZ/MMWR, Mogadishu, Somalia. Descriptors: Africa; East Africa; Hiran Somali Republic; hydrology; Maxaas Somalia; rural environment; Somali Republic; water resources; water supply; water wells; wells. Database: GeoRef. Accession Number: 1998-014259.

Saarberg Interplan, International (III). 1985. "Well n. GE 40 E, GE 51 E, Garbaharey, Gedo." Somalia: SI/GTZ/MMWR, Mogadishu, Somalia. Descriptors: Africa; East Africa; Garbaharey Somali Republic; Gedo; hydrology; Somali Republic; water resources; water supply; water wells; wells. Database: GeoRef. Accession Number: 1998-014260.

Saarberg Interplan, International (III). 1985. "Well n. BA 53, Geliyo, Bakool." Somalia: SI/GTZ/MMWR, Mogadishu, Somalia. Descriptors: Africa; Bakool; East Africa; Geliyo Somali Republic; hydrology; Somali Republic; water resources; water supply; water wells; wells. Database: GeoRef. Accession Number: 1998-014261.

Saarberg Interplan, International (III). 1985. "Well n. GE 42, Ceel Guduud, Gedo." Somalia: SI/GTZ/MMWR, Mogadishu, Somalia. Descriptors: Africa; Ceel Guduud Somali Republic; East Africa; Gedo; hydrology; Somali Republic; water resources; water supply; water wells; wells. Database: GeoRef. Accession Number: 1998-014262.

Saarberg Interplan, International (III). 1985. "Well n. GA 9, Galhareeri, Galgaduud." Somalia: SI/GTZ/MMWR, Mogadishu, Somalia. Descriptors: Africa; East Africa; Galgaduud; Galharreri Somali Republic; hydrology; rural environment; Somali Republic; water resources; water supply; water wells; wells. Database: GeoRef. Accession Number: 1998-014263.

Saarberg Interplan, International (III). 1985. "Well n. GE 41, Ceel Cadde, Gedo." Somalia: SI/GTZ/MMWR, Mogadishu, Somalia. Descriptors: Africa; Ceel Cadde Somali Republic; East Africa; Gedo; hydrology; rural environment; Somali Republic; water resources; water supply; water wells; wells. Database: GeoRef. Accession Number: 1998-014264.

Saarberg Interplan, International (III). 1985. "Well n. GE47, Kadija Haji, Gedo." Somalia: SI/GTZ/MMWR, Mogadishu, Somalia. Descriptors: Africa; East Africa; Gedo; hydrology; Kadija Haji Somalia; rural environment; Somali Republic; water resources; water supply; water wells; wells. Database: GeoRef. Accession Number: 1998-014265.

Saarberg Interplan, International (III). 1985. "Well n. HI 3, Dabayoodle, Hiran." Somalia: SI/GTZ/MMWR, Mogadishu, Somalia. Descriptors: Africa; Dabayoodle Somali Republic; East Africa; Hiran Somali Republic; hydrology; rural environment; Somali Republic; water resources; water supply; water wells; wells. Database: GeoRef. Accession Number: 1998-014266.

Saarberg Interplan, International (III). 1985. "Well n. GE 40 A, GE 51 A, Garbaharey, Gedo." Somalia: SI/GTZ/MMWR, Mogadishu, Somalia. Descriptors: Africa; East Africa; Garbaharey Somali Republic; Gedo; hydrology; rural environment; Somali Republic; water resources; water supply; water wells; wells. Database: GeoRef. Accession Number: 1998-014267.

Saarberg Interplan, International (III). 1985. "Well n. GE 50/1-3, Ged Weyne, Gedo." Somalia: SI/GTZ/MMWR, Mogadishu, Somalia. Descriptors: Africa; East Africa; Ged Weyne Somali Republic; Gedo; hydrology; rural environment; Somali Republic; water resources; water supply; water wells; wells. Database: GeoRef. Accession Number: 1998-014268.

Saarberg Interplan, International (III). 1985. "Well n. HI 1, Budga Kosar, Hiran." Somalia: SI/GTZ/MMWR, Mogadishu, Somalia. Descriptors: Africa; Budga Kosar Somali Republic; East Africa; Hiran Somali Republic; hydrology; rural environment; Somali Republic; water resources; water supply; water wells; wells. Database: GeoRef. Accession Number: 1998-014269.

Saarberg Interplan, Saudi Arabia. 1985. "Saudi Arabia Programme for Rural Water Supply in Somalia-Well File, Well n. BA 30, Gudo, Bakool." Saudi Arabia: Descriptors: Africa; Bakool Somali Republic; East Africa; ground water; Gudo Somali Republic; rural environment; Somali Republic; water resources; water supply; water wells; wells. Database: GeoRef. Accession Number: 2000-012061.

Saarberg Interplan, Saudi Arabia. 1985. "Saudi Arabian Programme for Rural Water Supply in Somalia-Well File, Well n. BA 28, Bioley, Bakool." Saudi Arabia: Descriptors: Africa; Bakool Somali Republic; Bioley Somali Republic; East Africa; ground water; rural environment; Somali Republic; water resources; water supply; water wells; wells. Database: GeoRef. Accession Number: 2000-012062.

Saarberg Interplan, Saudi Arabia. 1985. "Saudi Arabian Programme for Rural Water Supply in Somalia-Well File, Well, n. BA 29, Lowi Arjek, Bakool." Saudi Arabia: Descriptors: Africa; Bakool Somali Republic; East Africa; ground water; Lowi Arjek Somali Republic; rural environment; Somali Republic; water resources; water supply; water wells; wells. Database: GeoRef. Accession Number: 2000-012063.

Saarberg Interplan, Saudi Arabia. 1985. "Saudi Arabian Programme for Rural Water Supply in Somalia-Well File, Well n. BA 31 A, Ceel Garras, Bakool." Saudi Arabia: Descriptors: Africa; Bakool Somali Republic; Ceel Garas Somali Republic; East Africa; ground water; rural environment; Somali Republic; water resources; water supply; water wells; wells. Database: GeoRef. Accession Number: 2000-012064.

Saarberg Interplan, Saudi Arabia. 1985. "Saudi Arabian Programme for Rural Water Supply in Somalia-Well File. Well n. BA 31, Ceel Garras, Bakool." Saudi Arabia:

Descriptors: Africa; Bakool Somali Republic; Ceel Garas Somali Republic; East Africa; ground water; rural environment; Somali Republic; water resources; water supply; water wells; wells. Database: GeoRef. Accession Number: 2000-012065.

Saarberg Interplan, Saudi Arabia. 1985. "Saudi Arabian Programme for Rural Water Supply in Somalia-Well File, Well n. BA 32, Babaweyne, Bakool." Saudi Arabia: Descriptors: Africa; Babaweyne Somali Republic; Bakool Somali Republic; East Africa; ground water; rural environment; Somali Republic; water resources; water supply; water wells; wells. Database: GeoRef. Accession Number: 2000-012066.

Saarberg Interplan, Saudi Arabia. 1985. "Saudi Arabian Programme for Rural Water Supply in Somalia-Well File, Well n. BA 33, Rabdurre, Bakool." Saudi Arabia: Descriptors: Africa; Bakool Somali Republic; East Africa; ground water; Rabdurre Somali Republic; rural environment; Somali Republic; water resources; water supply; water wells; wells. Database: GeoRef. Accession Number: 2000-012067.

Saarberg Interplan, Saudi Arabia. 1985. "Saudi Arabian Programme for Rural Water Supply in Somalia-Well File, Well n. BA 34, Gaboodo, Bakool." Saudi Arabia: Descriptors: Africa; Bakool Somali Republic; East Africa; Gaboodo Somali Republic; ground water; rural environment; Somali Republic; water resources; water supply; water wells; wells. Database: GeoRef. Accession Number: 2000-012068.

Sabul, Ali A. 1978. Water Quality of Shebelle River. Somalia: Univ. Naz. Somala, Fac. Agraria, Mogadishu, Somalia. Descriptors: Africa; East Africa; geochemistry; hydrochemistry; hydrology; rivers and streams; Shebelle River; Somali Republic; surface water; water quality. Database: GeoRef. Accession Number: 1998-014293.

Sacchi, Rosalino and Zanferrari, Adriano. 1987. "Notes on some Shear Zones of Northern Somalia." Journal of African Earth Sciences. Volume 6, Issue 3, Pages 323-326. Descriptors: Africa; basement; displacements; ductility; East Africa; faults; northern Somalia; plate collision; plate tectonics; shear zones; Somali Republic; Somalia Republic; structural geology; tectonics; tectonophysics; thrust faults. Notes: References: 10; illustrations incl. sects., sketch maps. Abstract: Low-angle thrusts, displaying a well developed, stretching lineation, and west to south-west vergence, are reported from the basement of northern Somalia, and interpreted as the extreme evolution of a (Upper Proterozoic) phase of folding. This is seen as a late event, roughly coeval with gabbro emplacement, and later than the main metamorphism of the basement complex. Thrusting took place when the gabbros were still at a high temperature, as shown by 'hot' metamorphic assemblages within the zones of ductile shear. Development of abundant pegmatite and of some muscovite granite also took place, probably triggered by gabbro emplacement.Tectonic style suggests that here we may be dealing with the continuation of the collision zone with the East Gondwana Plate, recently recognized by some researchers near the eastern margin of the Saudi Arabian shield. ISSN: 0731-7247.

Sacchi, Rosalino, Zanferrari, Adriano, Kinnaird, Judith A. and Van Horn, Francoise D. 1985. "Notes on some Precambrian Terrains of N. Somalia; 13th Colloquium of African Geology; Abstracts --13 (Super e) Colloque De Geologie Africaine; Resumes." Publication Occasionnelle- Centre International Pour La Formation Et Les Echanges Géologiques = Occasional Publication- International Center for Training and Exchanges in the Geosciences. Mar. Volume 3, Pages 106-107. Descriptors: Africa; Arabian Peninsula; Arabian Shield; Asia; East Africa; faults; folds; gabbros; Hargeisa-

Berbera; igneous rocks; Mozambique; northern Somalia; plutonic rocks; Precambrian; Saudi Arabia; Somali Republic; structural geology; tectonics. ISSN: 0769-0541.

Sagri M, Abbate E and Bruni P. 1989. "Deposits of Ephemeral and Perennial Lakes in the Tertiary Daban Basin (Northern Somalia)." Palaeogeogr., Palaeoclimatol., Palaeoecol. Volume 70, Issue 1-3, Pages 225-233. Descriptors: Geographical Abstracts: Sediments and sedimentary processes- transport; Eocene; Oligocene; clastic deposit; lateral transition; lacustrine sediment; rifting; Tertiary. Notes: Somalia- Berbera- Daban Basin. Abstract: The Daban Basin is filled with 2700 m of Middle Eocene to Oligocene clastic deposits. From the bottom of the sequence upward six sedimentary environments are distinguished: restricted lagoon, delta, lagoon, alluvial plain, ephemeral and perennial lakes. A general vertical trend from an ephemeral, saline lake to perennial fresh-water lake has been recognized. The Daban Basin was a rapidly subsiding depression during the period that rifting was taking place in the adjacent Gulf of Aden. Lacustrine deposition was controlled by the paleogeographic and paleoclimatic changes induced by this event. ISSN: 0031-0182.

Said, Mohamed. 1987. Mudug Uranium Deposits; GeoSom 87; Geology of Somalia and Surrounding Regions; Late Abstracts. Somalia: Somali Natl. Univ., Dep. Geol., Mogadishu, Somalia. GeoSom 87; Geology of Somalia and Surrounding Regions, Mogadishu. Somalia Conference: Nov. 24-Dec. 1, 1987. Descriptors: Africa; anomalies; calcrete; carbonate rocks; carnotite; clastic sediments; Dusa Mareb; East Africa; economic geology; metal ores; mineral exploration; Mirrig; Mudug; radioactivity; reserves; sedimentary rocks; sediments; Somali Republic; uranium ores; vanadates; Wabo. OCLC Accession Number: 34175029.

Sainlos, J. C. and United Nations Environment Programme, Nairobi, Kenya. 1987. "Sectoral Report on Marine Environmental Legislation; Coastal and Marine Environmental Problems of Somalia; Annexes." International (III): 1987. Volume: 84, pages: 189-215. Descriptors: Africa; conservation; East Africa; environmental geology; impact statements; legislation; marine environment; Somali Republic. Notes: References: 9; ISSN: 1014-8647.

Saji, N. H. and Goswami, B. N. 1997. "Intercomparison of the Seasonal Cycle of Tropical Surface Stress in 17 AMIP Atmospheric General Circulation Models." Clim. Dyn. Aug. Volume 13, Issue 7-8, Pages 561-585. Descriptors: Ocean circulation; Ocean-atmosphere system; Air-water interface; Tropical oceanography; Simulation; Surface temperature; Wind stress; Atmospheric motion; Convergence; Equatorial Indian Ocean; South Pacific; Equatorial Pacific; Somali Dem. Rep. Marine. Notes: TR: CS0111768. Abstract: The mean state of the tropical atmosphere is important as the nature of the coupling between the ocean and the atmosphere depends nonlinearly on the basic state of the coupled system. The simulation of the annual cycle of the tropical surface wind stress by 17 atmospheric general circulation models (AGCMs) is examined and intercompared. The models considered were part of the Atmospheric Model Intercomparison Project (AMIP) and were integrated with observed sea surface temperature (SST) for the decade 1979-1988. Several measures have been devised to intercompare the performance of the 17 models on global tropical as well as regional scales. Within the limits of observational uncertainties, the models under examination simulate realistic tropical area-averaged zonal and meridional annual mean stresses. This is a noteworthy improvement over older generation low resolution models which were noted for their simulation of surface stresses considerably weaker than the observations. The models also simulate realistic magnitudes of the spatial distribution of the annual mean surface stress field and are seen to reproduce realistically its observed spatial pattern. Similar features are observed in the simulations of the annual variance field. The models perform well over almost all the tropical regions apart from a few. Of these, the simulations over Somali are interesting. Over this region, the models are seen to underestimate the annual mean zonal and meridional stresses. There is also wide variance between the different models in simulating these quantities. Large model-to-model variations were also seen in the simulations of the annual mean meridional stress field over equatorial Indian Ocean, south central Pacific, north east Pacific and equatorial eastern Pacific oceans. It is shown that the systematic errors in simulating the surface winds are related to the systematic errors in simulating the Inter-Tropical Convergence Zone (ITCZ) in its location and intensity. Weaker than observed annual mean southwesterlies simulated by most models over Somali is due to weaker than observed southwesterlies during the Northern Hemisphere summer. This is related to the weaker than observed land precipitation simulated by most models during the Northern Hemisphere summer. The diversity in simulation of the surface wind over Somali and equatorial Indian ocean is related to the diversity of AGCMs in simulating the precipitation zones in these regions. Database: Oceanic Abstracts. ISSN: 0930-7575.

Salad, Mohamed Khalief. 1977. Somalia: A Bibliographical Survey. Westport, Conn: Greenwood Press. Descriptors: 468. Notes: Somalia- Bibliography. Somalie, République démocratique de- Bibliographie. ISBN: 0837194806: LCCN: 76-51925.

Salad Hersi, O. and Hilowle Mohamed, A. 2000. "Stratigraphy and Petroleum Prospects of Northern Somalia; 2000 AAPG Eastern Section Meeting; Abstracts." AAPG Bull. Sep. Volume 84, Issue 9, Pages 1392. Descriptors: Adigrat Formation; Africa; Ahl Mado Basin; Ahl Mado Group; Auradu Formation; basin analysis; basins; Berbera Basin; Cenozoic; clastic rocks; East Africa; Karkar Formation; Marib-Hajar Basin; Mesozoic; natural gas; northern Somalia; petroleum; petroleum exploration; sandstone; Say'un-Al Masila Basin; sedimentary basins; sedimentary rocks; Somali Republic; stratigraphic units; Taleh Formation; Yesomma Sandstone. Abstract: The sedimentary cover of Northern Somalia includes post-Triassic continental and marine strata which accumulated in basins related to the disintegration of the Gondwanaland. Among these, the Berbera and Ahl Mado basins are the most important basins stratigraphically and hydrocarbon potential. Sedimentation in both basins begins with a Jurassic continental sandstone (Adigrat Formation) overlain by interbedded units of shallow marine limestones and shales (Bihendula sequence) in the Berbera Basin, and limestone-dominated strata with minor shale and sandstone interbeds (Ahl Mado Group) in the Ahl Mado Basin. The Cretaceous section, unconformable with the Jurassic sequence, is mainly continental (Yesomma Sandstone) in the Berbera Basin, but becomes shallow-marine, sandy to pure limestone with subordinate sandstone and shale (Tisje Formation) in the Ahl Mado Basin. By the end of the Cretaceous Period, a westward marine transgression permitted shallow-marine, Paleocene-lower Eocene limestone (Auradu Formation) deposition throughout northern Somalia. This is succeeded by thick anhydrite strata (Taleh Formation) overlain by Middle to Late Eocene shallow-marine limestone (Karkar Formation). The later is the youngest stratigraphic unit straddling the Gulf of Aden. Younger strata of syn- and post-rifting, continental to shallow-marine origin are confined in discrete basins along the coast of the gulf. Based on published and unpublished data, the geology of these basins proves that oil and gas have been generated with favorable reservoirs, as well as structural and stratigraphic traps. Moreover, continuation of these basins across the gulf, matching the hydrocarbon-producing Marib-Hajar and Say'un-Al Masila basins of Yemen, raises the hydrocarbon prospect of northern Somalia. ISSN: 0149-1423.

Salvadei, G. 1913. L'Alto Corso Del Fiume Uebi Scebeli. Upper Course of the Shebelle River. Italy: L'Oltremare, Rome, Italy. Descriptors: Africa; East Africa; fluvial features; hydrology; rivers; rivers and streams; Shebelle River; Somali Republic; surface water. Database: GeoRef. Accession Number: 1998-014305.

Samantar, Mohamed Said. 1994. The Conditions for Successful Management. England: University of Sussex (United Kingdom). Descriptors: Somalia. Abstract: This research project addresses a fundamental problem of development management in pastoral communities. I found that some scholars acknowledge that cooperation is indeed a possibility for better pastoral common resource management, while others are steadfast in their belief that nothing of the sort can take place. I also found in the literature that there are different solutions to the problems of risk and uncertainty which depend on the nature of the institution (with rules set up from outside and inside) attempting to solve the problem. This thesis deals with two models of solving this kind of problem, namely, the Prisoners' Dilemma solution and Assurance Problem solution. The important analytical difference which leads to different policy implications is that cooperation is reached in the former (under various adjustments) by adopting contrasting strategies for individual and social optima, while in the latter the same strategies are adopted. Therefore, in the Prisoners' Dilemma individual optimum is sacrificed for social optimum; in the Assurance Problem the two are compatible. This compatibility is more pronounced in the Pure Coordination game. Coming to the reality of the Somali pastoral situations, I found that the conditions for successful cooperation in pastoral CPRs are: (1) the existence of homogeneous groups; (2) the existence of a high-risk customary grazing area; (3) the existence of well-defined rules about resource use (xeer); (4) the existence of a flexible cooperative strategy among members of different groups. Moreover, I found that the conditions for the preservation of CPRs are: (1) the restoration of weakened customary institutions for pastoral communities; (2) xeer should be put in a written form, improved where necessary, and should constitute the basis for the formulation of national laws; (3) regional representatives of central authority should be elected locally and should come posted under the authority of the council of elders; (4) range and range water development should be based on the customary CPRs. Notes: D. Phil. URL: http://proquest.umi.com/pqdweb?did=744207391&Fmt=7&clientId=65345&RQT=309&VName=PQD

Samantar, Mohamed Said and Food and Agriculture Organization of the United Nations. 1989. A Study on Drought Induced Migration and its Impact on Land Tenure and Production in the Inter-Riverine Region of Somalia. Rome: Food and Agriculture Organization of the United Nations. Descriptors: 70; Colonization and settlement- Somalia; Migration- Somalia; Land tenure- Somalia; Agriculture and state- Somalia; Policy evaluation- Somalia; Aridity and arid regions- Somalia. Notes: Includes bibliographical references (p. 68-70). OCLC Accession Number: 34074359.

Samatar A. 1988. "The State, Agrarian Change and Crisis of Hegemony in Somalia." Rev. Afr. Polit. Econ. Volume 43, Pages 26-41. Descriptors: Social, economic and political aspects of rural change; peasantry; state; agrarian change. Abstract: Argues that while pre-capitalist production and political and cultural life are not necessarily advantageous to development, on the whole Africa's current agony is primarily attributable to cumulative colonial and post-colonial state policies which have directly contributed to the emergence of hybrid, moribund socio-economic structures incapable of lessening the vulnerabilities of the peasantry. The resulting deterioration of rural life continues to undermine the hegemony, i.e. the intellectual and moral leadership, of state classes, and bodies ill for any current or future collective struggle against underdevelopment. ISSN: 0305-6244.

Samatar A.I. 1993. "Structural Adjustment as Development Strategy? Bananas, Boom, and Poverty in Somalia." Economic Geography. Volume 69, Issue 1, Pages 25-43. Descriptors: International Aid and Investment; foreign investment; poverty effect; banana production; development strategy; structural adjustment. Abstract: The World Bank and the International Monetary Fund have become the most powerful macroeconomic development strategists in the Third World. Structural adjustment program (SAP) is the code term for their main strategy. One of SAP's objectives is to induce a business climate attractive to investors in Africa. This study evaluates Somalia's banana industry and associated foreign investment in the 1980s. The analysis shows that foreign investment modernized banana production and increased exports, but did not improve the starvation wages of plantation workers. Moreover, since nearly 75% of the earnings from exports leave the country, such investment does not enhance Somalia's capital accumulation fund. ISSN: 0013-0095.

Sameroone. 2005. "Is there Gold in Somalia?" MineralCollecting. Org. Article Title: Is there Gold in Somalia? Issue: January 22, 2004, Pages: July 14, 2006. Descriptors: Somalia; gold; mining. Abstract: No doubt, Somalia has precious metals including gold, zinc, silver but I strongly suggest a detailed investigation for gold and sulfides metallic over meta-sedimentary and meta-volcanic rocks has to be performed. The investigation should be included a detailed geological mapping using Landsat imagery and GIS. Unfortunately, the above geological tasks can not be carried out unless the current chronic instability and violence in Somali halt and Somalis bring peace and flexible solutions. However, the growing Somali higher institutions can establish geosciences faculty and take responsibilities of collecting and archiving all previous geological data about the nation. They can also carry out and organize workshops and seminars on the Somali minerals, water and environmental issues in close contact and cooperation with international universities, organizations and local NGO's. URL: http://www.mineralcollecting.org/

Samuelsson, Gunnar, Farah, Mohamed Hussein and Claeson, Per, et al. 1993/1. "Inventory of Plants used in Traditional Medicine in Somalia. IV. Plants of the Families Passifloraceae-Zygophyllaceae." Journal of Ethnopharmacology. Volume 38, Issue 1, Pages 1-29. Descriptors: medicinal plants; Somalia; traditional medicine. Notes: ID: 2366. Abstract: Thirty-seven plants are listed, which are used by traditional healers in the central and southern parts of Somalia. For each species are listed: the botanical name with synonyms, collection number, vernacular name, medicinal use, preparation of remedy and dosage. Results of a literature survey are also reported including medicinal use, substances isolated and pharmacological effects. Three plants which should have been included in Part I of the series have been added and some corrections to that paper have been made. With these additions the series comprises 180 different plant species, distributed in 59 plant families.

Samuelsson, Gunnar, Farah, Mohamed Hussein and Claeson, Per, et al. 1992/9. "Inventory of Plants used in Traditional Medicine in Somalia. III. Plants of the Families Lauraceae-Papilionaceae." Journal of Ethnopharmacology. Volume 37, Issue 2, Pages 93-112. Descriptors: medicinal plants; Somalia; traditional medicine. Abstract: Thirty-five plants are listed, which are used by traditional healers in the central and southern parts of Somalia. For each species are listed: the botanical name with synonyms, collection number, vernacular name, medicinal use, preparation of remedy and dosage. Results of a literature survey are also reported including medicinal use, substances isolated and pharmacological effects.

Samuelsson, Gunnar, Farah, Mohamed Hussein and Claeson, Per, et al. 1992/8. "Inventory of Plants used in Traditional Medicine in Somalia. II. Plants of the Families Combretaceae to Labiatae." Journal of Ethnopharmacology. Volume 37, Issue 1, Pages 47-70. Descriptors: medicinal plants; Somalia; traditional medicine. Abstract: Fifty-nine plants are listed, which are used by traditional healers in the central and southern parts of Somalia. For each species are listed: the botanical name with synonyms, collection number, vernacular name, medicinal use, preparation of remedy and dosage. Results of a literature survey are also reported including medicinal use, substances isolated and pharmacological effects.

Samuelsson, Gunnar, Farah, Mohamed Hussein and Claeson, Per, et al. 1991. "Inventory of Plants used in Traditional Medicine in Somalia. I. Plants of the Families Acanthaceae-Chenopodiaceae." Journal of Ethnopharmacology. Volume 35, Issue 1, Pages 25-63. Descriptors: medicinal plants; Somalia; traditional medicine. Abstract: Thirty-eight plants are listed, which are used by traditional healers in the central and southern parts of Somalia. For each species are listed: the botanical name with synonyms, collection number, vernacular name, medicinal use, preparation of remedy and dosage. Results of a literature survey are also reported including medicinal use. substances isolated and pharmacological effects.

Sanderson, I. 1985. Livestock Water Development in Central Rangeland of Somalia. Somalia: NRA, Mogadishu, Somalia. Descriptors: Africa; development; East Africa; prairies; rangelands; Somali Republic; water management; water resources. Database: GeoRef. Accession Number: 1998-014306.

Sartoni, G. 1976. "Researches on the Coast of Somalia. the Shore and Dune of Sar Uanle. 6.A Study of the Benthonic Algal Flora." Monit. Zool. Ital., Suppl. Volume 7, Issue 2, Pages 115-143. Descriptors: Benthos; Check lists; New records; Geographical distribution; Algae; Somali Dem. Rep., Sar Uanle; Marine. Notes: Records keyed from 1976 ASFA printed journals. TR: 1976. Abstract: The 69 spp found in a recent study of the benthonic algae at Sar Uanle, South of Chisimaio, in southern Somalia are listed. Many spp are reported for the 1st time for Somalia because there have been few studies of this region of the western Indian Ocean. A considerable number of the spp found are widely distributed in the Indo-Malaysian area. Database: ASFA: Aquatic Sciences and Fisheries Abstracts.

Sartoni, G. 1974. "Contribution to the Study of the Marine Algae of Sar Uanle (Southern Somali)." G. Bot. Ital. Volume 108, Issue 6, Pages 281-303. Descriptors: Biocoenosis; Check lists; Geographical distribution; New records; Marine ecology;

Algae; Somali Dem. Rep., Sar Uanle; Marine. Notes: Records keyed from 1976 ASFA printed journals. TR: 1976. Abstract: Some preliminary researches have been carried out on the algal flora of Sar Uanle, lying south of Chisimaio in southern Somalia. The material first collected for these researches, that includes 69 spp, appears mainly new for the Somalia coasts. This is due to the fact that the bibliography concerning this part of the western Indian ocean is very scarce. The presence of a remarkable percentage of spp that are widely distributed in the Indo-Malaysian area, has been found through the results of these enquiries. Database: ASFA: Aquatic Sciences and Fisheries Abstracts.

Sassi, F. P. and Arush, M. A. 1982. Lineamenti Geologici Della Somalia. Mogadiscio: Universitá Nazionale Somala. Descriptors: 328, 2; Geology- Somalia; Geology, Stratigraphic. Notes: In Italian or English. OCLC Accession Number: 13685383.

Sassi, F. P., Gatto, G. O. and Visona, D. 1983. "Present Status of Knowledge on the Crystalline Basement of North Somalia." Quaderni Di Geologia Della Somalia (Mogadiscio). Volume 7, Descriptors: Africa; areal geology; basement; crystalline rocks; East Africa; northern Somalia; Somali Republic. LCCN: 93093377.

Sassi, F. P. and Visona, D. 1993. Relics of Granulitic Mineral Assemblages in the Northern Somali Basement. Firenze; Istituto Agronomico L'Oltremar; 1993. Relazioni E Monografie Agrarie Subtropicali E Tropicali Nuova Serie113 A, pages: 83-90. International Meeting; 1st- 1987 Nov: Mogadishu. Notes: Geology of Somalia and surrounding regions: Geology and mineral resources of Somalia and surrounding regions. OCLC Accession Number: 34175029.

Sassi, F. P. Visona, D. Ferrara, G. and Gatto, G. O. 1993. The Crystalline Basement of Northern Somalia: Lithostratigraphy and Sequence of Events. Firenze; Istituto Agronomico L'Oltremar; 1993. Relazioni E Monografie Agrarie Subtropicali E Tropicali Nuova Serie113 A, pages: 3-40. International Meeting; 1st- 1987 Nov: Mogadishu. Notes: Geology of Somalia and surrounding regions: Geology and mineral resources of Somalia and surrounding regions. OCLC Accession Number: 34175029.

Schiarini, P. 1911. "Il Giuba Da Lugh a Bardera. Juba River from Luuq to Bardhere." Bollettino Della Societa Geografica Italiana. Volume 1911, Descriptors: Africa; Baardheere Somali Republic; East Africa; Gedo; geography; Juba River; Luuq Somalia; rivers and streams; Somali Republic. ISSN: 0037-8755.

Schlee G. 2003. "Redrawing the Map of the Horn: The Politics of Difference." Africa. Volume 73, Issue 3, Pages 343-368. Descriptors: regional politics; territorial delimitation; conflict management; geopolitics. Notes: Additional Info: South Africa; References: 39; Geographic: Africa- (East) Somalia Ethiopia Eritrea. Abstract: The paper examines the changing shapes of territories in the Horn of Africa and the discourses which legitimize these different shapes. It starts with the 'Horn' itself, the different ways to delineate it, and the interests behind these. Then Eritrea, Ethiopia, Somalia, and the de facto independent Somaliland are discussed and their justifications for being examined. These justifications are found not to follow the same pattern. The criteria for inclusion or exclusion of populations or territories differ and form a rich reservoir for future conflict. On a lower level, that of regional states comprised in a major unit, the Oromo of Ethiopia, the largest ethnic group in the Horn of Africa, are discussed in some detail. Accounts about how the Oromo have come to be and who is to be regarded as an Oromo are found to be mutually conflicting. In the last part, international and transnational relations in the Horn of Africa are looked at. Major groupings cross-cutting state boundaries are formed by states forming alliances with ethnic movements, opposition forces or warlords in neighboring states or ex-states, against other states or spheres of power. Publicity of such alliances is kept low and few efforts seem to be made to give them an ideological basis or historical justifications. The logic followed in these cases seems to be simply that the enemy of an enemy is a potential friend. ISSN: 0001-9720.

Schlueter, T. 1993. Geological Development and Economic Significance of Lacustrine Phosphate Deposits in Northern Tanzania. Firenze; Istituto Agronomico L'Oltremar; 1993. Relazioni E Monografie Agrarie Subtropicali E Tropicali Nuova Serie113 B, pages: 607-614. International Meeting; 1st- 1987 Nov: Mogadishu. Notes: Geology of Somalia and surrounding regions: Geology and mineral resources of Somalia and surrounding regions. OCLC Accession Number: 34175029.

Scholz U. 1993. "Landwirtschaftliche Probleme an Der Agronomischen Trockengrenze- Der Regenfeldbau in Somalia." Translated title: "Agricultural problems at the Agronomy drying border- the building of rain fields in Somalia." Giessener Beitrage Zur Entwicklungsforschung, Reihe I (Symposien). Volume 20, Pages 1-14. Descriptors: Agriculture; Crop and livestock production- general; agricultural production; animal husbandry; climatic problem; rain- fed agriculture; production problem; developing country; rainfed agriculture; farming system; sorghum production. Abstract: Apart from political disturbances, the arid climate constitutes the biggest handicap to the agricultural production of Somalia. The most adequate farming system is nomadic animal husbandry. Plant production is restricted to less than 2% of the total land area being split into two different farm types: irrigated farming concentrated along the two major rivers of the country, Shabelle and Juba, and rainfed farming located in the interriverine area between these two rivers. Rainfed agriculture deserves special attention due to its particular sensitivity towards the arid climatic conditions. The system is characterized by permanent mono-cultivation of a single crop (sorghum), restriction to a specific soil (vertisol), absence of any capital inputs, little labor input, low yields and the integration of animal husbandry as a secondary enterprise. The two biggest problems are the erratic water supply causing repeated crop failures and the restricted fodder supply at the end of the dry seasons. Ecologically rain fed farming has been proving sustainable without visible damage to the natural environment, but economically the system seems to offer only limited potential for the future development of both the local people and the whole country.

Scholz U. 1990. "Problems of Rainfed Agriculture in Southern Somalia." Geomethodica. Volume 15, Pages 47-70. Descriptors: Crop and livestock production-general; agricultural practice; developing country; water shortage; farming system; crop failure; rainfed agriculture; labor constraint; sorghum; agricultural policy; fodder supply. Notes: Somalia. Abstract: Apart from nomadic animal husbandry and irrigated farming, rainfed agriculture constitutes one of the three major farming systems of Somalia. Most of it is located in the interriverine area between the two rivers Shabelle and Juba. The present study examines the major problems of this system, tries to identify the causes and discusses strategies of the local people to overcome these constraints. The biggest problem is the high risk of cultivation including repeated crop failures caused by the erratic water supply. Other problems include: low yields caused by pests, diseases, and the absence of any yield raising production means; restricted availability of labor force which is entirely based on family labor; irregular fodder supply for the animals caused by the long dry seasons; and the poor role of the government whose interests are mainly concerned with irrigated but not rainfed agriculture.

Schott, Annette. 1991. "Map of the Democratic Republic of Somalia." Karlsruhe, Germany: Fachhochschule Karlsruhe, Studiengang Kartographie. Descriptors: Government publication; Local government publication. Notes: Description: 1 map; color; 52 x 45 cm., folded to 23 x 11 cm. Geographic: Somalia- Maps. Scale 1:3,000,000. Notes: Alternate title: Somalia 1:3 000 000. Responsibility: Diplomarbeit ... von Annette Schott unter der Leitung von Prof. H.J. Zylka. LCCN: 93-684223; OCLC Accession Number: 28844080.

Schottenloher, R. 1939. "Die Cercer, Garumullata, Und Nordlichen Arussi-Berglander Auf Der Somali-Hochscholle." Translated title: "Cercer, Garamullata, and the Northern Arussi Hills in the Somali Highlands." Petermann's Mitteilungen. Volume 1939, Descriptors: Africa; areal geology; Arussi; Cercer; East Africa; Ethiopia; Garamullata; Somali Highlands; Somali Republic. ISSN: 0323-7699.

Schrank E. 1990. "Upper Cretaceous Coal-Bearing Sediments in Northern Somalia: Short Note on Palynological Age and Palaeoenvironment." Berliner Geowissenschaftliche Abhandlungen, Reihe A. Volume 120, Issue 2, Pages 633-638. Descriptors: Micropaleontology; Cretaceous; pollen; spore; palaeoenvironment; Maastrichtian; tropical humid forest. Notes: Somalia. Abstract: Cores from three shallow boreholes in the Hed-Hed area yielded predominantly terrestrial microfloras including typical members of the Late Cretaceous Palmae province of Africa and northern South America. The angiosperm pollen indicates a Maastrichtian age. Common palms associated with probably hygrophile forms such as ferns, club mosses and hepatic mosses may have developed under humid conditions in a tropical coastal lowland with rivers, lakes and swamps. This interpretation is also supported by the presence of Ctenolophonaceae which are nowadays trees in tropical humid forests. ISSN: 0172-8784.

Schreck, Carl J. and Semazzi, Fredrick H. M. 2004. "Variability of the Recent Climate of Eastern Africa." Int. J. Climatol. Volume 24, Issue 6, Pages 681-701. Descriptors: Climatic variability; Empirical orthogonal functions; Satellite-rain gage data combination; Rainfall-atmospheric circulation relationships; Correlation analysis; Global warming; El Nino-Southern Oscillation event-climate relationships; Rainfall anomalies; Climate; Climatology; Rainfall; El Nino phenomena; Southern Oscillation; Sea surface; Surface temperature; Water temperature; Temperature anomalies; Ocean-atmosphere system; Teleconnections; Tropical meteorology; Winds; Africa, East; Burundi; Djibouti; Eritrea; Ethiopia; Kenya; Rwanda; Somalia; Sudan; Uganda; Tanzania; Indian Ocean; Congo, Rep. IS, Equatorial Pacific; Marine. Notes: TR: CS0414963. Abstract: The primary objective of this study is to investigate the recent variability of the eastern African climate. The region of interest is also known as the Greater Horn of Africa (GHA), and comprises the countries of Burundi, Djibouti, Eritrea, Ethiopia, Kenya, Rwanda, Somalia, Sudan, Uganda, and Tanzania. The analysis was based primarily on the construction of empirical orthogonal functions (EOFs) of gauge rainfall data and on CPC Merged Analysis of Precipitation (CMAP) data, derived from a combination of rain-gauge observations and satellite estimates. The investigation is based on the period 1961-2001 for the 'short rains' season of eastern Africa of October through to December. The EOF analysis was supplemented by projection of National Centers for Environmental

Prediction wind data onto the rainfall eigenmodes to understand the rainfall-circulation relationships. Furthermore, correlation and composite analyses have been performed with the Climatic Research Unit globally averaged surface-temperature time series to explore the potential relationship between the climate of eastern Africa and global warming. The most dominant mode of variability (EOF1) based on CMAP data over eastern Africa corresponds to el Nino-southern oscillation (ENSO) climate variability. It is associated with above-normal rainfall amounts during the short rains throughout the entire region, except for Sudan. The corresponding anomalous low-level circulation is dominated by easterly inflow from the Indian Ocean, and to a lesser extent the Congo tropical rain forest, into the positive rainfall anomaly region that extends across most of eastern Africa. The easterly inflow into eastern Africa is part of diffluent outflow from the maritime continent during the warm ENSO events. The second eastern African EOF (trend mode) is associated with decadal variability. In distinct contrast from the ENSO mode pattern, the trend mode is characterized by positive rainfall anomalies over the northern sector of eastern Africa and opposite conditions over the southern sector. This rainfall trend mode eluded detection in previous studies that did not include recent decades of data, because the signal was still relatively weak. The wind projection onto this mode indicates that the primary flow that feeds the positive anomaly region over the northern part of eastern Africa emanates primarily from the rainfall- deficient southern region of eastern Africa and Sudan. Although we do not assign attribution of the trend mode to global warming (in part because of the relatively short period of analysis), the evidence, based on our results and previous studies, strongly suggests a potential connection. ISSN: 1097-0088.

Schrope, M. 2001. "Pirates Attack US Research Ship Off Somalia." Nature. Volume 413, Issue 6852, Pages 97. ISSN: 0028-0836.

Schulte, L. and Berg, D. 1985. "Hydrogeologische Und Numerische Untersuchungen Zur Trinkwasserversorgung in Beled Weyn (Somalia)." Translated title: "Hydrogeological and Numerical Studies of Drinking Water in Beled Weyn, Somalia." Bbr. Brunnenbau, Bau von Wasserwerken, Rohrleitungsbau. Volume 36, Issue 7, Pages 263-267. Descriptors: Africa; Beled Weyn; data processing; digital simulation; drinking water; East Africa; electrical conductivity; ground water; hydrogeology; hydrometry; infiltration; numerical analysis; pressure meters; pump tests; representative basins; Shelli-Fluss; Somali Republic; surface water; surveys; water supply; water table. Notes: References: 6; illustrations incl. 2 tables, sketch maps; Reference includes data from Geoline, Bundesanstalt fur Geowissenschaften und Rohstoffe, Hanover, Germany. ISSN: 0340-3874.

Schwarz, R. A. 1983. The Somali Groundwater Development Project; the Community Participation Process; Monitoring, Evaluation and Training. Somalia: Descriptors: Africa; development; East Africa; ground water; monitoring; Somali Republic; water management; water resources. Database: GeoRef. Accession Number: 1998-014317.

Schwarz, Ronald A. and Vu Thi, Pierrette. 1987. The Social, Economic and Environmental Impact of Water Systems in Somalia: The Bay Region and the Central Rangeland Region. Training Resources and Information Networks. Descriptors: 1 v. (various paginations); Water-supply, Rural- Somalia- Bay Region; Water-supply, Rural-Somalia- Central Rangeland Region; Rural development projects- Somalia- Bay Region;

Rural development projects- Somalia- Central Rangeland Region. Notes: Cover title. "Draft under revision. For IDA use only"--Handwritten note on cover. "Selected sections of report only"--Leaf 5. OCLC Accession Number: 62140973.

Sclater, J. G., Grindlay, N. R., Madsen, J. A. and Rommevaux-Jestin, C. 2005. "Tectonic Interpretation of the Andrew Bain Transform Fault: Southwest Indian Ocean." Geochemistry Geophysics Geosystems. SEP 27. Volume 6, Pages Q09K10. Abstract: [1] Between 25° E and 35° E, a suite of four transform faults, Du Toit, Andrew Bain, Marion, and Prince Edward, offsets the Southwest Indian Ridge (SWIR) left laterally 1230 km. The Andrew Bain, the largest, has a length of 750 km and a maximum transform domain width of 120 km. We show that, currently, the Nubia Somalia plate boundary intersects the SWIR east of the Prince Edward, placing the Andrew Bain on the Nubia Antarctica plate boundary. However, the overall trend of its transform domain lies 10° clockwise of the predicted direction of motion for this boundary. We use four transform-parallel multibeam and magnetic anomaly profiles, together with relocated earthquakes and focal mechanism solutions, to characterize the morphology and tectonics of the Andrew Bain. Starting at the southwestern ridge-transform intersection, the relocated epicenters follow a 450-km-long, 20-km-wide, 6-km-deep western valley. They cross the transform domain within a series of deep overlapping basins bounded by steep inward dipping arcuate scarps. Eight strike-slip and three dip-slip focal mechanism solutions lie within these basins. The earthquakes can be traced to the northeastern ridge-transform intersection via a straight, 100-km-long, 10-km-wide, 4.5-km-deep eastern valley. A striking set of seismically inactive NE-SW trending en echelon ridges and valleys, lying to the south of the overlapping basins, dominates the eastern central section of the transform domain. We interpret the deep overlapping basins as two pull-apart features connected by a strike-slip basin that have created a relay zone similar to those observed on continental transforms. This transform relay zone connects three closely spaced overlapping transform faults in the southwest to a single transform fault in the northeast. The existence of the transform relay zone accounts for the difference between the observed and predicted trend of the Andrew Bain transform domain. We speculate that between 20 and 3.2 Ma, an oblique accretionary zone jumping successively northward created the en echelon ridges and valleys in the eastern central portion of the domain. The style of accretion changed to that of a transform relay zone, during a final northward jump, at 3.2 Ma.

Scott, Richard. 2005. "Piracy Attacks Rise off Somalia." Jane's Navy International. Issue AUG, Pages 2. Descriptors: Military engineering; Naval vessels; Military operations; Societies and institutions. Abstract: The International Chamber of Commerce's International Maritime Bureau (IMB) has warned of an upward spiral in acts of piracy off the war-ravaged east African state of Somalia. The IMB is very concerned about the recent increase in piracy activities by armed Somali militia. The IMB's Piracy Reporting Centre has confirmed that armed pirates in speedboats and gunboats are frequently opening fire on ships, seeking to hijack the vessel and hold the crew to ransom. The IBM issued an initial warning to the heightened piracy threat in June 2005, recommending that ships remain at least 50 miles offshore. ISSN: 1358-3719.

Seagull Exploration. 1978. Petroleum Exploration Guide, East Africa: Sudan, Ethiopia, Somalia, Kenya, Tanzania, Mozambique, Madagascar, South Africa. Seagull Exploration. Descriptors: 1 case; Petroleum- Geology- Africa, Eastern; Petroleum-

Geology- Sudan; Petroleum- Geology- Ethiopia; Petroleum- Geology- Somalia; Petroleum- Geology- Kenya; Petroleum- Geology- Tanzania; Petroleum- Geology- Mozambique; Petroleum- Geology- Madagascar; Petroleum- Geology- South Africa. Notes: Case includes: one volume of text (234 p.) + assorted maps. OCLC Accession Number: 14248400.

Seideman, Tony. 1993. "The Logistics of Salvation: Managing the Relief Effort in Somalia Means Getting Products into a Country where Infrastructure and Government Are Virtually Non-Existent." Distribution. Volume 92, No. 4 (Apr. 1993), Pages 46-48. Available from Northwestern University Transportation Library through interlibrary loan or document delivery.

Senesi N. 1980. "Valutazione Della Qualita Delle Acque Per Uso Irriguo e Problemi Connessi Alla Loro Utilizzazione Nell'Agricoltura Somala." Translated title: "Appraisal of the Quality of Waters for use in Irrigation and Problems Connected to their use in Somalia's Agriculture." Riv. Agric. Subtrop. Trop. Volume 74, Issue 1-2, Pages 17-37. Descriptors: Hydrology; Meteorology And Climatology; water quality; irrigation; Shabelle; Juba rivers; agriculture; Somalia. Notes: Special Feature: 3 figs, 8 tables, 13 refs. Abstract: The evaluation of water quality for irrigation from Shabelle and Juba rivers and from deep wells is considered essential for a correct development of agriculture in Somalia. This examines separately for each 1 of the 3 cited potential problems, its general cause and its effects on soil status and crop yields, successively to evaluate water quality for irrigation to predict the eventual arising of a potential problem, and to suggest available management alternatives to correct or prevent its occurrence. ISSN: 0035-6026.

Senni, L. 1929. "La Valorizzazione Del Giuba." Translated title: "Exploitation of Juba River." L'Agricoltura Coloniale. Volume 1929, Descriptors: Africa; East Africa; exploitation; fluvial features; geography; Juba River; rivers; Somali Republic; water resources. ISSN: 0394-2945.

Serrazanetti, R. M. 1931. "L'Irrigazione Dei Paesi Caldi (Somalia)." Translated title: "Irrigation of the Hot Countries, Somalia." L'Agricoltura Coloniale. Volume 1931, Descriptors: Africa; agriculture; East Africa; irrigation; Somali Republic; water management; water resources; water supply. ISSN: 0394-2945.

Sgavetti, Maria, Ferrari, M. Carla and Chiari, Roberto. 1995. "Stratigraphic Correlation by Integrating Photostratigraphy and Remote Sensing Multispectral Data: An Example from Jurassic-Eocene Strata, Northern Somalia." AAPG Bull. November. Volume 79, Pages 1571-1589. Descriptors: Petroleum geology/Somalia; Aerial photography; Correlation (Geology); Applied Science & Technology. Notes: Bibliography; Illustration; Map; Diagram. Abstract: Photostratigraphic and multispectral data were integrated for the study of Jurassic-Eocene strata in northern Somalia. Physical surfaces with chronostratigraphic significance were identified from aerial photographs, enabling photostratigraphic logs and stratal patterns and correlations to be established. Satellite multispectral data analysis was used to identify image facies based on the absorption characteristics of different lithologies. Lithologies were determined by direct comparison with selected sections measured and described in the field. The preliminary stratigraphic framework and photostratigraphic maps can contribute to existing data and provide the basis for further field studies. ISSN: 0149-1423.

Sharp, T. W., DeFraites, R. F., Thornton, S. A., Burans, J. P. and Wallace, M. R. 1995. "Illness in Journalists and Relief Workers Involved in International Humanitarian Assistance Efforts in Somalia, 1992-93." J. Travel Med. Jun 1. Volume 2, Issue 2, Pages 70-76. Abstract: Background: Journalists and relief workers participating in international relief efforts in Somalia following the intervention of outside armed forces in late 1992, were faced with a number of threats to their health. Principally these threats were from endemic infectious diseases and trauma. Methods: In-patient, emergency clinic, and laboratory records of U.S. military field hospitals, which provided the only available sophisticated medical care in Somalia during most of the study period (December 15, 1992, to February 15, 1993), were reviewed to determine the number of workers evaluated and the causes of their illnesses. In addition, two questionnaire surveys were conducted to elucidate risk factors for illness in these groups. Results: One hundred and thirty-eight journalists and relief workers, primarily from Europe and North America, were evaluated at a hospital for a variety of common travel-associated health problems, including diarrhea (33%), acute respiratory infection (21%), other febrile illnesses (11%), hepatitis (2%), major trauma (6%), and minor trauma (13%). Documented infectious disease pathogens included Plasmodium falciparum (7 cases), Shigella sp (3 cases), enterotoxigenic Escherichia coli (ETEC) (3 cases), dengue virus-2 (2 cases), and hepatitis E virus (3 cases). Two relief workers were killed by gunshot wounds. In the questionnaire surveys of 104 journalists and 98 relief workers, 84% of respondents reported that they had received some pre-travel medical advice, but only 70% sought a medical consultation in person. Thirty-four percent were not receiving a recommended antimalarial chemoprophylaxis regimen, and only 10% obtained a fluoroquinolone antimicrobial drug for self treatment of diarrhea. Sixty-four percent of both groups combined, reported having had diarrhea, and 26% experienced a nondiarrheal febrile illness. Sixty-eight percent reported that their work performance was adversely affected by illness. In multivariate logistic regression analyses, factors associated with an increased risk of diarrhea were age < 35 years (OR 1.5, 95% CI 1.1-1.9); residence in Somalia for more than 21 days (OR 1.7, 95% CI 1.3-2.1); and regular consumption of local food and water (OR 3.8, 95% CI 3.4-4.2). Factors associated with nondiarrheal febrile illness were age < 35 years (OR 1.4, 95% CI 1.1-1.8); residence in Somalia for more than 21 days (OR 1.8, 95% CI 1.4-2.2); and not having had an in-person pretravel medical consultation (OR 2.0, 95% CI 1.5-3.0). Conclusions: These data indicate that journalists and relief workers who traveled to Somalia in response to the massive humanitarian crisis themselves experienced substantial health problems. Improved pretravel medical preparation might prevent or limit illness in these unique groups and improve the efficiency of future disaster response efforts. (J Travel Med 2:70-76, 1995). ISSN: 1195-1982.

Sharp, T. W., Thornton, S. A. and Wallace, M. R., et al. 1995. "Diarrheal Disease among Military Personnel during Operation Restore Hope, Somalia, 1992-1993." Am. J. Trop. Med. Hyg. Volume 52, Issue 2, Pages 188-193. Descriptors: diarrhea; military personnel; man; Somalia. Abstract: The potential for widespread diarrheal disease was regarded as a substantial threat to U.S. troops participating in the early phases of Operation Restore Hope in Somalia. Outpatient surveillance of 20,859 U.S. troops deployed during the first eight weeks, however, indicated that a mean of only 0.8% (range 0.5-1.2%) of personnel sought care for diarrhea each week, and in three epidemiologic surveys, < 3% of troops reported experiencing a diarrheal illness per week.

Despite these low overall attack rates, diarrhea accounted for 16% of 381 hospital admissions and 20% of 245 patients admitted with a temperature greater than or equal to 38.5 degree C. Sixty-one specimens were obtained from inpatients and 52 were obtained from outpatients. Shigella sp. were isolated from 33%, enterotoxigenic Escherichia coli from 16%, Giardia lamblia from 4%, and rotavirus from 1% of 113 stool samples obtained from inpatient (61) outpatient (52) troops with diarrhea. Bacterial isolates obtained in Somalia were resistant to doxycycline (78%), ampicillin (54%), and sulfamethoxazole (49%), but uniformly sensitive to ciprofloxacin. With the exception of 10 Shigella sonnei isolates that were linked epidemiologically to one eating facility, bacterial pathogens occurred sporadically and demonstrated a wide variation of serotypes and antibiotic sensitivity patterns. Additionally, three of 11 paired sera collected from persons with nausea, vomiting, and watery diarrhea demonstrated a four-fold or greater increase in titer to Norwalk virus antibody. These data indicate that large outbreaks of diarrheal disease did not occur; however, highly drug-resistant enteric bacteria, and to a lesser extent viral and parasitic pathogens, were important causes of morbidity among U.S. troops in Somalia. ISSN: 0002-9637.

Shepherd G. 1992. "Forest Policies, Forest Politics." Overseas Development Institute, London; ODI Agricultural Occasional Paper, 13. Descriptors: 86 pages; Planning; social forestry; forestry project; institutional constraint; political aspect. Abstract: Social forestry projects are held in tension between the rural environment in which villagers and foresters must cooperate and the political, economic and legal environment created by the nation state. A good deal of attention has been given to local level issues in social forestry but it is also important to consider ways in which national level institutions help or hinder such projects. This book introduces the political and institutional context of forestry policy in an overview chapter. This is followed by four case studies drawn from papers first published in ODI's Social Forestry Network. These examine institutional constraints on forestry policy and woodland management in Mali, urban-rural conflict over wood fuel use in Nigeria, and the "reality of the commons' in Somalia. Notes: In: Forest policies, forest politics (1992) p. 86p; Geographic: Nigeria; Somalia; Mali. ISBN: 0850031680.

Shepherd G. 1989. "The Reality of the Commons: Answering Hardin from Somalia." Development Policy Review. Volume 7, Issue 1, Pages 51-63. Descriptors: natural resource; commons; pastoralism; common property rights; rural people. Notes: Somalia- Bay region. Abstract: Reports on work undertaken in the Bay Region of southern Somalia, during which it became clear that the assertion and denial of common property rights to bushland were at the heart of the energy issues which were the original field of study. It also describes the early stages of research aimed at returning common property rights to rural people who have lost them, and thereby ensuring the survival of resources which would otherwise be lost as well. ISSN: 0950-6764.

Shepherd G. Shepherd G. 1992. "The Reality of the Commons: Answering Hardin from Somalia." ODI Agricultural Occasional Paper, 13. Descriptors: 73-86; Economic; property rights; bushland; common property; land management; developing country; rural area. Abstract: It is still too readily assumed by field practitioners that there is and perhaps can be no successful management of lands held in common. The opposite proposition is argued here, that the real tragedy comes when the commons are thrown open and unrestricted exploitation allowed. This article reports on work undertaken in the

Bay region of southern Somalia between January and April 1988, during which it became clear that the assertion and denial of common property rights to bushland were at the heart of the energy issues which were the original field of study. It also describes the early stages of research aimed at returning common property rights to rural people who have lost them, and thereby ensuring the survival of resources which would otherwise be lost as well. Notes: In: Forest policies, forest politics (1992) p. 73-86; Geographic: Somalia- Bay Region. OCLC Accession Number: 1023360.

Shetye, Satish R. 1986. "Model Study Of The Seasonal Cycle of the Arabian Sea Surface Temperature." J. Mar. Res. Volume 44, Issue 3, Pages 521-542. Descriptors: Oceanography- Temperature Measurement; Thermodynamics; Seawater, Thermal Gradients; Fluid Dynamics; Meteorology- Climatology. Abstract: The annual variation of the SST along a zonal strip from the coast of Somalia to the southwest coast of India was simulated using available data (monthly-mean heat and momentum fluxes across the air-sea interface, surface advective field, etc.) as input to a Kraus-Turner mixed-layer model. Three cases were examined: influence of surface fluxes; the effects of surface fluxes and vertical advection; and effect of horizontal advection. The model forced with the surface heat and momentum fluxes alone simulated reasonably well the SST variability throughout the year except during the May-August (southwest monsoon) cooling phase. The model was found to be inadequate to handle the coastal areas during this phase. Over the open-sea regime the performance of the model was better; and, it improved when the influence of advection was included. ISSN: 0022-2402.

Shiddo, S. A., Mohamed, A. Aden and Akuffo, H. O., et al. 1995. "Visceral Leishmaniasis in Somalia: Prevalence of Markers of Infection and Disease Manifestations in a Village in an Endemic Area." Transactions of the Royal Society of Tropical Medicine and Hygiene. Volume 89, Issue 4, Pages 361-365. Descriptors: leishmaniasis; visceral; kala-azar; Leishmania donovani; prevalence; pathology; Somalia. Abstract: Prevalence and disease manifestations of visceral leishmaniasis (VL) were studied in a Somali village in an area which has long been known to be endemic for VL. Demographic data were collected from 102 households, comprising 438 inhabitants. Clinical examination was performed of 306 individuals, 72% of the 426 eligible persons. Of these, 276 (90%) agreed to give blood and 246 (80%) to be skin tested with leishmanin. Leishmanin reactions were positive; in 26% anti-Leishmania antibodies were detected in 11%, and splenomegaly was recorded in 14% (23% of those who were seropositive). Malaria was hypoendemic and therefore unlikely to be responsible for more than 10% of the cases with splenomegaly. Three of the seropositive villagers with splenomegaly complained of feeling ill. The remaining 91 sero- and/or leishmanin-positive individuals had no complaint regarding their health and had not experienced any long period of illness. There was a slight over-representation of males in the group of sero- and/or leishmanin-positive villagers, possibly due to a gender-associated difference in exposure to the parasite. Among the patients with clinical VL treated at Mogadishu hospitals during 1989 and 1990, the male/female ratio was 3[middle dot]3:1, which may indicate a selection of male patients for hospital care. Most patients were [les]15 years old, suggesting that the highest risk of becoming clinically ill was among children.

Shields T. 1993. "Biting the Hand that Feeds." Ceres. Volume 25, Issue 2, Pages 38-40. Descriptors: Water; food production; ethnic discrimination; clan violence. Notes: Somalia. Abstract: The worst of Somalia's clan violence was visited upon two mainly agricultural groups, the Cushitic Rahanwein and non-Cushitic Bantu, who between them provided Somalia with much of the pre-war food security it enjoyed. No matter how much food aid is delivered, the Somalis' selective decimation of these two groups may have helped doom the nation to chronic food shortages for years to come.

Shoham, Jeremy; Rivers, John; Payne, Philip; London School of Hygiene and Tropical Medicine; Centre for Human Nutrition and Save the Children Fund (Great Britain). 1989. Hartisheik. London: Centre for Human Nutrition, London School of Hygiene and Tropical Medicine. Descriptors: 27 pages; Refugees- Somalia; Refugees-Ethiopia; Refugee camps; Food supply; Malnutrition; Water supply; Management; Monitoring. Notes: Named Corp: United Nations. Office of the High Commissioner for Human Rights. UNHCR. Geographic: Hartisheik (Ethiopia: Refugee camp); Notes: "June 1989". OCLC Accession Number: 70139822.

Siad, M. F. 1954. "L'Aqua Vita Della Somalia, Il Bacino Di Coriolei." Translated title: "Water Vitality of Somalia, Qoryaleh Basin." Quaderni Di Geologia Della Somalia (Mogadiscio). Volume 48, Descriptors: Africa; East Africa; Qoryaleh Basin; Qoryaleh Somalia; Shebeellaha Hoose; Somali Republic; water resources. LCCN: 93093377.

"Siebens to Test Potential of New Basin in Western Indian Ocean." 1977. Oil Gas J. 10 Oct. Volume 75, Issue 42, Pages 170-171. Descriptors: basins; collision; economic geology; evolution; genesis; Indian Ocean; movement; natural gas; petroleum; petroleum exploration; plate tectonics; rifting; Somali Basin; West Indian Ocean. ISSN: 0030-1388.

Sighinolfi, G. P. Aberra, G. Gorgoni, G. and Valera, R. 1993. Distribution of Precious Metals in the Tulu Dimtu Ultramafic Body (Welega, Ethiopia). Firenze; Istituto Agronomico L'Oltremar; 1993. Relazioni E Monografie Agrarie Subtropicali E Tropicali Nuova Serie113 B, pages: 595-606. International Meeting; 1st- 1987 Nov: Mogadishu. Notes: Geology of Somalia and surrounding regions: Geology and mineral resources of Somalia and surrounding regions. OCLC Accession Number: 34175029.

Silvestri, Alfredo. 1945. "Sull'Esistenza Del Cretaceo Superiore Nella Somalia." Bollettino Della Societa Geologica Italiana. Descriptors: Cretaceous; historical geology; Italian Somaliland; Mesozoic; upper. pages 13-15, 1947. B., v. 64; Abstract: Records the occurrence of fossiliferous upper Cretaceous beds in Italian Somaliland. ISSN: 0037-8763.

Simoncelli, D. 1936. "La Valorizzazione Dell'Impero; Il Problema Minerario." Translated title: "Exploitation of the Empire; Mining Problems." Rivista Italiana De Scienze Economiche. Volume 8, Pages 521-545. Descriptors: Africa; East Africa; exploitation; mining; mining geology; rock mechanics; sedimentology; soil mechanics; Somali Republic. Database: GeoRef.

Simons, A. 1996. "Losing Mogadishu: Testing U.S. Policy in Somalia by Jonathan Stevenson." Armed Forces and Society. Volume 23, Issue 1, Pages 128. ISSN: 0095-327X.

Sinclair Somal Company, Mogadishu, Somalia. 1968. Subsurface Geologic Map of Central and Southern Somalia Showing Structure on Top of Middle Jurassic, Scale 1:100,000 [Modified]. Somalia: MRMI, Mogadishu, Somalia. Descriptors: Africa; East Africa; geologic maps; Jurassic; maps; Mesozoic; Middle Jurassic; Somali Republic; =; southern Somali Republic; central Somali Republic. Database: GeoRef. Accession Number: 1998-014340.

Singer, A., Stahr, K. and Zarei, M. 1998. "Characteristics and Origin of Sepiolite (Meerschaum) from Central Somalia." Clay Miner. Volume 33, Issue 2, Pages 349-362. Descriptors: Clay minerals; Composition; Lithology; Morphology; Limestone; Geomorphology; X ray diffraction analysis; Differential thermal analysis; Infrared spectroscopy. Abstract: Nearly pure sepiolite clay crops out in a playa-like depression near El Bur, Central Plateau region of Somalia. The deposit is associated with the Lower to Mid-Eocene Taleh Formation that includes, besides limestone, dolomite and gypsiferous marls, extensive anhydrite and various evaporites, primarily gypsum. The material was examined by X ray diffraction, differential thermoanalysis, infrared spectroscopy and electron microscopy. The material consists of well-crystallized sepiolite, accompanied in some layers to minor calcite and traces of quartz and halite. Lithology and geomorphology indicate a lacustrine, closed basin evaporative environment of formation for this deposit. ISSN: 0009-8558.

Sipkes J. 2003. "Mending a Broken Infrastructure: UNDP Assistance for Somalia." GIM International. Volume 17, Issue 3, Pages 70-71. Descriptors: Regional And Spatial Development And Planning; GIS, Remote Sensing; spatial analysis; mapping; civil service; development project. Notes: Somalia. Abstract: More than a decade of hostilities between various political and military factions has led to a total breakdown in civil administration in Somalia. This situation has in turn led the United Nations to offer assistance to groups within the country trying to recreate properly functioning institutions. For this purpose a United Nations Development Programme (UNDP) Office for Somalia has been set up, temporarily located in the Kenyan capital Nairobi. GIM International visited the UNDP Data and Information Management Unit (DIMU) to focus on the spatial analysis component of their work. ISSN: 1566-9076.

"Sketch Map of Somaliland." 1907. London. Notes: Description: 1 map; 50 x 39 cm. Geographic: Somalia- Maps. Scale 1:3,000,000. Cartgrph Code: Category of scale: a Constant ratio linear horizontal scale: 3000000; Notes: To accompany: Official history of the operations in Somaliland, 1901-4. Other Titles: Official history of the operations in Somaliland. OCLC Accession Number: 54334132.

Smith, A. D. 1896. "Expedition through Somaliland to Lake Rudolph." The Geographical Journal. Volume 8, Pages 120-137, 221-239. Descriptors: Africa; areal geology; East Africa; East African Lakes; expeditions; Lake Turkana; Somali Republic. ISSN: 0016-7398.

Smith, A. Donaldson. 1897. Through Unknown African Countries; the First Expedition from Somaliland to Lake Lamu. International (III): Publisher unknown, International (III). Descriptors: Africa; areal geology; East Africa; field trips; maps; Somali Republic; topographic maps; topography. Notes: 1 plate. Database: GeoRef. Accession Number: 1998-011359.

Smith, S. L. 1986. "Is the Somali Current a Biological River in the Northwestern Indian Ocean." Brookhaven National Lab., Upton, NY. Volume: BNL38942, CONF86031731, pages: 15. Descriptors: Communities; Population Dynamics; Productivity; Species Diversity; Upwelling; Indian Ocean; Water Currents; Zooplankton; Environmental Effects; ERDA/520100; ERDA/580500; Somali current; Environmental pollution and control Water pollution and control; Ocean technology and engineering Biological oceanography; Ocean technology and engineering Dynamic oceanography. Abstract: Calculations of diversity, equitability, average rank, and average percentage of total numbers show that the addition of zooplanktonic taxa such as Calanus carinatus arising in the upwelling areas during the southwest monsoon does significantly increase diversity and equitability of the community. Rank and percentages of total numbers, however, show that the response is not entirely due to the presence of Calanoides carinatus during the southwest monsoon. Another important change in the zooplanktonic community was the decline in the abundance of Oithona spages in upwelling areas during the southwest monsoon. (ERA citation 12:021127). Funder: Department of Energy, Washington, DC. Note: Symposium on marine biology of the Arabian Sea, Karachi, Pakistan, 29 Mar 1986. NTIS Accession Number: DE87003233XSP.

Smith, S. L. 1982. "The Northwestern Indian Ocean during the Monsoons of 1979: Distribution, Abundance, and Feeding of Zooplankton." Deep-Sea Res. Volume 29, Issue 11A, Pages 1331-1353. Descriptors: zooplankton; monsoons; upwelling; abundance; geographical distribution; Calanoides carinatus; Somalia; feeding behavior; Marine. Abstract: The upwelling regions, which are associated with a reproductively active population of the large-bodied Calanoides carinatus, are the primary features affecting distributions of zooplankton during the southwest monsoon and the main difference between monsoons. The ontogenetic migration of C. carinatus is essentially an annual life-history strategy and therefore on the same temporal scale as the reversals in the monsoonal winds and associated upwelling. The ability of C. carinatus to ingest readily the diatoms that dominate the upwelling regions and to store lipid is crucial to its dominance of the areas of upwelling both in numbers and biomass. ISSN: 0198-0149.

Smout, I. K. and Brown, D. A. 1984. Drainage Trial On Saline Land In Somalia. Fort Collins, CO: Int Commission on Irrigation & Drainage, New Delhi, In. pages: 219-243. Transactions- Twelfth International Congress on Irrigation and Drainage. Descriptors: Soils- Drainage; Sugar Cane; Agriculture- Somalia; Land Reclamation; Clay; Irrigation. Abstract: Field Trials on an abandoned sugar cane field on heavy clay (vertisol) soils provided data for designing irrigation and drainage systems for reclamation and salinity control. The previous irrigation system of small basins without any drainage was replaced by 300 m long furrows with surface drainage. Various subsurface drainage treatments were installed with buried drains at different spacings and with different filters. Sugar cane was planted, and irrigation, drainage, salinity and crop growth were monitored. The trials showed that leaching of salts and a reduction in water logging are technically feasible on these soils. The best drained plots showed considerable improvement in crop root distribution and cane growth and yield. Inspection of the buried plastic drain pipes after use revealed no problems of soil entry and confirmed the standard recommendation that drain filters are not required on such heavy soils. Based on the trials, development of 1,500 ha of the Estate was recommended, with 300 m furrows and buried plastic pipe drains 2 m deep at 33 m spacing.

Societa Africana d'Italia, Naples, Italy. 1911. "Osservazioni Idrometriche Del Fiume Uebi Scebeli Presso Afgoi Compiute Dall'Ufficio Per Lo Studio Della Linea Mogadiscio-Afgoi Negli Anni 1908-1909." Translated title: "Hydrometric Observation of Shebelle River Near Afgoi Completed at the Office for the Study of the Mogadishu-Afgoi Line in the Years of 1908-1909." Bollettino Della Societa Africana d'Italia. Volume 1-2, Descriptors: Afgoi Somali Republic; Africa; Benadir; East Africa; gauging; hydrology; measurement; Mogadishu Somali Republic; rivers and streams; Somali Republic; stream flow; surface water. ISSN: 0392-1468.

Societa Agricola Italo Somala, Mogadishu, Somalia. 1931. "Questioni Riguardanti l'Interramento Del Fiume Uebi Scebeli a Monte Delle Opere Di Derivazione e Dei Canali Di Irrigazione e l'Impianto Di Prosciugamento." Translated title: "Questions about Shebelle River Flow Near the Diversion and Irrigation Canals and Drainage System." Somalia: Descriptors: Africa; drainage basins; East Africa; hydrology; irrigation; rivers and streams; Shebelle River; Somali Republic; water resources; water use. Database: GeoRef. Accession Number: 2000-010226.

Societa Agricola Italo Somalo, Mogadishu, Somalia. 1930. Il Corso Dello Uebi Scebeli Da Genale Ad Haway e l'Uebi Gofca. Translated title: Shebelle River Course from Genale to Haway and Gofca River. Somalia: SAIS, Mogadishu, Somalia. Descriptors: Africa; East Africa; fluvial features; Genale Somali Republic; Gofca River; Haway Somalia; hydrology; rivers; rivers and streams; Somali Republic; surface water. Database: GeoRef. Accession Number: 1998-014299.

Societa Agricola Italo Somalo, Mogadishu, Somalia. 1930. L'Opera Della SAIS in Somalia. Translated title: SAIS Works in Somalia. Somalia: SAIS, Mogadishu, Somalia. Descriptors: Africa; agriculture; drainage basins; East Africa; hydrology; SAIS; Societa Agricola Italo Somalo; Somali Republic; water management. Database: GeoRef. Accession Number: 1998-014300.

Societa Agricola Italo Somalo, Mogadishu, Somalia. 1930. Planimetria Del Comprensorio Irriguo. Translated title: Planimetry of Well-Watered Territory. Somalia: SAIS, Mogadishu, Somalia. Descriptors: Africa; agriculture; East Africa; hydrology; irrigation; planimetry; Somali Republic; water supply. Database: GeoRef. Accession Number: 1998-014301.

Societa Agricola Italo Somalo, Mogadishu, Somalia. 1930. Sezioni e Portate Del Giuba. Translated title: Sections and Juba River Course. Somalia: SAIS, Mogadishu, Somalia. Descriptors: Africa; East Africa; fluvial features; hydrology; Juba River; rivers; rivers and streams; Somali Republic; surface water. Database: GeoRef. Accession Number: 1998-014302.

Societa Industriali Sapie, Rome, Italy. 1939. L'Industria Miniere Italiana e d'Otremare. Translated title: Mining Industry of Italy and Overseas Italian Territories. Italy: Societa Industriali Sapie, Rome, Italy. Descriptors: Africa; East Africa; Ethiopia; Europe; Italy; mineral resources; mining; mining geology; Somali Republic; Southern Europe. Notes: Volume 13. Database: GeoRef. Accession Number: 2000-010206.

SOGREAH, Grenoble, France. 1983. "Northwest Region Agricultural Development Project; Hydrogeology." France: Volume: 16, Descriptors: Africa; agriculture; East Africa; hydrology; northwestern Somalia; Somali Republic; water resources; water supply. Notes: Final report. Database: GeoRef. Accession Number: 1998-014344.

SOGREAH, Grenoble, France. 1983. "Northwest Region Agriculture Development Project; Climatology." France: Volume: 2, Descriptors: Africa; agriculture; climate; development; East Africa; irrigation; northwestern Somalia; Somali Republic; water resources; water supply. Notes: Technical report. Database: GeoRef. Accession Number: 1998-014345.

SOGREAH, Grenoble, France. 1981. "Northwest Region Agricultural Development Project; Hydrology." France: Volume: 3, Descriptors: Africa; agriculture; development; East Africa; feasibility studies; hydrology; irrigation; northwestern

Somalia; Somali Republic; water resources; water supply. Notes: Draft technical report. Database: GeoRef. Accession Number: 1998-014343.

"Soil Map: British Somaliland Protectorate. Based on Geological Map ... 1933 with Additions & Approximations." 1945. S.l: s.n.; Somalia. Descriptors: Soils- Somalia- Maps. Notes: Description: 1 map. Geographic: Somalia- Maps. Scale 1:1,000,000; conic projection. Notes: Date of information 1944. Shows: international boundaries, soils. "GP no. 2587." Coverage: Northern Somalia. OCLC Accession Number: 48603943.

"Somali Fishing Industry has Potential for Growth." 1982. Mar. Fish. Rev. Volume 44, Issue 12, Pages 25-26. Descriptors: International Development Abstracts. Somalia. Notes: Special Feature: map. Abstract: Somalia is one of the least developed countries in Africa. It has also suffered from food shortages caused by a severe drought during the early 1970s. The Somali Ministry of Fisheries and the Coastal Development Agency (CDA) want to increase the fisheries catch to help alleviate the country's food shortage. Somali fisheries are still largely undeveloped and play a minor role in the nation's economy. However, they could become important. ISSN: 0090-1830.

Somali Republic, Ministry of Mineral and Water Resources, Somalia. 1981. Energy Resources of Somali Democratic Republic; Proceedings of an International Conference on Long-Term Energy Resources. United States: Pitman Publ. Co., Boston, MA, United States. International Conference on Long-Term Energy Resources, Montreal, PQ. Canada Conference: Nov. 26-Dec. 7, 1979. Descriptors: Africa; East Africa; economic geology; energy sources; exploration; production; resources; Somali Republic. Notes: References: 5; 1 table. ISBN: 0273085344.

"Somalia." 1978. Annual Review of Mining. June. Volume 1978, Pages 486-487. Descriptors: 1977; Africa; deposits; East Africa; economic geology; economics; metals; mineral resources; production; review; Somali Republic. Database: GeoRef.

Somalia. Camera di Commercio, Industria e Agricoltura. Sezione e Mostre. 1958. "Bibliografia Somalia." Scula Tipografica Missione Cattolica, Mogadiscio. 135 pages. Pref. di Ariberto Forlani. "Part one, monographs, part two periodical articles; each arranged by subject. For books, gives only author, title, palce, date and paging; for articles, author, title and title of periodical, year and paging."[14] OCLC: 16576510.

Somalia; a Country Study. 1982. United States: American University, Foreign Area Stud., Washington, DC, United States. Descriptors: Africa; areal geology; East Africa; economics; history; philosophy; policy; politics; regional; Somali Republic; urbanization. Notes: Edition: 3. First published in 1977; individual papers are not cited separately; illustrations incl. tables. Database: GeoRef. Accession Number: 1983-032529.

Somalia Aid Coordination Body and Water and Environmental Sanitation Committee. 1998. Strategic Framework for Co-Ordinated Approaches to Promotion of Water and Environmental Sanitation Development in Somalia, 1 January 1999-31 December 2003: Report of the Proceedings and Outcomes of a Strategic Planning Workshop, Naro Moru, Kenya, 06-10 July 1998. Descriptors: 50 leaves; Water resources development- Somalia; Sanitation- Somalia. Accession Number: 53130336.

Somalia, CARS, Somalia. 1967. "Shebelle River Water Quality 1965/1966." Somalia: Descriptors: Africa; East Africa; geochemistry; hydrochemistry; hydrology;

---

[14] Anglemyer, Mary. 1970. Natural Resources: A Selection of Bibliographies." Second edition. Washington, DC: Engineer Agency for Resource Inventories. EARI Development Research Series. Report #3. Page 81.

Shebelle River; Somali Republic; southern Somali Republic; surface water; water quality. Database: GeoRef. Accession Number: 1998-015608.

Somalia, CARS, Somalia. 1965. "Irrigation Plans, Designs and Fundamentals." Somalia: Descriptors: Africa; design; development; East Africa; hydrology; irrigation; Somali Republic; surface water; water management. Database: GeoRef. Accession Number: 1998-015606.

Somalia, CITACO, Mogadishu, Somalia. 1975. Sablale and Kurtunwarey Agricultural Resettlement and Irrigation Scheme. Somalia: CITACO, Mogadishu, Somalia. Descriptors: Africa; agriculture; East Africa; irrigation; Kurtunwarey; Sablale; Shabeellaha Hoose Somali Republic; Shabellaha Hoose; Somali Republic; water resources; water supply. Database: GeoRef. Accession Number: 1998-017920.

Somalia, CITACO, Mogadishu, Somalia. 1974. Development Project Planned in Lower Juba and Constrains to their Implementation. Somalia: CITACO, Mogadishu, Somalia. Descriptors: Africa; development; East Africa; Juba River basin; land use; regional planning; Somali Republic; water resources. Database: GeoRef. Accession Number: 1998-017917.

Somalia, CITACO, Mogadishu, Somalia. 1973. Indagine Tecnico-Economica Per Un Sistema Di Allontanamento Delle Acque Di Rifiuto Della Citta' Di Mogadishu. Technical-Economic Investigation on a System of Waste Water Leaving Mogadishu. Somalia: CITACO, Mogadishu, Somalia. Descriptors: Africa; Benadir Somali Republic; East Africa; hydraulics; Mogadishu Somali Republic; Somali Republic; waste disposal; waste water. Database: GeoRef. Accession Number: 1998-017916.

"Somalia- Energy Situation in 1985." 1986. Bundesstelle für Aussenhandelsinformation, Cologne (Germany). Dec. Volume: NP7770290, pages: 12. Descriptors: Somalia; Imports; Petroleum; Power Generation; Electric Power; Energy Supplies; ERDA/293000; ERDA/012000; Foreign technology; Energy policy; Energy Policies regulations and studies; Energy use supply and demand; Energy Heating and cooling systems; Business and economics Foreign industry development and economics. Abstract: The energy situation of Somalia is reviewed on the basis of relevant data. Data on the country's national energy policy are followed by an outline of trends in energy sources and electric power generation. Key figures are presented on the country's external trade. (ERA citation 13:000975). Notes: In German. BfAI-Marktinformation. Reihe MI-BS. No. 29.534.86.273. NTIS Accession Number: DE87770290XSP.

"Somalia- Energy Situation 1983/1984." 1985. Bundesstelle für Aussenhandelsinformation, Cologne (Germany, F.R.). May. Volume: NP6770007, pages: 22. Descriptors: Electric Power; Energy Supplies; Exports; Imports; Petroleum; Petroleum Products; Power Generation; Somalia; Energy Policy; ERDA/292000; ERDA/293000; Foreign technology; Energy conversion non propulsive Conversion techniques; Energy use supply and demand; Energy Policies regulations and studies; Business and economics International commerce marketing and economics. Abstract: The energy situation of Somalia is reviewed on the basis of relevant data. Some remarks on the country's national and international energy policy are followed by an outline of trends in energy sources and electric power generation. (ERA citation 10:051543). Notes: In German. NTIS Accession Number: DE86770007XSP.

"Somalia: [Mineral Resources]." 1990. S.l: s.n. Descriptors: Mines and mineral resources- Somalia- Maps. Notes: Description: 1 map; color; 23 x 17 cm. Scale [ca.

1:7,000,000]; (E 410--E 510/N 120--S 10). Notes: Also shows undeveloped mineral resources. Includes location map. LCCN: 95-682840. OCLC Accession Number: 33164629.

"Somalia: A Tragedy." 1992. The Lancet. Volume 339, Issue 8798, April 11, 1992. Pages 924-924.

Somalia, Ministero di Lavori Pubblici SDR, Mogadishu, Somalia. 1965. Progetto Per Lo Studio Dei Corsi d'Acqua Temporanei Da Sbarrare Con Dighe in Terra o in Calcostruzzo. Translated title: Project for the Study of the Temporary Water Course to be Blocked with an Earth Dam Or Concrete. Somalia: Min Lav. Pubbl., Mogadishu, Somalia. Descriptors: Africa; construction; dams; earth dams; East Africa; gravity dams; hydraulics; Somali Republic; water resources; waterways. Database: GeoRef. Accession Number: 1999-067963.

Somalia, Ministero di Lavori Pubblici SDR, Mogadishu, Somalia. 1965. Rapporto Sulla Possibilita Di Valorizzazione Delle Fasce Rivierasche Del Giuba. Translated title: Report on the Possibility of Exploitation of the Coastal Part of Juba. Somalia: Min. Lav. Pubbl., Mogadishu, Somalia. Descriptors: Africa; East Africa; exploitation; hydrology; Juba River; Somali Republic; water resources. Database: GeoRef. Accession Number: 1999-067964.

Somalia, Ministero di Lavori Pubblici SDR, Mogadishu, Somalia. 19??. Schema Di Valorizzazione Delle Zone Del Basso Schebelli Da Audegle a Maringubai. Translated title: Exploitation Scheme of the Zone of the Lower Shebelle from Audegle to Maringubai. Somalia: Min. Lav. Pubbl., Mogadishu, Somalia. Descriptors: Africa; agriculture; Audegle; East Africa; hydrology; irrigation; Maringubai; Shabeellaha Hoose; Shebelle River; Somali Republic; water resources; water supply. Database: GeoRef. Accession Number: 1999-067965.

Somalia, Ministry of Agriculture, Mogadishu, Somalia. 1971. Scebeli River Water Salinity Measurements; Electrical Conductivity micromhs/cm 1965-1970. Somalia: Ministry of Agriculture, Mogadishu, Somalia. Descriptors: Africa; East Africa; electrical conductivity; hydrology; measurement; salinity; Shebelle River; Somali Republic; surface water. Database: GeoRef. Accession Number: 1999-067967.

Somalia, Ministry of Mining and Water Resources, Somalia. 1971. Hydrogeological Map. United States: U. N. Dev. Progr., New York, United States. Descriptors: Africa; East Africa; ground water; hydrogeologic maps; hydrogeology; hydrology; maps; Somali Republic; surveys. Database: GeoRef. Accession Number: 1987-047725.

Somalia, Ministry of Planning and Coordination, Mogadishu, Somalia. 1974. Five Year Development Programme, 1974-1978. Somalia: Min. Planning and Coordination, Mogadishu, Somalia. Descriptors: Africa; development; East Africa; land use; programs; Somali Republic; water resources. Database: GeoRef. Accession Number: 1999-067970.

Somalia, Ministry of Planning, Mogadishu, Somalia. 1965. Conclusion and Recommendations on Development of Water Resources and Irrigated Land Along Tra Schebeli Flood Plain Region Summarized from FAO. Somalia: Ministry of Planning & Coordination, Mogadishu, Somalia. Descriptors: Africa; agriculture; development; East Africa; irrigation; land use; Shebelle River; Somali Republic; water resources. Notes: Special fund agric. and water survey report. Database: GeoRef. Accession Number: 1999-067969.

Somalia, Ministry of Public Works, Mogadishu, Somalia. 1966. A Report on Development of Water Resources in Somali Republic. Somalia: Min. Public Works, Mogadishu, Somalia. Descriptors: Africa; development; East Africa; Somali Republic; water resources. Database: GeoRef. Accession Number: 1999-067971.

Somalia, Ministry of Public Works, Mogadishu, Somalia. 1966. Water Regulations. Somalia: Min. Publ. Works, Mogadishu, Somalia. Descriptors: Africa; East Africa; hydrology; regulations; Somali Republic; water management; water resources. Database: GeoRef. Accession Number: 1999-067972.

Somalia, National Water Centre, Mogadishu, Somalia. 1987. Field Manual for the Maintenance and Operation of National Groundwater Monitoring Network. Somalia: National Water Centre, Mogadishu, Somalia. Descriptors: Africa; climate; East Africa; field studies; ground water; monitoring; National Groundwater Monitoring Network; networks; Somali Republic. Notes: FAO/UNDP, Field document no. 2. Database: GeoRef. Accession Number: 1999-067992.

Somalia, State Planning Commission, Mogadishu, Somalia. 1979. Bibliography of Selected References on Water of Somalia. Somalia: State Planning Commission, Mogadishu, Somalia. Descriptors: Africa; bibliography; East Africa; Somali Republic; water resources. Database: GeoRef. Accession Number: 1998-014359.

Somalia, State Planning Commission, Mogadishu, Somalia. 1978. Schebelle River Development, Prospect and Problems. Somalia: State Planning Commission, Mogadishu, Somalia. Descriptors: Africa; development; East Africa; hydrology; rivers and streams; Shebelle River; Somali Republic; water management; water resources. Database: GeoRef. Accession Number: 1998-014358.

Somalia, State Planning Commission, Mogadishu, Somalia. 1976. Summary of Mogadishu Water Agency Project. Somalia: State Planning Commission, Mogadishu, Somalia. Descriptors: Africa; Benadir; East Africa; government agencies; Mogadishu Somali Republic; Mogadishu Water Agency; programs; Somali Republic; water resources. Database: GeoRef. Accession Number: 1998-014357.

Somalia. Geological Survey Dept. 1960-. "Annual Report of the Geological Survey Department of the Ministry of Industry and Commerce." Descriptors: Geology-Somalia; Dewey: 556.77 Preceding Title: Somaliland, British. Geological Survey Dept. 1953-1960. Annual report. Notes: Apr. 1960- Mar. 19; v.; 22 cm; Notes: Report year ends Mar. 31.

Somalia and Food and Agriculture Organization of the United Nations. 1984. Compendium of Agricultural Development Projects in Somalia, as at October 1984. Mogadishu, Somalia: Somali Democratic Republic: FAO. Descriptors: 110 leaves; Agricultural development projects- Somalia- Directories. Notes: "October 1984." Responsibility: SDR/FAO. LCCN: 90-980903; OCLC Accession Number: 30700984.

Somalia and Guddiga Qorshaynta Qaranka. 1979. Bibliography of Selected References on Waters of Somalia, October 1979. Mogadiscio: UNDP/UNESCO SOM/76/009 Project Documentation Centre, State Planning Commission. Descriptors: 17 pages, 2 leaves; Hydrology- Somalia- Bibliography; Water-supply- Somalia-Bibliography. Notes: Includes index. LCCN: 80-980133. OCLC Accession Number: 8763015.

"Somalia- Energiewirtschaft 1985. (Somalia- Energy Situation 1985)." 1986. Bundesstelle für Aussenhandelsinformation, Cologne (Germany). Dec. pages: 9.

Descriptors: Somalia; Petroleum; Imports; Power generation; Energy supplies; Electric power; Energy policy; Foreign technology; Energy use supply and demand; Energy Policies regulations and studies; Energy Fuels; Energy Electric power production; Business and economics Foreign industry development and economics. Abstract: The energy situation of Somalia is reviewed on the basis of relevant data. Data on the country's national energy policy are followed by an outline of trends in energy sources and electric power generation. Key figures are presented on the country's external trade. Notes: In German. BfAI-Marktinformation. Reihe MI-BS, no. 29.534.86.273. NTIS Accession Number: TIBB8711471XSP.

"Somalia- Energiewirtschaft 1986. (Somalia- Energy Situation 1986)." 1988. Bundesstelle für Aussenhandelsinformation, Cologne (Germany, F.R.). Feb. pages: 11. Descriptors: Petroleum; Natural gas; Coal; Electric power; Power generation; Exports; Imports; Energy supplies; Energy policy; Somalia; Foreign technology; Energy use supply and demand; Energy Electric power production; Energy Fuels; Energy Policies regulations and studies; Business and economics Foreign industry development and economics. Abstract: The energy situation of Somalia is reviewed on the basis of relevant data. Data on the country's national and international energy policy are followed by an outline of trends in energy sources and electric power generation. Key figures are presented on the country's external trade and balance of payments. Notes: In German, BfAI-Berichte ueber die Energiewirtschaft des Auslandes, no. 29.018.88.273. NTIS Accession Number: TIBB8882383XSP.

"Somalia- Energiewirtschaft 1987. (Somalia- Energy Situation 1987)." 1989. Bundesstelle für Aussenhandelsinformation, Cologne (Germany). Feb. pages: 16. Descriptors: Energy supplies; Petroleum; Electric power; Power generation; Exports; Imports; Energy policy; Somalia; Foreign technology; Energy Policies regulations and studies; Energy use supply and demand; Energy Fuels; Business and economics Foreign industry development and economics. Abstract: The energy situation of Somalia is reviewed on the basis of relevant data. Data on the country's national energy policy are followed by an outline of trends in energy sources and electric power generation. Key figures are presented on the country's external trade and balance of payments. Notes: In German, BfAI-Berichte ueber die Energiewirtschaft des Auslandes, no. 29.008.89.273. NTIS Accession Number: TIBB8981699XSP.

"Somalia- Energiewirtschaft 1988. (Somalia- Energy Situation 1988)." 1989. Bundesstelle für Aussenhandelsinformation, Cologne (Germany). Dec. pages: 12. Descriptors: Petroleum; Natural gas; Coal; Electric power; Power generation; Exports; Imports; Energy supplies; Energy policy; Somalia; Developing countries; Developing country application; Foreign technology; Energy Policies regulations and studies; Energy use supply and demand; Energy Fuels; Business and economics Foreign industry development and economics. Abstract: The energy situation of Somalia is reviewed on the basis of relevant data. Data on the country's national and international energy policy are followed by an outline of trends in energy sources and electric power generation. Key figures are presented on the country's external trade and balance of payments. Notes: In German. BfAI-Berichte ueber die Energiewirtschaft des Auslandes, no. 29.029.89.273. NTIS Accession Number: TIBB9080742XSP.

Somalia Geological Map. 1983. United States: United Nations Development Program, New York, NY, United States. Descriptors: Africa; areal geology; East Africa;

geologic maps; maps; Somali Republic. Database: GeoRef. Accession Number: 1987-047724.

Somalia; Guddiga Qorshaynta Qaranka and the United Nations Development Programme. 1979. Bibliography of Selected References on Waters of Somalia, October 1979. Mogadiscio: UNDP/UNESCO SOM/76/009 Project Documentation Centre, State Planning Commission. Descriptors: 17, 2 leaves; Hydrology- Somalia- Bibliography; Water-supply- Somalia- Bibliography. Notes: Includes index. Cover title. Reproduction: Microfilm. New York: New York Public Library, 1982. 1 microfilm reel; 35 mm. (MN *ZZ-20235). LCCN: 80-980133; OCLC Accession Number: 23788031.

Somalia; Italy and Ministero degli affari esteri. 1987. Geology of Somalia and Surrounding Regions: International Meeting: Mogadishu, 23 Nov.-30 Nov. 1987. Mogadishu: Somali National University. Descriptors: 1 v. Geology- Somalia- Congresses; Geology- Africa- Congresses; Geology- Somalia- Guidebooks. Abstract: General information, program, list of participants- Abstracts- Excursion A guidebook- Excursion B guidebook- Excursion C guidebook- The free-air and Bouguer maps of Somalia by A.S. Dorre and A. Rapolla. Notes: Somalia- Guidebooks. Notes: Cover title. Folded ill. inserted. Includes bibliographical references. LCCN: 89-150371; OCLC Accession Number: 20391584; 34175029.

Somalia; Ministry of Jubba Valley Development and USAID/Somalia. 1989. Final Evaluation of Jubba Development Analytical Studies Project: (USAID Project 649-0134): Provisional Report. Mogadishu: Ministry of Jubba Valley Development, Somali Democratic Republic: United States Agency for International Development, Somalia. Descriptors: 1 v. (various pagings); Regional planning- Juba River Valley (Ethiopia and Somalia); Water resources development- Juba River Valley (Ethiopia and Somalia); Land use- Juba River Valley (Ethiopia and Somalia)- Planning. Notes: "August 17, 1988." Other Titles: Jubba Development Analytical Studies final evaluation report. OCLC Accession Number: 44817146.

Somalia; Ministry of Jubba Valley Development and USAID/Somalia. 1989. Final Evaluation of Jubba Development Analytical Studies Project: (USAID Project 649-0134): Report. Mogadishu: Ministry of Jubba Valley Development, Somali Democratic Republic: United States Agency for International Development, Somalia. Descriptors: 1 v. (various pagings); Regional planning- Juba River Valley (Ethiopia and Somalia); Water resources development- Juba River Valley (Ethiopia and Somalia); Land use- Juba River Valley (Ethiopia and Somalia)- Planning. Notes: Cover title: Jubba Development Analytical Studies final evaluation report. "August 1, 1989." Other Titles: Jubba Development Analytical Studies final evaluation report. Library of Congress; LCCN: 90-980231; OCLC Accession Number: 29311907.

Somalia; National Refugee Commission and Office of the United Nations High Commissioner for Refugees. 1985. Farjano Settlement Project: Land Evaluation. Cambridge England: Sir M MacDonald & Partners Ltd. Descriptors: 1 v. (various pagings); Agricultural development projects- Somalia; Land use, Rural- Somalia; Soil moisture- Somalia- Measurement. Notes: At head of title: Somali Democratic Republic, National Refugee Commission, United Nations High Commissioner for Refugees. "April 1985." Five folded maps in pockets. Includes bibliographical references. National Agricultural Library. OCLC Accession Number: 23581801.

Somalia and Research Unit for Emergencies and Rural Development. 1982. A Preliminary Survey of the Nugal. Mogadishu Somalia: The Unit. Descriptors: 95 leaves, 8 leaves of plates; Water resources development- Somalia- Nugaal; Rural development- Somalia- Nugaal. Notes: Nugaal (Somalia)- Social conditions. Nugaal (Somalia)- Rural conditions. Notes: Cover title: A Preliminary survey of the Nugal region. Other Titles: Preliminary survey of the Nugal region. LCCN: 85-980502; OCLC Accession Number: 18716338.

"Somalia Upper-Air Climatic Atlas." 1993. Air Force Environmental Technical Applications Center Scott AFB, Il: Air Force Environmental Technical Applications Center Scott AFB, Il. January 1993. Volume: Ada259841, Page(S): 130 Pages(S). Descriptors: *Somalia, *Climate, *Upper Atmosphere, Atmospheres, Weather Forecasting, Tables (Data). Abstract: These statistical summaries for the upper atmosphere over Somalia are based on 1980-90 twice-daily gridded analyses produced by the European Centre for Medium Range Weather Forecasts (ECMWF). The ECMWF is a multi-national organization responsible for making global forecasts for the European community. Their data assimilation system uses multivariate optimal interpolation, which is able to use observations with differing error characteristics and spatial resolutions. The data is interpolated to a 2.5 by 2.5 degree grid in the horizontal and to the mandatory pressure levels in the vertical. For more information, see 'A Global Three-dimensional Multivariate Statistical Interpretation Scheme', by A.C. Lorenc, Monthly Weather Review, 1981, Vol. 109, pages 701-721. The tables in this atlas were prepared by the USAF Environmental Technical Applications Center's Operations Projects Branch, USAFETAC/DOC, in support of Operation RESTORE HOPE. The atlas provides regional analyses for Northern, Central, and Southern Somalia, each of which is defined on the map, opposite. Each table (two tables per page) provides monthly upper-air climatology at a specified grid point (every 2.5 degrees of latitude and longitude). The tables give D-value (in feet), mean temperature and dew point (in degrees centigrade), mean wind speed and percent occurrence frequency of direction (in knots and to 8-points of the compass), and mean geopotential heights (in feet) for 15 mandatory pressure levels (1,000 mb-10 mb) plus the surface. Except for wind direction frequency, all values have been rounded to the nearest whole number. Notes: APPROVED FOR PUBLIC RELEASE. Database: DTIC. DTIC Accession Number: ADA2598415XSP.

Somalia, Wasaaradda Beeraha and Food and Agriculture Organization of the United Nations. 1986. "Puntland Journal of Agriculture: A Ministry of Agriculture Publication." Descriptors: Agriculture- Somalia- Periodicals. Notes: Oct. 1986-; v. ill., maps; 28 cm; Notes: Title from cover. At head of title: FAO. Library of Congress; LCCN: sn 89-16064.

Somalia Water Supply and Sewerage Sector Study. 1977. United States: World Bank, Washington, DC, United States. Descriptors: Africa; East Africa; pollution; protection; sewage; Somali Republic; water resources; water supply. Database: GeoRef. Accession Number: 1998-015715.

"Somalia's Farmers Tap into Higher Yields." 1988. New Scientist. February 4. Volume 117, Pages 46. Descriptors: Agriculture- Somalia; Sorghum; Soil conservation; General Science; Applied Science & Technology. Notes: Illustration. ISSN: 0262-4079.

Somaliland Field Force. 1903-1904. "Somaliland Field Force. Triangulation Carried Out and Areas Mapped Under Different Scales. Sheet Nos. 1-25." Somaliland

Field Force; Somalia. Descriptors: Manuscript (mss). Notes: Description: 25 maps + index. Geographic: Somalia- Maps. Notes: Manuscript. Includes Nogal valley reconnaissance at 1:126,720- Plane table reconnaissance survey at 1:253,440- Berbera at 1:10,560- road traverse Wadamago- Olesan at 1:63,360. Scales vary. Coverage: Somalia. Other Titles: Triangulation carried out and areas mapped under different scales. Nogal valley reconnaissance. Plane table reconnaissance survey. Berbera. Road traverse Wadamago. Olesan. OCLC Accession Number: 48597670.

Somaliland Field Force. 1902. "Somaliland Expedition: May to October 1902." Somaliland Field Force; Somalia; Northern. Descriptors: Manuscript. Notes: Description: 1 map. Geographic: Somalia- Maps, Manuscript. Notes: Manuscript. Relief shown as hachures. Shows: tracks, routes of the Somaliland Field Force, wells. Scale 1:950,400. Coverage: Northern Somalia. OCLC Accession Number: 48603704.

Somaliland Field Force. 1901. "A Rough Sketch of the 'Mullah's Country'." Somaliland Field Force; Somalia; Northern. Notes: Description: 1 map. Geographic: Somalia- Maps, Manuscript. Scale ca. 1:400,000. Notes: Manuscript. Relief shown as form lines. Shows: tracks, routes of the Somaliland Field Force. General Info: Coverage: Northern Somalia. OCLC Accession Number: 48603541.

Somaliland Oil Exploration Company. 1954. A Geological Reconnaissance of the Sedimentary Deposits of the Protectorate of British Somaliland. Descriptors: Africa; Cenozoic; Cretaceous; East Africa; faults; geologic maps; Jurassic; maps; Mesozoic; petroleum; possibilities; Sedimentary deposits; Somali Republic; Tertiary. Abstract: An account of the development and tectonic relations of Jurassic to Tertiary formations overlying the faulted Precambrian basement of British Somaliland. In addition to the known seep at Dagah Shabell, other indications of oil have been found in Jurassic deposits of the region. The oil of the seep rises along slip planes of a shatter zone in faulted arenaceous beds of the middle Daban series (Oligo-Miocene). Notes: 41 pages, illustrations (including colored geological map), London, Crown Agents for the Colonies. Database: GeoRef Accession Number: 1954-004828; USGS Library.

Somaliland Protectorate. Military Government. 1944-1945. Report on the General Survey of British Somaliland. Colonial Development and Welfare Act. Economic Survey and Reconnaissance. Berbera? Government Press, B.M.A. Somalia. Includes folded maps, diagrams. USGS Library.

Somaliland Protectorate Geological Survey, Hargeisa, Somalia. 1962. Annual Report of the Geological Survey Department. Somalia: Geological Survey of Somaliland Protectorate, Hargeisa, Somalia. Descriptors: annual report; cartography; Geological Survey of Somaliland Protectorate; government agencies; maps; report; survey organizations. Database: GeoRef. Accession Number: 1998-018211; USGS Library.

Somaliland Protectorate Geological Survey, Somalia. 1962. Annual Report for Period April 1961-March 1962. Somalia: Min. Ind. Comm., Mogadishu, Somalia. Descriptors: annual report; Geological Survey Somaliland Protectorate; government agencies; report; survey organizations. Database: GeoRef. Accession Number: 1998-017933.

Somaliland Protectorate Geological Survey, Somalia. 1960. Hargeisa Water Supply; Investigation; a Report to the Crown Agents for Overseas Governments and Administrations. United Kingdom: Consult. Engin. Howard Humphreys & Sons (East Africa), Westminster, United Kingdom. Descriptors: Africa; East Africa; Hargeisa

Somali Republic; Somali Republic; Waqooyi Galbeed; water resources; water supply. Database: GeoRef. Accession Number: 1998-017932.

Somaliland, British. Survey Department. 1946. "Gazetteer. British Somaliland and Grazing Areas." Pages 45. Descriptors: Somaliland- Description and Travel-Gazetteer; Names, Geographical- Somaliland. Notes: First published March 1945; revised March 1946.

Sommavilla E and Turrini M.C. 2003. "Freshwater- the Basis of an Economic Development and Democracy Plan: A Case Study for Merka (Somalia)." IAHS-AISH Publication. Issue 286, Pages 322-331. Descriptors: Water Resources: Economic; water resource; water management; economic development; democracy. References: 6; Geographic: Somalia; East Africa; Sub-Saharan Africa. Abstract: The political instability in Africa, caused by droughts, is discussed. Effort is diverted from water research and management to conflicts. Loss of soil fertility forces many people to migrate, causing them to lose their cultural roots and destabilizing other areas with an already unstable situation. Lack of water and soil degradation make many people dependent upon international emergency aid. It is possible to escape from the tragic cycle of events by properly using the money given by private donors through small scale projects implemented by the communities themselves without anybody else making profit. ISSN: 0144-7815.

Sommavilla, E. 1978. "Un Territorio (Ad Esempio La Somalia) Visto Da Un Geologo." Translated title: "Territory (Example in Somalia) seen from its Geology." Cilmi Iyo Farsamo; Wargeyska Jaamacadda Ummadda Soomaaliyeed. Volume 2, Descriptors: Africa; areal geology; East Africa; Somali Republic. Database: GeoRef.

Sommavilla, E. 1977. "Geologia Strutturale Della Somalia. Structural Geology of Somalia." Quaderni Di Geologia Della Somalia (Mogadiscio). Volume 1, Pages 60-93. Descriptors: Africa; East Africa; Somali Republic; structural geology. LCCN: 93093377.

Sommavilla, E. Mase, G. Marchesi, S. and Ali Kassim, M. 1993. A Geological Model for Ground Water Research in the Shebeele Valley (Somalia). Firenze; Istituto Agronomico L'Oltremar; 1993. Relazioni E Monografie Agrarie Subtropicali E Tropicali Nuova Serie113 B, pages: 671-686. International Meeting; 1st- 1987 Nov: Mogadishu. Notes: Geology of Somalia and surrounding regions: Geology and mineral resources of Somalia and surrounding regions. Descriptors: Africa; aquifers; development; East Africa; ground water; Shabelle River; Somali Republic; surface water; water quality; water resources; water wells; wells. Notes: References: 13; 1 table, block diag., sects., sketch maps. OCLC Accession Number: 34175029.

Sommavilla, E. Sacdiya, C. Hussein Salad, M. and Ibrahim Mohamed, F. 1993. Neotectonic and Geomorphological Events in Central Somalia. Firenze; Istituto Agronomico L'Oltremar; 1993. Relazioni E Monografie Agrarie Subtropicali E Tropicali Nuova Serie113 A, pages: 389-396. International Meeting; 1st- 1987 Nov: Mogadishu. Notes: Geology of Somalia and surrounding regions: Geology and mineral resources of Somalia and surrounding regions. OCLC Accession Number: 34175029.

Sommavilla, E. and Turrini, M. C. 2004. Freshwater-the Basis of an Economic Development and Democracy Plan: A Case Study for Merka (Somalia). Wallingford; IAHS; 2004. Iahs Publication Volume: 286, pages: 322-332. 2003 Dec: Rome, Italy. Notes: International symposium on "the basis of civilization- water science?" sponsor: International association of hydrological sciences. ISBN: 1901502570.

Sommavilla, Elio and Turrini, Maria Chiara. 2003. "Freshwater- the Basis of an Economic Development and Democracy Plan: A Case Study for Merka (Somalia)." IAHS-AISH Publication. Issue 286, Pages 322-331. Descriptors: Water supply; Economic and social effects; Planning; Groundwater; Geomorphology; Project management; Rivers; Drought; Research and development management. Abstract: The political instability in Africa, caused by droughts, is discussed. Effort is diverted from water research and management to conflicts. Loss of soil fertility forces many people to migrate, causing them to lose their cultural roots and destabilizing other areas with an already unstable situation. Lack of water and soil degradation make many people dependent upon international emergency aid. It is possible to escape from the tragic cycle of events by properly using the money given by private donors through small scale projects implemented by the communities themselves without anybody else making profit. ISSN: 0144-7815.

Sommer, Corinna; Schneider, Wolfgang; Poutiers, Jean-Maurice and Food and Agriculture Organization of the United Nations. 1996. The Living Marine Resources of Somalia. Rome: Food and Agriculture Organization of the United Nations. Descriptors: 376, xxxii leaves of plates; Fishery resources- Somalia; Marine fishes- Somalia; Marine invertebrates- Somalia. Notes: "M-43"--T.p. verso. Includes bibliographical references (p. 342) and index. ISBN: 9251037426; ISSN: 1020-4547; LCCN: 97-161973.

South Africa. Survey Company. 1941. "Brava- Mogadiscio (Mogadishu) Road." EAF 243 to EAF 247. First ed. Scale: 1:37 000. [Nairobi]: [Survey Directorate, East Africa Command], 1941. 5 maps. Note(s): Sheets 1 to 5. Subjects: Somalia- Maps and charts 1941. Added Name: Great Britain. War Office. General Staff. Geographical Section. 5 sheets. British Library: Maps MOD EAF 243 to EAF 247.

South Africa. Survey Company. 1941. "Bulo Erillo- Bardera road." EAF 265 to EAF 268. First ed. Scale: 1:39 500. [Nairobi]: [Survey Directorate, East Africa Command], 1941. Maps. Note(s): Sheets 1 to 4. Subjects: Somalia- Maps and charts 1941. Added Name: Great Britain. War Office. General Staff. Geographical Section. 1 sheet. British Library: Maps MOD EAF 265 to EAF 268.

South Africa. Survey Company. 1941. "Dif- Afmadu road." EAF 155. First ed. Scale: 1:50 000. [Nairobi]: [Survey Directorate, East Africa Command], 1941. Maps. Note(s): Originally classified Secret. Subjects: Somalia- Maps and charts 1941. Added Name: Great Britain. War Office. General Staff. Geographical Section. 1 sheet. British Library: Maps MOD EAF 155

South Africa. Survey Company. 1941. "Gelib (Jelib)- Brava road." EAF 240 to EAF 242. First ed. Scale: 1:37000. [Nairobi]: [Survey Directorate, East Africa Command], 1941. 3 maps. Note(s): Sheets 1 to 3. Subjects: Somalia- Maps and charts 1941. Added Name: Great Britain. War Office. General Staff. Geographical Section. 3 sheets. British Library: Maps MOD EAF 240 to EAF 242.

South Africa. Survey Company. 1941. "Lower Giuba (Juba), 1:25 000 series." EAF 192 to EAF 210. First ed. Scale: 1:25 000. [Nairobi]: [Survey Directorate, East Africa Command], 1941. 19 maps. Subjects: Somalia- Maps and charts 1941. Added Name: Great Britain. War Office. General Staff. Geographical Section. 19 sheets. British Library: Maps MOD EAF 192 to EAF 210.

South Africa. Survey Company. 1941. "Mogadiscio (Mogadishu) & environs." EAF 275 & EAF 276. Scale: 1:25 000. [Nairobi]: [Survey Directorate, East Africa

Command], 1941. 3 maps. Note(s): Sheets 1 & 2. Subjects: Somalia- Maps and charts 1941. Added Name: Great Britain. War Office. General Staff. Geographical Section. 3 sheets. British Library: Maps MOD EAF 275 and EAF 276.

South Africa. Survey Company. 1941. "Mogadiscio (Mogadishu)." EAF 189. First ed. Scale: 1:14 300. [Nairobi]: [Survey Directorate, East Africa Command], 1941. Maps. Note(s): Originally classified Secret. Overprinted with defences. Subjects: Somalia- Maps and charts 1941. Added Name: Great Britain. War Office. General Staff. Geographical Section. 1 sheet. British Library: Maps MOD EAF 189.

Spears I.S. 2003. "Reflections on Somaliland and Africa's Territorial Order." Rev. Afr. Polit. Econ. Volume 30, Issue 95, Pages 89-98. Descriptors: Political; Conflict, protest, human rights; governance approach; territorial delimitation; independence. Notes: Additional Info: United Kingdom; References: 22; Geographic: Somalia-Somaliland. Abstract: This article examines the arguments for and against reforming the African state system in order to create more viable and peaceful states. It argues that while such a process has the potential to be enormously disruptive, selective recognition of some 'state-within-states', such as Somaliland, does offer promising approaches to more effective governance and more viable and coherent states. ISSN: 0305-6244.

Spigno, A. 1931. "Pozzi Tubolari Nel Letto Dell'Uebi Scebeli Per Rifornimento Di Acqua Potabile." Translated title: "Tube Wells in the Shebelle River Bed to Provide Drinking Water." Somalia: Descriptors: Africa; drinking water; East Africa; Shebelle River; Somali Republic; water resources; water supply; water wells. Database: GeoRef. Accession Number: 2000-012527.

Spraul, G. L. 1959. African Horn Basin. United States: Amerada Petroleum Corporation, Tulsa, OK, United States. Descriptors: Africa; areal geology; East Africa; Horn of Africa; Somali Republic. Database: GeoRef. Accession Number: 1998-014352.

Staab, Gordon E. 1989. "Construction of Port and Facilities on Horn of Africa-Lessons Learned." Journal of Construction Engineering Management. Volume 115, Issue 1, Pages 53-69. Descriptors: Ports and Harbors- Construction; Construction Industry--Quality Control; Port Structures--Costs; Civil Engineering--Project Management. Abstract: Sea trade to and from ports all over the world tends to make the Horn of Africa a focal point for waterborne traffic of the world. The assurance of the highest possible quality per dollar spent on port construction is of utmost importance to the host country (owner). Use of the principles and insights, which were learned during performance of quality assurance function on port construction project in Somalia and are shared herein, will maximize the probability that the quality of work produced is that which was designed. Proven-in-action success techniques related to production are divulged which will help the contractor realize expeditious, economical field operations and solve construction problems before they occur by preparing himself in advance to deal with them. ISSN: 0733-9364.

Stager, Curt. 1990. "Africa's Great Rift." National Geographic. May. Volume 177, Issue 5, Pages 2-41. Descriptors: Africa; African Plate; East Africa; East African Rift; ecosystems; faults; Great Rift Valley; igneous activity; plate tectonics; popular geology; rift zones; Somali Plate; tectonophysics; volcanism. Notes: illustrations incl. sketch map. ISSN: 0027-9358.

Standard Vacuum Oil Company, International (III). 1962. Geology of the Somalia Basin. Somalia: MRMI, Mogadishu, Somalia. Descriptors: Indian Ocean; marine geology; Somali Basin. Database: GeoRef. Accession Number: 1998-014353.

Stanford's Geographical Estabt. 1905. "Part of the Somali Protectorate." London: E. Stanford. Notes: Description: 1 map: color; 42 x 45 cm. Geographic: Somalia- Maps. Scale 1:1,000,000. Notes: Relief shown by landforms. Includes location map. Other Titles: Somali Protectorate. OCLC Accession Number: 54619564.

Stanley Associates Engineering, Canada. 1983. "Final Report for Development of Zeila District Water Supply Somalia; Project Reformation Mission." Canada: Descriptors: Africa; development; East Africa; Somali Republic; water resources; water supply; Woqooyi Galbeed Somali Republic; Zeila. Database: GeoRef. Accession Number: 1998-014355.

Stapleton, C. K. 1981. "Water Supply for Refugee Camps in Somalia." Somalia: Descriptors: Africa; East Africa; Somali Republic; water resources; water supply. Notes: Consultancy report. Database: GeoRef. Accession Number: 1998-014356.

Stefanini, G. 1938. "Cenni Sulle Localita Fossilifere Eoceniche Della Somalia. Notes on Eocene Fossil Localities in Somalia." Palaeontographia Italica. Volume 32, Pages 13-47. Descriptors: Cenozoic; Eocene; historical geology; Italian Somaliland; Paleogene; Tertiary. Notes: sup. 3; 12 figs. Abstract: An account of the stratigraphy of the Eocene deposits of Somaliland, with lists of fossils. ISSN: 0373-0972.

Stefanini, G. 1937. "Cenni Sulle Localita Fossilifere Oligoceniche e Mioceniche Della Somalia." Translated title: "Notes on Oligocene and Miocene Fossil Localities of Somalia." Palaeontographia Italica. Volume 32, Pages 1-24. Descriptors: Cenozoic; historical geology; Invertebrata; Italian Somaliland; paleontology; Tertiary. Notes: suppl. 2; 21 figs. (incl. sketch maps); Abstract: Fossil horizons in the Oligocene and Miocene of Italian Somaliland. ISSN: 0373-0972.

Stefanini, G. 1936. "Sull'Esistenza Del Giurassico Nella Valle Del Fafan (Ogaden)." Atti Della Societa Toscana Di Scienze Naturali Residente in Pisa. Processi Verbali. Volume 45, Issue 5, Pages 36-37. Descriptors: Fafan valley; historical geology; Italian Somaliland; Jurassic; Mesozoic; Ogaden. Reference includes data from Bibliography and Index of Geology Exclusive of North America, Geological Society of America, Boulder, CO, United States. Abstract: Notes on a small Jurassic fauna collected in the Fafan valley, Ogaden, Italian Somaliland. ISSN: 0365-7477.

Stefanini, G. 1936. "Il Problema Idraulico in Somalia (Replica Alla Lettera Di S. Caterini)." Translated title: "Hydraulic Problem in Somalia; Reply to the Letter of S. Caterini." L'Agricoltura Coloniale. Volume 1936, Descriptors: Africa; East Africa; hydraulics; irrigation; Somali Republic; water resources. ISSN: 0394-2945.

Stefanini, G. 1936. Saggio di una Carta Geologica Dell'Eritrea, Della Somalia e Dell'Ethiopia, Scala 1:2,000,000. Translated title: "Assay of a Geologic Map of Eritrea, Somalia and Ethiopia, Scale 1:2,000,000." Italy: C.N.R., Rome, Italy. 179 pages, illustrations, 2 folded maps in separate pocket. Bibliography page 151-172. Descriptors: Africa; areal geology; East Africa; Eritrea; Ethiopia; mapping; Somali Republic. USGS Library.

Stefanini, G. 1933. "Notizie Sulle Formazioni Plioceniche e Pleistoceniche Della Somalia." Translated title: "News of the Pleistocene Pliocene Formations of Somalia." Palaeontographia Italica. Volume 32, Pages 55-56. Descriptors: Cenozoic; historical geology; Italian Somaliland; Neogene; Pleistocene; Pliocene; Pliocene and Pleistocene; Quaternary; Somaliland (Italian); Tertiary. Notes: suppl. 1; 8 figs. (incl. sketch maps); Abstract: On the stratigraphy and conditions of deposition of the Pliocene and Pleistocene deposits of Italian Somaliland. ISSN: 0373-0972.

Stefanini, G. 1932. "Carta Geologica Dell'Eritrea, Della Somalia e Dell'Ethiopia, Scala 1:2,000,000." Translated title: "Geologic Map of Eritrea, Somalia, and Ethiopia, Scale 1:2,000,000." Italy: C.N.R., Ist. Geogr. Militare, Firenze, Italy. Descriptors: Africa; areal geology; East Africa; Eritrea; Ethiopia; geologic maps; maps; Somali Republic. Database: GeoRef. Accession Number: 1998-014374.

Stefanini, G. 1929. "Successione Ed Eta Della Serie Di Lugh Nella Somalia Italiana." Translated title: "Succession and State of Lugh Series of Italian Somalia." Report of the ...Session- International Geological Congress, Vol.15. Volume 2, Pages 223-238. Descriptors: Africa; East Africa; International Geological Congress; lithofacies; Lugh Series; sedimentary rocks; Somali Republic; stratigraphy. Notes: IGC, International Geological Congress. ISSN: 1023-3210.

Stefanini, G. 1928. "Sull'Esistenza Di Terreni Giurassici Nella Migiurtinia Settentrionale." Translated title: "Existence of Jurassic Terrains in Northern Migiurtinia." Atti Della Societa Toscana Di Scienze Naturali Residente in Pisa, Memorie. Volume 40, Descriptors: Africa; Bari; East Africa; historical geology; Jurassic; Mesozoic; Migiurtinia; northern Migiurtinia; Nugaal; Somali Republic. ISSN: 0365-7108.

Stefanini, G. 1925. "Primi Risultati Geologici Della Missione Della R. Societa' Geografica in Somalia." Translated title: "First Geologic Results of the Mission of R. Geographic Society in Somalia." Atti Accademia Nazionale Dei Lincei, Classe Di Scienze Fisiche Matematiche e Naturali, Rendiconti. Volume 6, Descriptors: Africa; areal geology; East Africa; expeditions; Somali Republic. ISSN: 0001-4435.

Stefanini, G. 1925. "Primi Risultati Geologici Della Missione Della R. Societa Geografica in Somalia, 1924." Translated title: "First Geologic Results of the Mission of the R. Geographic Society in Somalia, 1924." Atti Della Accademia Nazionale Dei Lincei, Rendiconti, Classe Di Scienze Fisiche, Matematiche e Naturali. Volume 1, Descriptors: Africa; areal geology; East Africa; Somali Republic. ISSN: 0392-7881.

Stefanini, G. 1925. "Risulti Geografici Di Una Missione Nella Somalia Settentrionale Italiana 1924." Translated title: "Geographical Results of a Mission in Northern Italian Somalia in 1924; International Geographic Congress." Proceedings- International Geographical Congress. Volume 1925, Descriptors: Africa; areal geology; East Africa; field trips; geography; northern Somali Republic; Somali Republic. Database: GeoRef.

Stefanini, G. 1925. "Sûr La Constitution Géologique De La Somalie Italiane Meridionale." Translated title: "Geology of Southern Italian Somalia; 13th International Geological Congress." Proceedings of the ...International Geological Congress. Volume 13, Descriptors: Africa; areal geology; East Africa; Somali Republic; southern Somali Republic. Database: GeoRef.

Stefanini, G. 1922. In Somalia; Note e Impressioni Di Viaggio. Translated title: In Somalia; Notes and Impressions of a Trip. Italy: Le Monnier, Florence, Italy. Descriptors: Africa; areal geology; East Africa; field studies; Somali Republic. Database: GeoRef. Accession Number: 1998-010430.

Stefanini, G. 1921. Il Problema Idraulico Della Somalia. Translated title: Hydraulic Problems of Somalia. Italy: Agricoltura Coloniale, Firenze, Italy. Descriptors: Africa; East Africa; Somali Republic; water resources; water supply. Database: GeoRef. Accession Number: 1998-014368.

Stefanini, G. 1919. Le Risorse Idrauliche Della Somalia Italiana. Translated title: Hydraulic Resources of Italian Somalia. Italy: Atti Congr. Naz. Colon., Italy. Descriptors: Africa; East Africa; Italian Somalia; Somali Republic; water resources. Database: GeoRef. Accession Number: 1998-014367.

Stefanini, G. 1916. Informazioni Ed Osservazioni Sul Regime Delle Sorgenti Di Baidoa. Translated title: Information and Observation on the Regime of the Springs of Baidoa. Italy: Missione Stefanini-Paoli, Firenze, Italy. Descriptors: Africa; Baidoa Somali Republic; Bay; East Africa; observations; Somali Republic; springs; water resources. Notes: Part 2, 7. Database: GeoRef. Accession Number: 1998-014365.

Stefanini, G. 1916. Studi Geologici e Idrografici. Translated title: Geologic Study and Hydrographics. Italy: Missione Stefanini-Paoli, Firenze, Italy. Descriptors: Africa; areal geology; East Africa; hydrographs; Somali Republic; waterways. Notes: Part 2. Database: GeoRef. Accession Number: 1998-014366.

Stefanini, G. 1914. "I Problemi Geografici Della Somalia Meridionale e Le Nuove Carte Dell'Istituto Geografico Militare." Translated title: "Geographic Problems of Southern Somalia and the New Map of the Military Geographic Institute." Rivista Geografica Italiana. Volume 1914, Descriptors: Africa; East Africa; geography; mapping; Somali Republic; southern Somali Republic. ISSN: 0035-6697.

Stefanini, G. 1913. "Osservazioni Geologiche Nella Somalia Italiana Meridionale (Nota Preventiva). Translated title: Geologic Observation in Southern Italian Somalia, Preventive Note." Bollettino Della Societa Geologica Italiana. Volume 32, Descriptors: Africa; areal geology; East Africa; Italian Somalia; Somali Republic; southern Somali Republic. ISSN: 0037-8763.

Stefanini, G. 1912. Schizzo Geologico Della Zona Mogadiscio, Mahaddei, Uanle Uen, Bur Acaba, Baidoa, Scale 1:500,000. Geologic Sketch of the Mogadishu Area, Mahadai, Wanle Weyne, Bur Hacaba, Baidoa, Scale 1:500,000. Italy: G. Giardi, Firenze, Italy. Descriptors: Africa; Baidoa Somali Republic; Benadir; Bur Hacaba Somali Republic; East Africa; geologic maps; Mahadai Somalia; maps; Mogadishu Somali Republic; Shabeellaha Dhexe; Somali Republic; Wanle Weyne Somalia. Database: GeoRef. Accession Number: 1998-014360.

Stefanini, G. 1912. Carta Geologica Della Somalia Italiana Meridionale, Scala 1:500,000. Translated title: Geologic Map of Southern Italian Somalia, Scale 1:500,000. Italy: G. Giardi, Firenze, Italy. Descriptors: Africa; areal geology; East Africa; geologic maps; Italian Somalia; maps; Somali Republic; southern Somali Republic. Database: GeoRef. Accession Number: 1998-014361.

Stefanini, G. and Corni, G. 1937. "Caratteri Geologici, Morfologici, e Fisici Della Somalia." Translated title: "Geologic Character, Morphology, and Physical Geography of Somalia." Italy: Ed. Arte Storia, Milan, Italy. Descriptors: Africa; areal geology; East Africa; geography; geomorphology; Somali Republic. Database: GeoRef. Accession Number: 1998-014388.

Stefanini, G. and Ferrara, A. 1924. "Stato Attuale Degli Studi Sul Terreno e Della Cartografia Geoagrologica Nell'Africa Orientale Italiana (Somalia, Eritrea)." Translated title: "Actual State of the Study on Terrain and Geo-Agricultural Cartography in Italian East Africa; Somalia, Eritrea." Giornale Di Geologia. Volume 19, Descriptors: Africa; agriculture; areal geology; cartography; East Africa; Eritrea; geography; Italian East Africa; Somali Republic. ISSN: 0017-0291.

Stefanini, G. and Puccioni, N. 1926. "Notize Preliminari Sui Principali Risultati Della Missione Della R. Societa Geografica in Somalia (1924)." Translated title: "Preliminary Note on the Principal Results of the R. Societa Geografica Mission in Somalia, 1924." Bollettino Della Societa Geografica Italiana. Volume 63, Pages 12-76. Descriptors: Africa; areal geology; East Africa; field studies; geography; Somali Republic. ISSN: 0037-8755.

Stefanini, Giuseppe. 1945. "Geological Survey. Italian, British, French Somaliland." S.l: s.n.; Somalia. Descriptors: Geology- Somalia- Maps. Description: 1 map. Geographic: Somalia- Maps. Scale 1:2,000,000. Cartgrph Code: Category of scale: a Constant ratio linear horizontal scale: 2000000; Notes: Date of information ca. 1933. Shows: grid, geological information, mineral deposits. "GP 2557." General Info: Coverage: Somalia. OCLC Accession Number: 48600872.

Stefanini, Giuseppe. 1933. Saggio Di Una Carta Geologica Dell'Eritrea, Della Somalia e Dell'Etiopia Alla Scala Di 1:2,000,000. Translated title: Explanatory Text for a Geologic Map of Eritrea, Somalia and Ethiopia at a Scale of 1:2,000,000. Italy: Consiglio Naz. Ricerche, Com. Geol, Firenze, Italy. Descriptors: Africa; bibliography; Cenozoic; Cretaceous; East Africa; Ethiopia; geologic maps; Geological map; historical geology; igneous and volcanic rocks; igneous rocks; Jurassic; maps; Mesozoic; Somali Republic; Tertiary; text for; Triassic. Abstract: Explanatory text, accompanied by geological maps, of the geology of Eritrea, Abyssinia, and Somaliland. Notes: 18 figs. (incl. sketch maps), 6 pls., 2 colored geol. maps. Database: GeoRef. Accession Number: 1933-003163.

Stefanini, Guiseppe and Paoli. 1916. "Ricerche Idrogeologiche, Botaniche Ed Entomologiche Fatte Nella Somalia Italiana Meridionale (1913)." Translated title: "Hydrogeological, Botanical and Entomological Studies done in Southern Italian Somalia." Relazioni e Monografie Agrario-Coloniali. Volume 7, Pages 255. Descriptors: Africa; agriculture; areal geology; Arthropoda; East Africa; hydrology; Insecta; Invertebrata; Mandibulata; Plantae; Somali Republic. 255 pages; illustrations incl. 33 plates, sketch maps. At head of title: "...Missione Stefanini-Paoli." USGS Library.

Stern W. 1988. "Entwicklungspolitische Massnahmen Im Bereich Mobiler Viehwirtschaft in Nord-Somalia." Translated title: "Measures of development policy within the range mobile cattle economics in north Somalia." Erde. Volume 119, Issue 4, Pages 235-242. Descriptors: mobile cattle keeping; Northern Range Development Project; grazing; cooperative ranching Species Term: Bos taurus; Somalia. Notes: Somalia (North). Abstract: Mobile cattle-keeping brings in money and is an important activity providing food, but little thought is given to assisting cattle keepers, eg in the Northern Range Development Project. Hence measures to improve grazing and make co-operative ranches have failed. But land is being enclosed, tending to over-pasturing in the land remaining. ISSN: 0013-9998.

Stern, H. P. 1984. The Development of Water Supplies in the Haud Area of the North West Region; Somalia. International (III): Descriptors: Africa; development; East Africa; Haud Somalia; Mudug; Somali Republic; water management; water resources; water supply. Database: GeoRef. Accession Number: 1998-014396.

Stern, Robert J. 2002. "Crustal Evolution in the East African Orogen: A Neodymium Isotopic Perspective." Journal of African Earth Sciences. Volume 34, Issue 3-4, Pages 109-117. Descriptors: East African Orogen; Nd isotopes; Neoproterozoic; Crustal evolution. ISSN: 1464-343X.

Stewart, J. A. B. 1956. "Geology of the Shalau Area." Report of the Geological Survey, Somaliland Protectorate. Volume K/12, Descriptors: Africa; areal geology; East Africa; northern Somalia; Sanaag; Shalau Somalia; Somali Republic. Database: GeoRef.

Stewart, J. A. B. 1955. "Geology of the Laferug Area." Report of the Geological Survey, Somaliland Protectorate. Volume K/1, Descriptors: Africa; areal geology; East Africa; Laferug Somalia; Somali Republic; Woqooyi Galbeed Somali Republic. Database: GeoRef.

Stewart, J. A. B. 1954. "The Geology of the Country around Las Dureh." Report of the Geological Survey, Somaliland Protectorate. Volume K/2, Descriptors: Africa; areal geology; East Africa; Las Dureh Somalia; Somali Republic; Woqooyi Galbeed Somali Republic. Database: GeoRef.

Stewart, J. A. B. Great Britain. Directorate of Overseas Surveys; British Somaliland and Geological Survey. 1960. "Plan of Dalan Cassiterite Prospect." Tolworth, Surrey: Directorate of Overseas Surveys. Descriptors: Geology- Somalia- Dalan- Maps. Notes: Description: 1 map: color; 44 x 56 cm. Scale 1:400. Cartgrph Code: Category of scale: a Constant ratio linear horizontal scale: 400; Notes: Published by Directorate of Overseas Surveys for Geological Survey, Somaliland Protectorate. Illustration A to Appendix 1 of report on Geology of Las Khoreh-Elayu area. Accompanied by "Dalan cassiterite prospect. Pit and trench diagrams." Illustrations B & C to Appendix 1. OCLC Accession Number: 54878035.

Stewart, J. A. B. and Greenwood, J. E. G. W. 1957. The Basement Geology of the Inda Ad Area, Erigavo District. Somalia: Geol. Surv. Somalil. Prot., Hargeisa, Somalia. Descriptors: Africa; areal geology; basement; East Africa; Erigavo Somali Republic; Inda Ad Somali Republic; Sanaag; Somali Republic. Database: GeoRef. Accession Number: 1998-014400.

Stigand, C. H. 1910. To Abyssinia through an Unknown Land; an Account of a Journey through Unexplored Regions of British East Africa by Lake Rudolf to the Kingdom of Menelek. United Kingdom: Publisher unknown, United Kingdom. Descriptors: Africa; areal geology; East Africa; East African Lakes; Ethiopia; exploration; field trips; history; Lake Turkana; maps; Somali Republic; topographic maps; topography. Notes: 32 plates. Database: GeoRef. Accession Number: 1998-011360.

Stilwell, J. 1976. "Oil, Gas Deposits seen Likely in Somalia." Oil Gas J. 02 Aug. Volume 74, Issue 31, Pages 165. Descriptors: Africa; East Africa; economic geology; gas; natural; natural gas; petroleum; petroleum exploration; possibilities; Somali Republic; stratigraphic traps; traps. Notes: sketch map. ISSN: 0030-1388.

Stipp, H. E., Jolly, J. H. and Petkof, B., et al. 1976. "The Mineral Industry of Other African Areas." Minerals Yearbook. US Geological Survey. Volume 1973, Vol. 3, Pages 1061-1126. Descriptors: 1973; Africa; Benin; Botswana; Burkina Faso; Burundi; Cameroon; Central Africa; Central African Republic; Chad; Congo; Djibouti; East Africa; economic geology; economics; Equatorial Guinea; Ethiopia; Fernando Po; Gambia; Guinea; Indian Ocean; Indian Ocean Islands; industry; Ivory Coast; Lesotho;

Madagascar; Malawi; Mali; Mauritania; Mauritius; metals; mineral resources; Niger; nonmetals; petroleum; production; Rwanda; Sahara; Senegal; Somali Republic; Southern Africa; Sudan; Swaziland; Toga; West Africa; Zimbabwe. Notes: tables. ISSN: 0076-8952.

Stock, J. H. 1982. "Researches on the Coast of Somalia. Shallow-Water Pycnogonida." Suppl. Volume 17, Pages 183-190. Descriptors: taxonomy; new species; geographical distribution; animal morphology; Pycnogonida; Somali Dem. Rep. range extension; Ammothella vanninni; Hannonia typica; Marine. Abstract: Five species of shallow-water Pycnogonida are recorded from Somalia. One species, Ammothella vanninii, is new to science. Of three others, the known range is considerably extended. The palp of Hannonia typica Hoek, 1881 is pluri-segmented instead of one-segmented as often assumed. Database: Aquatic Sciences and Fisheries Abstracts.

Stock, S. 1959. "Water Supply and Geology of Gabitec-Arapsio Area in the Northern Somaliland." Report of the Geological Survey, Somaliland Protectorate. Volume 1959, Descriptors: Africa; areal geology; East Africa; Gabitec-Arapsio; hydrogeology; northern Somalia; Somali Republic; water resources; water supply. Database: GeoRef.

Stock, S. 1959. "Water Supply and Geology of the Hand Plateau." Report of the Geological Survey, Somaliland Protectorate. Volume 1959, Descriptors: Africa; East Africa; Hand Plateau; hydrogeology; northern Somalia; Somali Republic; water resources; water supply. Database: GeoRef.

Stock, S. 1947. "Report on the Geology and Mineral Resources of North-Eastern Corner of Somaliland Protectorate." Report of the Geological Survey, Somaliland Protectorate. Volume 1947, Descriptors: Africa; areal geology; Somali Republic; East Africa; mineral resources; northern Somalia. Database: GeoRef.

Stoll, H. M., Ziveri, P., Geisen, M., Probert, I. and Young, J. R. 2002. "Potential and Limitations of Sr/Ca Ratios in Coccolith Carbonate: New Perspectives from Cultures and Monospecific Samples from Sediments." Philos. Transact A. Math. Phys. Eng. Sci. Apr 15. Volume 360, Issue 1793, Pages 719-747. Abstract: The Sr/Ca ratio of coccoliths was recently proposed as a potential indicator of past growth rates of coccolithophorids, marine algae, which play key roles in both the global carbonate and carbon cycles. We synthesize calibrations of this proxy through laboratory culture studies and analysis of monospecific coccolith assemblages from surface sediments. Cultures of coccolithophorids Helicosphaera carteri, Syracosphaera pulchra and Algirospira robusta confirm a 1-2% increase in Sr/Ca per degrees C previously identified in Emiliania huxleyi and Gephyrocapsa oceanica. This effect is not due merely to increases in growth rate with temperature and must be considered in palaeoceanographic studies. In light-limited cultures of E. huxleyi, Calcidiscus leptoporus and G. oceanica at constant temperature, coccolith Sr/Ca ratios vary by 10% across the range of possible growth and calcification rates for a given species. Among different species under similar culture conditions, Sr/Ca ratios vary by 30%. Although the highest ratios are in the cells with highest calcification and organic carbon fixation rates, at lower rates there is much scatter, indicating that different mechanisms control interspecific and intraspecific coccolith Sr/Ca variations. In field studies in the Equatorial Pacific and Somalia coastal region, coccolith Sr/Ca correlates with upwelling intensity and productivity. A more dynamic response is observed in larger coccoliths like C. leptoporus (23-55% variation in

Sr/Ca) than in smaller coccoliths of G. oceanica or Florisphaera profunda (6-15% variation in Sr/Ca). This response suggests that, despite temperature effects, coccolith Sr/Ca has potential as an indicator of coccolithophorid productivity. If the variable Sr/Ca response of different species accurately reflects their variable productivity response to upwelling (and not different slopes of Sr/Ca with productivity), coccolith Sr/Ca could provide useful data on past changes in coccolith ecology. The mechanism of coccolith Sr/Ca variations remains poorly understood but is probably more closely tied to biochemical cycles during carbon acquisition than to chemical kinetic effects on Sr incorporation in the calcite coccolith crystals. ISSN: 1364-503X.

Stommel, H. and Frassetto, R. 1968. "The Time of Appearance of Cold Water off Somalia." Proc. Natl. Acad. Sci. U. S. A. Jul. Volume 60, Issue 3, Pages 750-751. ISSN: 0027-8424.

Stringall, Stephen C. Broyles, Delores; International Travel Maps; ITMB Publishing Ltd and TerraCarta Media Group. . 2000. "Somalia: Scale 1:1,117,000." Vancouver, B.C: International Travel Maps, a division of ITMB Pub. Descriptors: Roads- Somalia- Maps. Notes: Description: 1 map; color; 99 x 69 cm. folded to 25 x 11 cm. Geographic: Somalia- Maps. Scale 1:1,117,000; Universal Transverse Mercator proj. (E 40045'--E 51015'/N 13045'--S 01045'). Notes: Relief shown by gradient tints and spot heights. Panel title. "Copyright ... ITMB." Includes index. Includes inset maps of Somali's climate and distribution of Somali people. Other Titles: International Travel Maps Somalia: scale 1:1,117,000; ISBN: 0783426410.

Stringall, Stephen C. Broyles, Delores; TerraCarta Media Group and International Travel Maps. 2000. "Somalia: Scale 1:1,117,000." Vancouver, B.C: International Travel Maps, a division of ITMB Pub. Descriptors: Roads- Somalia- Maps. Notes: Description: 1 map; color; 100 x 69 cm., folded to 25 x 11 cm. Geographic: Somalia- Maps, Tourist. Scale 1:1,117,000; (E 41000'--E 51020'/N 14000'--S 1030'). Notes: Relief shown by gradient tints and spot heights. Panel title. Includes text, color ill., index, index map, and 2 ancillary maps. Other Titles: International Travel Maps Somalia: scale 1:1,117,000; Responsibility: International Travel Maps; by Stephen C. Stringall and by Delores Broyles of TerraCarta Media Group. ISBN: 0921463685.

Study of Private Sector Participation in Somalia's Water Resources Development Industry; Main Report. 1986. Somalia: LBI/MMWR/WDA, Mogadishu, Somalia. Descriptors: Africa; development; East Africa; Somali Republic; water resources. Database: GeoRef. Accession Number: 1999-004729.

Study of Private Sector Participation in Somalia's Water Resources Development Industry; Main Report, Appendix 1. 1986. Somalia: LBI/MMWR/WDA, Mogadishu, Somalia. Descriptors: Africa; development; East Africa; Somali Republic; water resources. Database: GeoRef. Accession Number: 1999-004730.

Study of Private Sector Participation in Somalia's Water Resources Development Industry; Main Report; Appendix A-H. 1986. Somalia: LBI/MMWR/WDA, Mogdishu, Somalia. Descriptors: Africa; development; East Africa; industry; Somali Republic; water resources. Database: GeoRef. Accession Number: 1999-004725.

Stumpfl, E. F. Cerny, P. and Novak, M. 1995. Special Volume Lepidolite 200 Symposium. Austria: Volume: 55, 1-3, Descriptors: Africa; cell dimensions; Central Europe; collecting; crystal chemistry; crystal structure; differentiation; East Africa; electron probe; Europe; hydrothermal alteration; Iberian Peninsula; igneous rocks;

metasomatism; mica group; mineral localities; Nigeria; optical properties; paragenesis; pegmatite; petrology; plutonic rocks; Pyrenees; sheet silicates; silicates; Slovakia; Somali Republic; Southern Africa; Southern Europe; Spain; spectroscopy; Sudan; Swaziland; thermodynamic properties; United States; water-rock interaction; West Africa. Notes: Individual papers are cited separately; References: dissem. illustrations incl. 49 tables, sects., geol. sketch maps. ISSN: 0930-0708.

Subrahmanyam, B., Babu, V. Ramesh, Murty, V. S. N. and Rao, L. V. Gangadhara. 1996. "Surface Circulation off Somalia and Western Equatorial Indian Ocean during Summer Monsoon of 1988 from Geosat Altimeter Data." Int. J. Remote Sens. Volume 17, Issue 4, Pages 761. ISSN: 0143-1161

Suckert, E. 1960. "Il Problema Dell'Irrigazione in Somalia." Translated title: "Irrigation Problem in Somalia." Riv. Agric. Subtrop. Trop. Volume 54, Pages 4-9. Descriptors: Africa; East Africa; irrigation; Somali Republic; water resources. ISSN: 0035-6026.

"Sustaining Soldier Health and Performance in Somalia: Guidance for Small Unit Leaders." 1992. Natick, MA; United States: Army Research Inst. of Environmental Medicine. Dec. Volume: USARIEMTN931, page(s): 75. Descriptors: Somalia; Performance (Human); Military medicine; Preventive medicine; Deployment; Deserts; Handbooks; Hazards; Heat; Hydration; Infectious diseases; Morale; Nutrition; Physiology; Psychology; Sanitation; Health care facilities; Military operations; Military stress. Abstract: This handbook is intended as a guide and reference for U.S. military small unit commanders and NCOs. It includes pointers for sustainment of health and performance throughout predeployment, deployment, operations and redeployment. It addresses a broad range of important health issues, includes nutrition, hydration, managing work and environmental exposure, avoiding disease hazards, and maintaining morale.... Operation RESTORE HOPE, Somalia, Desert, Heat, Infectious disease, Environmental medicine, Water, Field nutrition and sanitation, Operational stress and hazards, Work, Rest, Physiology, Psychology. Database: NTIS. Accession Number: ADA2593028.

Swift, J. 1977. "Pastoral Development in Somalia; Herding Cooperatives as a Strategy Against Desertification and Famine." United States: Westview Press, Boulder, Colo., United States. Descriptors: Africa; arid environment; changes; climate; controls; desertification; deserts; East Africa; ecology; economics; ecosystems; environmental geology; eolian features; geomorphology; human ecology; land use; landform evolution; possibilities; processes; reclamation; Somali Republic. Notes: References: 18; tables, sketch map. Database: GeoRef. Accession Number: 1979-001785.

Szekield, A K. and Delaware Univ., Newark. Coll. of Marine Sciences. 1973. "Dynamics of Plankton Populations in Upwelling Areas (Dynamics of Plankton Populations in Upwelling Waters Along Arabian Coast, Somali Coast, and Northwest Coast of Africa)." Volume: E73-10325; NASA-CR-130780; Pagination 6P, Descriptors: Africa; Imagery; Plankton; Saudi Arabia; Somalia; Upwelling Water; Chlorophylls; Dust; Earth Resources Program; Microorganisms; Ph; Temperature Measurement. Abstract: There are no author-identified significant results in this report. Notes: Contract: NAS5-21784. Database: CSA Technology Research Database. Accession Number: N73-18357 (AH).

T

Takar, A.A, Dobrowolski, J.P and Thurow, T.L. 1990. "Influence of Grazing, Vegetation Life-Form, and Soil Type on Infiltration Rates and Interrill Erosion on a Somalian Rangeland." J. Range Manage. Volume 43, Issue 6, Pages 486-490. Descriptors: watershed management; grazing; vegetation; soil type; interrill erosion; infiltration rate Species Term: Somalia; Isoptera. Notes: Somalia. Abstract: Infiltration rate and interrill erosion on the sand site were significantly greater than on the clay site regardless of cover type or season. The clay site was dominated by annual forbs which rapidly decomposed. The sand site had greater annual and perennial grass cover which decomposed slower than forbs, providing longer and perhaps better protection from raindrop impact energy. Three growing seasons of livestock exclusion did not significantly increase soil cover on shrub interspaces; consequently, infiltration rates and interill erosion remained similar to the communally grazed sites. Interspace cover left by livestock was instead removed by termites and other microorganisms. Restricted ability of livestock to graze beneath thorny shrubs and increased phytomass from shrub leaf-fall resulted in a greater accumulation of cover and litter beneath shrubs, which aided infiltration on clay sites, regardless of season. ISSN: 0022-409X.

Takar, Aden A. 1988. Effects of Grazing Intensity, Vegetation Life-Form, and Soil Type on Infiltration Rate and Sediment Production on a Somali Rangeland. Descriptors: 81 leaves; Soil erosion- Somalia. Notes: Includes bibliographical references (leaves 75-81). Dissertation: Thesis (M.S.)--Utah State University. Dept. of Range Science, 1988. OCLC Accession Number: 19034475.

Talley, Michael J. 2003. "The Feasibility of Djibouti as an Intermediate Staging Base for U.S. Land Force Operations in the Middle East." Army Command and General Staff Coll, School of Advanced Military Studies, Fort Leavenworth, Ks: Army Command And General Staff College, School of Advanced Military Studies, Fort Leavenworth, KS. 2003. Volume: ADA416073, pages: 63 Pages. Descriptors: Military Operations; Middle East; Warfare; Global; Land Warfare; Policies; Political Science; Impact; Economics; Military Doctrine; Regions; History; Somalia; Staging; Geography; Terrorism; East Africa; Sudan. Abstract: As the United States sustains the Global War on Terrorism and the prospect of invading Iraq looms imminent, there remains a pressing need for viable intermediate staging bases in the Middle East to conduct decisive military operations. Recent policy changes and attitudes by the region's predominantly Muslim inhabitants have limited the US's choices for staging operations within the U.S. Central Command area of responsibility. Djibouti's geography, infrastructure, and capacity to accommodate a sizable military formation make it an ideal operations and logistics hub. The country's pro-Western stance and support for United Nations peace support initiatives and humanitarian civic actions illustrate the potential for a cooperative arrangement involving U.S. forces. Djibouti may offer the regional combatant commander a significant advantage through greater operational reach and increased flexibility in operational design. Destroying Hamm's strongholds in Sudan, neutralizing oppressive warlords in Somalia, or preparing to conduct offensive operations against rogue nations are all likely scenarios for U.S. land forces, and Djibouti may be the optimal launch pad. The study provides an overview of Djibouti and the Horn of Africa region by examining its historical background, socio-economic structure, political system, religious and cultural idiosyncrasies and the impact on military operations. It also defines optimal ISB standards and discusses the country's ability to support military operations. Finally, the study analyzes current regional disparities that affect U.S. and global interests and the likelihood for U.S. military intervention. Notes: The original document contains color images. Database: DTIC. URL: http://handle.dtic.mil/100.2/ADA416073.

Taylor, Hall. 1948. "Geology of Ethiopia being Developed by Sinclair in Exploring Concession." Oil Gas J. Volume 47, Issue 15, Pages 48-49. Descriptors: Africa; East Africa; economic geology; Ethiopia; historical geology; petroleum; possibilities; Somali plateau. Notes: 51, 5 figs. Abstract: The Somali plateau, southeastern Ethiopia, is composed of a sedimentary sequence of sands, limestones, shales, and gypsiferous layers ranging from Jurassic to Eocene in age. Both potential source beds and reservoir rocks are present, indicating the possibility of petroleum reserves. ISSN: 0030-1388.

Taylor, Hall. 1948. "Mesozoic Geology of the Somali-Ethiopia Area." Geological Society of America Bulletin. Volume 59, Issue 12, Part 2, Pages 1357. Descriptors: Africa; Cretaceous; East Africa; Ethiopia; historical geology; Jurassic; Mesozoic; Somali plateau. Abstract: The Somali-Ethiopia plateau area, east Africa, is largely underlain by Mesozoic sandstones, shales, limestones, and evaporites, forming a sedimentary section over 5000 feet thick. ISSN: 0016-7606.

Technische Universität Berlin. 1990. "Satellite Image Map Somalia 1:50,000." Berlin: Technische Universität Berlin; Somalia. Notes: Description: maps. Geographic: Remote-sensing images- Somalia. Scale 1:50,000; Gauss conformal proj. Category of scale: a Constant ratio linear horizontal scale: 50000; Notes: Relief shown as heights. Shows: grid, international and administrative boundaries, relief, roads (2 categories), tracks, paths, airports, telegraph and electricity lines. Coverage: Somalia. OCLC Accession Number: 48598661.

Technital S.p.A and Overseas Branch. 1975-1980. Juba River Valley Development Study. Rome: Technital S.p.A., Overseas Branch. Descriptors: 6 v. Economic development projects- Juba River Valley (Ethiopia and Somalia); Irrigation- Juba River Valley (Ethiopia and Somalia); Agriculture- Economic aspects- Juba River Valley (Ethiopia and Somalia); Geology- Juba River Valley (Ethiopia and Somalia); Hydrology- Juba River Valley (Ethiopia and Somalia). Abstract: v. 1. Summary report- v. 2. Human and social environment- v. 3. Geology, hydrology, and hydraulic engineering- v. 4. Agriculture and livestock sector- v. 5. Associated development. Economic analysis of the schemes- v. 6. Programme of all the measures. Short and medium term measures (1976-1986). Notes: Somalia- Economic conditions- 1960- Somalia- Social conditions- 1960-; Notes: At head of title: Democratic Republic of Somalia. Includes bibliographies. Library of Congress, LCCN: 85-980480; OCLC Accession Number: 12978992.

Technital, Rome, Italy. 1978. "Juba River Valley Development Study." Italy: Descriptors: Africa; development; East Africa; Juba River; Somali Republic; water resources. Notes: Final additional report. Database: GeoRef. Accession Number: 1998-014415.

Technital, Rome, Italy. 1975. Juba River Valley Development Study. Italy: Technital, Rome, Italy. Descriptors: Africa; development; East Africa; hydrology; Juba River; Somali Republic; water resources. Notes: 6 volumes. Database: GeoRef. Accession Number: 1998-014414.

Technoexport, Moscow, Russian Federation. 1976. "Program of Geological Investigations in the Somali Democratic Republic for 1976-1979." Russian Federation: Descriptors: Africa; areal geology; East Africa; field studies; Somali Republic; surveys. Database: GeoRef. Accession Number: 1998-014422.

Tedesco, Zammarano V. 1933. Tra Uebi e Giuba. Translated title: Between Shebelle River and Juba River. Italy: T. C. I., Le Vie d'Italia, Milan, Italy. Descriptors: Africa; East Africa; geography; hydrology; Juba River; rivers and streams; Shebelle River; Somali Republic. Database: GeoRef. Accession Number: 1998-014425.

Tedesco, Zammarano V. 1924. "Esplorazione Del Basso Uebi." Translated title: "Exploration of Lower Shebelle River." Bollettino Della Societa Geografica Italiana. Volume 3/4, 5/6, Descriptors: Africa; East Africa; exploration; fluvial features; geography; lower Shebelle River; rivers; Shebelle River; Somali Republic; surface water; water resources. ISSN: 0037-8755.

Teilhard de Chardin, P. 1930. "Études Géologiques en Éthiopie, Somalie et Arabie Méridionale." Translated title: "Geologic Studies in Ethiopia, Somalia and Southern Arabia." Mémoires De La Société Géologique de France, Nouvelle Série. Mémoire 14, 155 pages, color maps, bibliography pages 154-155. Descriptors: Africa; Arabian Peninsula; areal geology; Asia; East Africa; Ethiopia; Somali Republic; southern Arabian Peninsula. ISSN: 0249-7549.

Teilhard de Chardin, P., Lamare, P., Dreyfuss, M., Lacroix, A. and Basse, E. 1930. "Études Géologiques En Ethiopie, Somalie Et Arabie Meridionale." Translated title: "Geologic Studies in Ethiopia, Somalia and Southern Arabia." Memoires De La Société Géologique De France. Volume 6, Pages 1-83. Descriptors: Africa; Arabian Peninsula; areal geology; Asia; East Africa; Ethiopia; Somali Republic. ISSN: 0369-2027.

Teklay, M., Kroner, A., Mezger, K. and Oberhansli, R. 1998/2. "Geochemistry, Pb---Pb Single Zircon Ages and Nd---Sr Isotope Composition of Precambrian Rocks from Southern and Eastern Ethiopia: Implications for Crustal Evolution in East Africa." Journal of African Earth Sciences. Volume 26, Issue 2, Pages 207-227. Notes: ID: 2310. Abstract: Geochemical and isotope data for granitoid rocks from southern and eastern Ethiopia delineate the presumed margin of the Pan-African juvenile terrain of the Arabian-Nubian Shield against an older crustal segment of unknown origin extending from eastern Ethiopia to northern Somalia. Granitoids from southern Ethiopia have higher Na_2O and Na_2O/K_2O and lower Cr and Ni than granitoids with comparable SiO_2 values from eastern Ethiopia. In southern Ethiopia three periods of magmatism are identified on the basis on single zircon $207Pb/206Pb$ evaporation ages, namely at ~850, ~750-700 and ~650-550 Ma, and these correlate well with events documented from other parts of Ethiopia and the Arabian-Nubian Shield. The initial [epsilon]Nd(700 Ma) and [epsilon]sr(700 Ma) values range from -1.2 to +3.2 and from -13.4 to + 3.7, respectively, which precludes any significant contribution from much older continental crust in the generation of these rocks. Neodymium mean crustal residence ages, based on a depleted mantle model, range from 0.96 to 1.26 Ga. These data support the interpretation that southern Ethiopia constitutes part of the Arabian-Nubian Shield. In contrast, granitoids from eastern Ethiopia show geochemical features of S-type granites. In eastern Ethiopia Pal aeo-Neoproterozoic zircon ages (781-2489 Ma) are found. Initial [epsilon]Nd(700 Ma) and [epsilon]sr(700 Ma) values range from-4.3 to -18.3 and + 33.3 to + 99.8, respectively. Neodymium mean crustal residence ages range from 1.62 to 2.88 Ga. These data, in comparison to the western and southern parts of Ethiopia, are indicative of considerable reworking of pre-Pan-African crust. Variations in age, Sr---Nd isotope ratios and chemistry of the granitoids on a regional scale also suggest the existence of two separate basement terrains between southern and eastern Ethiopia, which may be separated by a tectonic line now concealed by Phanerozoic rocks. This tectonic line may represent a major tectonic boundary between the juvenile Arabian-Nubian Shield in the west and a pre-Pan-African gneissic terrain to the east, thus delineating the eastern margin of the Arabian-Nubian Shield. ISSN: 1464-343X.

Terreni, R. 1955. "Il Giuba Ed i Suoi Problemi." Translated title: "Juba River and its Problems." L' Universo. Volume 6, Descriptors: Africa; East Africa; hydrology; Juba River; Somali Republic; surface water. ISSN: 0042-0409.

Tesfaye, Samson, Harding, David J. and Kusky, Timothy M. 2003. "Early Continental Breakup Boundary and Migration of the Afar Triple Junction, Ethiopia." Geological Society of America Bulletin. September. Volume 115, Issue 9, Pages 1053-1067. Descriptors: Rifts (Geology)-East Africa; Plate tectonics-Africa; Geology-Ethiopia; Tertiary period; General Science; Applied Science & Technology. Notes: Bibliography; Graph; Illustration; Map; Table. Abstract: Migration of the Africa-Somalia-Arabia, or Afar, rift-rift-rift triple junction was studied. Analysis of structural features was used to infer the modern location of the Afar triple junction in the Lake Abhe area of Ethiopia, and the position of accommodation zones and geomorphic considerations were used to deduce the location of the paleotriple junction. The results indicate that the Afar triple junction has migrated around 1.5° (around 160 km) in a north-northeast direction relative to the African (Nubian) plate. This amount of migration is less than the 200 km migration predicted by plate-kinematic analysis. This discrepancy indicates either that the rate of spreading was slower than the present, roughly 1.6 cm/yr during the early phase of rifting, or that tectonic activity in the triple junction was initiated in the later early Miocene rather than in the Oligocene-Miocene. ISSN: 0016-7606.

Thomas, Abby; Institute for Development Anthropology (Binghamton, N.Y.); United States and Agency for International Development. 1985. A Soil and Water Conservation Project in Two Sites in Somalia: Seventeen Years Later. Descriptors: 20 l. Soil conservation- Somalia; Bunding- Somalia; Water conservation- Somalia; Applied anthropology. Notes: "(Draft submitted to PPC/E, AID Washington, in about February 1985; published, with some editorial changes, in August 1985 as AID Project Impact Evaluation Report No. 62)". OCLC Accession Number: 61727891.

Thomas, B. and Bowden, P. 2002. Horn of Africa: Special Issue. Amsterdam, Netherlands: Elsevier Science. Descriptors: 93 pages; Geology- Africa, Northeast; Geology- Somalia; Geology- Ethiopia. Notes: Text and 4 maps on 4 folded leaves issued in slip case. "Pergamon"--Cover. Includes bibliographical references. ISSN: 0899-5362.

Thompson V.B. 1995. "The Phenomenon of Shifting Frontiers: The Kenya-Somalia Case in the Horn of Africa, 1880s-1970s." Journal of Asian & African Studies. Volume 30, Issue 1-2, Pages 1-40. Descriptors: regional identity; shifting frontier; border change; historical studies. Notes: Kenya; Somalia. Abstract: That the imperial powers created borders and frontiers which did not previously exist in Africa is a myth of recent creation often perpetuated by Euro-centric writers. Borders and boundaries had existed over time. Frontiers shifted according to the strength or weakness of forces against those on whom the newcomers encroached. The Kenya-Somali border with which this paper is concerned is not an isolated entity, both countries sharing before and after the European partition of Africa (as they still do) common borders with Ethiopia, Somaliland, Uganda and the Sudan. These regions experienced trans-frontier movements before and after partition sometimes intermittent, and at other times if not frequent movements of nomadic people. Frontiers continued to shift throughout the colonial period and often these frontiers were shifting human frontiers. But the official perceptions of the Somalis in this frontier region of Kenya and emphasis on their being different from others, assisted in the formulation of the Somali-received tradition of themselves as a people distinct from other Africans in the zone. Despite these British colonial attitudes to the Somalis, as with other African peoples, they mirrored ambivalence thus perceiving them as people with sterling qualities and equally visualizing them as a nuisance.

Thompson, A. B. 1943. "Water Supply of British Somaliland." The Geographical Journal. Volume 101, Descriptors: Africa; East Africa; Somali Republic; water resources; water supply. ISSN: 0016-7398.

Thompson, J. D. 1977. "Ocean Deserts and Ocean Oases." United States: Westview Press, Boulder, Colo., United States. Descriptors: Africa; Arabian Peninsula; arid environment; Asia; changes; climate; desertification; deserts; East Africa; ecology; eolian features; geomorphology; landform description; landform evolution; marine environment; Namibia; North America; Northwest Africa; Peru; pollution; processes; productivity; salt water; Somali Republic; South America; Southern Africa; upwelling; water. Notes: References: 79; illustrations incl. sketch map. Database: GeoRef. Accession Number: 1979-001780.

Thomson A.M. 1983. "Somalia: Food Aid in a Long-Term Emergency." Food Policy. Volume 8, Issue 3, Pages 209-219. Descriptors: International Development Abstracts; Somalia; Food aid; Emergency relief programmes. Species Term: Somalia. Notes: Special Feature: 6 tables, 6 notes & refs. Abstract: From the mid-70s onwards, Somalia has suffered from a number of emergencies: drought, population disruptions, political conflict, and minor climatic problems. Examines the impact of continuous flows of food aid as a response to such emergencies, both in terms of the recipient government's attempts to stabilize the domestic economy and of donor governments' activities.

Thulin, M. 2005. "Three New Species of Chascanum (Verbenaceae) and Notes on the Genus in the Horn of Africa Region." Nordic Journal of Botany. Volume 23, Issue 5, Pages 513-517. Abstract: The new species Chascanian mixtum, from evergreen bushland and rocky slopes in north-western Somalia, C. elburense, from deciduous bushland mainly on sand in central Somalia, and C. glandulosum, from Acacia-Commiphora bushland on sandy soil in north-western Somalia, are described and illustrated. C. sessilifolium (Vatke) Moldenke is recircumscribed and neotypified with a specimen from the original type locality, and C. obovatum Sebsebe ssp. glaucum Sebsebe is placed in the synonymy of C. gillettii Moldenke. ISSN: 0107-055X.

Thulin, M. 2005. "Two New Species of Dyschoriste (Acanthaceae) from Somalia." Nordic Journal of Botany. Volume 23, Issue 5, Pages 519-522. Abstract: The new species Dyschoriste bayensis, from a granitic inselberg in southern Somalia, and D. miskatensis, from open bushland on limestone in north-eastern Somalia, are described and illustrated. ISSN: 0107-055X.

Thulin, M. 2005. "Three New Species of Heliotropium (Boraginaceae) from the Horn of Africa Region." Nordic Journal of Botany. Volume 23, Issue 5, Pages 527-532. Abstract: Heliotropium subspinosum sp. nov., from the coastal and subcoastal zones of Yemen, Oman and Somalia, is described and illustrated. This is a woody species with small leaves, and particularly the only 1-4-flowered inflorescences, the axes of which become subspinescent, are remarkable in the genus. H. laxum sp. nov., a slender shrublet from central Somalia, is described, and H. personatum sp. nov., a shrublet that is widespread and scattered, particularly on gypseous ground, in eastern Ethiopia, most parts of Somalia and in southern Yemen, is described and illustrated. ISSN: 0107-055X.

Thulin, M. 2005. "A New Species of Rhytidocaulon (Apocynaceae) from Somalia." Nordic Journal of Botany. Volume 23, Issue 5, Pages 533-534. Abstract: Rhytidocaulon baric:on, sp. nov., from open rocky limestone slopes in northeastern Somalia, is described and illustrated. ISSN: 0107-055X.

Thulin, M. 2005. "A New Species of Scutellaria (Lamiaceae) from Somalia." Nordic Journal of Botany. Volume 23, Issue 5, Pages 535-536. Abstract: The new species Scutellaria somalensis, from limestone slopes in north-eastern Somalia, is described and illustrated. ISSN: 0107-055X.

Thulin, M. 2005. "Notes on Convolvulus, Astripomoea, Ipomoea and Merremia (Convolvulaceae) from the Horn of Africa." Nordic Journal of Botany. Volume 23, Issue 5, Pages 629-640. Abstract: The six new species Convolvulus scopulatus, a shrub from gypseous semidesert coastal plains in northern Somalia, Astripomoea procera, a woody climber in Acacia-Commiphora bushland on sand in south-central Somalia, Ipomoea hiranensis, a prostrate or climbing shrublet in Acacia-Commiphora bushland in rocky places on sandstone in south-central Somalia, Ipomoea galhareriana, a woody climber in bushland on sand in central Somalia, Ipomoea pogonantha, a woody climber in bushland usually on red soil over limestone in eastern and southern Ethiopia, southern Somalia and northern Kenya, and Ipomoea corrugata, a herb with trailing stems in bushland on sandy soil in south-central Somalia, are described, and illustrations are provided for most of them. The new combinations Ipomoea ovatolanceolata (= I. adenoides var. ovatolanceolata) and Merremia obtusa (= M. ampelophylla subsp. obtusa) are made, and a new synonymy is provided for M, ampelophylla. Merremia ellenbeckii is neotypified and M. lobata is placed in synonymy. ISSN: 0107-055X.

Thulin, M. 2004. "Phyllanthus Xylorrhizus (Phyllanthaceae), a New Species from Somalia." Nordic Journal of Botany. Volume 23, Issue 4, Pages 385-387. Abstract: The new species Phyllanthus xylorrhizus, from limestone rocks along the coast of north-eastern Somalia, is described, illustrated, and compared with two other Somali endemics, P lunifolius and P spinosus. ISSN: 0107-055X.

Thulin, M. 2004. "Notes on Pluchea (Asteraceae) in Somalia and Ethiopia." Nordic Journal of Botany. Volume 22, Issue 6, Pages 659-665. Abstract: The new species Pluchea littoralis from limestone rocks and slopes facing the sea along the coast of north-eastern Somalia and P. lucens from rocky slopes on gypseous ground at 1500-1550 m altitude in northern Somalia, are described and illustrated. The new combinations P. kelleri (= Blumea kelleri) and P. somaliensis (= Blumea somaliensis) are made and a description is provided for P kelleri. P sarcophylla is resurrected and typified, and the circumscription of P. heterophylla is changed and an illustration of the species is provided. P. pinnatifida and Conyza stenodonta are treated as synonyms of P. arabica, and P. serra is placed in synonymy of Iphionopsis rotundifolia. ISSN: 0107-055X.

Thulin, M. 2002. "Notes on Withania (Solanaceae) in Somalia." Nordic Journal of Botany. Volume 22, Issue 4, Pages 385-389. Abstract: The identity of Withania reichenbachii is established, the name is neotypified, and the species is described and illustrated. The new combination W grisea (Hepper & Boulos) Thulin is made. The typification of W. somnifera is discussed, and it is concluded that the present typification can stand. A key to and a synopsis of the four species of Withania now recognized in Somalia are given. ISSN: 0107-055X.

Thulin, M. 2002. "New Species and Combinations in Kleinia (Asteraceae) from the Horn of Africa." Nordic Journal of Botany. Volume 22, Issue 4, Pages 419-426. Abstract: The new species Kleinia tuberculata, from open deciduous bushland on gypsum hills and gypseous limestone in northern and north-eastern Somalia, K. curvata, from evergreen bushland on limestone in northern Somalia, K. sabulosa, from deciduous bushland on sand in central Somalia, K. ogadensis, from deciduous bushland in eastern Ethiopia, K. gracilis, from dwarf bushland on limestone in north-eastern Somalia, and K. tortuosa, from alkaline plains in northern Somalia, are described, and for some of them illustrations are provided. The new combinations K. nogalensis and K. lunulata are made, and the recently described K. isabellae from eastern Ethiopia is placed in synonymy of K. lunulata. ISSN: 0107-055X.

Thulin, M. 2002. "Two New Species of Jatropha (Euphorbiaceae) from Somalia." Nordic Journal of Botany. Volume 22, Issue 4, Pages 427-430. Abstract: The new species Jatropha miskatensis, from limestone slopes of the Cal Miskaat Range in north-eastern Somalia, and J. marmorata, from bushland on shallow soil over limestone just south of the Nugaal valley in north-eastern Somalia, are described. An illustration is provided for J. miskatensis. ISSN: 0107-055X.

Thulin, M. 2002. "A New Species of Asparagus (Asparagaceae) from Somalia." Nordic Journal of Botany. Volume 22, Issue 4, Pages 431-433. Abstract: Asparagus brachiatus, a new unarmed species from limestone rocks in north-eastern Somalia, is described, illustrated, and compared with A. leptocladodius and A. africanus. ISSN: 0107-055X.

Thulin, M. 2002. "Cleome Socotrana (Capparaceae) and Allied Species in the Horn of Africa Region." Nordic Journal of Botany. Volume 22, Issue 2, Pages 215-218. Abstract: Four species of Cleome with small flowers, 6 stamens, petals without appendages, spreading to erect capsules, and hairy seeds, are recognized in the Horn of Africa region: C socotrana in the Socotra archipelago (Yemen), C. hadramautica, sp. nov., in southern Yemen, C. omanensis, comb. nov., in the Mahrah Region of Yemen and in Oman, and C. alhescens in northern Somalia. C. socotrana is lectotypified and a key to the species treated is provided. ISSN: 0107-055X.

Thulin, M. 2002. "A New Species and New Records of Helianthemum (Cistaceae) from the Horn of Africa Region." Nordic Journal of Botany. Volume 22, Issue 1, Pages 41-43. Abstract: Helianthemum speciosum, sp. nov., from gypsum outcrops in north-eastern Somalia, is described and illustrated. H. stipulation is recorded for the first time from Somalia and Yemen. ISSN: 0107-055X.

Thulin, M., Lavin, M., Pasquet, R. and Delgado-Salinas, A. 2004. "Phylogeny and Biogeography of Wafira (Leguminosae): A Monophyletic Segregate of Vigna Centered in the Horn of Africa Region." Syst. Bot. October-December. Volume 29, Issue 4, Pages 903-920. Abstract: Evidence from chloroplast trnK and nuclear ribosomal ITS sequences and morphological data reveals that the monotypic legume genus Wajira is nested within a clade comprising the species of Vigna subgen. Macrorhynchus. This Wajira-containing clade is basally branching in a larger clade that contains many of the genera traditionally referred to as tribe Phaseoleae subtribe Phaseolinae. Wajira is thus recircumscribed to include Vigna subgen. Macrorhynchus. Given the heterogeneity of floral morphology of its constituent species, Wajira is apomorphically diagnosed by woody stems and a pollen brush that comprises an introrse linear array of unicellular hairs. This recircumscribed genus now comprises five species, one of which is described as new, Wajira danissana. Three species require new nomenclatural combinations, Wajira grahamiana, Wajira praecox, and Wajira virescens. Wajira albescens, W danissana, W praecox, and W virescens are woody climbers that are each narrowly distributed in the and Somalia-Masai region characterized by sparse ground cover not subjected to seasonal burning. Wajira grahamiana has a thick woody subterranean rootstock that resprouts stems, and is widespread in the Sudano-Zarnbezian Region, southern India, and Sri Lanka, where grasslands subjected to seasonal burning predominate. This species is resolved in all phylogenetic analyses as derived from within the Somalia-Masai clade. An evolutionary rates analysis of trnK sequences suggests that the Wajira stem clade diverged from its closest relatives just over 10 million years ago, the extant diversification of the genus began around 6-7 million years ago, and Wajira grahamiana attained its widespread distribution during the last 2 million years.

Thurow T.L. 1989. "Decomposition of Grasses and Forbs in Coastal Savanna of Southern Somalia." Afr. J. Ecol. Volume 27, Issue 3, Pages 201-206. Descriptors: General Microbial Ecology; savannah; decomposition; termite; litter. Notes: Somalia. Abstract: Litter decomposition was rapid (95% disappearance in 1 yr) with no significant difference between grass and forb decomposition rates. Standing-dead decomposition rate was significantly slower (50% disappearance in 1 yr) with no significant difference between grass and forb decomposition rates. Rate of entry into the litter layer was significantly greater for forbs. Weight loss was not statistically correlated with rainfall received among sampling dates, in part because termites were an active decomposition agent throughout the dry season. ISSN: 0141-6707.

Tileston, F. M. 1964. Irrigation Development; Central Agricultural Research Station, Afgooye. United States: U. S. Agency for International Development, United States. Descriptors: Afgooye Somali Republic; Africa; agriculture; Benadir; development; East Africa; irrigation; Somali Republic; water management; water resources; water supply. Database: GeoRef. Accession Number: 1998-014433.

Toole, Michael J. 1993/7/24. "Military Role in Humanitarian Relief in Somalia." The Lancet. Volume 342, Issue 8865, Pages 190-191.

"Topographical Sketch Map ... British Somaliland." 1944. S.l: s.n.; Somalia; Northern. Notes: Description: 1 map. Geographic: Somalia- Maps, Topographic. Scale 1:1,180,000; conic projection; Cartgrph Code: Category of scale: a Constant ratio linear horizontal scale: 1180000; Notes: Relief shown as contours (interval 100 feet) and gradient tints. Shows: international boundaries, railways. "GP no. 2577." General Info: Coverage: Northern Somalia. OCLC Accession Number: 48603950.

Torr S.J, Parker A.G and Leigh-Browne G. 1989. "The Responses of Glossina Pallidipes Austen (Diptera: Glossinidae) to Odour-Baited Traps and Targets in Somalia." Bull. Entomol. Res. Volume 79, Issue 1, Pages 99-108. Descriptors: General Microbial Ecology; trap; Diptera Species Term: Glossina pallidipes; Somalia; Glossinidae; Diptera; Glossina (proper). Abstract: F3 traps caught 3 times as many tsetse as a biconical trap. The catch was increased by 1.6 times by releasing a mixture of 4-methylphenol (at 0.2 mg/h) and 3-n-propylphenol (0.04 mg/h), and by 4 times by releasing a mixture of acetone (500 mg/h), octenol (0.5 mg/h) and the 2 phenols. ISSN: 0007-4853.

Tozzi, R. 1961. "I Sistemi Tradizionali Dell'Agricoltura Irrigua in Somalia." Translated title: "Traditional System of Agricultural Irrigation in Somalia." Riv. Agric. Subtrop. Trop. Volume 1961, Descriptors: Africa; agriculture; East Africa; irrigation; Somali Republic; water resources; water supply. ISSN: 0035-6026.

Tozzi, R. 1960. "Contributo Allo Studio Del Regime Idraulico Del Fiume Scebeli Per Una Piu' Razionale Utilizzazione Delle Piene." Translated title: "Contribution to the Study of Hydraulic Regime of Shebelle River for Full Utilization." Riv. Agric. Subtrop. Trop. Volume 54, no.4-6, 7-9, Descriptors: Africa; East Africa; hydrology; Shebelle River; Somali Republic; surface water; utilization; water regimes; water use. ISSN: 0035-6026.

Tozzi, R. 1941. "Utilizzazione Agricola Delle Acque Del Giuba." Translated title: "Agricultural Utilization of Juba River Water." L'Agricoltura Coloniale. Volume 1941, Descriptors: Africa; agriculture; East Africa; irrigation; Juba River; Somali Republic; surface water; utilization; water management; water resources. ISSN: 0394-2945.

Tractionel, International (III). 1980. Fiume Scebeli. Shebelle River. Belgium: Fondo Europeo di Sviluppo, Brussels, Belgium. Descriptors: Africa; East Africa; hydrology; Shebelle River; Somali Republic; surface water. Database: GeoRef. Accession Number: 1998-014440.

Trashliev, S. 1977. "Lower Cretaceous Weathering and Prognosis for Finding Kaolin Deposits in Northwestern Somalia." Rudoobrazuvatelni Protsesi i Mineralni Nakhodista = Ore-Formation Processes and Mineral Deposits. Volume 6, Pages 27-37. Descriptors: Africa; ceramic materials; clastic sediments; Cretaceous; East Africa; economic geology; kaolin; Lower Cretaceous; Mesozoic; mineral deposits, genesis; northwest; prediction; sediments; Somali Republic; weathering. Notes: References: 4; illustrations incl. geol. sketch maps. ISSN: 0204-5311.

Trashliev, Stoyan. 1978. "Kaolin Occurrences in the Bur Area, Southwestern Somalia." Rudoobrazuvatelni Protsesi i Mineralni Nakhodista = Ore-Formation Processes and Mineral Deposits. Volume 9, Pages 15-26. Descriptors: Africa; Bur Galen; Bur Kananah; Bur region; Cenozoic; ceramic materials; chemical composition; chemically precipitated rocks; DTA data; East Africa; economic geology; gneisses; granites; humid environment; igneous rocks; kaolin deposits; metamorphic rocks; mineral composition; mineral deposits, genesis; Pleistocene; plutonic rocks; Precambrian; Quaternary; sedimentary rocks; Somali Republic; textures; weathering; weathering crust; X-ray data. References: 7. ISSN: 0204-5311.

Traversa G, Mohamed Ibrahim F, Morbidelli L, Nicoletti M and Brotzu P. 1986. "Preliminary Report on the Age and Nature of Basaltic Volcanism in Southern and Northern Somalia." Periodico Di Mineralogia. Volume 55, Issue 2-3, Pages 233-254. Descriptors: Volcanology; basaltic volcanism; radiometric age determinations; tholeiitic affinity. Notes: Somalia. Abstract: Starting from the Upper Oligocene, two mainly basaltic volcanic cycles of differing geodynamic significance developed in Somalia. The first, of Oligo-Miocene age (34-21 my), may be related to the Ethiopian trap basalts. The second, of Plio-Quaternary age (8.4-0.9 my), is connected with the opening of the Gulf of Aden. Radiometric age determinations have shown that the age of volcanism decreases from SE to NW in the Ethiopian-Somali Plateau, and that, in the southern edge of the Gulf of Aden, development of volcanic activity was more or less continuous from 8.4 to 0.9 my ago. Preliminary petrographic, petrochemical and mineral chemistry research has shown the predominantly transitional nature and tholeiitic affinity of the basaltic magmatism, of both Oligo-Miocene and Plio-Quaternary age. ISSN: 0369-8963.

Tribovillard N.-P, Lallier-Verges E, Caulet J.P, Riviere M and Ouahdi R. 1994. "L'Absence d'Un Enrichissement Marque En Matiere Organique Dans Des Sediments Deposes En Contexte d'Upwelling: Le Courant De Somalie." Bulletin- Société Géologique De France. Volume 165, Issue 1, Pages 65-75. Descriptors: Sedimentology; Sedimentary Geology; Chemical Oceanography; upwelling; Quaternary; organic matter; productivity; bottom water; phytoplankton; sediment chemistry. Notes: Somalia; Somalia Coast; Indian Ocean; Indian Ocean- Somalia Upwelling. Abstract: The first core is located in the upwelling zone where productivity is very high in surficial waters and where bottom waters are normally oxygenated. The second core is located at the edge of this upwelling-influenced domain, where bottom waters are oxygen depleted and slightly salt enriched. Both cores contain sediments of late Quaternary isotope stages 5 and 6. The emphasis has been put on the organic-matter content and on the trace-elements known to be related with sediments rich in organic matter (Cu, Zn, V, Mo) or with productivity (P, Ba).

Tribovillard N.-P, Tremblay P, Caulet J.P, Vergnaud-Grazzini C and Moureau N. 1996. "Lack of Organic Matter Accumulation on the Upwelling-Influenced Somalia Margin in a Glacial-Interglacial Transition." Mar. Geol. Volume 133, Issue 3-4, Pages 157-182. Descriptors: Oceans; Oceans; Depositional environments; organic matter; primary productivity; glacial/interglacial transition; upwelling; palaeoproductivity; glacial-interglacial transition. Notes: Indian Ocean- Somalia Margin. Abstract: Emphasis is placed upon the transition from isotopic Stage 6 to Stage 5, because the end of the glacial Stage 6 is characterised by intensified upwelling and primary productivity in the Somalia margin, whereas Stage 5 experienced a marked decrease in productivity. The results suggest that: the lack of marine organic matter (OM) in two settings favorable to organic carbon storage may be due to the nature of primary productivity. Coccolithophorid production would not be prone to OM accumulation, in contrast to the contribution of OM by diatoms or, particularly, naked phytoplankton; Ba is not an accurate palaeo-productivity marker in this region; the period of intensified upwelling in Stage 6 as well as the Stage 6-5 transition left no geochemical imprint on the sediments. ISSN: 0025-3227.

Tsui, A. O., Ragsdale, T. A. and Shirwa, A. I. 1991. "The Settlement of Somali Nomads." Genus. January-June. Volume 47, Issue 1-2, Pages 131-150. Descriptors: Africa; Africa South of the Sahara; Africa, Eastern; Age Distribution; Age Factors; Agriculture; Animal Population Groups; Conservation of Natural Resources; Data Collection; Demography; Developing Countries; Economics; Educational Status; Emigration and Immigration; Employment; Environment; Family Characteristics; Health

Manpower; Methods; Middle East; Natural Disasters; Population; Population Characteristics; Population Dynamics; Public Policy; Research; Sampling Studies; Sex Distribution; Sex Factors; Social Class; Socioeconomic Factors; Somalia; Transients and Migrants; Water Supply; Agricultural Workers; Animal Resources; Arab Countries; Community Surveys; Demographic Factors; Development Policy; Drought; Eastern Africa; Economic Factors; Family And Household; Households; Human Resources; Labor Force; Literacy; Macroeconomic Factors; Methodological Studies; Migrants; Migration; Natural Resources; Nomads; Policy; Research Methodology; School Age Population; Settlement And Resettlement; Socioeconomic Status; Studies; Surveys. Abstract: During the April 1973-June 1975 drought in Somalia, the government settled 100,000 nomads over 5 years in 3 agricultural (Dujuma, Sablaale, and Kurtunwary) and 3 fishing settlements (Brava, Adale, and Eil). They had earlier sought relief from the drought at some 20 relief camps. In 1982, the Ministry of National Planning and the Settlement Development Agency conducted a household survey in 4 of the 6 settlements (2059 households). Considerable problems occurred during the survey thus the data must be interpreted with caution. Nevertheless this survey provided 1 of the few sources on nomadic settlement conditions. 47.5% of the population in the settlement areas were children 15 years old. Fewer middle aged men than women lived in the settlement areas (9% vs. 22%). Males tended to be more literate and/or in school than females (74% vs. 50% and 64% vs. 43% respectively). Despite the disparity, the researchers found these proportions considerable and encouraging. Women headed many households (47.25%), especially in Adale (61%). Presumably many of the husbands returned to their pastoral ways. Other adult relatives and older children often lived in women headed households and provided support for farming, fishing, and other economic activities. Most respondents were satisfied with settled life and felt it would be permanent. Further 70-90% of respondents wanted their sons and daughters to be civil servants while 0-8% wanted them to be herders. 78-87% of respondents lost all livestock during the drought while only 2-10% acquired livestock after the drought. Since livestock provided considerable wealth in relation to incomes from agriculture and fishing and since nomads tended to be inexperienced in these new occupants, they underwent an extreme adjustment to settled life. In conclusion, the resettlement program had mixed successes. ISSN: 0016-6987.

Tucker M.R. 1984. "Possible Sources of Outbreaks of the Armyworm, Spodoptera Exempta (Walker) (Lepidoptera: Noctuidae), in East Africa at the Beginning of the Season." Bull. Entomol. Res. Volume 74, Issue 4, Pages 599-607. Descriptors: Ecological Abstracts; armyworm; Kenya; Tanzania; Somalia; Malawi; Zambia; Mozambique Species Term: Spodoptera exempta; Mythimna unipuncta; Noctuidae; Lepidoptera. Abstract: Eastern Kenya and E Tanzania were the most likely source areas for moths. Immigration from S Somalia into Kenya, and from Malawi, Zambia or N Mozambique into S Tanzania was possible in a few years. ISSN: 0007-4853.

Turner, W. R. 1978. "Hydrogeological Assessment; Trans Juba Livestock Project." International (VVV): Descriptors: Africa; East Africa; hydrogeology; hydrology; Juba River; livestock; Somali Republic; surface water; water resources; water use. Database: GeoRef. Accession Number: 1998-015675.

Turyahikayo, G. R. 1987. "Wave Characteristics off the East African Coast." Kenya J. Sci. (A Phys. Chem. Sci.). Volume 8, Issue 1-2, Pages 33-58. Descriptors:

Water waves; Equatorial easterlies; Monsoons; Tropical oceanography; Winds; Wind data; Wind-driven currents; Wave data; Climatic data; Water currents; Africa, East; Somali Dem. Rep., Kenya; Marine; Brackish. Notes: TR: KE0000065. Abstract: Sea and swell data collected by voluntary observing ships (VOS) plying the waters off the East African coast during January 1976-May 1977 and 1979-1982 are analysed. Monthly/seasonal wave characteristics in the region bounded by latitudes 10 degree S and 10 degree N, longitude 52 degree E and the East African coastline are discussed. June/July was found to be roughest period while March/April and November were the calmest. The ships encountered mainly sea and swell waves of low and moderate significant heights, respectively, most of the time although very rough and high waves in excess of 6.0 m significant height were observed occasionally during the north-east monsoon and the south-west/south-east trades seasons. The roughest area was found to be off the north-east Somali coast during the south-west monsoon winds. The maximum of the average height of the 10% highest waves observed was 10.2 m in June. The distribution of wave indicates that despite the dominance of low sea and moderate swell waves, rough waves can still be encountered in most of the area during both monsoons. (DBO). ISSN: 0250-8257.

U

United Nations. 2004. "Somalia." Department of Peacekeeping Operations. Cartographic Department. Map Number3690, Revision #6. July 2004. See: http://www.un.org/Depts/Cartographic/map/profile/somalia.pdf

United Nations Institute on Training and Research (UNITAR), Operational Satellite Applications Programme (UNOSAT). 2006. "Reported Incidents of Pirate Attacks & Hijackings off the Coast of Somalia, January 2005 through March 2006." See: http://unosat.web.cern.ch/unosat/freeproducts/east_and_horn_of_africa/UNOSAT_horn_piracy28mar06_small.jpg

United Nations Mineral and Groundwater Project, New York, NY, United States. 1972. Interim Report on Project Results and Conclusions; Phase II. United States: U. N. Min. Groundwater Proj., New York, NY, United States. Descriptors: Africa; Benadir; East Africa; hydrology; Mogadishu Somali Republic; Somali Republic; southern Somali Republic; water resources. Database: GeoRef. Accession Number: 1996-069195.

United States Agency for International Development, United States. 1984. "Comprehensive Groundwater Development Project." United States: Descriptors: Africa; development; East Africa; ground water; Somali Republic; water resources. Notes: Proposal for extension. Database: GeoRef. Accession Number: 1998-014442.

United States Agency for International Development, United States. 1980. "Bay Region Agricultural Development Project." United States: Descriptors: Africa; agriculture; Bay Somali Republic; development; East Africa; Somali Republic; water resources; water supply. Notes: Project document. Database: GeoRef. Accession Number: 1998-014441.

United States. Bureau of Mines. 1974. "The Mineral Industry of Other African Areas." United States: U. S. Bur. Mines, Washington, D. C. Descriptors: 1972; Africa; Benin; Botswana; Burkina Faso; Burundi; Cameroon; Central Africa; Central African Republic; Chad; Congo; data; Djibouti; East Africa; economic geology; economics; Ethiopia; Gambia; Guinea; Guinea-Bissau; Indian Ocean Islands; Ivory Coast; Lesotho;

Madagascar; Malawi; Mali; Mauritania; Mauritius; mineral resources; Niger; North Africa; petroleum; production; regional; Rwanda; Senegal; Somali Republic; Southern Africa; Sudan; Swaziland; Togo; West Africa; Western Sahara; Zimbabwe. Database: GeoRef. Accession Number: 1975-015347.

United States. Agency for International Development, Washington, DC, United States. 1979. Somalia; a Comprehensive Ground Water Project. United States: Agency for International Development, Washington, DC, United States. Descriptors: Africa; East Africa; ground water; Somali Republic; water resources. Database: GeoRef. Accession Number: 1996-067504.

Ufficio geologico d'Italia. 1936. Bibliografia Geologica Italiana Per Gli Anni 1915-1933: Africa Orientale Italiana. Roma: Istituto poligrafico dello Stato, Libreria. Descriptors: 116; Geology- Ethiopia- Bibliography; Geology- Somalia- Bibliography. Notes: "II Supplemento al volume LXI del Bollettino del R. Ufficio geologico d'Italia." Includes index. Other Titles: Bollettino del R. Ufficio geologico d'Italia. Supplemento. OCLC Accession Number: 11721844.

Ugolini, A. and Chelazzi, G. 1978. "Researches on the Coast of Somalia, the Shore and the Dune of Sar Uanle. 16. Notes on Cypraeidae (Mollusca Gastropoda)." Monit.Zool.Ital., Suppl., 10(5), 85-103. (1978). Descriptors: community composition; geographical distribution; Cypraeidae; Somali Dem. Rep. Marine. Notes: TR: IR7813592. Abstract: The authors describe 26 species of Cypraeidae (Mollusca, Gastropoda) collected on the coast of Somalia at Mogadiscio, Gesira, Merca, Brava, Bender Mtoni and Sar Uanle. Database: ASFA: Aquatic Sciences and Fisheries Abstracts.

UNICEF. 1982. Refugee Water Supply Project, Somali Democratic Republic: Technical Report. Mogadiscio: United Nations Children's Fund. Descriptors: 21, xxxx leaves, 14 leaves of plates; Wells- Somalia; Water-supply- Somalia. Notes: Includes bibliographical references (p. 21). Library of Congress; LCCN: 85-980490; OCLC Accession Number: 13120932.

UNICEF, Geneva, Switzerland. 1982. "Provision of Water Supply for Refugee Camps in Somalia; Drilling and Well Construction; Description and Summary Report." Switzerland: Descriptors: Africa; construction; drilling; East Africa; Somali Republic; water resources; water supply; water wells; wells. Database: GeoRef. Accession Number: 1998-015688.

UNICEF, Mogadishu, Somalia. 1984. "Problems and Prospects of Water Supply in the Haud Area of the North West Region." Somalia: Descriptors: Africa; East Africa; exploration; Haud Somali Republic; northern Somalia; Somali Republic; water resources; water supply. Notes: Discussion report. Database: GeoRef. Accession Number: 1998-015690.

UNICEF, Mogadishu, Somalia. 1982. "Refugee Water Supply Project; Somali Democratic Republic." Somalia: Descriptors: Africa; East Africa; Somali Republic; water resources; water supply. Notes: Technical report. Database: GeoRef. Accession Number: 1998-015689.

United Liberation Front of Western Somalia. 1973. "Map of Somali Territory Occupied by the Ethiopian Government for United Liberation Front of Western Somalia. [Kharitat Al-Aradi Al-Sumaliyah Al-Muhtallah Min Qibal Al-Ihtital Al-Habashi." N.p. Descriptors: Military occupation- Maps. Notes: Description: color map; 33 x 47 cm.

Scale 1:1,500,000. Geographic: Ethiopia- Boundaries- Somalia- Maps. Somalia-Boundaries- Ethiopia- Maps. Notes: English and Arabic. Accompanied by text, in Arabic. [16] pages. Library of Congress; LCCN: 74-692234; OCLC Accession Number: 5400357.

United Nations. 1993. United Nations Relief and Rehabilitation Programme for Somalia, Covering the Period 1 March-31 December 1993. New York: United Nations. Descriptors: 60; Humanitarian assistance- Somalia; Emergency relief; Rehabilitation; Decentralized government; Women; Agriculture; Repatriation; Food supply; Primary health care; Water-supply; Sanitation. Abstract: Includes statistics. Notes: Named Corp: United Nations. Unified Task Force Somalia. UN. Notes: "11 March 1993". OCLC Accession Number: 70065618.

United Nations. 1970. Mineral and Groundwater Survey, Somalia- United Nations Development Programme. United States: New York, united nations. Descriptors: Africa; East Africa; geophysical surveys; ground water; mineral deposits, genesis; Somali Republic; surveys. Reference includes data from Geophysical Abstracts, U. S. Geological Survey, Reston, VA, United States. Database: GeoRef. Accession Number: 1971-065341.

United Nations. 1955. "Frontier Area between Ethiopia and Trust Territory of Somaliland (Under Italian Administration)." New York: United Nations. Descriptors: Government publication; International government publication. Notes: Description: 1 map; color; 51 x 39 cm. Geographic: Ethiopia- Boundaries- Somalia- Maps. Somalia-Boundaries- Ethiopia- Maps. Scale 1:3,250,000; (E 400--E 510/N 130--S 20). Cartgrph Code: Category of scale: a Constant ratio linear horizontal scale: 3250000 Coordinates--westernmost longitude: E0400000 Coordinates--easternmost longitude: E0510000; Notes: Shows 1897, 1908, and 1950 boundaries. "Map no. 740." OCLC Accession Number: 54619427.

United Nations Development Programme. 1971. Mineral and Groundwater Survey (Phase II): Berbera Sulphur Deposits, Somalia. New York: The Programme. Descriptors: v. ill., maps; Sulphur; Mines and mineral resources- Somalia. OCLLC Accession Number: 41727473.

United Nations Development Programme. 1970. Mineral and Groundwater Survey, Somalia. U.N. Development Programme. Descriptors: Africa; East Africa; economic geology; ground water; mineral resources; regional; resources; Somali Republic. Notes: illustrations (incl. geol. maps), New York. Database: GeoRef. Accession Number: 1971-004941.

United Nations Development Programme. 1970. Mineral and Groundwater Survey, Somalia Interim Report. New York: United Nations. Descriptors: 133. Notes: Reproduction: Microfilm. Ann Arbor, Mich.: University Microfilms International, 1970? 1 microfilm reel; 35 mm. Responsibility: prepared for the government of the Somali Republic by the United Nations, as participating and executing agency for the United Nations Development Programme. LCCN: 89-892096.

United Nations Development Programme. 1968. Project: Mineral and Groundwater Survey; Report on Uranium, Thorium and Rare Earths at Alio Ghelle, Somalia. Ann Arbor, Mich., University Microfilms: Xerox. Descriptors: 49; Mineralogy-Somalia; Ore deposits- Somalia. Notes: Other Titles: Mineral and groundwater survey. Responsibility: Prepared for the Govt. of the Somali Republic by the United Nations acting as participating and executing agency for the United Nations Development Programme. Mogadiscio, 1968. OCLC Accession Number: 12347368.

United Nations Development Programme. 1968. United Nations Development Programme Project Mineral and Groundwater Survey: Mogadiscio: International Atomic Energy Agency. Descriptors: 49 leaves. Notes: Reproduction: Microfilm. Ann Arbor, Mich.: University Microfilms International, 1970? 1 microfilm reel; 35 mm. Other Titles: Mineral and groundwater survey. Report on uranium, thorium, and rare earths at Alio Ghelle, Somalia. Responsibility: report on uranium, thorium, and rare earths at Alio Ghelle, Somalia prepared for the government of the Somali Republic by the United Nations, acting as participating and executing agency for the United Nations Development Programme. LCCN: 89-892097; OCLC Accession Number: 20825135.

United Nations Development Programme. Mineral and Groundwater Survey (Phase II): Berbera Sulphur Deposits, Somalia. New York: United Nations. Descriptors: 14. Notes: Reproduction: Photocopy. Ann Arbor, University Microfilms, [n.d.]. Entry: 19840525; Update: 20020524. Accession Number: 10776357.

United Nations Development Programme, New York, NY, United States. 1975. "Mineral and Groundwater Survey; Phase III, Geochemical and Geological Investigation in the Northern Province [Modified]." United States: Volume: DP/UN/SOM-71-523-2, Descriptors: Africa; East Africa; geochemical methods; geochemical surveys; ground water; mineral exploration; mineral resources; Somali Republic; surveys; water resources. Notes: Technical report. Database: GeoRef. Accession Number: 1998-015686.

United Nations Development Programme, New York, NY, United States. 1975. "Mineral and Groundwater Survey; Phase III, 1972-74, Project Finding and Recommendation." United States: Descriptors: Africa; East Africa; ground water; mineral resources; Somali Republic; surveys; water resources. Database: GeoRef. Accession Number: 1998-015687.

United Nations Development Programme, New York, NY, United States. 1973. "Mineral and Groundwater Survey; Phase II, Goundwater in Somali Democratic Republic [Modified]." United States: Volume: 3, Descriptors: Africa; East Africa; ground water; mineral resources; northern Somalia; Somali Republic; surveys; water resources. Notes: Technical report. Database: GeoRef. Accession Number: 1998-015685.

United Nations Development Programme, New York, NY, United States. 1972. The First Geological Seminar. Somalia: Min. Min. water Res., Mogadishu, Somalia. Descriptors: Africa; areal geology; East Africa; Somali Republic; symposia. Database: GeoRef. Accession Number: 1998-015684.

United Nations Development Programme, New York, NY, United States. 1972. "Mineral and Groundwater Survey; Phase II, 1969-1971, Interim Report on Project Results and Conclusion." United States: Descriptors: Africa; East Africa; ground water; mineral resources; Somali Republic; surveys; water resources. Notes: Interim report. Database: GeoRef. Accession Number: 1998-015683.

United Nations Development Programme, New York, NY, United States. 1970. "Mineral and Groundwater Survey Project Somalia; Phase 1, 1964-1968." United States: Volume: DP/SF/UN/37 SOM, Descriptors: Africa; East Africa; ground water; mineral resources; Somali Republic; surveys; water resources. Notes: Interim report. Database: GeoRef. Accession Number: 1998-015682.

United Nations Development Programme, New York, NY, United States. 1967. Mineral and Groundwater Survey Project. Somalia: Min. Ind. e Comm., Geol. Dep., Mogadishu, Somalia. Descriptors: Africa; East Africa; exploration; ground water; mineral resources; Somali Republic; surveys; water resources. Database: GeoRef. Accession Number: 1998-015679.

United Nations Development Programme and Programme of Education for Emergencies and Reconstruction (Unesco). 2004. "An Atlas for Somalis." UNESCO PEER. Descriptors: Atlas. Notes: Description: 1 atlas (64 p.); color maps; 30 x 42 cm. Geographic: Somalia- Maps. Scale varies. Cartgrph Code: Category of scale: z; Other Titles: Atlaska Soomaalida; Responsibility: prepared and designed by the United Nations Development Programme's Data and Information Management Unit, Geographic Information System Lab, in collaboration with the United Nations Educational, Scientific and Cultural Organization Programme of Education for Emergencies and Reconstruction (UNESCO PEER). OCLC Accession Number: 61240461.

United Nations Environment Programme, Nairobi (Kenya). 1989. A Coast in Common: An Introduction to the Eastern African Action Plan. Nairobi (Kenya): UNEP. UNEP, Nairobi (Kenya)pages: 40. Descriptors: Biodiversity; Species diversity; Mangrove swamps; Coral reefs; Ecosystem management; Coastal zone management; Resource management; Marine fisheries; Kenya, Coast; Mozambique; Tanzania; Somalia; Madagascar; Indian Ocean, Mascarene Is., Reunion; Indian Ocean, Mauritius; Indian Ocean, Seychelles; Indian Ocean, Southwest; Marine. Abstract: The Eastern African region covers four coastal countries along the East African coast (Kenya, Mozambique, Tanzania and Somalia), one large island state (Madagascar), three smaller archipelagic states (Comores, Mauritius and Seychelles), and the territories of France in the southwest Indian Ocean (La Reunion). The environment here defies generalization, and encompasses several biogeographic provinces. Ecotypes include coastal dry forests, coastal dunes, coastal floodplains, fresh and brackish water marshes, mangrove forests, coral reefs, reef-back lagoons, sandy beaches and seabird rookeries (sea cliffs and nearshore islands). These areas function as essential habitat for local species including fish and migratory birds, as shoreline stabilizers, and as buffers against coastal erosion. The coast of Eastern Africa is bathed by the great current systems of the Indian Ocean, which vary greatly with the seasonal monsoons. The Indian Ocean has particularly narrow continental shelves along this coast, and thus lower biological productivity than many coastal regions. The coast is rich in varieties and numbers of marine life forms, however. Extensive and highly diverse coral reefs fringe its narrow shelves shores. Species-rich mangroves with their commercially important oysters, crabs and mullet abound near river estuaries and along the coasts, particularly those of Mozambique, Tanzania, Kenya and southern Somalia. The region's people are dependent to a significant extent on coastal resources. Fisheries rely on the trawlable inter-reef areas and the species-rich mangroves with their commercially important oysters, crabs and mullet. Coastal ecosystems are important economically for tourism and recreation. Notes: TR: EP0300056. ISBN: 9280712373.

United Nations Environment Programme, Nairobi, Kenya. 1987. "Coastal and Marine Environmental Problems of Somalia." International (III): 1987. Volume: 84, pages: 23. Descriptors: Africa; coastal environment; conservation; East Africa;

ecosystems; environment; environmental geology; impact statements; marine environment; Somali Republic. Notes: References: 21; ISSN: 1014-8647.

United Nations Environment Programme, Nairobi, Kenya. 1987. "Coastal and Marine Environmental Problems of Somalia; Annexes." International (III): 1987. Volume: 84, pages: 215. Descriptors: Africa; concepts; East Africa; environmental geology; marine environment; Somali Republic. Notes: Individual papers within scope are cited separately; illustrations; ISSN: 1014-8647.

United Nations, Development Program, Mineral and Groundwater Survey Project. 1979. Map of Mineral Deposits and Occurrences Somali Republic. Somalia: Descriptors: Africa; East Africa; economic geology; economic geology maps; maps; mineral resources; Somali Republic. Database: GeoRef. Accession Number: 1987-044354.

United Nations, Food and Agriculture Organisation, Mogadishu, Somalia. 1969. Conclusions and Recommendations on Development of Water Resources and Along the Schebelli Flood Plain Region. Somalia: FAO, Mogadishu, Somalia. Descriptors: Africa; development; East Africa; floodplains; fluvial features; Shebelle River; Somali Republic; southern Somali Republic; surface water; water resources. Database: GeoRef. Accession Number: 1998-016852.

United Nations, Food and Agriculture Organisation, Rome, Italy. 1981. Programme Development Mission; Somalia; Mission Findings and Recommendations. Italy: Descriptors: Africa; development; East Africa; programs; Somali Republic; water resources. Database: GeoRef. Accession Number: 1998-016854.

United Nations, Food and Agriculture Organisation, Rome, Italy. 1968. Agricultural Water Survey; Vol. III, Landforms and Soils. Italy: FAO, Rome, Italy. Descriptors: Africa; agriculture; East Africa; landforms; morphology; soils; Somali Republic; surveys; water resources. Database: GeoRef. Accession Number: 1998-016847.

United Nations, Food and Agriculture Organisation, Rome, Italy. 1968. Agricultural Water Survey; Vol. II, Water Resources. Italy: FAO, Rome, Italy. Descriptors: Africa; agriculture; East Africa; Somali Republic; surveys; water resources. Database: GeoRef. Accession Number: 1998-016848.

United Nations, Food and Agriculture Organisation, Rome, Italy. 1967. Agricultural and Water Survey; Vol. 1, General. Italy: FAO, Rome, Italy. Descriptors: Africa; agriculture; East Africa; Somali Republic; surveys; water resources. Database: GeoRef. Accession Number: 1998-016846.

United Nations, Food and Agriculture Organisation, Rome, Italy and United Nations, Food and Agriculture Organisation, Rome, United States. 1977. "Water use in Irrigated Agriculture." Italy: FAO, Rome, Italy. Descriptors: Africa; agriculture; East Africa; irrigation; Somali Republic; water resources; water use. Database: GeoRef. Accession Number: 1998-016853.

United Nations; Somalia and United Nations Development Programme. 1968. United Nations Development Programme Project: Mineral and Groundwater Survey; Report on Uranium, Thorium and Rare Earths at Alio Ghelle, Somalia. Mogadiscio: Descriptors: 49 l. 28 cm. Uranium ores- Somalia; Thorium ores- Somalia; Rare earths- Somalia. Notes: Responsibility: Prepared for the Government of the Somali Republic by the United Nations acting as participating and executing agency for the United Nations Development Programme. LCCN: 73-156061.

United Nations and United Nations Development Programme. 1970. Mineral and Groundwater Survey, Somalia: Interim Report Prepared for the Government of the Somali Republic. New York: United Nations. Descriptors: 133; Mines and mineral resources- Somalia; Groundwater- Somalia; Water-supply- Somalia. Notes: Photoreproduction. Ann Arbor, Mich. University Microfilms. Includes bibliographical references (p. 127-130). Responsibility: by the United Nations, participating and executing agency for the United Nations Development Programme. OCLC Accession Number: 5573507.

United States Board on Geographic Names; United States; Dept. of the Interior and Division of Geography. 1978. Ethiopia, Eritrea and the Somalilands: Official Standard Names Approved by the United States Board on Geographic Names. Washington, DC: U.S. Govt. Print. Off. Descriptors: 498. Notes: Ethiopia- Gazetteers. Eritrea- Gazetteers. Somalilands- Gazetteers. Responsibility: prepared in the Division of Geography, Department of the Interior, August, 1950. OCLC Accession Number: 16776318.

United States. Army Map Service. 1968. "Harar." Washington, D.C: The Service. Descriptors: Government publication; National government publication. Notes: Description: 1 map; color; 47 x 68 cm. Geographic: Ethiopia- Maps. Somalia- Maps. Scale 1,000,000; Lambert Conformal Conic proj. (E 42000'--E 48000'/N 12000'--N 8000'). Cartgrph Code: Category of scale: 1000000 Coordinates--westernmost longitude: 0420000 Coordinates--easternmost longitude: 0480000; Notes: Legend includes populated places, railroads, boundaries, roads, and other landmarks. Relief shown by contours, spot heights, and color. Includes a chart of gradient tints in meters and feet. Includes parts of Ethiopia and Somalia. Prime meridian: Greenwich. Other Titles: Ethiopia. Somalia. Responsibility: prepared by the Army Map Service (AMS), Corps of Engineers, U.S. Army. OCLC Accession Number: 12594430.

United States. Army Map Service. 1964. "Belet Uen." Washington, D.C: The Service. Notes: Description: 1 map; color; 47 x 69 cm. Geographic: Africa- Maps. Somalia- Maps. Scale 1:1,000,000; International Map Proj. (E 42000'--E 4800'/N 8000'--N 4 00'). Cartgrph Code: Category of scale: 1000000 Coordinates--westernmost longitude: 0420000 Coordinates--easternmost longitude: 0480000; Notes: Legend includes towns, railroads, boundaries, roads, and other landmarks. Includes a chart of altitude tints in meters and feet. Other Titles: Africa. Somalia. Responsibility: prepared by the Army Map Service (AMS), Corps of Engineers, U.S. Army. OCLC Accession Number: 12609254.

United States. Army Map Service. 1956. "Harar." Washington, D.C: The Service. Descriptors: Government publication; National government publication. Notes: Description: 1 map; color; 44 x 66 cm. Geographic: Ethiopia- Maps, Topographic. Djibouti- Maps, Topographic. Somalia- Maps, Topographic. Scale 1:1,000,000; Lambert conformal conic proj. Standard parallels N 8040' and N 11020'; (E 42000'--E 48000'/N 12000'--N 8000'). Cartgrph Code: Category of scale: a Constant ratio linear horizontal scale: 1000000 Coordinates--westernmost longitude: E0420000 Coordinates--easternmost longitude: E0480000; Notes: "Heights are in meters and are referred to mean sea level." Relief shown by gradient tints, spot heights, and pictorial relief. Shows boundaries, highways and roads, railways, rivers and water features, and other details. Compiled in 1961. Includes index to adjoining sheets, index to boundaries, reliability diagram, and glossary. Standard map series designation: Series 1301. Other Titles: Margin title; Harar, Ethiopia; Somalia; French Somaliland; Series 1301; Responsibility: printed by Army Map Service, Corps of Engineers, U.S. Army. OCLC Accession Number: 35841029.

United States. Army Map Service. 1956. "Djibouti." Washington: The Service. Descriptors: Government publication; National government publication. Notes: Description: 1 map; color; 45 x 63 cm. Geographic: Djibouti- Maps, Topographic. Ethiopia- Maps, Topographic. Somalia- Maps, Topographic. Tadjoura, Gulf of (Djibouti)- Bathymetric maps. Scale 1:250,000; Transverse Mercator proj. (E 42000'--E 43030'/N 12000'00"--N 11000'00"). Cartgrph Code: Category of scale: a Constant ratio linear horizontal scale: 250000 Coordinates--westernmost longitude: E0420000 Coordinates--easternmost longitude: E0433000; Notes: "Contour interval 100 meters." Relief shown by contours and spot heights. Depth shown by isolines. Shows boundaries, roads, trails, rivers and water features, and other details. "Compiled in 1956 ..." "Blue numbered lines indicate the 10,000 meter universal transverse Mercator grid, zone 38, Clarke 1880 spheroid. Red numbered ticks inside the neatline indicate the 10,000 meter East Africa Belt J grid, Clarke 1880 spheroid." Includes glossary, reliability diagram, and location diagram. Standard map series designation: Series Y502. Other Titles: Margin title; Djibouti, French Somaliland; British Somaliland; Ethiopia; Series Y502; Responsibility: prepared by the Army Map Service (AMS), Corps of Engineers, U.S. Army. OCLC Accession Number: 48561331.

United States. Army Map Service. 1950. "Mogadiscio." Washington, D.C: The Service. Descriptors: Government publication; National government publication. Notes: Description: 1 map; color; 44 x 66 cm. Geographic: Somalia- Maps, Topographic. Scale 1:1,000,000; [conic] proj. of the International Map [of the World]; (E 42000'--E 48000'/N 4000'--N 0000'). Notes: Relief shown by spot heights. Shows boundaries, highways and roads, railways, rivers and water features, and other details. "Reproduced by GSGS in 1946 from second edition, E.A.F. 1627, 1945." "British Crown copyright reserved. Reproduced with the permission of His Britannic Majesty's Stationery Office." Includes index to adjoining sheets, boundary diagram, and metric conversion table. Standard map series designation: Series 1301. "NA 38." "2-50." Other Titles: Series 1301; Title in margin; Africa 1:1,000,000; Responsibility: printed by Army Map Service, Corps of Engineers. OCLC Accession Number: 55002658.

United States. Army Map Service; Great Britain; Army and East African Force. 1959. "Obbia." Washington, D.C: The Service. Descriptors: Government publication; National government publication. Notes: Description: 1 map; color; 44 x 45 cm. Geographic: Somalia- Maps, Topographic. Scale 1:1,000,000; [conic proj. of the] International Map [of the World.] Central meridian E 510; (E 48000'--E 52000'/N 8000'--N 4000'). Notes: "Elevations are in meters and are referred to mean sea level." Relief shown by gradient tints, spot heights, and pictorial relief. Shows boundaries, highways and roads, rivers and water features, and other details. This edition reprinted, with revisions by AMS, from map published: London: Published by D. Survey, War Office and Air Ministry, 1956. "Reprint of third edition, E.A.F. 1176, 1942. Reproduced by War Office, 1946." "British Crown copyright reserved. Reproduced with the permission of Her Britannic Majesty's Stationery Office." Includes index to adjoining sheets, boundary diagram, and altitude tints table. Standard map series designation: Series 1301. Other

Titles: Series 1301; Responsibility: printed by Army Map Service, Corps of Engineers. OCLC Accession Number: 35677561.

United States. Army Map Service; Great Britain; Army and East African Force. 1959. "Belet Uen." Washington, D.C: The Service. Descriptors: Government publication; National government publication. Notes: Description: 1 map; color; 44 x 66 cm. Geographic: Ethiopia- Maps, Topographic. Somalia- Maps, Topographic. Scale 1:1,000,000; [conic proj. of the] International Map [of the World]; (E 42000'--E 48000'/N 8000'--N 4000'). Cartgrph Code: Category of scale: a Constant ratio linear horizontal scale: 1000000 Coordinates--westernmost longitude: E0420000 Coordinates--easternmost longitude: E0480000; Notes: "Elevations are in meters and feet and are referred to mean sea level." Relief shown by gradient tints, spot heights, hachures, and pictorial relief. Shows boundaries, highways and roads, railways, rivers and water features, and other details. This edition reprinted, with revisions by AMS, from map published: London: Published by D. Survey, War Office and Air Ministry, 1956. "Reprint from second edition, E.A.F. 1177, 1945. Reproduced by War Office, 1946." "British Crown copyright reserved. Reproduced with the permission of Her Britannic Majesty's Stationery Office." Includes index to adjoining sheets, boundary diagram, and metric conversion table. Standard map series designation: Series 1301. "Territorial boundaries are shown as at September 1939. They have not been demarcated in all instances, and their representation is not to be regarded as official." Other Titles: Series 1301; Responsibility: printed by Army Map Service, Corps of Engineers. OCLC Accession Number: 35676079.

United States. Army Map Service; Great Britain; Army and East African Force. 1956. "Mogadiscio." Washington, D.C: The Service. Descriptors: Government publication; National government publication. Notes: Description: 1 map; color; 44 x 66 cm. Geographic: Somalia- Maps, Topographic. Scale 1:1,000,000; [conic] proj. [of the] International Map [of the World]; (E 42000'--E 48000'/N 4000'--N 0000'). Notes: "Heights in metres and feet." Relief shown by gradient tints, spot heights, and pictorial relief. Shows boundaries, highways and roads, railways, rivers and water features, and other details. "Reprint of second edition, E.A.F. 1627, 1945. Reproduced by War Office, 1946. Printed by No. 1 SPC. RE., 1956." "British Crown copyright reserved. Reproduced with the permission of Her Britannic Majesty's Stationery Office." Includes index to adjoining sheets, boundary diagram, and metric conversion table. Standard map series designation: Series 1301. "Stock no. 1301XNA38." Other Titles: Margin title; Mogadiscio, Somali Republic; Series 1301; Responsibility: printed by Army Map Service, Corps of Engineers. OCLC Accession Number: 35731607.

United States. Army Map Service; Great Britain; Army and East African Force. 1943. "Marsabit." Washington, D.C: The Service. Descriptors: Government publication; National government publication. Notes: Description: 1 map; color; 45 x 67 cm. Geographic: Ethiopia- Maps, Topographic. Kenya- Maps, Topographic. Somalia- Maps, Topographic. Scale 1:1,000,000; [polyconic] proj. of the International map [of the World]; (E 36000'--E 42000'/N 4000'--N 0000'). Notes: "Heights in metres and feet." Relief shown by altitude tints, spot heights, and pictorial relief. Shows boundaries, highways and roads, railways, airports, rivers and water features, and other details. Reprint. Originally published: [London]: EAF, 1943. "Reprinted from first edition, E.A.F. 1178, 1943. Reproduced by War Office, 1946. Printed by no. 1 SPE, RE, 1956."

"Edition advanced to be concurrent with edition 2-GSGS--AMS-3." "British Crown copyright reserved. Reproduced with the permission of Her Britannic Majesty's Stationery Office." Includes index to adjoining sheets, and boundary diagram. Standard map series designation: Series 1301. Other Titles: Margin title; Marsabit, Africa; Series 1301; Responsibility: printed by Army Map Service, Corps of Engineers. OCLC Accession Number: 35801819.

United States. Army Map Service; Great Britain; Army; East African Force; Great Britain and Directorate of Military Survey. 1956. "Nairobi." Washington, D.C. The Service. Descriptors: Government publication; National government publication. Notes: Description: 1 map; color; 45 x 78 cm. Geographic: Kenya- Maps, Topographic. Somalia- Maps, Topographic. Tanzania- Maps, Topographic. Scale 1:1,000,000; [conic] proj. of the International map [of the World]; (E 360--E 430/S 00--S 40). Notes: "Heights in metres and feet." Relief shown by gradient tints, and spot heights. Legend includes towns, railroads, boundaries, roads, and other landmarks. Reprint. Originally published: London: D. Survey War Office and Air Ministry, 1956." "Reprint of first edition, E.A.F. 1943. Reproduced by War Office, 1946." "British Crown copyright reserved. Reproduced with the permission of His Britannic Majesty's Stationery Office." "Territorial boundaries are shown as at September 1939." Includes index to adjoining sheets, and boundary diagram. Standard map series designation: Series 1301. Other Titles: Margin title; Nairobi, Africa; Series 1301; Responsibility: printed by Army Map Service, Corps of Engineers, U.S. Army. OCLC Accession Number: 34844337.

United States. Army Map Service; Great Britain; Army and Middle East Drawing and Reproduction. 1945. "Alula." Washington, D.C: The Service. Descriptors: Government publication; National government publication. Notes: Description: 1 map; color; 44 x 45 cm. Geographic: Somalia- Maps, Topographic. Scale 1:1,000,000; [conic] proj. of the International Map [of the World.] Central meridian E 510; (E 48000'--E 52000'/N 12000'--N 8000'). Notes: "Elevations are in meters and feet and are referred to mean sea level." Relief shown by gradient tints, spot heights, and hachures. Shows boundaries, highways and roads, rivers and water features, and other details. Reprint of a map copied by GSGS, that was originally published in 1945. "Reprinted from first edition, M.D.R. 1/12290, 1945." "British Crown copyright reserved. Reproduced with the permission of Her Britannic Majesty's Stationery Office." Includes index to adjoining sheets, boundary diagram, and metric conversion table. Standard map series designation: Series 1301. Other Titles: Margin title; Alula, East Africa; Series 1301; Entry: 19970808; Update: 20050111. Accession Number: 37438832.

United States. Central Intelligence Agency. 1992, 2002. Somalia Summary Map.

Agricultural Land Use and Natural Resources from Somalia Summary Map, CIA 2002 (84K). See:
http://www.lib.utexas.edu/maps/africa/somalia_nat_res_2002.jpg

Agricultural Land Use and Natural Resources from Somalia Summary Map, CIA 1992 (89K). See:
http://www.lib.utexas.edu/maps/africa/somalia_ag92.jpg

Ethnic Groups from Somalia Summary Map, CIA 2002 (103K). See:
http://www.lib.utexas.edu/maps/africa/somalia_ethnic_grps_2002.jpg

Ethnic Groups from Somalia Summary Map, CIA 1992 (119K). See:
http://www.lib.utexas.edu/maps/africa/somalia_ethnic92.jpg

Population from Somalia Summary Map, CIA 2002 (83K). See:
http://www.lib.utexas.edu/maps/africa/somalia_pop_2002.jpg
Population from Somalia Summary Map, CIA 1992 (91K). See:
http://www.lib.utexas.edu/maps/africa/somalia_pop92.jpg

    United States. Defense Mapping Agency. 1996. "Indian Ocean, Africa--East Coast, Somalia, Raas Garmaal to Raas Cabaad." Bethesda, Md. Riverdale, MD: The Agency; NOAA Distribution Branch, N/CG33, National Ocean Service, distributor. Descriptors: Government publication; National government publication. Notes: Description: 1 map; color; 109 x 74 cm. Geographic: Indian Coast (Somalia) Indian Ocean- Navigation. Scale 1:300,000 at lat. 05000'; Mercator proj. (E 49003'00"--E 51003'00"/N 8055'00"--N 6000'00"). Notes: Depths shown by isolines and soundings. Relief shown by contours, spot heights, and hachures. "From Italian and French charts to 1962." "Copyright 1996 by the United States Government. No copyright claimed under Title 17 U.S.C." "Soundings in meters." Inset: Qooriga Neegro. Scale 1:50,000. Includes source diagram. Other Titles: Raas Garmaal to Raas Cabaad; Responsibility: prepared and published by the Defense Mapping Agency. OCLC Accession Number: 34735744.

    United States. Defense Mapping Agency. 1996. "Africa--East Coast, Somalia-Kenya-Tanzania, Webi Jubba to Zanzibar." Bethesda, Md. Riverdale, MD: The Agency; NOAA Distribution Branch, N/CG33, National Ocean Service, distributor. Descriptors: Government publication; National government publication. 1 map; color; 82 x 63 cm. Geographic: Indian Coast (Somalia) Indian Coast (Kenya) Indian Coast (Tanzania) Indian Ocean- Navigation. Scale 1:971,600 at lat. 3030'S; Mercator proj. (E 38000'--E 43030'/S 0000'--S 7010'). Notes: Depths shown by isolines and soundings. Relief shown by spot heights. "From U.S. and British charts to 1992." "Copyright 1996 by the United States Government. No copyright claimed under Title 17 U.S.C." "Soundings in meters." Includes source diagram and index to next larger scale charts. Other Titles: Webi Jubba to Zanzibar; Responsibility: prepared and published by the Defense Mapping Agency. OCLC Accession Number: 35589545.

    United States. Defense Mapping Agency. 1996. "Africa--East Coast, Plans in Somalia." Bethesda, Md. Riverdale, MD: The Agency; NOAA Distribution Branch, N/CG33, National Ocean Service, distributor. Descriptors: Harbors- Somalia- Kismaayo-Maps; Harbors- Somalia- Buur Gaabo- Anchorage; Government publication; National government publication. Notes: Description: 2 maps on 1 sheet; color; 49 x 65 cm. and 44 x 65 cm., sheet 109 x 76 cm. Geographic: Indian Coast (Somalia); Scale 1:35,000; Mercator proj. (E 42027'18"--E 42039'30"/S 0019'00"--S 0028'20"). Scale 1:35,000; Mercator proj. (E 41044'48"--E 41057'00"/S 1009'38"--S 1017'50"). Notes: Depths shown by isolines and soundings. Relief shown by contours and spot heights. "Copyright 1996 by the United States Government. No copyright claimed under Title 17 U.S.C." "Soundings in meters." Components: A. Qooriga Kismaayo. From Italian, French, and U.S. charts to 1982- B. Buur Gaabo anchorage. From a 1942 Italian chart. Includes source diagram and tidal information. Other Titles: Plans in Somalia; Responsibility: prepared and published by the Defense Mapping Agency. OCLC Accession Number: 36285624.

    United States. Defense Mapping Agency. 1995. "Africa--East Coast, Somalia, Hobyo to Kismaayo." Bethesda, Md. Riverdale, MD: The Agency; NOAA Distribution Branch, N/CG33, National Ocean Service, distributor. Descriptors: Government publication; National government publication. Notes: Description: 1 map; color; 80 x 107 cm. Geographic: Indian Coast (Somalia) Indian Ocean- Navigation. Scale 1:973,000 at lat. 2030'; Mercator proj. (E 41040'00"--E 51000'00"/N 6000'00"--S 1000'00"). Notes: Depths shown by isolines and soundings. Relief shown by spot heights. "From Italian and U.S. Navy surveys to 1986." "Copyright 1995 by the United States Government. No copyright claimed under Title 17 U.S.C." "Soundings in meters." Other Titles: Hobyo to Kismaayo; Responsibility: prepared and published by the Defense Mapping Agency. OCLC Accession Number: 34489388.

United States. Defense Mapping Agency. 1995. "Africa--East Coast, Somalia, Cadale (Itala) to Baraawe." Bethesda, Md. Riverdale, MD: The Agency; NOAA Distribution Branch, N/CG33, National Ocean Service, distributor. Descriptors: Government publication; National government publication. Notes: Description: 1 map; color; 74 x 108 cm. Geographic: Indian Coast (Somalia) Indian Ocean- Navigation. Scale 1:300,000 at lat. 5000'; Mercator proj. (E 44000'00"--E 46055'00"/N 2055'00"--N 0055'00"). Notes: Depths shown by isolines and soundings. Relief shown by form lines and spot heights. "From Italian charts and U.S. Navy surveys to 1985." "Copyright 1995 by the United States Government. No copyright claimed under Title 17 U.S.C." "Soundings in meters." Includes source diagram. Other Titles: Cadale (Itala) to Baraawe; Responsibility: prepared and published by the Defense Mapping Agency. OCLC Accession Number: 33890847.

United States. Defense Mapping Agency. Hydrographic/Topographic Center. 1993. "Somalia City Graphic 1:12,500. Muqdisho." Bethesda, MD: Defense Mapping Agency, Hydrographic/Topographic Center. Descriptors: Government publication; National government publication. Notes: Description: 1 map on 2 sheets; color; 118 x 180 cm., on sheets 146 x 108 and 137 x 82 cm. Geographic: Mogadishu (Somalia)- Maps. Mogadishu (Somalia)- Maps, Topographic. Scale 1:12,500; Transverse Mercator proj. Notes: Relief shown by contours. Includes indexes of features and streets, sheet index, location map, and boundary diagram. "NIMA REF NO. Y921XMUQDIS01." ""NIMA REF NO. Y921XMUQDIS02." Other Titles: Muqdisho; Standard map series designation; Y921; Responsibility: prepared ... by the Defense Mapping Agency, Hydrographic/Topographic Center. OCLC Accession Number: 53502081.

United States. Defense Mapping Agency. Hydrographic/Topographic Center. 1992. "Somalia City Graphic 1:15,000. Dhuusamarreeb (Dusa Marreb)." Bethesda, MD: Defense Mapping Agency Hydrographic/Topographic Center. Descriptors: Government publication; National government publication. Notes: Description: 1 map; color; 49 x 49 cm., on sheet 74 x 56 cm. Geographic: Dhuusamarreeb (Somalia)- Maps. Scale 1:15,000; Transverse Mercator proj. Notes: Includes location and boundary maps. "NIMA ref. no. Y921DHUUSAMARR." "Printed by DMAHTC 12-92." "Copyright 1992 by the United States government." Other Titles: Standard map series designation; Series Y921; Dhuusamarreeb (Dusa Marreb); Responsibility: prepared and published by the Defense Mapping Agency, Hydrographic/Topographic Center. OCLC Accession Number: 56908345.

United States. Defense Mapping Agency. Hydrographic/Topographic Center. 1992. "Somalia City Graphic 1:15,000. Baydhabo (Baidoa)." Bethesda, MD: Defense Mapping Agency Hydrographic/Topographic Center. Descriptors: Government publication; National government publication. 1 map; color; 51 x 50 cm., on sheet 75 x

56 cm. Geographic: Baydhabo (Somalia)- Maps. Scale 1:15,000; Transverse Mercator proj. Notes: Includes location and boundary maps. "NIMA ref. no. Y921XBAYDHABO." "Map information as of 1992." "Printed by DMAHTC 12-92." "Reprinted by NIMA 4-01." Other Titles: Standard map series designation; Series Y921; Baydhabo (Baidoa); Responsibility: prepared and published by the Defense Mapping Agency, Hydrographic/Topographic Center. Baydhabo (Baidoa) Series Y921, Edition 1-DMA, Somalia City Graphic, Original scale 1:15,000, U.S. National Imagery and Mapping Agency, 1992, reprinted 2001 (1.3MB) See: http://www.lib.utexas.edu/maps/africa/baydhabo_1992.jpg. OCLC Accession Number: 56892929.

United States. Defense Mapping Agency. Hydrographic/Topographic Center. 1992. "Somalia City Graphic 1:15,000. Waajid." Bethesda, MD: Defense Mapping Agency Hydrographic/Topographic Center. Descriptors: Government publication; National government publication. Notes: Description: 1 map; color; 49 x 50 cm., on sheet 74 x 56 cm. Geographic: Waajid (Somalia)- Maps. Scale 1:15,000; Transverse Mercator proj. Notes: Includes location and boundary maps. "NIMA ref. no. Y921WAAJID." "Copyright 1992 by the United States government." "Reprinted by NIMA 6-01." Other Titles: Standard map series designation; Series Y921; Waajid. Prepared and published by the Defense Mapping Agency, Hydrographic/Topographic Center. Waajid. Series Y921, Edition 1-DMA, Somalia City Graphic, Original scale 1:15,000, U.S. National Imagery and Mapping Agency, published 1992, reprinted 2001 (1.8MB) See: http://www.lib.utexas.edu/maps/africa/waajid_1992.jpg. OCLC Accession Number: 56908295.

United States. Defense Mapping Agency. Hydrographic/Topographic Center. 1992. "Somalia City Graphic 1:15,000. Marka (Merka)." Bethesda, MD: Defense Mapping Agency Hydrographic/Topographic Center. 1 map; color; 50 x 50 cm., on sheet 72 x 65 cm., folded to 18 x 33 cm. Geographic: Shabeellaha Hoose Region (Somalia)- Maps. Scale 1:15,000; Transverse Mercator proj. (E 44044'00"--E 44048'00"/N 01045'00"--N 01041'00"). Notes: Title from upper margin. "Prepared and published by the Defense Mapping Agency Hydrographic/Topographic Center." Includes inset map of Shalaamboot Airfield, glossary, and ancillary locator map. "DMA stock no Y921XMARKA, ed. no. 001." Other Titles: Standard map series designation; Series Y921; Marka (Merka); OCLC Accession Number: 50216366.

United States. Defense Mapping Agency. Hydrographic/Topographic Center. 1992. "Somalia City Graphic 1:20,000. Kismaayo." Bethesda, MD: Defense Mapping Agency Hydrographic/Topographic Center; (2000 printing). Descriptors: Government publication; National government publication. 1 map; color; 49 x 79 cm., on sheet 66 x 83 cm. Geographic: Kismaayo (Somalia)- Maps. Scale 1:20,000; Transverse Mercator proj. Notes: Includes location and boundary maps. Includes index. "NIMA ref. no. Y921XKISMAAYO." "Compiled in 1984 from best available source materials." "Printed by DMAHTC 12-92." "Copyright 1992 by the United States government." "Reprinted by NIMA 12-00." Other Titles: Standard map series designation; Series Y921; Kismaayo; Responsibility: prepared and published by the Defense Mapping Agency, Hydrographic/Topographic Center. Kismaayo Series Y921, Edition 3-DMA, Somalia City Graphic, Original scale 1:20,000, U.S. National Imagery and Mapping Agency, compiled 1984, published 1992, reprinted 2000 (9.8MB). See:

http://www.lib.utexas.edu/maps/africa/kismaayo_1992.jpg; OCLC Accession Number: 56892804.

United States. Defense Mapping Agency. Hydrographic/Topographic Center. 1991. "Indian Ocean, Somalia--East Coast, Mudqdisho (Mogadishu)." Washington, D.C; Denver, CO: The Center; USGS Branch of Distribution, Box 25286, distributor. Descriptors: Harbors- Somalia- Mogadishu- Maps; Nautical charts- Indian Ocean; Government publication; National government publication. Notes: Description: 1 map; color; 71 x 106 cm. Geographic: Indian Ocean- Navigation. Scale 1:15,000; Mercator proj. (E 45017'00"--E 42025'30"/S 2002'30"--S 1056'45"). Cartgrph Code: Notes: Depths shown by isolines and soundings. "From U.S. Navy surveys in 1985 and an Italian chart to 1941." "Copyright restrictions from the country of origin continue to exist." "Soundings in meters." Includes source diagram. Other Titles: Muqdisho (Mogadishu); Responsibility: prepared and published by the Defense Mapping Agency Hydrographic/Topographic Center. OCLC Accession Number: 26018778.

United States. Defense Mapping Agency. Hydrographic/Topographic Center. 1990. "Africa--East Coast, Somalia, Gulf of Aden, Approaches to Berbera." Washington, D.C; Denver, CO: The Center; USGS Branch of Distribution, Box 25286, distributor. Descriptors: Harbors- Somalia- Berbera- Maps; Government publication; National government publication. Notes: Description: 2 maps on 1 sheet; color; 54 x 75 cm., sheet 123 x 86 cm. Geographic: Aden, Gulf of- Navigation. Scale 1:15,000; Mercator proj. (E 44056'00"--E 45001'00"/N 10028'00"--N 10024'00"). Scale 1:75,000; Mercator proj. (E 44046'21"--E 45017'00"/N 10044'00"--N 8022'00"). Notes: Depths shown by isolines and soundings. Relief shown by contours and spot heights. "From U.S. Navy surveys to 1985, and British Admiralty and Italian charts to 1989." "Copyright restrictions of the country of origin continue to exist." "Soundings in meters." Includes source diagram. Other Titles: Approaches to Berbera. Responsibility: prepared and published by the Defense Mapping Agency Hydrographic/Topographic Center. OCLC Accession Number: 26200969.

United States. Defense Mapping Agency. Hydrographic/Topographic Center. 1984. "Africa-East Coast, Somalia, Cadale (Itala) to Baraawe." Washington, D.C: The Center; DMA, DVCP, distributor. Descriptors: Nautical charts- Somalia; Government publication; National government publication. Notes: Description: 1 map; color; 74 x 108 cm. Geographic: Indian Ocean- Navigation. Scale 1:300,000 at lat. 5000'; Mercator proj. (E 44000'00"--E 46055'00"/N 2055'00"--N 0055'00"). Notes: "OMEGA." Depths shown by isolines and soundings. Relief shown by spot heights. "Soundings in meters." From Italian charts to 1941: 818, 819, 805, and 833. Copyright restrictions of the country of origin continue to exist. Inset: Mugdisho (Mogadishu). Scale 1:20,000. Includes compilation diagram. Other Titles: Cadale (Itala) to Baraawe. Responsibility: prepared and published by the Defense Mapping Agency Hydrographic/Topographic Center. OCLC Accession Number: 11749983.

United States. Defense Mapping Agency. Hydrographic/Topographic Center. 1978. "Africa--East Coast, Somalia, Obbia to Kisimaayo." Washington, D.C: Defense Mapping Agency, Hydrographic/Topographic Center. Descriptors: Nautical charts- Indian Ocean; Government publication; National government publication. Notes: Description: 1 map; color; 80 x 107 cm. Geographic: Indian Ocean- Navigation. Indian Coast (Africa)- Navigation. Indian Coast (Somalia)- Navigation. Scale 1:973000;

Mercator proj. (E 41040'00"--E 51000'00"/N 6000'00"--S 1000'00"). Notes: Depths shown by isolines and soundings. Relief shown by contours and spot heights. "OMEGA." "From an Italian survey in 1938 and 1939." "Soundings in fathoms." Includes glossary and fathoms to meters conversion table. Other Titles: Obbia to Kisimaayo; Responsibility: prepared and published by the Defense Mapping Agency, Hydrographic/Topographic Center. OCLC Accession Number: 62094375.

United States. Defense Mapping Agency. Hydrographic/Topographic Center. 1978. "Indian Ocean--Arabian Sea, Africa--Arabian Peninsula, Gulf of Aden and Adjacent Coasts." Washington, D.C: The Center. Descriptors: Nautical charts- Aden, Gulf of; Indian Coast (Somalia)- Navigation; Nautical charts- Indian Ocean; Government publication; National government publication. 1 map; color; 81 x 129 cm. Geographic: Aden, Gulf of- Navigation. Indian Ocean- Navigation. Scale 1:945,200 at lat 14000'; Mercator proj. (E 44055'--E 56010'/N 17018'--N 10020'). Notes: "From British, French and Italian surveys to 1940. With additions and corrections to 1951." "OMEGA." "Soundings in fathoms." Depths shown by isolines and soundings. "Heights in feet." Relief shown by hachures and spot heights. "Title at right foot of chart: Gulf of Aden. Inset: Bandar Raysut. Scale 1:12,000. Other Titles: Gulf of Aden. Gulf of Aden and adjacent coasts. Bandar Raysut. Responsibility: prepared and published by the Defense Mapping Agency Hydrographic/Topographic Center. OCLC Accession Number: 31674450.

United States. Defense Mapping Agency. Hydrographic/Topographic Center. 1977. "Indian Ocean, Africa--East Coast, Cabo Guardafui to Baía De Lourenço Marques, Including Mozambique Channel and Madagascar." Washington, D.C: The Center. Descriptors: Nautical charts- Indian Ocean; Government publication; National government publication. Notes: Description: 1 map; color; 118 x 65 cm. Geographic: Indian Coast (Kenya)- Navigation. Indian Coast (Madagascar)- Navigation. Indian Coast (Mozambique)- Navigation. Indian Coast (Somalia)- Navigation. Indian Coast (Tanzania)- Navigation. Indian Coast (Africa)- Navigation. Indian Ocean- Navigation. Scale 1:3,776,000 at lat. 70; Mercator proj. (E 32015'--E 54030'/N 13000'--S 26030'). Notes: "From various sources to 1958, corrected from Notice to mariners through NM 27, 1977." "Soundings in fathoms." Depths shown by isolines and soundings. "Heights in feet." Relief shown by spot heights. Title at right foot of chart: Cabo Guardafui to Baía de Lourenço Marques. Other Titles: Cabo Guardafui to Baía de Lourenço Marques. Mozambique Channel and Madagascar. Responsibility: prepared and published by the Defense Mapping Agency Hydrographic/Topographic Center. OCLC Accession Number: 31673351.

United States. Defense Mapping Agency. Hydrographic/Topographic Center; United States and National Imagery and Mapping Agency. . 2000. "Somalia City Graphic 1:15,000: Beledweyne." Bethesda, MD: The Agency. Descriptors: Government publication; National government publication. Notes: Description: 1 map; color; 49 x 50 cm., on sheet 74 x 57 cm. Geographic: Beledweyne (Somalia)- Maps. Scale 1:15,000; Transverse Mercator proj. (E 45010'30"--E 45014'30"/N 4046'--N 4042'). Notes: "This sheet falls within NB 38-15, 1501, 1:250,000 and NB 38-115, Y630, 1:100,000." "Grid 1,000 meter UTM: Zone 38 ..." "This map is red-light readable." Includes location and boundaries maps and glossary. "NIMA REF. NO. Y921XBELEDWYNE." "NSN 7643014033781." Other Titles: Beledweyne; Standard map series designation; Series

Y921; Responsibility: prepared ... by Defense Mapping Agency. Hydrographic/Topographic Center. OCLC Accession Number: 49864519.

United States. Defense Mapping Agency. Hydrographic/Topographic Center; United States and National Imagery and Mapping Agency. 1999. "Somalia City Graphic 1:15,000: Baardheere." Bethesda, MD: The Agency. Descriptors: Government publication; National government publication. Notes: Description: 1 map; color; 49 x 50 cm., on sheet 70 x 56 cm. Geographic: Baardheere (Somalia)- Maps. Scale 1:15,000; Transverse Mercator proj. (E 42021'--E 42019'/N 2023'--N 2019'). Notes: "This sheet falls within NA 38-5, 1501, 1:250,000 and NA 38-049, Y630, 1:100,000." "Grid 1,000 meter UTM: Zone 38 ..." Includes location and boundaries maps and glossary. "NIMA REF. NO. Y921XBAARDHEERE." "NSN 7643014033646." "This map is red-light readable." Other Titles: Baardheere; Standard map series designation; Series Y921; Responsibility: prepared ... by Defense Mapping Agency. Hydrographic-Topographic Center. OCLC Accession Number: 49864560.

United States. Defense Mapping Agency. Hydrographic/Topographic Center; United States and National Imagery and Mapping Agency. 1999. "Somalia City Graphic 1:15,000: Jalalaqsi." Bethesda, MD: The Agency. Descriptors: Government publication; National government publication. Notes: Description: 1 map; color; 49 x 49 cm., on sheet 74 x 56 cm. Geographic: Jalalaqsi (Somalia)- Maps. Scale 1:15,000; Transverse Mercator proj. (E 43034'--E 45038'/N 3025'--N 3021'). Notes: "This sheet falls within NA 38-07, 1501, 1:250,000 and NA 38-020, Y630, 1:100,000." "Grid 1,000 meter UTM: Zone 38 ..." Includes location and boundaries maps and glossary. "NIMA REF. NO. Y921XJALALAQSI." "NSN 7643014033643." "This map is red-light readable." Other Titles: Jalalaqsi; Standard map series designation; Series Y921; Responsibility: prepared ... by Defense Mapping Agency. Hydrographic/Topographic Center. OCLC Accession Number: 49864508.

United States. Defense Mapping Agency. Hydrographic/Topographic Center; United States and National Imagery and Mapping Agency. 1999. "Somalia City Graphic 1:15,000. Jalalaqsi." Bethesda, MD: Defense Mapping Agency Hydrographic and Topographic Center. Descriptors: Cities and towns- Somalia- Maps. Notes: Description: 1 map; color; 50 x 50 cm., on sheet 75 x 56 cm., folded to 19 x 28 cm. Scale 1:15,000; Transverse Mercator proj. (E 45 34'00--E 45 38'00 /N 03 25'00--N 03 21'00). Notes: Title from upper margin. "Prepared and published by the Defense Mapping Agency Hydrographic/Topographic Center." "This sheet falls within NA 38-07, 1501, 1:250,000 and NA 38-020, Y630, 1:100,000." Includes locator map. "NIMA ref. no. Y921XJALALAQSI." "NSN 7643014033643." Other Titles: Standard map series designation; Series Y921; Jalalaqsi; Entry: 20020523; Update: 20050115. OCLC Accession Number: 49857799.

United States. Defense Mapping Agency. Hydrographic/Topographic Center; United States and National Imagery and Mapping Agency. 1992. "Somalia City Graphic 1:15,000: Xudder (Oddur)." Bethesda, MD: The Agency. Descriptors: Government publication; National government publication. Notes: Description: 1 map; color; 49 x 49 cm., on sheet 74 x 56 cm. Geographic: Xudder (Somalia)- Maps. Xudder (Somalia)- Maps, Topographic. Scale 1:15,000; Transverse Mercator proj. (E 43051'30"--E 43055'30"/N 04009'30"--N 04005'00"). Notes: "This sheet falls within NB 38-14, 1501, 1:250,000 and NB 38-136, Y630, 1:100,000." "Grid 1,000 meter UTM: Zone 38 ..."

Includes location and boundaries maps, notes, and glossary. "Printed by DMAHTC 12-92." "This map is red-light readable." Other Titles: Xudder (Oddur); Standard map series designation; Series Y921; Responsibility: prepared ... by Defense Mapping Agency. Hydrographic/Topographic Center. OCLC Accession Number: 49864489.

United States. Defense Mapping Agency. Hydrographic/Topographic; United States and National Imagery and Mapping Agency. 1998. "Somalia City Graphic 1:15,000: Marka (Merka)." Washington, D.C. The Agency. Descriptors: Government publication; National government publication. Notes: Description: 1 map; color; 49 x 50 cm., on sheet 74 x 78 cm. Geographic: Marka (Somalia)- Maps. Marka (Somalia)- Maps, Topographic. Scale 1:15,000; Transverse Mercator proj. (E 44044'--E 44048'/N 1045'--N 1041'). Notes: "Map information as of 1992." "Reprinted by NIMA 12-1998." "This sheet falls within NA 38-10, 1501 1:250,000 and NA 38-78, Y630, 1:100,000." Includes inset, location map and glossary. Other Titles: Marka (Merka); Standard map series designation; Series Y921; Responsibility: prepared ... by Defense Mapping Agency. Hydrographic/Topographic Center. Accession Number: 49607023.

United States. Defense Mapping Agency. Hydrographic/Topographic; United States and National Imagery and Mapping Agency. 1992. "Somalia City Graphic 1:15,000: Luuq." Washington, D.C. The Agency. Descriptors: Government publication; National government publication. Notes: Description: 1 map; color; 49 x 50 cm., on sheet 74 x 57 cm. Geographic: Luuq (Somalia)- Maps. Luuq (Somalia)- Maps, Topographic. Scale 1:15,000; Transverse Mercator proj. (E 42031'--E 42035'/N 3050'--N 3046'). Notes: "This sheet falls within NC 38-1, 1501, 1:250,000 and NA 38-2, Y630, 1:100,000." "Printed by DMAHTC 12-92." "Grid 1,000 meter UTM: Zone 38." Includes location map and glossary. Other Titles: Luuq; Standard map series designation; Series Y921; Responsibility: prepared ... by Defense Mapping Agency. Hydrographic/Topographic Center. OCLC Accession Number: 49606933.

United States. Defense Mapping Agency. Hydrographic/Topographic; United States and National Imagery and Mapping Agency. 1992. "Somalia City Graphic 1:15,000: Boosaaso (Bender Cassim)." Washington, D.C. The Agency. Descriptors: Government publication; National government publication. Notes: Description: 1 map; color; 49 x 49 cm., on sheet 76 x 59 cm. Geographic: Boosaaso (Somalia)- Maps. Boosaaso (Somalia)- Maps, Topographic. Scale 1:15,000; Transverse Mercator proj. (E 49008'--E 49012'/N 11018'30"--N 11014'30"). Notes: "This sheet falls within NC 39-1, 1501, 1:250,000 and NC 39-27, 1:100,000." "Printed by DMAHTC 12-92." "Grid 1,000 meter UTM: Zone 39." Includes location map and glossary. Other Titles: Boosaaso (Bender Cassim); Standard map series designation; Series Y921; Responsibility: prepared ... by Defense Mapping Agency. Hydrographic/Topographic Center. OCLC Accession Number: 49606971.

United States. Defense Mapping Agency. Hydrographic/Topographic; United States and National Imagery and Mapping Agency. 1992. "Somalia City Graphic 1:15,000: Gaalkacyo (Galcaio)." Washington, D.C. The Agency. Descriptors: Government publication; National government publication. Notes: Description: 1 map; color; 49 x 49 cm., on sheet 74 x 57 cm. Geographic: Gaalkacyo (Somalia)- Maps. Gaalkacyo (Somalia)- Maps, Topographic. Scale 1:15,000; Transverse Mercator proj. (E 47024'30"--E 47028'30"/N 6048'00"--N 6044'00"). Notes: "This sheet falls within NC 38-8, 1501, 1:250,000 and NB 38-47, Y630, 1:100,000." "Printed by DMAHTC 12-92."

"Grid 1,000 meter UTM: Zone 38." Includes location map and glossary. Other Titles: Gaalkacyo (Galcaio); Standard map series designation; Series Y921; Responsibility: prepared ... by Defense Mapping Agency. Hydrographic/Topographic Center. OCLC Accession Number: 49606960.

United States. Defense Mapping Agency. Hydrographic/Topographic; United States and National Imagery and Mapping Agency. 1984. "Somalia City Graphic 1:12,500: Caluula (Alula)." Washington, D.C. The Agency. Descriptors: Government publication; National government publication. Notes: Description: 1 map; color; 47 x 66 cm., on sheet 57 x 74 cm. Geographic: Caluula (Somalia)- Maps. Caluula (Somalia)- Maps, Topographic. Scale 1:12,500; Transverse Mercator proj. (E 50043'30"--E 50048'00"/N 11059'30"--N 11056'20"). Notes: "Printed by DMAHTC 9-84." "Grid 1,000 meter UTM: Zone 39." Includes location map and glossary. Other Titles: Caluula (Alula); Standard map series designation; Series Y921; Responsibility: prepared ... by Defense Mapping Agency. Hydrographic/Topographic Center. OCLC Accession Number: 49606982.

United States. Dept. of State. Office of the Geographer. 1979. "Djibouti-Somalia Boundary." Pages 5. Descriptors: LC: JX4111. Geographic: Djibouti- Boundaries- Somalia. Somalia- Boundaries- Djibouti.; Note(s): Includes bibliographical references.; Responsibility: Office of the Geographer, Bureau of Intelligence and Research, Department of State. LOC.

United States. Dept. of State. Office of the Geographer. 1975. "Ethiopia-Somalia Boundary." Pages 12. Descriptors: LC: DT382.3.S55; Dewey: 916.3. Geographic: Ethiopia- Boundaries- Somalia. Somalia- Boundaries- Ethiopia.; Responsibility: U.S. Dept. of State. Office of the Geographer. LOC.

United States, Dept. of State and Office of the Geographer. 1968. "French Territory of Afars and Issas-Somalia Boundary." Pages 2. Descriptors: Dewey: 916.771. Geographic: Djibouti- Boundaries- Somalia. Somalia- Boundaries- Djibouti.; Note(s): "Prepared by the Geographer, Office of Strategic and Functional Research, Bureau of Intelligence and Research, Department of State." Includes bibliographical references. LOC.

United States. Embassy (Somalia). 1987. "Mogadishu City Maps." Wellingborough, Northants, Great Britain: Collets; Process Print. Descriptors: Atlas. Notes: Description: 1 atlas (12 p.); color maps; 26 x 11 cm. Geographic: Mogadishu (Somalia)- Maps. Scale not given. Notes: Includes index. Library of Congress; LCCN: 90-114777.

United States. National Imagery and Mapping Agency. 1999. "Africa--East Coast, Somalia, Raas Garmaal to Raas Binna." Bethesda, Md. Riverdale, Md: The Agency; NOAA Distribution Division, N/ACC3, National Ocean Service, distributor. Descriptors: Government publication; National government publication. Notes: Description: 1 map; color; 111 x 76 cm. Geographic: Indian Ocean- Navigation. Indian Coast (Somalia); Scale 1:300,000 at lat. 05000'; Mercator proj. (E 50018'00"--E 52020'00"/N 11020'00"--N 8022'00"). Notes: Depths shown by isolines and soundings. Relief shown by contours, hachures, and spot heights. "From various sources to 1999." Shipping list no. 2000-2005-S. "Copyright 1999 by the United States Government. No copyright claimed under Title 17 U.S.C." "Soundings in meters." Includes source diagram. Other Titles:

Raas Garmaal to Raas Binna; Responsibility: prepared and published by the National Imagery and Mapping Agency. OCLC Accession Number: 43946664.

United States. National Imagery and Mapping Agency. 1998. "Africa--East Coast, Somalia-Kenya, Manda Island to Kismaayo." Bethesda, Md. Riverdale, MD: The Agency; NOAA Distribution Division, N/ACC3, National Ocean Service, distributor. Descriptors: Government publication; National government publication. Notes: Description: 1 map; color; 74 x 108 cm. Geographic: Indian Coast (Somalia) Indian Coast (Kenya) Indian Ocean- Navigation. Scale 1:300,000 at lat. 50; Mercator proj. (E 40052'56"--E 43047'56"/S 0020'16"--S 2020'16"). Notes: Depths shown by isolines and soundings. Relief shown by spot heights. "From Italian and British Admiralty charts to 1966." "Copyright 1998 by the United States Government. No copyright claimed under Title 17 U.S.C." "Soundings in meters." Includes source diagram. Other Titles: Manda Island to Kismaayo. OCLC Accession Number: 39182650.

United States. National Imagery and Mapping Agency. 1998. "Indian Ocean, Africa--East Coast, Somalia, Kismaayo to Baraawe." Bethesda, Md. Riverdale, MD: The Agency; NOAA Distribution Division, N/ACC3, National Ocean Service, distributor. Descriptors: Government publication; National government publication. Notes: Description: 1 map; color; 74 x 108 cm. Geographic: Indian Coast (Somalia) Indian Ocean- Navigation. Scale 1:300,000 at lat. 50; Mercator proj. (E 42025'00"--E 45020'00"/N 1010'00"--S 0050'00"). Notes: Depths shown by isolines and soundings. Relief shown by spot heights. "From Italian charts to 1955." "Copyright 1998 by the United States Government. No copyright claimed under Title 17 U.S.C." "Soundings in meters." Includes source diagram. Other Titles: Kismaayo to Baraawe. OCLC Accession Number: 39182659.

United States. National Imagery and Mapping Agency. 1998. "Indian Ocean, Africa--East Coast, Somalia, Raas Cusbad to Raas Cabaad (Ras Assuad to Ras Auad)." Bethesda, Md. Riverdale, MD: The Agency; NOAA Distribution Division, N/ACC3, National Ocean Service, distributor. Descriptors: Government publication; National government publication. Notes: Description: 1 map; color; 74 x 108 cm. Geographic: Indian Coast (Somalia) Indian Ocean- Navigation. Scale 1:300,000 at lat. 50; Mercator proj. (E 47050'00"--E 50045'00"/N 6020'00"--N 4020'00"). Notes: Depths shown by isolines and soundings. Relief shown by contours, spot heights, and hachures. "From Italian charts to 1940." "Copyright 1998 by the United States Government. No copyright claimed under Title 17 U.S.C." "Soundings in meters." Includes compilation diagram. Other Titles: Raas Cusbad to Raas Cabaad (Ras Assuad to Ras Auad). OCLC Accession Number: 39182664.

United States. National Imagery and Mapping Agency. 1998. "Africa--East Coast, Somalia, Raas Xaafuun to Hobyo." Bethesda, Md. Riverdale, MD: The Agency; NOAA Distribution Division, N/ACC3, National Ocean Service, distributor. Descriptors: Government publication; National government publication. Notes: Description: 1 map; color; 69 x 109 cm. Geographic: Indian Coast (Somalia) Indian Ocean- Navigation. Scale 1:964,515 at lat. 80; Mercator proj. (E 46030'00"--E 56000'00"/N 11000'00"--N 5000'00"). Notes: Depths shown by isolines and soundings. Relief shown by hachures. "From U.S. charts and Italian surveys to 1996." "Copyright 1998 by the United States Government. No copyright claimed under Title 17 U.S.C." "Soundings in meters."

Includes source diagram and index to next larger scale charts. Other Titles: Raas Xaafuun to Hobyo. OCLC Accession Number: 39656065.

United States. National Imagery and Mapping Agency. 1997. "Indian Ocean, Africa--East Coast, Somalia, Cadale to Raas Cusbad (Itala to Ras Assuad)." Bethesda, Md. Riverdale, MD: The Agency; NOAA Distribution Division, N/ACC3, National Ocean Service, distributor. Descriptors: Government publication; National government publication. Notes: Description: 1 map; color; 74 x 108 cm. Geographic: Indian Coast (Somalia) Indian Ocean- Navigation. Scale 1:300,000 at lat. 5000'; Mercator proj. (E 46010'00"--E 49005'00"/N 4035'00"--N 2035'00"). Notes: Depths shown by isolines and soundings. Relief shown by contours, spot heights, and hachures. "From Italian charts to 1941." "Copyright 1997 by the United States Government. No copyright claimed under Title 17 U.S.C." "Soundings in meters." Includes compilation diagram. Other Titles: Cadale to Raas Cusbad; Cadale to Raas Cusbad (Itala to Ras Assuad). OCLC Accession Number: 39025286.

United States. Office of Geography and United States Board on Geographic Names. 1950. Ethiopia, Eritrea, and the Somalilands; Official Standard Names Approved by the United States Board on Geographic Names. Washington: Descriptors: 498; Geografische namen. Notes: Ethiopia- Gazetteers. Eritrea- Gazetteers. Somaliland- Gazetteers. Library of Congress; LCCN: 73-10041; OCLC Accession Number: 6188069.

United States. Army; Corps of Engineers and Mediterranean Division. 1966. Chisimaio Port Facilities: Investigation of Retaining Wall Repair and Wave Action Damage Moles 1 and 2. Livorno, Italy: US Army Engineer Division, Mediterranean. Descriptors: 20, 34 leaves, 1 folded leaf of plates; Harbors- Somalia- Design and construction; Retaining walls; Water waves. Notes: Somalia. OCLC Accession Number: 7484894.

United States; Defense Mapping Agency and Hydrographic/Topographic Center. 1984. "Indian Ocean, Africa-East Coast, Somalia, g. Raas Garmaal to g. Raas Cabaad." Washington, D.C: Defense Mapping Agency Hydrographic/Topographic Center. Descriptors: Nautical charts- Somalia; Government publication; National government publication. Notes: Description: 1 map; color; 109 x 74 cm. Geographic: Indian Ocean- Navigation. Scale 1:300,000 at lat. 05000'; Mercator proj. (E 49003'00"--E 51003'00"/N 8055'00"--N 6000'00"). Notes: "OMEGA." From Italian and French charts to 1962: Italian chart 815 (1:300,000 ed. 1940, corr. 1955), Italian chart 816 (1:300,000 ed. 1940, corr. 1959), French chart 6267 (1:421,670 edited and corrected 1962). "Copyright restrictions of the country of origin continue to exist." Depths shown by isolines and soundings. Relief shown by contours, hachures, and spot heights. "Soundings in meters." Includes compilation diagram. Inset: Qooriga Neegro (Scale 1:50,000. From an Italian chart in 1940). Other Titles: G. Raas Garmaal to g. Raas Cabaad. Raas Garmaal to g. Raas Cabaad. Entry: 19850228; Update: 20041120 National Library Cataloging: U.S. Government Printing Office (GPO). Accession Number: 11745540.

United States; Defense Mapping Agency and Topographic Center. 1973. "Somalia and the French Territory of the Afars and Issas." S.1: DMATC. Descriptors: Ethnology- Somalia- Maps; Government publication; National government publication. Notes: Description: 1 map; color; 47 x 34 cm. Geographic: Somalia- Maps. Somalia- Population- Maps. Somalia- Economic conditions- Maps. Scale 1:3,480,000; (E 420--E 500/N130--S 1036'). Notes: Includes location and comparative size diagram and maps of

"Population," "Economic activity," and "Ethnic groups." OCLC Accession Number: 63613156.

United States; Office of Geography and United States Board on Geographic Names. 1963. Ethiopia, Eritrea, and the Somalilands: Official Standard Names Approved by the United States Board on Geographic Names. Washington: U.S. G.P.O. Descriptors: 498. Notes: Ethiopia- Gazetteers. Eritrea- Gazetteers. Djibouti- Gazetteers. OCLC Accession Number: 14952080.

"U.S. Helicopter Quick Reaction Force Defends U.N. Peace Keepers in Somalia." 1993. Vertiflite. Volume 39, Issue 5, Pages 18. ISSN: 0042-4455.

Università di Siena and Università nazionale della Somalia. 1994. "Geological Map of the Gedo and Bakool Region, Southern Somalia." Siena: Dipartimento di Scienze della Terra Università di Siena; Centrooffset. Descriptors: Geology- Somalia- Maps. Notes: Description: 1 map; color; 71 x 101 cm. Scale 1:250,000. Notes: Relief shown by spot heights. Includes location map; index map to 1:200,000 and 1:100,000 topographic sheets; tectonic sketch map; 2 color cross-sections; general map of area; stratigraphic chart of the Gedo, Bakool and Bay regions; tectonic sketch map of upper Jubba valley; and 3 maps showing the "Geomorphological evolution of the upper Jubba valley." Responsibility: by Ali Kassim M., Carmignani L., Fantozzi P.L. & Conti P.; Dipartimento di Scienze della Terra Università di Siena; Somali National University. OCLC Accession Number: 58676272.

Università di Siena; Università nazionale della Somalia and Università di Padova. 1994. "Geological Map of the Bay Region, Southern Somalia." Siena: Dipartimento di Scienze della Terra Università di Siena; Centrooffset). Descriptors: Geology- Somalia- Maps. Notes: Description: 1 map; color; 60 x 89 cm. Scale 1:250,000. Notes: Relief shown by spot heights. Includes location map; index map to 1:200,000 and 1:100,000 topographic sheets; tectonic sketch map; 1 color cross-section; and index to mapping. Responsibility: by Abdirahman H.M., Abdirahim M.M., Ali Kassim M., Bakos F., Carmignani L., Conti P., Fantozzi P.L. & Sassi F.P.; Dipartimento di Scienze della Terra Università di Siena; Somali National University; Dipartimento di Mineralogia et Petrologia Università di Padova. OCLC Accession Number: 58676285.

Università nazionale della Somalia. 1986-1988. Geophysical Survey in Sablaale Area: WEBI Shabeele Plain. Mogadishu: The University. Descriptors: 18, 46; Groundwater- Shebeli River Watershed (Ethiopia and Somalia); Groundwater- Somalia; Water chemistry- Shebeli River Watershed (Ethiopia and Somalia); Water chemistry- Somalia; Geophysical well logging. Notes: Cover title. Responsibility: Somali National University. LCCN: 88-151657; OCLC Accession Number: 22120042.

Università nazionale della Somalia and Facoltà di Geologia. 1900s-. "Quaderni Di Geologia Della Somalia." Translated title: "Quaternary Geology of Somalia." Descriptors: Geology- Somalia- Periodicals; LC: WMLC 93/1719; QE339.S6. Notes: Frequency: Quarterly; Formerly: Annually; v.; 24 cm; Notes: Description based on: Magg. 1977. LCCN: sf 93-93377; sn 88-22011.

Università nazionale della Somalia and Food and Agriculture Organization of the United Nations. 1991. Report of a Sub-Regional Seminar on the Dynamics of Pastoral Land and Resource Tenure in the Horn of Africa: Mogadishu, 8-11 April 1990, Organized by Somali National University. Rome: Food and Agriculture Organization of the United Nations. Descriptors: 64 pages; Aridity and arid regions- East Africa;

Pastoralism- East Africa; Pastoralism- Somalia; Resource management- East Africa; Resource management- Somalia; Land tenure- Africa, East; Land tenure- Somalia. Notes: Cover title. Includes bibliographical references. OCLC Accession Number: 28532308.

Universita Nazionale Somala, Divisione della Geologia, Mogadishu, Somalia. 1986. Farjanno Groundwater Investigations; Final Report. Somalia: Universita Nazionale Somala, Mogadishu, Somalia. Descriptors: Africa; East Africa; Farjanno Somali Republic; ground water; Shabeellaha Hoose Somali Republic; Somali Republic; water resources. Database: GeoRef. Accession Number: 1998-016778.

"Unrest and Restrictive Terms Limit Abundant Potential." 1993. World Oil. August. Volume 214, Pages 96+. Descriptors: Petroleum industry- Africa; Applied Science & Technology. Abstract: Part of a special issue on the outlooks for the world's oil- and gas-producing regions. Reforms in northern Africa produced good local results, putting pressure on officials in other areas to overhaul and simplify contracts. Meanwhile, operators on the west coast, cautious about exploration, continue to develop existing fields. However, deepwater exploration and production is moving forward. The oil and gas industry's prospects in Egypt, Libya, Tunisia, Algeria, Morocco, Nigeria, Cameroon, Gabon, Congo, Angola, South Africa, Chad, Equatorial Guinea, Ethiopia, Côte D'Ivoire, Kenya, Mozambique, Namibia, Niger, Senegal, Somalia, and Sudan are discussed, and exploration and production efforts in Benin, Madagascar, Seychelles, Tanzania, Zaire, and Zimbabwe are detailed. ISSN: 0043-8790.

Unruh J.D. 1995. "The Relationship between Indigenous Pastoralist Resource Tenure and State Tenure in Somalia." GeoJournal. Volume 36, Issue 1, Pages 19-26. Descriptors: Policy; Social, economic and political aspects of rural change; developing country; policy effect; environmental impact; agricultural policy; resources conflict; pastoralism; crop production; land tenure; state-pastoralist relations; tenure system. Notes: Somalia. Abstract: In Somalia the transient resource rights and resource use arrangements that are critical to transhumant pastoralism were ignored in the formulation of the national tenure regime which favored crop cultivation. The results were increased land degradation, resource use conflicts, declines in pastoral production, and impacts on Somali clan alliances which in many cases regulate rational resource access and use. Somalia possesses the greatest proportion of pastoralists in Africa. Transhumant pastoralism, as the most widespread agricultural enterprise in the country, will play a critical role in food production for the foreseeable future. The relationship between indigenous pastoralist tenure and state imposed tenure has decreased the ability of pastoralism to reproduce itself, thereby compromising the rational utilization of very large areas of rangeland interior, which have very few alternative uses. ISSN: 0343-2521.

Unruh J.D. 1993. "An Acacia-Based Design for Sustainable Livestock Carrying Capacity on Irrigated Farmlands in Semi-Arid Africa." Ecol. Eng. Volume 2, Issue 2, Pages 131-148. Descriptors: Livestock; Evolution And Palaeoecology; Crop and livestock production- general; sustainable agriculture; livestock production; developing country; semi-arid area; irrigated land; acacia plantation; land-use design; livestock; farmland; fodder; livestock carrying capacity; integrated farming system; fodder tree; acacia; irrigated agriculture; pastoralism Species Term: Acacia; Somalia; Faidherbia albida; Animalia; Acacia albida. Notes: Somalia. Abstract: Multiple land-use designs are becoming increasingly necessary in semi-arid Africa as growing populations focus numerous production systems onto spatially limited arable lands. With data gathered in

Somalia, this study considers a design in which fodder-producing trees and irrigated agriculture could be integrated. Following a discussion of the benefits of such an integration, this analysis focuses on a quantitative examination of potential livestock carrying capacity from Acacia albida trees in an irrigated area. Comparisons are made with observed livestock numbers in order to determine if such a design could accommodate the seasonal influx of nomadic herds. The land-use elements that comprise carrying capacity are themselves examined to see which elements might be managed to offset or take advantage of those which are not easily managed. ISSN: 0925-8574.

Unruh J.D. 1991. "Nomadic Pastoralism and Irrigated Agriculture in Somalia. Utilization of Existing Land use Patterns in Designs for Multiple Access of "High Potential' Areas of Semi-Arid Africa." GeoJournal. Volume 25, Issue 1, Pages 91-108. Descriptors: Agriculture; Crop and livestock production- general; river basin; production system; multiple land use; food production; semiarid area; developing country; agricultural practice; land-use pattern; nomads; pastoralism; irrigation system; resources management; irrigated agriculture; nomadic pastoralism. Notes: Somalia. Abstract: The persistent interplay of food production problems, land degradation, and social and climatic difficulties on the Horn of Africa result in recurring famines in spite of vast sums of money spent on agricultural development. As land resources- which undergird both social and production systems in Africa- become increasingly degraded, development efforts, especially in problematic areas, need to become part of comprehensive resource use programs that take into account the existing regional land use ecology. Designs which disrupt the ecology of established land uses can lead to extensive degradation because such uses are linked to wider areas; and the effects of such disruption can ultimately threaten the viability of the proposed schemes themselves. While African agriculture had traditionally met greater food needs by expanding the area under cultivation and irrigation, the increasing scarcity of new high quality arable land means that multiple use of "high potential' areas will become a priority. This paper describes a multiple land use in a "high potential' river basin of Somalia, in the context of the existing use pattern involved in irrigated agriculture and nomadic pastoralism. The spatial and temporal access and use of resources are analyzed, and recommendations made for improving the integration of these production systems. ISSN: 0343-2521.

Unruh, J. D. 1991. "Nomadic Pastoralism and Irrigated Agriculture in Somalia." GeoJournal. Volume 25, Issue 1, Pages 91. ISSN: 0343-2521.

Upatham, E. S., Koura, M., Ahmed, M. D. and Awad, A. H. 1981. "Studies on the Transmission of Schistosoma Haematobium and the Bionomics of Bulinus (Ph.) Abyssinicus in the Somali Democratic Republic." Ann. Trop. Med. Parasitol. Feb. Volume 75, Issue 1, Pages 63-69. Descriptors: Animals; Bulinus, parasitology, physiology; Demography; Ecology; Geography; Schistosoma haematobium; Schistosomiasis- epidemiology- transmission; Somalia; Water Supply. ISSN: 0003-4983.

"USAF Somalia Focus Shifts to Food, Water." 1993. Aviation Week & Space Technology. January 11. Volume 138, Pages 23. Descriptors: Airplanes in relief work; Somalia History, 1991- (Civil War); Relief work; United States Air Force, Air Mobility Command; Applied Science & Technology; Readers' Guide (Current Events); Business; Aircraft; Public order and safety; National security; Aircraft Manufacturing. Abstract: Now that the United States is nearing the end of its deployment of 24,000 troops in Somalia, the U.S. Air Force Air Mobility Command is shifting its focus to moving large amounts of food, water, and utility items. In the first 18 days of Operation Restore Hope, AMC flew 126 million ton-miles, providing supplies for Somalis and moving a Marine expeditionary unit and most of an Army division to and within Somalia. ISSN: 0005-2175.

Usoni, Luigi. 1952. "Giacimenti Ferriferi Della Somalia." Report of the ...Session- International Geological Congress. Volume 1, Pages 183-185. Descriptors: economic geology; iron; Italian Somaliland; metals; mineral deposits, genesis. Notes: IGC, International Geological Congress; 19th, Algeria, Symposium . . . fer; geol. sketch map (under separate cover); Abstract: Describes a lenticular bed of limonite in lower Eocene deposits of Migiurtinia, Italian Somaliland, believed to have been formed by chemical precipitation from circulating ground waters of meteoric origin, and magnetite-ilmenite ores in quartz gangue, associated with granitic masses in southern Somaliland. Extracted from a study of the mineral resources of Italian east Africa. ISSN: 1023-3210.

Usoni, Luigi. 1952. "Risorse Minerarie Dell'Africa Orientale; Eritrea-Etiopia-Somalia." Translated title: "Mineral Resources of Eastern Africa- Eritrea- Ethiopia-Somalia." Ministerio Africa Italiana, Ispettorato Gen. Miner. Jandi Sapi, Rome. Descriptors: Africa; East Africa; economic geology; Eritrea-Ethiopia-Italian Somaliland; Ethiopia; Italian Somaliland; mines and mineral resources; mineral waters; springs. Notes: illustrations (incl. maps), 553 pages, bibliography pages 521-532. Abstract: A monographic study of the mineral resources of Eritrea, Ethiopia, and Italian Somaliland; a section on mineral springs is included. LCCN: 53-23196.

Utke A, Huth A, Matheis G and Hawa H.H. 1990. "Geological and Structural Setting of the Maydh and Inda Ad Basement Units in Northeast Somalia." Berliner Geowissenschaftliche Abhandlungen, Reihe A. Volume 120, Issue 2, Pages 537-550. Descriptors: Regional Geology; basement; tectonics; metamorphism; magmatism; sedimentation. Notes: Somalia. Abstract: Controlled by Phanerozoic tectonics, the basement in northern Somalia outcrops along the Gulf of Aden. The last reactivation is recorded in the late Proterozoic (Pan-African), accompanied by the intrusion of syn- to posttectonic granitoids and gabbros. The Inda Ad Group is flanked to the west by a greenschist series named the Maydh "greenstone belt'. The series is tentatively believed to represent a rift basin with little oceanic crust. The contact to the metasediments of the Inda Ad Group is marked by metaconglomerates and -greywackes, indicating a strong relief in the western part of the basin. An E-W compression lead to folding and greenschist metamorphism of both units. I-type granitoids intruded the series after their deformation. ISSN: 0172-8784.

V

Vail, J. R. 1988. "Tectonics and Evolution of the Proterozoic Basement of Northeastern Africa." Federal Republic of Germany: Friedr. Vieweg & Sohn, Braunschweig, Federal Republic of Germany. Descriptors: Africa; amphibolite facies; Arabian Peninsula; Arabian Shield; Asia; basement; Bayuda Desert; East Africa; Egypt; evolution; facies; gneisses; granite gneiss; granulites; IGCP; igneous rocks; Kenya; metamorphic rocks; metasedimentary rocks; Mozambique Belt; North Africa; northeastern Africa; Nubian Shield; ophiolite; Pan-African Orogeny; plate collision; Precambrian; Proterozoic; Sinai Egypt; Somali Republic; structural geology; Sudan; tectonics; upper Precambrian; volcanic rocks. Notes: SP: IGCP, International Geological

Correlation Programme; IGCP Project No. 215; References: 100; geol. sketch maps. ISBN: 3528063254.

Vail, J. R. 1990. "Research in Sudan, Somalia, Egypt and Kenya. Edited by Berliner Geowissenschaftliche Abhandlungen, Section A, Volumes 120.1 and 120.2 1072 p., 1990. ISSN 0172-8784; ISBN 3-927541-15-X and 16-X." Journal of African Earth Sciences. Volume 11, Issue 1-2, Pages 231-232. ISSN: 1464-343X.

Valette, J. N. 1975. "(Geochemical Study of Lac Abhe (TFAI))." Rapages P.-V. Reun., Comm. Int. Explor. Sci. Mer. Mediterr., Monaco. Volume 23, Issue 3, Pages 111 pages. Descriptors: Sediment analysis; Water analysis; Somali Dem. Rep., Affambo L; Freshwater. Notes: Records keyed from 1977 ASFA printed journals. TR: 1977. Abstract: During a TFAI survey, 100 water and 100 sediment samples were collected at 50 stns. The results show that numerous phenomena must be taken into consideration to explain the chemistry of the waters and sediments of lac Abhe. If the influence of the Aouche is of prime importance, the effects of volcanism must be recalled. The leaching of basalts plays an important role and undoubtedly there is a chemical modification of the waters, while underground (due to reheating and exchanges). All studies seem to support the hypothesis that there is a connection between the waters of the Aouche and those of the northern zones and probably those of the chimneys sector. There would be an important charging of the water in the North, near lac Afanbo, in the alluvial deposits of the Aouche, which dominate the plain of lac Abhe for {approx} 100 m (Lopoukhine 1973). The waters of the Aouche would infiltrate under the Dama Ale, and would be reheated there, leaching the overhanging rocks, then they would gush forth through fissures at the level of the sources. The role of volcanism is important but the main physico-chemical characteristics of the lake are produced by the confinement and intense evaporation of the waters. One finds at lac Abhe, the same phenomena seem in the north of Chad, in the region of Kannem, and at lac Nexton, in East Africa. The geochemical peculiarities affect the lakes supplied by streams with endoreic flow, in the desert type zones, where evaporation greatly exceeds the supplies. Database: ASFA: Aquatic Sciences and Fisheries Abstracts.

Van Blarcom B, Knudsen O and Nash J. 1993. "The Reform of Public Expenditures for Agriculture." World Bank Discuss. Pap. Volume 216, Pages 96p. Descriptors: International Aid and Investment; Agricultural policy and theory; structural adjustment; government expenditure; agricultural finance; developing country; policy reform; agriculture. Notes: India Ghana Mexico Somalia Venezuela. Abstract: The study examines questions about public spending levels in agriculture and the composition of that spending. It provides evidence of the effect of structural adjustment and stabilization programs in the 1980s. The case studies cover India, Ghana, Mexico, Somalia and Venezuela. The study's particular focus is on reform of government expenditure for agriculture. It describes a number of issues, with lessons and best practices gleaned from structural adjustment experience and from recent public expenditure reviews. On some of these issues the discussion indicates how reform of agricultural spending policy relates to larger economic policy questions in structural adjustment programs. Other issues are related to budgetary processes rather than policies, and how reforms in these processes should be supportive of adjustment efforts. A concluding section lays out an agenda for future data collection efforts and research. ISSN: 0259-210X.

van Herwaarden, G. J. 1993. Assessment of the 1991/92 Drought in Somalia using FAO-ARTEMIS Information. Graz, Austria: Publ by Environmental Research Inst of Michigan, Ann Arbor, MI. Pages: 679. Proceedings of the 25th International Symposium on Remote Sensing and Global Environmental Change, Apr 4-8 1993. FAO Library.

van Linschoten, Jan Huygen; Blaeue, Willem Janszoon; Survey of Kenya; British Museum and Trustees. 1961. "Part of a Map of the East Coast of Africa, J.H. Van Linschoten, 1596 A.D.; Part of a Map of Aethiopia, Willem Blaueu, 1662 A.D." Nairobi: Survey of Kenya. Descriptors: Maps- Facsimile; Government publication; National government publication. Notes: Description: 2 maps on 1 sheet; 17 x 27 cm. and 18 x 27 cm., sheet 45 x 48 cm. Geographic: Kenya- Maps- Early works to 1800. Somalia- Maps- Early works to 1800. Tanzania- Maps- Early works to 1800. Indian Coast (Africa)- Maps- Early works to 1800. Scale not given. Notes: Relief shown pictorially. "Reproduced by kind permission of the Trustees of the British Museum, London." Both maps show portions of the coastal areas of present-day Somalia, Kenya, and Tanzania. Includes ill. of native animals on both and of sea creatures on the van Linschoten map. "43." Original: Atlas of Kenya. Other Titles: Part of a map of Aethiopia, Willem Blaueu, 1662 A.D. OCLC Accession Number: 39686208.

Vannini, M. 1982. "Notes on Somalian Species of the Genus Hypocolpus (Decapoda, Brachyura, Xanthidae) with the Description of a New Species." Crustaceana. Volume 42, Issue 1, Pages 101-105. Descriptors: geographical distribution; new species; taxonomy; Somali Dem. Rep. Pacific Ocean Southwest; morphology (organisms); morphology; Somalia Coast; Hypocolpus diverticulatus; Hypocolpus perfectus; Hypocolpus rugosus stenocoelus; Hypocolpus; Marine. Abstract: During the Somalian expeditions of the Centro di Faunistica ed Ecologia Tropicali of the C.N.R. (Director, Prof. L. Pardi) from 1970 to 1979, various specimens of Hypocolpus were collected by the author, on the southern coast of Somalia: Hypocolpus diverticulatus (Strahl), Sar Uanle, 20 km S. of Chisimaio, 2 males and 1 female (Museum of Zoology, Florence, no. 606), Hypocolpus perfectus Guinot, Sar Uanle, 3 males (Museum of Zoology, Florence, no. 607), and Hypocolpus rugosus stenocoelus Guinot, Gesira, 20 km S. of Mogadiscio, 2 males (Museum of Zoology, Florence, no. 608). The last two species were previously known only from the original description by Guinot-Dumortier (1960), whose specimens originated from the southern coast of Madagascar and from Mauritius, respectively. H. perfectus was described by Guinot-Dumortier (1960) after three females; so far the male was unknown. Therefore a figure of the male gonopod of the species is provided. ISSN: 0011-216X.

Vannini, M. 1976. "Researches on the Coast of Somalia. the Shore and the Dune of Sar Uanle. 8. Notes on Atelecyclidae and Portunidae (Decapoda, Brachyura)." Monit.Zool.Ital.Suppl. Volume 8, Issue 2, Pages 119-127. Descriptors: Check lists; Biological surveys; Biological collections; Coasts; New records; Geographical distribution; Portunidae; Atelecyclidae; Somali Dem. Rep., Sar Uanle; Marine. Notes: Records keyed from 1977 ASFA printed journals. 1977. Abstract: From a collection of crabs from Sar Uanle (southern Somalia), a single sp of Atelecyclidae and 13 spp of Portunidae (of which only 3 have been previously recorded in Somalia) are listed. Database: ASFA: Aquatic Sciences and Fisheries Abstracts.

Varlamoff, Nicolas. 1973. "Quelques Remarques Sûr La Tectonique Cassante Des Plates-Formes Africaines Et Les Mineralisations Connexes." Translated title: "The "Breaking" Tectonics of the African Platforms and their Associated Mineralizations." France: Masson & Cie, Paris. Descriptors: actinides; Africa; Burkina Faso; East Africa; economic geology; Egypt; exploration; gold ores; granites; igneous rocks; Ivory Coast; Mauritania; metal ores; metals; metasomatism; mineral exploration; molybdenum ores; Niger; Nigeria; North Africa; ore deposits; plutonic rocks; polymetallic ores; Precambrian; programs; regional; Sahara; Somali Republic; tectonics; thorium; uranium ores; west; West Africa. Database: GeoRef. Accession Number: 1974-030759.

Venema, Siebren C. 1983. "Fishery Resources in the North Arabian Sea and Adjacent Waters." Deep-Sea Research, Part A: Oceanographic Research Papers. Volume 31, Issue 6-8A Pt A, Pages 1001-1018. Descriptors: Fisheries; Oceanography- Ecology; Marine Biology. Abstract: In this paper the estimates of standing biomass, and the potential yields derived there from, obtained through all these surveys and other assessments including a few examples of classical stock assessments, will be compared with actual landings, as reported in the FAO Statistical Yearbook to assess sustainable levels of exploitation and the possibilities for development. Attention is also paid to the relationship between climatic variability in oceanographic phenomena and fluctuations in the abundance of living resources, in particular tuna. This review includes the Arabian Sea from the Maldives to Somalia, the Gulfs between Iran and the Arabian Peninsula, the Gulf of Aden and to a limited extent the Red Sea. The fish resources will be discussed by major surveys or areas, while tunas, crustaceans and cephalopods will be dealt with separately. Refs.

Venter, Al J. 1999. "Arabia's curse." Middle East. October 1999, Issue 294, pages 35, 3p, 2c, 1bw. Subject Terms: Drug Abuse; Drugs Of Abuse; Qat. Geographic Terms: Middle East; Somalia. Abstract: Focuses on the proliferation of illegal drugs in the Middle East. Social costs of qat addiction; Existence of a `qat mafia' in Somalia and Kenya. Database: Military & Government Collection. ISSN: 0305-0734. Persistent link to this record: http://search.epnet.com/login.aspx?direct=true&db=mth&an=2382499&loginp age=Login.asp&scope=site

Venter, H. J. T., Thulin, M. and Verhoeven, R. L. 2006. "Two New Species of Cryptolepis (Apocynaceae: Periplocoideae) from Somalia, North-East Africa." South African Journal of Botany. Volume 72, Issue 1, Pages 139-143. Descriptors: Africa (Somalia); Apocynaceae; Cryptolepis spp. nov. Notes: ID: 2230. Abstract: Two new species, Cryptolepis nugaalensis and Cryptolepis somaliensis are described. Both were discovered in arid environment in Somalia. Both belong to a unique group of xerophytic, mostly shrub-like species in the Horn of Africa, island of Socotra and southern Arabia, isolated from the rest of the genus.

Vernier, A. 1993. Aspects of Ethiopian Hydrogeology. Firenze; Istituto Agronomico L'Oltremar; 1993. Relazioni E Monografie Agrarie Subtropicali E Tropicali Nuova Serie113 B, pages: 687-698. International Meeting; 1st- 1987 Nov: Mogadishu. Notes: Geology of Somalia and surrounding regions: Geology and mineral resources of Somalia and surrounding regions. OCLC Accession Number: 34175029.

Vestal, W. J. and Searcy, W. P. 1989. "Seismic Reflection Profiles- Somali Basin Data Report. USNS Wilkes 1977-1979. Technical Rept. FY1977-FY1979." Naval Oceanographic Office, NSTL Station, MS. Dec. Volume: NOOTR300, pages: 63.

Descriptors: Indian Ocean; Seismic reflection; East Africa; Seismic data; Ocean basins; Somali basin; AGS 33 vessel; Ocean technology and engineering Marine geophysics and geology (47E). Abstract: The Somali Basin and contiguous sedimentary basins comprise an inverted L-shaped area of about 1,300,000 square miles in the western Indian Ocean, bounded by the African coasts of Somalia, Kenya, and Tanzania on the west, the Carlsberg Ridge on the north and east, and the northwest coast of the Malagasi Republic (Madagascar) and the Amirante Ridge-trench complex on the south-west. During the period 1977 to 1979, the USNS WILKES collected over 20,000 miles of reflection seismic data in the Somali Basin, along with bathymetry, magnetics, and other data. Photographic reproductions of much of the ten-second seismic analogs are reproduced in this volume. Emphasis has been placed on data in the main part of the basin where survey tracks were concentrated. DTIC Accession Number: ADA2260503XSP.

Vigny, C., Asfaw L.M, Huchon P, Ruegg J.-C and Khanbari K. 2006. "Confirmation of Arabia Plate Slow Motion by New GPS Data in Yemen." Journal of Geophysical Research B: Solid Earth. February 4. Volume 111, Pages Article Number: B02402. Descriptors: Plate tectonics; Arabian plate; triple junction; plate tectonics; plate motion; GPS Species Term: Somalia. Notes: Additional Info: United States; References: Number: 38; Geographic: Yemen Middle East Asia Eurasia. Abstract: During the last 10 years, a network of about 30 GPS sites was measured in Djibouti, East Africa. Additional points were also measured in Yemen, Oman, Ethiopia, Iran, and on La Réunion island. Merged with data from the available International GPS Service permanent stations scattered on the different plates in the area (Eurasia, Anatolia, Africa, Arabia, Somalia), this unique data set provides new insight on the current deformation in the Africa-Somalia-Arabia triple junction area and on the Arabian plate motion. Here we show that coherent motions of points in Yemen, Bahrain, Oman, and Iran allow us to estimate a geodetically constrained angular velocity for the Arabian plate (52.59° N, 15.74 ° W, 0.461°;/Myr in ITRF2000). This result differs significantly from earlier determinations and is based upon our vectors in Yemen. They provide new additional data and better geometry for angular velocity determination. Combined with the African and Somalian motions, this new angular velocity results in predicted spreading rates in the Red Sea and the Gulf of Aden which are 15-20% lower than those measured from oceanic magnetic anomalies and thus averaged over the last 3 Myr. With respect to Eurasia, the geodetic motion of Arabia is also about 30% slower than predicted by NUVEL-1A. On the basis of the kinematic results presented here and on other evidence for a similar slower geodetic rate of the Indian plate, we suggest that the whole collision zone between Africa, Arabia, India on one hand and Eurasia on the other hand has slowed down in the last 3 Myr. ISSN: 0148-0227.

Viola, G., Andreoli, M., Ben-Avraham, Z., Stengel, I. and Reshef, M. 2005. "Offshore Mud Volcanoes and Onland Faulting in Southwestern Africa: Neotectonic Implications and Constraints on the Regional Stress Field." Earth and Planetary Science Letters. February 28. Volume 231, Issue 1-2, Pages 147-160. Abstract: Recently discovered mud volcanoes in the Orange Basin, offshore southwestern Africa, denote the existence of neotectonic faults in the submerged continental shelf. Interpretation of seismic lines perpendicular to the trend of the alignment of the mud volcanoes shows flower structures, diagnostic of strike-slip faulting along a N/NNW direction. Analysis at the regional scale of onland neotectonic features in southwestern Africa shows that recent faulting occurred both in central Namibia and Namaqualand, South Africa and that it created both N/NNW- and NW-trending lineaments. It is proposed that the newly discovered offshore neotectonic activity and the onland structures described in this paper represent the structural expression of the same stress field. These structures form a set of conjugate transtensive faults, which constrain the regional horizontal greatest compressive stress in a NW/NNW direction. Such stress orientation, also supported by in situ stress measurements, defines the so-called Wegener stress anomaly, the predominant present-day stress field of southwest Africa. The Wegener anomaly is incompatible with the stress orientation required by plate-scale tectonic constraints, mainly in the form of recently published GPS motion values for the African plate. ISSN: 0012-821X.

Visona, D. 1993. The Daba Shabeli Gabbro-Syenite Complex: An Element of the Gabbro Belt in the Northern Somali Basement. Firenze; Istituto Agronomico L'Oltremar; 1993. Relazioni E Monografie Agrarie Subtropicali E Tropicali Nuova Serie113 A, pages: 59-82. International Meeting; 1st- 1987 Nov: Mogadishu. Notes: Geology of Somalia and surrounding regions: Geology and mineral resources of Somalia and surrounding regions. OCLC Accession Number: 34175029.

Visona, D., Abdullahi, Hayder M., Hawa Hersi, H., Ibrahim Hersi, A. and Said, Ahmed A. 1983. "Nuovi Dati Di Campagna Sul Basamento Cristallino Della Somalia Settentrionale; Escursione Settembre-Novembre)." Translated title: "New Data of Countryside on the Crystalline Basement of Northern Somalia; Excursion of September-November." Quaderni Di Geologia Della Somalia (Mogadiscio). Volume 7, Descriptors: Africa; areal geology; basement; crystalline rocks; East Africa; expeditions; northern Somalia; Somali Republic. LCCN: 93093377.

Visona, D. and Said, Ahmed A. 1987. "Note Di Campagna Sul Basamento Della Somalia Settentrionale; Escursione 18 Ottobre-1 Novembere 1986. Note of the Countryside on the Basement of Northern Somalia; Excursion of 18 October-1 November, 1986." Quaderni Di Geologia Della Somalia (Mogadiscio). Volume 9, Descriptors: Africa; areal geology; basement; East Africa; expeditions; northern Somalia; Somali Republic; surveys. LCCN: 93093377.

Vitale, Charles S. 1967. Bibliography on the Climate of the Somali Republic. Silver Spring, Md: U.S. Environmental Data Service. Descriptors: 56. Notes: Somalia-Climate- Bibliography. Includes indexes. Arranged chronologically with a subject index. OCLC Accession Number: 3702417.

Voigt B, Gabriel B, Lassonczyk B and Ghod M.M. 1990. "Quaternary Events at the Horn of Africa." Berliner Geowissenschaftliche Abhandlungen, Reihe A. Volume 120, Issue 2, Pages 679-694. Descriptors: Landforms; Oceans; tufa; sinter dam; monsoon; aridity; climate oscillation; terrace; Quaternary. Abstract: Geomorphological, sedimentological and pedological indications show that the ecological conditions have changed several times during the Quarternary in northern Somalia. The amplitudes of climatic oscillations seem not to have been as high as in the Saharan desert belt. A sequence of five terraces near Candala yields no soil with genetic characteristics of strong humidity. The mostly widespread Quaternary accumulations are torrentially deposited coarse gravels. But on the other hand there are also several terraces, which partly consist of laminated fine clastics, fossil-bearing, lacustrine marl and flat tufa deposits. These sediment facies suggest pluvial environment conditions. ISSN: 0172-8784.

Voigt, B., Ghod, M. M. and Gabriel, B. 1990. "Karin-Formation in NordSomalia Als Spaetpleistozaener Klimazeuge." Translated title: "The Karin Formation of Northern Somalia as a Late Pleistocene Climate Indicator." Zentralblatt Fuer Geologie Und Palaeontologie, Teil I: Allgemeine, Angewandte, Regionale Und Historische Geologie. Volume 1990, Issue 4, Pages 437. Descriptors: Africa; carbonate rocks; Cenozoic; chemically precipitated rocks; clastic rocks; clastic sediments; East Africa; evaporites; fluvial sedimentation; fluviolacustrine sedimentation; fresh-water sedimentation; Karin Formation; lacustrine sedimentation; middle Pleistocene; paleoclimatology; paleoecology; Paleosols; Pleistocene; Quaternary; sedimentary rocks; sedimentation; sediments; shingle; soils; Somali Republic; terraces; travertine; upper Pleistocene. Notes: Presented at the Kolloquium Afrikagruppe deutscher Geowissenschaftler, Bonn, 1989; Reference includes data from Geoline, Bundesanstalt fur Geowissenschaften und Rohstoffe, Hanover, Germany. ISSN: 0340-5109.

"Volcanism in French Somaliland." 1940. Nature (London). May. Volume 145, Issue 3682, Pages 828-829. Descriptors: Africa; Djibouti; East Africa; Extinct volcanism; Fumaroles, French Somaliland; historical geology; igneous rocks; thermal waters. Abstract: A review of a paper by E. Aubert de la Ruee in Bulletin Volcanologique (Assoc. Volcanologie, Union Geod.) s. 2, t. 5, p. 71-108, 10 figs., 13 pls. (incl. map), 1939, describing the volcanic rock series, fumaroles, and thermal springs of French Somaliland. ISSN: 0028-0836.

W

Wallace, Mark R., Sharp, Trueman W. and Smoak, Bonnie, et al. 1996. "Malaria among United States Troops in Somalia." The American Journal of Medicine. Volume 100, Issue 1, Pages 49-55. Abstract: United States military personnel deployed to Somalia were at risk for malaria, including chloroquine-resistant Plasmodium falciparum malaria. This report details laboratory, clinical, preventive, and therapeutic aspects of malaria in this cohort. The study took place in US military field hospitals in Somalia, with US troops deployed to Somalia between December 1992 and May 1993. Centralized clinical care and country-wide disease surveillance facilitated standardized laboratory diagnosis, clinical records, epidemiologic studies, and assessment of chemoprophylactic efficacy.: Forty-eight cases of malaria occurred among US troops while in Somalia; 41 of these cases were P falciparum. Risk factors associated with malaria included: noncompliance with recommended chemoprophylaxis (odds ratio [OR] 2.4); failure to use bed nets (OR 2.6); and failure to keep sleeves rolled down (OR 2.2). Some patients developed malaria in spite of mefloquine (n = 8) or doxycycline (n = 5) levels compatible with chemoprophylactic compliance. Five mefloquine failures had both serum levels >=650 ng/mL and metabolite:mefloquine ratios over 2, indicating chemoprophylactic failure. All cases were successfully treated, including 1 patient who developed cerebral malaria. P falciparum malaria attack rates were substantial in the first several weeks of Operation Restore Hope. While most cases occurred because of noncompliance with personal protective measures or chemoprophylaxis, both mefloquine and doxycycline chemoprophylactic failures occurred. Military or civilian travelers to East Africa must be scrupulous in their attention to both chemoprophylaxis and personal protection measures.

Walters, Kenneth R., Sr. and Arnold, Richard D. 1992. "Horn of Africa Climate and Weather- Executive Summary." Air Force Environmental Technical Applications

Center, Scott Afb, Il: Air Force Environmental Technical Applications Center, Scott AFB, IL. December 1992. 36 Pages(s). Descriptors: Climatology, East Africa, Ethiopia, Kenya, Somalia, Sudan, Weather, Yemen, Monsoons, Atmospheric Precipitation. Abstract: Provides a brief executive summary of annual weather and climatology for the region generally known as the 'Horn of Africa,' an area that comprises Somalia, Yemen, Djibouti, Ethiopia, Sudan, and Kenya.... Climatology Weather, Summary, Horn of Africa, Somalia, Yemen, Djibouti, Ethiopia, Sudan, Kenya. DTIC: ADA260152,.

Walters, K. R. and Arnold, R. D. 1992. "Horn of Africa Climate and Weather, Executive Summary. Technical Note." Air Force Environmental Technical Applications Center, Scott AFB, IL. Dec. Volume: USAFETACTN92006, pages: 36. Descriptors: Ethiopia; Kenya; Somalia; Sudan; Weather; Yemen; Monsoons; Atmospheric precipitation; Climatology; East Africa; Djibouti; Restore Hope operation; Atmospheric sciences Meteorological data collection analysis and weather forecasting. Abstract: Provides a brief executive summary of annual weather and climatology for the region generally known as the 'Horn of Africa,' an area that comprises Somalia, Yemen, Djibouti, Ethiopia, Sudan, and Kenya.... Climatology Weather, Summary, Horn of Africa, Somalia, Yemen, Djibouti, Ethiopia, Sudan, Kenya. DTIC Accession Number: ADA2601524XSP.

Warden, A. J. 1962. "Geology of Hargeisa-Bulhar-Laferug Area." Report of the Geological Survey, Somaliland Protectorate. Volume AJW/20, Descriptors: Africa; areal geology; Bulhar Somali Republic; East Africa; Hargeisa Somali Republic; Laferug Somalia; Somali Republic; Woqooyi Galbeed Somali Republic. Revision of JEM/4 and 7. Database: GeoRef.

Warden, A. J. 1962. "Somali Republic; Annual Report of the Geological Survey Department of the Ministry of Industry and Commerce for Period April 1961-March 1962." Descriptors: Africa; East Africa; exploration; Geological Survey; mineral resources; report; Somali; Somali Republic; Surveys, reports. Notes: Hargeisa; Abstract: Includes summarized data on the results of geologic mapping and mineral exploration. Database: GeoRef.

Warden, A. J. 1959. "The Geology of the Basement Rocks to the North and North-West of Bihendula, Berbera District." Report of the Geological Survey, Somaliland Protectorate. Volume AJW/14, Descriptors: Africa; areal geology; basement; Berbera Somali Republic; Bihendula; East Africa; Somali Republic; Woqooyi Galbeed Somali Republic. Database: GeoRef.

Warden, A. J. 1959. "The Geology of the Basement Rocks in the Area West of Manjah-Asseh and Hudiso, Berbera District." Report of the Geological Survey, Somaliland Protectorate. Volume AJW/12, Descriptors: areal geology; basement; Berbera Somali Republic; Hudiso; Manja-Asseh; Woqooyi Galbeed Somali Republic. Database: GeoRef.

Warden, A. J. 1957. "Report on the Las Khoren Inlier of Inda Ad Series and Contiguous Younger Sediments, Erigavo District." Report of the Geological Survey, Somaliland Protectorate. Volume 1957, Descriptors: Africa; areal geology; East Africa; Erigavo Somali Republic; Inda Ad Series; Las Khoren Inlier; regional; Sanaag; sediments; Somali Republic. Database: GeoRef.

Warden, A. J. and Daniels, John L. 1982. "Evolution of the Precambrian of Northern Somalia; Programme and Abstracts of the First Symposium I.G.C.P. 164; Pan-

African Crustal Evolution in the Arabian Nubian Shield." F.E.S. Research Series. Volume 13, Pages 105-106. Descriptors: Africa; amphibolite facies; basement; crystalline rocks; East Africa; evolution; facies; folds; gneisses; Gondwana; granulite facies; intrusions; metamorphic rocks; metamorphism; metasedimentary rocks; Mozambique Belt; northern Somali Republic; orogeny; Pan-African Orogeny; petrology; plate collision; Precambrian; Proterozoic; sedimentation; Somali Republic; structural geology; upper Precambrian; volcanism. Notes: IGCP, International Geological Correlation Programme; IGCP Project No. 164. Database: GeoRef.

Warden, A. J. and Daniels, J. L. 1982. "Evolution of the Precambrian of Northern Somalia." Precambrian Research. Volume 16, Issue 4, Pages A42-A43.

Warden, A. J. and Daniels, John L. 1962. "The Precambrian of the Northern Somali Republic." Report of the Geological Survey, Somaliland Protectorate. Volume AJW/19-JLD/27, Descriptors: Africa; areal geology; East Africa; Precambrian; Somali Republic. Database: GeoRef.

Warden, A. J. and Horkel, A. D. 1984. "The Geological Evolution of the NE-Branch of the Mozambique Belt (Kenya, Somalia, Ethiopia)." Mitteilungen Der Oesterreichischen Geologischen Gesellschaft. Dec. Volume 77, Pages 161-184. Descriptors: Africa; East Africa; Ethiopia; faults; interpretation; Kenya; metamorphic rocks; Mozambique Belt; rift zones; Somali Republic; Somalia Republic; structural analysis; structural geology; systems. Notes: References: 61; illustrations incl. 4 tables, sect., geol. sketch maps. ISSN: 0251-7493.

"Warning Ignored as Famine Hits Somalia again." 1987. New Scientist. May 21. Volume 114, Pages 30. Descriptors: Famines; Somalia- Climate; Droughts; General Science; Applied Science & Technology. ISSN: 0262-4079.

Warsame A.A. 2001. "How a Strong Government Backed an African Language: The Lessons of Somalia." International Review of Education. Volume 47, Issue 3-4, Pages 341-360. Descriptors: Cultural; Culture and Development; language; cultural identity; military government. Notes: References: 18; Geographic: Somalia. Abstract: Although Somali is the mother tongue of over 95 per cent of the population of Somalia, when the country received independence in 1960 it took English, Italian, and Arabic as its official languages. Because of controversy involving technical, religious and political questions, no script for the Somali language could be agreed upon, either in the colonial era or in parliamentary years, 1960-1969. The consequences of this non-decision were considerable for Somali society. However, when the authoritarian military rulers came to the power in the early 1970s, they made a final decision in regard to script. They also issued a decree to the effect that Somali was to be the language of political and administrative discourse in the Somali Republic. That act marked the beginning of the restoration of cultural and linguistic rights for Somali society. This article examines how Somalia, under a strong and totalitarian regime, was able to promote its language. The article also presents an overview of the organization and the implementation of the literacy campaigns carried out in Somalia, as well as some notes on planning and the theoretical framework behind the campaigns. ISSN: 0020-8566.

Water Development Agency, Mogadishu, Somalia. 1983. "Development of Small Water Reservoirs in Somalia by Means of Underground Dams or Other Water Molding Structures." Somalia: Descriptors: Africa; dams; East Africa; ground-water dams;

reservoirs; Somali Republic; water management; water resources; water storage. Notes: Project proposal. Database: GeoRef. Accession Number: 1998-015703.

[Water in Somalia]. 1966. Descriptors: Developing Countries- Pictorial Works; Water Supply- Pictorial Works; Photograph; Picture. Abstract: People gathered around community wells, water troughs, several kinds of pumps, sub-surface reservoir, and dam. Shots of several kinds of water tanks. Shot of the PHS team. Notes: 11 photographs; Named Corp: United States. Public Health Service. Geographic: Somalia- Pictorial works. Notes: Title supplied by cataloger. Most photos have cropping marks in image area. Photos include captions. "230-980-21a." OCLC Accession Number: 27771819.

Watson R.M. 1981. Down Market Remote Sensing." Descriptors: 5-36; Remote Sensing, Gis And Mapping; range resource surveys; Somalia; oil pollution surveys; North Sea; forest inventory; Sudan; sodium fluoride prospecting; Kenya. Abstract: Some examples of down market remote sensing technologies are described for range resource surveys in Somalia, for oil pollution surveys in the North Sea, for forest inventory in Sudan, and for sodium flouride prospecting in Kenya. In: Matching remote sensing technologies and their applications. Proc. 9th conference, Remote Sensing Society, London 1981, (Remote Sensing Society, Reading) (1981) p. 5-36; Notes: Special Features: 8 figs, 25 refs. Update: A 01 JAN 1982. OCLC Accession Number: 0376112.

Watson, R. M. Nimmo, J. M. and Resource Management & Research. 1985. "Somali Democratic Republic, Southern Rangelands Survey." London: Resource Management & Research. Descriptors: Range management- Somalia- Maps; Range ecology- Somalia- Maps; Range plants- Somalia- Maps; Rangelands- Somalia- Maps; Rangelands- Somalia- Statistics; Atlas. Notes: Description: 1 atlas (4 v. in); ill., maps; 30-82 x 61 cm. Scale 1:1,000,000; (E 410--E 470/N 50--S 20). Scale 1:2,500,000; (E 410--E 470/N 50--S 20). Notes: "The survey was carried out from April 1983 to November 1984"--V. 1, pt. 1, Introd. "Updating and corrigenda; 30 April 1985" (v. 1, pt. 1/1, p. A-E). Pt. 2 of each v. consists of map plates with title: Somali Democratic Republic, southern rangelands survey, 1983/4, census results--maps. Includes bibliographical references. Responsibility: R.M. Watson, J.M. Nimmo. Library of Congress. Abstract: v. 1, pt. 1. Appendices- pt. 1/1. The static range resources of the southern rangelands, text- pt. 1/2. The static range resources of the southern rangelands, tables and figures- pt. 2. 1983/4 census results--maps- pt. 5/1-2. Summarized descriptions and index of monitoring sites- v. 2, pt. 1. [without special title]- pt. 2. 1983/4 census results--maps- v. 3, pt. 1. [without special title]- pt. 2. 1983/4 census results--maps- v. 4, pt. 1. The dynamic range resources of the southern rangelands, methods and some comments on the results. LCCN: 87-675503; 90-982157.

Weaver J.N, Brownfield M.E and Bergin M.J. 1990. "Coal in Sub-Saharan-African Countries Undergoing Desertification." Journal of African Earth Sciences. Volume 11, Issue 3-4, Pages 261-271. Descriptors: Economic Geology; Environment And Development; coal; peat; energy resources; energy; fuel; coal deposits; peat deposits. Notes: Senegal Mali Niger Benin Nigeria Cameroon Central African Republic Sudan Ethiopia Somalia Mauritania Africa- (sub-Sahara). Abstract: Coal has been reported in 11 of the 16 sub-Saharan countries discussed in this appraisal: Mauritania, Senegal, Mali, Niger, Benin, Nigeria, Cameroon, Central African Republic, Sudan, Ethiopia, and Somalia. No coal occurrences have been reported in Gambia, Togo, Burkina, Chad, and Djibouti but coal may be present within these countries because neighboring countries do contain coal-bearing rocks. Most of these countries are undergoing desertification or will in the near future. Nine sedimentary basins, completely or partially within the sub-Saharan region, have the potential of either coal and/or peat deposits of economic value. Peat occurs in the deltas, lower river, and interdunal basin areas of Senegal, Mauritania, and Sudan. ISSN: 0899-5362.

Weaver, J. N. and Landis, E. R. 1990. "Coal and Peat in the Sub-Saharan Region of Africa; Alternative Energy Options?" Natural Resources Forum. Volume 14, Issue 1, Pages 64-69. Descriptors: Africa; basins; Benue Valley; Chad Basin; Chari Basin; coal; distribution; economic geology; Gao Trough; Indian Ocean; Niger Basin; organic residues; peat; resources; sedimentary basins; sedimentary rocks; sediments; Senegal Basin; Somali Basin; Sudan Trough; Taoudenni; West Africa. Notes: Non-USGS publications with USGS authors; References: 12; illustrations incl. 1 table, geol. sketch maps. ISSN: 0165-0203.

Webersik C. 2005. "Fighting for the Plenty: The Banana Trade in Southern Somalia." Oxford Development Studies. Volume 33, Issue 1, Pages 81-97. Descriptors: Conflict, protest, human rights; Cultural; violence; agricultural trade; social conflict. References: Number: 29; Geographic: Somalia East Africa Sub-Saharan Africa Eastern Hemisphere World. Abstract: In this paper it is argued that economic interests by multinational corporations, local businessmen and faction leaders are significant elements in the perpetuation of civil violence in Somalia. This study examines the banana trade regime in southern Somalia in relation to conflict over export levies at the national level and farm land and water at the regional level. Small but influential groups come to have an economic interest in prolonged conflict. This viewpoint affirms that it can be misleading to associate war with complete collapse or breakdown of an economy- although it may certainly skew the development of an economy. Two further points arise in respect of such analyses. First, are the initial causes of violent conflict necessarily the same as the factors perpetuating this situation? Second, to what extent are more conventional explanations of conflict in Africa, such as ethnicity, religion and economic inequality, of relevance in this case? ISSN: 1360-0818.

Webersik C. 2004. "Differences that Matter: The Struggle of the Marginalized in Somalia." Africa. Volume 74, Issue 4, Pages 516-533. Descriptors: Political; Political economy, class and state analysis; ethnopolitics; political conflict; state role; social impact. Notes: References: 34; Geographic: Somalia East Africa Sub-Saharan Africa. Abstract: Somalia has been without a government for the past thirteen years. After the ousting of Siyaad Barre in 1991 observers were left with the question why a promising, even democratic, society sharing the same ethnicity, one religion, a common language and a predominantly pastoral culture was overtaken by a devastating civil war. Analysts stressed the significance of kinship and clan politics in the maintenance of sustained conflict. They argued that Somalia's state collapse must be placed in a historical context taking into consideration the cultural heritage of Somali society and the legacy of the colonial past. The purpose of the article is twofold: first, it seeks to explore an alternative explanation for the break down of Barre's dictatorial regime; and second, to analyze the social consequences of political and economic exclusion that followed the state collapse. The paper argues that Somalia's state failure can be explained by the unjust distribution of new sources of wealth in post-colonial Somalia. This modernization process was accompanied by violent clashes and continued insecurity. The breakdown of the former regime did not create a representative government. Instead, faction leaders fought for political supremacy at the cost of the lives of thousands of civilians. In the absence of a functioning government that could guarantee security and protection, clan loyalties gained importance. Clan affiliation became a condition of being spared from violence. Unjust distribution of pockets of wealth, such as the high-potential agricultural land in the riverine areas in southern Somalia, led to localized clashes. It will be argued that horizontal inequalities, or inequalities between groups, are based on both material and imagined differences. Somali faction leaders use these differences instrumentally, to maintain and to exercise power. Irrespective of the existence of invisible and physical markers, it is important to understand what existing social boundaries mean to their participants. A localized clan conflict in Lower Shabelle between the Jido and the Jareer clan families illustrates the consequences of social and economic exclusion. Groups who felt excluded from economic and political life, such as the Jareer, took up arms. Violence became a means of being heard and taken seriously in the current Somali peace talks in Kenya. ISSN: 0001-9720.

Weegar, A. A. 1964. Photogeologic Notes, Hiran Area and Ogaden Area. Somalia: Mobil Petroleum Company, Mogadishu, Somalia. Descriptors: Africa; areal geology; East Africa; geologic maps; Hiran Somali Republic; maps; Ogaden Ethiopia; photogeology; Somali Republic. Database: GeoRef. Accession Number: 1998-015704.

Weir, Joh. 1929. Jurassic Fossils from Jubaland, East Africa. Glasgow: Jackson, Wylie and Co. Descriptors: 4; Paleontology- Jurassic; Paleontology- Somalia- Jubbada Hoose; Geology, Stratigraphic- Jurassic; Geology- Somaliland. Half-title: Glasgow university publications. XIII. Bibliography: p. 57-59. Responsibility: collected by V.G. Glenday, and the Jurassic geology of Somaliland; by J. Weir ... with descriptions of Echinoidea by Dr. Ethel D. Currie ... and of corals by Mary Latham. Monographs of the Geological Department of the Hunterian Museum, Glasgow University, Volume III. LCCN: 29-18377. OCLC Accession Number: 5225099.

Weissel J.K, Pratson L.F and Malinverno A. 1994. "The Length-Scaling Properties of Topography." Journal of Geophysical Research. Volume 99, Issue B7, Pages 13,997-14,012. Descriptors: Landforms; topography; scaling property; digital elevation model; hypsometric curve; length scale. Notes: Somalia Ethiopia Saudi Arabia. Abstract: The scaling properties of synthetic topographic surfaces and digital elevation models (DEMs) of topography are examined by analyzing their "structure functions', i.e., the qth order powers of the absolute elevation differences: $hq(l) = E \{|h(x + l)- h(x)|q$. We find that the relation $h1(l)\≃ clH$ describes well the scaling behavior of natural topographic surfaces, as represented by DEMs gridded at 3 arc sec. Average values of the scaling exponent H between ~0.5 and 0.7 characterize DEMs from Ethiopia, Saudi Arabia, and Somalia over 3 orders of magnitude range in length scale l (~0.1-150 km). Differences in apparent topographic roughness among the three areas most likely reflect differences in the amplitude factor c. Hypsometric curves, which probably reflect the relative importance of tectonic and erosional processes in shaping topography, clearly show that statistical moments higher than the second are important in describing topographic surfaces. Scaling analysis is a valuable tool for assessing the quality and accuracy of DEM representations of the Earth's topography. ISSN: 0148-0227.

Welby, M. S. 1901. Twixt Sirdar and Menelik; an Account of a Year's Expeditions from Zeila to Cairo through Unknown Abyssinia. United Kingdom:

Publisher unknown, London, United Kingdom. Descriptors: Africa; areal geology; Cairo Egypt; East Africa; Egypt; Ethiopia; expeditions; North Africa; Somali Republic; Zeila Somalia. Database: GeoRef. Accession Number: 1999-004837.

Wendler, Ines, Zonneveld, Karin A. F. and Willems, Helmut. 2002. "Oxygen Availability Effects on Early Diagenetic Calcite Dissolution in the Arabian Sea as Inferred from Calcareous Dinoflagellate Cysts; from Process Studies to Reconstruction of the Palaeoenvironment; Advances in Palaeoceanography." Global Planet. Change. Nov. Volume 34, Issue 3-4, Pages 219-239. Descriptors: Africa; Arabian Sea; Asia; bioclastic sedimentation; calcite; carbonates; continental margin sedimentation; diagenesis; Dinoflagellata; early diagenesis; East Africa; ecology; geochemistry; hydrochemistry; Indian Ocean; Indian Peninsula; marine environment; marine sedimentation; marine sediments; modern; organic compounds; oxygen; Pakistan; palynomorphs; preservation; productivity; sediment-water interface; sedimentation; sedimentation rates; sediments; solution; Somali Republic. Notes: Includes appendix; References: 99; illustrations incl. 2 tables, sketch map. Abstract: In oceanic regions with high primary production, such as the Arabian Sea, the primary signals of proxies are often altered by diagenetic processes. The present study aims at assessing the effects of early diagenesis on calcareous dinoflagellate cysts, which represent a relatively new tool for reconstructing the paleoenvironmental conditions within the photic zone. For this purpose, surface sediment samples from within and below the oxygen minimum zone (OMZ) of the northeastern and southwestern Arabian Sea have been analyzed quantitatively for their calcareous dinoflagellate cyst content. The calculated cyst accumulation rates (ARs), the relative abundances and cyst fragmentation values were compared to bottom water oxygen (BWO) content and ARs of organic carbon at the sample positions. Different patterns were found in the northeastern and southwestern part of the Arabian Sea. In the SW, no relationship between cyst ARs and BWO is distinguishable, and the distribution of cyst ARs is thought to largely reflect primary cyst production. In the NE, much higher ARs of all species are found in samples from within the OMZ in comparison to samples from below it. This is interpreted to result from better calcite preservation within the OMZ, presumably due to reduced oxic degradation of organic matter. The differential drop of cyst ARs of the individual species at the lower boundary of the OMZ in the NE Arabian Sea, as well as the species-specific change in relative abundance and fragmentation, indicate different sensitivity to calcite dissolution of the different species. These results show that early diagenetic calcite dissolution can change both relative and absolute abundances of calcareous dinoflagellate cysts, which has to be considered if using them for paleoenvironmental reconstructions. Furthermore, it is shown that considerable calcite dissolution can occur above the carbonate saturation horizon in high productive areas. However, calcite preservation can be substantially increased, as soon as oxygen concentrations are too low for oxic degradation of OM. Under low oxic conditions (within and near the OMZ), the main factor controlling organic matter (OM) preservation appears to be BWO concentrations. Under higher oxygen levels (below approximately 1500 m depth in the NE Arabian Sea) there seems to be an increasing influence of bioturbation and sedimentation rate on the preservation of OM by controlling its oxygen exposure time. This study presents an example of a highly productive basin in which differences in early diagenetic processes can lead to the preservation of a signal that is either dominated by primary production (off Somalia) or by secondary alteration (off

Pakistan), although in both areas, an oxygen depleted zone is present. For estimating the effects of early diagenetic calcite dissolution in a sediment by metabolic CO (sub 2) (and probably by H (sub 2) S oxidation), not only the content of organic carbon but also other geochemical proxies for paleoredox-conditions have to be included for paleoenvironmental reconstructions. ISSN: 0921-8181.

"What about Piracy?" 2005. Shipages World Shipbuilding. Volume 206, Issue 4219, Pages 20-22. Descriptors: Security systems; Ships; Contracts; Water; Societies and institutions; Acoustic devices; Project management. Abstract: The November 2005 attack on the cruise ship 'Seabourn Spirit' off the coast of Somalia put piracy back into the media spotlight. Following this attack, the Somalian government has contracted an American private security company to provide anti-piracy operations in its waters. American Technology Corporation has reported a significant increase in orders for its Long Range Acoustic Devices (LRAD) following successful deployment in the Seabourn Attack. The International Maritime Organization's anti-piracy project is also focusing on the development of regional agreements on counter-piracy measures and has initiated a program of sub-regional meetings to promote regional action in the wider context of maritime security. ISSN: 0037-3931.

White Star Mining Company, International (III). 1970. "Report." International (III): Descriptors: Africa; areal geology; East Africa; Somali Republic. Database: GeoRef. Accession Number: 1998-015708.

White, W. C. 1955. "Preliminary Report on the Geology of the Dibrawein Area, Borama District." Report of the Geological Survey, Somaliland Protectorate. Volume 1955, Descriptors: Africa; areal geology; Borama Somali Republic; Dibrawein Somali Republic; East Africa; regional; Somali Republic; Woqooyi Galbeed Somali Republic. Database: GeoRef.

White, W. C. 1955. "A Preliminary Report on the Geology of the Western Borama District." Report of the Geological Survey, Somaliland Protectorate. Volume 1955, Descriptors: Africa; areal geology; Borama Somali Republic; East Africa; regional; Somali Republic; Woqooyi Galbeed Somali Republic. Database: GeoRef.

White, W. C. and Daniels, John L. 1970. Abdal Qadr. United Kingdom: Directorate Overseas Surv., United Kingdom. Descriptors: Abdal Qadr; Africa; areal geology; East Africa; geologic maps; maps; Somali Republic. Notes: Latitude: N103000, N110000 Longitude: E0430000, E0423000. Database: GeoRef. Accession Number: 1989-007758.

White, W. C. and Daniels, John L. 1970. Gocti. United Kingdom: Directorate Overseas Surv., United Kingdom. Descriptors: Africa; areal geology; East Africa; geologic maps; Gocti; maps; Somali Republic. Notes: Latitude: N100000, N103000 Longitude: E0430000, E0423000. Database: GeoRef. Accession Number: 1989-007770.

White, W. C. and Daniels, John L. 1970. Borama. United Kingdom: Directorate Overseas Surveys, United Kingdom. Descriptors: Africa; areal geology; Borama; East Africa; geologic maps; maps; Somali Republic. Notes: Latitude: N092230, N100000 Longitude: E0433000, E0430000. Database: GeoRef. Accession Number: 1989-007771.

White, W. C. Mason, J. E. and Daniels, John L. 1971. Bawn. United Kingdom: Directorate Overseas Surv., United Kingdom. Descriptors: Africa; areal geology; Bawn; East Africa; geologic maps; maps; Somali Republic. Notes: Latitude: N100000, N103000 Longitude: E0433000, E0430000. Database: GeoRef. Accession Number: 1989-007769.

White, W. C. Mason, J. E. and Stewart, J. A. B. 1970. Silil. United Kingdom: Directorate Overseas Surv., United Kingdom. Descriptors: Africa; areal geology; East Africa; geologic maps; maps; Silil; Somali Republic. Notes: Latitude: N103000, N110000 Longitude: E0432000, E0430000. Database: GeoRef. Accession Number: 1989-007757.

Whiteman A. 1982. "East African Basins: Reserves, Resources and Prospects." Descriptors: 51-99; International Development Abstracts, volume 74; petroleum geology; S.Africa; Mozambique; Madagascar; Tanzania; Kenya; Somalia. Abstract: A detailed descriptive account of the petroleum geology and prospects of the E African basin, summarizing resources of the following countries: S Africa, Mozambique, Madagascar, Tanzania, Kenya and Somalia. Notes: In: Petroleum exploration strategies in developing countries. Proc. meeting, The Hague, 1981, (Graham & Trotman, London, for UN Natural Resources & Energy Division) (1982) p. 51-99; Special Features: 15 figs, 8 tables, 19 refs. Update: A 01 JAN 1983. OCLC Accession Number: 0411120.

Whiteman, A. J. and Exploration Consultants Limited. 1977. Geology and Hydrocarbon Prospects of Eastern Africa: Cape of Good Hope to the Horn. Marlow, England: Exploration Consultants. Descriptors: 5 v. in 7; Petroleum- Geology- Africa, Eastern; Petroleum- Geology- Africa, Southern; Petroleum- Geology- Cape of Good Hope (South Africa: Cape); Petroleum- Geology- South Africa; Petroleum- Geology- Mozambique; Petroleum- Geology- Tanzania; Petroleum- Geology- Kenya; Petroleum- Geology- Somalia; Geology- Africa, Eastern; Geology- Africa, Southern; Geology- Cape of Good Hope (South Africa: Cape); Geology- South Africa; Geology- Mozambique; Geology- Tanzania; Geology- Kenya; Geology- Somalia. Notes: Cover title: Eastern Africa: South Africa to Somalia. Other Titles: Eastern Africa: South Africa to Somalia. Worldcat Accession Number: 14210706.

Wilhelmi F. 1990. "Methodische Probleme Der Erfassung Von Quelea Quelea (L.), Einem Schadvogel Im Getreideanbau Sud-Somalias." Translated title: "Methodical problems in the collection of Quelea Quelea, a bird harmful to the cultivation of grain." Geomethodica. Volume 15, Pages 141-170. Descriptors: Oil and chemical pollution; red-billed weaverbird; weaverbird; habitat selection. Species Term: Quelea quelea. Notes: Somalia. Abstract: The significance of topographical structures relevant for red-billed weaverbird habitat selected is emphasized. An approach to remote sensing for Somalia is presented, concerning the examination of dynamic vegetation structures.

Williams, M. A. J. Alsharhan, A. S. and Glennie, K. W. 1995. Evidence of Late Quaternary Monsoon Variability Along the Southern Margins of the Sahara, Including Ethiopia and Somalia; Abstracts of the International Conference on Quaternary Deserts and Climatic Change. United Arab Emirates (ARE): United Arab Emirates University, Desert and Marine Environment Center, Al Ain, United Arab Emirates (ARE). International Conference on Quaternary Deserts and Climatic Change, Al Ain. United Arab Emirates Conference: Dec. 9-11, 1995. Descriptors: Africa; C-13/C-12; carbon; Cenozoic; climate change; deserts; East Africa; eolian features; Ethiopia; fresh-water environment; Gastropoda; ground water; Holocene; IGCP; Invertebrata; isotope ratios; isotopes; Jubba Valley; last glacial maximum; levels; lower Holocene; Mollusca; monsoons; O-18/O-16; oxygen; paleohydrology; Pleistocene; Quaternary; Sahara; semi-arid environment; shells; Somali Republic; south-central Sahara; stable isotopes; stratigraphic boundary; terrestrial environment; upper Pleistocene. Notes: international geological correlation programme; IGCP project no. 349. Database: GeoRef. Accession Number: 2000-009008.

Willig, Dierk. 1994. "Wehrgeologen in Somalia." Translated title: "Military Geologists in Somalia." Nachrichten- Deutsche Geologische Gesellschaft. Volume 51, Pages 86-90. Descriptors: Africa; boreholes; East Africa; foundations; military geology; Somali Republic; water supply. Notes: References: 7; illustrations; Reference includes data from Geoline, Bundesanstalt fur Geowissenschaften und Rohstoffe, Hanover, Germany. ISSN: 0375-6262.

Wilpolt, R. H. and Simov, S. D. 1979. "Uranium Deposits in Africa; Uranium Deposits in Africa; Geology and Exploration; Proceedings of a Regional Advisory Group Meeting, Organized by the International Atomic Energy Agency." Panel Proceedings Series- International Atomic Energy Agency. Issue STI/PUB/509, Pages 3-20. Descriptors: Africa; African Platform; Algeria; Bakouma; Central Africa; Central African Republic; cratons; Cretaceous; East Africa; economic geology; evolution; Franceville Basin; Gabon; geology; granitic composition; igneous rocks; Iullemenden Basin; Karroo Supergroup; Katangan Orogeny; Mesozoic; metal ores; metamorphic rocks; mineralization; mobile belts; Mudugh; Namibia; Niger; North Africa; Nubian Sandstone; plutonic rocks; Precambrian; Rossing uranium deposit; sediments; Somali Republic; South Africa; Southern Africa; Transvaal South Africa; uranium ores; volcanic rocks; West Africa; Witwatersrand. Notes: References: 14; geol. sketch maps. ISBN: 9200410790.

Wilson, G. 1958. Groundwater Geology of Somalia. United States: UNDP, New York, NY, United States. Descriptors: Africa; East Africa; ground water; Somali Republic; water resources. Database: GeoRef Accession Number: 1998-015712.

"Wind Energy Applications and Training Symposium. Final Report. Progress Rept." 1990. American Wind Energy Association, Arlington, VA. Performer: American Solar Energy Society, Boulder, CO. Volume: CONF9009334 Summary, CONF9009335 Summary, CONF9009336 Summary, pages: 130. Descriptors: California; Cape Verde Islands; Chile; Costa Rica; Diesel Engines; Economic Analysis; Energy Source Development; Financing; Hybrid Systems; India; Indonesia; Leading Abstract; Maintenance; Mali; Meetings; Mexico; Operation; Pakistan; Philippines; Photovoltaic Power Supplies; Progress Report; Resource Potential; Somalia; Turkey; USA; Wind Power; Wind Turbine Arrays; Wind Turbines; Wind; Energy Miscellaneous energy conversion and storage; Administration and management Research program administration and technology transfer; Business and economics Foreign industry development and economics. Abstract: Sixteen representatives from 11 developing nations participated in the 1990 WEATS program. Consistent with previous symposia, the format included classroom-style training and field trip experiences to acquaint the participants with the history and progress to wind energy development in the US, technologically and economically. Brief presentations about wind energy development in all the countries represented were made by the participants. Several reports were prepared and presented along with slides for further explanation. The one-on-one symposium wrap-up session on the last day continues to be a good method of discovering what can be the next step in working with each country to development their wind energy potential. Seventeen papers have been indexed separately for inclusion on the data base. Notes: Funder: Department of Energy, Washington, DC. Wind energy applications and training symposium; Wind energy applications and training symposium; Wind energy applications and training symposium, San Francisco, CA; Amarillo, TX; Washington, DC, 19-28 Sep 1990; 19-28 Sep 1990; 19-28 Sep 1990. Sponsored by Department of Energy, Washington, DC. NTIS Accession Number: DE91006703XSP.

Windley, Brian F., Whitehouse, Martin J. and Ba-Bttat, Mahfood A. O. 1996. "Early Precambrian Gneiss Terranes and Pan-African Island Arcs in Yemen: Crustal Accretion of the Eastern Arabian Shield." Geology. February. Volume 24, Pages 131-134. Descriptors: Geology- Yemen Arab Republic; Gneiss; Island arcs; General Science; Applied Science & Technology. Notes: Bibliography; Illustration; Map. Abstract: Six Precambrian terranes in Yemen are identified and discussed in terms of the crustal evolution of the eastern Arabian Shield. The 2 island-arc terranes and 4 gneiss terranes provide important constraints on the tectonic framework of northeast Gondwana and the rate of Pan-African crustal growth. The 2 western gneiss terranes were correlated with the Asir and Afif terranes in Saudi Arabia. The Abas and Al-Mahfid gneiss terranes to the east yielded Sm-Nd model ages of 1.7 to 2.3 Ga and 1.3 to 2.7 Ga, respectively, and could not be correlated with terranes in Saudi Arabia. The Al-Mahfid gneiss terrane, bounded to the east by an island-arc terrane, is separated from the Abas gneiss terrane to the west by another island-arc terrane. This Pan-African island-arc terrane was obducted onto one or both of the gneiss terranes. The terranes in Yemen may be similar to terranes on the eastern margin of the Arabian Shield and in northern Somalia. ISSN: 0091-7613.

WoldeGabriel, Giday and Dickinson, William R. (chairperson). 1987. "Pre-Tertiary Tectonic Processes in the Development of the East African Rift System; Geological Society of America, 1987 Annual Meeting and Exposition." Abstracts with Programs- Geological Society of America. Volume 19, Issue 7, Pages 895. Descriptors: Afar; Africa; basalts; crystalline rocks; domes; East Africa; East African Rift; epeirogeny; Ethiopia; Ethiopian Rift; evolution; Gondwana; igneous rocks; Indian Ocean; lineaments; lithosphere; Red Sea; sea-level changes; Somali Republic; stress; structural geology; tectonics; terrains; transgression; uplifts; upwelling; volcanic rocks. ISSN: 0016-7592.

Wolfenden, E., Ebinger, C., Yirgu, G., Deino, A. and Ayalew, D. 2004. "Evolution of the Northern Main Ethiopian Rift: Birth of a Triple Junction." Earth & Planetary Science Letters. JUL 30. Volume 224, Issue 1-2, Pages 213-228. Abstract: Models for the formation of the archetypal rift-rift-rift triple junction in the Afar depression have assumed the synchronous development of the Red Sea-Aden-East African rift systems soon after flood basaltic magmatism at 31 Ma, but the timing of initial rifting in the northern sector of the East African rift system had been poorly constrained. The aims of our field, geochronology, and remote sensing studies were to determine the timing and kinematics of rifting in the 3rd arm, the Main Ethiopian rift (MER), near its intersection with the southern Red Sea rift. New structural data and 10 new SCLF Ar-40/Ar-39 dates show that extension in the northern Main Ethiopian rift commenced after 11 Ma, more than 17 My after initial rifting in the southern Red Sea and Gulf of Aden. The triple junction, therefore, could have developed only during the past 11 My, or 20 My after the flood basaltic magmatism. Thus, the flood basaltic magmatism and separation of Arabia from Africa are widely separated in time from the opening of the Main Ethiopian rift, which marks the incipient Nubia-Somalia plate boundary; triple junction formation is not a primary feature of break-up above the Afar mantle plume. The East African rift system appears to have propagated northward from the Mesozoic Anza rift system into the Afar depression to cut across Oligo-Miocene rift structures of the Red Sea and Gulf of Aden, in response to global plate reorganisations. Structural patterns reveal a change from 130°E-directed extension to 105°E-directed extension sometime in the interval 6.6 to 3 Ma, consistent with predictions from global plate kinematic studies. The along-axis propagation of rifting in each of the three arms of-the triple junction has led to a NE-migration of the triple junction since 11 Ma. ISSN: 0012-821X.

Wondem, Atalay Ayele. 1995. "Spatio-Temporal Variation of Seismic Energy and b-Value in the Horn of Africa; International Union of Geodesy and Geophysics; XXI General Assembly; Abstracts." International Union of Geodesy and Geophysics, General Assembly. Volume 21, Week B, Pages 405. Descriptors: Afar; Africa; b-values; Djibouti; earthquake prediction; earthquakes; East Africa; Ethiopia; geologic hazards; Global Seismic Hazard Assessment Program; GSHAP; Horn of Africa; programs; Red Sea region; risk assessment; seismic energy; seismic risk; seismology; Somali Republic; spatial variations; time variations. Database: GeoRef.

Wood, D. A. Phillips Petroleum Company Europe-Africa and Africa Exploration. 1982. A Regional Study of the Potential for Hydrocarbon Exploration in Africa: Including Map Showing the Distribution and Geological Characteristics of the Sedimentary Basins. London: Phillips Petroleum Co. Europe-Africa. Descriptors: 1 v. (loose-leaf); Petroleum- Geology- Benin; Petroleum- Geology- Botswana; Petroleum- Geology- Burundi; Petroleum- Geology- Central African Republic; Petroleum- Geology- Chad; Petroleum- Geology- Djibouti; Petroleum- Geology- Equatorial Guinea; Petroleum- Geology- Ethiopia; Petroleum- Geology- Guinea; Petroleum- Geology- Guinea-Bissau; Petroleum- Geology- Kenya; Petroleum- Geology- Lesotho; Petroleum- Geology- Liberia; Petroleum- Geology- Madagascar; Petroleum- Geology- Malawi; Petroleum- Geology- Mali; Petroleum- Geology- Mauritania; Petroleum- Geology- Mozambique; Petroleum- Geology- Niger; Petroleum- Geology- Senegal; Petroleum- Geology- Gambia; Petroleum- Geology- Sierra Leone; Petroleum- Geology- Somalia; Petroleum- Geology- South Africa; Petroleum- Geology- Swaziland; Petroleum- Geology- Tanzania; Petroleum- Geology- Togo; Petroleum- Geology- Uganda; Petroleum- Geology- Zambia; Petroleum- Geology- Zimbabwe. Worldcat Accession Number: 14223293.

Woodward, D. J. and Stockton, G. 1989. "Somalia: A Study of the Profitability of Somali Exports." Deloitte, Haskins and Sells, Washington, DC. Performer: Abt Associates, Inc., Washington, DC. Performer: Idaho Univ., Moscow. Postharvest Inst. for Perishables. Aug. Volume: AIDPNABD389, pages: 138. Descriptors: Agricultural economics; Profits; Livestock; Leather; Hides; Somalia; Exports; Government policies; Commodities; Bananas; Foreign exchange rate; Business and economics International commerce marketing and economics; Agriculture and food Agricultural economics. Abstract: In an effort to provide policy-related data on the effect of differing exchange rates on Somali exports, the present study: assesses the profitability of Somalia's major exports- livestock (sheep/goats, cattle, and camels), hides and skins, and bananas; presents an overview of Somalia's export sector and the major commodities involved; provides a detailed discussion of cost composition to point of export for each commodity/commodity group; reviews specific export markets (Saudi Arabia, North

Yemen, Egypt, and Italy) to establish actual prices realized for commodities (instead of minimum export prices); assesses issues surrounding Somalia's entry into each market; and details the results of the profitability analysis. The implications of the analysis for exchange rate policy are discussed, as are the linkages between exchange rate adjustment and export performance. Recommendations for ensuring the continuation of data collection and analysis are included. A major conclusion is that increased export profitability helps only the seller in the willing buyer/willing seller trade equation. Notes: Funder: Agency for International Development, Washington, DC. Office of Science and Technology. NT: Prepared in cooperation with Abt Associates, Inc., Washington, DC., and Idaho Univ., Moscow. Postharvest Inst. for Perishables. Sponsored by Agency for International Development, Washington, DC. Office of Science and Technology. NTIS Accession Number: PB90224585XSP.

World Bank. 1990. "Somalia Infrastructure Rehabilitation Project, Road Network." Washington, D.C. World Bank. Descriptors: Roads- Somalia- Maps. Notes: Description: 1 map; color; 19 x 16 cm. Scale [ca. 1:1,000,000]; (E 400--E 520/N 120--N 00). Notes: "February 1990." "IBRD-22016." Includes location map. LCCN: 95-681294; OCLC Accession Number: 32548486.

World Bank.1968. "Republic of Somalia. Highway System." I.B.R.D. Somalia. Descriptors: Roads- Somalia- Maps. Notes: Description: 1 map. Scale ca. 1:6,500,000. Notes: Shows: international and regional boundaries, roads (3 categories), roads under construction, airports, seaports. "IBRD 134OR4." General Info: Coverage: Somalia. OCLC Accession Number: 48602943.

World Health Organization/GTZ, Mogadishu, Somalia. 1984. Plan for Drinking Water Supply and Sanitation Sector in Somalia for the Decade 1981-1990 and for the Period 1984-1990 [Modified]. Somalia: WHO/GTZ, Mogadishu, Somalia. Descriptors: Africa; drinking water; East Africa; protection; sanitation; Somali Republic; water resources; water supply. Notes: In 2 volumes. Database: GeoRef. Accession Number: 1998-015711.

Wyllie, Bruce Kerr N. 1925. "Geology of a Portion of British Somaliland." Monographs of the Geological Department of the Hunterian Museum. Volume 1, Pages 9-10. Descriptors: Africa; areal geology; East Africa; northern Somalia; Somali Republic.

Wyllie, Bruce Kerr N. Smellie, W. R. Gregory, J. W. and Hunterian Museum (University of Glasgow). 1925. "The Collections of Fossils and Rocks from Somaliland: Made by Messrs. B.N.K. Wyllie, B. Sc., F.G.S. and W.R. Smellie, D. Sc., F.R.S. Ed, with an Account by them of Part of the Geology of Somaliland, an Introduction by J.W. Gregory." Glasgow: Maclehose, Jackson. Descriptors: vi, 180 pages; Geology- Somalia; Paleontology- Somalia; Rocks- Somalia- Catalogs and collections; Fossils- Somalia- Catalogs and collections. Monographs of the Geological Department of the Hunterian Museum (University of Glasgow), Number I. Notes: Includes index. WorldCat Accession Number: 57503174.

Wyte, Joey. 1993. "Bridging the Juba." Engineer. August 1993, Vol. 23 Issue 3, page 10, 2p, 2bw. Subject Terms: Military Engineering; United States. Army. Corps of Engineers. Geographic Terms: Juba River (Ethiopia & Somalia); Somalia. Abstract: Recounts how the 41st Engineer Battalion of the US Army Corps of Engineers accomplished their mission to bridge the Juba River, north of Kismayo in Somalia. Relocation to the port city of Kismayo; Equipment used; Organization of the company and attachments; Shipment of bridge sets to the place; Reconnaissance and rehearsals for bridge erection; Site preparation; Construction. ISSN: 0046-1989.

X, Y, Z

Xayder, Ali Salad; Abdulle, Ali Omar and Yussuff, Faarah Mohamud. 1987. The Exploration, Development and use of Water Resources in Somalia; GeoSom 87; International Meeting; Geology of Somalia and Surrounding Regions; Abstracts. Somalia: Somali Natl. Univ., Mogadishu, Somalia. GeoSom 87; Geology of Somalia and Surrounding Regions, Mogadishu. Somalia Conference: Nov. 23-30, 1987. Descriptors: Africa; drilling; East Africa; economic geology; ground water; hydrochemistry; hydrology; Somali Republic; surveys; water resources; water wells; wells. OCLC Accession Number: 34175029.

Xuseen Haybe Cige; Somalia and Machadka Shaqo-ku-Tababaridda Macallimiinta. 1984. Cimilada, Cimilo Gooreedka, Cimilo Goboleedyada Adduunka Iyo Dhigaha Iyo Loolalka. Mogadishu, Somali Democratic Republic: IITT. Descriptors: 51; Climatology- Study and teaching- Somalia. Notes: "Mogadisho, December, 1984." "T2." At head of title: Jamhuuriyadda Dimuqraadiga Soomaaliya, Wasaaradda Waxbarashada iyo Barbaarinta, Machadka Shaqo-ku-Tababaridda Macallimiinta = Somali Democratic Republic, Ministry of Education, Institute of In-service Teacher Training. Responsibility: waxa soo diyaariyay Xuseen Haybe Cige. OCLC Accession Number: 23665844.

Yates, J., Gupta, K. R. and Kazokas, William C. (chairperson). 1990. "India's Programme on Polymetallic Nodules; 22nd Annual Offshore Technology Conference; 1990 Proceedings; Bridging Four Decades." Proceedings- Offshore Technology Conference. Volume 22, Vol. 2, Pages 273-279. Descriptors: Arabian Basin; Asia; cadmium ores; cobalt ores; copper ores; deep-sea environment; development; economic geology; history; India; Indian Ocean; Indian Peninsula; lead ores; manganese ores; marine environment; metal ores; mineral exploration; molybdenum ores; nickel ores; nodules; oceanography; ore grade; policy; polymetallic ores; programs; resources; Somali Basin; zinc ores. Notes: OTC 6299; References: 17; 2 tables. ISSN: 0160-3663.

Yusuf, Mahad A. 1985. "Metodologie Statistiche in Idrologia; Analisi Della Portata Del Juba a Lugh." Translated title: "Statistical Methods in Hydrology; Analysis of the Course of Juba River to Lugh." Cilmi Iyo Farsamo; Wargeyska Jaamacadda Ummadda Soomaaliyeed. Volume 13, Descriptors: Africa; East Africa; fluvial features; Gedo Somali Republic; hydrology; Juba River; Lugh Somali Republic; methods; rivers; rivers and streams; Somali Republic; statistical analysis; surface water. Database: GeoRef.

Yusuf, Omar Shire; Ahmed, Osman Mohamed and Hassan, Abdirahman Dahir. 1987. Building Materials in the Banadir District; Problems of Quarry Exploitation and Environmental Protection. Somalia: Somali Natl. Univ., Mogadishu, Somalia. GeoSom 87; Geology of Somalia and Surrounding Regions, Mogadishu. Somalia Conference: Nov. 23-30, 1987. Descriptors: Africa; Banadir; construction materials; desertification; East Africa; economic geology; effects; mining geology; pollution; quarries; sands; Somali Republic. OCLC Accession Number: 34175029.

Yusuf, Omar Shire; Mortari, R. Hassan, Abdirahman Dahir and Ahmed, Osman Mohamed. 1987. Some Technical Characteristics of Benadir Sands. Somalia: Somali Natl. Univ., Mogadishu, Somalia. GeoSom 87; Geology of Somalia and Surrounding

Regions, Mogadishu. Somalia Conference: Nov. 23-30, 1987. Descriptors: Africa; age; Benadir; density; East Africa; engineering geology; foundations; grain size; penetration tests; properties; sands; soil mechanics; Somali Republic. OCLC Accession Number: 34175029.

Zaccarini, G. 1913. Carta Della Media e Bassa Goscia Al 50,000. Translated title: Map of the Middle and Lower Goscia at 1:50,000. Italy: Ministero dell'Africa Italiana, Rome, Italy. Descriptors: Africa; areal geology; Bay; Bur Somali Republic; East Africa; geologic maps; Gossia Somalia; maps; Somali Republic. Database: GeoRef. Accession Number: 1998-015721.

Zaccarini, G. 1912. Carta Della Regione Dell'Uebi Scebeli Presso Merca Al 50,000. Map of the Shebelle River Region Near Merca at 1:50,000. Italy: Servizio Cartografico del Ministero dell'Africa Italian, Rome, Italy. Descriptors: Africa; areal geology; East Africa; geologic maps; maps; Merca Somalia; Shebeellaha Hoose; Shebelle River; Somali Republic. Database: GeoRef. Accession Number: 1998-015720.

Zaccarini, G. 1910. Carta Della Somalia Al 200,000. Translated title: Map of Somalia at 1:200,000. Italy: Servizio Cartografico del Ministero dell'Africa Italiana, Rome, Italy. Descriptors: Africa; areal geology; East Africa; geologic maps; maps; Somali Republic. Database: GeoRef. Accession Number: 1998-015719.

Zambrano, R. 1980. "Metodo Gravimetrico Di Prospezione Geofisica; Applicazioni Alla Geologia, Alla Ricerca Mineraria, all'Ingeneria; Aspetti Interpretativi. Gravimetric Method of Geophysical Prospecting; Application to Geology, Mining Research, Engineering; Interpretation Aspects." Cilmi Iyo Farsamo; Wargeyska Jaamacadda Ummadda Soomaaliyeed. Volume 6, Descriptors: Africa; applications; East Africa; engineering geology; geophysical methods; geophysical surveys; gravity methods; mining; mining geology; Somali Republic; surveys. Database: GeoRef.

Zanettin, B. 1993. Stratigraphy, Magmatism and Structural Evolution of the Ethiopian Volcanic Province. Firenze; Istituto Agronomico L'Oltremar; 1993. Relazioni E Monografie Agrarie Subtropicali E Tropicali Nuova Serie113 A, pages: 279-310. International Meeting; 1st- 1987 Nov: Mogadishu. Notes: Geology of Somalia and surrounding regions: Geology and mineral resources of Somalia and surrounding regions. OCLC Accession Number: 34175029.

Zardi, O., Adorisio, E., Harare, Osman and Nuti, M. 1980. "Serological Survey of Toxoplasmosis in Somalia." Transactions of the Royal Society of Tropical Medicine and Hygiene. Volume 74, Issue 5, Pages 577-581. Abstract: Of 356 samples of human sera collected from native patients in two distinct zones of Somalia, 53 % were positive ([ges] 1:8) to the dye-test for Toxoplasma gondii antibodies. A significantly lower incidence (P T. gondii infection exists among the Somalian population.These differences were regarded as being due to climatic and geographical conditions rather than to diet or socio-economic conditions.

Zavattari, E. 1940. Dal Giuba Al Lago Rodolfo. from Juba to Lake Rudolf. Italy: Reale Accademia d'Italia, Rome, Italy. Descriptors: Africa; areal geology; East Africa; East African Lakes; Ethiopia; Juba River; Lake Turkana; Somali Republic. Database: GeoRef. Accession Number: 1998-015724.

Zeltner, Alexandra. 2000. Monsoonal Influenced Changes of Coccolithophore Communities in the Northern Indian Ocean; Alteration during Sedimentation and Record in Surface Sediments. Federal Republic of Germany: University of Tubingen, Tubingen,

Federal Republic of Germany. Descriptors: Africa; algae; Arabian Peninsula; Asia; Cenozoic; climate effects; Coccolithophoraceae; communities; East Africa; Indian Ocean; marine sedimentation; marine sediments; microfossils; monsoons; nannofossils; northern Indian Ocean; ocean basins; ocean circulation; Oman; phytoplankton; plankton; Plantae; Quaternary; sea water; seasonal variations; sedimentation; sediments; Somali Republic; suspended materials. Notes: illustrations incl. sects., geol. sketch maps. ISSN: 0937-373X.

Zhabina, N. N., Demidova, T. P. and Morozov, A. A. 1979. "Soyedineniya Sery v Somaliyskoy Kotlovine Indiyskogo Okeana." Translated title: "Sulfur Compounds in Sediments of the Somali Basin, Indian Ocean." Geokhimiya. Dec. Volume 1979, Issue 12, Pages 1868-1882. Descriptors: carbon; Eh; geochemistry; Indian Ocean; marine sediments; oceanography; organic carbon; organic compounds; organic materials; pH; pore water; pyrite; sediments; Somali Basin; sulfates; sulfides; sulfur. Notes: References: 21; illustrations incl. 2 tables. ISSN: 0016-7525.

Zhelnin, A. A., Kolchitskii, N. N., Lisogurskii, N. I., Petrichev, A. Z. And Petrova, L. I. 1981. "The Transport and Evolution of the Air and Water-Vapor Masses in the Troposphere Over the Indian Ocean." Meteorologicheskie Issledovaniia. Issue 24, Pages 32-38. Descriptors: Water vapor; Aircraft components; Somalia; Coastal environments; Troposphere; Oceans; Air Masses; Indian Ocean; Marine Meteorology; Aerology; Air Flow; Annual Variations; Atmospheric Moisture; Evolution (Development); Gas Transport; Mass Transfer; Quantitative Analysis. Notes: In Russian. Abstract: Data from aerological observations made during the Monsoon-77 expedition are used. It is shown that quantitative changes occur in the transport of air and water vapor over the Somali Current. It is pointed out that mass transport falls off near the equator and increases in moving to the north or south. In evaluating the transport of air and water vapor from the western to the eastern part of the sea, it is found that in June the quantity of air transported is comparable in both parts; the quantity of water vapor, however, falls off in moving eastward over the Arabian Sea. In July, the transported air mass is equal in both parts of the sea, whereas the quantity of water vapor is found to increase as it is transported from the African to the Indian coast. Database: CSA Technology Research Database.

Zhenfu, Y. 1983. "Report for Jalalaqsi Groundwater Exploration." Somalia: Descriptors: Africa; East Africa; exploration; ground water; Hiran Somali Republic; Jalalaqsi Somalia; Somali Republic; water resources. Database: GeoRef. Accession Number: 1998-015726.

Zollner, Douglas. 1986/10. "Sand Dune Stabilization in Central Somalia." Forest Ecology and Management. Volume 16, Issue 1-4, Pages 223-232. Abstract: As part of a reforestation project in refugee impacted areas of Somalia, a sand dune stabilization pilot project was started in June, 1983, in the Hiiraan region of central Somalia. Thirty ha of the 5.6 km2 of broken sands have been crosshatched with Commiphora (sp.) cuttings and then planted with trees.The Commiphora cuttings were obtained from nearby areas of natural vegetation. The 1.5 m long cuttings were then planted in dry sand in rows at 10 m intervals, closing to 5 m at the crest of the dune. The live micro-windbreaks formed reduced wind velocity and blowing sand. Perhaps more importantly, they excluded livestock, mainly goats and sheep, from the stabilized area.Eighteen tree/shrub species, grown in plastic pots in a small nursery, were planted between the micro-windbreaks at a

5 m x 4 m spacing. The seedlings were outplanted during the rainy season without supplementary watering.After planting, ground cover increased from 5% to 26% in 18 months. The new vegetation is more dense at the base of the dune, becoming sparser on the steep slopes and crest. These grasses and forbs were started from seed blown in by the wind and caught by the live fencing. They became established and grew because livestock have been excluded from the area.Prosopis juliflora out-performed all other tree species. The largest trees have branches 3 m long after 15 months with extensive root systems. Survival of the seedlings is high and can be attributed to a lens of water approximately 1 m below the surface of the dune (deeper at the crest). Seedling roots, initiated during the rainy season, when the entire soil profile is wet, were able to reach this lens of moist sand. ISSN: 0378-1127.

www.ingramcontent.com/pod-product-compliance
Lightning Source LLC
Chambersburg PA
CBHW080410270326
41929CB00018B/2966

* 9 7 8 1 7 8 0 3 9 1 8 5 4 *